AF353121

Functionally Relevant Macromolecular Interactions of Disordered Proteins

Functionally Relevant Macromolecular Interactions of Disordered Proteins

Special Issue Editor

Istvan Simon

MDPI • Basel • Beijing • Wuhan • Barcelona • Belgrade • Manchester • Tokyo • Cluj • Tianjin

Special Issue Editor
Istvan Simon
Institute of Enzymology, TTK
Hungary

Editorial Office
MDPI
St. Alban-Anlage 66
4052 Basel, Switzerland

This is a reprint of articles from the Special Issue published online in the open access journal *International Journal of Molecular Sciences* (ISSN 1422-0067) (available at: https://www.mdpi.com/journal/ijms/special_issues/macromolecular_interactions).

For citation purposes, cite each article independently as indicated on the article page online and as indicated below:

LastName, A.A.; LastName, B.B.; LastName, C.C. Article Title. *Journal Name* **Year**, *Article Number*, Page Range.

ISBN 978-3-03936-521-0 (Hbk)
ISBN 978-3-03936-522-7 (PDF)

© 2020 by the authors. Articles in this book are Open Access and distributed under the Creative Commons Attribution (CC BY) license, which allows users to download, copy and build upon published articles, as long as the author and publisher are properly credited, which ensures maximum dissemination and a wider impact of our publications.

The book as a whole is distributed by MDPI under the terms and conditions of the Creative Commons license CC BY-NC-ND.

Contents

About the Special Issue Editor

Istvan Simon was born in Budapest, Hungary, in 1947. He graduated as a physicist and habilitated in biology and in physics. He is a member of the Hungarian Academy of Sciences and currently a professor emeritus at the Research Centre of the Hungarian Academy of Sciences, where he has been since 1969. He turned his attention to computational analysis at Cornell University, where he spent several years in the group of Harold A. Scheraga. He continued his career in this field in Hungary, and pioneered computational protein structure research at the end of the 1970s. He has published 8 book chapters, 132 papers. The publications have been cited over 12,000 times, and he has twice been listed among the highly cited researchers according to the Web of Science. Together with his research group he has provided 16 databases and prediction servers on the World Wide Web. These include the prediction of "stabilization centers", i.e., residue pairs that are responsible for keeping a protein's structure intact, the prediction of disulfide-forming cysteines (CYSREDOX), and a number of top-cited transmembrane prediction algorithms (DAS, HMMTOP, and PDBTM). Recently, his group has uncovered the statistical thermodynamics forming the background of protein disorder, and provided the corresponding prediction server, IUPred, followed by the prediction of functional regions of disordered proteins (ANCHOR). The strength of these methods is the groundbreaking discovery of principles underlying protein structure organization.

International Journal of
Molecular Sciences

Editorial

Macromolecular Interactions of Disordered Proteins

István Simon

Institute of Enzymology, RCNS, Lorand Eotvos Research Network, Center of Excellence of the Hungarian Academy of Sciences, Magyar Tudósok krt. 2., H-1117 Budapest, Hungary; simon.istvan@ttk.mta.hu

Received: 3 January 2020; Accepted: 10 January 2020; Published: 13 January 2020

Proteins are social beings. Especially disordered proteins like company, they hardly act without interacting with another macromolecule. In most known cases, these other macromolecules are other proteins, sometimes nucleic acids and very seldom something else. Disordered proteins are rather newcomers in protein science. The first papers on these proteins came out in the fourth quarter of the last century. What is more, they were hardly recognized before the great paper of Wright and Dyson published in J. Mol. Biol. in 1999 [1]. By now, it is well known that a large portion of all existing proteins are intrinsically disordered under physiological conditions. They perform vital roles in many living cells. For more than a decade, it was generally thought that disordered proteins or disorder parts of partially disordered proteins have different amino acid composition than folded proteins have and various prediction methods were developed based on this principle. Dosztanyi et al. [2,3] provided a physical background of a disorder prediction methods (IUPred) by estimating the lowest value of the sum of the pairwise interaction energies between residues from the amino acid sequences without considering structural information. This calculated energy per residue value for globular proteins was well separated from the ones calculated for disordered (unstructured) proteins know at that time. This principle of pair energy estimation applied in IUPred also worked well, when the method ANCHOR [4,5] was developed to predict binding site within disordered parts of protein by which the disordered protein bounds to a folded one and its structure is formed upon binding. Those segments of the disordered proteins are identified as binding site where in amino acids, considering together with the average composition of folded protein, exhibit low enough pairwise interaction energy to be stable. Recently however shreds of evidence were accumulated about the existence of a different type of disordered proteins [6]. It turned out that some disordered proteins can undergo coupled folding and binding without the involvement of an already folded protein, but by intra-acting with disordered proteins. This second protein can be the same as the first one (formation of homodimers) or can be different (formation of heterodimers. They can also form higher order oligomers. These proteins which can stabilize their structure via "mutual synergistic folding" have residue compositions similar to that of the folded globular water-soluble proteins. Their residue compositions are different from the composition of the traditional disordered proteins, which can only be stabilized on the surface of an already stabilized macromolecule, in most cases on the surface of a folded protein. These traditional disordered proteins can be named as "coupled folding and binding" protein. Recently the "mutual synergistic folding" proteins were collected in a database MFIB [7], the "coupled folding and binding" proteins were collected in a database DIBS [8] and the structural and functional properties of these two types of protein were compared [9]. Beside the large variation of protein-protein interactions, in the past decade, more and more examples are found, where disordered proteins interact with non-protein macromolecules in various forms [10]. There is also a very new phenomenon when proteins, including disordered ones, are involved in phase separation, which can be a weak but functionally important macromolecular interaction [11].

Paper of two merged special issues on the same topic: "Functionally relevant macromolecular interactions of disordered proteins" are summarized and listed in the order of their online publication date. Research, review and concept papers listed separately, starting with the oldest one in this Editorial.

1. Research

In the first paper of this series, a study on the effect of acetylation on the phase separation tendency of Tau protein was reported by Ferreon et al. It is well known that intrinsically disordered protein Tau is involved in Alzheimer's disease. Recently it was shown that Tau is capable of undergoing liquid-liquid phase separation, which involves weak protein-protein interactions and it is considered as an initiation of Tau aggregation observed in Alzheimer's disease. In this work, it was shown that acetylation disfavors phase separation and aggregation of Tau, therefore, acetylation prevents the toxic effects of liquid-liquid phase separation dependent aggregation [12].

Srivastava et al. tried to decipher RNA-recognition patterns of IDPs in the next paper. They analyzed the protein-RNA complexes which undergo disordered to ordered transition (DOT) during binding. The DOT region is small and positively charged, like the binding sites in globular proteins. However, for DOTs of IDP have significantly higher exposure to the water, than their counterpart in structured protein. These findings can help to develop tools for identifying DOT regions in RNA binding proteins [13].

Contreras et al. studied a Protein (LrtA protein of Synechocystis sp. PCC 6803) which is oligomeric and folded in solution, but the single-chain is only folded and stable in their N terminal half of the polypeptide (residues 1–100) while the other half (101–197) is very unstable and rather disordered with chameleonic sequence properties. While disordered protein which undergoes mutual synergistic folding upon binding to each other, this is a rather rear case, when it happens only with half of the protein. The other half remains folded before and after self-association [14].

The origin of the thermal stability of eukaryotic proteins was studied and compared with that of thermophilic and mesophilic proteins of prokaryotes by Alvarez-Ponce et al. The eukaryotic model system was *Arabidopsis thaliana* at 22 and 37 °C, and they compare both the amino acid compositions and levels of intrinsic disorder of heat-induced and heat-repressed proteins. Heat-induced proteins are enriched in intrinsically disordered regions and depleted in hydrophobic amino acids in contrast to thermophile prokaryotic proteins [15].

A decision-tree based meta server to predict disordered parts of proteins and their residues involved in binding motifs has been developed by Zhao and Xue. The meta server is based on four predictors: DisEMBL, IUPred, VSL2, and ESpritz. The meta server provides higher accuracy than each of these independent predictors [16].

Arvidsson and Wright applied a protein disorder approach characterizing differentially expressed genes analyzing cell adhesion regulated gene expression in lymphoma cells. They checked if predicted protein disorder was differentially associated with proteins encoded by differentially regulated genes in lymphoma cells. Intrinsic disorder protein properties were extracted from the Database of Disordered Protein Prediction (D^2P^2). They concluded that down-regulated genes in stromal cell-adherent lymphoma cells encode proteins that are characterized by elevated levels of disorder [17].

The co-evolution of IDPs and folded partner proteins was studied by checking their evolutionary couplings. Pancsa et al. pointed that due to the lack of strict structural constraints, IDPs undergo faster evolutionary changes than folded proteins, which makes the reliable identification and alignment of IDP homologs difficult. They demonstrated that partner binding imposes constraints on IDP sequences that manifest in detectable inter-protein evolutionary couplings. It brings hope that IDP–partner interactions could soon be successfully dissected through residue co-variation analysis [18].

A principal part of the physical bases of disordered proteins involved in mutual synergetic folding in homodimers has been uncovered by Magyar et al. The authors concluded that homodimer proteins have a larger solvent-accessible main-chain surface area on the contact surface of the subunits, when compared to globular homodimer proteins. The main driving force of the dimerization is the mutual shielding of the water-accessible backbones and the formation of extra intermolecular interactions [19].

Szabó et al. reported their finding that disordered parts of Mixed Lineage Leukemia 4 (MLL4) protein are capable of RNA binding. They explored the RNA binding capability of two; uncharacterized regions of MLL4; with the aim of shedding light to the existence of possible regulatory lncRNA interactions of the protein They demonstrated that both regions; one that contains a predicted RNA binding sequence and one that does not, are capable of binding to different RNA constructs in vitro [20].

A method to characterize the hydration of proteins based on evaluating two-component wide-line ^{1}H NMR signals is presented in the next paper. Tompa et al. also provided a description of key elements of the procedure conceived for the thermodynamic interpretation of such results. The results enable a quantitative description of the ratio of ordered and disordered parts of proteins, and the energy relations of protein–water bonds in aqueous solutions of the proteins [21].

Homma et al, studied the evolution rate structural domains (SDs) and intrinsically disordered regions (IDRs) of immune-related mammalian proteins. IDRs are generally subject to fewer constraints and evolve more rapidly than SDs. However, it turned out that for immune-related proteins in mammals, the evolution rates in SDs come close to those in IDRs [22].

Moosa, M.M. et al. applied direct single-molecule observation to study sequential DNA bending transitions by the SoxSox2 is a transcription factor which assumed to achieve its regulatory diversity via heterodimerization with partner transcription factors. However, single-molecule fluorescence spectroscopy suggests that Sox2 alone can modulate structural landscape of the DNA in a dosage-dependent manner [23].

In a paper which was a follow-up of the Contreras, L.M. et al. paper [14], Neira et al. reported a study on the structure of the C-terminal half (residues 102–191) of the LrtA protein of *Synechocystis* sp. PCC 6803 in separated form with various physical-chemical techniques. At physiological conditions isolated C-LrtA intervened in a self-association equilibrium, involving several oligomerization reactions. They concluded that C-LrtA was an oligomeric disordered protein [24].

Mishra et al. extended their one-bead-per-amino-acid model for intrinsically disordered proteins to account for phosphorylation in studying the effect of phosphorylation on nuclear pore complex selectivity. The simulations show that upon phosphorylation the transport rate of inert molecules increases, while that of nuclear transport receptors decreases. The models provide a molecular framework to explain how extensive phosphorylation decreases the selectivity of the nuclear pore complexes [25].

Walter et al. studied the hydrodynamic properties of the intrinsically disordered potyvirus genome-linked protein, VVPg), of the translation initiation factor, eIF4E, and of their binary complex (VPg)-eIF4E. N-terminal His tag decreased the conformational entropy of this intrinsically disordered region. A comparative study revealed the His tag contribution to the hydrodynamic behavior of proteins [26].

The role of intrinsically disordered linkers in the confinement of binding domains in enzyme actions was studied in the following paper. By statistical physical modeling Szabo et al. show that this arrangement results in processive systems, in which the linker ensures an optimized effective concentration around novel the binding site(s), favoring rebinding over full release of the polymeric partner. By analyzing 12 enzymes they suggest a unique type of entropic chain function of intrinsically disordered proteins, that may impart functional advantages on diverse enzymes in a variety of biological contexts [27].

Machulin et al. studied the contribution of repeats in ribosomal S1 proteins into the tendency for intrinsic disorder and flexibility within and between structural domains for all available UniProt S1 sequences. Using charge–hydrophobicity plot cumulative distribution function (CH-CDF) analysis they classified 53% of S1 proteins as ordered proteins, the remaining proteins were related to molten globule state. According to the FoldUnfold and IsUnstruct programs, relatively short flexible or disordered regions are predominant in the multi-domain proteins. Their results suggest that the ratio of flexibility in the separate domains is related to their roles in the activity and functionality of S1 [28].

The decrease of disorder level of p53-DBD upon interacting with the anticancer protein Azurin by mean of Raman spectroscopy was monitored by Signorelli et al. This technique was found to be suitable to elucidate the structural properties of intrinsically disordered proteins and was applied to investigate the changes in both the structure and the conformational heterogeneity of the DNA-binding domain (DBD) belonging to the intrinsically disordered protein p53 upon its binding to Azurin, an electron-transfer anticancer protein from *Pseudomonas aeruginosa*. The results show an increase of the secondary structure content of DBD concomitantly with a decrease of its conformational heterogeneity upon its binding to Azurin [29].

Structural and functional properties of a capsid protein of dengue and related flavivirus. Dengue, West Nile and Zika have very similar viral particle with an outer lipid bilayer containing two viral proteins in the nucleocapsid core were studied by Faustino et al. Using dengue virus capsid protein as the main model, the protein size, thermal stability, and function with its structure/dynamics features were correlated. Their findings suggest that the capsid protein interaction with host lipid systems leads to minor allosteric changes that may modulate the specific binding of the protein to the viral RNA [30].

Chan-Yao-Chong, et al. investigated the early steps of actin recognition of Neural Wiskott–Aldrich Syndrome Protein (N-WASP) domain V. Using docking calculations and molecular dynamics simulations, their study shows that actin is first recognized by the N-WASP domain V regions which have the highest propensity to form transient α –helices. The WH2 motif consensus sequences "LKKV" subsequently binds to actin through large conformational changes of the disordered domain V [31].

Mentes et al.'s paper is the follow-up of the Magyar's paper [19] of this collection. It reports the properties of heterodimer Mutual Synergistic Folding (MSF) proteins instead of homodimeric ones. The main driving force of the dimerization is the mutual shielding of the water-accessible backbones and the formation of extra intermolecular interactions just like in homodimers. However here shielding of the β-sheet backbones and the formation of a buried structural core along with the general strengthening of inter-subunit interactions together could be important factors [32].

Conformational ensembles of alpha-Synuclein were studied using single-molecule force spectroscopy and mass spectroscopy by Corti et al. This work applies single-molecule force spectroscopy to probe conformational properties of α-synuclein in solution and its conformational changes induced by ligand binding. This analysis provides support to the structural interpretation of charge-state distributions obtained by native mass spectrometry and helps defining the conformational components detected by single-molecule force spectroscopy [33].

The topic of the Mészáros et al. paper is closely related to the ones of the Mentes et al's. paper [32] and the Magyar et al. paper [19]. The authors report the sequence and structure properties of protein complexes formed by disordered proteins via Mutual Synergistic Folding (MSF). A method is presented which differences in binding strength, subcellular localization, and regulation are encoded in the sequence and structural properties of proteins. It serves as a better representation of structures arising through this specific interaction mode [34].

Three Rett syndromes (RTT) treatment-related genes MECP2, CDKL5 and FOXG1 in silico by evolutionary classification and disordered region assessment were reported in this paper. Fahmi, M. et al. provided insight into the structural characteristics, evolution and interaction landscapes of those three proteins. They also reported the disordered structure properties and evolution of those proteins which may provide valuable information for the development of therapeutic strategies of RTT [35].

2. Review

Sánchez-López et al. report about the structural determinants of the N-terminus of the prion protein and the effect of binding copper ions in their review. They discuss the current knowledge of how mutations can impact the copper-binding properties of prion protein both in health and disease progression [36].

In their review paper Ciemny et al. first introduced the technique of Monte Carlo and other simulation for predicting possible structures of unstructured protein, protein-peptide complexes and unfolded states of globular proteins. They presented several case studies on various disordered proteins. They also proposed the use of the CABS coarse-grained model with Monte Carlo sampling scheme. They also show that CABS can be combined with the use of experimental data too [37].

The literature information on the alteration of disordered proteins in neurodegenerative and other diseases reported in this review. Martinelli et al. discussed how the misfolded proteins can be involved in Alzheimer's, Parkinson's and other diseases. The most common form of misfolding IDPs is the formation of neurotoxic amyloid plaques. The review discusses important special cases of beta-amyloid, alpha-synuclein, tau etc. They also show drug candidates for later use in to treatment of diseases caused by misfolded IDPs [38].

This is the first review from the Greb-Markiewicz lab in which Kolonko and Greb-Markiewicz summarized our current knowledge on helix-loop-helix/Per-ARNT-SIM (bHLH–PAS) proteins, considering their structures and intrinsic disorder nature based on NMR and X-ray analysis. Currently, all determined structures comprise only selected domains (bHLH and/or PAS), while parts of proteins, comprising their long C-termini, have not been structurally characterized yet since these regions appear to be intrinsically disordered. These intrinsically disordered parts contribute a lot to the flexibility and function of these proteins [39].

The second review from the same lab is the paper of Tarczewska and Greb-Markiewicz which is a follow-up publication of the review paper of Kolonko and Greb-Markiewicz [39], the currently available information on "The Significance of the Intrinsically Disordered Regions for the Function of the BHLH Transcription Factors" is reported. Their aim was to emphasize the significance of existing disordered regions within the helix–loop–helix (bHLH) transcription factors for their functionality [40].

Finally, in the last review paper of this collection, Owen and Shewmaker summarized our current knowledge on "The Role of Post-Translational Modification in the Phase Transition of Intrinsically Disordered Proteins". They pointed that intrinsically disordered regions are critical to the liquid–liquid phase separation that facilitates specialized cellular functions and discuss how post-translational modifications of intrinsically disordered protein segments can regulate the molecular condensation of macromolecules into functional phase-separated complexes [41].

3. Concept

In his concept paper, Rikkerin, E.H.A. considers the "first line response" role of disordered protein in the protection against pathogens and disease. He presents several examples of how disorder and post-translational changes can play in the response of organisms to the stress of a changing environment. He proposes that some disordered proteins enable organisms to sense and react rapidly as the first line responds [42].

Funding: Hungarian Research and Developments Fund OTKA K115698.

Conflicts of Interest: The author declares no conflict of interest.

References

1. Wright, P.E.; Dyson, H.J. Intrinsically unstructured proteins: Re-assessing the protein structure-function paradigm. *J. Mol. Biol.* **1999**, *293*, 321–331. [CrossRef]
2. Dosztányi, Z.; Csizmók, V.; Tompa, P.; Simon, I. The pairwise energy content estimated from amino acid composition discriminates between folded and intrinsically unstructured proteins. *J. Mol. Biol.* **2005**, *347*, 827–839. [CrossRef]
3. Dosztanyi, Z.; Csizmok, V.; Tompa, P.; Simon, I. IUPred: Web server for the prediction of intrinsically unstructured regions of proteins based on estimated energy content. *Bioinformatics* **2005**, *21*, 3433–3434. [CrossRef]
4. Mészáros, B.; Simon, I.; Dosztányi, Z. Prediction of protein binding regions in disordered proteins. *PLoS Comput. Biol.* **2009**, *5*, e1000376.
5. Dosztányi, Z.; Mészáros, B.; Simon, I. ANCHOR: Web server for predicting protein binding regions in disordered proteins. *Bioinformatics* **2009**, *25*, 2745–2746. [CrossRef]
6. Demarest, S.J.; Martinez-Yamout, M.; Chung, J.; Chen, H.; Xu, W.; Dyson, H.J.; Evans, R.M.; Wright, P.E. Mutual synergistic folding in recruitment of CBP/p300 by p160 nuclear receptor coactivators. *Nature* **2002**, *415*, 549. [CrossRef]

7. Fichó, E.; Reményi, I.; Simon, I.; Mészáros, B. MFIB: A repository of protein complexes with mutual folding induced by binding. *Bioinformatics* **2017**, *33*, 3682–3684. [CrossRef]

8. Schad, E.; Fichó, E.; Pancsa, R.; Simon, I.; Dosztányi, Z.; Mészáros, B. DIBS: A repository of disordered binding sites mediating interactions with ordered proteins. *Bioinformatics* **2018**, *34*, 535–537. [CrossRef]

9. Mészáros, B.; Dobson, L.; Fichó, E.; Tusnády, G.E.; Dosztányi, Z.; Simon, I. Sequential, Structural and Functional Properties of Protein Complexes Are Defined by How Folding and Binding Intertwine. *J. Mol. Biol.* **2019**, *43*, 4408–4428.

10. Fuxreiter, M.; Simon, I.; Bontos, S. Dynamic protein-DNA recognition: Beyond what can be seen. *Trends Biochem. Sci.* **2011**, *36*, 415–423. [CrossRef]

11. Boeynaems, S.; Tompa, P.; Van den Bosch, L. Phasing in on cell cycle. *Cell Division* **2018**, *13*. Article Number: 1. [CrossRef]

12. Ferreon, J.C.; Jain, A.; Choi, K.-J.; Tsoi, P.S.; MacKenzie, K.R.; Jung, S.Y.; Ferreon, A.C. Acetylation Disfavors Tau Phase Separation. *Int. J. Mol. Sci.* **2018**, *19*, 1360. [CrossRef]

13. Srivastava, A.; Ahmad, S.; Gromiha, M.M. Deciphering RNA-Recognition Patterns of Intrinsically Disordered Proteins. *Int. J. Mol. Sci.* **2018**, *19*, 1595. [CrossRef]

14. Contreras, L.M.; Sevilla, P.; Cámara-Artigas, A.; Hernández-Cifre, J.G.; Rizzuti, B.; Florencio, F.J.; Muro-Pastor, M.I.; García de la Torre, J.; Neira, J.L. The Cyanobacterial Ribosomal-Associated Protein LrtA from *Synechocystis* sp. PCC 6803 Is an Oligomeric Protein in Solution with Chameleonic Sequence Properties. *Int. J. Mol. Sci.* **2018**, *19*, 1857.

15. Alvarez-Ponce, D.; Ruiz-González, M.X.; Vera-Sirera, F.; Feyertag, F.; Perez-Amador, M.A.; Fares, M.A. Arabidopsis Heat Stress-Induced Proteins Are Enriched in Electrostatically Charged Amino Acids and Intrinsically Disordered Regions. *Int. J. Mol. Sci.* **2018**, *19*, 2276. [CrossRef]

16. Zhao, B.; Xue, B. Decision-Tree Based Meta-Strategy Improved Accuracy of Disorder Prediction and Identified Novel Disordered Residues Inside Binding Motifs. *Int. J. Mol. Sci.* **2018**, *19*, 3052. [CrossRef]

17. Arvidsson, G.; Wright, A.P.H. A Protein Intrinsic Disorder Approach for Characterising Differentially Expressed Genes in Transcriptome Data: Analysis of Cell-Adhesion Regulated Gene Expression in Lymphoma Cells. *Int. J. Mol. Sci.* **2018**, *19*, 3101. [CrossRef]

18. Pancsa, R.; Zsolyomi, F.; Tompa, P. Co-Evolution of Intrinsically Disordered Proteins with Folded Partners Witnessed by Evolutionary Couplings. *Int. J. Mol. Sci.* **2018**, *19*, 3315. [CrossRef]

19. Magyar, C.; Mentes, A.; Fichó, E.; Cserző, M.; Simon, I. Physical Background of the Disordered Nature of "Mutual Synergetic Folding" Proteins. *Int. J. Mol. Sci.* **2018**, *19*, 3340. [CrossRef]

20. Szabó, B.; Murvai, N.; Abukhairan, R.; Schád, É.; Kardos, J.; Szeder, B.; Buday, L.; Tantos, Á. Disordered Regions of Mixed Lineage Leukemia 4 (MLL4) Protein Are Capable of RNA Binding. *Int. J. Mol. Sci.* **2018**, *19*, 3478. [CrossRef]

21. Tompa, K.; Bokor, M.; Tompa, P. The Melting Diagram of Protein Solutions and Its Thermodynamic Interpretation. *Int. J. Mol. Sci.* **2018**, *19*, 3571. [CrossRef]

22. Homma, K.; Anbo, H.; Noguchi, T.; Fukuchi, S. Both Intrinsically Disordered Regions and Structural Domains Evolve Rapidly in Immune-Related Mammalian Proteins. *Int. J. Mol. Sci.* **2018**, *19*, 3860. [CrossRef]

23. Moosa, M.M.; Tsoi, P.S.; Choi, K.-J.; Ferreon, A.C.M.; Ferreon, J.C. Direct Single-Molecule Observation of Sequential DNA Bending Transitions by the Sox2 HMG Box. *Int. J. Mol. Sci.* **2018**, *19*, 3865. [CrossRef]

24. Neira, J.L.; Giudici, A.M.; Hornos, F.; Arbe, A.; Rizzuti, B. The C Terminus of the Ribosomal-Associated Protein LrtA Is an Intrinsically Disordered Oligomer. *Int. J. Mol. Sci.* **2018**, *19*, 3902. [CrossRef]

25. Mishra, A.; Sipma, W.; Veenhoff, L.M.; Van der Giessen, E.; Onck, P.R. The Effect of FG-Nup Phosphorylation on NPC Selectivity: A One-Bead-Per-Amino-Acid Molecular Dynamics Study. *Int. J. Mol. Sci.* **2019**, *20*, 596. [CrossRef]

26. Walter, J.; Barra, A.; Doublet, B.; Céré, N.; Charon, J.; Michon, T. Hydrodynamic Behavior of the Intrinsically Disordered Potyvirus Protein VPg, of the Translation Initiation Factor eIF4E and of their Binary Complex. *Int. J. Mol. Sci.* **2019**, *20*, 1794. [CrossRef]

27. Szabo, B.; Horvath, T.; Schad, E.; Murvai, N.; Tantos, A.; Kalmar, L.; Chemes, L.B.; Han, K.-H.; Tompa, P. Intrinsically Disordered Linkers Impart Processivity on Enzymes by Spatial Confinement of Binding Domains. *Int. J. Mol. Sci.* **2019**, *20*, 2119. [CrossRef]

28. Machulin, A.; Deryusheva, E.; Lobanov, M.; Galzitskaya, O. Repeats in S1 Proteins: Flexibility and Tendency for Intrinsic Disorder. *Int. J. Mol. Sci.* **2019**, *20*, 2377. [CrossRef]

29. Signorelli, S.; Cannistraro, S.; Bizzarri, A.R. Raman Evidence of p53-DBD Disorder Decrease upon Interaction with the Anticancer Protein Azurin. *Int. J. Mol. Sci.* **2019**, *20*, 3078. [CrossRef]

30. Faustino, A.F.; Martins, A.S.; Karguth, N.; Artilheiro, V.; Enguita, F.J.; Ricardo, J.C.; Santos, N.C.; Martins, I.C. Structural and Functional Properties of the Capsid Protein of Dengue and Related *Flavivirus*. *Int. J. Mol. Sci.* **2019**, *20*, 3870. [CrossRef]

31. Chan-Yao-Chong, M.; Durand, D.; Ha-Duong, T. Investigation into Early Steps of Actin Recognition by the Intrinsically Disordered N-WASP Domain V. *Int. J. Mol. Sci.* **2019**, *20*, 4493. [CrossRef]

32. Mentes, A.; Magyar, C.; Fichó, E.; Simon, I. Analysis of Heterodimeric "Mutual Synergistic Folding"-Complexes. *Int. J. Mol. Sci.* **2019**, *20*, 5136. [CrossRef]

33. Corti, R.; Marrano, C.A.; Salerno, D.; Brocca, S.; Natalello, A.; Santambrogio, C.; Legname, G.; Mantegazza, F.; Grandori, R.; Cassina, V. Depicting Conformational Ensembles of α-Synuclein by Single Molecule Force Spectroscopy and Native Mass Spectroscopy. *Int. J. Mol. Sci.* **2019**, *20*, 5181. [CrossRef]

34. Mészáros, B.; Dobson, L.; Fichó, E.; Simon, I. Sequence and Structure Properties Uncover the Natural Classification of Protein Complexes Formed by Intrinsically Disordered Proteins via Mutual Synergistic Folding. *Int. J. Mol. Sci.* **2019**, *20*, 5460. [CrossRef]

35. Fahmi, M.; Yasui, G.; Seki, K.; Katayama, S.; Kaneko-Kawano, T.; Inazu, T.; Kubota, Y.; Ito, M. In Silico Study of Rett Syndrome Treatment-Related Genes, *MECP2*, *CDKL5*, and *FOXG1*, by Evolutionary Classification and Disordered Region Assessment. *Int. J. Mol. Sci.* **2019**, *20*, 5593. [CrossRef]

36. Sánchez-López, C.; Rossetti, G.; Quintanar, L.; Carloni, P. Structural Determinants of the Prion Protein N-Terminus and Its Adducts with Copper Ions. *Int. J. Mol. Sci.* **2019**, *20*, 18. [CrossRef]

37. Ciemny, M.P.; Badaczewska-Dawid, A.E.; Pikuzinska, M.; Kolinski, A.; Kmiecik, S. Modeling of Disordered Protein Structures Using Monte Carlo Simulations and Knowledge-Based Statistical Force Fields. *Int. J. Mol. Sci.* **2019**, *20*, 606. [CrossRef]

38. Martinelli, A.H.S.; Lopes, F.C.; John, E.B.O.; Carlini, C.R.; Ligabue-Braun, R. Modulation of Disordered Proteins with a Focus on Neurodegenerative Diseases and Other Pathologies. *Int. J. Mol. Sci.* **2019**, *20*, 1322. [CrossRef]

39. Kolonko, M.; Greb-Markiewicz, B. bHLH–PAS Proteins: Their Structure and Intrinsic Disorder. *Int. J. Mol. Sci.* **2019**, *20*, 3653. [CrossRef]

40. Tarczewska, A.; Greb-Markiewicz, B. The Significance of the Intrinsically Disordered Regions for the Functions of the bHLH Transcription Factors. *Int. J. Mol. Sci.* **2019**, *20*, 5306. [CrossRef]

41. Owen, I.; Shewmaker, F. The Role of Post-Translational Modifications in the Phase Transitions of Intrinsically Disordered Proteins. *Int. J. Mol. Sci.* **2019**, *20*, 5501. [CrossRef] [PubMed]

42. Rikkerink, E.H.A. Pathogens and Disease Play Havoc on the Host Epiproteome—The "First Line of Response" Role for Proteomic Changes Influenced by Disorder. *Int. J. Mol. Sci.* **2018**, *19*, 772. [CrossRef] [PubMed]

© 2020 by the author. Licensee MDPI, Basel, Switzerland. This article is an open access article distributed under the terms and conditions of the Creative Commons Attribution (CC BY) license (http://creativecommons.org/licenses/by/4.0/).

International Journal of
Molecular Sciences

Article

Acetylation Disfavors Tau Phase Separation

Josephine C. Ferreon [1,*]**, Antrix Jain** [2]**, Kyoung-Jae Choi** [1]**, Phoebe S. Tsoi** [1]**,
Kevin R. MacKenzie** [1,3]**, Sung Yun Jung** [4] **and Allan Chris Ferreon** [1,*]

[1]   Department of Pharmacology and Chemical Biology, Baylor College of Medicine, Houston, TX 77030, USA;
     Kyoungjae.Choi@bcm.edu (K.-J.C.); Phoebe.Tsoi@bcm.edu (P.S.T.); Kevin.MacKenzie@bcm.edu (K.R.M.)
[2]   Advanced Technology Cores, Baylor College of Medicine, Houston, TX 77030, USA; antrixj@bcm.edu
[3]   Department of Pathology and Immunology, Baylor College of Medicine, Houston, TX 77030, USA
[4]   Department of Biochemistry and Molecular Biology, Baylor College of Medicine, Houston, TX 77030, USA;
     syjung@bcm.edu
*   Correspondence: josephine.ferreon@bcm.edu (J.C.F.); allan.ferreon@bcm.edu (A.C.F.);
     Tel.: +1-713-798-1756 (J.C.F.); +1-713-798-1754 (A.C.F.)

Received: 8 April 2018; Accepted: 2 May 2018; Published: 4 May 2018

Abstract: Neuropathological aggregates of the intrinsically disordered microtubule-associated protein
Tau are hallmarks of Alzheimer's disease, with decades of research devoted to studying the protein's
aggregation properties both in vitro and in vivo. Recent demonstrations that Tau is capable of
undergoing liquid-liquid phase separation (LLPS) reveal the possibility that protein-enriched phase
separated compartments could serve as initiation sites for Tau aggregation, as shown for other
amyloidogenic proteins, such as the Fused in Sarcoma protein (FUS) and TAR DNA-binding
protein-43 (TDP-43). Although truncation, mutation, and hyperphosphorylation have been shown
to enhance Tau LLPS and aggregation, the effect of hyperacetylation on Tau aggregation remains
unclear. Here, we investigate how the acetylation of Tau affects its potential to undergo phase
separation and aggregation. Our data show that the hyperacetylation of Tau by p300 histone
acetyltransferase (HAT) disfavors LLPS, inhibits heparin-induced aggregation, and impedes access
to LLPS-initiated microtubule assembly. We propose that Tau acetylation prevents the toxic effects
of LLPS-dependent aggregation but, nevertheless, contributes to Tau loss-of-function pathology by
inhibiting Tau LLPS-mediated microtubule assembly.

Keywords: intrinsically disordered protein; membrane-less organelle; neurodegenerative disease;
p300 HAT acetylation; post-translational modification; protein aggregation; Tau fibrillation

1. Introduction

Tau inclusions are key components of neurofibrillary tangles (NFTs) a recurring pathological
feature for several neurodegenerative diseases including Alzheimer's disease (AD) [1–4]. There are
two prevailing hypotheses on the mechanism of Tau pathology linked to protein misfolding and
aggregation, both of which are not mutually exclusive [5–7]. One is that Tau has intrinsic aggregation
motifs that enable fibrillation, leading to gain-in-toxic function(s) [8–11] exacerbated by the inability
of the cellular degradation machinery to remove misfolded or aggregated Tau [12,13]. Another
pathological mechanism is that aggregation-promoting Tau accumulation stems from loss-of-normal
function(s). Tau is essential for microtubule dynamics and stability; impairment of this function linked
to Tau sequestration into aggregates results in neuronal loss [14]. Alterations in protein sequence
and structure (such as truncations, mutations, or post-translational modifications) contribute to both
Tau loss-of-normal function and gain-in-toxic dysfunction by affecting the protein's ability to bind
microtubules or propensity to misfold and aggregate [6,7,11].

Tau is an intrinsically disordered protein (IDP), a class of proteins characterized by a high degree of
structural flexibility, conformational heterogeneity and binding promiscuity. Often, these properties not

only allow for complex functions involving networks of interactions, but also facilitate dysfunctions as a result of misfolding or aggregation [15–17]. Tau is rich in serine/threonine (S/T) and lysine (K) residues, and is known to undergo post-translational modifications (PTMs), such as phosphorylation, acetylation, ubiquitination, and sumoylation. These PTMs are linked to both Tau function and pathology [18]. Phosphorylation is known to modulate Tau's ability to promote microtubule assembly. Abnormal Tau hyperphosphorylation, however, results in fibrillation, as evidenced by hyperphosphorylated Tau being the primary component of NFTs [1]. At least 20 phosphorylation sites [14] and 23 acetylation sites [19–21] have been reported for Tau.

Tau is a macromolecular polyampholyte consisting of negatively-charged N- and C-terminal domains, and a positively-charged central Proline-rich (P) domain with microtubule binding regions (MTBR, R1–R4; Figure 1A). An increase in negative charge via phosphorylation or removal of lysine positive charge by acetylation can have significant effects on Tau function (microtubule assembly/stabilization) and dysfunction (Tau aggregation). Tau acetylation can be mediated by p300/CREB-binding protein (CBP) HAT [19–23], and reports indicate that Tau, itself, has intrinsic acetyltransferase activity [19,20,22]. In fact, hyperacetylated Tau (Ac-Tau) has been used as a diagnostic marker for AD [20,21,24]. Although the role of hyperphosphorylation in facilitating Tau pathological aggregation is not debated, there are conflicting reports on the role of hyperacetylation in Tau pathology. Several groups have found acetylated Tau in pathological inclusions in vivo [20,21], as well as in co-deposits with hyperphosphorylated Tau [21]. However, Cook et al. report that acetylation at key Tau motifs ($K_{259/353}$IGS) can be protective through the inhibition of phosphorylation of a nearby serine that otherwise would promote aggregation [23]. In addition, observations from in vitro experiments are contradictory: Cohen et al. report that acetylation accelerates Tau heparin-induced fibrillation [20], whereas others indicate that acetylation inhibits Tau filament assembly [19,23].

Liquid-liquid phase separation (LLPS) has recently gained attention as a physical mechanism for proteins to self-assemble into compartments termed membrane-less organelles [25–28]. The LLPS-mediated enrichment of proteins into membrane-less organelles, such as stress granules [29], provides "hotbeds" or seeds for protein aggregation [30–32]. LLPS was shown to initiate the aggregation of several neurodegenerative disease-associated proteins, such as FUS [30], TDP-43 [31,33], and hnRNPA1 [34]. Recently, Tau LLPS has been implicated in both the functional role of Tau in promoting microtubule assembly [35] and dysfunction in initiating Tau self-interaction and fibrillation [36–38]. Although it has been shown that hyperphosphorylation accelerates Tau LLPS and aggregation, consistent with hyperphosphorylated Tau's abundance in pathological inclusions [36,37], there are no reports on the role of acetylation on Tau phase separation and LLPS-mediated aggregation. Since Tau LLPS is expected to be strongly influenced by electrostatics, here, we investigate the role of acetylation in driving LLPS and determine if this role is consistent with the current hypothesis that LLPS can initiate and mediate Tau aggregation.

2. Results and Discussion

2.1. p300-Mediated Acetylation of Tau

Hyperacetylated Tau (Ac-Tau) was prepared from wt Tau using p300 HAT (see Section 3). Tau acetylation was verified by Western blot against acetyl-lysines (Figure 1B) and mass spectrometry (Figure 1C–E). We identified 15 acetylation sites (99% sequence coverage) using tandem mass spectrometry (Figure 1C; Section 3), including K148 near the Tau N-terminal domain; K163, K174, K190, K224, K234, and K240 in the P1–2 regions; K254, K280, K281, K290, and K311 in the microtubule binding region (MTBR, R1–R4); and, K375, K385, and K395 in the P3 region (Figure 1A,C). Figure 1D,E shows the representative MS/MS spectra (VQIINK$_{280}$K$_{281}$ and VQIVYK$_{311}$, respectively). Fragmented b (red) and y (blue) ions from low energy collisions in mass spectrometer are marked. The b and y ions refer to the peaks corresponding to the prefix ions observed sequentially in the spectrum with each prefix offset from the previous by the mass of an amino acid. A 42.016 Da increase in mass difference due to lysine acetylation is included in the sequential mass difference to construct the peptide sequence. The acetylation sites that were identified are consistent with previous reports [19–21]. Notably, K280/K281 and K311, which are in the hexapeptide aggregation motifs VQIINK$_{280}$K$_{281}$ and VQIVYK$_{311}$, were found to be acetylated (Figure 1D,E). These motifs play critical roles in Tau interaction with negatively-charged microtubules and with polyanions such as heparin [9]. If we assume complete acetylation, we expect the theoretical pI for full-length Tau to change from 8.2 to 5.5 (Prot pi web tool; https://www.protpi.ch). Such modifications can significantly alter Tau electrostatic properties, with 50% acetylation already corresponding to a pI of 6.2. Interestingly, the observed sites of acetylation are concentrated in the positively-charged central region of Tau (Figure 1A).

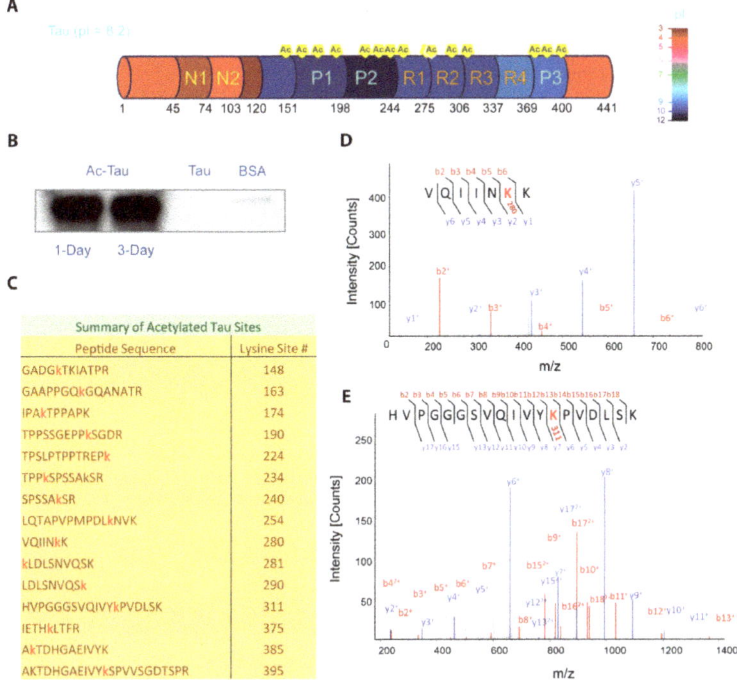

Summary of Acetylated Tau Sites	
Peptide Sequence	Lysine Site #
GADGkTKIATPR	148
GAAPPGQkGQANATR	163
IPAkTPPAPK	174
TPPSSGEPPkSGDR	190
TPSLPTPPTREPk	224
TPPkSPSSAkSR	234
SPSSAkSR	240
LQTAPVPMPDLkNVK	254
VQIINkK	280
kLDLSNVQSK	281
LDLSNVQSk	290
HVPGGGSVQIVYkPVDLSK	311
IETHkLTFR	375
AkTDHGAEIVYK	385
AKTDHGAEIVYkSPVVSGDTSPR	395

Figure 1. Tau hyperacetylation. (**A**) Domain organization of Tau. Protein segments are color-coded to reflect their respective pIs using the provided color palette. (**B**) In vitro Tau acetylation was verified by Western blot against acetyl-lysine (left to right lanes: acetylated Tau after one- and three-day reaction incubations, and negative controls using wild-type Tau and BSA, respectively). (**C**) Tau lysine acetylation sites (red) identified by mass spectrometry. (**D,E**) MS/MS spectra of peptides that contain the hexapeptide aggregation motifs VQIINK$_{280}$K$_{281}$ and VQIVYK$_{311}$, showing acetylation at K280 and K311, respectively (shown in red). The 'b' ions (shown in red) represent fragment peaks generated from the amino to carboxyl terminus. The 'y' ions (shown in blue) represent fragment peaks generated from the carboxyl to amino terminus. The suffix numbers represent the corresponding number of amino acids. See Section 3 for details.

2.2. Acetylation Changes Tau Phase Behavior

LLPS has recently been observed for wt and hyperphosphorylated Tau, and truncation mutant (K18) [35–38]. At low salt conditions (5 mM sodium phosphate, pH 7.8), we observed near-instantaneous formation of wt Tau droplets (Figure 2A). Subsequent fusions indicate the liquid nature of the wt Tau droplets. We characterized the protein concentration (2.5–20 µM) and salt concentration (0–250 mM NaCl) dependencies of wt Tau LLPS. Consistent with the literature [37,38], higher salt concentrations disfavor LLPS and higher protein concentrations favor LLPS (Figure 2E). For the case of Ac-Tau, we observed a dramatic reduction in droplet formation (Figure 2B,F). Similar results were also observed when LLPS experiments were performed with wt Tau or Ac-Tau in the presence of a crowding agent (10% PEG 8K, 200 mM NaCl, 10 mM acetate, 10 mM glycine, 10 mM sodium phosphate, pH 7.5; Figure 2C,D). Thus, independent of the presence or absence of crowding, the hyperacetylation of Tau disfavors LLPS.

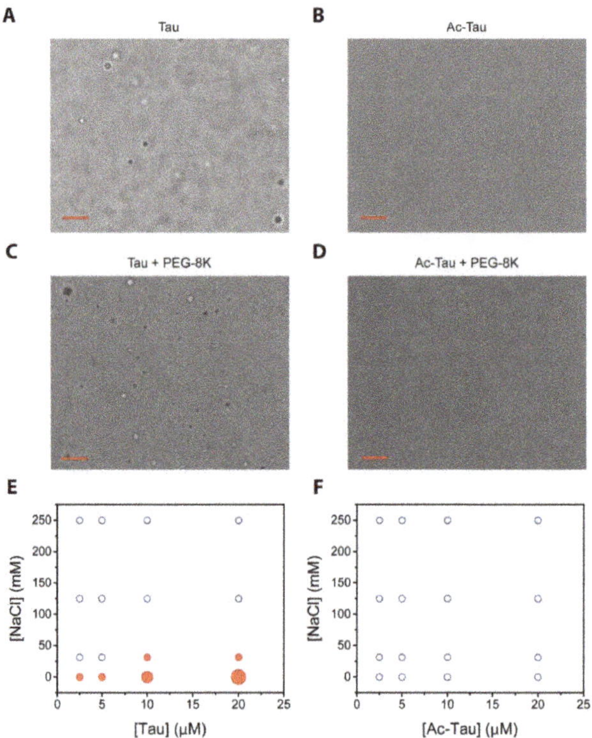

Figure 2. Acetylation disfavors Tau liquid-liquid phase separation (LLPS). Droplet formation of wt Tau versus hyperacetylated (Ac-Tau), respectively, in the absence (**A,B**) and presence of the crowding agent PEG-8K (**C,D**). Phase transition maps of Tau versus Ac-Tau (in the absence of crowders), respectively (**E,F**). The diagrams present protein and salt concentration dependencies of droplet formation. Blue open circles represent conditions of minimal droplet formation (<5 droplets/frame); small red solid circles represent 5–20 droplets, medium red solid circle 20–100 droplets, and large red circle >100 droplets. The bars in (**A–D**) represent 10 μm. See Section 3 for details.

Interestingly, even though both hyperphosphorylation and hyperacetylation decrease the overall pI of Tau, the two PTMs seem to have opposite effects on LLPS. In contrast to LLPS enhancement by hyperphosphorylation [36,37], hyperacetylation clearly disfavors Tau LLPS (Figure 2). Further experiments performed in identical or comparable conditions using the same Tau constructs, full-length or otherwise, are needed for a clear and direct comparison of LLPS behaviors of hyperphosphorylated, hyperacetylated, and wt Tau proteins. Nevertheless, we think that hyperacetylation disfavors full-length Tau LLPS by neutralizing the lysine positive charges, thereby affecting opposite-charge attractions that help support Tau self- and mesoscale interactions. Our data also give direct support that electrostatics plays a major role in Tau LLPS.

2.3. Acetylation of Tau Inhibits Heparin-Induced Aggregation

Heparin has been widely used to induce and accelerate Tau aggregation [11]. Utilizing a truncated Tau construct, Ambadipudi et al. demonstrated that heparin promotes Tau fibrillation via LLPS [37]. Similarly, we observed that heparin induces LLPS of full-length wt Tau and facilitates subsequent protein aggregation (Figure 3A–C,F). In contrast, Ac-Tau failed to undergo heparin-induced LLPS in the same experimental conditions (Figure 3D). Additionally, Ac-Tau (relative to wt Tau) exhibited a dramatic decrease in the fibrillation rate as reported by Th T fluorescence (Figure 3F). Residues in the $VYINK_{280}K_{281}$ and $VQIVK_{311}$ regions of the Tau microtubule binding repeats (R1–R4), which we identified as Tau acetylation sites (Figure 1), are also known interaction sites for heparin [39]. Thus, the observed effects of acetylation on Tau heparin-induced aggregation can be attributed to the loss of binding to heparin.

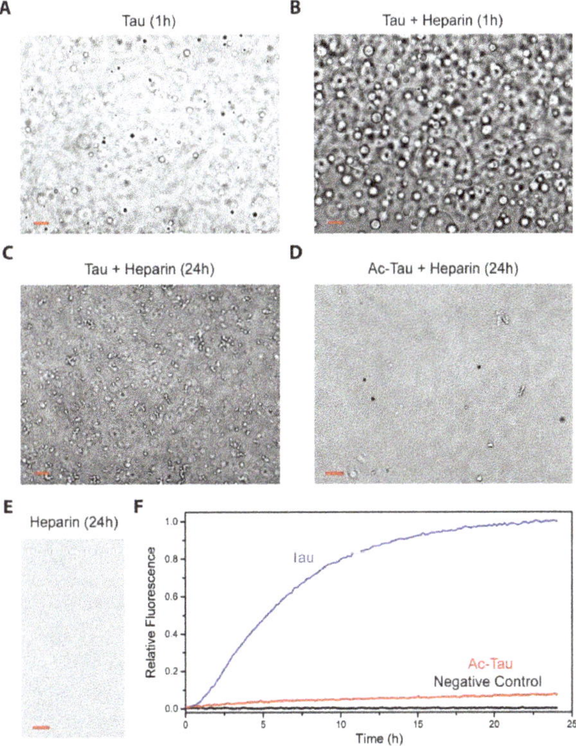

Figure 3. Acetylation disfavors heparin-induced Tau aggregation. (**A,B**) Heparin accelerates non-acetylated wt Tau LLPS. The presence of heparin (5 μM heparin: 20 μM Tau ratio) results in more droplets and larger fused droplets at the bottom of the dish. Droplet formation was not observed for Ac-Tau in the absence (Figure 2B,D) or presence of heparin (24 h, (**D**)). After 24 h incubation, an abundance of irregularly-shaped oligomers/aggregates was observed for wt Tau but not for Ac-Tau (**C,D**). No aggregation/LLPS was observed for the heparin control (**E**). Ac-Tau displayed minimal aggregation compared to wt Tau as reported by Th T fluorescence assay (**F**). Scale bars represent 10 μm. See Section 3 for details.

Although heparin accelerates wt Tau LLPS, it is unknown whether heparin is equally distributed in the Tau-rich and Tau-poor phases (which we think to be unlikely). LLPS, nevertheless, allows Tau to co-localize and thereby concentrate, with the Tau-rich condensed phase facilitating Tau aggregation nucleation and/or seeding.

2.4. Acetylation of Tau Prevents Access to LLPS-Mediated Microtubule Assembly

A recent report by Hernandez-Vega et al. suggests that Tau phase separated droplets (induced using the crowding agents PEG, Ficoll or dextran) can initiate microtubule assembly [35]. To assess LLPS-mediated microtubule assembly by wt Tau and Ac-Tau independent of crowding agents, we performed our phase separation experiments in low-salt conditions. After mixing rhodamine-labeled and unlabeled tubulin heterodimers with wt Tau, we observed an initial increase in solution turbidity. The ensuing dynamic microtubule assembly was visible by fluorescence microscopy within 1 h of incubation (Figure 4A). In contrast, Ac-Tau neither displayed turbidity nor detectable microtubule assembly up to 18 h of incubation (Figure 4B). Whereas previous studies have shown that acetylation reduces Tau's ability to bind to microtubules [20], our data clearly demonstrates that the failure of Ac-Tau to undergo LLPS affects its potential for microtubule assembly.

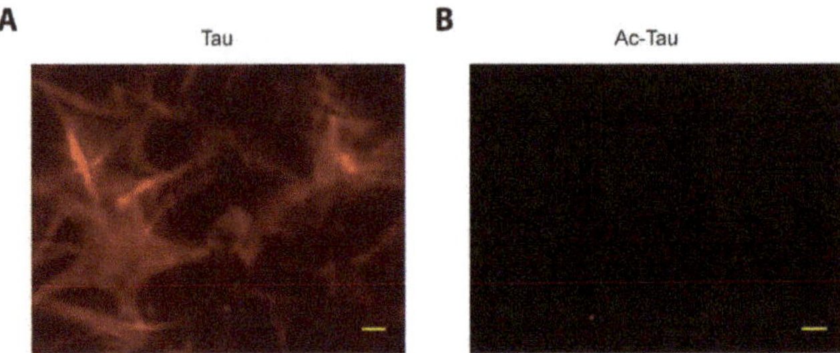

Figure 4. Tau acetylation prevents access to LLPS-mediated microtubule assembly. (**A,B**) Fluorescence microscopy images showing microtubule assembly from mixtures of rhodamine-labeled and unlabeled tubulin heterodimers with Tau or Ac-Tau, respectively. Scale bars represent 10 μm. See Section 3 for details.

Our in vitro data indicate that Ac-Tau is less prone to aggregation as compared to wt Tau. Cryo-EM structures of AD patient-derived filaments indicate that Tau residues 306–378 form the amyloid core [40]. The stable core is composed of several β-strands that pack intra- and inter-molecularly, with β1 ($_{306}$VYINK$_{311}$) in close proximity to β8 [40]. Our results show that in Ac-Tau, K311 (β1), and K375 (β8) are both acetylated; we speculate that this influences interactions within the amyloid core, and contributes to inhibition of Tau aggregation. Further experiments on the acetylation of the amyloid core residues will be needed to directly assess the effect of Tau acetylation on the amyloid structure.

Recent reports suggest that Tau aggregation is accelerated through LLPS [36,37]. Our data clearly show that acetylation decreases or abolishes Tau LLPS. Our findings are consistent with an LLPS-mediated model of aggregation (but do not prove whether such a mechanism is operative in vivo). Since acetylation reduces the propensity of Tau to undergo LLPS, we conclude that acetylation in vivo is unlikely to enhance or lead directly to condensation-mediated aggregation, in contrast to the demonstrated effect of hyperphosphorylation [36]. It is, however, possible that combinations of phosphorylations and acetylations can favor LLPS and/or aggregation; future experiments with Tau bearing homogeneous PTMs will be needed to address this conclusively.

Tau participates in microtubule formation and stabilization, and Tau LLPS has been shown as a mechanism by which a Tau-rich condensed phase can recruit tubulin dimers and facilitate their assembly [20]. Acetylation at key Tau sites that interfere with tubulin binding would affect this function, as would acetylation that disfavors partitioning of Tau into a Tau-rich phase. Thus, we speculate that the primary contribution of Tau acetylation to cellular dysfunction is not through a gain-of-function mechanism, such as toxic aggregation, but through a loss of physiologic function mechanism (i.e., reduced binding to tubulins/microtubules, and decreased LLPS-mediated initiation of microtubule assembly; Figure 5).

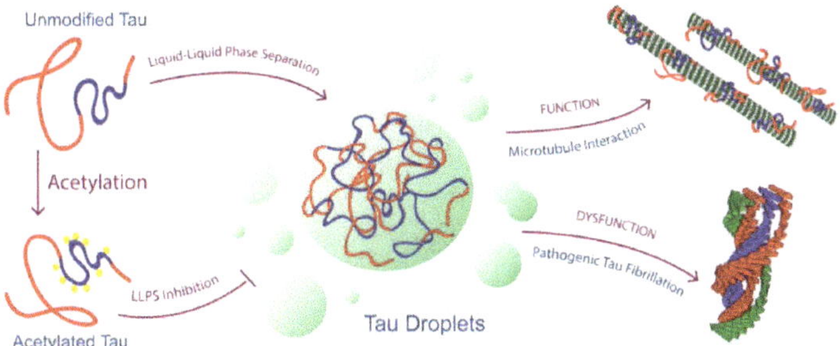

Figure 5. Model for Tau's loss of physiologic function and gain of pathologic dysfunction linked to its ability to undergo LLPS as modulated by acetylation.

Less direct effects on physiologic Tau function may also be important. Many of the same lysines (K254, K311, and K353) implicated as sites of ubiquitination [41] are also sites of acetylation and, thus, might be involved in evading the ubiquitin-lysosome proteasomal degradation machinery. Acetylation has been shown to inhibit Tau degradation by inhibiting its ubiquitination [21], and results in the accumulation of Tau, including hyperphosphorylated Tau. The presence of lysine deacetylase (SIRT1) has been shown to inhibit neuronal loss in an AD mouse model and deletion of SIRT1 results to pathologic levels of Tau in vivo [42]. Cross-talk between the different PTMs has also been reported. For example, hypoacetylation of Tau at key KIGS motifs in the *R1-4* regions increases vulnerability to hyperphosphorylation, which leads to filament aggregation [23]. Hyperphosphorylation of Tau has been reported to enhance Tau LLPS. However, other reports also show that hyperphosphorylation reduces microtubule assembly [14]. Thus, LLPS-mediated mechanisms by hyperphosphorylated Tau could be detrimental for both function and dysfunction pathways (Figure 5). We plan to carry out further experiments on hyperphosphorylated Tau to assess how this PTM of Tau can modulate microtubule assembly and protein aggregation, both in LLPS and non-LLPS conditions. Nevertheless, we speculate that the hyperacetylation of Tau is detrimental to Tau function, but not instrumental to LLPS-mediated Tau dysfunction (Figure 5). It would also be interesting to know the cross-talks between hyperphosphorylation and hyperacetylation in LLPS-mediated microtubule assembly and promotion of pathologic fibrils. Can hyperphosphorylated Tau also recruit hyperacetylated Tau into droplets? If so, this might explain the presence of hyperacetylated Tau in pathological inclusions of hyperphosphorylated Tau.

In conclusion, our data affirm the importance of electrostatics in Tau LLPS. Furthermore, we show that hyperacetylation disfavors Tau LLPS and, as a consequence, LLPS-facilitated aggregation. Finally, by preventing access to LLPS-mediated microtubule assembly and stabilization, hyperacetylation contributes to Tau dysfunction primarily through a loss-of-function mechanism.

3. Materials and Methods

3.1. Tau Expression and Purification

Wild-type (wt) Tau (2N4R isoform; 441 residues) plasmid (Addgene plasmid #16316, a gift from Peter Klein) was transformed into *Escherichia coli* BL21 star cells. Cells were grown at 37 °C in Terrific

Broth medium in the presence of kanamycin until the optical density at 600 nm (OD_{600}) reaches 0.8–1.0, then induced with 1 mM isopropyl β-D-1-thiogalactopyranoside (IPTG) and grown overnight at 18 °C.

wt Tau was purified using a similar procedure described by Barghorn et al. [43]. Briefly, wt Tau cell pellets were resuspended in 50 mM NaCl, 5 mM DTT, 50 mM sodium phosphate, pH 6.5, and supplemented with a protease inhibitor cocktail (GenDEPOT, Barker, TX, USA). The cells were lysed using a homogenizer (Avestin, Ottawa, ON, Canada). Additional salt was then added (for a final concentration of 450 mM NaCl) before the solution was incubated for 20 min in hot water (~80–90 °C). The supernatant was concentrated, diluted to a final salt concentration of 50 mM NaCl, and purified by FPLC (Bio-Rad, Hercules, CA, USA) using a salt gradient applied to a heparin sepharose HP column (GE, Marlborough, MA, USA). Fractions containing wt Tau were concentrated and further purified by reverse-phase HPLC (Agilent, Santa Clara, CA, USA), lyophilized, and stored at −80 °C until later use. Purified acetylated Tau (Ac-Tau) was prepared using reverse-phase HPLC after in vitro acetylation of wt Tau (see below).

3.2. p300 Histone Acetyltransferase (HAT) Domain Expression and Purification

Enzymatically-active p300 HAT was prepared as previously described [44]. Briefly, p300 HAT and Sir2 expression plasmids (generous gifts from Phillip Cole) were co-transformed into *E. coli* BL21 AI cells (Invitrogen, Carlsbad, CA, USA). Cells were grown at 37 °C in Terrific Broth medium until induction ($OD_{600} \approx 0.8$–1.0) with 1 mM IPTG, followed by overnight growth at 18 °C. Both proteins were purified using FPLC (Bio-Rad) with a Talon cobalt resin (GE) and a Q HP sepharose column (GE). Separate p300 HAT and Sir2 fractions were stored in −80 °C until later use. The final storage buffer for p300 HAT is ~150 mM NaCl, 125 mM TCEP, 25% (*v/v*) glycerol, 20 mM Tris, pH 8.

3.3. In Vitro p300 HAT-Mediated Acetylation Reactions

Acetylation of wt Tau by p300 HAT was performed by combining 500 μL of 86 μM purified wt Tau (dissolved in water), 200 μL of 15 μM p300 HAT, 25 μL of 10 mM acetyl-CoA (Sigma, Saint Louis, MO, USA), and 25 μL of 1 M Tris, pH 8. The acetylation reaction was allowed to proceed for three days at RT (unless stated otherwise).

3.4. Western Blot of Acetylated Tau

Tau acetylation was verified by western blot against acetyl-lysine. 100-ng samples of Ac-Tau, wt Tau and BSA were loaded on a 4–20% gradient SDS-PAGE gel (Mini-PROTEAN TGX Precast Gels, Bio-Rad). After electrophoresis, the gel was transferred to a polyvinylidene fluoride (PVDF) membrane using the Trans-Blot Turbo Transfer System, following the manufacturer's protocols (Bio-Rad). After incubation with 5% (*w/v*) nonfat milk in TBS-T (150 mM NaCl, 0.1% Tween-20, 20 mM Tris-HCl, pH 7.5) at RT for 2 h, the membrane was incubated with antibody against acetyl-lysine (1:100 in 1% nonfat milk/TBS-T; sc-32268, Santa Cruz Biotech, Dallas, TX, USA) overnight at 4 °C. The membrane was washed six times for 10 min with TBS-T and incubated with HRP-conjugated anti-mouse antibody (1:1000; #7076, Cell Signaling, Danvers, MA, USA) at RT for 30 min. The membrane was washed six times and developed with Clarity Western ECL Substrate according to the manufacturer's protocols (Bio-Rad). Chemiluminescent signals were measured using ChemiDoc MP Image System (Bio-Rad).

3.5. Mass Spectrometry of Acetylated Tau

Ac-Tau sample was boiled in 30 μL of 1× NuPAGE LDS sample buffer (Invitrogen) and subjected to SDS-PAGE (NuPAGE 10% Bis-Tris gel, Invitrogen) then visualized with Coomassie Brilliant blue-stain. The SDS-PAGE gel containing the band corresponding to Tau was excised, destained, and subjected to in-gel digestion using 100 ng trypsin (#T9600, GenDepot). The digested peptides were resuspended in 10 μL of 0.1% formic acid and subjected to a nanoHPLC-MS/MS system with an EASY-nLC 1200 coupled to Fusion Tribrid Orbitrap Lumos mass spectrometer (Thermo Fisher, Waltham, MA, USA). The peptides were loaded onto a Reprosil-Pur Basic C18 (1.9 μm, Dr. Maisch

GmbH, Ammerbuch-Entringen, Germany) pre-column of 2 cm × 100 μm size. The pre-column was switched in-line with an in-housed 50 mm × 150 μm analytical column packed with Reprosil-Pur Basic C18 equilibrated in 0.1% formic acid. The peptides were eluted using a 45-min discontinuous gradient of 4–28% acetonitrile/0.1% formic acid at a flow rate of 750 nL/min. The eluted peptides were directly electro-sprayed into mass spectrometer operated in the data-dependent acquisition mode acquiring fragmentation spectra of the top 30 strongest ions under direct control of Xcalibur software (4.0; Thermo Fisher). Parent MS spectrum was acquired in the Orbitrap with full MS range of 300–1400 m/z in the resolution of 120,000. CID fragmented MS/MS spectrum was acquired in ion-trap with rapid scan mode. Obtained MS/MS spectra were searched against the target-decoy human refseq database (June 2015 release, containing 73,637 entries) in Proteome Discoverer 1.4 interface (Thermo Fisher) with the Mascot algorithm (Mascot 2.4, Matrix Science, London, UK). Variable modifications of lysine and arginine acetylation, methionine oxidation, and N-terminal acetylation were allowed. The precursor mass tolerance was confined within 20 ppm with fragment mass tolerance of 0.5 Da and with a maximum of two missed cleavages allowed. The peptides identified in the Mascot results file were validated with a 5% false discover rate (FDR) and subjected to manual verification to confirm lysine acetylation.

3.6. Microscopy Imaging

Liquid-liquid phase separation (LLPS) experiments were performed using variable protein (wt Tau and Ac-Tau; 2.5–20 μM) and salt (0–250 mM NaCl) concentrations in 5 mM sodium phosphate buffer, pH 7.8. 10–20-μL drops were pipetted onto 35-mm glass bottom dishes (ibidi, Martinsried, Germany) and immediately monitored for droplet formation (with incubation time of 5 min). For heparin-induced LLPS experiments, heparin (8–25 kDa; Santa Cruz Biotech) was mixed with wt Tau or Ac-Tau at approximately 1:4 molar ratio (heparin:protein). LLPS experiments in the presence of crowding agent were carried out using 30 μM wt Tau or Ac-Tau in 10% PEG-8K, αβγ buffer (10 mM glycine, 10 mM acetate, 10 mM sodium phosphate, pH 7.5). Microscopy images were recorded at RT using an FL EVOS imaging system (Invitrogen).

3.7. Thioflavin T (Th T) Aggregation Assay

Heparin-induced wt Tau and Ac-Tau aggregation were detected following changes in Th T fluorescence using a Biotek Synergy H1 plate reader, employing 440 nm excitation and 480 nm emission wavelengths. Protein aggregation reactions were conducted using 20 μM wt Tau or Ac-Tau in 5 μM heparin, 10 μM Th T (GenDepot), 0.25 mM TCEP, 5 mM sodium phosphate, pH 7.8. Aggregation kinetics were monitored for ~24 h.

3.8. Microtubule Assembly Assay

The ability of wt Tau and Ac-Tau to promote microtubule assembly was investigated using fluorescence imaging. Rhodamine-labeled and unlabeled tubulin heterodimers (1:20 ratio; Cytoskeleton, Inc., Denver, CO, USA) were mixed with wt Tau or Ac-Tau (9 μM tubulin heterodimers and 27.5 μM Tau) in a final buffer condition of 0.2 mM $MgCl_2$, 0.1 mM GTP, 50 μM EDTA, 9 mM sodium PIPES, pH 6.9. Microtubule formation was visually monitored using an FL EVOS fluorescence microscope (Invitrogen) starting from 10 min up to 18 h.

Author Contributions: J.C.F. and A.C.F. designed the experiments. J.C.F., K.J.C., and P.S.T. performed the microscopy imaging and molecular biology experiments. A.J. and S.Y.J. acquired and analyzed the mass spectrometry data. J.C.F., K.R.M., and A.C.F. wrote the manuscript.

Funding: This work was supported by laboratory startup funds from Baylor College of Medicine (J.C.F. and A.C.F.).

Acknowledgments: We thank Jin Wang for the use of the Biotek Synergy H1 instrument.

Conflicts of Interest: The authors declare no competing financial interests.

Abbreviations

Ac-Tau	Hyperacetylated Tau
AD	Alzheimer's disease
IDP	Intrinsically disordered protein
HAT	Histone acetyltransferase
LLPS	Liquid-liquid phase separation
PTMs	Post-translational modifications
wt	Wild-type

References

1. Buee, L.; Bussiere, T.; Buee-Scherrer, V.; Delacourte, A.; Hof, P.R. Tau protein isoforms, phosphorylation and role in neurodegenerative disorders. *Brain Res. Brain Res. Rev.* **2000**, *33*, 95–130. [CrossRef]
2. Dickson, D.W. Tau and synuclein and their role in neuropathology. *Brain Pathol.* **1999**, *9*, 657–661. [CrossRef] [PubMed]
3. Holtzman, D.M.; Morris, J.C.; Goate, A.M. Alzheimer's disease: The challenge of the second century. *Sci. Transl. Med.* **2011**, *3*, 77sr71. [CrossRef] [PubMed]
4. Ghetti, B.; Oblak, A.L.; Boeve, B.F.; Johnson, K.A.; Dickerson, B.C.; Goedert, M. Invited review: Frontotemporal dementia caused by microtubule-associated protein tau gene (MAPT) mutations: A chameleon for neuropathology and neuroimaging. *Neuropathol. Appl. Neurobiol.* **2015**, *41*, 24–46. [CrossRef] [PubMed]
5. Trojanowski, J.Q.; Lee, V.M. Pathological tau: A loss of normal function or a gain in toxicity? *Nat. Neurosci.* **2005**, *8*, 1136–1137. [CrossRef] [PubMed]
6. Feinstein, S.C.; Wilson, L. Inability of tau to properly regulate neuronal microtubule dynamics: A loss-of-function mechanism by which tau might mediate neuronal cell death. *Biochim. Biophys. Acta* **2005**, *1739*, 268–279. [CrossRef] [PubMed]
7. Iqbal, K.; Alonso Adel, C.; Chen, S.; Chohan, M.O.; El-Akkad, E.; Gong, C.X.; Khatoon, S.; Li, B.; Liu, F.; Rahman, A.; et al. Tau pathology in alzheimer disease and other tauopathies. *Biochim. Biophys. Acta* **2005**, *1739*, 198–210. [CrossRef] [PubMed]
8. Li, W.; Lee, V.M. Characterization of two VQIXXK motifs for tau fibrillization in vitro. *Biochemistry* **2006**, *45*, 15692–15701. [CrossRef] [PubMed]
9. Mukrasch, M.D.; Bibow, S.; Korukottu, J.; Jeganathan, S.; Biernat, J.; Griesinger, C.; Mandelkow, E.; Zweckstetter, M. Structural polymorphism of 441-residue tau at single residue resolution. *PLoS Biol.* **2009**, *7*, e34. [CrossRef] [PubMed]
10. Von Bergen, M.; Friedhoff, P.; Biernat, J.; Heberle, J.; Mandelkow, E.M.; Mandelkow, E. Assembly of tau protein into alzheimer paired helical filaments depends on a local sequence motif ((306)VQIVYK(311)) forming beta structure. *Proc. Natl. Acad. Sci. USA* **2000**, *97*, 5129–5134. [CrossRef] [PubMed]
11. Ramachandran, G.; Udgaonkar, J.B. Mechanistic studies unravel the complexity inherent in tau aggregation leading to alzheimer's disease and the tauopathies. *Biochemistry* **2013**, *52*, 4107–4126. [CrossRef] [PubMed]
12. Cripps, D.; Thomas, S.N.; Jeng, Y.; Yang, F.; Davies, P.; Yang, A.J. Alzheimer disease-specific conformation of hyperphosphorylated paired helical filament-tau is polyubiquitinated through Lys-48, Lys-11, and Lys-6 ubiquitin conjugation. *J. Biol. Chem.* **2006**, *281*, 10825–10838. [CrossRef] [PubMed]
13. Tai, H.C.; Serrano-Pozo, A.; Hashimoto, T.; Frosch, M.P.; Spires-Jones, T.L.; Hyman, B.T. The synaptic accumulation of hyperphosphorylated tau oligomers in alzheimer disease is associated with dysfunction of the ubiquitin-proteasome system. *Am. J. Pathol.* **2012**, *181*, 1426–1435. [CrossRef] [PubMed]
14. Tepper, K.; Biernat, J.; Kumar, S.; Wegmann, S.; Timm, T.; Hubschmann, S.; Redecke, L.; Mandelkow, E.M.; Muller, D.J.; Mandelkow, E. Oligomer formation of tau protein hyperphosphorylated in cells. *J. Biol. Chem.* **2014**, *289*, 34389–34407. [CrossRef] [PubMed]
15. Ferreon, A.C.; Moran, C.R.; Ferreon, J.C.; Deniz, A.A. Alteration of the alpha-synuclein folding landscape by a mutation related to parkinson's disease. *Angew. Chem. Int. Ed. Engl.* **2010**, *49*, 3469–3472. [CrossRef] [PubMed]
16. Ferreon, A.C.; Ferreon, J.C.; Wright, P.E.; Deniz, A.A. Modulation of allostery by protein intrinsic disorder. *Nature* **2013**, *498*, 390–394. [CrossRef] [PubMed]

17. Ferreon, J.C.; Lee, C.W.; Arai, M.; Martinez-Yamout, M.A.; Dyson, H.J.; Wright, P.E. Cooperative regulation of p53 by modulation of ternary complex formation with CBP/p300 and HDM2. *Proc. Natl. Acad. Sci. USA* **2009**, *106*, 6591–6596. [CrossRef] [PubMed]

18. Martin, L.; Latypova, X.; Terro, F. Post-translational modifications of tau protein: Implications for alzheimer's disease. *Neurochem. Int.* **2011**, *58*, 458–471. [CrossRef] [PubMed]

19. Kamah, A.; Huvent, I.; Cantrelle, F.X.; Qi, H.; Lippens, G.; Landrieu, I.; Smet-Nocca, C. Nuclear magnetic resonance analysis of the acetylation pattern of the neuronal tau protein. *Biochemistry* **2014**, *53*, 3020–3032. [CrossRef] [PubMed]

20. Cohen, T.J.; Guo, J.L.; Hurtado, D.E.; Kwong, L.K.; Mills, I.P.; Trojanowski, J.Q.; Lee, V.M. The acetylation of tau inhibits its function and promotes pathological tau aggregation. *Nat. Commun.* **2011**, *2*, 252. [CrossRef] [PubMed]

21. Min, S.W.; Cho, S.H.; Zhou, Y.; Schroeder, S.; Haroutunian, V.; Seeley, W.W.; Huang, E.J.; Shen, Y.; Masliah, E.; Mukherjee, C.; et al. Acetylation of tau inhibits its degradation and contributes to tauopathy. *Neuron* **2010**, *67*, 953–966. [CrossRef] [PubMed]

22. Cohen, T.J.; Friedmann, D.; Hwang, A.W.; Marmorstein, R.; Lee, V.M. The microtubule-associated tau protein has intrinsic acetyltransferase activity. *Nat. Struct. Mol. Biol.* **2013**, *20*, 756–762. [CrossRef] [PubMed]

23. Cook, C.; Carlomagno, Y.; Gendron, T.F.; Dunmore, J.; Scheffel, K.; Stetler, C.; Davis, M.; Dickson, D.; Jarpe, M.; DeTure, M.; et al. Acetylation of the kxgs motifs in tau is a critical determinant in modulation of tau aggregation and clearance. *Hum. Mol. Genet.* **2014**, *23*, 104–116. [CrossRef] [PubMed]

24. Min, S.W.; Chen, X.; Tracy, T.E.; Li, Y.; Zhou, Y.; Wang, C.; Shirakawa, K.; Minami, S.S.; Defensor, E.; Mok, S.A.; et al. Critical role of acetylation in tau-mediated neurodegeneration and cognitive deficits. *Nat. Med.* **2015**, *21*, 1154–1162. [CrossRef] [PubMed]

25. Brangwynne, C.P. Phase transitions and size scaling of membrane-less organelles. *J. Cell Biol.* **2013**, *203*, 875–881. [CrossRef] [PubMed]

26. Feric, M.; Vaidya, N.; Harmon, T.S.; Mitrea, D.M.; Zhu, L.; Richardson, T.M.; Kriwacki, R.W.; Pappu, R.V.; Brangwynne, C.P. Coexisting liquid phases underlie nucleolar subcompartments. *Cell* **2016**, *165*, 1686–1697. [CrossRef] [PubMed]

27. Kato, M.; Han, T.W.; Xie, S.; Shi, K.; Du, X.; Wu, L.C.; Mirzaei, H.; Goldsmith, E.J.; Longgood, J.; Pei, J.; et al. Cell-free formation of rna granules: Low complexity sequence domains form dynamic fibers within hydrogels. *Cell* **2012**, *149*, 753–767. [CrossRef] [PubMed]

28. Li, P.; Banjade, S.; Cheng, H.C.; Kim, S.; Chen, B.; Guo, L.; Llaguno, M.; Hollingsworth, J.V.; King, D.S.; Banani, S.F.; et al. Phase transitions in the assembly of multivalent signalling proteins. *Nature* **2012**, *483*, 336–340. [CrossRef] [PubMed]

29. Lin, Y.; Protter, D.S.; Rosen, M.K.; Parker, R. Formation and maturation of phase-separated liquid droplets by rna-binding proteins. *Mol. Cell* **2015**, *60*, 208–219. [CrossRef] [PubMed]

30. Patel, A.; Lee, H.O.; Jawerth, L.; Maharana, S.; Jahnel, M.; Hein, M.Y.; Stoynov, S.; Mahamid, J.; Saha, S.; Franzmann, T.M.; et al. A liquid-to-solid phase transition of the als protein fus accelerated by disease mutation. *Cell* **2015**, *162*, 1066–1077. [CrossRef] [PubMed]

31. Conicella, A.E.; Zerze, G.H.; Mittal, J.; Fawzi, N.L. Als mutations disrupt phase separation mediated by alpha-helical structure in the TDP-43 low-complexity C-terminal domain. *Structure* **2016**, *24*, 1537–1549. [CrossRef] [PubMed]

32. Li, Y.R.; King, O.D.; Shorter, J.; Gitler, A.D. Stress granules as crucibles of als pathogenesis. *J. Cell Biol.* **2013**, *201*, 361–372. [CrossRef] [PubMed]

33. Tsoi, P.S.; Choi, K.J.; Leonard, P.G.; Sizovs, A.; Moosa, M.M.; MacKenzie, K.R.; Ferreon, J.C.; Ferreon, A.C.M. The N-terminal domain of ALS-linked TDP-43 assembles without misfolding. *Angew. Chem. Int. Ed. Engl.* **2017**, *56*, 12590–12593. [CrossRef] [PubMed]

34. Molliex, A.; Temirov, J.; Lee, J.; Coughlin, M.; Kanagaraj, A.P.; Kim, H.J.; Mittag, T.; Taylor, J.P. Phase separation by low complexity domains promotes stress granule assembly and drives pathological fibrillization. *Cell* **2015**, *163*, 123–133. [CrossRef] [PubMed]

35. Hernandez-Vega, A.; Braun, M.; Scharrel, L.; Jahnel, M.; Wegmann, S.; Hyman, B.T.; Alberti, S.; Diez, S.; Hyman, A.A. Local nucleation of microtubule bundles through tubulin concentration into a condensed tau phase. *Cell Rep.* **2017**, *20*, 2304–2312. [CrossRef] [PubMed]

36. Wegmann, S.; Eftekharzadeh, B.; Tepper, K.; Zoltowska, K.M.; Bennett, R.E.; Dujardin, S.; Laskowski, P.R.; MacKenzie, D.; Kamath, T.; Commins, C.; et al. Tau protein liquid-liquid phase separation can initiate tau aggregation. *EMBO J.* **2018**, *37*, e98049. [CrossRef] [PubMed]

37. Ambadipudi, S.; Biernat, J.; Riedel, D.; Mandelkow, E.; Zweckstetter, M. Liquid-liquid phase separation of the microtubule-binding repeats of the alzheimer-related protein tau. *Nat. Commun.* **2017**, *8*, 275. [CrossRef] [PubMed]

38. Zhang, X.; Lin, Y.; Eschmann, N.A.; Zhou, H.; Rauch, J.N.; Hernandez, I.; Guzman, E.; Kosik, K.S.; Han, S. Rna stores tau reversibly in complex coacervates. *PLoS Biol.* **2017**, *15*, e2002183. [CrossRef] [PubMed]

39. Mukrasch, M.D.; Biernat, J.; von Bergen, M.; Griesinger, C.; Mandelkow, E.; Zweckstetter, M. Sites of tau important for aggregation populate {beta}-structure and bind to microtubules and polyanions. *J. Biol. Chem.* **2005**, *280*, 24978–24986. [CrossRef] [PubMed]

40. Fitzpatrick, A.W.P.; Falcon, B.; He, S.; Murzin, A.G.; Murshudov, G.; Garringer, H.J.; Crowther, R.A.; Ghetti, B.; Goedert, M.; Scheres, S.H.W. Cryo-em structures of tau filaments from alzheimer's disease. *Nature* **2017**, *547*, 185–190. [CrossRef] [PubMed]

41. Lee, M.J.; Lee, J.H.; Rubinsztein, D.C. Tau degradation: The ubiquitin-proteasome system versus the autophagy-lysosome system. *Prog. Neurobiol.* **2013**, *105*, 49–59. [CrossRef] [PubMed]

42. Kim, D.; Nguyen, M.D.; Dobbin, M.M.; Fischer, A.; Sananbenesi, F.; Rodgers, J.T.; Delalle, I.; Baur, J.A.; Sui, G.; Armour, S.M.; et al. Sirt1 deacetylase protects against neurodegeneration in models for alzheimer's disease and amyotrophic lateral sclerosis. *EMBO J.* **2007**, *26*, 3169–3179. [CrossRef] [PubMed]

43. Barghorn, S.; Davies, P.; Mandelkow, E. Tau paired helical filaments from alzheimer's disease brain and assembled in vitro are based on beta-structure in the core domain. *Biochemistry* **2004**, *43*, 1694–1703. [CrossRef] [PubMed]

44. Thompson, P.R.; Wang, D.; Wang, L.; Fulco, M.; Pediconi, N.; Zhang, D.; An, W.; Ge, Q.; Roeder, R.G.; Wong, J.; et al. Regulation of the p300 hat domain via a novel activation loop. *Nat. Struct. Mol. Biol.* **2004**, *11*, 308–315. [CrossRef] [PubMed]

 © 2018 by the authors. Licensee MDPI, Basel, Switzerland. This article is an open access article distributed under the terms and conditions of the Creative Commons Attribution (CC BY) license (http://creativecommons.org/licenses/by/4.0/).

 International Journal of *Molecular Sciences*

Article

Deciphering RNA-Recognition Patterns of Intrinsically Disordered Proteins

Ambuj Srivastava [1], Shandar Ahmad [2] and M. Michael Gromiha [1,*]

[1] Department of Biotechnology, Bhupat and Jyoti Mehta School of Biosciences,
    Indian Institute of Technology Madras, Chennai 600 036, Tamilnadu, India; ambuj.88.in@gmail.com
[2] School of Computational and Integrative Sciences, Jawaharlal Nehru University, New Delhi 110 067, India;
    shandar@jnu.ac.in
* Correspondence: gromiha@iitm.ac.in; Tel.: +91-44-2257-4138

Received: 12 April 2018; Accepted: 16 May 2018; Published: 29 May 2018

Abstract: Intrinsically disordered regions (IDRs) and protein (IDPs) are highly flexible owing to their lack of well-defined structures. A subset of such proteins interacts with various substrates; including RNA; frequently adopting regular structures in the final complex. In this work; we have analysed a dataset of protein–RNA complexes undergoing disorder-to-order transition (DOT) upon binding. We found that DOT regions are generally small in size (less than 3 residues) for RNA binding proteins. Like structured proteins; positively charged residues are found to interact with RNA molecules; indicating the dominance of electrostatic and cation-π interactions. However, a comparison of binding frequency shows that interface hydrophobic and aromatic residues have more interactions in only DOT regions than in a protein. Further; DOT regions have significantly higher exposure to water than their structured counterparts. Interactions of DOT regions with RNA increase the sheet formation with minor changes in helix forming residues. We have computed the interaction energy for amino acids–nucleotide pairs; which showed the preference of His–G; Asn–U and Ser–U at for the interface of DOT regions. This study provides insights to understand protein–RNA interactions and the results could also be used for developing a tool for identifying DOT regions in RNA binding proteins.

Keywords: intrinsically disorder proteins; disorder-to-order regions; protein–RNA interactions; unstructured proteins

---

1. Introduction

Intrinsically disordered proteins lack stable three-dimensional structures under physiological conditions and are known to perform important roles in several processes including signalling, enzymatic activity, and gene regulation [1,2]. To perform these functions, disordered regions interact with protein, RNA, DNA, and other small molecules to gain ordered structures [3,4]. Experimentally, interactions mediated by IDRs can be observed using NMR and X-ray crystallography. However, because of poor resolution, problems in crystallization, and high time and resource consumption, computational methods are necessary to identify disorder-mediated interactions [5,6].

Several methods have been developed for understanding the disorderness of proteins using sequence or structural information [7–10]. In addition, the transition of disorder-to-order regions in protein–protein interactions (PPI) is well studied experimentally and computationally [8–12]. For example, LMO4, a putative breast oncoprotein, interacts with various tandem LIM-domain containing proteins mediated by disordered regions [13]. BRCA1, a tumour suppressor protein, helps in binding with multiple protein and DNA partners by its central disorder region of ~1500 amino acids [14]. Recently, Papadakos et al. [15] showed that inducing intrinsic disorder in high-affinity protein–protein interactions reduces the affinity of binding.

Many proteins contain disordered regions and some of the regions attained ordered structures after binding to their cognate substrates, which are also known as MoRF (Molecular Recognition Features) segments [16,17]. Sugase et al. [18] have shown that folding and binding of IDPs or IDRs are coupled processes. Furthermore, binding partners are also shown to influence affinity and kinetics of binding. The flexibility of IDPs helps them to bind with multiple partners and have co-operative interactions [19]. Although induced fit and conformational selection processes are proposed explanations for the coupling of folding and binding, the exact model which is preferred by IDPs is not known [11,20].

The dynamics of the RNA molecule makes it more amenable to interact with disorder-mediated protein–RNA interactions [21]. The recognition of the protein–RNA complex has been experimentally studied using EMSA, yeast-3-hybrid assay, pull-down assay and CLIP [22,23]. On the other hand, plenty of tools have been developed to identify binding sites in RNA-binding proteins [24–32]. All these methods use the information in their sequence to compute the feature and/or evaluate the performance. Recently, Peng and Kurgan [33] developed a webserver for prediction of disorder-mediated interactions in RNA, DNA and protein–protein complexes. However, the knowledge for understanding the mechanisms or factors responsible for binding of disordered region with RNA has not yet been completely explored.

In this work, we constructed a dataset for protein–RNA complexes (provided in supplementary information), which are involved in disorder-to-order transitions. Utilizing the dataset, we analyzed the number and size of DOT regions in protein–RNA complexes, preference of residues involved in binding in DOT regions, secondary structure, solvent accessibility, pair preference at the interface, preference in different secondary structures of RNA, and interaction energy between protein and RNA DOT and non-DOT regions at the interface.

2. Results and Discussion

Our dataset contains a total of 23,452 and 2412 residues in non-ribosomal and ribosomal protein–RNA complexes. Among them, 1175 (5%) and 155 (6.4%) residues are found to be in DOT regions in non-ribosomal and ribosomal complexes, respectively. The residues binding with RNA are obtained by using 3.5 and 6 Å distance cut-offs and similar trends are obtained. Therefore, we have presented the results with 3.5 Å and those for 6 Å are shown in supplementary material.

2.1. Number of DOT Regions in Protein–RNA Complexes and Length of DOT Regions

The variation in the number of DOT regions in non-ribosomal and ribosomal complexes is shown in Figure 1. We observed that most of the complexes have less than three DOT regions (88% in non-ribosomal and 100% in ribosomal complexes). Most non-ribosomal proteins have one DOT region, whereas ribosomal proteins have mostly two or more DOT regions. In addition, at most eight DOT regions per complex are found in our dataset.

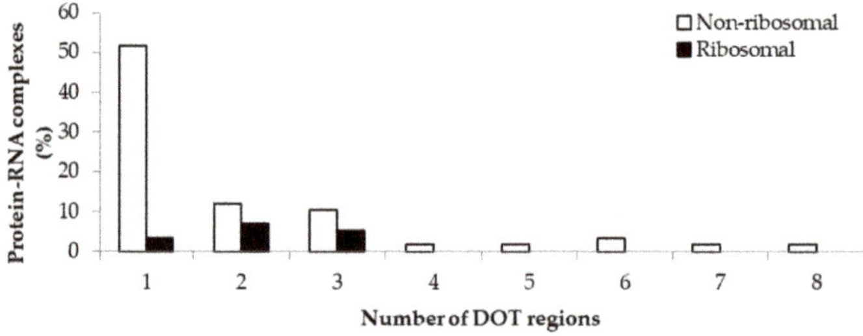

Figure 1. Percentage of protein–RNA complexes containing different number of DOT regions.

Further, we analysed the length of each DOT region in non-ribosomal and ribosomal protein–RNA complexes, which shows that most DOT regions are short, as shown in Figure 2. In both non-ribosomal and ribosomal complexes, more than 70% of DOT regions have three to 10 residues and very few (only 5) regions have a length of more than 50 residues. This leads to a speculation that only a small conformational change might be required for bringing shape complementarity in protein–RNA complexes and these small DOT regions help in obtaining the same.

a. Non-ribosomal complexes

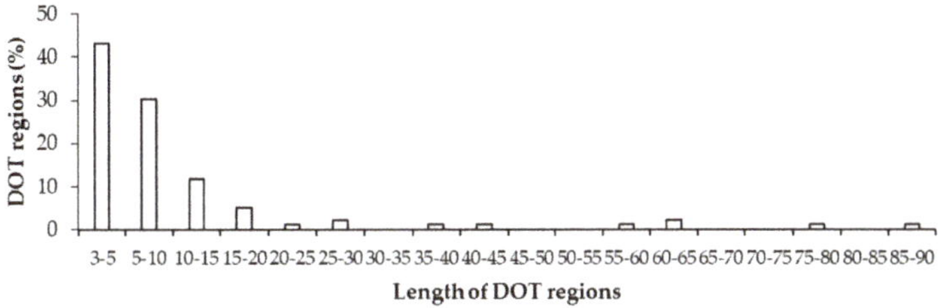

b. Ribosomal

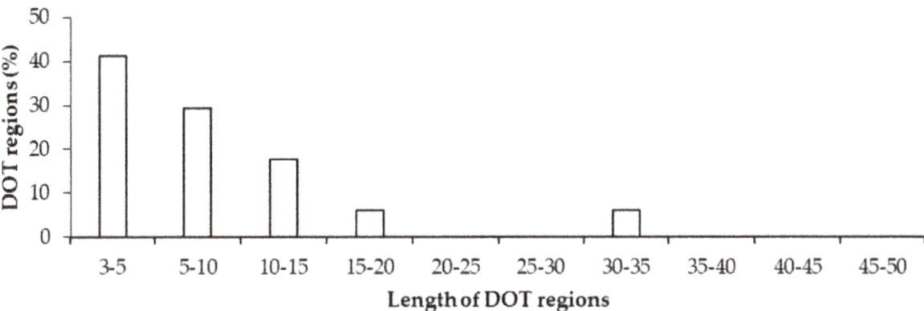

Figure 2. Length distribution of DOT regions in protein–RNA complexes in (**a**) non-ribosomal and (**b**) ribosomal complexes.

2.2. Binding Frequency of Residues at DOT Regions

The binding frequencies of residues in DOT regions using 3.5 Å (NR3.5 and RB3.5) and 6 Å (NR6 and RB6) distance cut-offs are shown in Figure 3 and Figure S2, respectively. We observed that among all positively charged residues (Arg, Lys and His), Arg and Lys have high preference for binding in both NR3.5 (Figure 3a) and NR6 (Figure S2a) datasets. Interestingly, only eight and 13 among 20 residues are observed in binding DOT regions at RB3.5 (Figure 3b) and RB6 (Figure S2b) datasets, and Arg has the highest frequency of binding. Cys, Met, and Trp in DOT regions are not involved in binding with RNA, whereas in ordered complexes 0.97%, 4.52%, and 5.54% of Cys, Met, and Trp are involved in binding, respectively. The comparison of binding site residues in DOT regions and the whole protein showed an expected presence of 1.5% and 2.7% of Met and Trp, respectively, in the interface of the DOT region. These results showed that the non-occurrence of Met and Trp at the interface of the DOT regions is statistically significant.

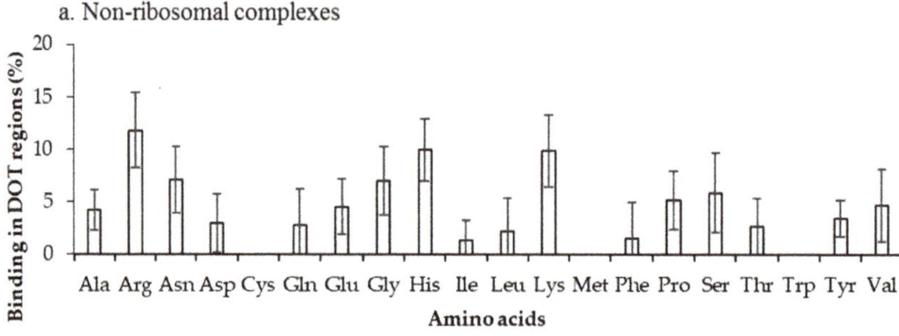

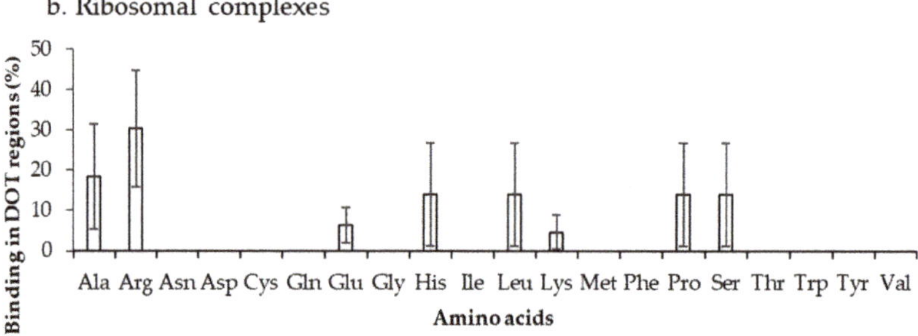

Figure 3. Amino acid frequency of binding in the DOT region for (**a**) non-ribosomal and (**b**) ribosomal protein–RNA complexes.

We have computed the preference of binding of residues in DOT regions by dividing the number of residues in DOT regions with the total number of binding residues, and the results are presented in Figure 4 and Figure S3 for 3.5 Å (NR3.5 and RB3.5 datasets) and 6 Å (NR6 and RB6 datasets), respectively. In Figure 4a, high frequency of Arg, Gly, Lys, and Ser (z-score > 1) is observed for the NR3.5 dataset, which suggests that these residues are more probable to contact DOT regions with respect to all residues in contact with RNA. However, for the NR6 dataset (Figure S3a), the result is only consistent for Lys, and two other residues (Glu and Pro) show high binding frequency. In ribosomal protein complexes with 3.5 Å and 6 Å, Ala & Glu, and Glu & Tyr have high frequencies, respectively (Figure 4b and Figure S3b).

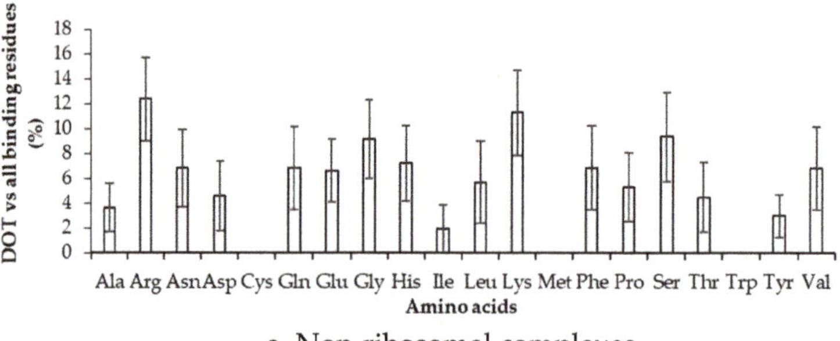

Figure 4. *Cont.*

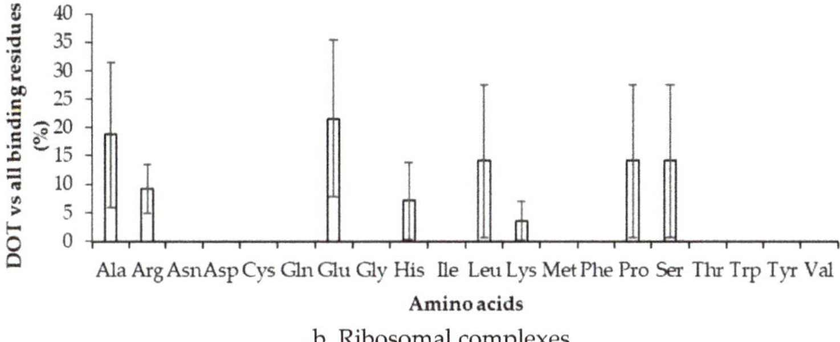

Figure 4. Frequency of DOT regions by contact residues for (**a**) non-ribosomal and (**b**) ribosomal complexes.

2.3. Binding Propensity of Residues at DOT Region

Propensity is calculated by normalizing the binding frequency of residues in DOT regions with the overall frequency of the respective residues to be in a protein, using Equation (3). This can measure the bias in binding of residues in DOT regions, independent of their count in DOT regions. We have calculated the propensity of amino acids to be in DOT regions using distance cut-offs of 3.5 Å and 6 Å and the results are shown in Figure 5 and Figure S4, respectively. In the NR3.5 (Figure 5a) dataset, His, Arg, Asn, Gln, Phe, and Tyr have high propensity of binding, whereas in ribosomal proteins (RB3.5 dataset; Figure 5b), only His showed a high propensity. In the NR6 (Figure S4a), His has high propensity, whereas Asn, His and Tyr have high propensity in the RB6 (Figure S4b) dataset. Similarly, high propensity for binding is observed for positively charged residues along with Tyr and Phe in protein–RNA complexes [34]. On the other hand, among all charged residues only Arg has high tendency to bind with DOT regions in protein–protein complexes [35]. Furthermore, non-specific interactions occurred frequently in protein–protein complexes, which is not a common trend in the binding residues of DOT regions in protein–RNA complexes. Therefore, we can infer that the preferred residues at DOT regions are specific in protein–RNA complexes and, especially, charged interactions are important in DOT regions for binding with RNA.

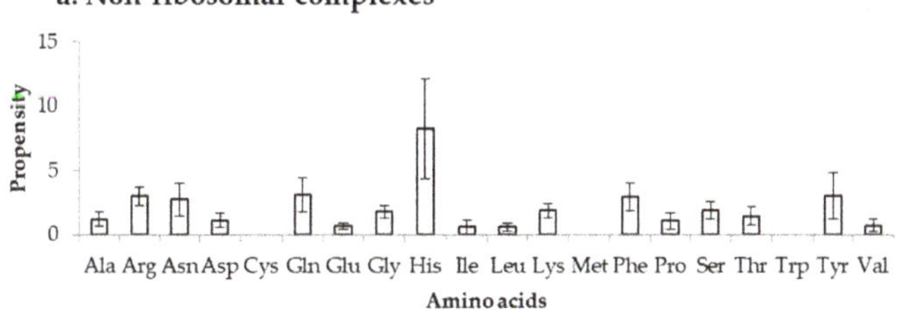

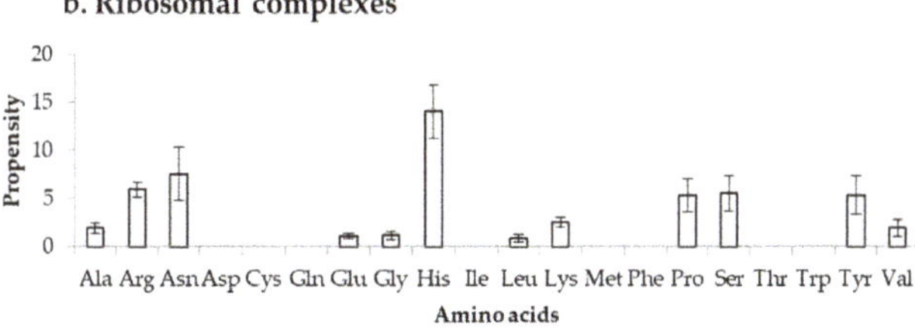

Figure 5. Propensity for amino acids in (**a**) non-ribosomal and (**b**) ribosomal complexes.

2.4. Comparison of Frequency of Binding in the DOT Region and Other Residues of a Protein

To estimate the difference between binding in DOT regions and other part of proteins, we calculated the binding frequency of amino acids in these regions, as shown in Figure 6 and Figure S5. Amino acids significantly differ in their binding with RNA in DOT regions and in the complete protein (p-value for the mean is less than 0.01). In non-ribosomal proteins, when the 3.5 Å cut-off is considered, nonpolar and aromatic residues mostly have high frequency values in the DOT regions than in the overall protein. All the frequencies are observed to be significant when statistical analysis is performed for the bootstrapped sample of the frequencies (p-value is less than 0.01). Residues such as His, Phe, and Leu are found to have a more than 3-fold increase in the frequency of binding in the DOT regions than in other parts of the proteins. A similar trend is observed in the NR6 dataset (Figure S5).

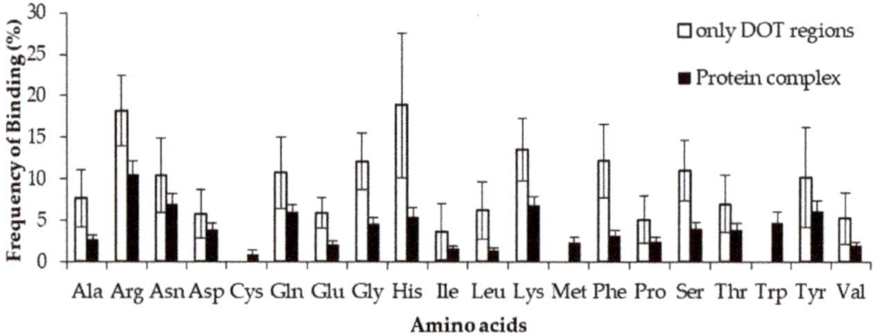

Figure 6. Binding frequency of each amino acid in the DOT region and in the overall protein for non-ribosomal complexes using the 3.5 Å cut-off.

2.5. Amino Acid Contact Frequency with Nucleotides

We have also analysed amino acid contacts with each nucleotide in non-ribosomal complexes using 3.5 Å and 6 Å distance cut-offs for contacting residues and the results are shown in Figure 7 and Figure S6. In the 3.5 Å distance criterion, Arg and Lys have a high frequency to bind with nucleotides. Arg and Lys are observed to have the most and least binding frequencies with Guanine and Uracil, respectively. Whereas in the 6 Å criterion, almost the same frequency of binding is observed for Arg and Lys with Adenine, Guanine and Cytosine nucleotides; least binding was observed in the Uracil nucleotide. When compared with the results presented for ordered protein–DNA and protein–RNA complexes in our earlier works, Arg, Lys, Trp, and Tyr were favoured by RNA and Arg was selected by DNA-binding proteins together with Guanine in DNA and Uracil in RNA–protein complexes [36].

Figure 7. Normalized amino acid nucleotide contact frequency in non-ribosomal protein–RNA complexes at 3.5 Å.

2.6. Secondary Structure of DOT and RNA-Interacting DOT Residues

The secondary structures of DOT residues are quantified to study the bias of residues to have a specific secondary structure in binding and non-binding regions and data are presented in Table 1 and Table S1 for NR3.5 and NR6 datasets, respectively. In the NR3.5 dataset, all the DOT residues have lower and higher preference in sheet (15%) and other structure class (8%), respectively. Interestingly, in DOT residues, binding with RNA molecules, strand-forming residues have a higher preference (15.2%) as compared to helical (8.6%) and other regions (8.9%).

Table 1. Secondary structure of all DOT residues and residues binding with RNA in DOT regions in the NR3.5 dataset.

Secondary Structure	Number of Binding Residues in DOT Regions (N_{idt})	Number of Residues in DOT Region (N_d)	Relative Binding in DOT Regions (%)
Helix	25 (22.12)	288 (24.51)	8.6
Sheet	22 (19.47)	145 (12.34)	15.2
Others (coil, turn, bend)	66 (58.41)	742 (63.15)	8.9

Percentage is mentioned in the parenthesis. Relative binding in DOT regions are calculated by $N_{idt}/N_d \times 100$.

2.7. Relative Solvent Accessibility of DOT Residues

The spatial arrangement of DOT residues is further explored by solvent accessibility calculation and the result is shown in Table 2. Comparison of RASA of DOT regions and complete protein–RNA complex revealed that in DOT regions, solvent accessibility of every amino acid is more than that of other amino acids of a protein. As expected, charged residues have low fold difference (1.18 to 1.28) in RASA in DOT regions and the complete protein. However, most hydrophobic residues (Ala, Cys, Ile, Leu, Met, Phe, Tyr, and Val) have about 1.8 to 2 folds higher RASA in DOT regions than the complete protein, Met has the highest difference. On the other hand, the mean solvent accessibility of DOT regions of proteins is 44 Å^2, which is similar to the average RASA of binding DOT regions (43 Å^2) of protein–protein complexes [17].

Table 2. Relative average solvent accessibility (RASA) of DOT residues and all residues in non-ribosomal protein–RNA complexes.

Amino Acids	RASA in DOT Regions	RASA in Complete Protein	Fold Difference
Ala	44.743	23.305	1.920
Arg	52.168	40.822	1.278
Asn	63.583	42.721	1.488
Asp	56.552	43.811	1.291
Cys	22.32	11.426	1.953
Gln	47.805	38.988	1.226
Glu	53.688	47.838	1.122
Gly	51.599	35.272	1.463
His	43.229	35.372	1.222
Ile	26.618	14.692	1.812
Leu	32.007	16.374	1.955
Lys	58.529	49.520	1.182
Met	41.13	20.391	2.017
Phe	32.481	17.519	1.854
Pro	56.463	38.230	1.477
Ser	54.574	34.622	1.576
Thr	49.852	31.074	1.604
Trp	21.571	19.029	1.134
Tyr	44.579	24.752	1.801
Val	31.433	17.405	1.806

2.8. Number of Residues in Contact with Nucleotides in the DOT Region and in Entire Protein

Among 1175 residues in DOT regions in our dataset, only 96 (8.17%) and 268 (22.81%) are in contact with nucleotides in the NR3.5 and the NR6 dataset, respectively. Almost all the residues have a similar tendency of binding with nucleotides in proteins, ranging between 20% to 29%, as shown in Table 3 and Table S2. However, the number of nucleotides interacting with DOT residues is somewhat different, that is, the range of interaction is 18 to 33%. The DOT residues are more likely to bind with Guanine (20.4%), followed by Cytosine and Uracil, than to binding with Adenine (13.1%).

Table 3. Number of interaction of nucleotides with DOT residues and with complete protein at 3.5 Å.

Nucleotides	Number of Nucleotide in Contact with DOT Regions (N_{idt})	Number of Nucleotides in Contact with Any Residue of Proteins (N_{prot})	Relative Contact in DOT Regions (%)
A	18 (18.75)	137 (25.66)	13.1
C	26 (27.08)	131 (24.53)	19.8
G	32 (33.33)	157 (29.40)	20.4
U	20 (20.83)	109 (20.41)	18.3

Percentage is mentioned in the parenthesis. Relative contact in DOT regions are calculated by $N_{idt}/N_{prot} \times 100$.

2.9. Secondary Structure of Nucleotides Interacting with DOT Residues

Further, we have classified the nucleotides based on location and contacts with DOT residues and preference of amino acids in a protein and the results are presented in Table 4 and Table S3. Among all secondary structures formed by nucleotides, unpaired bases are most likely to bind with DOT residues. Specifically, we observed that A and U in unpaired regions prefer to interact with DOT residues, whereas C and G in unpaired and base-paired positions interact with DOT residues with a similar preference. G and C also interact with DOT residues in pseudoknot secondary structure, whereas A and U are least likely to exist in pseudoknot form when bound to DOT regions.

Table 4. Preference of nucleotides in different secondary structures to bind with DOT residues.

Nucleotides	Secondary Structure	Number of Nucleotide in Contact with DOT Regions (N_{idt})	Number of Nucleotides in Contact with Any Residue of Proteins (N_{prot})	Relative Contact in DOT Regions (%)
A	Unpaired	12 (12.50)	106 (19.56)	11.01
A	Basepaired	6 (6.25)	30 (5.54)	20.00
A	Pseudoknot	0 (0)	0 (0)	0
C	Unpaired	8 (8.33)	70 (12.92)	11.42
C	Basepaired	17 (17.71)	59 (10.89)	28.81
C	Pseudoknot	1 (1.04)	5 (0.92)	20.00
G	Unpaired	16 (16.67)	87 (16.05)	18.39
G	Basepaired	15 (15.63)	71 (13.10)	21.13
G	Pseudoknot	1 (1.04)	4 (0.74)	25.00
U	Unpaired	15 (15.63)	81 (14.94)	18.51
U	Basepaired	5 (5.21)	29 (5.35)	17.24
U	Pseudoknot	0 (0)	0 (0)	0
All	Unpaired	51 (53.13)	344 (63.47)	14.83
All	Basepaired	43 (44.79)	189 (34.87)	22.75
All	Pseudoknot	2 (2.08)	9 (1.66)	22.22

Percentage is mentioned in parenthesis. Relative contacts in DOT regions are calculated by $N_{idt}/N_{prot} \times 100$.

2.10. Interaction Energy of DOT Residues with Nucleotides

We have computed the interaction energy between amino acids and nucleotides in DOT and ordered regions at the binding interface and the results are presented in Table 5. Most of the amino acids have stronger interactions with nucleotides in ordered regions than DOT regions. However, we noticed that some combinations of amino acid–nucleotide pairs have favourable energy when interacting with DOT regions. For example, Arg, His, Ile, Leu, Val, and Phe interact with G, His, Ser, and Val with C, and Asn, Asp, Gly, Ile, Leu, and Ser with U. In addition, hydrophobic residues Ile, Leu, and Val have more favourable interactions with G at DOT regions than others. Since Arg and Lys are important for protein–RNA complex formation through electrostatic interactions these residues have stronger energies in ordered regions than DOT regions. On the other hand, His in the DOT region has favourable energy with G and C. These differences in energy could be important to understand the interactions between DOT regions and the RNA molecule, which might also be used to distinguish the RNA binding residues of proteins in DOT and other regions.

Table 5. Interaction energy between amino acids and nucleotides in DOT regions.

Amino Acids	A	G	C	U
Ala	−0.62 (−0.55)	−0.34 (−0.57)	−0.49 (−0.53)	−0.55 (−0.64)
Arg	−0.36 (−1.23)	**−1.15** (−0.83)	−0.89 (−0.95)	**−1.06** (−0.98)
Asn	−0.45 (−0.68)	−0.59 (−0.73)	−0.48 (−0.83)	**−1.85** (−0.82)
Asp	−0.75 (−0.74)	−0.39 (−0.79)	−0.19 (−0.56)	**−1.40** (−0.92)
Cys	0.00 (−0.87)	−0.01 (−0.03)	−0.03 (−1.10)	−0.63 (−1.13)
Gln	−0.15 (−0.87)	−0.57 (−0.74)	−0.08 (−0.84)	−0.36 (−0.71)
Glu	−0.72 (−0.80)	−0.41 (−0.64)	−0.43 (−0.62)	−0.68 (−0.59)
Gly	−0.28 (−0.47)	−0.37 (−0.69)	−0.58 (−0.57)	**−1.07** (−0.79)
His	−0.81 (−1.17)	**−2.13** (−1.41)	**−1.53** (−1.21)	−0.70 (−1.01)
Ile	−0.60 (−0.64)	**−1.63** (−0.80)	−0.54 (−0.50)	**−1.33** (−0.76)
Leu	−0.35 (−0.75)	**−1.19** (−0.50)	−0.42 (−0.49)	−0.54 (−0.41)
Lys	−0.74 (−0.76)	−0.86 (−0.83)	−0.66 (−0.90)	−0.83 (−0.83)
Met	−0.64 (−1.05)	−0.07 (−0.75)	−0.16 (−1.03)	−0.83 (−1.19)
Phe	−0.81 (−1.03)	**−1.12** (−0.89)	−0.54 (−1.32)	−0.24 (−1.42)
Pro	−0.88 (−0.83)	−0.60 (−0.88)	−0.62 (−0.91)	−0.69 (−1.00)
Ser	−0.79 (−0.77)	−0.29 (−0.56)	**−1.24** (−0.71)	**−1.41** (−0.68)
Thr	−0.39 (−0.68)	−0.66 (−0.56)	−0.67 (−0.64)	−0.53 (−1.00)
Trp	**−1.15** (−1.10)	0.00 (−1.64)	0.00 (−0.99)	0.00 (−1.34)
Tyr	**−1.53** (−1.36)	−1.11 (−1.42)	−0.63 (−1.05)	−0.16 (−1.09)
Val	−0.38 (−0.70)	−0.66 (−0.53)	−0.64 (−0.53)	−0.52 (−0.76)

Interaction energy for non-DOT residues is mentioned in the parenthesis. Amino acid–nucleotide pairs with favourable interaction energies in DOT regions are shown in bold.

We have compared the interaction energy of amino acid–nucleotide pairs in the interface of DOT and other regions and two typical examples are shown in Figure 8. We noticed a wide range of interactions such as stacking, cation-π, electrostatic, and van der Waals interactions at the interface. Most favourable energy is observed for Asn and His with U (−3.26 kcal/mol) and G (−5.44 kcal/mol), respectively, in DOT regions (Figure 8a). On the other hand, Arg and Phe have favourable energy with A (−8.49 kcal/mol) and C (−4.88 kcal/mol), respectively, in non-DOT regions.

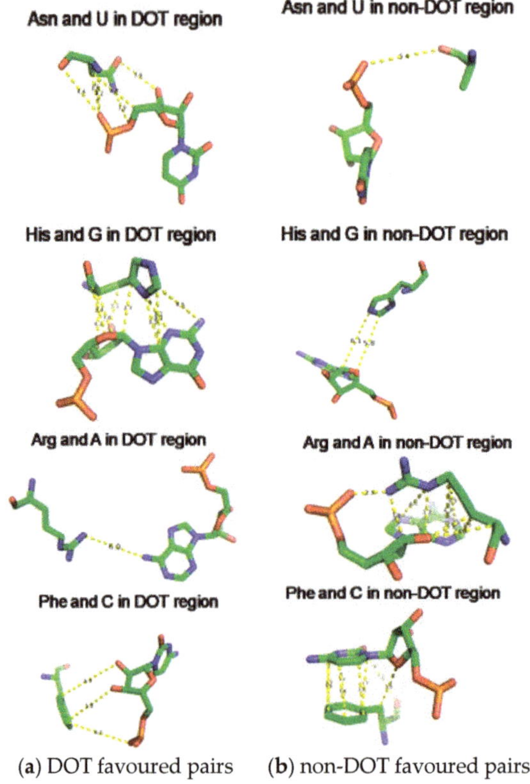

Figure 8. Amino acid showing (**a**) strong interaction in DOT and weak interaction in non-DOT regions and (**b**)weak interaction in DOT and strong interaction in non-DOT regions.

3. Materials and Methods

We adopted the following protocol to obtain a set of protein–RNA complexes with disorder-to-order transition (DOT) regions: (i) Downloaded the protein–RNA complexes from PDB and NDB databases (www.rcsb.org) [37–39]; (ii) Clustered all the protein–RNA complexes with 30% sequence identity cut-off using CD-Hit suite [40]; (iii) Performed BLAST search (using 99% identity cut-off) of protein sequences to obtain free proteins corresponding to each protein–RNA complex [41,42]. The free proteins have the same sequences as the protein part of protein–RNA complexes but crystallized without RNA. Note that free proteins contain unique PDB IDs, which is distinct from the protein–RNA complex; (iv) Disordered residues are obtained from missing residues information in the protein–RNA complex and free protein pairs by locating "REMARK 465" statement in the protein structure file; (v) DOT residues are isolated by comparing the disorder residues of free and protein–RNA complex pairs such that the residue is ordered in the protein–RNA complex but disordered in free protein. Note that only the regions having 3 or more continuous DOT residues are considered. The final dataset contains 101 DOT regions in 52 proteins and complete data are given in supplementary information. The representation of DOT and ordered region in a typical protein–RNA complex (PDB ID: 4H4K) is shown in Figure 9.

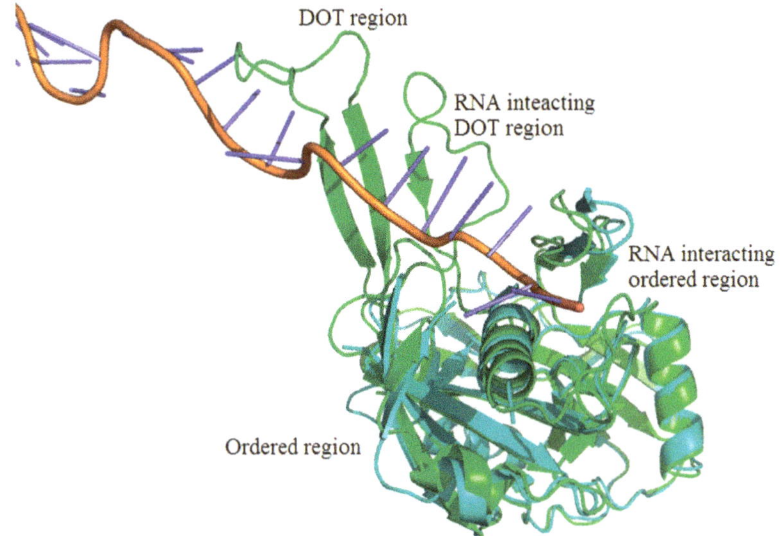

Figure 9. Representation of disorder-to-order mediated interactions. Free protein, RNA, and complex (CRISPR-Cas RNA Silencing Cmr Complex) are shown in cyan, orange and green, respectively. The PDB IDs are 4H4K:A (free protein), 3XIL:I (RNA of protein–RNA complex) and 3XIL:B (RNA-bound protein). The disorder-to-order transition (DOT) region can be clearly seen in green with a missing overlapping region of free protein.

3.1. Number of DOT Regions and Their Lengths

The number of DOT regions and their lengths are obtained by counting the number of non-consecutive and consecutive residues, respectively, using custom build python scripts.

3.2. DOT Residues in Contact with RNA

The residues in contact with RNA molecules are obtained by using distance cut-offs mentioned in literature, that is, 3.5 Å and 6 Å [43–45]. Binding residues in DOT regions are obtained by taking common residues in the DOT dataset and RNA contacting residues. We have classified protein–RNA complexes in non-ribosomal and ribosomal classes because of the difference in their interaction pattern, number of interacting amino acids, and residue bias in them [46]. Therefore, using the type of complex and distance cut-off for interacting residues, we divided protein–RNA complexes into four different datasets: (1) NR3.5: non-ribosomal complex with a contact distance of 3.5 Å; (2) RB3.5: ribosomal complex with a contact distance of 3.5 Å; (3) NR6: non-ribosomal complex with a contact distance of 6 Å; and (4) RB6: ribosomal complex with a contact distance of 6 Å.

We computed the frequency of each DOT residue involved in binding using the Equation (1).

$$\text{Frequency of binding residues in DOT region} = \frac{N_{ib}}{N_{id}} \tag{1}$$

where N_{ib}: number of ith residues binding in the DOT region and N_{id}: number of ith residues in DOT.

Moreover, the differences in the frequency of binding residues in DOT regions and in the protein complexes are obtained.

3.3. Frequency of Binding in DOT and Other Residues

We also computed the frequency of residues binding in DOT regions over all the binding residues by using Equation (2), an error bar is plotted using the bootstrap method by randomly re-sampling an equal sized data with a replacement 1000 times.

$$\text{Frequency of binding by contact residues} = \frac{N_{ibd}}{N_{ib}} \tag{2}$$

where N_{ibd}: number of ith residues binding in DOT region; N_{ib} is number of ith residues binding with RNA in complete protein.

3.4. Propensity of Binding Residues in DOT Region

The normalization of frequency of residues present in DOT regions by individual residue frequency provides the tendency of a residue in DOT regions. Accordingly, propensity values are calculated using the following equation:

$$\text{Propensity}(I) = \frac{N_{ibd}/N_{id}}{N_{ip}/N_{p}} \tag{3}$$

where Propensity (I): propensity of ith residue; N_{ibd}: number of ith residue binding in DOT region; N_{id}: number of ith residue in DOT regions; N_{ip}: number of ith residue in protein; N_{p}: number of residues in protein.

3.5. Boot Strap Sampling

To obtain the standard error in frequency and propensity calculations, bootstrap sampling is performed. In this technique all the protein–RNA complexes are sampled randomly and each sample contains complexes equal to the number of protein–RNA complexes. Therefore, each sample will have redundancy of some complexes and will be devoid of some complexes. In this manner, we have created 1000 samples on which the calculations are performed.

3.6. Relative Average Solvent Accessibility (RASA)

The DOT residues buriedness is analysed by the NACCESS [47] program and the RASA of each residue is calculated by using Equation (4).

$$\text{RASA} = \frac{A_{ibd}}{\sum_{i=1}^{n}(A_{ibd})} \tag{4}$$

where A_{ibd}: RASA of ith residue binding with RNA in DOT region; n: number of DOT residues in a protein–RNA complex.

3.7. Secondary Structure of Protein and RNA

Secondary structure of both proteins and RNA molecules are analysed by DSSP and DSSR programs, respectively [48,49]. The DSSR program gives dot bracket notation of secondary structure of RNA as shown in Figure S1, in which "." represents unpaired nucleotide, "(" or ")" represent paired bases, and "{" or "}" or "[" or "]" or "<" or ">" represent pseudoknot bases.

3.8. Binding Preference of Nucleotides for Amino Acids

The binding preference of nucleotide with DOT residues has been calculated by counting the occurrence of nucleotides–amino acid interacting pairs under the distance of 3.5 Å.

3.9. Interaction Energy between Amino Acids and Nucleotides at Binding Interface

The interaction energy of amino acids with nucleotides is computed using van der Waals and coulombs potential using AMBER force field [50]. It is given by

$$\text{Energy} = \sum \left[\left(\frac{A_{ij}}{r_{ij}^{12}} - \frac{B_{ij}}{r_{ij}^{6}} \right) + \frac{q_i q_j}{\varepsilon r_{ij}} \right] \tag{5}$$

where, $A_{ij} = \varepsilon_{ij}^* (R_{ij}^*)^{12}$ and $B_{ij} = 2\,\varepsilon_{ij}^* (R_{ij}^*)^6$; $R_{ij}^* = (R_i^* + R_j^*)$; and $\varepsilon_{ij}^* = (\varepsilon_i^* \varepsilon_j^*)^{1/2}$; R^* and ε^* van der Waals radius and well depth, respectively, and these parameters are obtained from Gromiha et al. [51]; q_i and q_j is the charge on atom i and j, respectively and R_{ij} is the distance separating atom i and j.

4. Conclusions

The analysis of DOT regions in protein–RNA complexes revealed that in each complex these regions are generally small in size. Electrostatic interactions are found to be important, with the involvement of positively charged residues (Arg, Lys and His) in DOT regions. Among nucleotide–amino acid pairs, guanine–Arg and uracil–Lys pairs are identified to be the most and the least preferred ones at the interface, respectively. Generally, nucleotides prefer to bind DOT regions than other regions of protein. Further, DOT regions are significantly more exposed to solvent than other residues of protein–RNA complexes. Specifically, hydrophobic residues have higher difference in RASA of DOT regions and complete proteins. DOT regions are preferred to form coils, turns, and bends than regular secondary structures such as helices and strands. On the RNA side, DOT residues prefer to bind unpaired A and U and paired regions of C and G. In pseudoknot condition, mostly C and G interact with DOT residues. The interaction energy calculations revealed the types of interactions and preferred amino acid-nucleotide pairs at the interface based on energy.

The frequencies and propensities obtained in the present study could be used for discriminating DOT binding residues from other residues. Further, the location of DOT binding residues based on solvent accessibility and secondary structure of protein and RNA along with energy calculations may help to understand the recognition mechanism.

We obtained the DOT regions by comparing 3D coordinates of the missing residues in protein–RNA complexes and their respective free proteins. This might be an under representation of DOT regions since the structures solved by crystallization often stabilize the residues and reduce the native disorder. Hence, the disordered residues having 3D coordinates in free proteins are not considered. The current study can further be refined with the availability of more numbers of protein–RNA complexes and the improvements in structure determination techniques. In addition, development of disorder specific databases for protein–nucleic acid complexes with large datasets could enhance the confidence level of the result reported in the present study.

Supplementary Materials: Supplementary materials can be found at http://www.mdpi.com/1422-0067/19/6/1595/s1.

Author Contributions: M.M.G. and S.A. conceived the project and designed experiments. A.S. constructed the dataset and performed the analysis. M.M.G, S.A. and A.S took part in discussions. A.S. drafted the manuscript. M.M.G and S.A. edited and refined the manuscript.

Acknowledgments: We thank the Department of Biotechnology, Indian Institute of Technology Madras for computational facilities. A.S. thank Ministry of Human Resource and Development (MHRD) for the fellowship. This project is partially supported by the Council of Scientific & Industrial Research (CSIR), Government of India to M.M.G. and S.A. (grant numbers: 37(1694)/17/EMR-II and 37(1695)/17/EMRII respectively). S.A. would like to acknowledge a grant from University for potential of excellence (UpoE-II) #270 and support DST-PURSE.

Conflicts of Interest: The authors declare no conflict of interest.

References

1. Habchi, J.; Tompa, P.; Longhi, S.; Uversky, V.N. Introducing protein intrinsic disorder. *Chem. Rev.* **2014**, *114*, 6561–6588. [CrossRef] [PubMed]
2. Fuxreiter, M.; Toóth-Petroóczy, A.; Kraut, D.A.; Matouschek, A.T.; Lim, R.Y.; Xue, B.; Kurgan, L.; Uversky, V.N. Disordered proteinaceous machines. *Chem. Rev.* **2014**, *114*, 6806–6843. [CrossRef] [PubMed]
3. Babu, M.M.; Van der, L.R.; de Groot, N.S.; Gsponer, J. Intrinsically disordered proteins: Regulation and disease. *Curr. Opin. Struct. Biol.* **2011**, *21*, 432–440. [CrossRef] [PubMed]
4. Wright, P.E.; Dyson, H.J. Intrinsically disordered proteins in cellular signaling and regulation. *Nat. Rev. Mol. Cell Biol.* **2015**, *16*, 18–29. [CrossRef] [PubMed]
5. Deller, M.C.; Kong, L.; Rupp, B. Protein stability: A crystallographer's perspective. *Acta Cryst. F* **2016**, *72*, 72–95. [CrossRef] [PubMed]
6. Johnson, D.E.; Xue, B.; Sickmeier, M.D.; Meng, J.; Cortese, M.S.; Oldfield, C.J.; Le Gall, T.; Dunker, A.K.; Uversky, V.N. High-throughput characterization of intrinsic disorder in proteins from the Protein Structure Initiative. *J. Struct. Biol.* **2012**, *180*, 201–215. [CrossRef] [PubMed]
7. Dosztányi, Z.; Csizmok, V.; Tompa, P.; Simon, I. IUPred: Web server for the prediction of intrinsically unstructured regions of proteins based on estimated energy content. *Bioinformatics* **2005**, *21*, 3433–3434. [CrossRef] [PubMed]
8. Jones, D.T.; Cozzetto, D. DISOPRED3: Precise disordered region predictions with annotated protein-binding activity. *Bioinformatics* **2014**, *31*, 857–863. [CrossRef] [PubMed]
9. Disfani, F.M.; Hsu, W.L.; Mizianty, M.J.; Oldfield, C.J.; Xue, B.; Dunker, A.K.; Uversky, V.N.; Kurgan, L. MoRFpred, a computational tool for sequence-based prediction and characterization of short disorder-to-order transitioning binding regions in proteins. *Bioinformatics* **2012**, *28*, i75–i83. [CrossRef] [PubMed]
10. Meng, F.; Uversky, V.N.; Kurgan, L. Comprehensive review of methods for prediction of intrinsic disorder and its molecular functions. *Cell. Mol. Life Sci.* **2017**, *74*, 3069–3090. [CrossRef] [PubMed]
11. Berlow, R.B.; Dyson, H.J.; Wright, P.E. Functional advantages of dynamic protein disorder. *FEBS Lett.* **2015**, *589*, 2433–2440. [CrossRef] [PubMed]
12. Basu, S.; Söderquist, F.; Wallner, B. Proteus: A random forest classifier to predict disorder-to-order transitioning binding regions in intrinsically disordered proteins. *J. Comput. Aided Mol. Des.* **2017**, *31*, 453–466. [CrossRef] [PubMed]
13. Deane, J.E.; Ryan, D.P.; Sunde, M.; Maher, M.J.; Guss, J.M.; Visvader, J.E.; Matthews, J.M. Tandem LIM domains provide synergistic binding in the LMO4: Ldb1 complex. *EMBO J.* **2004**, *23*, 3589–3598. [CrossRef] [PubMed]
14. Mark, W.Y.; Liao, J.C.; Lu, Y.; Ayed, A.; Laister, R.; Szymczyna, B.; Chakrabartty, A.; Arrowsmith, C.H. Characterization of segments from the central region of BRCA1: An intrinsically disordered scaffold for multiple protein–protein and protein–DNA interactions? *J. Mol. Biol.* **2005**, *345*, 275–287. [CrossRef] [PubMed]
15. Papadakos, G.; Sharma, A.; Lancaster, L.; Bowen, R.; Kaminska, R.; Leech, A.P.; Walker, D.; Redfield, C.; Kleanthous, C. Consequences of inducing intrinsic disorder in a high-affinity protein-protein interaction. *J. Am. Chem. Soc.* **2015**, *137*, 5252–5255. [CrossRef] [PubMed]
16. Fukuchi, S.; Amemiya, T.; Sakamoto, S.; Nobe, Y.; Hosoda, K.; Kado, Y.; Murakami, S.D.; Koike, R.; Hiroaki, H.; Ota, M. IDEAL in 2014 illustrates interaction networks composed of intrinsically disordered proteins and their binding partners. *Nucleic Acids Res.* **2014**, *42*, D320–D325. [CrossRef] [PubMed]
17. Vacic, V.; Oldfield, C.J.; Mohan, A.; Radivojac, P.; Cortese, M.S.; Uversky, V.N.; Dunker, A.K. Characterization of molecular recognition features, MoRFs, and their binding partners. *J. Proteome Res.* **2007**, *6*, 2351–2366. [CrossRef] [PubMed]
18. Sugase, K.; Dyson, H.J.; Wright, P.E. Mechanism of coupled folding and binding of an intrinsically disordered protein. *Nature* **2007**, *447*, 1021–1025. [CrossRef] [PubMed]
19. Shammas, S.L.; Travis, A.J.; Clarke, J. Allostery within a transcription coactivator is predominantly mediated through dissociation rate constants. *Proc. Natl. Acad. Sci. USA* **2014**, *111*, 12055–12060. [CrossRef] [PubMed]
20. Shammas, S.L.; Crabtree, M.D.; Dahal, L.; Wicky, B.I.; Clarke, J. Insights into coupled folding and binding mechanisms from kinetic studies. *J. Biol. Chem.* **2016**, *291*, 6689–6695. [CrossRef] [PubMed]

21. Dyson, H.J. Roles of intrinsic disorder in protein–nucleic acid interactions. *Mol. Biosyst.* **2012**, *8*, 97–104. [CrossRef] [PubMed]

22. Dey, B.; Thukral, S.; Krishnan, S.; Chakrobarty, M.; Gupta, S.; Manghani, C.; Rani, V. DNA–protein interactions: Methods for detection and analysis. *Mol. Cell. Biochem.* **2012**, *365*, 279–299. [CrossRef] [PubMed]

23. Popova, V.V.; Kurshakova, M.M.; Kopytova, D.V. Methods to study the RNA-protein interactions. *Mol. Biol.* **2015**, *49*, 472–481. [CrossRef]

24. Walia, R.R.; Caragea, C.; Lewis, B.A.; Towfic, F.; Terribilini, M.; El-Manzalawy, Y.; Dobbs, D.; Honavar, V. Protein–RNA interface residue prediction using machine learning: An assessment of the state of the art. *BMC Bioinform.* **2012**, *13*, 89. [CrossRef] [PubMed]

25. Kumar, M.; Gromiha, M.M.; Raghava, G.P. Prediction of RNA binding sites in a protein using SVM and PSSM profile. *Proteins* **2008**, *71*, 189–194. [CrossRef] [PubMed]

26. Terribilini, M.; Sander, J.D.; Lee, J.H.; Zaback, P.; Jernigan, R.L.; Honavar, V.; Dobbs, D. RNABindR: A server for analyzing and predicting RNA-binding sites in proteins. *Nucleic Acids Res.* **2007**, *35*, W578–W584. [CrossRef] [PubMed]

27. Wang, L.; Brown, S.J. BindN: A web-based tool for efficient prediction of DNA and RNA binding sites in amino acid sequences. *Nucleic Acids Res.* **2006**, *34*, W243–W248. [CrossRef] [PubMed]

28. Zhang, J.; Ma, Z.; Kurgan, L. Comprehensive review and empirical analysis of hallmarks of DNA-, RNA-and protein-binding residues in protein chains. *Brief. Bioinform.* **2017**, 1–19. [CrossRef] [PubMed]

29. Wang, L.; Huang, C.; Yang, M.Q.; Yang, J.Y. BindN+ for accurate prediction of DNA and RNA-binding residues from protein sequence features. *BMC Syst. Biol.* **2010**, *4*, S3. [CrossRef] [PubMed]

30. Yan, J.; Friedrich, S.; Kurgan, L. A comprehensive comparative review of sequence-based predictors of DNA-and RNA-binding residues. *Brief. Bioinform.* **2015**, *17*, 88–105. [CrossRef] [PubMed]

31. Tuszynska, I.; Bujnicki, J.M. DARS-RNP and QUASI-RNP: New statistical potentials for protein–RNA docking. *BMC Bioinform.* **2011**, *12*, 348. [CrossRef] [PubMed]

32. Wang, Y.; Guo, Y.; Pu, X.; Li, M. A sequence-based computational method for prediction of MoRFs. *RSC Adv.* **2017**, *7*, 18937–18945. [CrossRef]

33. Peng, Z.; Kurgan, L. High-throughput prediction of RNA, DNA and protein binding regions mediated by intrinsic disorder. *Nucleic Acids Res.* **2015**, *43*, e121. [CrossRef] [PubMed]

34. Kim, O.T.; Yura, K.; Go, N. Amino acid residue doublet propensity in the protein–RNA interface and its application to RNA interface prediction. *Nucleic Acids Res.* **2006**, *34*, 6450–6460. [CrossRef] [PubMed]

35. Mohan, A.; Oldfield, C.J.; Radivojac, P.; Vacic, V.; Cortese, M.S.; Dunker, A.K.; Uversky, V.N. Analysis of molecular recognition features (MoRFs). *J. Mol. Biol.* **2006**, *362*, 1043–1059. [CrossRef] [PubMed]

36. Fernandez, M.; Kumagai, Y.; Standley, D.M.; Sarai, A.; Mizuguchi, K.; Ahmad, S. Prediction of dinucleotide-specific RNA-binding sites in proteins. *BMC Bioinform.* **2011**, *12*, S5. [CrossRef] [PubMed]

37. Rose, P.W.; Prlić, A.; Altunkaya, A.; Bi, C.; Bradley, A.R.; Christie, C.H.; Costanzo, L.D.; Duarte, J.M.; Dutta, S.; Feng, Z.; Green, R.K. The RCSB protein data bank: Integrative view of protein, gene and 3D structural information. *Nucleic Acids Res.* **2017**, *45*, D271–D281. [PubMed]

38. Berman, H.M.; Olson, W.K.; Beveridge, D.L.; Westbrook, J.; Gelbin, A.; Demeny, T.; Hsieh, S.H.; Srinivasan, A.R.; Schneider, B. The nucleic acid database. A comprehensive relational database of three-dimensional structures of nucleic acids. *Biophys. J.* **1992**, *63*, 751–759. [CrossRef]

39. Coimbatore Narayanan, B.; Westbrook, J.; Ghosh, S.; Petrov, A.I.; Sweeney, B.; Zirbel, C.L.; Leontis, N.B.; Berman, H.M. The Nucleic Acid Database: New features and capabilities. *Nucleic Acids Res.* **2014**, *42*, D114–D122. [CrossRef] [PubMed]

40. Huang, Y.; Niu, B.; Gao, Y.; Fu, L.; Li, W. CD-HIT Suite: A web server for clustering and comparing biological sequences. *Bioinformatics* **2010**, *26*, 680–682. [CrossRef] [PubMed]

41. Boratyn, G.M.; Schäffer, A.A.; Agarwala, R.; Altschul, S.F.; Lipman, D.J.; Madden, T.L. Domain enhanced lookup time accelerated BLAST. *Biol. Direct.* **2012**, *7*, 12. [CrossRef] [PubMed]

42. Altschul, S.F.; Gish, W.; Miller, W.; Myers, E.W.; Lipman, D.J. Basic local alignment search tool. *J. Mol. Biol.* **1990**, *215*, 403–410. [CrossRef]

43. Gromiha, M.M. *Protein Bioinformatics: From Sequence to Function*; Academic Press: Cambridge, MA, USA, 2010.

44. Si, J.; Zhao, R.; Wu, R. An overview of the prediction of protein DNA-binding sites. *Int. J. Mol. Sci.* **2015**, *16*, 5194–5215. [CrossRef] [PubMed]

45. Nagarajan, R.; Gromiha, M.M. Prediction of RNA binding residues: An extensive analysis based on structure and function to select the best predictor. *PLoS ONE* **2014**, *9*, e91140. [CrossRef] [PubMed]
46. Ciriello, G.; Gallina, C.; Guerra, C. Analysis of interactions between ribosomal proteins and RNA structural motifs. *BMC Bioinform.* **2010**, *11*, S41. [CrossRef] [PubMed]
47. NACCESS, V2.1.1. *A Computer Program for Solvent Accessible Area Calculations*; Department of Biochemistry and Molecular Biology, University College London: London, UK, 1993.
48. Kabsch, W.; Sander, C. Dictionary of protein secondary structure: Pattern recognition of hydrogen-bonded and geometrical features. *Biopolymers* **1983**, *22*, 2577–2637. [CrossRef] [PubMed]
49. Lu, X.J.; Bussemaker, H.J.; Olson, W.K. DSSR: An integrated software tool for dissecting the spatial structure of RNA. *Nucleic Acids Res.* **2015**, *43*, e142. [CrossRef] [PubMed]
50. Cornell, W.D.; Cieplak, P.; Bayly, C.I.; Gould, I.R.; Merz, K.M.; Ferguson, D.M.; Spellmeyer, D.C.; Fox, T.; Caldwell, J.W.; Kollman, P.A. A second generation force field for the simulation of proteins, nucleic acids, and organic molecules. *J. Am. Chem. Soc.* **1995**, *117*, 5179–5197. [CrossRef]
51. Gromiha, M.M.; Yokota, K.; Fukui, K. Understanding the recognition mechanism of protein–RNA complexes using energy based approach. *Curr. Protein Pept. Sci.* **2010**, *11*, 629–638. [CrossRef] [PubMed]

© 2018 by the authors. Licensee MDPI, Basel, Switzerland. This article is an open access article distributed under the terms and conditions of the Creative Commons Attribution (CC BY) license (http://creativecommons.org/licenses/by/4.0/).

International Journal of
Molecular Sciences

Article

The Cyanobacterial Ribosomal-Associated Protein LrtA from *Synechocystis* sp. PCC 6803 Is an Oligomeric Protein in Solution with Chameleonic Sequence Properties

Lellys M. Contreras [1,†], Paz Sevilla [2,3], Ana Cámara-Artigas [4], José G. Hernández-Cifre [5], Bruno Rizzuti [6,*], Francisco J. Florencio [7], María Isabel Muro-Pastor [7], José García de la Torre [5] and José L. Neira [8,9,*,†]

[1]   Center for Environmental Biology and Chemistry Research, Facultad Experimental de Ciencias y Tecnología, Universidad de Carabobo, 2001 Valencia, Venezuela; lellyscontreras@gmail.com

[2]   Facultad de Farmacia, Departamento de Química Física II, Universidad Complutense de Madrid, 28040 Madrid, Spain; paz@farm.ucm.es

[3]   Instituto de Estructura de la Materia, IEM-CSIC, Serrano 121, 28006 Madrid, Spain

[4]   Department of Chemistry and Physics, Research Centre CIAIMBITAL, University of Almería- ceiA3, 04120 Almería, Spain; acamara@ual.es

[5]   Department of Physical Chemistry, University of Murcia, 30003 Murcia, Spain; jghc@um.es (J.G.H.-C.); jgt@um.es (J.G.d.l.T.)

[6]   CNR-NANOTEC, Licryl-UOS Cosenza and CEMIF.Cal, Department of Physics, University of Calabria, 87036 Rende, Italy

[7]   Instituto de Bioquímica Vegetal y Fotosíntesis, CSIC-Universidad de Sevilla, 41092 Seville, Spain; floren@us.es (F.J.F.); imuro@ibvf.csic.es (M.I.M.-P.)

[8]   Instituto de Biología Molecular y Celular, Universidad Miguel Hernández, 03202 Elche (Alicante), Spain

[9]   Instituto de Biocomputación y Física de Sistemas Complejos, Joint Units IQFR-CSIC-BIFI, and GBsC-CSIC-BIFI, Universidad de Zaragoza, 50009 Zaragoza, Spain

*   Correspondence: bruno.rizzuti@cnr.it (B.R.); jlneira@umh.es (J.L.N.); Tel.: +39-0984-49-6078 (B.R.); +34-96-6658475 (J.L.N.)

†   These authors contributed equally to this work.

Received: 18 May 2018; Accepted: 20 June 2018; Published: 24 June 2018

Abstract: The LrtA protein of *Synechocystis* sp. PCC 6803 intervenes in cyanobacterial post-stress survival and in stabilizing 70S ribosomal particles. It belongs to the hibernating promoting factor (HPF) family of proteins, involved in protein synthesis. In this work, we studied the conformational preferences and stability of isolated LrtA in solution. At physiological conditions, as shown by hydrodynamic techniques, LrtA was involved in a self-association equilibrium. As indicated by Nuclear Magnetic Resonance (NMR), circular dichroism (CD) and fluorescence, the protein acquired a folded, native-like conformation between pH 6.0 and 9.0. However, that conformation was not very stable, as suggested by thermal and chemical denaturations followed by CD and fluorescence. Theoretical studies of its highly-charged sequence suggest that LrtA had a Janus sequence, with a context-dependent fold. Our modelling and molecular dynamics (MD) simulations indicate that the protein adopted the same fold observed in other members of the HPF family (β-α-β-β-β-α) at its N-terminal region (residues 1–100), whereas the C terminus (residues 100–197) appeared disordered and collapsed, supporting the overall percentage of overall secondary structure obtained by CD deconvolution. Then, LrtA has a chameleonic sequence and it is the first member of the HPF family involved in a self-association equilibrium, when isolated in solution.

Keywords: conformational plasticity; disordered protein; folding; ribosomal protein; spectroscopy; protein stability

---

1. Introduction

The *lrtA* gene was first identified in *Synechococcus* sp. PCC 7002 as a sequence encoding a light-repressed protein [1], with a larger half-life in dark conditions than in the presence of light [2]. Although the exact functions of the LrtA protein are unknown, recent studies have shown that it is involved in post-stress survival in *Synechocystis* sp. PCC 6803, stabilizing the 70S ribosomal particles [3].

LrtA is related to other proteins, which are highly present among bacteria and associated with ribosomes. These proteins modulate ribosome activity to preserve their integrity and aid in cell survival during stress circumstances. Under these conditions, stalling of the protein synthesis, a major energy-consuming process in living cells, is downregulated, usually by proteins involved in ribosome inhibition. Reduction of translation activity is associated with: (i) dimerization of 70S particles to form the translationally inactive 100S disome (also known as hibernating ribosomes [4]), mediated by intermolecular interactions among proteins; or alternatively; (ii) interaction of canonical ribosomal proteins with the ribosome [5,6]. Among the most studied members of this protein family are two *Escherichia coli* proteins: YfiA (also known as PY or RaiA, ribosome associated inhibitor A); and YhbH (also known as HPF, hibernation promoting factor). YfiA is thought to inhibit translation indirectly, by modulating a more stringent proofreading mechanism involving 70S particles [7,8]. On the other hand, HPF stops translation by stabilizing 100S dimers [8–10]. Formation of 100S disomes is also mediated by other proteins, known as ribosome modulation factors (RMFs), in γ-proteobacteria species or double YfiA- and YhbH- knocked-out cells [10]. Phylogenetic analyses have shown that most bacteria have at least one of those HPF or YfiA homologues [10]. These homologues have been classified in three classes, based on the presence of a conserved domain and, in some cases, additional sequence extensions: long HPF, short HPF, and YFiA. The conserved domain has a β-α-β-β-β-α fold, with the two α-helices packed against one side of the four-stranded β-sheet [5,11]. According to its sequence, the LrtA from *Synechocystis* sp. PCC 6803 could be classified within the long HPF sub-family; in addition, it also bears similarity to the spinach plastid-specific ribosomal protein, which is present in the chloroplast stroma, either associated or unbound to the 30S ribosomal unit [3]. Although we have shown that LrtA stabilizes 70S particles [3], nothing is known about the conformation or stability of the isolated protein in solution.

In this work, we embarked in the characterization of the conformational stability and structure of LrtA from *Synechocystis* sp. PCC 6803 by using experimental and *in silico* approaches. Our results are the first characterization of the conformation and stability of a member of the long HPF subfamily. At physiological pH, LrtA was involved in a self-association equilibrium, as shown by hydrodynamic techniques. The protein acquired a native-like conformation around pH 6.0, as judged from intrinsic fluorescence (monitoring tertiary structure), ANS (8-anilino-1-naphthalene sulfonic acid) fluorescence (revealing hydrophobic solvent-exposed patches), CD (reporting secondary structure) and 1D ^{1}H NMR experiments. The MD simulations and analyses of sequence suggested that the protein had a Janus sequence, and its conformation was solvent-dependent, with an N-terminal region acquiring the fold of other members of the HPF family and the C-terminal region appearing disordered and collapsed. Therefore, LrtA is the first member of the HPF family with chameleonic features encoded in its sequence and shown to be involved in a self-association equilibrium when isolated in solution.

2. Results

2.1. Isolated LrtA Was Involved in a Self-Association Equilibrium in Solution

We first tried to elucidate the oligomerization state of the protein to identify the protein-concentration range where we must characterize the conformational stability of the protein.

To map the hydrodynamic properties of LrtA we used three hydrodynamic techniques: DOSY-NMR (diffusion ordered NMR spectroscopy), DLS (dynamic light scattering), size exclusion chromatography (SEC), glutaraldehyde cross-linking, and lifetime fluorescence measurements. Furthermore, we tried to measure the self-association of LrtA by using isothermal titration calorimetry

(ITC), but in all attempts, protein precipitated at the concentrations required to carry out the experiments. It is important to pinpoint the differences among the different hydrodynamic techniques used in this work. With NMR, we shall obtain information about the low-molecular weight species, whose overall rotational tumbling is very fast. By using DLS, we shall obtain information about the hydrodynamic parameters, assuming a spherical shape, for all the species (high or low molecular weight) present in solution, and we shall be able to see whether those hydrodynamic parameters are protein-concentration-dependent. By using SEC, we shall be able to monitor the elution volume of LrtA, which will depend on the molecular weight and the shape of the molecule, but that volume could be also affected by possible interactions with the column. Finally, by lifetime fluorescence measurements, we shall determine how the decay of the electronic excited states can be affected by: (i) the presence of conformational isomers; (ii) energy transfer among the eight Tyr residues in LrtA; or (iii) even transient electronic effects in collisional quenching [12,13]. To elucidate the oligomerization state of the protein, we also tried to carry out T_2 echo measurements estimating the averaged correlation time of the species present in solution from the amide region. However, at the times used (2.9 ms and 400 μs) in our echo experiments, most of the amide peaks disappeared, and only the proton resonances of the His ring (around 8.5 ppm) could be clearly measured, yielding a very long value for the T_2, unreliable to estimate the mobility of the backbone of the polypeptide chain.

The DOSY-NMR measurements at pH 8.0 yielded a translational diffusion coefficient (D) with a value of $(7.2 \pm 0.2) \times 10^{-7}$ cm^2 s^{-1} (Figure S1A). By taking into account the hydrodynamic radius, R_S, of dioxane (2.12 Å), and its D under our conditions ($(6.8 \pm 0.2) \times 10^{-6}$ cm^2 s^{-1}), the R_S (Stokes radius) estimated for LrtA was 20 ± 2 Å. We can compare this value with that theoretically determined for a polypeptide with the length of LrtA. The R value for an unsolvated, ideal, spherical molecule can be estimated from [14]: $R = \sqrt[3]{3M\overline{V}/4N_A\pi}$, where N_A is Avogadro's number, M is the molecular weight (22.717 kDa) and $\overline{V}$ the specific volume of LrtA (0.729 mL/g). The calculated R for LrtA is 18.7 Å, but since the hydration shell is 3.2 Å wide [15], the hydration radius would be 21.9 Å, which is similar to the R_S from DOSY-NMR. On the other hand, it has been shown that the R_S of a folded spherical protein can be approximated by [16]: $R_S = (4.75 \pm 1.11)N^{0.29}$, where N is the number of residues; in a 197-residue-long protein such as LrtA, this expression yields 21 ± 6 Å, similar to that determined by DOSY-NMR experiments. Therefore, by the DOSY-NMR measurements, we are only detecting a monomeric globular species of LrtA. In fact, the 1D ^{1}H-NMR spectrum of LrtA at pH 8.0 (with 500 mM NaCl) (Figure S2) corresponds to that of a well-folded protein with dispersed peaks in the methyl and amide regions; interestingly enough, the spectrum had down-field shifted H$_\alpha$ protons (between 5.0 and 6.0 ppm), suggesting the presence of residues involved in β-strands. It is interesting to note that, as possible higher-order molecular species could not be observed in the NMR spectrum due to their molecular weight (and therefore signal broadening), the presence of the majority of the amide protons belonging to possible disordered regions in the protein (see below, 2.4.) would not be observed in the amide region due to: (i) solvent hydrogen-exchange at pH 8.0; (ii) conformational exchange broadening; or (iii) overlapping with the signals from the well-folded region (as indicated by the largest increase of intensity around 8.3 ppm, Figure S2B). In addition, the alkyl resonances belonging to possible polypeptide disordered regions would be hindered by the rest of the methyl groups of the protein in the up-field shifted region. On the other hand, the 1D ^{1}H-NMR spectrum at pH 4.5 showed a smaller intensity and a poorer signal-to-noise ratio (due probably to the precipitation during sample preparation) than the spectrum at pH 8.0. In addition, the spectrum also showed broader peaks (and less intense, as it is evident by comparing Figure S3B with Figure S2B), and the absence of well-dispersed signals in the amide and methyl regions (Figure S3), as shown, for instance, by the lack of peaks around 0.3 and 0.5 ppm, which appeared at pH 8.0. The broader peaks at pH 4.5 (when compared to pH 8.0) could be due to the presence of uni-molecular conformational-exchange equilibria or, alternatively, to the presence of self-associated species. Thus, to elucidate whether at pH 4.5 there were concentration-dependent equilibria, we carried out far UV CD experiments at protein concentrations of 4.5 and 9.8 μM (in protomer units); these experiments (Figure S4) showed that the molar ellipticity and the shape of the spectra were protein-concentration-dependent. Therefore, these results suggest that the conformation of the protein and its self-associated features were different at the two pH values of 4.5 and 8.0.

Second, we measured the hydrodynamic features of LrtA by using DLS at several protein concentrations. Two peaks were identified in the size distribution analysis for the concentration of 68 μM (in protomer units): the first peak with $R_S = 39 \pm 5$ Å that accounts for the 97.5% of the protein in the solution and a second peak with $R_S = 409 \pm 200$ Å corresponding to a small amount of aggregates (Figure 1A). Taking into account that particle scattering intensity is proportional to the square of the molecular weight, a small percentage (in this case 2.5%) of protein aggregates dominates the intensity distribution, which can be misleading, and therefore, the results in Figure 1A are shown as size distribution by volume instead of intensity. That R_S, obtained from the first peak, corresponds to a molecular weight of 81 kDa, by using an empirical mass vs. size calibration curve in the instrument software. The experiments at different LrtA concentrations indicate that the R_S (obtained from the volume peak measurements of the first observed peak) varied with the protein concentration. These findings suggest the presence of a self-associated equilibrium at pH 8.0 (Figure 1B) involving the protein, as also described in other oligomeric proteins [17]. It is interesting to note that these oligomeric species should not be observed in NMR due to their large molecular weights [15].

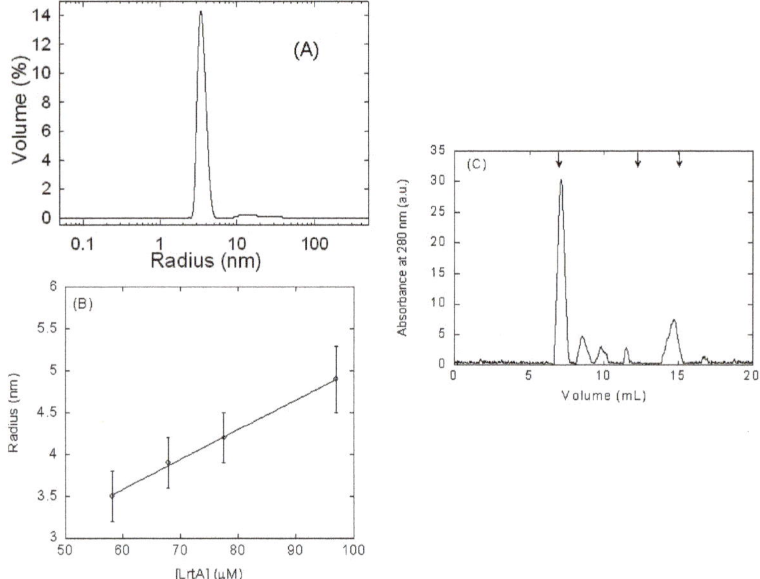

Figure 1. Hydrodynamic measurements of LrtA by dynamic light scattering (DLS) and size exclusion chromatography (SEC): (**A**) DLS measurements of the hydrodynamic radius RS of LrtA as a function of the percentage of the volume peak at 68 μM concentration (in protomer units). (**B**) Variation of the calculated R_S with LrtA concentration (in protomer units). Error bars are standard deviations from the fitting to a spherical shape. (**C**) SEC chromatogram of LrtA at 97 μM (in protomer units) at pH 8.0 (50 mM Tris) in 0.7 M NaCl; the arrows at the top indicate (from left to right) the elution volumes of blue dextran (7.1 ± 0.1 mL), albumin (12.1 ± 0.1 mL; 63.7 kDa), and bovine RNase A (15.1 ± 0.1 mL; 15.7 kDa) (the errors are standard deviations of three independent measurements). Chromatogram was baseline-corrected by UNICORN 5.01 software (GE Healthcare), and therefore, the origin of the sharpening observed in the peaks. Experiments were carried out at 25 °C.

Third, we also detected the presence of oligomeric species using glutaraldehyde cross-linking, which can react as monomer, but also as a heterogeneous polymer, involving accessible lysine residues. Our results (Figure S5) indicate that in the presence of a final concentration of 1% gutaraldehyde, there were dimers (which appear close to the band of the protein marker at 48 kDa) and other high-molecular-weight species with molecular weights larger than 210 kDa, at the top of the SDS-PAGE (sodium dodecyl sulfate polyacrylamide gel electrophoresis) gel lanes. These high molecular weight species could be due to the presence of cross-linked dimeric species.

Next, we used SEC in a Superose 12 10/300 GL, in buffer pH 8.0 (50 mM Tris) with 0.7 M NaCl to elucidate whether the protein behaved as an oligomer. We used such high concentrations of NaCl to avoid, as much as possible, any kind of protein-column interactions (as those we have observed to occur with other kind of matrix columns, see below Section 4). The protein markers used to calibrate the column were also loaded in the same buffer. At loading concentrations of 97 µM (in protomer units) of LrtA, the protein eluted as several peaks (errors are standard deviations from three independent measurements) at 7.3 ± 0.2, 8.5 ± 0.1, 9.8 ± 0.2, 11.5 ± 0.1, and 14.7 ± 0.2 mL (Figure 1C). These peaks, especially that at 7.3 mL, indicate that the protein behaved as an oligomer, with molecular weights larger than those of albumin (63.7 kDa) and bovine RNase A (15.7 kDa), although other higher-order molecular species of LrtA were present in solution. The peak at 14.7 mL could be due to the monomeric species, which was observed under these conditions. The other oligomeric species could be assigned to hexamers (11.5 mL) and dodecamers (9.8 mL), whereas the other two could be due to the presence of aggregates (as those species detected in DLS, see above); however, it is important to indicate that some of the peaks could be also due to protein-column interactions even in the presence of high NaCl concentration.

Finally, we measured the fluorescence lifetimes of LrtA at different protein concentrations. The experimental decay of the total protein fluorescence was best fit to bi-exponential functions (Table 1, Figure S6), and thus, two lifetimes were observed; attempts to fit the experimental data to more than two exponentials led to an increase in the χ^2. At any of the protein concentrations, the shortest lifetime corresponded to the largest amplitude (a_1), and it did not change with the protein concentration. Interestingly enough, the longest lifetime (as well as its amplitude, a_2) was concentration-dependent (Table 1). Furthermore, the $<\tau>$ showed also a concentration-dependence: going from a value of 6 ns (at the smallest concentration) to 1 ns at 98 µM. It is well-known that the intrinsic fluorescence lifetime of the first excited electronic singlet state does not change, but due to various quenching processes, changes in the environment around the fluorophores or even conformational changes in the molecular species, the measured lifetime is different from the intrinsic one [12,13]. Thus, even ruling out a possible fitting of the data to a monomer $\leftrightarrow$ oligomer equilibria because we are observing the lifetimes of eight Tyr residues, each of them with a different environment, we can conclude that the concentration-dependence observed in the $<\tau>$ (Table 1) should have its origin in association-dissociation events.

Table 1. Fluorescence lifetimes of LrtA (100 mM, phosphate buffer (pH 8.0), with 500 mM NaCl) at 25 °C [a].

Concentration (µM)	τ_1 (ns)	a_1	τ_2 (ns)	a_2	$<\tau>$ (ns)	χ^2
98	0.49 ± 0.08	1.9 ± 0.2	3.4 ± 0.1	0.072 ± 0.005	1.101	1.384
9.8	0.46 ± 0.07	2.0 ± 0.2	3.6 ± 0.2	0.048 ± 0.006	0.9622	1.528
7.8	0.54 ± 0.08	1.8 ± 0.2	2.7 ± 0.2	0.07 ± 0.01	0.9037	1.35
1.9	0.69 ± 0.09	0.71 ± 0.09	14.9 ± 0.8	0.0156 ± 0.0008	5.233	1.22
0.98	0.48 ± 0.07	1.0 ± 0.1	19.5 ± 0.4	0.0104 ± 0.0005	5.996	1.238

[a] Errors are from fitting to a bi-exponential function.

It could be thought that the detected self-associated species could be random-oligomers; although we cannot rule out the presence of aggregates (from the DLS, SEC, and glutaraldehyde cross-linking results), however, there are at least three pieces of evidence suggesting that the self-associated species do not oligomerize un-specifically: (i) the DLS results show a linear dependence with the concentration (Figure 1B); (ii) the protein-dependent, and almost exponential, variation of the life-times; and, (iii) the presence of bands at particular molecular weights in the glutaraldehyde experiments. Therefore, those results, together with experiments from (GdmCl) chemical denaturations (see below, Section 2.3.), must be due to the presence of self-associated equilibria, involving the regions around some of the eight Tyr residues.

2.2. LrtA Acquired a Native-Like Conformation Between pH 6.0 and 9.0

We analyzed the structure of LrtA at varying pH to find out in which interval the protein acquired a native-like conformation. To this end, we used several biophysical techniques, namely, intrinsic and ANS fluorescence, CD and NMR. We used intrinsic fluorescence to monitor changes in the tertiary structure around its eight Tyr residues. Furthermore, ANS fluorescence was used to monitor the burial of solvent-exposed hydrophobic patches. We acquired far-UV CD spectra to monitor the changes in secondary structure. Finally, we acquired 1D ^{1}H NMR spectra that show the presence of secondary and tertiary structure at physiological pH (see above, Section 2.1.). These spectra indicate (Figures S2 and S3) that the secondary and tertiary structures of the protein at pH 4.5 and 8.0 were completely different.

2.2.1. Fluorescence

Intrinsic Steady-State Fluorescence and Thermal Denaturations—The fluorescence spectrum of LrtA at physiological pH showed a maximum at 308 nm, as expected for a polypeptide chain with fluorescent Tyr residues. The pH-dependence of the intrinsic $<1/\lambda>$ showed two transitions (Figure 2A, left axis, filled circles). The first transition finished at pH 6.0, but we could not determine its pK_a due to the absence of an acidic baseline. This transition was probably due to the titration of some of the seventeen Glu and/or twelve Asp residues of the LrtA sequence [18,19], which can alter the environment around some of the Tyr residues. However, we cannot rule out that it could be also due to the titration of some of the eight naturally-occurring His residues, taking place at an unusually low pK_a. The second transition occurred at basic pH, starting at pH > 9.0, but, in this case, we could not determine the pK_a due to the absence of a baseline at the highest pH values. This transition was probably due to the titration of at least some of the eight Tyr residues in the sequence. Therefore, the changes observed in fluorescence as the pH was changed could be due to titrations of specific residues around the Tyr residues, or alternatively, to conformational changes involving those fluorescent amino acids.

Thermal denaturations of LrtA were carried out at several pH values with a protein concentration of 9.8 μM, in protomer units. At pH values larger than 6.0, we observed an irreversible broad transition (Figure 2C, left axis, blank circles). Below pH 6.0, we did not observe any sigmoidal behavior, and we did not observe any sigmoidal transition at pH 13.0 either (Figure S7A). It could be argued that as fluorescence is intrinsically temperature-sensitive [12,13], we are not monitoring the denaturation of the protein. However, it must be kept in mind that fluorescence temperature sensitivity is linear (as observed at low pH values, Figure S7A; or in the native and unfolded baselines of the curve shown in Figure 2C), but it is not sigmoidal as observed in the denaturations at pH 7.0, with a midpoint around 45 °C (Figure 2C), or at pH 8.4 (Figure S7A). We also carried out experiments at LrtA concentrations of 5 μM, in protomer units (Figure 2C), and denaturation was also irreversible. Therefore, irreversibility was not associated with the amount of protein used during thermal denaturations.

ANS-Binding—At low pH, the ANS fluorescence intensity at 480 nm was large and decreased as the pH was raised (Figure 2A, right axis, blank squares), suggesting that LrtA had solvent-exposed hydrophobic regions. We could not determine the pK_a of this titration due to the absence of an acidic baseline. The burial of solvent-exposed hydrophobic residues was complete at pH 6.0, as it happens with the transition observed by following the intrinsic fluorescence (see above). Since ANS reports on burial of hydrophobic surface, and therefore it monitors conformational changes, we must conclude that the protein had structural changes at acidic pH values; then, the variations monitored by intrinsic fluorescence (see above) at acidic pH must be associated with conformational changes due to the protonation of Asp and Glu residues (or the other amino acids described above).

In conclusion, our results indicate that, at low pH values, LrtA had solvent-exposed hydrophobic regions.

Solvent-Exposure of Tyr Residues Monitored by Iodide and Acrylamide Quenching—We carried out quenching experiments at pH 3.0, 7.0, and 11.0, because these are the three regions where we observed a different intrinsic fluorescence behavior of LrtA (Figure 2A). We used two quenching agents because of the charge effects probably occurring at extreme pH value with I^-. We have assumed that the fluorescence lifetimes of the self-associated protein (for a fixed protein concentration) did not change in the whole pH interval. The K_{sv} values for KI and acrylamide in the absence of denaturant were smaller than those measured in other proteins containing only Tyr residues [20,21] (Table 2). As a general trend, the K_{sv} values of LrtA in the presence of acrylamide were smaller at acidic pH values than at physiological or basic ones; these differences could be due to the presence of higher-order self-associated species at the acidic pH values (as suggested by the ANS results (Figure 2A) and the CD data at low pH (Figure S4)). Furthermore, these results indicate that the structure of LrtA underwent some conformational changes at acidic pH (in agreement with results from intrinsic and ANS fluorescence, Figure 2A, and the NMR results, Figures S2 and S3). In the presence of GdmCl (guanidine hydrochloride), the K_{sv} values were larger (either in KI or acrylamide) than those in the absence of denaturant (Table 2), suggesting that Tyr residues were more solvent-exposed.

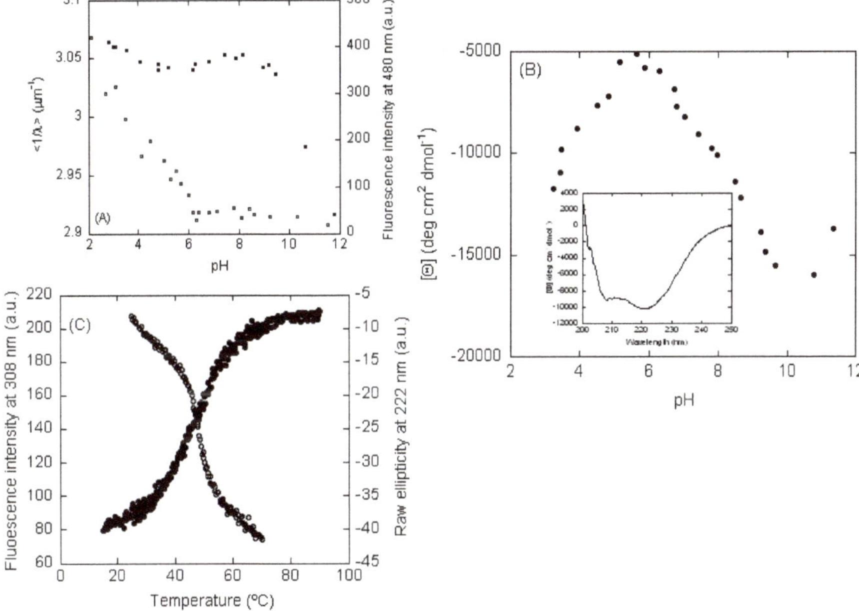

Figure 2. pH-denaturation of LrtA: (**A**) Intrinsic (left axis, filled circles) and ANS (right axis, blank squares) fluorescence of LrtA, as the pH was modified. (**B**) Changes in the [⊖] at 222 nm as the pH was varied (filled circles). Inset: far-UV CD spectrum of LrtA at 100 mM phosphate buffer (pH 8.0), with 500 μM NaCl at 25 °C. (**C**) Thermal denaturations followed by intrinsic fluorescence (left axis, blank circles) at pH 7.0 and 5 μM (in protomer units) of LrtA, and raw ellipticity at 222 nm at pH 7.0 and 9.8 μM, in protomer units (right axis, filled circles).

Table 2. Quenching parameters for LrtA under several conditions at 25 °C.

Solution Conditions	KI	Acrylamide	
	K_{sv} (M^{-1})	K_{sv} (M^{-1})	υ (M^{-1})
pH 3.0	- [a]	7.7 ± 0.5	2.1 ± 0.1
pH 7.0	1.63 ± 0.03	11 ± 1	2.2 ± 0.3
pH 11	0.78 ± 0.02	9.7 ± 0.5	0.5 ± 0.1
GdmCl (pH 7.0)	3.1 ± 0.1	22 ± 4	2.0 ± 0.3

[a] Not determined due to protein precipitation.

2.2.2. CD

The far-UV (ultraviolet) CD spectrum of LrtA at pH 8.0 had minima at 222 nm and 210 nm (Figure 2B inset), suggesting the presence of helix- or turn-like conformations. Decomposition of the far-UV CD spectrum at pH 8.0, by using the k2d algorithm, available online at the DICHROWEB site [22,23], yields a 30% of helical structure, 17% of β-sheet and 52% of random-coil. However, as LrtA has 8 Tyr, 4 Phe, and 14 His residues (eight naturally-occurring residues and six in the purification tail, see Section 4), we cannot rule out the absorbance of aromatic residues at this wavelength [24,25].

The molar ellipticity, [$\ominus$], at 222 nm showed a dumb-bell shape, with a maximum value at pH 6.0 (Figure 2B). These results suggest that there were changes in the secondary structure (or alternatively in the environment around aromatic residues [24,25]) above and below pH 6.0; interestingly enough, the changes at low pH mirrored those observed by intrinsic and ANS fluorescence (Figure 2A). Since the fluorescence results indicated that the environment around Tyr residues remained essentially unaltered until pH 9.0 (Figure 2A, filled circles), and taking into account the ANS-fluorescence (Figure 2A, blank circles), we can conclude that between pH 6.0 and 9.0, although the protein had a folded conformation (Figure S2A,B), either the secondary structure of the protein changed or, alternatively, the environment around some of the 4 Phe and 14 His residues in LrtA. We could not determine the pK_a values corresponding to the titrations at the two sides of the curve due to the absence of acidic and basic baselines, respectively.

As it happened with the thermal denaturations followed by fluorescence, the transitions followed by the ellipticity at 222 nm did not show any sigmoidal behavior below pH 6.0 (Figure S7B), but above that pH there was an irreversible broad transition (Figure 2C, right axis, filled circles).

To sum up, the spectroscopic probes (intrinsic and ANS fluorescence, CD, and NMR) indicate that LrtA acquired a native, with well-folded regions (Figure S2) from pH 6.0 to 9.0.

2.3. LrtA Showed an Irreversible Complex Unfolding Equilibrium

As the thermal denaturations were irreversible (either followed by fluorescence or CD, Figure 2C and Figure S7), we tried to determine the conformational stability of LrtA by using GdmCl-denaturations followed by fluorescence and CD (urea-denaturations followed by fluorescence did not show any sigmoidal behavior, Figure S8A, inset). The tendency in both refolding and unfolding curves, followed by fluorescence and CD, was the same (Figure S8); however, the refolding CD results indicate that the final ellipticity, acquired by the native state, was not the same as that in the unfolding experiments. Moreover, the fluorescence refolding curves indicate that the value of the <1/λ> was different to that in the unfolding ones, even though we used the same protein concentration. Then, we conclude that there was a hysteresis behavior, and chemical denaturations were also irreversible, as it could be expected for a protein composed of several domains (see below, Section 2.4.).

In addition, comparison of the unfolding CD and fluorescence results suggest that the unfolding of LrtA was not a simple two-state process, as the denaturation curves by both techniques were different. Whereas fluorescence curves showed two transitions, CD reported a sole one, whose apparent midpoint did not overlap with that of the fluorescence (Figure S8). Fluorescence denaturation curves at several protein concentrations (in the range from 1.9 to 19 µM (in protomer units)) indicate that the first transition monitored corresponded to a protein-concentration-dependent process (Figure S8B inset), as at high protein concentrations (19 µM) two transitions were observed, with apparent midpoints around 1.0 and 2.0 M GdmCl. This result further confirms that LrtA was an oligomeric protein.

2.4. Sequence Properties and Molecular Modeling of LrtA

The primary structure of LrtA possesses a relatively large fraction of charged residues, both acidic and basic. Due to their high hydrophilicity, these residues tend to hamper the hydrophobic collapse and increase disorder in the protein backbone. Predictors of local disorder [26–29] based on a variety of physical properties (Figure S9) were used to estimate the propensity of the protein sequence to fold. There is a consensus indicating the region around residues 100–130 is highly disordered, together with a few other residues at both protein termini. Since the experimentally determined radius of monomeric LrtA has a value close to that of a compact protein (see Section 2.1., DOSY-NMR results), this may indicate that the region 100–130 was either a long, disordered loop within a single domain protein or a coil region separating two distinct but spatially close domains.

The overall propensity of LrtA to fold into a well-structured protein was also explored by mapping its properties in terms of charge and hydropathy (Figure 3). In particular, Figure 3A compares the location of LrtA sequence within an Uversky diagram [30], which provides indication on the possibility that the protein belongs to the IDP (intrinsically disordered protein) class through the identification of a boundary hydropathy that separates folded and unfolded polypeptides. When the overall primary structure was considered, LrtA fell into the region of the diagram that is mostly populated by well-folded proteins [31], although also accessible to a few IDPs. However, when the sequence of LrtA was divided into two separate portions, they had distinct features. The first-half of LrtA sequence (residues 1–100) more distinctly belonged to the region occupied by ordered polypeptides. In contrast, the second-half (residues 100–191) fell in the region of the diagram that is populated by IDPs, although also accessible to some well-folded proteins. On the other hand, a Das–Pappu diagram [32] showed that LrtA should be considered a so-called 'Janus sequence' in between weak and strong polyampholytes (Figure 3B), independently whether the whole protein or just the two halves of its sequence are considered. This observation strongly suggests that the structure of LrtA is context-dependent, and may easily become more expanded/collapsed or structured/unstructured according to the environment (such as solution conditions or the presence of biomolecular partners). From our experimental results, LrtA had folded regions at pH 8.0 in aqueous solution (Figure S2 and Figure 2B, inset), although with a small stability, as suggested by thermal denaturations (Figure 2C and Figure S7A).

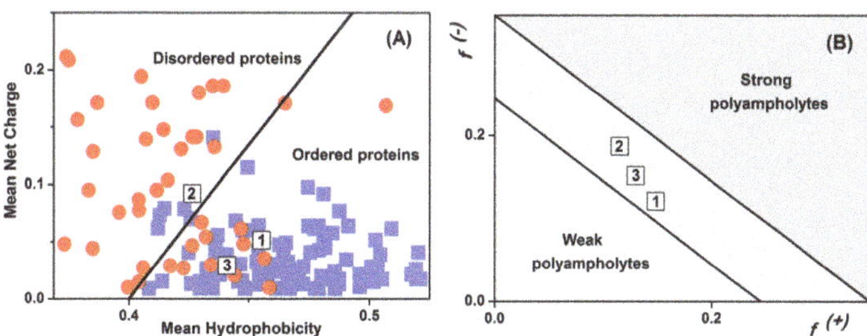

Figure 3. Location of LrtA in the diagram of state for charged polypeptides: Symbols "1", "2", and "3" indicate, respectively, the first two halves of the LrtA sequence (residues 1–100 and 101–191) and the whole protein. (**A**) Uversky plot based on the absolute mean net charge as a function of the mean scaled hydropathy, as obtained with PONDR [31]; well-folded (blue squares) and disordered proteins (red circles) are shown. (**B**) Das–Pappu plot based on the fraction f(+) and f(−) of positively and negatively charged residues, respectively [32].

With the aim of building a model for the secondary and tertiary structures of LrtA, the protein sequence was submitted to full-chain protein structure prediction servers [33–36]. A particularly interesting result was obtained by using I-TASSER [33], which is one of the most popular and accurate software for generating high-quality model predictions of tridimensional protein structures. The best models predicted by I-TASSER (Figure S10) all included a well-structured domain spanning the first 100 residues, followed by a collapsed and poorly-structured region. The well-structured domain consisted of two parallel α-helices, and a β-sheet formed by four anti-parallel β-strands. These models were remarkable because they predicted a degree of order in the structure of LrtA that is in reasonable agreement with our expectations based on the CD experimental results (see above, Section 2.2.2), whereas in most cases the algorithms tend to overestimate the amount of secondary structure when applied to intrinsically unfolded polypeptides. Furthermore, the absence of a defined folding topology for the second half of LrtA sequence is consistent with the theoretical predictions discussed above. It is worth mentioning that the C-terminal region of HPF of *S. aureus*, another member of the long sub-family of HPF, in EM (electron microscopy) preparations was folded [5], in contrast with our model; then, it seems that in LrtA from *Synechocystis* sp. PCC 6803, the C-terminal region has specific features, which might be related to protein function. Finally, the conformations predicted for the first-half of LrtA sequence were in common with those obtained with the other algorithms of structural modeling that we used (i.e., FALCON [34], SWISS-MODEL [35], and Robetta [36]), although details of the geometry and orientation of the α-helices and β-strands were in some models different. This was particularly intriguing, especially because a four-strand motif is typical of many RNA-binding proteins ([5] and references therein).

Our theoretical predictions are not difficult to reconcile with the findings show by NMR, at physiological pH, where the spectrum of LrtA was that of a folded molecule (Figure S2). In fact, the signals of the proton nuclei in the unfolded and folded halves of the protein had a different behavior. The amide protons of the unfolded half of the protein would appear between 8.0 and 8.5 ppm [37], where they would be probably obscured, although they should be sharper than the rest of the signals, by many of the amide resonances of the folded half (those of the residues connecting the α-helices and the β-strands); it is interesting to note, however, the presence of a higher intensity at 8.2–8.3 ppm (Figure S2B), which could be due to the sharper resonance of the unfolded region of LrtA. Furthermore, the majority, if not all, of the amide protons of the unfolded half will be broadened and exchanged with the solvent at pH 8.0 [37], as it has been observed to occur in other intrinsically disordered regions, when the pH is raised and even when the temperature is decreased at the highest explored pH [38]. However, we tried to acquire a 1D ^{1}H NMR spectrum at pH 6.9 (in the presence of 0.5 M NaCl) and 15 °C; under these conditions (Figure S11) some amide signals appearing between 7.8 and 8.3 became sharper, as expected for a disordered region that has a fast molecular tumbling. The methyl region, under these solvent conditions, was similar to that acquired at higher pH and temperature (Figure S2). In addition, all the methyl peaks corresponding to the side-chains of Val, Ile, and Leu residues of the disordered half of the protein under any of the conditions explored (pH 8.0, 20 °C or pH 6.9, 15 °C) would appear at basically the same chemical shifts as those of the corresponding folded region, i.e., around 0.8 ppm [37].

The models predicted by I-TASSER provided a static picture of LrtA that does not take into account its dynamics, which could be expected to be significant to determine the properties of such a chameleonic protein. Furthermore, the main difference between the predicted models and our experimental findings is the presence of a larger amount of β-structures in the former. Thus, we suspected that the structure predicted corresponded to the most stable structure that LrtA can assume, e.g., under ideal conditions in solution or when bound to a partner molecule, although, the experimental evidence (Figure 2C, Figures S7 and S8) suggests that this structure was not very stable. For those reasons, we used MD simulations to study the behavior of the protein structure both at room and high temperatures. The latter case corresponds to the simplest and most direct way to investigate the dynamics of a protein under non-native conditions [39], speeding up the sampling by overcoming the energetic barriers that restrain the structure in a given conformation.

The MD results showed that the region including residues 1–100 is stable and maintains its folding topology and structure when simulated at room temperature (Figure S12). In particular, as shown in Figure 4, Tyr19 and Tyr77 interact to fix the two α-helices, whereas the other two Tyr residues are on the opposite face of the protein (Figure 4B). In contrast, when the temperature was raised, Tyr19 and Tyr77 lost their coordination and the first N-terminal β-strand immediately started losing its anchoring with the rest of the protein and the β-sheet scaffold (Figure 4B), increasing the amount of coil and helical structure. The anchoring was not recovered in annealing runs performed by reducing back the temperature, unless they were started at the earliest step of the local unfolding process. This finding suggests that the folding of the N-terminal region of LrtA was possibly assisted by interactions with other biomolecules, which may include other monomers of LrtA or binding partners, such as RNA. In contrast, the rest of the protein was very stable, and did not lose its folding topology even under the most extreme simulation conditions (Figure 4C).

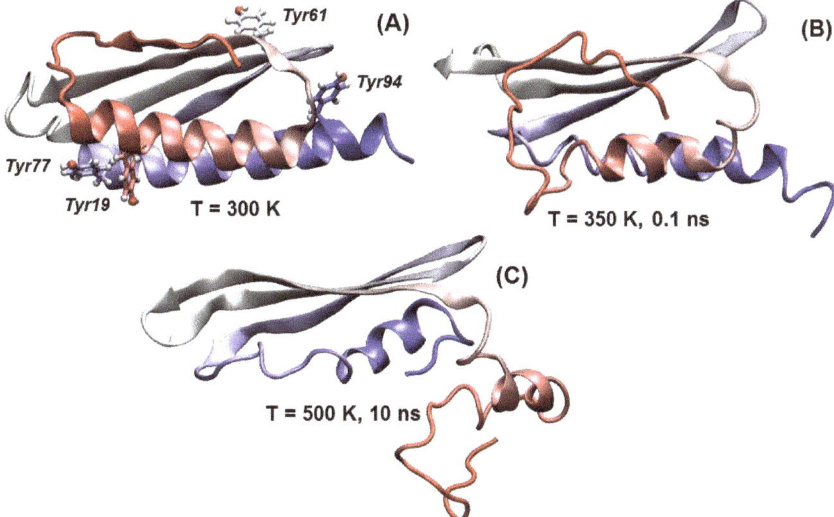

Figure 4. Dynamic behavior of the predicted folded domain of LrtA: The region comprising residues 1–100 of LrtA is shown in cartoon representation (colored from red (N terminus) to silver-white (mid-sequence regions) up to blue (C terminus of the domain)). (**A**) Structure at room temperature, with Tyr residues indicated. (**B**) Simulation under unfolding conditions: in the N-terminal region (indicated in red), the first β-strand loses it structure and coordination with the rest of the β-sheet. (**C**) Structure under extreme conditions: the folding topology is maintained. VMD [40] is used for the protein displays.

3. Discussion

LrtA seemed to acquire a native-like conformation from pH 6.0 to 9.0. Changes in secondary (far-UV CD) and tertiary (intrinsic fluorescence) structures, and in the burial of hydrophobic residues (ANS fluorescence) occurred concomitantly at low pH. Under acidic conditions (pH 3.0), species with a higher amount of secondary structure (as indicated by a larger (in absolute value) ellipticity, Figure 2B) appeared to be populated, although they had hindered solvent-accessibility towards I⁻ and acrylamide quenchers (Table 1). Therefore, at low pH, LrtA had non-native conformations with non-stable secondary and tertiary structures (as judged by the absence of a sigmoidal shape in thermal denaturation curves). These results are further supported by the NMR spectra acquired at low pH, where there was no dispersion of amide or methyl signals (Figure S3), suggesting the presence of conformations with an unfolded structure. Besides, the broadening observed in the methyl and amide regions suggested the presence of aggregation, which was further confirmed by the far-UV CD spectra at pH 4.5 at different protein concentrations. Then, the protein at acidic pH had a larger tendency to associate than at physiological pH. The increase of ANS fluorescence at low pH, indicating a large solvent-accessible hydrophobic surface area, may appear difficult to reconcile with the results of I⁻ or acrylamide quenching, suggesting a smaller solvent accessibility towards Tyr (Table 1). However, the larger amount of hydrophobic surface area (monitored by ANS), that became solvent-exposed at acidic pH values, could involve several of the Val, Leu, and Ile residues, which are highly abundant in LrtA (46 out of 197 amino acids).

LrtA had well-folded regions in the pH range from 6.0 to 9.0, as indicated by: (i) the sigmoidal curves in the thermal and chemical denaturations (fluorescence and far-UV CD) (Figure 2C, Figure S7 and S8); and (ii) the 1D ^{1}H-NMR spectrum at pH 7.0 and 8.0 (Figure S2). Moreover, the protein was capable of binding homogenous yeast RNA (from Sigma) with an affinity of ~1 μM (Figure S13), and therefore the purification protocol did not affect the conformational features of the protein. However, this structure was not highly rigid, as judged from the apparent thermal midpoint obtained from the irreversible denaturation curves (~40 °C, Figure 2C); these findings agree with the MD results in this work. As the secondary and tertiary structures of LrtA are only stable in a narrow thermal range, an increase in the temperature environment reduces the availability of well-folded and active protein, and therefore, a larger amount of protein is needed to carry out the cyanobacterial functions. The LrtA structure under native conditions, as suggested by the deconvolution of CD data, had a smaller percentage of α-helix structure than other HPF members (30% vs. 45%), as well as a lower percentage of β-sheet (17% vs. 27%). These experimental percentages were confirmed by the results of our MD simulations.

LrtA was an oligomeric protein at physiological pH. We showed that some of the Tyr residues seemed to be involved in the self-associating interface, as judged by the changes in the fluorescence lifetimes (Table 1) or the protein-concentration dependence of the curve denaturations midpoints (Figure S8B inset). In addition, our MD simulations at room temperature suggest that Tyr19 and Tyr77 in the folded domain of the protein were key in anchoring amino acids of the β-sheet, and the loss of such anchoring during the high-temperature simulations caused a partial disruption of the protein β-sheet. Then, Tyr residues were important for quaternary and secondary scaffolding in LrtA. Recently, it has been observed that, in EM preparations, the HPF of *S. aureus* (another member of the long HPF subfamily) is forming domain-swapped dimeric species [5]. Furthermore, crystals of the short HPF from *Vibrio cholerae* show the presence of dimers mediated by Co (II) anchoring residues of the β-sheets of two monomers [11], further pinpointing to the crucial importance of residues in the β-sheet for a possible quaternary arrangement of any HPF member. However, the importance of oligomerization for the function of all those proteins (included LrtA) remains to be elucidated, as it could be an adaptive mechanism of regulation to interact with other proteins or even with RNA.

4. Materials and Methods

4.1. Materials

Deuterium oxide and IPTG was obtained from Apollo Scientific (Stockport, UK). Sodium trimethylsilyl [2,2,3,3-^{2}H$_4$] propionate (TSP), imidazole, DNase, Trizma base and acid, yeast RNA, glutaraldehyde (25% *w/v* solution), ANS, deuterated acetic acid, its sodium salt and His-Select HF nickel resin were from Sigma-Aldrich (Madrid, Spain). The β-mercaptoethanol (β-ME) was from BioRad (Madrid, Spain). Triton X-100 and protein marker, PAGEmark-tricolor (G Biosciences) were from VWR (Barcelona, Spain). Dialysis tubing, with a molecular weight cut-off of 3500 Da, was from Spectrapor (Spectrum Laboratories, Breda, The Netherlands). Amicon centrifugal devices with a cut-off molecular weight of 3000 Da were from Millipore (Barcelona, Spain). Standard suppliers were used for all other chemicals. Water was deionized and purified on a Millipore system.

4.2. Protein Expression and Purification

Expression of LrtA was carried out in BL21(DE3) or C41 [41] strains with a final ampicillin concentration of 100 mg/mL at 37 °C. The cells were cultured in 1 L flasks. Protein expression was induced with a final concentration of 1.0 mM IPTG when the absorbance of the cell culture at 600 nm was 0.4–0.9, and the cells were grown for 15–16 h at 37 °C. Cells were harvested at 8000 rpm in a JA-10 rotor (Beckman Coulter, Miami, FL, USA) for 15 min. The pellet from 5 L of culture was re-suspended in 50 mL of buffer A (500 mM NaCl, 5 mM imidazole, 20 mM Tris buffer (pH 8), 0.1% Triton X-100 and 1 mM β-ME), supplemented with a tablet of Sigma Protease Cocktail EDTA-free and 2 mg of DNase (per 5 L of culture). After being incubated with gentle agitation at 4 °C for 10 min, cells were disrupted by sonication (Branson sonicator, 750 W, Richmond, VA, USA), with 10 cycles of 45 s at 55% of maximal power output and an interval of 15 s between the cycles. All the sonication steps and the interval waits were carried out in ice. The lysate was clarified by centrifugation at 18,000 rpm for 40 min at 4 °C in a Beckman JSI30 centrifuge with a JA-20 rotor (Beckman Coulter, Miami, FL, USA).

The clarified lysate from such first centrifugation did not contain a large amount of LrtA, and thus, we suspected that most of the protein was present in the cell debris precipitate. Therefore, the precipitate was treated with buffer A supplemented with 8 M urea and a tablet of Sigma Protease Cocktail EDTA-free and 2 mg of DNase (per 5 L of culture). The re-suspended sample was treated with another 10 cycles of sonication in ice, and the sample was clarified by centrifugation at 20,000 rpm for 30 min at 4 °C. LrtA was in the supernatant and was purified by immobilized affinity chromatography (IMAC). The supernatant was added to 5 mL of Ni-resin previously equilibrated in buffer A supplemented with 8 M urea. The mixture was incubated for 20 min at 4 °C, and afterwards, the lysate was separated from the resin by gravity. On-column refolding was carried out during the washing step with 20 mL of buffer B (20 mM Tris buffer (pH 8.0), 500 mM NaCl, 1 mM β-ME, and 20 mM imidazole); the protein was eluted by gravity from the column with buffer C (20 mM Tris buffer (pH 8.0), 500 mM NaCl, 1 mM β-ME, and 500 mM imidazole). The eluted LrtA was extensively dialyzed against buffer D (100 mM sodium phosphate buffer (pH 8.0) with 500 mM NaCl). Precipitate in the dialysis tubing after five dialysis steps in buffer D was removed by centrifugation at 20,000 rpm for 30 min at 4 °C. The final yield of protein was 4.5–6.5 mg/L of culture (with both cellular strains assayed), and the protein was 90–95% pure as judged by SDS gels (Figure S14). This purity percentage takes into account the possible contamination due to the presence of deoxyribonucleotides, as judged by their absorbance at 260 nm (see below).

We attempted to re-purify the protein recovered from IMAC by using gel filtration chromatography in a Superdex 16/600, 75 pg column (GE Healthcare, Barcelona, Spain) connected to an AKTA FPLC system (GE Healthcare) by monitoring the absorbance at 280 nm; nevertheless, the protein was bound to the column and did not come out within its bed volume. Binding to the column has been also observed during purification of the recombinant HPF from *Staphylococcus aureus* [5], another member of the long subfamily of HPF.

The eluted protein from IMAC showed absorbance at 260 nm, suggesting that it was probably contaminated with di- or tri-deoxyribonucleotides (from the cleavage of the DNase used during purification and even though sample was dialyzed against 500 mM of NaCl). Presence of deoxyribonucleotides has been also observed in the recombinant HPF from *S. aureus* after its purification [5]. We tried to remove the deoxyribonucleotides by using different concentrations of polyethylenimine (PEI), ranging from 0.2 to 1% (*v/v*) [42], but most of LrtA co-precipitated with the oligonucleotides. The total protein concentration, P_c (in mg/mL) was determined by using the expression [43]: $P_c = 1.55\ A_{280} - 0.75\ A_{260}$, where A_{280} and A_{260} are the absorbance of the dialyzed protein solution at 280 and 260 nm, respectively. However, it is important to note that the presence of deoxyribonucleotides did not affect the spectroscopic signals either of both fluorescence and far-UV

CD, because DNA is spectroscopically silent in fluorescence, and in the far-UV region of CD spectra between 210–240 nm, deoxyribonucleotides do not absorb [44,45]. It is also interesting to note that deoxyoligonucleotides in aqueous solution or intact DNA show a small or slightly positive ellipticity around 222 nm [45–49] where we have carried out the study of the CD biophysical properties of the protein (see Results section).

4.3. Fluorescence

Fluorescence spectra were collected on a Cary Varian spectrofluorimeter (Agilent, Foster City, CA, USA), interfaced with a Peltier, at 25 °C. LrtA concentration in the pH- or chemical-denaturation experiments was 9.8 μM (in protomer units). For experiments with ANS, a final probe concentration of 100 μM was added. A 1-cm-pathlength quartz cell (Hellma, Mullheim, Germany) was used.

In the pH-induced unfolding curves, the pH was measured after completion of the experiments with an ultra-thin Aldrich electrode in a Radiometer pH-meter (Madrid, Spain). The acids and salts used were: pH 2.0–3.0, phosphoric acid; pH 3.0–4.0, formic acid; pH 4.0–5.5, acetic acid; pH 6.0–7.0, NaH_2PO_4; pH 7.5–9.0, Tris acid; pH 9.5–11.0, Na_2CO_3; pH 11.5–13.0, Na_3PO_4. Chemical and pH denaturations were repeated three times with new samples at any of the concentrations assayed. Appropriate blank corrections were made in all spectra both in pH- and chemical-denaturation experiments.

For GdmCl-denaturation experiments the samples were prepared the day before from a 7 M GdmCl concentrated stock and left overnight to equilibrate; before experiments, samples were left to 25 °C for 1 h. For the refolding experiments, the sample was exchanged in 7 M GdmCl by using Amicon centrifugal devices; protein concentration was the same as in the unfolding experiments.

The emission intensity weighted average of the inverse wavelengths (also called the spectrum mass center, or the spectral average energy of emission), <1/λ>, was calculated as described [50]. Briefly, we define <1/λ> as: $\langle 1/\lambda \rangle = \sum_{1}^{n} \frac{1}{\lambda_i} I_i / \sum_{1}^{n} I_i$, where I_i is the intensity at wavelength λ_i. We shall report <1/λ> in units of μm^{-1}.

Steady-State Spectra—The experimental set-up for the intrinsic and ANS fluorescence pH-denaturation experiments has been described previously [50]. Briefly, protein samples were excited at 278 nm, for the intrinsic fluorescence, and 380 nm for the ANS experiments. In all cases, excitation and emission slits were 5 nm. The experiments were recorded between 300 and 400 nm (for the intrinsic fluorescence) and between 400 to 600 for the ANS experiments. The signal in all cases was acquired for 1 s and the increment of wavelength was set to 1 nm. For the chemical denaturations, following intrinsic fluorescence, several protein concentrations were used in the range from 1.9 to 19.6 μM (in protomer units) at 100 mM phosphate buffer (pH 7.0) and 50 mM NaCl.

Thermal Denaturations—Thermal denaturations of isolated LrtA at different pH values were carried out with the same experimental set-up described [50] and protein concentrations of 9.8 and 5 μM (in protomer units). Briefly, these experiments were performed at constant heating rates of 60 °C/h and an average time of 1 s. Thermal scans were collected at 308 nm after excitation at 278 nm from 25 to 95 °C and acquired every 0.2 ° C.

Fluorescence Quenching—Quenching by iodide and acrylamide was examined at different solution conditions, with an LrtA concentration of 9.8 μM (in protomer units): pH 3.5 (formic buffer, 50 mM), pH 7.0 (phosphate buffer, 50 mM), and pH 11.0 (boric buffer, 50 mM). Experiments were also carried out in the presence of 6 M GdmCl at pH 7.0 (50 mM, phosphate buffer). The experimental set-up for both quenchers was the same described above for the intrinsic fluorescence experiments. The data for KI were fitted to [12]

$$\frac{F_0}{F} = 1 + K_{sv}[KI] \tag{1}$$

where K_{sv} is the Stern-Volmer constant for collisional quenching; F_0 is the fluorescence intensity in the absence of KI; and F is that at any KI concentration. The range of KI concentrations explored was 0–0.7 M. For experiments with acrylamide, the data were fitted to [12]

$$\frac{F_0}{F} = (1 + K_{sv}[acrylamide])e^{v[acrylamide]} \tag{2}$$

where v is the dynamic quenching constant. Fittings to Equations (1) and (2) were carried out by using Kaleidagraph (Synergy software, Dubai, United Arab Emirates).

4.4. Fluorescence Lifetimes

Lifetimes were measured on an EasyLife V™ lifetime fluorometer (Madrid, Spain) with the stroboscopic technique, by using as excitation source a pulsed light of a diode LED operating at 278 nm. The number of channels used for each scan was 500, and the integration time was 1 s. Three scans were averaged in each experiment, and they were repeated twice at 25 °C in 50 mM phosphate buffer (pH 8.0) and 500 mM NaCl.

The experimental fluorescence decays ($D(t)$) were fitted to a sum of exponential functions: $D(t) = \sum_{i=1}^{n} a_i \exp(-t/\tau_i)$, where τ_i is the the lifetime of the electronic excited states of the fluorescent species present in solution, and a_i the pre-exponential factor of those electronic states. The pre-exponential factors can be interpreted not only in terms of the populations of the corresponding species, but also in terms of the radiative probability constants of Tyr residues, in the case of LrtA. We also determined the mean lifetime, <τ>, as: $<\tau> = \sum_{i=1}^{n} f_i\tau_i$, where the f_is are defined as: $f_i = a_i\tau_i/\sum_j a_j\tau_j$. The fitting procedure of the experimental fluorescence lifetime curves used an iterative method based on the Levenberg–Marquardt algorithm [51]. The temporal width of the excitation pulse, which distorted the observed decay, was taken into account through the instrument response function (IRF), which was determined by using a scattered solution of Ludox. Goodness of the fittings was tested by using a reduced χ^2, that was calculated by measuring the spectral noise at time t, and determining the measurement uncertainties [52].

4.5. CD

The far-UV CD spectra were collected on a Jasco J815 spectropolarimeter (Jasco, Tokyo, Japan) fitted with a thermostated cell holder, and interfaced with a Peltier unit, at 25 °C. The instrument was periodically calibrated with (+)-10-camphorsulphonic acid. Several protein concentrations of LrtA (4.5, 9.8 and 19.6 μM, in protomer units) were used to test for concentration-dependent changes in the shape and intensity of the steady-state spectrum at two different pH values; differences were not observed at these concentrations (9.8 and 19.6 μM) at pH 8.0. However, protein-concentration-dependent changes (in the concentration range of 4.5 and 9.8 μM, in protomer units) were observed in the shape and ellipticity at pH 4.5, suggesting that at this pH the protein showed a larger tendency to self-associate (see above, Section 2.1, Figure S4) than at pH 8.0. Molar ellipticity was calculated as described [44,45,50].

Steady-State Spectra—Experiments were acquired with the same experimental set-up described previously [50]. Typically, spectra were acquired at a scan speed of 50 nm/min with a response time of 2 s and averaged over six scans with a bandwidth of 1 nm. Spectra were corrected by subtracting the baseline in all cases. Protein concentrations were 9.8 μM (in protomer units) for pH- and chemical-denaturation experiments (for both unfolding and refolding). The refolding samples of LrtA were prepared as described above for fluorescence.

Thermal Denaturations—Thermal denaturations were carried out with the same experimental set-up described previously [50] and a protein concentration of 9.8 μM. Briefly, thermal denaturations were performed at a constant heating rate of 60 °C/h, a response time of 8 s, a band width of 1 nm, and acquired every 0.2 °C. Thermal scans were collected in the far-UV region at 222 nm from 25 to 85 °C.

4.6. NMR Spectroscopy

The NMR experiments were acquired at 20 °C or 15 °C, when stated, on a Bruker Avance DRX-500 spectrometer equipped with a triple resonance probe and z-pulse field gradients.

1D ^{1}H-NMR experiments—Homonuclear 1D-^{1}H-NMR experiments were performed with LrtA at a concentration of 30 μM (in protomer units) in 0.5 mL, in 100 mM phosphate buffer (pH 8.0 or pH 6.9) and 500 mM NaCl in H_2O/D_2O (90%/10%, v/v), uncorrected for deuterium isotope effects. The 1D-^{1}H spectra under these conditions were acquired with 16 K data points, with 2 K scans and a 6000 Hz spectral width (12 ppm), with the WATERGATE sequence [53]. Baseline correction and zero-filling were applied before processing. We also acquired a spectrum at pH 4.5 (100 mM acetate) with 64 K scans, while the other experimental parameters were as above. During buffer exchange in Amicon centrifugal devices the sample precipitated and we had to increase the number of scans to have a good signal-to-noise ratio. All spectra were processed and analyzed by using TopSpin 2.1 (Bruker GmbH, Karlsruhe, Germany). TSP was used as the external chemical shift reference, taking into account the pH-dependence of its resonance [15].

Translational Diffusion Measurements—The DOSY experiments at pH 8.0 and 4.5 were performed with the pulsed-gradient spin-echo sequence, as described previously [50], with 16 gradient strengths ranging linearly from 2 to 95% of the total power of the gradient unit. At both pH values, the duration of the pulse gradient was 2.5 ms, and the time between the two gradients was 150 ms. Samples were exchanged in D_2O buffer (100 mM phosphate buffer (pH 8.0, not corrected by isotope effects) or 100 mM deuterated acetate buffer (pH 4.5, not corrected by isotope effects) with 500 mM NaCl at both pH values) by using Amicon centrifugal devices during 4 to 6 h. Precipitation was observed after the buffer exchange (from pH 8.0 to pH 4.5), and the sample was centrifuged at 13,000 rpm for five minutes, before putting it into the NMR tube. Further and more severe precipitation was observed during the D_2O buffer exchange at pH 4.5 and attempts to acquire a DOSY by increasing the number of scans (until 8 K) at each pulse field gradient strength failed. The methyl region was used for intensity measurements in the experiment acquired at pH 8.0.

4.7. DLS

DLS measurements were performed at 25 °C in 100 mM phosphate buffer (pH 8.0) with 500 mM NaCl at different protein concentrations. Experiments were performed at fixed angle ($\theta = 173°$) in a Zetasizer nano instrument (Malvern Instrument Ltd., Malvern, UK) equipped with a 10-mW helium-neon laser ($\lambda = 632.8$ nm) and a thermoelectric temperature controller. Experiments were analyzed with Zetasizer software (Malvern Instrument Ltd., Malvern, UK), and they were carried out as described [50]. Each sample was measured 10 times with 10 runs of 30 s each. The Z-average size was obtained by fitting the autocorrelation function with the cumulants method. The R_S was calculated by applying the Stokes–Einstein equation: $D = kT/6\pi\eta R_S$, where k is the Boltzmann constant and T is the temperature (in K). Experiments were also carried out at pH 4.5 (100 mM acetic acid) and 500 mM NaCl, but the poor signal obtained precluded any reliable conclusion.

4.8. Molecular Modelling

Disorder propensity was estimated by using scoring functions according to independent algorithms [26–29], as calculated through submission to their respective web servers. Three-dimensional models of the protein were obtained using the on-line structure predictors I-TASSER [33], FALCON [34], SWISS-MODEL [35], and Robetta [36]. Folding stability in the LrtA structure predicted by I-TASSER was assessed in MD simulations at room temperature performed with the GROMACS package [54]. The AMBER ff99SB-ILDN force field [55] and TIP3P model [56] were used for the protein and water, respectively, and other simulation conditions were as previously described [57,58]. High temperature simulations were performed at 400 and 500 K, according to a protocol previously adopted [59].

4.9. SEC

SEC experiments were carried out in a Superose 12 10/300 GL column at pH 8.0 (50 mM Tris) and 0.7 M NaCl, connected to an AKTA FPLC (GE Healthcare), and monitoring the absorbance at 280 nm. Flow rate was 0.7 mL/min, and the protein was loaded in a volume of 100 µL at a concentration of 97 µM (in protomer units). Baselines of the chromatograms were strongly curved and therefore the UNICORN software 5.0.2 (GE Healthcare, Barcelona, Spain) was used for automatic baseline correction. The markers used to calibrate the column were albumin, bovine RNase A and blue dextran (GE Healthcare); the isolated markers were loaded in the column in the above described buffer in three independent measurements.

4.10. Glutaraldehyde Cross-Linking

Glutaraldehyde cross-linking was carried out at pH 8.0 (50 mM Tris buffer) in the presence of 0.5 M NaCl at room temperature. Sample volume was 200 µL. Protein concentration was 97 µM (in protomer units). The protocol used was that described previously [60], taking aliquots of a fresh 25% (*w/v*) glutaraldehyde solution (Sigma-Aldrich) to yield a final concentration of the cross-linker in the protein solution of 1%. Aliquots of 30 µL were extracted at defined times and the cross-linking reaction was stopped by adding an identical volume of SDS-loading buffer to the aliquot. The extracted samples were run in a 12% SDS-PAGE gel immediately. Experiments were repeated twice.

5. Conclusions

Our results show that LrtA from cyanobacteria, a member of the long HPF subfamily, shows a clear propensity to self-associate and has an apparent low conformational stability. The structure was predicted to be formed by two domains, one of which was well-folded, whereas the other was disordered. These conformational features, together with the presence of a relatively high number of charged residues, provide an overall structural plasticity.

Supplementary Materials: The following are available online at http://www.mdpi.com/1422-0067/19/7/1857/s1, There are 14 figures, showing: the intensity curve of the methyl groups from the DOSY experiment at pH 8.0 and 500 mM NaCl (Figure S1); the spectroscopic characterization of LrtA by NMR at pH 8.0 and 500 mM NaCl and 20 °C (Figure S2); the 1D 1H-NMR spectrum of LrtA at pH 4.5 and 500 mM NaCl and 20 °C (Figure S3); far-UV CD spectra of LrtA at pH 4.5 at two different concentrations (Figure S4); the SDS-PAGE gel with the glutaraldehyde cross-linking results (Figure S5); the fluorescence lifetime at pH 8.0 and 500 mM NaCl of LrtA at 0.98 µM (in protomer units) (Figure S6); the thermal denaturations followed by fluorescence at different pH values (Figure S7); the chemical (urea and GdmCl) denaturations of LrtA followed by fluorescence and CD (Figure S8); the predictions of disorder probability along the sequence (Figure S9); several models of the LrtA structure as obtained from homology modeling (Figure S10); the spectroscopic characterization of LrtA by NMR at pH 6.9 and 500 mM NaCl at 15 °C (Figure S11); the secondary structure of the protein obtained in MD simulations (Figure S12); the titration of LrtA to yeast RNA (Figure S13); and the SDS-gel of purified LrtA (Figure S14).

Author Contributions: Conceptualization, L.M.C., P.S., A.C.-A., J.G.H.-C., B.R., and J.L.N.; Methodology, L.M.C., P.S., A.C.-A., J.G.H.-C., B.R., and J.L.N.; Software, B.R.; Investigation, L.M.C., P.S., A.C.-A., J.G.H.-C., B.R., and J.L.N.; Resources, F.J.F. and M.I.M.-P.; Writing—Original Draft Preparation, L.M.C., P.S., A.C.-A., J.G.H.-C., B.R., F.J.F., M.I.M.-P., J.G.d.l.T., and J.L.N.; Writing—Review and Editing, B.R., A.C.-A., and J.L.N.; Supervision, J.G.d.l.T. and J.L.N.

Funding: This research was funded by Spanish Ministry of Economy and Competitiveness (CTQ2015-64445-R (to J.L.N.), BIO2016-78020-R to A.C.A., FIS2014-52212-R to P.S. and BIO2016-75634-P (to M.I.M.-P. and F.J.F.), with Fondo Social Europeo (E.S.F.)). J.G.T. and J.G.H.-C. We also like to express our gratitude for support from Fundación Séneca (Comunidad Autónoma de la Región de Murcia) (19353/PI/14). B.R. acknowledges kind hospitality and use of computational resources in the European Magnetic Resonance Center (CERM), Sesto Fiorentino (Florence), Italy. L.M.C. thanks support from Universidad Miguel Hernández for her sabbatical stay.

Conflicts of Interest: The authors declare no conflict of interest.

Abbreviations

ANS	8-anilino-1-naphthalene sulfonic acid
β-ME	β-mercaptoethanol
CD	circular dichroism
DLS	dynamic light scattering
DOSY	diffusion ordered spectroscopy
EM	electron microscopy
GdmCl	guanidine hydrochloride
HPF	hibernating promoting factor
IDP	intrinsically disordered protein
IMAC	immobilized affinity chromatography
IRF	instrument response function
ITC	isothermal titration calorimetry
MD	molecular dynamics
NMR	nuclear magnetic resonance
PEI	polyethylenimine
RMF	ribosome modulation factor
RNase	ribonuclease
SDS-PAGE	sodium dodecyl sulfate polyacrylamide gel electrophoresis
SEC	size exclusion chromatography
UV	ultraviolet

References

1. Tan, X.; Varughese, M.; Widger, W.R. A light-repressed transcript found in *Synechococcus* sp. PCC 7002 is similar to a chloroplast-specific small subunit ribosomal protein and to a transcription modulator protein associated with sigma 54. *J. Biol. Chem.* **1994**, *269*, 20905–20912. [PubMed]

2. Samartzidou, H.; Widger, W.R. Transcriptional and post-transcriptional control of mRNA from lrtA, a light-repressed transcript in *Synechococcus* sp. PC 7002. *Plant Physiol.* **1998**, *117*, 225–234. [CrossRef] [PubMed]

3. Galmozzi, C.V.; Florencio, F.J.; Muro-Pastor, M.I. The cyanobacterial ribosomal-associated protein LrtA is involved in post-stress survival in *Synechocystis* sp. PCC 6803. *PLoS ONE* **2016**. [CrossRef] [PubMed]

4. Yoshida, H.; Wada, A. The 100S ribosome: Ribosomal hibernation induced by stress. *Wiley Interdiscip. Rev. RNA* **2014**, *5*, 723–732. [CrossRef] [PubMed]

5. Khusainov, I.; Vicens, Q.; Ayupov, R.; Usachev, K.; Myasnikov, A.; Simonetti, A.; Validov, S.; Kieffer, B.; Yuspova, G.; Yusupov, M.; et al. Structures and dynamics of hibernating ribosomes from *Staphylococcus aureus* mediated by intermolecular interactions of HPF. *EMBO J.* **2017**, *36*, 2073–2087. [CrossRef] [PubMed]

6. Starosta, A.L.; Lasak, J.; Jung, K.; Wilson, D.N. The bacterial translation stress response. *FEMS Microbiol. Rev.* **2014**, *38*, 1172–1201. [CrossRef] [PubMed]

7. Agafonov, D.E.; Spirin, A.S. The ribosome-associated inhibitor A reduces translation errors. *Biochem. Biophys. Res. Commun.* **2004**, *320*, 354–358. [CrossRef] [PubMed]

8. Polikanov, Y.S.; Blaha, G.M.; Steitz, T.A. How hibernation factors RMF, HPF and YfiA turn off protein synthesis. *Science* **2012**, *336*, 915–918. [CrossRef] [PubMed]

9. Ueta, M.; Yoshida, H.; Wada, C.; Baba, T.; Mori, H.; Wada, A. Ribosome binding proteins YHbH and YfiA have opposite functions during 100S formation in the stationary phase of *Escherichia coli*. *Genes Cells* **2005**, *10*, 1103–1112. [CrossRef] [PubMed]

10. Ueta, M.; Ohniwa, R.L.; Yoshida, H.; Maki, Y.; Wada, C.; Wada, A. Role of HPF (hibernation promoting factor) in translational activity in *Escherichia coli*. *J. Biochem.* **2008**, *143*, 425–433. [CrossRef] [PubMed]

11. De Bari, H.; Berry, E.A. Structure of *Vibrio cholerae* ribosome hibernation factor. *Acta Crystallogr. Sect. F* **2013**, *69*, 228–236. [CrossRef] [PubMed]

12. Lakowicz, J.R. *Principles of Fluorescence Spectroscopy*, 2nd ed.; Plenum Press: New York, NY, USA, 1999.

13. Albani, J.R. *Principle and Applications of Fluorescence Spectroscopy*; Blackwell Publishing: Oxford, UK, 2007.

14. Cantor, C.R.; Schimmel, P.R. *Biophysical Chemistry*; W. H. Freeman: New York, NY, USA, 1980.

15. Cavanagh, J.F.; Wayne, J.; Palmer, A.G., III; Skelton, N.J. *Protein NMR Spectroscopy: Principles and Practice*, 1st ed.; Academic Press: San Diego, CA, USA, 1996.

16. Wilkins, D.K.; Grimshaw, S.B.; Receveur, V.; Dobson, C.M.; Jones, J.A.; Smith, L.J. Hydrodynamic radii of native and denatured proteins measured by pulse field gradient NMR techniques. *Biochemistry* **1999**, *38*, 16424–16431. [CrossRef] [PubMed]

17. Nobbmann, U.; Connah, M.; Fish, B.; Varley, P.; Gee, C.; Mulot, S.; Chen, J.; Zhou, L.; Lu, Y.; Sheng, F.; et al. Dynamic light scattering as a relative tool for assessing the molecular integrity and stability of monoclonal antibodies. *Biotechnol. Genet. Eng. Rev.* **2007**, *24*, 117–128. [CrossRef] [PubMed]

18. Pace, C.N.; Grimsley, G.R.; Scholtz, J.M. Protein ionizable groups: pK values and their contribution to protein stability and solubility. *J. Biol. Chem.* **2009**, *284*, 13285–13289. [CrossRef] [PubMed]

19. Grimsley, G.R.; Scholtz, J.M.; Pace, C.N. A summary of the measured pK values of the ionizable groups in folded proteins. *Protein Sci.* **2009**, *18*, 247–251. [PubMed]

20. Lux, B.; Gerard, D.; Laustriat, G. Tyrosine fluorescence of S8 and S15 *Escherichia coli* ribosomal proteins. *FEBS Lett.* **1977**, *80*, 66–70. [CrossRef]

21. Soengas, M.S.; Mateo, C.R.; Salas, M.; Acuña, A.U.; Gutiérrez, C. Structural features of φ29 single-stranded DNA-binding protein. Environment of tyrosines in terms of complex formation with DNA. *J. Biol. Chem.* **1997**, *272*, 295–302. [CrossRef] [PubMed]

22. Whitmore, L.; Wallace, B.A. Protein secondary structure analysis from circular dichroism spectroscopy: Methods and reference databases. *Biopolymers* **2008**, *89*, 392–400. [CrossRef] [PubMed]

23. Whitmore, L.; Wallace, B.A. DICHROWEB, an online server for protein secondary structure analyses from circular dichroism spectroscopic data. *Nucleic Acids Res.* **2004**, *32*, W668–W673. [CrossRef] [PubMed]

24. Woody, R.W. Circular dichroism. *Methods Enzymol.* **1995**, *246*, 34–71. [PubMed]

25. Kelly, S.M.; Jess, T.J.; Price, N.C. How to study proteins by circular dichroism. *Biochim. Biophys. Acta* **2005**, *1751*, 119–139. [CrossRef] [PubMed]

26. Dosztányi, Z.; Csizmók, V.; Tompa, P.; Simon, I. IUPred: Web server for the prediction of intrinsically unstructured regions of proteins based on estimated energy content. *Bioinformatics* **2005**, *21*, 3433–3434. [CrossRef] [PubMed]

27. Yang, Z.R.; Thomson, R.; McNeil, P.; Esnouf, R.M. RONN: The bio-basis function neural network technique applied to the detection of natively disordered regions in proteins. *Bioinformatics* **2005**, *21*, 3369–3376. [CrossRef] [PubMed]

28. Ishida, T.; Kinoshita, K. PrDOS: Prediction of disordered protein regions from amino acid sequence. *Nucleic Acids Res.* **2007**, *35*, W460–W464. [CrossRef] [PubMed]

29. Lobanov, M.Y.; Sokolovskiy, I.V.; Galzitskaya, O.V. IsUnstruct: Prediction of the residue status to be ordered or disordered in the protein chain by a method based on the Ising model. *J. Biomol. Struct. Dyn.* **2013**, *31*, 1034–1043. [CrossRef] [PubMed]

30. Uversky, V.; Gillespie, J.; Fink, A. Why are "natively unfolded" proteins unstructured under physiological conditions? *Proteins* **2000**, *41*, 415–427. [CrossRef]

31. Dunker, A.K.; Lawson, J.D.; Brown, C.J.; Williams, R.M.; Romero, P.; Oh, J.S.; Oldfield, C.J.; Campen, A.M.; Ratliff, C.M.; Hipps, K.W.; et al. Intrinsically disordered proteins. *J. Mol. Graph. Model.* **2001**, *19*, 26–59. [CrossRef]

32. Das, R.K.; Pappu, R.V. Conformations of intrinsically disordered proteins are influenced by linear sequence distributions of oppositely charged residues. *Proc. Natl. Acad. Sci. USA* **2013**, *110*, 13392–13397. [CrossRef] [PubMed]

33. Yang, J.; Yan, R.; Roy, A.; Xu, D.; Poisson, J.; Zhang, Y. The I-TASSER Suite: Protein structure and function prediction. *Nat. Methods* **2015**, *12*, 7–8. [CrossRef] [PubMed]

34. Wang, C.; Zhang, H.; Zheng, W.M.; Xu, D.; Zhu, J.; Wang, B.; Ning, K.; Sun, S.; Li, S.C.; Bu, D. FALCON@home: A high-throughput protein structure prediction server based on remote homologue recognition. *Bioinformatics* **2016**, *32*, 462–464. [CrossRef] [PubMed]

35. Biasini, M.; Bienert, S.; Waterhouse, A.; Arnold, K.; Studer, G.; Schmidt, T.; Kiefer, F.; Cassarino, T.G.; Bertoni, M.; Bordoli, L.; et al. SWISS-MODEL: Modelling protein tertiary and quaternary structure using evolutionary information. *Nucleic Acids Res.* **2014**, *42*, W252–W258. [CrossRef] [PubMed]

36. Kim, D.E.; Chivian, D.; Baker, D. Protein structure prediction and analysis using the Robetta server. *Nucleic Acids Res.* **2004**, *32*, W526–W531. [CrossRef] [PubMed]

37. Wüthrich, K. *NMR of Proteins and Nucleic Acids*; Wiley and Sons: New York, NY, USA, 1986.

38. Pantoja-Uceda, D.; Neira, J.L.; Saelices, L.; Robles-Rengel, L.; Florencio, F.J.; Muro-Pastor, M.I.; Santoro, J. Dissecting the binding between glutamine synthetase and its tow natively unfolded protein inhibitors. *Biochemistry* **2016**, *55*, 3370–3382. [CrossRef] [PubMed]

39. Rizzuti, B.; Daggett, V. Using simulations to provide the framework for experimental protein folding studies. *Arch. Biochem. Biophys.* **2013**, *531*, 128–135. [CrossRef] [PubMed]

40. Humphrey, W.; Dalke, A.; Schulten, K. VMD: Visual molecular dynamics. *J. Mol. Graph. Model.* **1996**, *14*, 33–38. [CrossRef]

41. Miroux, B.; Walker, J.E. Over-production of proteins in *Escherichia coli*: Mutant hosts that allow synthesis of some membrane proteins and globular proteins at high levels. *J. Mol. Biol.* **1996**, *260*, 289–298. [CrossRef] [PubMed]

42. Burgess, R.R. Protein precipitation techniques. *Methods Enzymol.* **2009**, *463*, 331–342. [PubMed]

43. Dunn, M.J. Initial planning: Determination of total protein concentration. In *Protein Purification Methods*; Harris, E.L.V., Angal, S., Eds.; Oxford University Press: Oxford, UK, 1995; pp. 10–20.

44. Neira, J.L. Fluorescence, circular dichroism and mass spectrometry as tools to study virus structure. *Subcell. Biochem.* **2013**, *68*, 177–202. [PubMed]

45. Tinoco, I., Jr.; Sauer, K.; Wang, J.C. *Physical Chemistry: Principles and Applications in Biological Sciences*, 3rd ed.; Prentice-Hall: New York, NY, USA, 1995; pp. 585–588.

46. Bokma, J.T.; Johnson, W.C., Jr.; Blok, J. CD of the Li-salt of DNA in ethanol/water mixtures: Evidence for the B- to C-form transition in solution. *Biopolymers* **1987**, *26*, 893–909. [CrossRef] [PubMed]

47. Jin, X.; Johnson, W.C., Jr. Comparison of base inclination in Ribo-AU and Deoxyribo-AT polymers. *Biopolymers* **1995**, *36*, 303–312. [CrossRef] [PubMed]

48. Charnavets, T.; Nunvar, J.; Necasova, I.; Völker, J.; Breslauer, K.J.; Schneider, B. Conformational diversity of the single-stranded DNA from bacterial repetitive extragenic palindromes: Implications for the DNA recognition elements of transposases. *Biopolymers* **2015**, *103*, 585–596. [CrossRef] [PubMed]

49. Atkins, P.; De Paula, J. *Physical Chemistry for the Life Sciences*, 1st ed.; W. H. Freeman: Oxford, UK, 2006; pp. 565–566.

50. Neira, J.L.; Hornos, F.; Bacarizo, J.; Cámara-Artigas, A.; Gómez, J. The monomeric species of the regulatory domain of Tyrosine Hydroxylase has a low conformational stability. *Biochemistry* **2016**, *55*, 6209–6220. [CrossRef] [PubMed]

51. Bevington, P.R.; Robinson, K.D. *Data Reduction and Error Analysis for the Physical Sciences*, 3rd ed.; McGraw-Hill: New York, NY, USA, 2003.

52. James, D.R.; Siemiarczuk, A.; Ware, W.R. Stroboscopic optical boxcar technique for the determination of fluorescence lifetimes. *Rev. Sci. Instrum.* **1992**, *63*, 1710–1716. [CrossRef]

53. Piotto, M.; Saudek, V.; Sklenar, V. Gradient-tailored excitation for single-quantum NMR spectroscopy of aqueous solutions. *J. Biomol. NMR* **1993**, *2*, 661–665. [CrossRef]

54. Abraham, M.J.; Murtola, T.; Schulz, R.; Páll, S.; Smith, J.C.; Hess, B.; Lindahl, J. GROMACS: High performance molecular simulations through multi-level parallelism from laptops to supercomputers. *SoftwareX* **2015**, *1–2*, 19–25. [CrossRef]

55. Lindorff-Larsen, K.; Piana, S.; Palmo, K.; Maragakis, P.; Klepeis, J.L.; Dror, R.O.; Shaw, D.E. Improved side-chain torsion potentials for the Amber ff99SB protein force field. *Proteins* **2010**, *78*, 1950–1958. [CrossRef] [PubMed]

56. Jorgensen, W.L.; Chandrasekhar, J.; Madura, J.D.; Impey, R.W.; Klein, M.L. Comparison of simple potential functions for simulating liquid water. *J. Chem. Phys.* **1983**, *79*, 926–935. [CrossRef]

57. Evoli, S.; Guzzi, R.; Rizzuti, B. Molecular simulations of β-lactoglobulin complexed with fatty acids reveal the structural basis of ligand affinity to internal and possible external binding sites. *Proteins* **2014**, *82*, 2609–2619. [CrossRef] [PubMed]

58. Rizzuti, B.; Bartucci, R.; Sportelli, L.; Guzzi, R. Fatty acid binding into the highest affinity site of human serum albumin observed in molecular dynamics simulation. *Arch. Biochem. Biophys.* **2015**, *579*, 18–25. [CrossRef] [PubMed]

59. Guglielmelli, A.; Rizzuti, B.; Guzzi, R. Stereoselective and domain-specific effects of ibuprofen on the thermal stability of human serum albumin. *Eur. J. Pharm. Sci.* **2018**, *112*, 122–131. [CrossRef] [PubMed]
60. Rudolph, R.; Böhm, G.; Lilie, H.; Jaenicke, R. Folding proteins. In *Protein Function: A Practical Approach*, 2nd ed.; Creighton, T.E., Ed.; Oxford University Press: Oxford, UK, 1997; pp. 57–99.

© 2018 by the authors. Licensee MDPI, Basel, Switzerland. This article is an open access article distributed under the terms and conditions of the Creative Commons Attribution (CC BY) license (http://creativecommons.org/licenses/by/4.0/).

International Journal of
Molecular Sciences

Article

Arabidopsis Heat Stress-Induced Proteins Are Enriched in Electrostatically Charged Amino Acids and Intrinsically Disordered Regions

David Alvarez-Ponce [1,2,*], Mario X. Ruiz-González [2,3], Francisco Vera-Sirera [2], Felix Feyertag [1], Miguel A. Perez-Amador [2] and Mario A. Fares [2,3,†]

[1]   Biology Department, University of Nevada, Reno, NV 89557, USA; ffeyertag@unr.edu
[2]   Instituto de Biología Molecular y Celular de Plantas, CSIC-UPV, 46022 Valencia, Spain;
      marioxruizgonzalez@gmail.com (M.X.R.-G.); fravesi@ibmcp.upv.es (F.V.-S.);
      mpereza@ibmcp.upv.es (M.A.P.-A.)
[3]   Smurfit Institute of Genetics, University of Dublin, Trinity College, Dublin 2, Ireland
*   Correspondence: dap@unr.edu; Tel.: +1-(775)-682-5735
†   Posthumous author.

Received: 9 July 2018; Accepted: 31 July 2018; Published: 3 August 2018

Abstract: Comparison of the proteins of thermophilic, mesophilic, and psychrophilic prokaryotes has revealed several features characteristic to proteins adapted to high temperatures, which increase their thermostability. These characteristics include a profusion of disulfide bonds, salt bridges, hydrogen bonds, and hydrophobic interactions, and a depletion in intrinsically disordered regions. It is unclear, however, whether such differences can also be observed in eukaryotic proteins or when comparing proteins that are adapted to temperatures that are more subtly different. When an organism is exposed to high temperatures, a subset of its proteins is overexpressed (heat-induced proteins), whereas others are either repressed (heat-repressed proteins) or remain unaffected. Here, we determine the expression levels of all genes in the eukaryotic model system *Arabidopsis thaliana* at 22 and 37 °C, and compare both the amino acid compositions and levels of intrinsic disorder of heat-induced and heat-repressed proteins. We show that, compared to heat-repressed proteins, heat-induced proteins are enriched in electrostatically charged amino acids and depleted in polar amino acids, mirroring thermophile proteins. However, in contrast with thermophile proteins, heat-induced proteins are enriched in intrinsically disordered regions, and depleted in hydrophobic amino acids. Our results indicate that temperature adaptation at the level of amino acid composition and intrinsic disorder can be observed not only in proteins of thermophilic organisms, but also in eukaryotic heat-induced proteins; the underlying adaptation pathways, however, are similar but not the same.

Keywords: temperature response; protein thermostability; salt bridges; intrinsically disordered proteins

---

1. Introduction

Proteins of thermophilic prokaryotes (those adapted to high temperatures) exhibit several distinctive features that increase their thermostability. One of the most consistent observations in thermophile proteins is an enrichment in salt bridges [1,2]. Salt bridges consist of electrostatic interactions among amino acid residues with positive (Lys and Arg) and negative (Glu and Asp) charges, and their contribution to increasing the stability of thermophilic bacteria was first proposed by Perutz and Raidt [3]. In addition, compared with proteins of mesophiles (adapted to intermediate temperatures) and psychrophiles (adapted to low temperatures), thermophile proteins tend to exhibit more disulfide bonds and non-covalent interactions, including hydrogen bonds, and hydrophobic interactions, features that also tend to increase protein stability by linking together distant parts of the amino acid sequence [4,5]. These structural trends have an impact on the amino acid composition of thermophilic proteomes: the proteins of thermophilic bacteria tend to be enriched in charged amino acids and depleted in polar ones such as Ser, Thr, Asn, and Gln [6–12].

A few studies in prokaryotes have also shown that thermophile proteins are depleted in intrinsically disordered regions (IDRs), i.e., regions that lack a defined three-dimensional structure [13–15]. This observation is consistent with the fact that high temperatures induce disorder, but in contrast with the fact that IDRs confer thermoresistance [16–18].

Much less is known about how eukaryotic proteomes adapt to high temperatures. Some studies have suggested that the same biases in amino acid composition observed in thermophilic prokaryotes can be observed in thermophilic fungi (compared to other fungi; ref. [19]) and endothermic vertebrates (compared to ectothermic vertebrates; ref. [20]). In agreement with this notion, comparison of the orthologous proteins of two closely related fish, *Pachycara brachycephalum* (from Antarctica) and *Zoarces viviparous* (from a temperate zone) revealed an excess of Ser and a reduction of Glu and Asn in the cold-adapted species [21]. To our knowledge, the relationship between temperature and intrinsic disorder has not been investigated in eukaryotic proteomes.

Protein adaptation to high temperatures is expected to be observed not only in the proteins of thermophilic organisms, but also in some of the proteins of any mesophilic organism. When an organism is exposed to high temperatures, a subset of its proteins is overexpressed, whereas others are repressed (heat-induced and heat-repressed proteins, respectively, e.g., ref. [22]). As heat-induced function at relatively high temperatures, we hypothesize that they should be similar to those of thermophilic organisms.

Plants represent particularly suitable models to test this hypothesis, as they are sessile organisms that cannot escape from their environment, and they lack the effective thermoregulation mechanisms exhibited by homeotherms. Therefore, plants are expected to have developed adaptations to cope with heat stress [23]. To test our hypothesis, we grew *Arabidopsis thaliana* plants under normal (22 °C) and heat stress conditions (37 °C), and measured gene expression levels. Proteins overexpressed under heat stress were enriched in electrostatically charged amino acids and depleted in polar and hydrophobic amino acids. However, in contrast with our expectations, these proteins were also enriched in IDRs. These results indicate that *Arabidopsis* heat-induced proteins exploit some, but not all the same mechanisms as thermophile proteins to cope with high temperatures.

2. Results

2.1. Proteins That Are Overexpressed at High Temperatures Are Enriched in Electrostatically Charged Amino Acids and Depleted in Polar and Hydrophobic Amino Acids

We grew *Arabidopsis* plants at 22 and 37 °C for 24 h, and performed microarray analyses to measure gene expression levels at the beginning of the experiment ($E_{0,22}$ = expression at time 0 and 22 °C) and at the end of the experiment ($E_{24,22}$ and $E_{24,37}$). $E_{0,22}$ strongly correlated with $E_{24,22}$ (Spearman's rank correlation coefficient, $\rho = 0.991$, $p < 10^{-200}$; Figure 1) supporting the robustness of our gene expression measures—the small differences between gene expression at both time points could be due to differences in gene expression during development and to measurement errors. The correlation between $E_{24,22}$ and $E_{24,37}$ was weaker ($\rho = 0.897$, $p = 10^{-200}$; Figure 2), highlighting the effect of heat stress on the expression of many genes.

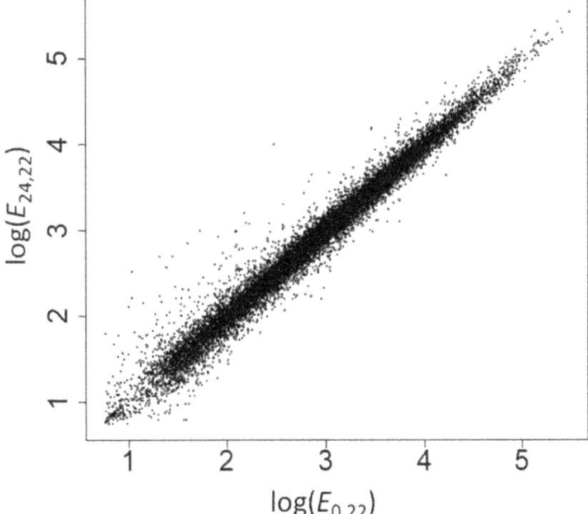

Figure 1. Correlation between gene expression levels at 22 °C at time 0 and at time 24 h.

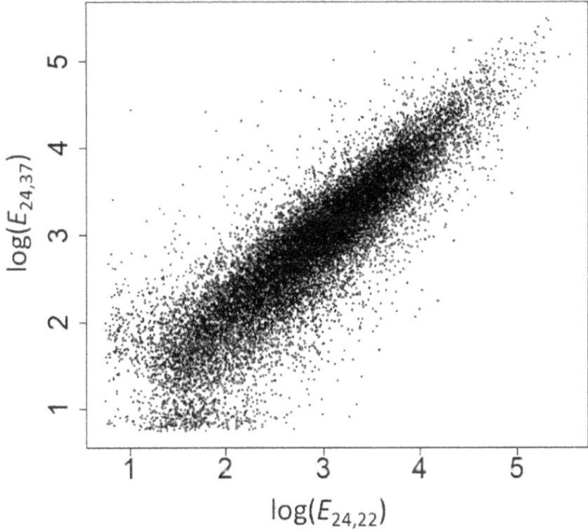

Figure 2. Correlation between gene expression levels at 22 °C at time 24 h and at 37 °C at time 24 h.

For each gene with available probes (n = 20,491), we computed a response to heat stress (R) as the logarithm in base 2 of the ratio of expression levels at 37 and 22 °C (following formula 1). Genes with $R > 0$ are overexpressed at high temperatures, and genes with $R < 0$ are repressed. Genes with $R > 1$ (strongly overexpressed) are enriched in Gene Ontology biological processes "protein refolding", "protein folding", "chaperone cofactor-dependent protein refolding", "chaperone-mediated protein folding", "de novo posttranslational protein folding", "de novo protein folding", "cellular response to heat", "response to heat", "response to temperature stimulus", and "heat acclimation". They are also enriched in molecular functions "misfolded protein binding", "heat shock protein binding", "protein binding involved in protein folding", and "unfolded protein binding" (Tables S1–S3).

We observed a positive correlation between R and the fraction of charged amino acids (ρ = 0.146, $p = 2.47 \times 10^{-98}$), and negative correlations between R and both the fraction of polar ($\rho = -0.076$,

$p = 1.72 \times 10^{-27}$) and hydrophobic ($\rho = -0.084$, $p = 4.08 \times 10^{-33}$) amino acids (Figure 3). We next computed the correlation between R and the frequency of each amino acid separately. The correlation was significantly positive for all four charged amino acids (Arg, Asp, Glu, and Lys), negative for all hydrophobic amino acids (significant for Gly, Ile, Phe, Pro, and Val), except Met (for which the correlation was non-significantly positive), and negative for all polar amino acids (significant for Asn, Ser, Thr, Trp and Tyr), except for Gln, for which the correlation was significantly positive (Table 1). All these correlations remained significant after controlling for multiple testing (Table 1).

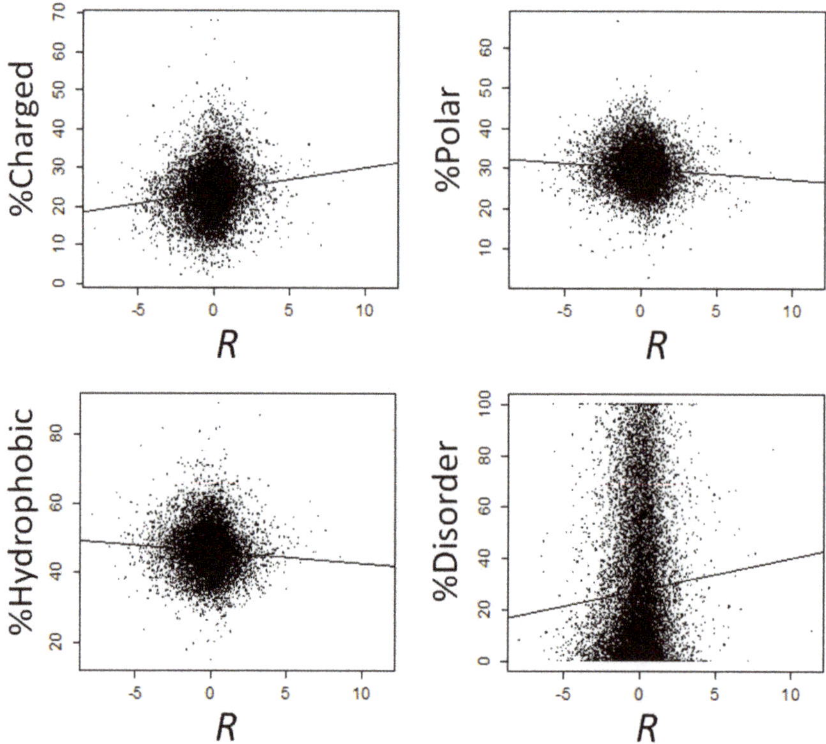

Figure 3. Correlations between response to high temperature (R) and the fraction of charged, polar, hydrophobic and disordered amino acids. Lines represent regression lines.

Table 1. Correlations between amino acid frequencies and response to high temperature.

Type	Amino Acid	No Control			Controlling for GC Content			Controlling for $E_{24,22}$			Controlling for $E_{24,37}$		
		ρ	p-Value	q-Value	ρ	p-Value	q-Value	ρ	p-Value	q-Value	ρ	p-Value	q-Value
Charged	Arg	0.075	1.31×10^{-26}	4.37×10^{-26}	0.068	2.05×10^{-22}	5.86×10^{-22}	0.068	**0.013**	**0.015**	0.079	1.02×10^{-29}	3.40×10^{-29}
	Asp	0.104	1.62×10^{-50}	1.62×10^{-49}	0.105	9.95×10^{-52}	9.95×10^{-51}	0.106	7.16×10^{-53}	7.16×10^{-52}	0.095	1.84×10^{-42}	1.23×10^{-41}
	Glu	0.118	5.48×10^{-64}	1.10×10^{-62}	0.122	5.60×10^{-69}	1.12×10^{-67}	0.115	2.61×10^{-61}	5.22×10^{-60}	0.115	2.60×10^{-61}	5.20×10^{-60}
	Lys	0.082	8.23×10^{-32}	4.12×10^{-31}	0.100	9.58×10^{-47}	4.79×10^{-46}	0.081	2.27×10^{-31}	9.08×10^{-31}	0.079	8.76×10^{-30}	3.40×10^{-29}
	Total	0.146	2.47×10^{-98}		0.155	3.61×10^{-111}		0.145	1.04×10^{-97}		0.140	1.65×10^{-90}	
Polar	Asn	−0.025	3.86×10^{-4}	**0.001**	−0.018	**0.011**	**0.015**	−0.044	4.17×10^{-10}	8.34×10^{-10}	0.005	0.433	0.433
	Cys	−0.011	0.127	0.158	−0.009	0.187	0.208	−0.034	1.54×10^{-6}	2.57×10^{-6}	0.026	2.07×10^{-4}	3.19×10^{-4}
	Gln	0.046	3.20×10^{-11}	7.11×10^{-11}	0.053	2.36×10^{-14}	5.24×10^{-14}	0.046	6.79×10^{-11}	1.51×10^{-10}	0.044	3.95×10^{-10}	7.90×10^{-10}
	His	−0.010	0.134	0.158	−0.009	0.210	0.221	−0.024	**0.001**	**0.001**	0.010	0.146	0.154
	Ser	−0.036	2.25×10^{-7}	4.09×10^{-7}	−0.042	2.36×10^{-9}	4.72×10^{-9}	−0.052	1.00×10^{-13}	2.86×10^{-13}	−0.012	0.092	0.102
	Thr	−0.099	1.10×10^{-45}	7.33×10^{-45}	−0.100	9.24×10^{-47}	4.79×10^{-46}	−0.098	1.12×10^{-44}	7.47×10^{-44}	−0.096	2.75×10^{-43}	2.75×10^{-42}
	Trp	−0.033	2.26×10^{-6}	3.77×10^{-6}	−0.036	2.50×10^{-7}	4.55×10^{-7}	−0.039	2.11×10^{-8}	3.84×10^{-8}	−0.022	**0.002**	**0.002**
	Tyr	−0.024	**0.001**	**0.001**	−0.016	**0.021**	**0.026**	−0.025	3.72×10^{-4}	4.96×10^{-4}	−0.021	**0.003**	**0.004**
	Total	−0.076	1.72×10^{-27}		−0.072	1.11×10^{-24}		−0.102	9.48×10^{-49}		−0.034	9.25×10^{-7}	
Hydrophobic	Ala	−0.008	0.280	0.311	−0.020	**0.004**	**0.006**	0.027	1.32×10^{-4}	1.89×10^{-4}	−0.060	1.50×10^{-17}	3.75×10^{-17}
	Gly	−0.054	1.40×10^{-14}	3.50×10^{-14}	−0.066	1.99×10^{-21}	4.98×10^{-21}	−0.028	5.46×10^{-5}	8.40×10^{-5}	−0.092	1.17×10^{-39}	5.85×10^{-39}
	Ile	−0.045	1.01×10^{-10}	2.02×10^{-10}	−0.035	5.63×10^{-7}	9.38×10^{-7}	−0.052	1.55×10^{-13}	3.88×10^{-13}	−0.033	2.91×10^{-6}	5.29×10^{-6}
	Leu	−0.004	0.547	0.547	−0.004	0.533	0.533	−0.016	**0.021**	**0.023**	0.015	**0.029**	**0.034**
	Met	0.006	0.387	0.407	0.014	**0.042**	**0.049**	−0.001	0.942	0.942	0.017	**0.017**	**0.021**
	Phe	−0.075	1.04×10^{-26}	4.16×10^{-26}	−0.070	9.79×10^{-24}	3.26×10^{-23}	−0.084	2.59×10^{-33}	1.30×10^{-32}	−0.056	1.36×10^{-15}	3.02×10^{-15}
	Pro	−0.060	8.03×10^{-18}	2.29×10^{-17}	−0.074	1.85×10^{-26}	7.40×10^{-26}	−0.052	8.41×10^{-14}	2.80×10^{-13}	−0.070	7.88×10^{-24}	2.25×10^{-23}
	Val	−0.017	**0.012**	**0.017**	−0.024	**0.001**	**0.001**	−0.006	0.370	0.390	−0.033	3.30×10^{-6}	5.50×10^{-6}
	Total	−0.084	4.08×10^{-33}		−0.096	1.31×10^{-43}		−0.064	2.88×10^{-20}		−0.109	2.73×10^{-55}	

p-values and q-values shown in bold face represent significant tests at $\alpha = 0.05$ or $q = 0.05$.

Next, we compared the amino acid composition of proteins encoded by genes that are overexpressed ($R > 0$, n = 10,728) vs. proteins encoded by genes that are repressed ($R < 0$, n = 9763) at 37 °C. Overexpressed proteins were enriched in charged amino acids (median percent in overexpressed proteins: 24.32%; median percent in repressed proteins: 23.20%; Mann-Whitney's U test, $p = 1.90 \times 10^{-66}$) and depleted in both polar (median percent in overexpressed proteins: 29.54%; median percent in repressed proteins: 30.04%; $p = 2.53 \times 10^{-20}$) and hydrophobic (median percent in overexpressed proteins: 45.77%; median percent in repressed proteins: 46.43%; $p = 6.56 \times 10^{-21}$) amino acids. In almost perfect agreement with our correlation analyses, proteins encoded by overexpressed genes were significantly enriched in Arg, Asp, Gln, Glu, and Lys, and significantly depleted in Asn, Gly, Ile, Phe, Pro, Ser, Thr, and Trp (Table 2).

Table 2. Amino acid frequencies in overexpressed ($R > 0$) and repressed ($R < 0$) proteins at high temperatures.

Type	Amino Acid	Median Overexpressed (%)	Median Repressed (%)	p-Value	q-Value
Charged	Arg	5.43	5.19	$\mathbf{8.06 \times 10^{-21}}$	$\mathbf{4.61 \times 10^{-20}}$
	Asp	5.36	5.10	$\mathbf{1.60 \times 10^{-36}}$	$\mathbf{6.40 \times 10^{-35}}$
	Glu	6.61	6.15	$\mathbf{8.28 \times 10^{-44}}$	$\mathbf{6.62 \times 10^{-42}}$
	Lys	6.33	6.06	$\mathbf{1.20 \times 10^{-21}}$	$\mathbf{8.00 \times 10^{-21}}$
	Total	24.32	23.20	$\mathbf{1.90 \times 10^{-66}}$	
Polar	Asn	4.08	4.12	**0.017**	**0.024**
	Cys	1.59	1.60	**0.043**	0.060
	Gln	3.27	3.16	$\mathbf{7.77 \times 10^{-8}}$	$\mathbf{1.88 \times 10^{-7}}$
	His	2.11	2.10	0.204	0.244
	Ser	8.79	8.96	$\mathbf{2.27 \times 10^{-7}}$	$\mathbf{5.19 \times 10^{-7}}$
	Thr	4.90	5.13	$\mathbf{5.31 \times 10^{-34}}$	$\mathbf{1.42 \times 10^{-32}}$
	Trp	1.07	1.11	$\mathbf{4.75 \times 10^{-4}}$	**0.001**
	Tyr	2.65	2.68	0.132	0.163
	Total	29.54	30.04	$\mathbf{2.53 \times 10^{-20}}$	
Hydrophobic	Ala	6.32	6.30	0.889	0.889
	Gly	6.18	6.41	$\mathbf{2.77 \times 10^{-10}}$	$\mathbf{8.21 \times 10^{-10}}$
	Ile	5.12	5.23	$\mathbf{1.87 \times 10^{-7}}$	$\mathbf{4.40 \times 10^{-7}}$
	Leu	9.24	9.27	0.675	0.720
	Met	2.38	2.37	0.399	0.449
	Phe	4.08	4.28	$\mathbf{1.55 \times 10^{-18}}$	$\mathbf{7.75 \times 10^{-18}}$
	Pro	4.54	4.71	$\mathbf{2.56 \times 10^{-12}}$	$\mathbf{8.53 \times 10^{-12}}$
	Val	6.67	6.68	0.178	0.215
	Total	45.77	46.43	$\mathbf{6.56 \times 10^{-21}}$	

p-values correspond to the Mann-Whitney's U test. p-values and q-values shown in bold face represent significant tests at $\alpha = 0.05$ or $q = 0.05$.

Similar results were obtained when using a more stringent threshold to classify genes as overexpressed ($R > 2$, n = 826) or repressed ($R < -2$, n = 1214) at 37 °C. Overexpressed proteins are enriched in charged amino acids (median percent in overexpressed proteins: 25.30%; median percent in repressed proteins: 22.54%; $p = 1.50 \times 10^{-26}$) and depleted in both polar (median percent in overexpressed proteins: 29.74%; median percent in repressed proteins: 30.17%; $p = 3.20 \times 10^{-8}$) and hydrophobic (median percent in overexpressed proteins: 45.20%; median percent in repressed proteins: 47.24%; $p = 6.04 \times 10^{-11}$) amino acids. More specifically, overexpressed proteins are significantly enriched in Arg, Asp, Gln, Glu, and Lys, and significantly depleted in Asn, Cys, Gly, His, Ile, Phe, Pro, Thr, Trp, and Tyr (Table 3).

Table 3. Amino acid frequencies in highly overexpressed ($R > 2$) and highly repressed ($R < -2$) proteins at high temperatures.

Type	Amino Acid	Median Overexpressed (%)	Median Repressed (%)	*p*-Value	*q*-Value
Charged	Arg	5.26	4.80	**7.82×10^{-9}**	**1.04×10^{-7}**
	Asp	5.51	4.95	**1.62×10^{-12}**	**4.32×10^{-11}**
	Glu	6.92	5.92	**1.31×10^{-17}**	**1.05×10^{-15}**
	Lys	6.78	6.17	**1.78×10^{-7}**	**1.78×10^{-6}**
	Total	25.30	22.54	**1.50×10^{-26}**	
Polar	Asn	4.04	4.29	**2.81×10^{-4}**	**0.001**
	Cys	1.66	1.69	**0.031**	**0.045**
	Gln	3.13	2.94	**2.87×10^{-4}**	**6.57×10^{-4}**
	His	2.03	2.12	**0.023**	**0.035**
	Ser	8.47	8.41	0.780	0.810
	Thr	4.95	5.26	**3.52×10^{-6}**	**1.56×10^{-5}**
	Trp	1.05	1.15	**0.035**	**0.050**
	Tyr	2.57	2.86	**6.09×10^{-5}**	**1.87×10^{-4}**
	Total	29.74	30.17	**3.20×10^{-8}**	
Hydrophobic	Ala	6.11	6.12	0.867	0.878
	Gly	6.05	6.50	**5.49×10^{-5}**	**1.76×10^{-4}**
	Ile	5.25	5.48	**0.001**	**0.002**
	Leu	9.01	9.17	0.215	0.292
	Met	2.46	2.52	0.321	0.395
	Phe	4.12	4.65	**9.55×10^{-13}**	**3.82×10^{-11}**
	Pro	4.31	4.62	**9.34×10^{-5}**	**2.58×10^{-4}**
	Val	6.76	6.84	0.294	0.386
	Total	45.20	47.24	**6.04×10^{-11}**	

p-values correspond to the Mann-Whitney's *U* test. *p*-values and *q*-values shown in bold face represent significant tests at $\alpha = 0.05$ or $q = 0.05$.

2.2. The Amino Acid Composition of Heat-Induced Proteins Is Not due to Covariation of Amino Acid Composition with GC Content, Gene Expression Levels, or Subcellular Location

We considered whether our results could be affected by confounding factors. First, GC content is known to affect amino acid composition [24], and R significantly correlates with GC content ($\rho = 0.088$, $p = 9.76 \times 10^{-37}$). Combined, these correlations alone might potentially explain the observed trends. To discard this possibility, we computed partial correlations between R and the frequency of each amino acid, while controlling for GC content, with very similar results. The correlation continued to be significantly positive for charged amino acids and significantly negative for polar and hydrophobic ones (Table 1). More specifically, the correlation was significantly positive for Arg, Asp, Gln, Glu, and Lys and significantly negative for Asn, Gly, Ile, Phe, Pro, Ser, Thr, Trp, Tyr, and Val. Both the negative correlation between R and Ala frequency and the positive correlation between R and Met frequency, which were initially not significant, became significant after controlling for GC content (Table 1).

Second, highly expressed proteins resemble proteins from thermophiles in their amino acid composition [25], and expression levels correlate with R (expression level at 22 °C: $\rho = -0.156$, $p = 4.88 \times 10^{-112}$; expression level at 37 °C: $\rho = 0.241$, $p = 1.18 \times 10^{-268}$). To discard the potential confounding effects of expression levels, we computed partial correlations between R and the frequency of each amino acid, while controlling for expression levels, again with very similar results. When controlling for expression levels at 22 °C, R correlated positively with the frequencies of Ala, Arg, Asp, Gln, Glu, and Lys and negatively with the frequencies of Asn, Cys, Gly, His, Ile, Leu, Phe, Pro, Ser, Thr, Trp, and Tyr. When controlling for expression levels at 37 °C, R correlated positively with the frequencies of Arg, Asp, Cys, Gln, Glu, Leu, Lys, and Met and negatively with the frequencies of Ala, Gly, Ile, Phe, Pro, Thr, Trp, Tyr, and Val. In both cases, the positive correlations between R and the frequency charged amino acids and the negative correlations between R and the frequencies of polar and hydrophobic amino acids remained significant (Table 1).

Proteins locating to different parts of the cell differ in their amino acid compositions and in their response to heat stress ([26,27]; Table 4). To discard subcellular location as a confounding factor, we analyzed the correlation between R and the amino acid composition separately for proteins locating to 10 different subcellular compartments (Table 5). The correlation between R and the fraction of charged amino acids was positive in nine of the compartments, which represents a significant departure from the 50% expected at random (one-tailed binomial test, $p = 0.011$). The correlation was significantly positive for the cytosol, the plastid (the compartments with the higher number of known/inferred proteins), and the mitochondrion. The correlation between R and the fraction of hydrophobic amino acids was negative in eight of the compartments (one-tailed binomial test, $p = 0.055$), significantly negative in the plastid and the mitochondrion, and significantly positive in the nucleus. The correlation between R and the fraction of polar amino acids was negative in half of the compartments, and significantly negative in the cytosol and the nucleus. These results suggest that the enrichment of heat-induced proteins in charged amino acids and their depletion in hydrophobic amino acids are not a byproduct of covariation of both R and amino acid composition with subcellular location. The lack of significance in most of the individual correlations is probably due to the low number of proteins for which location information is available, ranging from 720 for the plastid to 63 in the peroxisome (Table 4), which is expected to greatly reduce the statistical power of our compartment-specific analyses. However, we note an exception: among nuclear proteins R exhibits a significantly positive correlation with the percent of hydrophobic residues (Table 5).

2.3. Proteins That Are Overexpressed at High Temperatures Are Highly Disordered

For each *Arabidopsis* protein, we computed the percentage of amino acids that belong to IDRs using IUPred [28]. This percentage correlates positively with R ($\rho = 0.059$, $p = 4.93 \times 10^{-17}$; Figure 3). Genes that are overexpressed at 37 °C ($R > 0$) encode proteins that are more disordered than those that are repressed ($R < 0$), with median disorder percent of 19.19% and 16.51% for induced and repressed genes, respectively (Mann-Whitney's U test, $p = 2.01 \times 10^{-35}$). The differences are more solid when comparing genes that are strongly overexpressed at 37 °C ($R > 2$) vs. those that are strongly repressed ($R < -2$), with percentages of median disorder of 21.54% and 11.51% for induced and repressed genes, respectively (Mann-Whitney's U test, $P = 2.03 \times 10^{-23}$).

In agreement with previous works [29,30], we found a positive correlation between GC content and the percent of disordered residues ($\rho = 0.044$, $p = 2.84 \times 10^{-10}$). In addition, GC content positively correlates with R ($\rho = 0.088$, $p = 9.76 \times 10^{-37}$), making it possible that the positive correlation between R and disorder might be due to the covariation of both parameters with GC content. The correlation between R and disorder, however, is significant, even after controlling for GC content ($\rho = 0.055$, $p = 3.44 \times 10^{-15}$).

Likewise, intrinsic disorder positively correlates with expression levels (at 22 °C: $\rho = 0.040$, $p = 1.03 \times 10^{-8}$; and at 37 °C: $\rho = 0.072$, $p = 7.75 \times 10^{-25}$), in agreement with previous results in *Escherichia coli* [31], but in contrast with observations in yeasts [32,33]. Disorder, however, significantly correlates with R after controlling for expression levels (at 22 °C: $\rho = 0.066$, $p = 4.64 \times 10^{-21}$; and at 37 °C: $\rho = 0.043$, $p = 1.03 \times 10^{-9}$).

Both intrinsic disorder and R substantially vary among proteins locating to different subcellular compartments (Table 4), thus raising the possibility that covariation of both factors with subcellular location may account for the observed enrichment of stress-induced proteins in IDRs. We analyzed the correlation between intrinsic disorder and R separately for proteins locating to 10 different subcellular compartments. The correlation was positive for eight of the tissues (significantly positive for the cytosol, endoplasmic reticulum, and the vacuole) and significantly negative for the nucleus and the plasma membrane (Table 5). These results indicate that the positive correlation between disorder and R, while generalized, does not apply to proteins locating to all compartments.

Table 4. Amino acid composition, intrinsic disorder and response to heat stress of proteins locating to different subcellular locations.

Subcellular Location	n	Median Charged Amino Acids (%)	Median Polar Amino Acids (%)	Median Hydrophobic Amino Acids (%)	Median Intrinsic Disorder (%)	Median R
Cytosol	633	25.46	26.74	47.31	15.64	0.131
Endoplasmic reticulum	163	24.12	27.22	48.68	10.11	0.147
Extracellular	197	18.94	32.87	48.43	8.61	−0.296
Golgi	375	23.20	29.37	47.38	14.22	0.108
Mitochondrion	286	23.12	27.63	49.26	14.93	0.261
Nucleus	446	26.50	29.16	43.92	42.73	0.406
Peroxisome	63	23.16	26.47	50.00	10.61	−0.207
Plasma membrane	343	22.21	28.63	48.73	14.97	−0.195
Plastid	720	23.33	27.65	48.91	15.28	−0.190
Vacuole	81	21.14	28.24	49.75	8.12	−0.008

Table 5. Correlations between amino acid frequencies and response to high temperature among proteins of different subcellular locations.

Subcellular Location	Correlation R-Charged Amino Acids		Correlation R-Polar Amino Acids		Correlation R-Hydrophobic Amino Acids		Correlation R-Intrinsic Disorder	
	ρ	p-Value	ρ	p-Value	ρ	p-Value	ρ	p-Value
Cytosol	0.171	$\mathbf{1.54 \times 10^{-5}}$	−0.142	$\mathbf{3.27 \times 10^{-4}}$	−0.069	0.082	0.123	**0.002**
Endoplasmic reticulum	0.061	0.437	−0.015	0.847	−0.112	0.155	0.226	**0.004**
Extracellular	0.054	0.452	0.021	0.765	−0.073	0.309	0.068	0.346
Golgi	−0.046	0.370	0.068	0.191	−0.009	0.866	0.073	0.156
Mitochondrion	0.124	**0.036**	0.060	0.312	−0.125	**0.034**	0.065	0.272
Nucleus	0.020	0.681	−0.119	**0.012**	0.102	**0.031**	−0.207	$\mathbf{1.09 \times 10^{-5}}$
Peroxisome	0.016	0.902	−0.104	0.416	0.080	0.535	0.237	0.062
Plasma membrane	0.064	0.254	−0.017	0.750	−0.004	0.947	−0.154	**0.004**
Plastid	0.137	$\mathbf{2.20 \times 10^{-4}}$	0.007	0.859	−0.110	**0.003**	0.062	0.095
Vacuole	0.184	0.099	0.082	0.466	−0.189	0.091	0.266	**0.017**

p-values shown in bold face represent significant tests at $\alpha = 0.05$.

3. Discussion

We show that *Arabidopsis* proteins whose expression levels increase at high temperatures (heat-induced proteins) are enriched in charged amino acids, and depleted in polar and hydrophobic amino acids, compared to heat-repressed proteins. The enrichment of heat-induced proteins in charged amino acids and the depletion in polar amino acids are trends that mirror those observed in the proteins of thermophilic prokaryotes. The observed enrichment of heat-induced proteins in electrostatically charged amino acids was expected, as such amino acids can engage in salt bridges, which usually increase protein thermostability [1–3]—it should be noted, nonetheless, that not all charged amino acids participate in salt bridges, and that not all salt bridges increase thermostability [34]. However, the depletion of heat-induced proteins in hydrophobic amino acids was not expected, as the proteins of thermophilic prokaryotes are usually enriched in such amino acids (e.g., ref. [35]).

Despite the overall observed trends (heat-induced proteins being enriched in charged amino acids and depleted in polar and hydrophobic amino acids), not all amino acids vary according to these rules. In particular, the frequencies of Cys (polar), His (polar), Ala (hydrophobic), Leu (hydrophobic), and Met (hydrophobic) do not correlate significantly with R, and Gln (a polar amino acid) is more frequent in heat-induced proteins than in heat-repressed ones (Table 1). The enrichment of heat-induced proteins in Gln is surprising, given its tendency to undergo deamination at high temperatures [36].

We show that the observed overall trends are not due to heat-induced genes/proteins being different in terms of expression levels, GC content or subcellular location. When controlling for these factors, however, the direction of the correlations for certain amino acids change (Table 1). Thus, the observed trends in amino acid composition are likely the result of adaptation of heat-induced and heat-repressed *Arabidopsis* proteins to high and low temperatures, respectively.

Burra et al. [13] predicted that the proteins of thermophilic prokaryotes should be enriched in IDRs, as intrinsically disorder proteins are often resistant to high temperatures [16–18]. However, contradicting their predictions, they observed that thermophiles often are depleted in IDRs, which may compensate for the disorder induced by temperature. Similar observations were made in both another proteome-level analysis [15] and an analysis of FlgM proteins from bacteria adapted to different temperatures [14]. In agreement with Burra et al.'s prediction, we observed that *Arabidopsis* heat-induced proteins are enriched in IDRs. Our results suggest that there are different ways in which ordered/disordered regions can promote thermostability.

The correlations described in the current work are moderate, albeit statistically significant. Several scenarios may account for the weakness of the correlations. First, amino acid composition and protein intrinsic disorder may be affected by factors other than temperature. Second, the difference between the temperatures used in this study (22 vs. 37 °C) is small compared to the differences between the optimal temperatures of psychrophiles, mesophiles, and thermophiles. Third, certain plant genes may have changed their patterns of response to heat stress during the recent evolutionary history of *Arabidopsis*. i.e., certain genes that are currently heat-induced may have been heat-repressed in the past, and certain genes that are currently heat-repressed may have been heat-induced in the past. As amino acid and disorder adjustment to temperature is expected to take a relatively long amount of time, such switches in expression profiles may have limited the adaptation of proteomes to temperatures. Fourth, the adaptability of plant proteomes to temperatures may be more limited than that of prokaryotic proteomes, e.g., due to the higher complexity of protein-protein interaction networks and the smaller effective population size of plants [37].

In summary, the amino acid composition of heat-induced proteins in *Arabidopsis* mirrors to some extent, but not completely, that of the proteomes of thermophilic prokaryotes. This indicates that protein adaptation to high temperatures takes place partly through similar molecular mechanisms in prokaryotes and eukaryotes. Our observations also indicate that adaptation of proteins at the level of amino acid composition and protein intrinsic disorder can be detected not only when comparing the proteomes of species adapted to very different temperatures, but also among the proteins of the same species with different temperature response profiles. These observations expand our view of how eukaryotic proteomes adapt to different temperatures.

4. Materials and Methods

4.1. Plant Material, Growth Conditions, and Experimental Design

Arabidopsis thaliana Columbia ecotype seeds were sterilized with 70% ethanol for 20 min, 2.5% sodium hypochlorite (commercial bleach) with 0.05% Triton X-100 for 10 min, and finally, four washes with sterile dH_2O. Seeds were placed onto Whatmann paper in Murashige and Skoog (MS) medium plates (Duchefa, Haarlem, The Netherlands). Plates were kept in the dark at 4 °C for 96 h for stratification, and incubated during 8 h in light at 22 °C to promote germination. Plates were transferred to darkness at 22 °C for 72 h. At this moment plates were either kept at 22 °C or transferred to 37 °C. Seedlings were harvested at 0 and 24 h with four biological replicates. Samples were frozen in liquid nitrogen and stored at −80 °C.

4.2. Microarray Analysis

Total RNA was extracted using the RNeasy Plant Mini Kit (Qiagen, Hilden, Germany), and RNA integrity was tested with the 2100 Bioanalyzer (Agilent, Santa Clara, CA, USA). Transcriptome analyses were carried out according to Minimum Information About a Microarray Experiment (MIAME) guidelines. We used the Agilent *Arabidopsis* (V4) Gene Expression 4 × 44K Microarray in a one-color experimental design. The microarray contained 43,803 probes (60-mer oligonucleotides). Four biological replicates were analyzed for each treatment (time points 0 and 24 h at 22 °C and 24 h at 37 °C).

Half a µg of RNA was amplified and labeled with the Agilent Low Input Quick Amp Labeling Kit. To assess the labeling and hybridization efficiencies we used an Agilent Spike-In Kit. Hybridization and slide washing were performed with the Gene Expression Hybridization Kit (Agilent) and Gene Expression Wash Buffers (Agilent), respectively. Then, slides were scanned at 5 µm resolution in an Agilent G2565AA microarray scanner, and image files were analyzed with the Feature Extraction software 9.5.1. We used the GeneSpring 12.1 software (Agilent) to perform the interarray analyses. To ensure a high-quality data set we removed control features, and selected only features for which the 'IsWellAboveBG' parameter was one in at least three out of four biological replicates (31,921 features from 43,803). Our microarray data sets have been submitted to the Gene Expression Omnibus database (accession number: GSE116592).

A new gene annotation of probes in the microarray was carried out using BLASTN searches (https://blast.ncbi.nlm.nih.gov/Blast.cgi), using the sequences of each probe as query against the Arabidopsis genome annotation in The Arabidopsis Information Resource (TAIR; www.arabidopsis.org), version 10. BLAST results for each probe were filtered with a minimum E-value of 9.9×10^{-6}, a minimum sequence identity of 98% between probe and transcript, and a minimum overlap of the 75% of the probe sequence length. Probes matching multiple genes were not considered. Results for this gene annotation are quite similar to those obtained in similar analyses performed by TAIR (ftp://ftp.arabidopsis.org/Microarrays/Agilent/).

4.3. Gene Overexpression/Repression Analysis

For each probe and experimental condition (three conditions: 0 h at 22 °C, 24 h at 22 °C, and 24 h at 37 °C), expression levels were averaged across the four biological replicates. For those genes that mapped to more than one probe, expression levels were averaged across all probes. As a result, a single expression level was obtained for each gene and experimental condition.

For each gene with available probes ($n = 20{,}491$), the response (R) of its expression to heat stress was computed as:

$$R = \log_2 \frac{E_{24,37}}{E_{24,22}} \qquad (1)$$

where $E_{24,37}$ is expression level at 37 °C at 24 h, and $E_{24,22}$ is expression level at 22 °C at 24 h. R takes positive values for genes that are overexpressed at 37 °C compared to 22 °C, and negative values for those that are repressed.

4.4. Protein and Gene Sequence Analysis

All *Arabidopsis* protein sequences were obtained from Ensembl Plants [38] (assembly: TAIR10). For each gene encoding multiple proteins (alternative splicing isoforms), the longest protein was selected for analysis. For each protein, the frequency of each amino acid was computed by dividing the number of occurrences of the amino acid by the length of the protein. GC content of each gene was retrieved from Ensembl Plants' Biomart [38,39]. For each protein, the most likely subcellular location was retrieved from the SUBA4 database [40]. The consensus location was used. Only proteins located to a single compartment were used in compartment-specific analyses.

4.5. Prediction of Protein Intrinsic Disorder

Protein intrinsic disorder prediction was carried out using IUPred [28] for regions of disorder of at least 30 amino acids ("long" option). IUPred predicts tendency for polypeptide chains to be intrinsically disordered or ordered by analyzing the composition of amino acids within a window of 30 consecutive amino acids. It does so by utilizing an energy predictor matrix to estimate the tendency for pairs of amino acids to form strong stabilizing connections, the underlying assumption being that globular proteins form strong stabilizing contacts whereas structurally disordered proteins lack this capacity. IUPred reports a disorder score for each residue ranging from 0 to 1, conferring complete order to disorder, respectively. In this study, we used a threshold of >0.4 to calculate the proportion of amino acids within each protein that were likely to be in disordered regions.

4.6. Statistical Analyses

Statistical analyses were conducted using R [41]. Partial correlation analyses were conducted using the R function pcor.test [42]. Tests repeated on all 20 amino acids were corrected for multiple testing using the Benjamini-Hochberg approach [43].

Supplementary Materials: Supplementary materials can be found at http://www.mdpi.com/1422-0067/19/8/2276/s1.

Author Contributions: M.X.R.-G., F.V.-S., M.A.P.-A. and M.A.F. conceived, designed, and conducted the experiments and microarray analyses. D.A.-P. conceived and designed the bioinformatics analyses. D.A.-P. and F.F. conducted the bioinformatics analyses. D.A.-P. and M.X.R.-G. wrote the paper. All authors contributed to editing the manuscript, read, and approved the manuscript, except M.A.F., who is a posthumous author.

Funding: D.A.-P. and F.F. were supported by funds from the University of Nevada, Reno, and by pilot grants from Nevada INBRE (P20GM103440) and the Smooth Muscle Plasticity COBRE from the University of Nevada, Reno (5P30GM110767-04), both funded by the National Institute of General Medical Sciences (National Institutes of Health). M.X.R.-G. and M.A.F. were supported by grants from Science Foundation Ireland (12/IP/1637) and the Spanish Ministerio de Economía y Competitividad, Spain (MINECO-FEDER; BFU201236346 and BFU2015-66073-P) to MAF. MXRG was supported by a JAE DOC fellowship from the MINECO, Spain. F.V.-S. and M.A.P.-A. were supported by grant BIO2014-55946-P from MINECO-FEDER.

Acknowledgments: D.A.-P., M.X.R.-G., F.V.-S., F.F. and M.A.P.-A. dedicate this work to the memory of M.A.F. Current address of F.F.: Structural Genomics Consortium; Target Discovery Institute, Nuffield Department of Medicine, University of Oxford, Oxford OX3 7FZ, United Kingdom.

Conflicts of Interest: The authors declare no conflict of interest. The funders had no role in the design of the study; in the collection, analyses, or interpretation of data; in the writing of the manuscript, and in the decision to publish the results.

Abbreviations

IDR	Intrinsically Disordered Region
MS	Murashige and Skoog
TAIR	The Arabidopsis Information Resource
BLAST	Basic Local Alignment Search Tool

References

1. Karshikoff, A.; Ladenstein, R. Ion pairs and the thermotolerance of proteins from hyperthermophiles: A 'traffic rule' for hot roads. *Trends Biochem. Sci.* **2001**, *26*, 550–557. [CrossRef]
2. Strop, P.; Mayo, S.L. Contribution of surface salt bridges to protein stability. *Biochemistry* **2000**, *39*, 1251–1255. [CrossRef] [PubMed]
3. Perutz, M.; Raidt, H. Stereochemical basis of heat stability in bacterial ferredoxins and in haemoglobin A2. *Nature* **1975**, *255*, 256–259. [CrossRef] [PubMed]
4. Argos, P.; Rossmann, M.G.; Grau, U.M.; Zuber, H.; Frank, G.; Tratschin, J.D. Thermal stability and protein structure. *Biochemistry* **1979**, *18*, 5698–5703. [CrossRef] [PubMed]
5. Beeby, M.; D O'Connor, B.; Ryttersgaard, C.; Boutz, D.R.; Perry, L.J.; Yeates, T.O. The genomics of disulfide bonding and protein stabilization in thermophiles. *PLoS Biol.* **2005**, *3*, e309. [CrossRef] [PubMed]
6. Farias, S.T.; Bonato, M. Preferred amino acids and thermostability. *Genet. Mol. Res.* **2003**, *2*, 383–393. [PubMed]
7. Haney, P.J.; Badger, J.H.; Buldak, G.L.; Reich, C.I.; Woese, C.R.; Olsen, G.J. Thermal adaptation analyzed by comparison of protein sequences from mesophilic and extremely thermophilic *Methanococcus* species. *Proc. Natl. Acad. Sci. USA* **1999**, *96*, 3578–3583. [CrossRef] [PubMed]
8. Kreil, D.P.; Ouzounis, C.A. Identification of thermophilic species by the amino acid compositions deduced from their genomes. *Nucleic Acids Res.* **2001**, *29*, 1608–1615. [CrossRef] [PubMed]
9. Tekaia, F.; Yeramian, E.; Dujon, B. Amino acid composition of genomes, lifestyles of organisms, and evolutionary trends: A global picture with correspondence analysis. *Gene* **2002**, *297*, 51–60. [CrossRef]
10. Zeldovich, K.B.; Berezovsky, I.N.; Shakhnovich, E.I. Protein and DNA sequence determinants of thermophilic adaptation. *PLoS Comput. Biol.* **2007**, *3*, e5. [CrossRef] [PubMed]
11. Chakravarty, S.; Varadarajan, R. Elucidation of determinants of protein stability through genome sequence analysis. *FEBS Lett.* **2000**, *470*, 65–69. [CrossRef]
12. Cambillau, C.; Claverie, J.-M. Structural and genomic correlates of hyperthermostability. *J. Biol. Chem.* **2000**, *275*, 32383–32386. [CrossRef] [PubMed]
13. Burra, P.V.; Kalmar, L.; Tompa, P. Reduction in structural disorder and functional complexity in the thermal adaptation of prokaryotes. *PLoS ONE* **2010**, *5*, e12069. [CrossRef] [PubMed]
14. Wang, J.; Yang, Y.; Cao, Z.; Li, Z.; Zhao, H.; Zhou, Y. The role of semidisorder in temperature adaptation of bacterial FlgM proteins. *Biophys. J.* **2013**, *105*, 2598–2605. [CrossRef] [PubMed]
15. Vicedo, E.; Schlessinger, A.; Rost, B. Environmental pressure may change the composition protein disorder in prokaryotes. *PLoS ONE* **2015**, *10*, e0133990. [CrossRef] [PubMed]
16. Galea, C.A.; High, A.A.; Obenauer, J.C.; Mishra, A.; Park, C.-G.; Punta, M.; Schlessinger, A.; Ma, J.; Rost, B.; Slaughter, C.A. Large-scale analysis of thermostable, mammalian proteins provides insights into the intrinsically disordered proteome. *J. Proteome Res.* **2008**, *8*, 211–226. [CrossRef] [PubMed]
17. Tsvetkov, P.; Myers, N.; Moscovitz, O.; Sharon, M.; Prilusky, J.; Shaul, Y. Thermo-resistant intrinsically disordered proteins are efficient 20S proteasome substrates. *Mol. Biosyst.* **2012**, *8*, 368–373. [CrossRef] [PubMed]
18. Galea, C.A.; Nourse, A.; Wang, Y.; Sivakolundu, S.G.; Heller, W.T.; Kriwacki, R.W. Role of intrinsic flexibility in signal transduction mediated by the cell cycle regulator, p27^{Kip1}. *J. Mol. Biol.* **2008**, *376*, 827–838. [CrossRef] [PubMed]

19. Van Noort, V.; Bradatsch, B.; Arumugam, M.; Amlacher, S.; Bange, G.; Creevey, C.; Falk, S.; Mende, D.R.; Sinning, I.; Hurt, E. Consistent mutational paths predict eukaryotic thermostability. *BMC Evol. Biol.* **2013**, *13*, 7. [CrossRef] [PubMed]

20. Wang, G.-Z.; Lercher, M.J. Amino acid composition in endothermic vertebrates is biased in the same direction as in thermophilic prokaryotes. *BMC Evol. Biol.* **2010**, *10*, 263. [CrossRef] [PubMed]

21. Windisch, H.S.; Lucassen, M.; Frickenhaus, S. Evolutionary force in confamiliar marine vertebrates of different temperature realms: Adaptive trends in zoarcid fish transcriptomes. *BMC Genom.* **2012**, *13*, 549. [CrossRef] [PubMed]

22. Albanese, V.; Yam, A.Y.-W.; Baughman, J.; Parnot, C.; Frydman, J. Systems analyses reveal two chaperone networks with distinct functions in eukaryotic cells. *Cell* **2006**, *124*, 75–88. [CrossRef] [PubMed]

23. Berry, J.; Bjorkman, O. Photosynthetic response and adaptation to temperature in higher plants. *Annu. Rev. Plant Physiol.* **1980**, *31*, 491–543. [CrossRef]

24. Sueoka, N. Correlation between base composition of deoxyribonucleic acid and amino acid composition of protein. *Proc. Natl. Acad. Sci. USA* **1961**, *47*, 1141–1149. [CrossRef] [PubMed]

25. Cherry, J.L. Highly expressed and slowly evolving proteins share compositional properties with thermophilic proteins. *Mol. Biol. Evol.* **2010**, *27*, 735–741. [CrossRef] [PubMed]

26. Nakashima, H.; Nishikawa, K. The amino acid composition is different between the cytoplasmic and extracellular sides in membrane proteins. *FEBS Lett.* **1992**, *303*, 141–146. [PubMed]

27. Nakashima, H.; Nishikawa, K. Discrimination of intracellular and extracellular proteins using amino acid composition and residue-pair frequencies. *J. Mol. Biol.* **1994**, *238*, 54–61. [CrossRef] [PubMed]

28. Dosztanyi, Z.; Csizmok, V.; Tompa, P.; Simon, I. IUPred: Web server for the prediction of intrinsically unstructured regions of proteins based on estimated energy content. *Bioinformatics* **2005**, *21*, 3433–3434. [CrossRef] [PubMed]

29. Peng, Z.; Uversky, V.N.; Kurgan, L. Genes encoding intrinsic disorder in eukaryota have high GC content. *Intrinsically Disord. Proteins* **2016**, *4*, e1262225. [CrossRef] [PubMed]

30. Yruela, I.; Contreras-Moreira, B. Genetic recombination is associated with intrinsic disorder in plant proteomes. *BMC Genom.* **2013**, *14*, 772. [CrossRef] [PubMed]

31. Paliy, O.; Gargac, S.M.; Cheng, Y.; Uversky, V.N.; Dunker, A.K. Protein disorder is positively correlated with gene expression in *Escherichia coli*. *J. Proteome Res.* **2008**, *7*, 2234–2245. [CrossRef] [PubMed]

32. Singh, G.P.; Dash, D. How expression level influences the disorderness of proteins. *Biochem. Biophys. Res. Commun.* **2008**, *371*, 401–404. [CrossRef] [PubMed]

33. Yang, J.R.; Liao, B.Y.; Zhuang, S.M.; Zhang, J. Protein misinteraction avoidance causes highly expressed proteins to evolve slowly. *Proc. Natl. Acad. Sci. USA* **2012**, *109*, E831–840. [CrossRef] [PubMed]

34. Hendsch, Z.S.; Tidor, B. Do salt bridges stabilize proteins? A continuum electrostatic analysis. *Protein Sci.* **1994**, *3*, 211–226. [CrossRef] [PubMed]

35. Zhou, X.-X.; Wang, Y.-B.; Pan, Y.-J.; Li, W.-F. Differences in amino acids composition and coupling patterns between mesophilic and thermophilic proteins. *Amino Acids* **2008**, *34*, 25–33. [CrossRef] [PubMed]

36. Catanzano, F.; Barone, G.; Graziano, G.; Capasso, S. Thermodynamic analysis of the effect of selective monodeamidation at asparagine 67 in ribonuclease a. *Protein Sci.* **1997**, *6*, 1682–1693. [CrossRef] [PubMed]

37. Charlesworth, B. Fundamental concepts in genetics: Effective population size and patterns of molecular evolution and variation. *Nat. Rev. Genet.* **2009**, *10*, 195–205. [CrossRef] [PubMed]

38. Bolser, D.; Staines, D.M.; Pritchard, E.; Kersey, P. Ensembl Plants: Integrating tools for visualizing, mining, and analyzing plant genomics data. *Methods Mol. Biol.* **2016**, *1374*, 115–140. [PubMed]

39. Kasprzyk, A.; Keefe, D.; Smedley, D.; London, D.; Spooner, W.; Melsopp, C.; Hammond, M.; Rocca-Serra, P.; Cox, T.; Birney, E. EnsMart: A generic system for fast and flexible access to biological data. *Genome Res.* **2004**, *14*, 160–169. [CrossRef] [PubMed]

40. Hooper, C.M.; Castleden, I.R.; Tanz, S.K.; Aryamanesh, N.; Millar, A.H. SUBA4: The interactive data analysis centre for arabidopsis subcellular protein locations. *Nucleic Acids Res.* **2016**, *45*, D1064–D1074. [CrossRef] [PubMed]

41. R Core Team. R: A language and environment for statistical computing. R Foundation for Statistical Computing. Available online: http://www.R-project.org/ (accessed on 31 October 2014).

42. Kim, S. Ppcor: An R package for a fast calculation to semi-partial correlation coefficients. *Commun. Stat. Appl. Methods* **2015**, *22*, 665. [CrossRef] [PubMed]
43. Benjamini, Y.; Hochberg, Y. Controlling the false discovery rate: A practical and powerful approach to multiple testing. *J. R. Stat. Soc. B* **1995**, *57*, 289–300.

 © 2018 by the authors. Licensee MDPI, Basel, Switzerland. This article is an open access article distributed under the terms and conditions of the Creative Commons Attribution (CC BY) license (http://creativecommons.org/licenses/by/4.0/).

 International Journal of
Molecular Sciences

Article

Decision-Tree Based Meta-Strategy Improved Accuracy of Disorder Prediction and Identified Novel Disordered Residues Inside Binding Motifs

Bi Zhao and Bin Xue *

Department of Cell Biology, Microbiology and Molecular Biology, School of Natural Sciences and Mathematics, College of Arts and Sciences, University of South Florida, Tampa, FL 33620, USA; bizhao@mail.usf.edu
* Correspondence: binxue@usf.edu; Tel.: +1-813-974-6008

Received: 20 August 2018; Accepted: 4 October 2018; Published: 7 October 2018

Abstract: Using computational techniques to identify intrinsically disordered residues is practical and effective in biological studies. Therefore, designing novel high-accuracy strategies is always preferable when existing strategies have a lot of room for improvement. Among many possibilities, a meta-strategy that integrates the results of multiple individual predictors has been broadly used to improve the overall performance of predictors. Nonetheless, a simple and direct integration of individual predictors may not effectively improve the performance. In this project, dual-threshold two-step significance voting and neural networks were used to integrate the predictive results of four individual predictors, including: DisEMBL, IUPred, VSL2, and ESpritz. The new meta-strategy has improved the prediction performance of intrinsically disordered residues significantly, compared to all four individual predictors and another four recently-designed predictors. The improvement was validated using five-fold cross-validation and in independent test datasets.

Keywords: meta strategy; dual threshold; significance voting; decision tree based artificial neural network; protein intrinsic disorder

---

1. Introduction

Intrinsically disordered proteins (IDPs) and intrinsically disordered regions (IDRs) play critical functions in many biological processes [1–7]. Among all the possible molecule mechanisms for the functions of IDPs/IDRs, a major one is that IDPs/IDRs physically interact with their partners through either conformational search or induced fit [8–10]. Eventually, due to them having structural flexibility, IDPs/IDRs may bind to the partners with low-affinity but high-specificity [11–15], and thus regulate the downstream biological processes. Clearly, to characterize the dynamic process of the interaction and the mechanism of regulation, the exact locations of those intrinsically disordered amino acids (IDAAs) involved in the interaction need to be determined. However, high-accuracy experimental methods for the detection of IDPs/IDRs/IDAAs are time-consuming and cost-inefficient. Besides, high-through experimental identification of disordered residues, although having attracted a lot of attention and approaches have been widely scouted [16], is still challenging and the methods are not currently available.

Consequently, many computational tools have been developed to identify IDPs/IDRs/IDAAs and associated molecular interactions. The Protein Data Bank (PDB) [17], while being used in the majority of cases for the three-dimensional structures of biomolecules, does contain information on residues with missing coordinates. These residues are interpreted as IDAAs. Furthermore, PDB also contains the structure of molecular complexes, which frequently provides information of molecular interactions involving IDPs/IDRs. DisProt [18], which is the first database of IDPs/IDRs, not only collects IDPs/IDRs/IDAAs, but also integrates the information of molecular partners.

Similarly, IDEAL [19], another database of IDPs, incorporates the interaction networks of IDPs in the database. DisBind [20], DIBS [21], and MFIB [22] are three recently developed databases for IDPs/IDRs based molecular interactions. These databases can be used to search for IDAA/IDR/IDP, or to develop computational predictors for various purposes. In fact, both PDB and DisProt are frequently used for the development of disorder predictors. In addition, PDB contains complex structures formed between a short IDR and another protein. Many of these short IDRs are known as MoRFs (Molecular Recognition Features) [23]. MoRFs are the very first type of IDRs found in molecular interaction. Based on this discovery, many MoRF related predictors have developed, such as: MoRF [24], MoRFpred [25], MFSPSSMpred [26], MoRFchibi [27], MoRFPred-plus [28], and OPAL [29], among many others. Furthermore, many other predictors have been developed for the general binding site/regions within IDPs/IDRs, e.g., ANCHOR [30], SLiMpred [31], PepBindPred [32], DISOPRED3 [33], IUPred2A/ANCHOR2 [34], etc.

All these computational tools provide information on protein intrinsic disorder for different aspects. Databases are collections of experimentally observed examples; predictors can be used to analyze novel sequences. Disorder predictors identify the location and, to some extent, the scale of flexibility of IDRs/IDAAs; binding motif predictors spot the location of binding regions; other types of predictors may provide information on various structural features and functional roles. Frequently, the outputs of disorder predictors are used as input for other predictors to improve the prediction accuracy [32,35–40]. Clearly, accurate identification of IDAAs is very important for studies associated with protein structure, intrinsic disorder, interaction, and function. Therefore, improving the prediction accuracy of protein intrinsic disorder predictors is always desirable, though also a real challenge at present time. Furthermore, improving the prediction accuracy of IDAAs has other important impacts on basic science. With more and more IDPs/IDRs being discovered, our knowledge on the actual content of protein intrinsic disorder in nature is still elusive. Part of the reason is that the accuracy of existing computational tools is still not able to meet the requirements. Therefore, developing high-accuracy predictors is still in urgent need. In addition, it could also be expected that when developing new predictors, novel computational strategies could be innovated, and thus, make a much broader impact.

In our previous studies on the development of intrinsic disorder predictors [41,42], as well as studies by many other researchers [43–47], meta-strategy has been demonstrated to have multiple advantages over individual predictors that adopt a single computational strategy in the prediction. One oversimplified but straightforward explanation for the success of meta-strategy is that meta-strategy is able to combine the strengths of all individual predictors, and thus improve the prediction accuracy. Nonetheless, a direct integration of multiple individual predictors may not improve the prediction accuracy significantly [48,49], however, further integration of various data pre-processing techniques will. Data pre-processing, such as angle-shift technique in protein dihedral angle prediction, was used in artificial neural network based predictor and improved the accuracy remarkably [50]. A combination of non-linear transformation and principal component analysis-based dimension reduction together with meta-strategy was used to improve the prediction accuracy of miRNAs [48]. With these proofs, it is expected that other novel techniques can also be used to improve the prediction accuracy of protein intrinsic disorder. In this project, dual-threshold value and two-step significance voting were integrated into a decision-tree based neural network to improve the prediction accuracy of IDAAs.

2. Results

2.1. Prediction Performance of Component Predictors

The ROC (Receiver Operating Characteristic) curves of four component predictors was presented in Figure 1A. The AUC (Area Under the Curve) for DisEMBL, IUPred, VSL2, and ESpritz are 0.78, 0.82, 0.84, and 0.88, respectively. The balanced accuracy (Acc-b) of these four predictors at their default settings are: 68%, 76%, 77%, and 73%, accordingly. In Figure 1B, the overlap and coverage between every two predictors were analyzed for the positive samples (disordered residues) and negative samples (structured residues). Here, overlap stands for the ratio of true-positive (or true-negative) predictions made by both predictors over the total number of positive (or negative) samples, and coverage is defined as the ratio of correct predictions made by either predictors over the total number of samples. Clearly, the overlap of positive samples between predictors normally ranges from ~30% to 50%; however, the number for the overlap between IUpred and VSL2 went up to ~65%. In terms of coverage, the numbers were in the range from ~60% to ~80%. For negative samples, the overlap was from ~70% to ~90%, and the coverage was normally higher than ~90%. The highest values of coverage, as shown by bars at the most right-hand side of both panels, were ~85% and 97% for positive and negative samples, respectively. These two values may outline the theoretical uplimits of combining these four predictors.

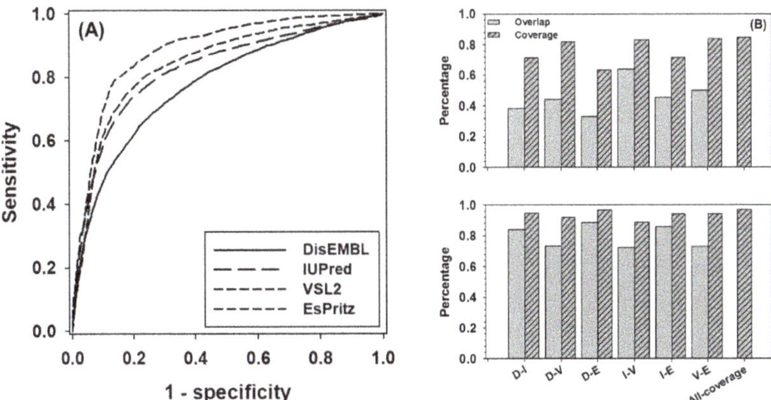

Figure 1. Prediction performance of four component predictors, including DisEMBL, IUPred, VSL2, and ESpritz. (**A**) ROC curves of four component predictors. The ROC curves were obtained by using the default settings of these predictors. (**B**) The pairwise overlap (gray bars) and coverage (dashed bars) for true positive predictions (upper panel) and true negative predictions (lower panel) between each pair of predictors. Axis shows pairs of predictors as follows: D-DisEMBL, I-IUPred, V-VSL2, and E-ESpritz. All-coverage on *x*-axis stands for the maximum coverage of all predictors.

2.2. Use Information Gain to Choose Threshold Values

To use the new meta-strategy, threshold values of the decision tree need to be determined first. Other than using the method based on the distribution of positive samples and negative samples as a function of prediction score [49], the information gain of all the component predictors in the dataset was analyzed and compared to the distribution of positive samples and negative samples, as shown in Figure 2. The curves of information gain can be roughly characterized by a single-peak distribution, and the location of peaks was, roughly, on the right-hand side of the cross-point where the ratio of positive samples surpassed negative samples. More specifically, the locations of peaks for DisEMBL, IUPred, VSL2, and ESpritz were around 0.5, 0.52, 0.64, and 0.26, respectively. By notation, the locations of the peaks provide a rough estimation of the threshold values, which can be used to maximally partition positive samples and negative samples. Clearly, these values can hardly be determined by the analysis of distribution of positive and negative samples.

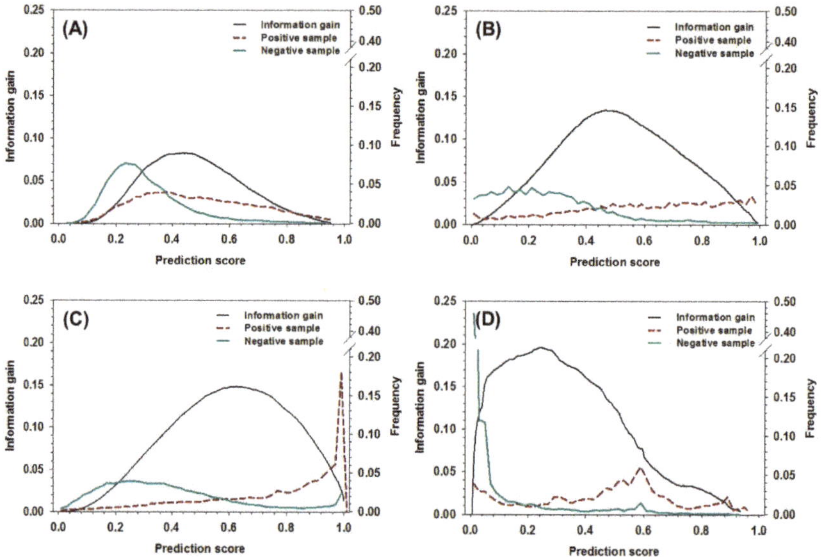

Figure 2. The distribution of information gain, positive sample, and negative samples as a function of prediction score for (**A**) DisEMBL, (**B**) IUPred, (**C**) VSL2, and (**D**) ESpritz. The *x*-axis shows the prediction score, the *y*-axis on the left shows the values of information gain, and the *y*-axis on the right shows the fractions of positive samples and negative samples at different prediction scores, respectively.

2.3. Performance of the New Predictor

Table 1 shows the prediction performance of the new meta-strategy compared to the component predictors, as well as another four recently-developed predictors under five-fold cross-validation. In brief, the performance of new meta-strategy developed in this project was obviously better than others. In terms of accuracy (Acc), balanced accuracy (Acc-b), Matthews Correlation Coefficient (MCC), F1 score, Area Under ROC Curve (AUC_ROC), and Under Precision-Recall Curve (AUC_PR), the new prediction strategy achieved 84.2%, 83.1%, 0,635, 0.744, 0.899, and 0.788, respectively, and was ranked at the first place among all eight different predictors. The new meta-strategy was ranked at the second place on sensitivity (Sens), with one percentage point behind VSL2. With regard to specificity (Spec), the new strategy was inferior to the predictors ESpritz (94%), DisEMBL (91.4%), AUCpreD (90.9%), IUPred2 (87.7%), and IUPred (87.4%). Regardless, it should be noted the Sens values of these predictors are at least 15 percentage points lower than the new meta-strategy.

Table 1. Prediction performance of the new strategy under five-fold cross-validation, in comparison with four component predictors, another four recently-designed predictors.

	DisEMBL	IUPred	VSL2	Espritz	PONDR-FIT	MFDp2	IUPred2	AUCpreD	This Work
Sens	0.440 ± 0.008	0.650 ± 0.003	0.817 ± 0.004	0.514 ± 0.009	0.713 ± 0.004	0.777 ± 0.004	0.640 ± 0.003	0.592 ± 0.006	0.807 ± 0.012
Spec	0.914 ± 0.002	0.874 ± 0.004	0.736 ± 0.003	0.939 ± 0.002	0.859 ± 0.004	0.859 ± 0.004	0.877 ± 0.004	0.909 ± 0.002	0.856 ± 0.007
Acc	0.779 ± 0.003	0.810 ± 0.003	0.759 ± 0.002	0.818 ± 0.004	0.817 ± 0.004	0.836 ± 0.003	0.810 ± 0.003	0.819 ± 0.003	0.842 ± 0.003
Acc-b	0.677 ± 0.006	0.762 ± 0.002	0.776 ± 0.002	0.726 ± 0.004	0.786 ± 0.003	0.818 ± 0.003	0.759 ± 0.002	0.751 ± 0.003	0.831 ± 0.004
MCC	0.410 ± 0.007	0.529 ± 0.006	0.504 ± 0.004	0.521 ± 0.007	0.561 ± 0.007	0.614 ± 0.006	0.526 ± 0.006	0.535 ± 0.006	0.635 ± 0.006
F1	0.531 ± 0.006	0.660 ± 0.003	0.658 ± 0.003	0.616 ± 0.006	0.689 ± 0.004	0.729 ± 0.003	0.657 ± 0.004	0.651 ± 0.005	0.744 ± 0.004
AUC_ROC	0.776 ± 0.004	0.823 ± 0.001	0.841 ± 0.003	0.886 ± 0.003	0.857 ± 0.003	0.879 ± 0.002	0.822 ± 0.001	0.869 ± 0.003	0.899 ± 0.004
AUC_PR	0.607 ± 0.007	0.675 ± 0.007	0.656 ± 0.020	0.752 ± 0.006	0.696 ± 0.004	0.629 ± 0.006	0.657 ± 0.004	0.716 ± 0.007	0.788 ± 0.010

Note Bene. The measures of predictor performance include: sensitivity (Sens), specificity (Spec), accuracy (Acc), balanced accuracy (Acc-b), Matthews Correlation Coefficient (MCC), F1 score, Area Under ROC Curve (AUC_ROC), and Area Under Precision-Recall Curve (AUC_PR). The highest value in each of these measures is in bold and highlighted (red).

The performance of all these nine predictors was also assessed using the independent dataset as shown in Table 2. By comparing the data of Tables 1 and 2, it is obvious that although the numbers have fluctuations, the overall levels and trends of all the measures of prediction performance are essentially the same.

Table 2. Prediction performance of all nine predictors in the independent dataset.

	DisEMBL	IUPred	VSL2	Espritz	PONDR-FIT	MFDp2	IUPred2	AUCpreD	This Work
Sens	0.454	0.656	0.82	0.529	0.728	0.78	0.647	0.609	0.811 ± 0.007
Spec	0.915	0.872	0.735	0.932	0.856	0.857	0.87	0.908	0.856 ± 0.006
Acc	0.784	0.811	0.759	0.818	0.82	0.835	0.811	0.823	0.844 ± 0.003
Acc-b	0.684	0.764	0.777	0.731	0.792	0.819	0.761	0.759	0.834 ± 0.001
MCC	0.424	0.532	0.507	0.521	0.569	0.615	0.53	0.53	0.639 ± 0.003
F1	0.544	0.663	0.659	0.622	0.696	0.729	0.66	0.66	0.747 ± 0.002
AUC_ROC	0.779	0.824	0.841	0.888	0.857	0.88	0.822	0.872	0.9 ± 0.002
AUC_PR	0.617	0.673	0.642	0.754	0.695	0.622	0.672	0.72	0.789 ± 0.005

Note Bene. The new strategy was optimized five times under five-fold cross-validation. Therefore, the performance was also tested in the independent test dataset five times. The results shown in the table is the average of all five times. The highest value in each of these measures is in bold and highlighted (red).

The performance of this new meta-strategy, as well as other predictors, for twenty types of amino acids was analyzed using balanced accuracy in Figure 3. Overall, the new meta-strategy has the highest Acc-b values in fifteen types of residues. The new meta-strategy was also ranked first together with the more recent predictor MFDp2 for residues P and Q. However, the new meta-strategy was ranked at the second position for C, N, and Y, with several percentage points behind MFDp2.

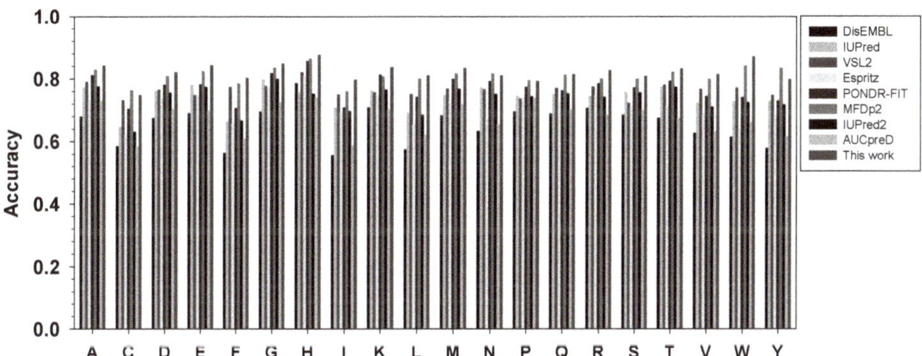

Figure 3. Comparison of balanced accuracy for twenty types of amino acids. The *x*-axis shows amino acid types in the alphabetic order, while the *y*-axis shows the value of balanced accuracy. For each type of amino acid, the predictors from left to right are: DisEMBL, IUPred, VSL2, ESpritz, PONDR-FIT, MFDp2, IUPred2A, and AUCpreD.

The balanced accuracies of all predictors for terminal residues were also analyzed in Figure 4. Obviously, the accuracy is location and predictor dependent. For many predictors, the closer to the termini, the lower the accuracy. For N-terminal residues, IUPred, ESpritz, MFDp2, and IUpred2 achieved ~67% balanced accuracy, which was also largely location independent. For DisEMBL, VSL2, and AUCpreD, the balanced accuracies increased gradually from ~55% to ~65% in the window from the 5th to the ~15th residues and then kept similar accuracy afterwards. The newly designed meta-strategy had a lower balanced accuracy of ~52% for the first several residues. The accuracy then increased gradually to ~63% at the 25th residue. PONDR-FIT, a more recently developed predictor, was the least accurate predictor for N-terminal residues, especially in the range from the 10th to the 20th residues where its accuracy was 2–5 points lower than the new strategy. For C-terminal residues, the patterns of accuracy were different from N-terminal residues. First, the balanced accuracy was higher in general than N-terminal residues by several percentage points. Second, although the accuracies of predictors were still either location-independent or location-dependent, the values of accuracies were highly diversified. AUCpreD, MFDp2, IUPred, IUPred2, and ESpritz made location-independent predictions for C-terminal residues, however, the accuracy of these predictors spread from ~74% to 68%, accordingly. DisEMBL, VSL2, and PONDR-FIT's accuracy increased gradually from ~55 to62% at the 5th residue to ~67% at the 20th residue. The accuracy of the newly designed strategy for C-terminal residues was at the lower-end for the first several terminal residues, though increased consistently and achieved the highest balanced accuracy for residues at the ~20th position.

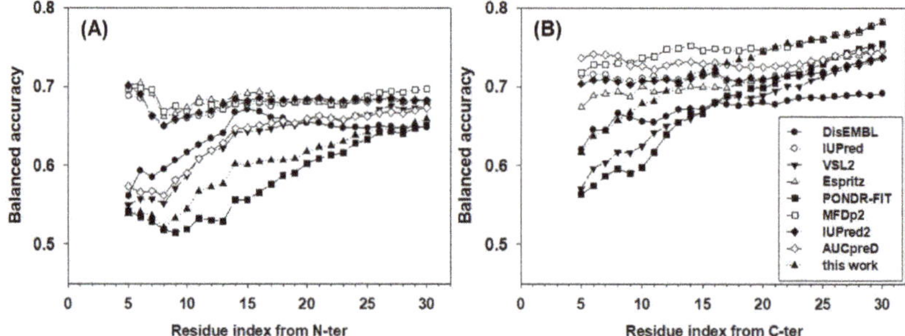

Figure 4. Balanced accuracy of (**A**) N-terminal and (**B**) C-terminal residues. The *x*-axis shows the distance from the first (N-terminal) or the last (C-terminal) residue. The analysis starts at the fifth residue on both N- and C-termini. The *y*-axis shows the value of the balanced accuracy.

With these observations, all the samples were regrouped into three new datasets each containing the first 25 N-terminal residues, the first 25 C-terminal residues, and the middle region, respectively. The meta-strategy was re-trained in three different datasets separately. The prediction performance of all predictors in all three regions under five-fold cross-validation was compared and analyzed in Table 3. Evidently, compared to the results in Figure 4, the prediction accuracy of terminal residues improved substantially. More specifically, the values of improvement of accuracy, balanced accuracy, F1, MCC in N-ter, Mid, and C-ter datasets ranged from 1 to 5 percentage points. For sensitivity and specificity, since many other predictors were trained to maximize either sensitivity or specificity, the new meta-strategy was normally not able to compete with them.

Table 3. Comparison of prediction performance under five-fold cross-validation of eight predictors, as well as the new strategy trained for N-terminal, middle region, and C-terminal residues. The highest value in each of these measures is in bold and highlighted (red).

		DisEMBL	IUPred	VSL2	Espritz	PONDR-FIT	MFDp2	IUPred2	AUCpreD	This Work
N-ter	Sens	0.553 ± 0.009	0.541 ± 0.017	0.782 ± 0.011	0.582 ± 0.010	**0.837 ± 0.004**	0.782 ± 0.009	0.539 ± 0.015	0.748 ± 0.011	0.829 ± 0.023
	Spec	0.741 ± 0.016	0.841 ± 0.020	0.524 ± 0.024	0.789 ± 0.012	0.405 ± 0.020	0.582 ± 0.039	**0.842 ± 0.020**	0.590 ± 0.038	0.572 ± 0.049
	Acc	0.614 ± 0.003	0.639 ± 0.012	0.698 ± 0.005	0.650 ± 0.004	0.697 ± 0.006	0.718 ± 0.012	0.638 ± 0.014	0.697 ± 0.011	**0.746 ± 0.014**
	Acc-b	0.647 ± 0.004	0.691 ± 0.011	0.653 ± 0.010	0.686 ± 0.004	0.621 ± 0.011	0.682 ± 0.020	0.691 ± 0.014	0.669 ± 0.016	**0.701 ± 0.020**
	MCC	0.277 ± 0.009	0.364 ± 0.021	0.308 ± 0.019	0.349 ± 0.009	0.265 ± 0.024	0.361 ± 0.038	0.363 ± 0.027	0.330 ± 0.029	**0.410 ± 0.035**
	F1	0.660 ± 0.007	0.669 ± 0.015	0.778 ± 0.006	0.692 ± 0.008	0.789 ± 0.007	0.789 ± 0.007	0.668 ± 0.014	0.770 ± 0.010	**0.815 ± 0.013**
Middle	Sens	0.387 ± 0.009	0.682 ± 0.005	0.820 ± 0.004	0.481 ± 0.010	**0.663 ± 0.004**	0.777 ± 0.005	0.672 ± 0.005	0.539 ± 0.007	0.807 ± 0.013
	Spec	0.927 ± 0.001	0.877 ± 0.004	0.751 ± 0.004	0.948 ± 0.002	0.888 ± 0.004	0.875 ± 0.005	**0.880 ± 0.004**	0.927 ± 0.001	0.877 ± 0.006
	Acc	0.801 ± 0.004	0.831 ± 0.004	0.767 ± 0.003	0.839 ± 0.004	0.835 ± 0.004	0.852 ± 0.005	0.831 ± 0.004	0.836 ± 0.003	**0.861 ± 0.004**
	Acc-b	0.657 ± 0.005	0.780 ± 0.003	0.786 ± 0.003	0.715 ± 0.005	0.776 ± 0.003	0.826 ± 0.004	0.776 ± 0.003	0.732 ± 0.003	**0.842 ± 0.005**
	MCC	0.376 ± 0.009	0.544 ± 0.005	0.497 ± 0.006	0.506 ± 0.008	0.546 ± 0.006	0.616 ± 0.008	0.540 ± 0.007	0.510 ± 0.006	**0.643 ± 0.008**
	F1	0.477 ± 0.007	0.628 ± 0.003	0.622 ± 0.005	0.583 ± 0.008	0.653 ± 0.004	0.711 ± 0.004	0.650 ± 0.004	0.606 ± 0.005	**0.731 ± 0.006**
C-ter	Sens	0.584 ± 0.014	0.615 ± 0.017	**0.847 ± 0.016**	0.609 ± 0.019	0.828 ± 0.017	0.771 ± 0.016	0.598 ± 0.016	0.681 ± 0.013	0.790 ± 0.018
	Spec	0.787 ± 0.021	0.838 ± 0.023	0.586 ± 0.014	0.857 ± 0.015	0.615 ± 0.017	0.743 ± 0.015	**0.845 ± 0.019**	0.796 ± 0.019	0.769 ± 0.021
	Acc	0.686 ± 0.013	0.727 ± 0.005	0.715 ± 0.007	0.734 ± 0.009	0.720 ± 0.008	0.757 ± 0.007	0.723 ± 0.007	0.739 ± 0.009	**0.780 ± 0.009**
	Acc-b	0.685 ± 0.012	0.726 ± 0.006	0.716 ± 0.007	0.733 ± 0.007	0.721 ± 0.009	0.757 ± 0.007	0.722 ± 0.008	0.739 ± 0.008	**0.780 ± 0.009**
	MCC	0.379 ± 0.026	0.465 ± 0.014	0.448 ± 0.015	0.482 ± 0.015	0.453 ± 0.018	0.514 ± 0.014	0.459 ± 0.018	0.481 ± 0.018	**0.560 ± 0.018**
	F1	0.649 ± 0.012	0.691 ± 0.008	0.747 ± 0.008	0.694 ± 0.009	0.746 ± 0.008	0.759 ± 0.006	0.682 ± 0.012	0.722 ± 0.007	**0.781 ± 0.006**

The performance of all the predictors were then tested in CASP10 dataset and then compared to DISOPRED3, which is one of the two best predictors in CASP10 competition (see Appendix A for more details). In brief, DISOPRED3 and AUCpreD have very similar performance and are better than other predictors on multiple measures, such as specificity, accuracy, MCC, F1, and AUC-ROC. PONDR-FIT achieved the highest balanced accuracy. The new meta-strategy has the highest sensitivity. In addition to the whole dataset analysis, the per-sequence accuracy was also analyzed. The balanced accuracy of PONDR-FIT, MFDp2, AUCpreD, and the new meta-strategy in CASP10 dataset was compared in Figure 5A. All the symbols above the diagonal line represent sequences with higher accuracy when predicted using PONDR-FIT, MFDp2, or AUCpreD, and vice versa. For symbols in the dashed circle, the prediction accuracies of the compared four predictors are all not satisfactory. Symbols in dashed box constitute another group of sequences of which the prediction accuracy of the new meta-strategy is much higher than the other three predictors. For pair-wise comparison between predictors, there are more open circles above the diagonal line, more triangles under the diagonal line, and similar numbers of filled circles on both sides of the diagonal line. Thus, PONDR-FIT (open circles) has better per-sequence prediction performance in the CASP10 dataset. The new meta-strategy and AUCpreD achieved similar results on per-sequence prediction performance. Since the new meta-strategy also made a very low-accuracy prediction on some of the sequences, analyzing the potential reasons could be beneficial. For this purpose, the per-sequence balanced accuracy, fraction of experimentally validated IDAAs per sequence, and the length of each sequence were analyzed in Figure 5B. In this figure, it is apparent that sequences with a very low fraction of experimentally validated IDAAs have very low accuracy. Therefore, the fraction of IDAAs is a critical factor for the performance of the new meta-strategy.

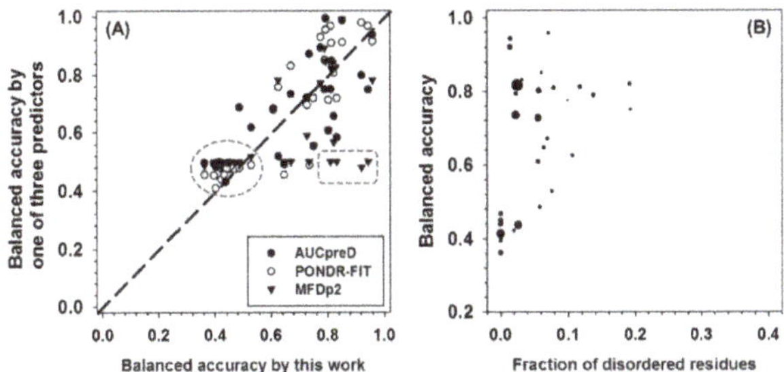

Figure 5. (**A**) Comparison of per-sequence balanced accuracy among AUCpreD (filled circle), PONDR-FIT (open circle), MFDp2 (filled triangle), and this work on sequences in the CASP10 test dataset. The reasons for selecting these predictors are: (1) they are developed in recent years; (2) they have higher performance on some of the accuracy measures; (3) for simplicity of visualization, only four predictors were selected. The *x*-axis shows the per-sequence balanced accuracy of this work, and the *y*-axis shows the per-sequence accuracy of the other three predictors. (**B**) Per-sequence balanced accuracy of this work (*y*-axis) as a function of the fraction of experimentally validated intrinsically disordered amino acids (IDAAs) (*x*-axis). The size of the symbol is proportional to the length of the sequence.

3. Discussion

Intrinsically disordered proteins play critical roles in biomolecular interaction and signaling; therefore, identifying these residues is crucial for the subsequent analysis and biological studies of the functions and mechanisms. Many experimental techniques have been designed for characterizing these residues. Nonetheless, these techniques are normally time-consuming and/or cost-inefficient. Besides, these techniques may not be appropriate for proteomic studies, although many new approaches are under development [16,51,52]. Therefore, using computational tools to predict intrinsically disordered residues becomes practical, especially for novel protein sequences. Under this situation, using high-accuracy predictors is essential. However, as shown in the previous analysis, the current levels of prediction accuracy of many disordered predictors still have a lot of room for improvement.

There are multiple ways to improve the accuracy of machine learning based techniques. Tuning the list of input features is often the first trial. Recently, deep learning and meta-strategy have also been applied to improve the prediction accuracy. Our previous studies and the studies of other groups [41–47] a direct application of meta-strategy may not lead to the improvement of prediction accuracy, although it has been demonstrated that meta-strategy has many advantages [48]. In these cases, novel data processing techniques are very helpful [48,49]. Therefore, in this project, a dual-threshold was employed; two-step voting with different accuracy stringency was also integrated in the pipeline, based on the analysis of information gain. These techniques eventually contributed remarkably to the improvement of prediction accuracy. The outcomes of this new strategy demonstrate that: (1) integrating lower-accuracy predictors is able to produce higher-accuracy output; (2) the improvement of prediction performance of meta-strategy is significant and impressive, compared to individual predictors and other state-of-the-art predictors, including deep-learning based predictors; (3) the meta-strategy has well-balanced results for sensitivity and specificity, and therefore, is able to achieve higher values on other evaluation quantities, such as F1, MCC, etc.; (4) the meta-strategy provides novel ideas on the renovation of existing predictors.

Many data-processing techniques could be integrated into the meta-strategy. In this project, dual-threshold and two-step significance voting were designed and were critical for the improvement of prediction performance. Dual-threshold refers to true prediction and false prediction having different threshold values. By using dual-threshold, it is possible to control the increase of false positive rate and false negative rate. Two-step voting is a technique to use two sets of threshold values at two steps. At the first step, a set of more stringent threshold values are used, and at the second step less-stringent threshold values are used. In this way, the results from the first step have higher reliability than the second step. Significance-voting is another very useful technique complementary to the well-known majority-voting. When using majority-voting, the number of predictors making true predictions and the number of predictors making false predictions competes to determine the final results. In the application of significance-voting, the Euclidean distance of a prediction score from the corresponding threshold value is calculated, then the sum of distances of predictors making true predictions is compared to that of predictors making false predictions. Clearly, this technique is also beneficial for reducing the prediction error. For majority-voting based strategy, overlap is a critical measurement. However, in significance-voting based predictor, although overlap is still very important, coverage plays a more critical role. In addition, results from majority-voting and from significance-voting predictors have different preferences. Majority-voting is strong in selecting part of the true predictions that have very high confidence. However, significance-voting is able to pick up additional true predictions that cannot be identified by majority-voting.

When selecting individual predictors, overlap and coverage between a pair of predictors or among multiple predictors can be calculated and used to check the similarity of two predictors, and to evaluate whether the combination of these two predictors is able to improve final prediction accuracy. If the two predictors have extremely high overlap and very low coverage, these two predictors are very similar to each other in terms of the predictive results, and vice versa. Evidently, these two types of situations need to be avoided in most cases when selecting the component predictors. Normally, the selected component predictors should have a reasonably level of overlap and a higher level of coverage. The values of coverage also provide an estimation on the maximum values of true-positive and true-negative predictions by combining a pair or several predictors.

It should also be noted that most experimental work aiming at IDAA validation is focused on in vitro approaches, and consequently, the corresponding data analysis and computational strategies are also focused on in vitro data. Regardless, the in vitro foldability of amino acid residues could be very different from in vivo environment [53]. Therefore, novel ideas to develop large-scale in vivo conformational assays are also urgently needed. In fact, novel in vivo labeling strategies of IDAAs have been proposed [53]. It is hopeful that these in vivo techniques or at least the data of in vivo studies will be eventually incorporated into novel predictors of in vivo foldability of IDAA.

4. Materials and Methods

DisProt v7.0 and PDB (Protein DataBank) were combined to build the dataset of disordered residues. DisProt contains over 800 protein sequences, in which the IDAAs/IDRs have been identified using various experimental techniques, such as X-ray, NMR, circular dichroism (CD) spectrometry, proteolysis, etc. For all the DisProt sequences, IDAAs have already been annotated. PDB sequences were extracted using the PISCES server [54]. All the PDB structures in the list have 2.5 angstrom or better resolution and 30% or less sequence identity. Then, 20% of the PDB sequences were randomly selected for further analysis. The missing residues in these PDB sequences were assigned as IDAAs, while all other residues were determined to be structured residues. All the extracted sequences from both DisProt and PDB were further filtered using CD-HIT [55] to remove sequences with 30% or higher sequence identity. Finally, there are 312 protein sequences, containing 30,140 disordered residues and 75,945 structured residues. All the sequences with X-ray structures in CASP10 [56] were also extracted. These sequences were each aligned with all the sequences in the above-mentioned main dataset to check the sequence identity. Only sequences with 30% or lower sequences identity were kept to make the second independent test dataset. This second independent test dataset has 35 sequences.

The infrastructure of the meta-strategy is shown in Figure 6. The prediction results of DisEMBL [57], IUPred [58], VSL2 [59], and ESpritz [60], were used as input. The major reasons for choosing these four predictors are as follows: (1) these predictors were designed using very different strategies. DisEMBL uses artificial neural networks. IUPred uses knowledge-based interaction potential. VSL2 uses neural networks on sequences of different lengths. ESpritz applied bidirectional recursive neural network (BRNN) and was trained separately on N-terminal, C-terminal, and the general sequences; (2) they achieved relatively higher prediction accuracy; (3) these predictors have standalone versions. These four scores were then fed into a decision-tree based artificial neural network (DBann) to make the final prediction. The DBann combines four specific techniques including dual-threshold, significance-voting, two-step selection [49], and two-hidden-layer Artificial Neural Network (ANN). Dual-threshold is a technique using different threshold values for true prediction and false predictions. Significance-voting is complementary to majority-voting by calculating the Euclidean distance of prediction scores to their corresponding threshold values and then comparing the distances of true predictions and false predictions to make selections. For example, when two predictors make true predictions and another two predictors make false predictions, comparing the number of true predictions (N_T) and the number of false predictions (N_P) may have limited usage. In this case, comparing the sum of distances from true thresholds value (d_T) and the total distance from false threshold values (d_F) provides more useful information of the relative significance of true predictions and false predictions. Two-step selection uses two sets of dual-threshold values together with significance-voting as follows: (1) use more stringent values as the first-step threshold values for both true predictions and false predictions; (2) select less stringent values as the second-step threshold values for both true predictions and false predictions; (3) if the numbers of predictors for true prediction and false prediction are equal in the first-step, second-step examination will be performed. If the numbers are still the same, the significance voting will be carried out; (4) based on the results of the above-mentioned comparison, the predictive results of individual predictors will be encoded differently. The encoded predictive results will then be fed into the two-hidden-layer ANN, which is a fully connected ANN and has ten and two nodes in the input and output layers, respectively, as well as twenty nodes in both hidden layers. The activation function for all the nodes is hyperbolic tangent function. In addition, in the output layer, the output was further transformed using a soft matrix function.

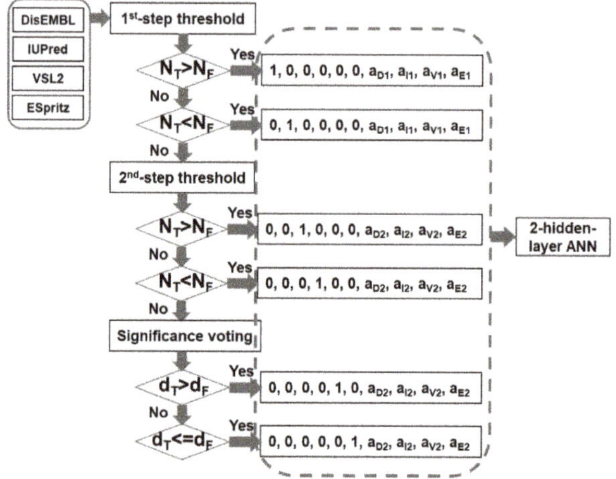

Figure 6. Infrastructure of the new meta-strategy. N_T and N_F are the numbers of predictors making true prediction and false prediction, respectively. "a1" and "a2" are the differences of prediction score from the 1st-step threshold and the 2nd-step threshold values, respectively. The letter subscripts represent DisEMBL (D), IUPred(I), VSL2(V), and ESpritz(E), accordingly. "d_T" and "d_F" are Euclidean distances of prediction scores from their corresponding threshold values for true predictions and false predictions, accordingly.

All the selected sequences were grouped into two datasets. One contains a randomly-selected 20% of all the samples and was set as the independent test dataset, while the other, containing the rest 80% of the samples, was designated as the training and validation dataset. The ratios of positive samples (disordered residues) to negative samples (structured residues) in two datasets are roughly the same. The training and validation dataset was further split into five subsets for five-fold cross-validation. In brief, three out of five subsets were used to train the predictor, the forth subset was used to prevent overfitting, and the last one was used to validate the final prediction performance. By using the different subsets for training, preventing overfitting, and validation, the aforementioned process was repeated five times. The final prediction performance was the average of all five times in the validation subsets. The trained predictors were also evaluated in the independent test dataset.

The performance of predictors was assessed using Sensitivity (Sens), Specificity (Spec), Accuracy (Acc), balanced accuracy (Acc-b, the average of sensitivity and specificity), F1 score (F1), Matthews Correlation Coefficient (MCC), Area Under ROC Curve (AUC, or AUC_ROC), and Area Under precision-recall Curve (AUC_PR) under five-fold cross-validation and in independent datasets. The performance of newly designed predictor was compared to four component predictors (DisEMBL, IUPred, VSL2, and ESpritz), as well as another four recently developed predictors, including PONDR-FIT [42], MFDp2 [61], IUPred2A [34], and AUCpreD [62].

Information Gain (IG) was calculated as a function of predictive score as follows:

$$IG(x) = \sum_{i=1,2} p_i \log_2 p_i - \sum_{j=1,2} f_j(x) \sum_{k=1,2} p_{j,k} \log_2 p_{j,k} \tag{1}$$

In which, p_i is the fraction of positive ($i = 1$) or negative ($i = 2$) samples in the dataset; "x" is the threshold prediction score to split the dataset into two groups; $f_j(x)$ is the fraction of samples with prediction score higher than the threshold ($j = 1$) or the fraction of samples with prediction score lower than the threshold ($j = 2$); and $p_{j,k}$ refers to the fraction of positive samples ($k = 1$) or negative samples ($k = 2$) in the j-th group.

Author Contributions: Conceptualization, B.X.; Methodology, B.X. and B.Z.; Software, B.Z.; Validation, B.Z.; Formal Analysis, B.X. and B.Z.; Writing, B.X.

Funding: This research received no external funding.

Acknowledgments: The authors acknowledge with thanks the usage of DisEMBL, IUPred, VSL2, ESpritz, PONDR-FIT, MFDp2, IUPred2A, AUCpreD, and DISOpred3.

Conflicts of Interest: The authors declare no conflict of interest.

Abbreviations

IDP	Intrinsically disordered protein
IDR	Intrinsically disordered region
IDAA	Intrinsically disordered amino acid
ANN	Artificial neural network
IG	Information gain
Sens	Sensitivity
Spec	Specificity
Acc	Accuracy
Acc-b	Balanced accuracy
MCC	Mathew's correlation coefficient
AUC-ROC	Area under ROC curve
AUC-PR	Area under precision-recall curve

Appendix A

Table A1. Comparison of prediction accuracy of in the CASP10 test dataset.

	Sen	Spec	Acc	Acc-b	MCC	F1	AUC_ROC	AUC_PR
Disembl	0.379	0.954	0.929	0.666	0.286	0.318	0.754	0.295
IUPred	0.168	0.958	0.924	0.563	0.122	0.161	0.618	0.175
VSL2	0.612	0.811	0.803	0.712	0.214	0.213	0.774	0.275
Espritz	0.512	0.921	0.903	0.716	0.298	0.316	0.815	0.404
DISOPRED3	0.362	0.993	0.966	0.678	0.495	0.481	0.860	0.495
PONDRFIT	0.586	0.929	0.914	0.758	0.362	0.374	0.830	0.358
MFDp2	0.325	0.975	0.947	0.650	0.322	0.349	0.778	0.352
IUPred2	0.164	0.959	0.924	0.561	0.119	0.158	0.616	0.170
AUCpreD	0.425	0.984	0.960	0.705	0.465	0.481	0.863	0.501
This work	0.629	0.840	0.831	0.734	0.249	0.245	0.793	0.305

Note Bene. DISOPRED3 is one of the two best predictors in CASP10 competition and therefore was used in the comparison. The other one and its web server were not available when the study was carried out. The highest value in each of these measures is in bold and highlighted (red).

References

1. Dunker, A.K.; Silman, I.; Uversky, V.N.; Sussman, J.L. Function and structure of inherently disordered proteins. *Curr. Opin. Struct. Biol.* **2008**, *18*, 756–764. [CrossRef] [PubMed]

2. Uversky, V.N.; Dunker, A.K. Multiparametric analysis of intrinsically disordered proteins: Looking at intrinsic disorder through compound eyes. *Anal. Chem.* **2012**, *84*, 2096–2104. [CrossRef] [PubMed]

3. Csermely, P.; Sandhu, K.S.; Hazai, E.; Hoksza, Z.; Kiss, H.J.; Miozzo, F.; Veres, D.V.; Piazza, F.; Nussinov, R. Disordered proteins and network disorder in network descriptions of protein structure, dynamics and function: Hypotheses and a comprehensive review. *Curr. Protein Pept. Sci.* **2012**, *13*, 19–33. [CrossRef] [PubMed]

4. Tompa, P. Intrinsically disordered proteins: A. 10-year recap. *Trends Biochem. Sci.* **2012**, *37*, 509–516. [CrossRef] [PubMed]

5. Uversky, V.N.; Dave, V.; Iakoucheva, L.M.; Malaney, P.; Metallo, S.J.; Pathak, R.R.; Joerger, A.C. Pathological unfoldomics of uncontrolled chaos: Intrinsically disordered proteins and human diseases. *Chem. Rev.* **2014**, *114*, 6844–6879. [CrossRef] [PubMed]

6. Fuxreiter, M.; Toth-Petroczy, A.; Kraut, D.A.; Matouschek, A.; Lim, R.Y.; Xue, B.; Kurgan, L.; Uversky, V.N. Disordered proteinaceous machines. *Chem. Rev.* **2014**, *114*, 6806–6843. [CrossRef] [PubMed]

7. Wright, P.E.; Dyson, H.J. Intrinsically disordered proteins in cellular signalling and regulation. *Nat. Rev. Mol. Cell Biol.* **2015**, *16*, 18–29. [CrossRef] [PubMed]

8. Follis, A.V.; Hammoudeh, D.I.; Wang, H.; Prochownik, E.V.; Metallo, S.J. Structural rationale for the coupled binding and unfolding of the c-Myc oncoprotein by small molecules. *Chem. Biol.* **2008**, *15*, 1149–1155. [CrossRef] [PubMed]

9. Wright, P.E.; Dyson, H.J. Linking folding and binding. *Curr. Opin. Struct. Biol.* **2009**, *19*, 31–38. [CrossRef] [PubMed]

10. Schulenburg, C.; Hilvert, D. Protein conformational disorder and enzyme catalysis. *Top Curr. Chem.* **2013**, *337*, 41–67. [CrossRef] [PubMed]

11. Dunker, A.K.; Garner, E.; Guilliot, S.; Romero, P.; Albrecht, K.; Hart, J.; Obradovic, Z.; Kissinger, C.; Villafranca, J.E. Protein disorder and the evolution of molecular recognition: Theory, predictions and observations. *Pac. Symp. Biocomput.* **1998**, *3*, 473–484.

12. Uversky, V.N. Intrinsic disorder-based protein interactions and their modulators. *Curr. Pharm. Des.* **2013**, *19*, 4191–4213. [CrossRef] [PubMed]

13. Dogan, J.; Gianni, S.; Jemth, P. The binding mechanisms of intrinsically disordered proteins. *Phys. Chem. Chem. Phys.* **2014**, *16*, 6323–6331. [CrossRef] [PubMed]

14. Liu, Z.; Huang, Y. Advantages of proteins being disordered. *Protein Sci.* **2014**, *23*, 539–550. [CrossRef] [PubMed]

15. Teilum, K.; Olsen, J.G.; Kragelund, B.B. Globular and disordered-the non-identical twins in protein-protein interactions. *Front. Mol. Biosci.* **2015**, *2*, 40. [CrossRef] [PubMed]

16. Minde, D.P.; Dunker, A.K.; Lilley, K.S. Time, space, and disorder in the expanding proteome universe. *Proteomics* **2017**, *17*. [CrossRef] [PubMed]

17. Berman, H.M.; Westbrook, J.; Feng, Z.; Gilliland, G.; Bhat, T.N.; Weissig, H.; Shindyalov, I.N.; Bourne, P.E. The Protein Data Bank. *Nucleic Acids Res.* **2000**, *28*, 235–242. [CrossRef] [PubMed]

18. Piovesan, D.; Tabaro, F.; Micetic, I.; Necci, M.; Quaglia, F.; Oldfield, C.J.; Aspromonte, M.C.; Davey, N.E.; Davidovic, R.; Dosztanyi, Z.; et al. DisProt 7.0: A major update of the database of disordered proteins. *Nucleic Acids Res.* **2017**, *45*, D219–D227. [CrossRef] [PubMed]

19. Fukuchi, S.; Sakamoto, S.; Nobe, Y.; Murakami, S.D.; Amemiya, T.; Hosoda, K.; Koike, R.; Hiroaki, H.; Ota, M. IDEAL: Intrinsically Disordered proteins with Extensive Annotations and Literature. *Nucleic Acids Res.* **2012**, *40*, D507–D511. [CrossRef] [PubMed]

20. Yu, J.F.; Dou, X.H.; Sha, Y.J.; Wang, C.L.; Wang, H.B.; Chen, Y.T.; Zhang, F.; Zhou, Y.; Wang, J.H. DisBind: A database of classified functional binding sites in disordered and structured regions of intrinsically disordered proteins. *BMC Bioinform.* **2017**, *18*, 206. [CrossRef] [PubMed]

21. Schad, E.; Ficho, E.; Pancsa, R.; Simon, I.; Dosztanyi, Z.; Meszaros, B. DIBS: A repository of disordered binding sites mediating interactions with ordered proteins. *Bioinformatics* **2018**, *34*, 535–537. [CrossRef] [PubMed]

22. Ficho, E.; Remenyi, I.; Simon, I.; Meszaros, B. MFIB: A repository of protein complexes with mutual folding induced by binding. *Bioinformatics* **2017**, *33*, 3682–3684. [CrossRef] [PubMed]

23. Cheng, Y.; Oldfield, C.J.; Meng, J.; Romero, P.; Uversky, V.N.; Dunker, A.K. Mining alpha-helix-forming molecular recognition features with cross species sequence alignments. *Biochemistry* **2007**, *46*, 13468–13477. [CrossRef] [PubMed]

24. Malhis, N.; Gsponer, J. Computational identification of MoRFs in protein sequences. *Bioinformatics* **2015**, *31*, 1738–1744. [CrossRef] [PubMed]

25. Disfani, F.M.; Hsu, W.L.; Mizianty, M.J.; Oldfield, C.J.; Xue, B.; Dunker, A.K.; Uversky, V.N.; Kurgan, L. MoRFpred, a computational tool for sequence-based prediction and characterization of short disorder-to-order transitioning binding regions in proteins. *Bioinformatics* **2012**, *28*, i75–i83. [CrossRef] [PubMed]

26. Fang, C.; Noguchi, T.; Tominaga, D.; Yamana, H. MFSPSSMpred: Identifying short disorder-to-order binding regions in disordered proteins based on contextual local evolutionary conservation. *BMC Bioinform.* **2013**, *14*, 300. [CrossRef] [PubMed]

27. Malhis, N.; Jacobson, M.; Gsponer, J. MoRFchibi SYSTEM: Software tools for the identification of MoRFs in protein sequences. *Nucleic Acids Res.* **2016**, *44*, W488–W493. [CrossRef] [PubMed]

28. Sharma, R.; Bayarjargal, M.; Tsunoda, T.; Patil, A.; Sharma, A. MoRFPred-plus: Computational Identification of MoRFs in Protein Sequences using Physicochemical Properties and HMM profiles. *J. Theor. Biol.* **2018**, *437*, 9–16. [CrossRef] [PubMed]

29. Sharma, R.; Raicar, G.; Tsunoda, T.; Patil, A.; Sharma, A. OPAL: Prediction of MoRF regions in intrinsically disordered protein sequences. *Bioinformatics* **2018**, *34*, 1850–1858. [CrossRef] [PubMed]

30. Dosztanyi, Z.; Meszaros, B.; Simon, I. ANCHOR: Web server for predicting protein binding regions in disordered proteins. *Bioinformatics* **2009**, *25*, 2745–2746. [CrossRef] [PubMed]

31. Mooney, C.; Pollastri, G.; Shields, D.C.; Haslam, N.J. Prediction of short linear protein binding regions. *J. Mol. Biol.* **2012**, *415*, 193–204. [CrossRef] [PubMed]

32. Khan, W.; Duffy, F.; Pollastri, G.; Shields, D.C.; Mooney, C. Predicting binding within disordered protein regions to structurally characterised peptide-binding domains. *PLoS ONE* **2013**, *8*, 72838. [CrossRef] [PubMed]

33. Jones, D.T.; Cozzetto, D. DISOPRED3: Precise disordered region predictions with annotated protein-binding activity. *Bioinformatics* **2015**, *31*, 857–863. [CrossRef] [PubMed]

34. Meszaros, B.; Erdos, G.; Dosztanyi, Z. IUPred2A: Context-dependent prediction of protein disorder as a function of redox state and protein binding. *Nucleic Acids Res.* **2018**, *46*, W329–W337. [CrossRef] [PubMed]

35. Li, B.Q.; Cai, Y.D.; Feng, K.Y.; Zhao, G.J. Prediction of protein cleavage site with feature selection by random forest. *PLoS ONE* **2012**, *7*, e45854. [CrossRef] [PubMed]

36. Zhao, X.; Dai, J.; Ning, Q.; Ma, Z.; Yin, M.; Sun, P. Position-specific analysis and prediction of protein pupylation sites based on multiple features. *BioMed Res. Int.* **2013**, *2013*, 109549. [CrossRef] [PubMed]

37. Tretyachenko, V.; Vymetal, J.; Bednarova, L.; Kopecky, V., Jr.; Hofbauerova, K.; Jindrova, H.; Hubalek, M.; Soucek, R.; Konvalinka, J.; Vondrasek, J.; et al. Random protein sequences can form defined secondary structures and are well-tolerated in vivo. *Sci. Rep.* **2017**, *7*, 15449. [CrossRef] [PubMed]

38. Hu, J.; Li, Y.; Zhang, Y.; Yu, D.J. ATPbind: Accurate Protein-ATP Binding Site Prediction by Combining Sequence-Profiling and Structure-Based Comparisons. *J. Chem. Inf. Model.* **2018**, *58*, 501–510. [CrossRef] [PubMed]

39. Basu, S.; Soderquist, F.; Wallner, B. Proteus: A random forest classifier to predict disorder-to-order transitioning binding regions in intrinsically disordered proteins. *J. Comput. Aided Mol. Des.* **2017**, *31*, 453–466. [CrossRef] [PubMed]

40. Klausen, M.S.; Jespersen, M.C.; Nielsen, H.; Jensen, K.K.; Jurtz, V.I.; Soenderby, C.K.; Sommer, M.O.A.; Winther, O.; Nielsen, M.; Petersen, B.; et al. NetSurfP-2.0: Improved prediction of protein structural features by integrated deep learning. *BioRxiv* **2018**. [CrossRef]

41. Xue, B.; Oldfield, C.J.; Dunker, A.K.; Uversky, V.N. CDF it all: Consensus prediction of intrinsically disordered proteins based on various cumulative distribution functions. *FEBS Lett.* **2009**, *583*, 1469–1474. [CrossRef] [PubMed]

42. Xue, B.; Dunbrack, R.L.; Williams, R.W.; Dunker, A.K.; Uversky, V.N. PONDR-FIT: A meta-predictor of intrinsically disordered amino acids. *Biochim. Biophys. Acta* **2010**, *1804*, 996–1010. [CrossRef] [PubMed]

43. Schlessinger, A.; Punta, M.; Yachdav, G.; Kajan, L.; Rost, B. Improved disorder prediction by combination of orthogonal approaches. *PLoS ONE* **2009**, *4*, e4433. [CrossRef] [PubMed]

44. Hirose, S.; Shimizu, K.; Noguchi, T. POODLE-I: Disordered region prediction by integrating POODLE series and structural information predictors based on a workflow approach. *In Silico Biol.* **2010**, *10*, 185–191. [CrossRef] [PubMed]

45. Kozlowski, L.P.; Bujnicki, J.M. MetaDisorder: A meta-server for the prediction of intrinsic disorder in proteins. *BMC Bioinform.* **2012**, *13*, 111. [CrossRef] [PubMed]

46. Huang, Y.J.; Acton, T.B.; Montelione, G.T. DisMeta: A meta server for construct design and optimization. *Methods Mol. Biol.* **2014**, *1091*, 3–16. [CrossRef] [PubMed]

47. Mizianty, M.J.; Stach, W.; Chen, K.; Kedarisetti, K.D.; Disfani, F.M.; Kurgan, L. Improved sequence-based prediction of disordered regions with multilayer fusion of multiple information sources. *Bioinformatics* **2010**, *26*, i489–i496. [CrossRef] [PubMed]

48. Xue, B.; Lipps, D.; Devineni, S. Integrated Strategy Improves the Prediction Accuracy of miRNA in Large Dataset. *PLoS ONE* **2016**, *11*, e0168392. [CrossRef] [PubMed]

49. Zhao, B.; Xue, B. Improving prediction accuracy using decision-tree-based meta-strategy and multi-threshold sequential-voting exemplified by miRNA target prediction. *Genomics* **2017**, *109*, 227–232. [CrossRef] [PubMed]

50. Xue, B.; Dor, O.; Faraggi, E.; Zhou, Y. Real-value prediction of backbone torsion angles. *Proteins* **2008**, *72*, 427–433. [CrossRef] [PubMed]

51. Aebersold, R.; Mann, M. Mass-spectrometric exploration of proteome structure and function. *Nature* **2016**, *537*, 347–355. [CrossRef] [PubMed]

52. Mann, M. Origins of mass spectrometry-based proteomics. *Nat. Rev. Mol. Cell Biol.* **2016**, *17*, 678. [CrossRef] [PubMed]

53. Minde, D.P.; Ramakrishna, M.; Lilley, K.S. Biotinylation by proximity labelling favours unfolded proteins. *BioRxiv* **2018**. [CrossRef]

54. Wang, G.; Dunbrack, R.L., Jr. PISCES: Recent improvements to a PDB sequence culling server. *Nucleic Acids Res.* **2005**, *33*, W94–W98. [CrossRef] [PubMed]

55. Li, W.; Godzik, A. Cd-hit: A fast program for clustering and comparing large sets of protein or nucleotide sequences. *Bioinformatics* **2006**, *22*, 1658–1659. [CrossRef] [PubMed]

56. Monastyrskyy, B.; Kryshtafovych, A.; Moult, J.; Tramontano, A.; Fidelis, K. Assessment of protein disorder region predictions in CASP10. *Proteins* **2014**, *82*, 127–137. [CrossRef] [PubMed]

57. Linding, R.; Jensen, L.J.; Diella, F.; Bork, P.; Gibson, T.J.; Russell, R.B. Protein disorder prediction: Implications for structural proteomics. *Structure* **2003**, *11*, 1453–1459. [CrossRef] [PubMed]

58. Dosztanyi, Z.; Csizmok, V.; Tompa, P.; Simon, I. IUPred: Web server for the prediction of intrinsically unstructured regions of proteins based on estimated energy content. *Bioinformatics* **2005**, *21*, 3433–3434. [CrossRef] [PubMed]

59. Peng, K.; Radivojac, P.; Vucetic, S.; Dunker, A.K.; Obradovic, Z. Length-dependent prediction of protein intrinsic disorder. *BMC Bioinform.* **2006**, *7*, 208. [CrossRef] [PubMed]

60. Walsh, I.; Martin, A.J.; Di Domenico, T.; Tosatto, S.C. ESpritz: Accurate and fast prediction of protein disorder. *Bioinformatics* **2012**, *28*, 503–509. [CrossRef] [PubMed]

61. Mizianty, M.J.; Peng, Z.; Kurgan, L. MFDp2: Accurate predictor of disorder in proteins by fusion of disorder probabilities, content and profiles. *Intrinsically Disord. Proteins* **2013**, *1*, e24428. [CrossRef] [PubMed]

62. Wang, S.; Ma, J.; Xu, J. AUCpreD: Proteome-level protein disorder prediction by AUC-maximized deep convolutional neural fields. *Bioinformatics* **2016**, *32*, i672–i679. [CrossRef] [PubMed]

© 2018 by the authors. Licensee MDPI, Basel, Switzerland. This article is an open access article distributed under the terms and conditions of the Creative Commons Attribution (CC BY) license (http://creativecommons.org/licenses/by/4.0/).

International Journal of
Molecular Sciences

Article

A Protein Intrinsic Disorder Approach for Characterising Differentially Expressed Genes in Transcriptome Data: Analysis of Cell-Adhesion Regulated Gene Expression in Lymphoma Cells

Gustav Arvidsson and Anthony P. H. Wright *

Clinical Research Center, Department of Laboratory Medicine, Karolinska Institutet, Huddinge SE 141 57, Sweden; gustav.arvidsson@ki.se
* Correspondence: anthony.wright@ki.se

Received: 30 August 2018; Accepted: 4 October 2018; Published: 10 October 2018

Abstract: Conformational protein properties are coupled to protein functionality and could provide a useful parameter for functional annotation of differentially expressed genes in transcriptome studies. The aim was to determine whether predicted intrinsic protein disorder was differentially associated with proteins encoded by genes that are differentially regulated in lymphoma cells upon interaction with stromal cells, an interaction that occurs in microenvironments, such as lymph nodes that are protective for lymphoma cells during chemotherapy. Intrinsic disorder protein properties were extracted from the Database of Disordered Protein Prediction (D^2P^2), which contains data from nine intrinsic disorder predictors. Proteins encoded by differentially regulated cell-adhesion regulated genes were enriched in intrinsically disordered regions (IDRs) compared to other genes both with regard to IDR number and length. The enrichment was further ascribed to down-regulated genes. Consistently, a higher proportion of proteins encoded by down-regulated genes contained at least one IDR or were completely disordered. We conclude that down-regulated genes in stromal cell-adherent lymphoma cells encode proteins that are characterized by elevated levels of intrinsically disordered conformation, indicating the importance of down-regulating functional mechanisms associated with intrinsically disordered proteins in these cells. Further, the approach provides a generally applicable and complementary alternative to classification of differentially regulated genes using gene ontology or pathway enrichment analysis.

Keywords: intrinsic disorder; intrinsic disorder prediction; intrinsically disordered region; protein conformation; transcriptome; RNA sequencing; Microarray; differentially regulated genes; gene ontology analysis; functional analysis

---

1. Introduction

Genome-wide approaches to identify genes that are differentially expressed under different conditions of interest have become a standard approach to investigating mechanisms involved in biological processes. The analysis pipeline used in such studies generally leads quickly to some form of gene ontology analysis, in order to identify biological functions that are associated with the differentially regulated genes. A complementary approach would be to analyse differentially regulated genes in relation to predicted conformational properties of the proteins they encode but such an approach has not been reported.

Characterisation of differentially regulated genes in relation to predicted or known conformational properties of the proteins they encode would be of interest in the light of recent discoveries showing overall relationships between conformational properties and different types of protein functionality or mechanism of action [1]. For example, the catalytic domains of enzymes are generally ordered globular conformations while transcription factors are characterised by a preponderance of intrinsic disorder leading to ensembles of many alternative conformational forms [2]. It is now clear that about half the proteins in eukaryotes contain at least one extended (>30 amino acid residues) intrinsically disordered region (IDR) and some proteins are completely disordered [3].

Interestingly, IDRs occur more frequently in regulatory proteins and disease-related proteins [4,5]. We recently identified genes that are differentially expressed in mantle cell lymphoma (MCL) cells that adhere to stromal cells with which they are co-cultured compared to non-adherent MCL cells in the same culture [6]. The differentially regulated gene set defined in this in vitro model system showed substantial overlap with genes that are differentially regulated in the lymph node microenvironment of MCL and chronic lymphoblastic leukaemia (CLL) patients. Retention of lymphoma cells in microenvironments is thought to lead to minimal residual disease, in which a subpopulation of cancer cells receives survival signals from normal cells in microenvironments, thus allowing them to survive during treatment and to subsequently cause disease relapse. In vitro, minimal residual disease is mimicked by cell adhesion mediated drug resistance whereby, for example, lymphoma cells residing in close proximity to stromal cells manifest an enhanced level of resistance to cytostatic drugs [7]. Thus, differentially regulated genes in co-cultured adherent lymphoma cells are likely to represent processes important for cell adhesion mediated drug resistance and minimal residual disease.

In our recent study, we identified 1050 genes that were differentially regulated in MCL cells adhered to stromal cells compared to non-adherent MCL cells in the same co-culture. The four main functional themes characterised by the differentially regulated gene set were cell adhesion, anti-apoptosis and B-cell signalling/immune-modulation, associated with up-regulated genes in adherent cells, as well as early mitotic processes, associated with down-regulated genes [6]. Here we test whether the differentially regulated gene set or its subsets encode proteins that differ in IDR properties compared to non-regulated genes.

2. Results

To determine whether there might be a difference in the frequency of IDRs (defined as predicted IDRs $\geq$ 30 amino acid residues in length) in proteins encoded by adhesion-regulated genes (adsu, n = 1009) compared to other proteins (nadsu, n = 17,612), we calculated the percentage of IDRs in adhesion-related proteins for each IDR predictor (Figure 1, blue line) and compared it to the proportion of genes in the adhesion-regulated gene set (5.4%, Figure 1, red line). For all predictors, the proportion of predicted IDRs associated with the adhesion gene set exceeded the frequency expected based on the proportion of proteins in the set. For many predictors, Figure 1 also shows a tendency towards a larger number of longer IDRs in proteins encoded by the adhesion gene set at the expense of shorter IDRs.

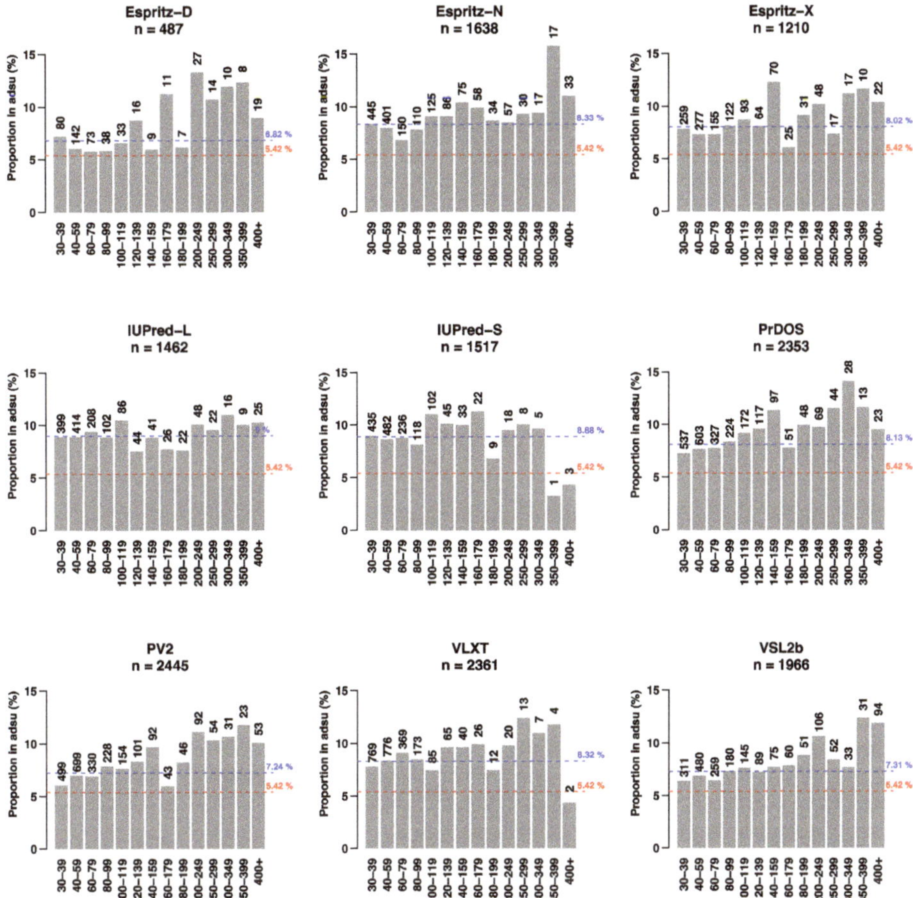

Figure 1. Enrichment of intrinsically disordered regions (IDRs) in proteins encoded by genes that are differentially expressed in lymphoma cells upon adhering to stromal cells. The number (*n*) of IDRs (≥30 residues) for each predictor is shown as well as how the detected IDRs are distributed in relation to length. The number of IDRs in each size category is shown. The blue line shows the percentage of all IDRs encoded by adhesion-related genes (adsu) and non-adhesion-related genes (nadsu) that are associated with the adsu set, while the red line shows the percentage expected if IDRs are equally distributed between the adsu and nadsu sets.

To determine whether the enhanced frequency of IDRs in proteins encoded by adhesion-regulated genes was significant, we used a resampling approach to test whether the IDR frequency associated with the 1009 adhesion-regulated genes lay outside the distribution of frequencies generated by 1000-fold resampling of 1009 genes from the control gene set (*n* = 17,612). A *z*-score and associated *p*-value was generated for data from each predictor. As shown in Table 1 (adsu vs. nadsu), the enrichment of IDRs in proteins encoded by adhesion-regulated genes (adsu) was significant for all predictors.

Table 1. Intrinsically disordered regions are enriched in proteins encoded by down-regulated genes in lymphoma cells upon adherence to stromal cells.

Gene Set Comparison	Espritz-D [#]	Espritz-N [#]	Espritz-X [#]	IUPred-L [#]	IUPred-S [#]	PrDOS [#]	PV2 [#]	VLXT [#]	VSL2b [#]
Adsu vs. Nadsu									
IDR number in adsu	487	1638	1210	1462	1517	2353	2445	2361	1966
IDR number in nadsu *	382	1036	796	847	892	1520	1794	1492	1431
Adjusted *p*-value	2.38×10^{-8}	1.32×10^{-27}	2.81×10^{-23}	4.34×10^{-27}	1.32×10^{-27}	4.24×10^{-34}	2.85×10^{-19}	2.75×10^{-25}	8.29×10^{-19}
Adsu_Down vs. Adsu_Up									
IDR number in adsu_down	276	1072	760	983	1025	1507	1508	1572	1216
IDR number in adsu_up *	171	458	363	387	397	758	683	639	606
Adjusted *p*-value	3.51×10^{-78}	$<1.00 \times 10^{-99}$	$<1.00 \times 10^{-99}$	$<1.00 \times 10^{-99}$	$<1.00 \times 10^{-99}$	$<1.00 \times 10^{-99}$	$<1.00 \times 10^{-99}$	$<1.00 \times 10^{-99}$	$<1.00 \times 10^{-99}$

* mean of 1000 resamples of n proteins encoded by genes in nadsu or adsu_up, where n = the number of genes in adsu or adsu_down, respectively. Abbreviations: adsu (adhesion-regulated genes); nadsu (non-adhesion-regulated genes); adsu_down (down-regulated adsu); adsu_up (up-regulated adsu). [#] Predictors of intrinsic disorder that appear in the D^2P^2 database.

Next, we tested whether the enrichment of IDRs associated with the adhesion-regulated gene set could be ascribed to subsets of the adhesion-regulated genes. Comparison of proteins encoded by genes manifesting a greater degree of regulation (fold change $\geq$ 1.3) relative to the remaining regulated genes showed fewer IDRs in more highly regulated genes compared to less highly regulated genes for all predictors and with lower levels of significance compared to the comparison of regulated and non-regulated genes (data not shown). Thus, there is an enrichment of IDRs in adhesion-regulated genes but the enrichment is not related to the extent of their regulation. Comparison of the up-regulated subset (adsu_up, change >1) relative to the down-regulated subset (adsu_down, change <1), on the other hand, showed an enhanced enrichment of IDRs in proteins encoded by the adsu_down subset compared to the enhancement levels in Figure 1, with high levels of significance (Table 1, adsu_down vs. adsu_up). Thus, the enrichment in IDRs in proteins encoded by adhesion-regulated genes is mainly associated with proteins encoded by down-regulated genes.

We next investigated whether the length of IDRs in proteins encoded by adsu genes tends to be longer than in other proteins (nadsu). As expected, IDR length is not normally distributed, as indicated by the consistently higher value of the mean compared to the median (Table 2), as well as tests of normality (data not shown). Thus, a Mann–Whitney test was used to test the significance of differences in IDR length between groups. Table 2 shows that some predictors (notably PV2, PrDOS and VSL2b) predict longer IDRs in proteins encoded by adsu genes that in other proteins (nadsu), but for other predictors the difference is less significant or lacking in statistical support. IDRs encoded by adsu_down genes were significantly longer than IDRs encoded by adsu_up genes for all predictors. Thus, IDRs in proteins encoded by genes that are down-regulated in adherent cells tend to be both more frequent and longer than IDRs in other proteins.

Table 2. Intrinsically disordered regions tend to be longer in proteins encoded by down-regulated genes in lymphoma cells upon adherence to stromal cells.

Gene Set Comparison	Espritz-D	Espritz-N	Espritz-X	IUPred-L	IUPred-S	PrDOS	PV2	VLXT	VSL2b
Adsu vs. Nadsu									
Median (mean) IDR length (adsu)	64 (125)	57 (97)	66 (98)	54 (87)	52 (67)	61 (98)	61 (90)	48 (62)	74 (128)
Median (mean) IDR length (nadsu)	61 (98)	56 (91)	62 (93)	54 (86)	51 (68)	55 (86)	56 (82)	47 (61)	66 (107)
Adjusted *p*-value *	6.04×10^{-2}	7.33×10^{-2}	$\mathbf{2.50 \times 10^{-2}}$	6.02×10^{-1}	5.10×10^{-1}	$\mathbf{3.75 \times 10^{-11}}$	$\mathbf{4.97 \times 10^{-9}}$	$\mathbf{2.42 \times 10^{-2}}$	$\mathbf{4.97 \times 10^{-9}}$
Adsu_Down vs. Adsu_Up									
Median (mean) IDR length (adsu_down)	68.5 (154)	60 (104)	76 (108)	56 (93)	53 (70)	64 (107)	65 (96)	49 (65)	79.5 (142)
Median (mean) IDR length (adsu_up)	59 (89)	54 (86)	58 (83)	50 (75)	50 (62)	58 (85)	57 (80)	46 (57)	67 (104)
Adjusted *p*-value *	$\mathbf{1.24 \times 10^{-3}}$	$\mathbf{5.26 \times 10^{-3}}$	$\mathbf{2.14 \times 10^{-7}}$	$\mathbf{5.08 \times 10^{-3}}$	$\mathbf{1.66 \times 10^{-2}}$	$\mathbf{3.38 \times 10^{-4}}$	$\mathbf{1.42 \times 10^{-5}}$	$\mathbf{1.73 \times 10^{-3}}$	$\mathbf{7.05 \times 10^{-5}}$

* Mann–Whitney test; bold text = $p < 0.05$. Abbreviations: adsu (adhesion-regulated genes); nadsu (non-adhesion-regulated genes); adsu_down (down-regulated adsu); adsu_up (up-regulated adsu).

We next addressed how IDRs are distributed among the proteins encoded by the adsu_down gene set in relation to proteins associated with the adsu_up and nadsu gene sets (Table 3).

Table 3. Distribution of IDRs predicted by VSL2b in proteins encoded by genes that are differentially regulated in lymphoma cells upon interaction with stromal cells.

Gene Set Comparison	Number of Proteins	Number (%) of Completely Disordered Proteins	Number (%) of Proteins with IDR	Median Percent IDR Per Protein (All Proteins)	Median Percent IDR Per Protein (IDR-Containing Proteins)
adsu_down	445	19 (4.3)	367 (82.5)	38.1	48.1
adsu_up	556	11 (2)	370 (66.5)	18.1	37.6
nadsu	17,459	476 (2.7)	11,248 (64.4)	17.8	37.6

adsu_down (down-regulated adhesion-regulated genes); adsu_up (up-regulated adhesion-regulated genes); nadsu (non-adhesion-regulated genes).

The proportion of completely disordered proteins was higher for the adsu_down sets than for proteins encoded by the other gene sets, as was the proportion of proteins containing at least one IDR. The median proportion of the protein sequences that were predicted as IDR was higher for the adsu_down group, irrespective of whether all proteins were considered or only proteins containing IDRs. Table 3 shows data for the VSL2b predictor but other predictors generally produced a similar result, especially PrDOS and PV2. The frequency of IDR-containing proteins with different IDR proportions for the different gene sets is compared graphically in Figure 2A.

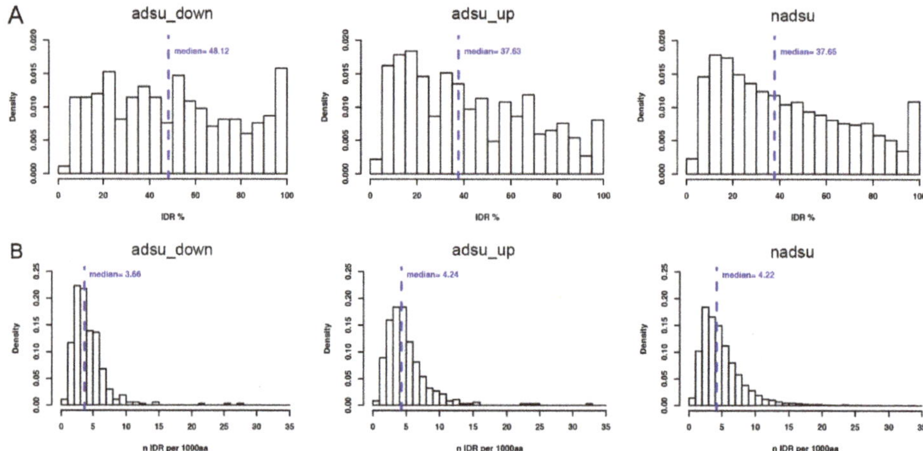

Figure 2. Relative frequency distributions of proportion of IDR per protein and length-normalized number of IDRs per protein for proteins encoded by adsu_down genes in relation to adsu_up and nadsu genes. IDR predictions were made using VSL2b. (**A**) Relative frequency distributions (Density) of IDR-containing proteins with different percent IDR content. The median position and value are shown in blue. (**B**) Relative frequency distributions (Density) of numbers of IDRs per IDR-containing protein, normalized for differences in protein length (IDR number per 1000 amino acid residues). The median position and value are shown in blue.

For proteins encoded by nadsu and adsu_up, the relative frequency declines progressively as the proportion of IDR per protein increases. Contrastingly, a more even distribution of relative frequencies is seen for adsu_down proteins, with relatively fewer low-IDR content proteins and an increased proportion of high-IDR content proteins. Interestingly, the protein length-normalized number of IDRs per protein is somewhat lower for proteins encoded by adsu_down genes, compared to adsu_up and nadsu genes (Figure 2B). Thus, the greater IDR content of adsu_down encoded genes tends to be associated with fewer and longer IDRs when only IDR-containing proteins are analyzed.

To further investigate differences in IDR lengths between groups, we plotted the length of the longest IDR in each protein as a function of protein length to compare adsu_down and adsu_up encoded proteins (Figure 3).

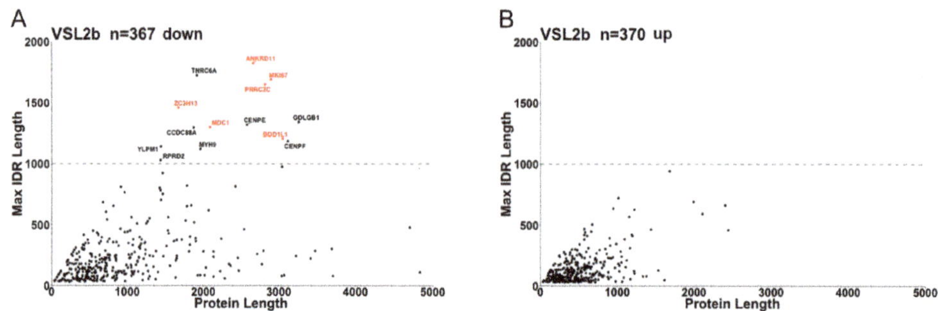

Figure 3. Comparison of proteins encoded by down- or up-regulated adhesion-regulated genes with regard to longest IDR length per protein and protein length. IDR-containing proteins encoded by (**A**) down-regulated adhesion-regulated genes (adsu_down) and (**B**) up-regulated adhesion-regulated genes (adsu_up) are shown. Of the 14 proteins in (**A**) for which the maximum IDR length is greater than 1000 residues (above dotted line), 6 proteins (red text) were also found in the sets of 14 proteins with the longest IDRs predicted by the PV2 and PrDOS predictors.

adsu_down encoded proteins are characterized by both longer protein length and longer length of the longest IDR (VSL2b). There are 14 adsu_down encoded proteins with IDRs longer than 1000 residues and these are also among proteins with the longest IDRs for most other predictors (notably PrDOS and PV2). The IDR score profiles for the 6 proteins that are reproducibly found in the top 14 proteins with longest IDRs by the VSL2b, PV2 and PrDOS predictors (red text in Figure 3A) are shown in Figure 4.

Consistent with Figure 3A, most of the proteins are predicted to be disordered throughout most of their length. Some contain extended regions with close to maximal intrinsic disorder scores (e.g., ZC3H13), while others are characterized by fluctuating levels of intrinsic disorder (e.g., MKI67). Some proteins contain both patterns in different regions of the protein (e.g., BOD1L1). Many of the proteins have short regions that are predicted to be ordered and that could correspond to folded protein domains. The different types of predicted conformation could inform about mechanisms involved in the function of proteins encoded by down-regulated genes in relation to up-regulated genes (see Discussion).

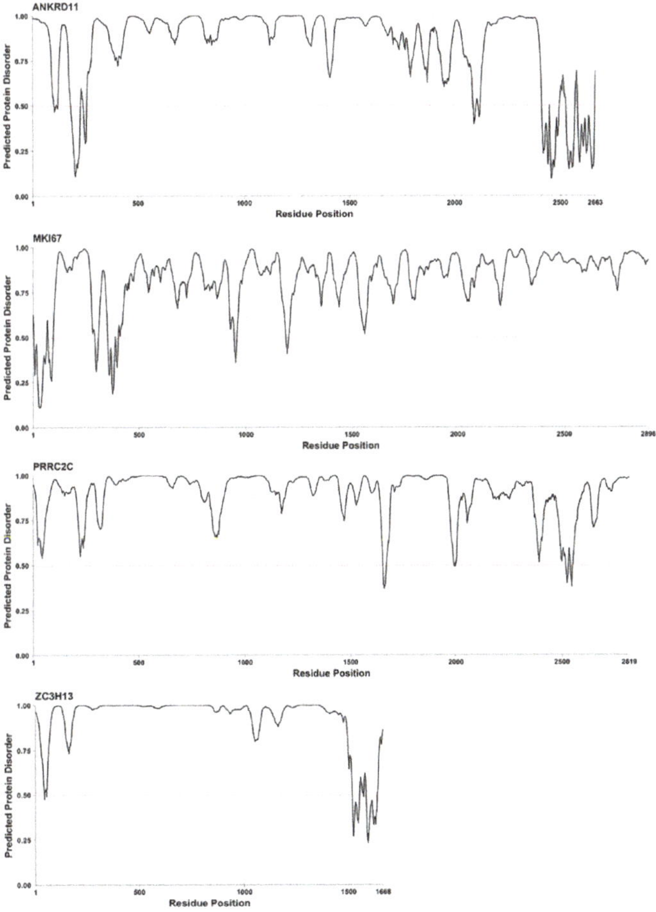

Figure 4. *Cont.*

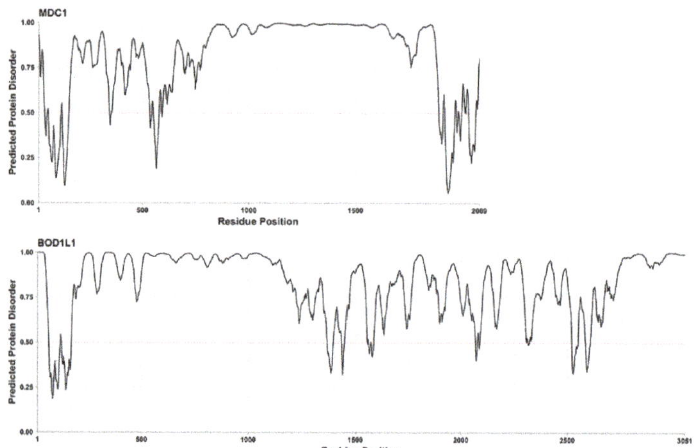

Figure 4. Examples of proteins with long IDRs. Proteins that are reproducibly found by the VSL2b, PV2 and PrDOS predictors in the set of 14 proteins with the longest predicted IDRs (red text in Figure 3A) are shown. The residue-by-residue intrinsic disorder score (VSL2b) is plotted as a function of residue number throughout the length of the respective proteins. The horizontal gridline at a score of 0.5 distinguishes regions predicted to be ordered (<0.5) or intrinsically disordered (>0.5).

3. Discussion

The main finding of this work is that proteins encoded by genes that are down-regulated in lymphoma cells upon adhering to stromal cells, typically found in microenvironments that increase cancer-cell survival, tend to have more frequent and longer regions of predicted intrinsically disordered conformation than proteins encoded by up-regulated genes or other expressed genes in the same cells. Our previous work has shown that many proteins encoded by down-regulated genes in adherent cells are involved in early stages of mitosis [6]. The present results complement this observation by suggesting that proteins encoded by the down-regulated gene set tend to function by mechanisms that are associated with intrinsically disordered regions. A secondary finding is that many of the proteins encoded by down-regulated genes are larger than proteins encoded by up-regulated genes.

Intrinsically disordered protein regions can be broadly divided into regions that are always disordered and disordered regions that form one or more ordered conformations in particular molecular environments, such as during coupled binding and folding interactions with partner proteins [8]. Some IDRs have been shown to bind partners in the disordered state via multi-valent interactions, mediated by short linear motifs that are distributed along the length of the IDR [5,9–11]. However, IDRs have other functions in addition to interaction with partners. One such function is mediation of phase transitions in cells that allow for compartmentalization of cellular regions in so-called "membrane-less organelles" that include nucleoli, nuclear speckles, P-bodies and chromatin [12–16]. These kinds of functional mechanisms might be associated with the IDRs that have consistently close-to-maximal prediction scores over extended regions of proteins encoded by down-regulated genes, as exemplified by some of the proteins in Figures 3 and 4.

The clearest example of a protein that is predicted to be maximally disordered throughout most of the protein sequence is ZC3H13. Interestingly, ZC3H13 is part of the WTAP complex, which is involved in RNA splicing and processing and is localized in nuclear speckles [17]. It is likely that such speckles result from phase transition processes and it is possible that the disordered region of ZC3H13 is important for speckle formation or ZC3H13 localization to the speckle. In fact, many documented types of so-called proteinaceous membrane-less organelles are located in the nucleus and include chromatin in addition to nuclear speckles, nucleoli and many other bodies [14]. The MDC1 protein

(Figures 3 and 4) contains a central region predicted to be completely disordered, flanked by less disordered/structured regions, which are known to mediate binding to several partner proteins at chromatin regions containing double-stranded DNA breaks [18]. Thus, MDC1 has been regarded as a "scaffold" protein responsible for spreading of DNA-repair factors over the damaged chromatin region and it is tempting to speculate that the central disordered region could play a role in phase-transitions. Other proteins in Figures 3 and 4 that have extensive regions predicted to be completely disordered and that work in a chromatin environment are YLPM1, involved in regulating telomerase activity, and BOD1L1, a protein that protects stalled DNA replication forks.

MKI67 is predicted to be disordered (with varying score) throughout almost its entire length (see Figure 4). Interestingly, MKI67 orchestrates formation of the perichromosomal layer, which coats the condensed chromosomes during mitosis in order to prevent chromosome aggregation [19]. In mitotic mammalian cells, the nuclear membrane and nucleolus are broken down and nucleolar proteins including the known phase-transition proteins, Nucleophosmin and Fibrillarin, that drive nucleolus formation in interphase cells [20], are also found in the mitotic perichromosomal layer. This fact, taken together with the RNA-binding activity associated with MKI67, suggests that the perichromosomal layer may be formed by phase transition phenomena. Interestingly, higher expression of MKI67 is a negative prognostic marker for MCL patients [21].

In the IDR class that conditionally adopts ordered conformations in some molecular contexts, the ordered conformations are characterized by varying degrees of "fuzziness", defined as the existence of a heterogeneous range of ordered conformations in the context of, for example, interaction with a single partner [22]. Many proteins that conditionally adopt ordered conformations contain pre-structure motifs (PreSMos), defined as short protein regions within IDRs that have a weak propensity for secondary structure formation leading to formation of unstable secondary structure elements in a minority sub-population of IDR-containing proteins [23]. PreSMos become stabilized during coupled binding and folding, and form part of the folded protein conformation that is seen in complexes with partner proteins. Protein regions encoded by down-regulated genes that show alternating sub-regions of higher and lower intrinsic disorder scores might correspond to these kinds of IDR since the short regions with lower intrinsic disorder scores may represent PreSMos. The CENPE and CENPF proteins are characterized by disordered regions interspersed with regions with lower disorder scores that could represent regions containing PreSMos. This would be consistent with the multiple interactions made by these proteins within the kinetochore structure that binds to the centromeric chromatin of chromosomes during mitosis. TNRC6A is a member of the GW182 family of scaffold proteins that are important for organization of proteins needed for RNA-mediated gene silencing and are found in P-bodies that are formed by a phase transition process [24].

Although somewhat speculative, the preceding sections suggest mechanisms by which some of the large proteins with large amounts of intrinsic disorder might contribute the propagation of lymphoma cells in suspension as well as how their down-regulation could lead to reduced proliferation of lymphoma cells adhered to stromal cells. Reduced proliferation is known to increase the survival of cancer cells during chemotherapy, which primarily targets proliferating cells [7,25]. Further, the cell cycle arrest that occurs in adherent MCL cells [26] would be expected to reduce the need for apoptotic responses and we previously showed that adherence to stromal cells is associated with up-regulation of anti-apoptotic genes [6].

We have shown that predicted intrinsic disorder can be used to interrogate proteins encoded by transcriptome data and that identification of gene sets encoding proteins with characteristic predicted disorder properties can provide information relevant for understanding the mechanisms underlying the functionality of groups of proteins. This approach complements the commonly used gene ontology analysis approach, which primarily gives information about the cellular components or processes that are characteristic for the function of protein sets. Both approaches provide information that can be used for hypothesis building and the design of further experiments.

In this work, we have only analyzed predicted protein disorder as a conformational characteristic. There are other predictors that could be used to expand the approach in the future and new predictors are continuously being developed as more is learned about how protein functionality is coupled to the conformational flexibility of proteins. Examples are the s2D predictor [27], which predicts secondary structure elements in relation to random coil regions, and Dynamine [28], which predicts the rigidity of the peptide backbone throughout protein sequences, as well as the ANCHOR [29] and MoRFpred [30] predictors, which predict protein interaction sites. More recently developed predictors include prediction of protein regions involved in phase transitions [31], prediction of decomposed residue-by-residue solvation free energy [32] and prediction of residue-by-residue compactness/secondary structure [33]. Thus, it is easy to see that a battery of predictors could be used to reveal many different conformational aspects of protein sets encoded by groups of differentially regulated genes identified in transcriptome data. Databases like the Database of Disordered Protein Prediction (D^2P^2) [34] or the more recently developed MobiDB [35], which contain collections of prediction data from different sources, will be useful tools for this purpose.

4. Materials and Methods

4.1. Data

Human protein regions predicted to be disordered and related data were downloaded from the publically available D^2P^2 database (available online: http://d2p2.pro/search/build) on 11 September 2017. Default options were used for the download except that "Genome" was set to "Homo sapiens 63_37" and the "Limit to" option was set to "all". The downloaded data contained all predicted IDRs detected in a total of 917,132 features for each of 9 different intrinsic disorder predictors (Espritz_Disprot, Espritz_NMR, Espritz_Xray, IUPred_long, IUPred_short, PV2, PrDOS, VL-XT, and VSL2b). See the D^2P^2 website (available online: http://d2p2.pro) or [34] for details. Mean fold-change transcriptome values for 1050 genes that show significantly altered transcript levels when Jeko-1 mantle lymphoma cells adhere to MS-5 stromal cells were taken from a recently published study from our group [6].

4.2. Data Analysis

Data were imported into and analysed using the *R* statistical programming platform (version 3.4.3, https://cran.r-project.org) [36] using packages shipped with the standard version, together with the following additional packages: data.table [37], nortest [38], formattable [39], org.Hs.eg.db [40].

To match gene expression data to IDR data for proteins in the D^2P^2 data set, it was first necessary to match an ENSEMBL protein id (from the EMSEMBL database, http://www.ensembl.org/index.html) to each of the genes identified in the RNAseq experiment. This was done by matching entries in the RNAseq data with entries in the org.Hs.eg.db annotation database from which fields for ENSEMBL protein id (ENSEMBLPROT) and gene name (SYMBOL) were extracted and appended to the RNAseq data using ENTREZID as a common key. 18,686 of 23,445 entries in the RNAseq data set were matched and also had identical gene names. This set was used in the further analysis. The annotation for the vast majority of the non-matched genes indicated that they represented non-protein-coding genes, putative protein encoding genes or pseudogenes. 1009 of the 1050 adhesion regulated genes were matched to an ENSEMBL protein id and a control set of 17,612 genes that were not shown to be regulated by adhesion were uniquely matched. Entries for which "SEQID" in the D^2P^2 IDR data matched the ENSEMBL protein id in sets or subsets of the adhesion-regulated genes or non-regulated genes were used for analysis of the sets or subsets. Only D^2P^2 entries for IDRs $\geq$ 30 amino acid residues were used and data for the 9 different IDR predictors were extracted from the database and analysed separately.

Differences in IDR number between test sets and control sets were evaluated statistically by z-scores and associated p-values calculated from the measured test value compared to the mean of

1000 control values, calculated from 1000 re-samples (with replacement) randomly selected from the control data. The size of the control re-samples was the same as the size of the test set. Differences in IDR length between test sets and control sets were evaluated statistically using a Mann–Whitney test, a non-parametric test appropriate for non-normally distributed data. p-values were adjusted for multiple testing using the false discovery rate method.

Author Contributions: Conceptualization, G.A. and A.P.W.; Methodology, G.A. and A.P.W.; Formal Analysis, G.A. and A.P.W.; Investigation, G.A. and A.P.W.; Data Curation, G.A. and A.P.W.; Writing-Original Draft Preparation, A.P.W.; Writing-Review & Editing, G.A. and A.P.W.; Visualization, G.A. and A.P.W.; Supervision, A.P.W.; Project Administration, A.P.W.; Funding Acquisition, A.P.W.

Funding: This research was funded by the Swedish Cancer Society and the Swedish Research Council.

Conflicts of Interest: The authors declare no conflict of interest.

References

1. Das, R.K.; Ruff, K.M.; Pappu, R.V. Relating sequence encoded information to form and function of intrinsically disordered proteins. *Curr. Opin. Struct. Biol.* **2015**, *32*, 102–112. [CrossRef] [PubMed]

2. Arai, M.; Sugase, K.; Dyson, H.J.; Wright, P.E. Conformational propensities of intrinsically disordered proteins influence the mechanism of binding and folding. *Proc. Natl. Acad. Sci. USA* **2015**, *112*, 9614–9619. [CrossRef] [PubMed]

3. van der Lee, R.; Buljan, M.; Lang, B.; Weatheritt, R.J.; Daughdrill, G.W.; Dunker, A.K.; Fuxreiter, M.; Gough, J.; Gsponer, J.; Jones, D.T.; et al. Classification of intrinsically disordered regions and proteins. *Chem. Rev.* **2014**, *114*, 6589–6631. [CrossRef] [PubMed]

4. Uversky, V.N.; Oldfield, C.J.; Dunker, A.K. Intrinsically disordered proteins in human diseases: Introducing the d2 concept. *Annu. Rev. Biophys.* **2008**, *37*, 215–246. [CrossRef] [PubMed]

5. Marasco, D.; Scognamiglio, P.L. Identification of inhibitors of biological interactions involving intrinsically disordered proteins. *Int. J. Mol. Sci.* **2015**, *16*, 7394–7412. [CrossRef] [PubMed]

6. Arvidsson, G.; Henriksson, J.; Sander, B.; Wright, A.P. Mixed-species rnaseq analysis of human lymphoma cells adhering to mouse stromal cells identifies a core gene set that is also differentially expressed in the lymph node microenvironment of mantle cell lymphoma and chronic lymphocytic leukemia patients. *Haematologica* **2018**, *103*, 666–678. [CrossRef] [PubMed]

7. Medina, D.J.; Goodell, L.; Glod, J.; Gelinas, C.; Rabson, A.B.; Strair, R.K. Mesenchymal stromal cells protect mantle cell lymphoma cells from spontaneous and drug-induced apoptosis through secretion of b-cell activating factor and activation of the canonical and non-canonical nuclear factor kappab pathways. *Haematologica* **2012**, *97*, 1255–1263. [CrossRef] [PubMed]

8. Dyson, H.J.; Wright, P.E. Intrinsically unstructured proteins and their functions. *Nat. Rev. Mol. Cell Biol.* **2005**, *6*, 197–208. [CrossRef] [PubMed]

9. Huang, C.; Rossi, P.; Saio, T.; Kalodimos, C.G. Structural basis for the antifolding activity of a molecular chaperone. *Nature* **2016**, *537*, 202–206. [CrossRef] [PubMed]

10. Scognamiglio, P.L.; Di Natale, C.; Leone, M.; Poletto, M.; Vitagliano, L.; Tell, G.; Marasco, D. G-quadruplex DNA recognition by nucleophosmin: New insights from protein dissection. *Biochim. Biophys. Acta* **2014**, *1840*, 2050–2059. [CrossRef] [PubMed]

11. Pauwels, K.; Lebrun, P.; Tompa, P. To be disordered or not to be disordered: Is that still a question for proteins in the cell? *Cell. Mol. Life Sci.* **2017**, *74*, 3185–3204. [CrossRef] [PubMed]

12. Shin, Y.; Brangwynne, C.P. Liquid phase condensation in cell physiology and disease. *Science* **2017**, *357*, eaaf4382. [CrossRef] [PubMed]

13. Chong, P.A.; Forman-Kay, J.D. Liquid-liquid phase separation in cellular signaling systems. *Curr. Opin. Struct. Biol.* **2016**, *41*, 180–186. [CrossRef] [PubMed]

14. Darling, A.L.; Liu, Y.; Oldfield, C.J.; Uversky, V.N. Intrinsically disordered proteome of human membrane-less organelles. *Proteomics* **2018**, *18*, e1700193. [CrossRef] [PubMed]

15. Mitrea, D.M.; Kriwacki, R.W. Phase separation in biology; functional organization of a higher order. *Cell Commun. Signal.* **2016**, *14*, 1. [CrossRef] [PubMed]

16. Toretsky, J.A.; Wright, P.E. Assemblages: Functional units formed by cellular phase separation. *J. Cell. Biol.* **2014**, *206*, 579–588. [CrossRef] [PubMed]

17. Ping, X.L.; Sun, B.F.; Wang, L.; Xiao, W.; Yang, X.; Wang, W.J.; Adhikari, S.; Shi, Y.; Lv, Y.; Chen, Y.S.; et al. Mammalian wtap is a regulatory subunit of the rna n6-methyladenosine methyltransferase. *Cell Res.* **2014**, *24*, 177–189. [CrossRef] [PubMed]

18. Nagy, Z.; Kalousi, A.; Furst, A.; Koch, M.; Fischer, B.; Soutoglou, E. Tankyrases promote homologous recombination and check point activation in response to dsbs. *PLoS Genet.* **2016**, *12*, e1005791. [CrossRef] [PubMed]

19. Booth, D.G.; Earnshaw, W.C. Ki-67 and the chromosome periphery compartment in mitosis. *Trends Cell. Biol.* **2017**, *27*, 906–916. [CrossRef] [PubMed]

20. Feric, M.; Vaidya, N.; Harmon, T.S.; Mitrea, D.M.; Zhu, L.; Richardson, T.M.; Kriwacki, R.W.; Pappu, R.V.; Brangwynne, C.P. Coexisting liquid phases underlie nucleolar subcompartments. *Cell* **2016**, *165*, 1686–1697. [CrossRef] [PubMed]

21. Katzenberger, T.; Petzoldt, C.; Holler, S.; Mader, U.; Kalla, J.; Adam, P.; Ott, M.M.; Muller-Hermelink, H.K.; Rosenwald, A.; Ott, G. The ki67 proliferation index is a quantitative indicator of clinical risk in mantle cell lymphoma. *Blood* **2006**, *107*, 3407. [CrossRef] [PubMed]

22. Fuxreiter, M.; Tompa, P. Fuzzy complexes: A more stochastic view of protein function. *Adv. Exp. Med. Biol.* **2012**, *725*, 1–14. [PubMed]

23. Lee, S.H.; Kim, D.H.; Han, J.J.; Cha, E.J.; Lim, J.E.; Cho, Y.J.; Lee, C.; Han, K.H. Understanding pre-structured motifs (presmos) in intrinsically unfolded proteins. *Curr. Protein Pept. Sci.* **2012**, *13*, 34–54. [CrossRef] [PubMed]

24. Yao, B.; Li, S.; Chan, E.K. Function of gw182 and gw bodies in sirna and mirna pathways. *Adv. Exp. Med. Biol.* **2013**, *768*, 71–96. [PubMed]

25. Kurtova, A.V.; Balakrishnan, K.; Chen, R.; Ding, W.; Schnabl, S.; Quiroga, M.P.; Sivina, M.; Wierda, W.G.; Estrov, Z.; Keating, M.J.; et al. Diverse marrow stromal cells protect cll cells from spontaneous and drug-induced apoptosis: Development of a reliable and reproducible system to assess stromal cell adhesion-mediated drug resistance. *Blood* **2009**, *114*, 4441–4450. [CrossRef] [PubMed]

26. Lwin, T.; Hazlehurst, L.A.; Dessureault, S.; Lai, R.; Bai, W.; Sotomayor, E.; Moscinski, L.C.; Dalton, W.S.; Tao, J. Cell adhesion induces p27kip1-associated cell-cycle arrest through down-regulation of the scfskp2 ubiquitin ligase pathway in mantle-cell and other non-hodgkin b-cell lymphomas. *Blood* **2007**, *110*, 1631–1638. [CrossRef] [PubMed]

27. Sormanni, P.; Camilloni, C.; Fariselli, P.; Vendruscolo, M. The s2d method: Simultaneous sequence-based prediction of the statistical populations of ordered and disordered regions in proteins. *J. Mol. Biol.* **2015**, *427*, 982–996. [CrossRef] [PubMed]

28. Cilia, E.; Pancsa, R.; Tompa, P.; Lenaerts, T.; Vranken, W.F. The dynamine webserver: Predicting protein dynamics from sequence. *Nucleic Acids Res.* **2014**, *42*, W264–W270. [CrossRef] [PubMed]

29. Meszaros, B.; Erdos, G.; Dosztanyi, Z. Iupred2a: Context-dependent prediction of protein disorder as a function of redox state and protein binding. *Nucleic Acids Res.* **2018**, *46*, W329–W337. [CrossRef] [PubMed]

30. Disfani, F.M.; Hsu, W.L.; Mizianty, M.J.; Oldfield, C.J.; Xue, B.; Dunker, A.K.; Uversky, V.N.; Kurgan, L. Morfpred, a computational tool for sequence-based prediction and characterization of short disorder-to-order transitioning binding regions in proteins. *Bioinformatics* **2012**, *28*, i75–i83. [CrossRef] [PubMed]

31. Vernon, R.M.; Chong, P.A.; Tsang, B.; Kim, T.H.; Bah, A.; Farber, P.; Lin, H.; Forman-Kay, J.D. Pi-pi contacts are an overlooked protein feature relevant to phase separation. *eLife* **2018**, *7*, e31486. [CrossRef] [PubMed]

32. Chong, S.-H.; Lee, C.; Kang, G.; Park, M.; Ham, S. Structural and thermodynamic investigations on the aggregation and folding of acylphosphatase by molecular dynamics simulations and solvation free energy analysis. *J. Am. Chem. Soc.* **2011**, *133*, 7075–7083. [CrossRef] [PubMed]

33. Konrat, R. The protein meta-structure: A novel concept for chemical and molecular biology. *Cell. Mol. Life Sci.* **2009**, *66*, 3625–3639. [CrossRef] [PubMed]

34. Oates, M.E.; Romero, P.; Ishida, T.; Ghalwash, M.; Mizianty, M.J.; Xue, B.; Dosztányi, Z.; Uversky, V.N.; Obradovic, Z.; Kurgan, L.; et al. D^{2}p^2: Database of disordered protein predictions. *Nucleic Acids Res.* **2013**, *41*, D508–D516. [CrossRef] [PubMed]

35. Piovesan, D.; Tabaro, F.; Paladin, L.; Necci, M.; Micetic, I.; Camilloni, C.; Davey, N.; Dosztanyi, Z.; Meszaros, B.; Monzon, A.M.; et al. Mobidb 3.0: More annotations for intrinsic disorder, conformational diversity and interactions in proteins. *Nucleic Acids Res.* **2018**, *46*, D471–D476. [CrossRef] [PubMed]
36. R Core Team. *R: A Language and Environment for Statistical Computing*; R Foundation for Statistical Computing: Vienna, Austria, 2008.
37. Dowle, M.; Srinivasan, M. Data.Table: Extension of 'Data.Frame'. R Package Version 1.10.4-3. 2017. Available online: https://cran.r-project.org/package=data.table (accessed on 4 October 2018).
38. Gross, J.; Ligges, U. Nortest: Tests for Normality. R Package Version 1.0-4. 2015. Available online: https://cran.r-project.org/package=nortest (accessed on 4 October 2018).
39. Ren, K.; Russell, K. Formattable: Create 'Formattable' Data Structures. R Package Version 0.2.0.1. 2016. Available online: https://cran.r-project.org/package=formattable (accessed on 4 October 2018).
40. Carlson, M. Org.Hs.Eg.Db: Genome Wide Annotation for Human. R Package Version 3.5.0. 2017. Available online: http://bioconductor.org/packages/org.Hs.eg.db/ (accessed on 4 October 2018).

© 2018 by the authors. Licensee MDPI, Basel, Switzerland. This article is an open access article distributed under the terms and conditions of the Creative Commons Attribution (CC BY) license (http://creativecommons.org/licenses/by/4.0/).

International Journal of
Molecular Sciences

Article

Co-Evolution of Intrinsically Disordered Proteins with Folded Partners Witnessed by Evolutionary Couplings

Rita Pancsa [1,*,†], Fruzsina Zsolyomi [2,†] and Peter Tompa [1,3,*]

[1] Research Centre for Natural Sciences of the Hungarian Academy of Sciences, Institute of Enzymology, 1117 Budapest, Hungary

[2] Department of Biochemistry and Molecular Biology, Faculty of Medicine, University of Debrecen, 4032 Debrecen, Hungary; zsolyomi.fruzsina@med.unideb.hu

[3] Center for Structural Biology, Flanders Institute for Biotechnology (VIB), Vrije Universiteit Brussel, 1050 Brussels, Belgium

* Correspondence: pancsa.rita@ttk.mta.hu (R.P.); peter.tompa@vub.be (P.T.); Tel.: +36-1-382-6705 (R.P.); +32-2-629-1924 (P.T.)

† These authors contributed equally to this work.

Received: 30 September 2018; Accepted: 22 October 2018; Published: 25 October 2018

Abstract: Although improved strategies for the detection and analysis of evolutionary couplings (ECs) between protein residues already enable the prediction of protein structures and interactions, they are mostly restricted to conserved and well-folded proteins. Whereas intrinsically disordered proteins (IDPs) are central to cellular interaction networks, due to the lack of strict structural constraints, they undergo faster evolutionary changes than folded domains. This makes the reliable identification and alignment of IDP homologs difficult, which led to IDPs being omitted in most large-scale residue co-variation analyses. By preforming a dedicated analysis of phylogenetically widespread bacterial IDP–partner interactions, here we demonstrate that partner binding imposes constraints on IDP sequences that manifest in detectable interprotein ECs. These ECs were not detected for interactions mediated by short motifs, rather for those with larger IDP–partner interfaces. Most identified coupled residue pairs reside close (<10 Å) to each other on the interface, with a third of them forming multiple direct atomic contacts. EC-carrying interfaces of IDPs are enriched in negatively charged residues, and the EC residues of both IDPs and partners preferentially reside in helices. Our analysis brings hope that IDP–partner interactions difficult to study could soon be successfully dissected through residue co-variation analysis.

Keywords: intrinsically disordered; disordered protein; structural disorder; correlated mutations; co-evolution; evolutionary couplings; residue co-variation; interaction surface; residue contact network

---

1. Introduction

Protein sequences provide rich information on structural and functional constraints in the form of residue co-variation in evolution. With the rapid expansion of available sequence data and computational power, and improvements in global statistical approaches [1], the problem of transitive residue correlations (false positive correlations observed for residues that do not actually contact each other in space) could be largely overcome and the analysis of evolutionary couplings (ECs) between protein residues has achieved important breakthroughs [2,3]. Several groups have demonstrated that the analysis of sequence co-variation can be efficiently used for predicting protein structures [2–10], including transmembrane proteins [11,12], defining evolutionary units within proteins [13], and identifying contacting residues of interaction partners [14,15], and interacting subunits of larger complexes [14]. The reason why co-variation analysis does not (yet) provide the ultimate solution to the sequence-based prediction of protein structures is that it requires large, good-quality alignments of sufficiently diverse sequences [16], restricting its applicability to phylogenetically widespread and reasonably conserved proteins.

Intrinsically disordered proteins/regions (IDPs/IDRs) lack well-defined 3D structures, rather, they exist and function as ensembles of rapidly interconverting conformers [17–20]. The conformational variability and adaptability, extended interaction surface, various embedded interaction motifs [21,22], and post-translational modification sites [23] of IDPs make structural disorder indispensable in regulatory [24], complex-assembly [25], and scaffolding [26,27] functions. IDPs are central to cellular interaction networks [28,29] and are frequently associated with human diseases [30].

IDPs mostly interact with their partners through eukaryotic/short linear motifs (ELMs/SLiMs) comprised of a few specificity-determining residues embedded into a disordered sequence environment that ensures the right positioning of their mostly hydrophobic, crucial interaction residues [21,22,31–33]. SLiM-mediated interactions are frequently switched on and off by alternative splicing [31,34–36], and are frequently rewired in evolution [31]. As IDPs lack a well-defined tertiary structure and their functional modules are restricted to a few critical residues, most of their sequences are under limited structural and/or functional constraints. As compared to folded domains, this inherent freedom leads to increased rates [37] and altered types [38] of residue changes in evolution, hampering both the identification and correct alignment of homologous IDPs. Despite their indisputable importance, true IDPs (with the exception of ribosomal proteins [15]) and their interactions were not subject to high-scale residue co-variation analyses [6,7,9,11,12,14,15] until very recently. Although disordered regions have less co-varying residues than folded domains [39], Toth-Petroczy and colleagues have recently refined their method, EVfold, to predict potential structured states of IDPs through detecting ECs within their chains [40]. The prevalent phenomenon of induced folding or disorder-to-order transition [41] in IDP-partner recognition, however, suggests that the (partly) structured states of IDPs are encoded not only in their own sequence but also in that of the partner. It follows then that co-evolutionary signals in the interface of the IDP and its partner may actually be just as pronounced as those within a protein fold or between two folded interaction partners.

Therefore, to see if functional constraints originating from partner binding resulted in detectable interprotein ECs of (at least for certain phylogenetically wide-spread) IDPs and to characterize the respective protein pairs, interfaces, and their co-evolved residue pairs, we performed a targeted screen to identify such cases.

2. Results

Even though bacterial proteomes are relatively poor in IDPs/IDRs [42–44], we restricted our search to bacterial IDP–partner interactions to ensure a sufficient number and diversity of orthologous sequences for residue co-variation analysis. Although there are already a few hundred eukaryotic genomes available, they are not evenly distributed among phylogenetic groups (i.e., a large fraction of them are from mammals) and thus do not show enough sequence diversity on the level of proteins. This is well supported by the fact that recent large-scale residue co-variation analyses have all focused on bacterial protein complexes [9,14,15]. Therefore, we analyzed the 42 bacterial IDPs bound to their folded partners available in the Database of disordered binding sites (DIBS) [45] because there the structural states of the constituent protein chains are backed by experimental evidence. After getting rid of 4 redundant structures and one where the IDP chain was only four residues long, the phylogenetically fairly wide-spread 19 complexes (those with >130 sequences for the IDP in Pfam database 31 [46] full alignments) were selected for co-variation analysis (Supplementary Tables S1 and S2). This way, species–specific complexes, such as virulence factors, certain toxin–antitoxin, effector–chaperone and effector–immunity protein pairs could be counter-selected instead of needlessly occupying co-evolution analysis servers. The sequences were trimmed for the interacting regions/domains or extended to reach the minimum length of 30 residues required for Gremlin analysis as indicated in Table 1, and the Gremlin [15] and EVcomplex [14] servers were used to identify interprotein ECs (See Methods for further details). These methods perform co-variation analyses along similar lines. They first prepare a so-called paired alignment for the protein pair provided, which means that they detect the closest homolog of each of the two provided proteins in each analyzed proteome, and build an alignment wherein the interacting sequences are linked together and filtered for similarity. If the resulting paired alignment has enough sequences, they proceed with detecting co-varying residue pairs (evolutionary couplings, ECs) therein. They use somewhat different approaches to score the pairs of residues based on evolutionary co-variation. From the outputs we exclusively take interprotein ECs, where the coupled residues come from the two different proteins because those provide clues on the interaction. Gremlin could be successfully run on 13 complexes (Table 1), while it stopped in 6 cases due to the alignments being insufficient for analysis. For 7 of the 13 successfully analyzed complexes, Gremlin detected coupled interprotein residue pairs with a scaled score $\geq$1.3 and probability in the top 12% ($p > 0.88$), hereafter referred to as ECs. For EVcomplex, interprotein ECs with an EVcomplex score >0.9 have been accepted as ECs. EVcomplex also identified ECs for 7/19 complexes (Supplementary Table S1). Both the identified complexes and the detected ECs showed a good overlap between the two methods. Since Gremlin ran the sequence search on a more up-to-date sequence database, it obtained better sequence coverage values and consequently identified more ECs than EVcomplex for almost all the analyzed protein pairs. Based on their residue–residue distances, almost all Gremlin EC pairs fell spatially close, implying that they could be correctly identified co-varying pairs. Thus, we decided to continue the analysis and show the results for Gremlin ECs. The ECs were then checked by PDBe PISA [47] in order to see if they are at the interface (IF) and if they engage in physical interaction (hydrogen bonds or salt bridges; see Table 1).

2.1. ECs Were Detected Outside the Sequence Ranges with PDB Coordinates, but Visible ECs Reside on the Interfaces

Gremlin identified 31 ECs in 7 complexes (Supplementary Tables S1 and S3). Intriguingly, for 9 of these co-varying residue pairs, at least one of the constituent residues (mostly the IDP residue) fell outside the segments with PDB coordinates (invisible ECs; marked with N/A in Supplementary Table S3). These invisible ECs make up a remarkably high percentage of all identified ECs. Based on their Gremlin scores and probabilities, they do not seem to be mistakenly predicted pairs as they often have higher scores and probabilities than interface residue pairs with several atomic contacts. Of the 10 invisible EC residues, 5 were present in the respective PDB sequences but had no atomic coordinates, while the other 5 were not included in the constructs used for structure determination. High occurrence of these invisible EC residues suggests that the IDP segments used for structure determination and interaction analyses are often too short, lacking residues that could still form important contacts with the partner. There are actually several instances where the contribution to binding of segments that are disordered in the partner-bound state has been proven, such as binding of CREB to CBP [48] and binding of SF1 to U2Af65 [49], which forms the basis of the concept of fuzzy interactions [50].

All except one of the 22 visible IDP EC residues (Table 1, Figure 1) are interface residues according to PDBe PISA, and 12 of them are assigned as forming a direct H-bond or salt bridge (or both) with the partner (Table 1), although not necessarily with their corresponding EC pair (Supplementary Table S3). These values already support the validity of the predicted ECs, nevertheless, we decided to investigate them in more detail.

Table 1. Residue co-variation analysis of IDP–partner interactions with broad phylogenetic spread.

Complex	Folded Partner			IDP			Gremlin Analysis Results		
DIBS ID-PDB ID (Chains Partner(s) I IDP)	IF Area (Å^2)	Gene Name	Uniprot AC_region	Gene Name	Uniprot AC_Region	Length	# of Seq. in PFAM 31 Full	Coverage (seq/res)	ECs by Gremlin/by EVComplex/ Gremlin IDP ECs on IF/in Bonds [a]
DI4200001-3B1K (BG I D)	1016	gap2	Q9R6W2_78-215	cp12	Q6BBK3_46-75	30	507	0.76	1/0/1/1
DI2200001-3HPW (AB I C)	1483	ccdB	P62554_1-101	ccdA	P62552_37-72	36	395	4.06	7/0/6(1 inv [b])/2
DI2200002-5CQX (AB I C)	1128	mazF	P0AE70_1-111	mazE	P0AE72_53-82	30	4349	3.00	9/1/5(4 inv)/5
DI2200004-3M91 (AC I B)	1065	mpa	P9WQN5_46-96	pup	P9WHN5_21-64	44	511	2.19	5/0/2(2 inv)/1
DI2200006-3M4W (AC I E)	2321	rseB	P0AFX9_220-318	rseA	P0AFX7_125-195	71	253	1.67	2/0/2/1
DI1200004-1SC5 (A I B)	1641	fliA	O67268_1-236	flgM	O66683_1-88	88	1450	3.43	3/10/3/1
DI1210003-2A7U (B I A)	611	atpH	P0ABA5_1-134	atpA	P0ABB3_1-30	30	14,854	8.34	4/3/2(2 inv)/1
DI2200005-1SUY (AB I C)	1588	kaiA	Q79V62_177-283	kaiC	Q79V60_485-518	34	2739	0.85	0/0/0/0
DI1200003-1QFN (A I B)	732	grxA	P68688_1-85	nrdA	P00452_732-761	30	7563	1.59	0/0/0/0
DI1200001-1R1R (A I D)	851	nrdA	P00452_335-729	nrdB	P69924_347-376	30	4425	0.78	0/1/0/0
DI1200006-5D0O (D I C)	2072	bamD	P0AC02_1-245	bamC	P0A903_30-85	56	726	0.38	0/0/0/0
DI1210006-4Z0U (B I E)	358	rnhA	A7ZHV1_1-155	ssb	P0AGE0_149-178	30	8735	0.31	0/0/0/0
DI1210004-3C94 (A I B)	320	sbcB	P04995_13-355	ssb	A0A0H3GL04_145-174	30	8735	0.11	0/0/0/0

[a] In this column, bonds were assigned to the IDP EC residues based on PDBe PISA H-bond and salt bridge annotations. [b] The locations, bonds, and distances of invisible EC pairs could not be analyzed as they reside outside the sequence ranges with PDB coordinates. The numbers of such invisible EC pairs (inv) are indicated in brackets in the last column along with interface (IF) pairs, since they provide the explanation for the difference between the number of identified ECs and interface ECs.

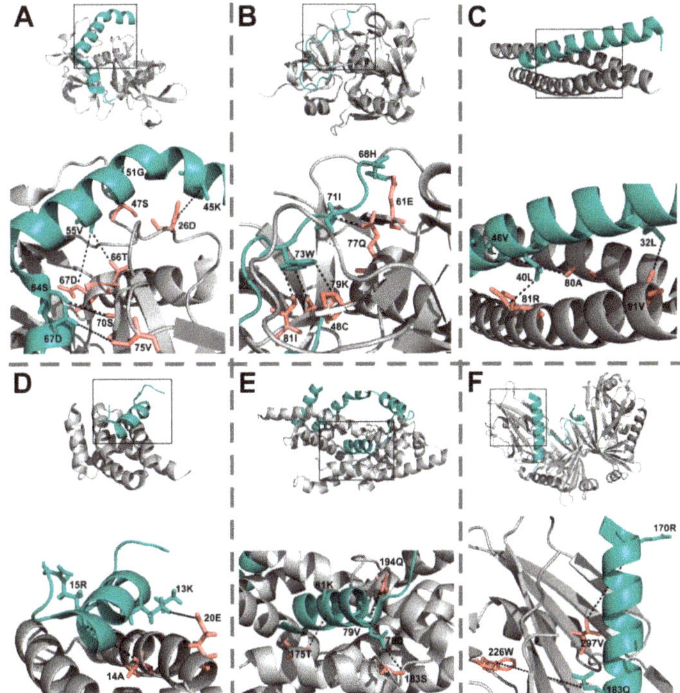

Figure 1. High-scoring interprotein ECs indicating IDP-partner co-evolution. The six protein complexes with multiple high-scoring interprotein ECs predicted by Gremlin are introduced: (**A**) PDB complex 3HPW; (**B**) 5CQX; (**C**) 3M91; (**D**) 2A7U; (**E**) 1SC5; (**F**) 3M4W. IDPs are depicted in teal cartoon style, while the partners are in light grey. Rectangles indicate the part of the complexes which are depicted as detailed interaction maps below. In the interaction maps, the EC side chains are depicted as sticks colored teal in the IDP and salmon in the partner. The interprotein ECs have their respective residue side chains labelled and are connected by dashed lines. For the few pairs forming direct H-bonds or salt bridges according to PISA, the connecting lines are doubled. The 183Q-226W IDP-partner EC pair on the F panel had a distance 17.5Å, thus it was handled as an outlier. None of the depicted complexes represent SLiM-mediated interactions, as we could not identify any ECs for those (see Section 2.5 for further details).

2.2. EC Pairs Have More Atomic Contacts between Them than Other Pairs

Interchain residue–residue atomic contacts have been obtained from the Protein Contacts Atlas (PCA) [51] portal for the 34 X-Ray structures. Among EC-containing complexes, 2A7U is an NMR structure so it was not included in the following residue-contact analysis. From the contact lists between the IDP chain and the relevant partner chains, the numbers of atomic contacts between each contacting residue pair have been obtained, the residue pairs have been grouped as ECs, EC residue/no-EC residue pairs and no-EC residue/no-EC residue pairs, and the numbers of atomic contacts for these three groups have been compared. Although only 7 of the 20 visible ECs of X-ray structures had atomic contacts in PCA, with a median of 5, ECs had significantly more contacts between them than contacting residue pairs in the other two groups (Figure 2A). A likely reason behind EC/no-EC pairs having significantly less contacts than no-EC/no-EC pairs is that in the EC/no-EC dataset the contacting residues of EC residues were skimmed. Their respective EC pairs were not taken into account in this dataset; still, they had less space to make contacts with other, no-EC residues. We also wanted to see how many contacts EC residues have, regardless of their partner residue. To this end, we added up the total number of atomic contacts (with all contacting partner residues) for all residues in the contact lists for the IDPs and partners separately. These values were then compared between EC and non-EC residues. We did not find significant differences neither for IDPs nor for their partners (Supplementary Figure S1). Thus, we can claim that although the individual residues of IDP-partner ECs do not have more atomic contacts than other residues, ECs taken as pairs do have more atomic contacts with each other than other residue pairs.

2.3. ECs Are Significantly Closer to Each Other than Randomly Paired IDP–Partner Interface Residues

We were also interested in the distance distribution of ECs. For the pairs with PCA atomic contacts, we took the shortest atomic contact as residue–residue distance, while for the rest of the visible ECs (including the ones in the NMR structure 2A7U) we calculated the minimal residue–residue distances using PyMOL (Supplementary Table S3). We compared the distance distribution so obtained to an equivalent reference distance distribution of randomly selected and subsequently paired IDP and partner residues picked from the same interfaces and found that ECs are highly significantly closer in space than the randomly paired interchain interface residue pairs (Figure 2B). The sampling of randomly paired interface residues was carried out 100 times and their distances were consistently significantly larger than those of ECs ($p < 0.01$). The descriptive distribution features (minimum, 1st quantile, mean, median, 3rd quantile, maximum) of the 100 samples have shown an average standard deviation of 2.31 which we interpret as low variation among the random samples. 18/22 visible ECs were closer than 8 Å, two were between 8 and 10 Å far, and only two had >10 Å distances (Figure 2C). The latter two were assigned as outliers by R, they probably represent mistakenly identified pairs (or have larger distances due to other reasons [52]) and thus have been handled as outliers and excluded from distance comparisons (Figure 2B). One of the outliers is the 183Q-226W IDP-partner EC pair in the RseA/RseB (PDB: 3M4W) complex as indicated in Figure 1. The other outlier is the sole EC pair (74D-133Y) identified in the CP12/Glyceraldehyde-3-phosphate dehydrogenase (PDB: 3B1K) complex. Therefore, the latter complex is left without any reliable ECs, and thus it is not shown in Figure 1.

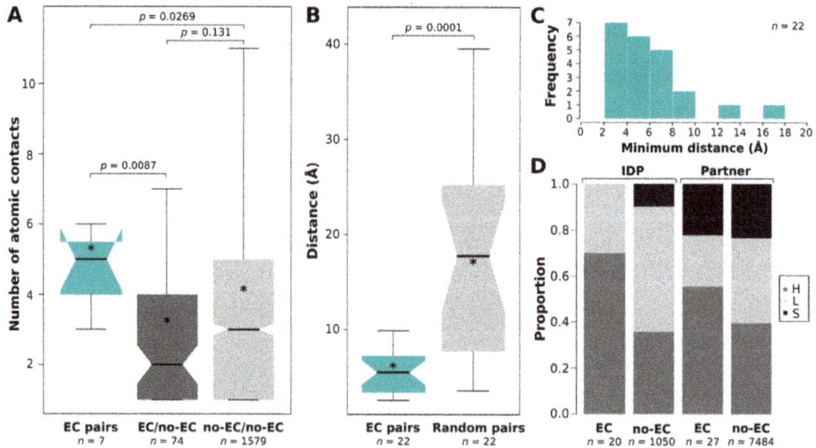

Figure 2. Atomic contacts and distances of ECs. (**A**) The numbers of atomic contacts connecting EC pairs according to PCA compared to those of other contacting residue pairs. Pairwise comparisons were done with Mann–Whitney U test with the corresponding p-values indicated; (**B**) the minimum distances between visible EC pairs are compared to those of randomly selected and paired IDP–partner interface residues; (**C**) histogram showing the distribution of minimum distances between visible ECs; (**D**) the proportions of main secondary structure element types (H—helix, S—strand, L—loop) are compared between EC and no-EC residues for IDPs and partners separately. In the boxplots stars (*) indicate the average values of the datasets.

2.4. EC Residues Preferentially Occur in Helices in Both IDPs and Partners

To assess the structural preferences of EC residues, we have computed DSSP secondary structure assignments for the complexes. The EC residues preferentially resided in helices for both IDPs ($p = 0.003$) and partners ($p = 0.039$) compared to other residues (Figure 2D). Also, the detected interprotein ECs typically cluster in one alpha helix of the IDPs, although in most cases the studied chains did not have more: in anti-sigma-28 factor FlgM they cluster in helix4 (Figure 1E), while in RseA in the longer helix of the two (Figure 1F). Furthermore, ECs have also been detected in the toxin-antitoxin MazF-MazE pair, where the IDP MazE does not fold into a regular secondary structure on the surface of the MazF dimer (Figure 1B).

2.5. General Trends Observed for the Protein Complexes with High-Scoring Interprotein ECs

The seven protein pairs yielding high-scoring ECs mostly had >2 Gremlin sequence/residue alignment coverage values, while the ones that stopped running or did not provide high-scoring ECs tended to have values <1 (Table 1). The coverage values did not correlate with the number of IDP homologs in PFAM. Longer-interacting IDP regions with wider phylogenetic spread (for the corresponding phylogenetic groups see Supplementary Table S1) had a higher chance for sufficient coverage values, while short IDP chains that required significant extensions for reaching the minimum length of 30 residues did not have a good chance for >1 coverage values even if mediating a phylogenetically widely conserved interaction. Therefore, the complexes where the IDPs interact with their partners in the typical IDP manner—through ELMs/SLiMs or molecular recognition features (MoRFs)—namely the enolase- and PNPase-interacting motifs of RNase E [53], the SspB-interacting region of the N-terminal part of RseA, and the C-terminal interaction motif of single-stranded DNA-binding protein (SSB) recruiting diverse partner proteins [54,55], did not have enough PFAM sequences for analysis or, despite a vast amount of sequences in PFAM, did not show any high-scoring ECs. The lack of a sufficient number or diversity of detectable homologs for these important interaction motifs/regions could be due to their short length, the relatively small fraction of the actual specificity-determining residues and the fast evolutionary turnover of the surrounding other residues [31].

In the case of the interactions mediated by the SSB C-terminal motif, that are among the phylogenetically most widespread ones, the lack of detected ECs could be attributed to different reasons. The C-terminal motif is 9 residues in length, so it had to be extended by 21 residues from the poorly conserved SSB linker region to reach a total length of 30 residues. The lack of conservation in the linker segment dilutes the information in the motif, whereas the multitude of interaction partners simultaneously restricting the evolution of the motif [55,56] leads to a complete lack of sequence variation in most of its residues, which could both contribute to the lack of detected ECs.

While EC residues showed a strong preference for interface helices, only 58.8% of ELM instances form secondary structure elements, with only 16.2% of them being mostly helical and 7.6% being partially helical according to a large-scale analysis of the eukaryotic linear motif (ELM) database [31]. Although our dataset only contains a few bacterial short linear motifs, which are not part of the ELM database, they show a similar distribution among secondary structure types as proposed for their eukaryotic counterparts. The SSB C-terminal motif does not form a secondary structure with any of its 5 partners, while the SspB binding motif of RseA forms two very short helices. The PNPase- and enolase-binding motifs of RNase E were not subjected to EC analysis because they did not have the sufficient amount of PFAM sequences, but the former binds through beta sheet augmentation, while the latter forms a short helix. Thus, the identified SLiMs do not show a preference for helical conformation, and by mostly spanning only 4–9 residues, even the helix-forming ones are not long enough to form extended helical structures on the surface of binding partners, which could also contribute to the lack of detected ECs. In all, the complete lack of ECs for SLiM-mediated interactions regardless of their phylogenetic spread definitely represents a major limitation of applying residue co-variation-based approaches for the analysis of IDPs.

Although multisubunit enzymes often contain predicted disordered chains that occupy a completely extended conformation in the complex, and such cases would be perfect candidates for IDP-partner co-evolution analysis, the structural features of these subunits are rarely analyzed on their own. Due to this reason, our DIBS-derived dataset only has 4/37 permanent protein complexes, while most of them depict transient interactions (Supplementary Table S1). The seven complexes with identified ECs contained both permanent (1) and transient (6) complexes in similar fractions as seen for the overall dataset, so permanence of the interactions does not seem to largely affect their co-evolution patterns at first glance. However, it is interesting to note that all the complexes with identified ECs had IDP–partner interface areas >1000 Å^2, except for the sole complex with the permanent interaction of two subunits of the ATP synthase that only had 610.5 Å^2. This might imply that, in permanent protein complexes, which are phylogenetically widely distributed, co-evolution of interacting subunits is so prominent that it can be detected even if the corresponding interaction surfaces are relatively small.

2.6. ECs Could Not Be Identified for Very Small Interfaces with no H-Bonds or Salt Bridges

By comparing the interfaces of the complexes with and without ECs (the latter group comprised all the 30 non-redundant complexes that did not provide high-scoring ECs by Gremlin) for their interface areas, H-bond and salt bridge densities, we did not find any significant difference (Figure 3). However, it became evident that ECs could not be detected for interfaces <600 Å^2 (for transient complexes this threshold rather seems as 1000 Å^2) in size, not even for exceptionally widely conserved interactions. ECs could also not be identified for interfaces without any assigned H-bonds or salt bridges (Figure 3C).

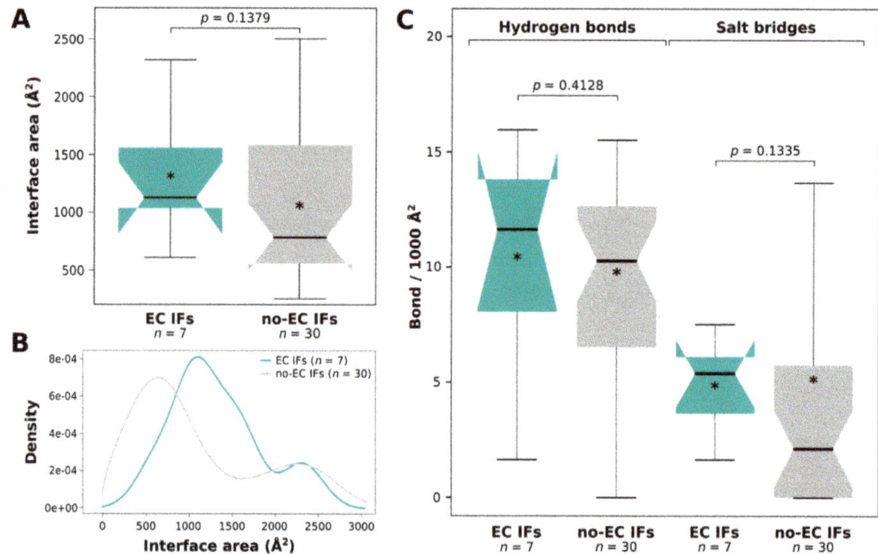

Figure 3. Comparison of interface areas and bonds for EC and no-EC interfaces. (**A**) Comparison of EC and non-EC interface areas with Mann–Whitney U test. (**B**) Density plot of the interfaces in the two groups. (**C**) Comparison of the normalized numbers of H-bonds and salt bridges between EC and non-EC interfaces with Mann–Whitney U test, with the corresponding *p*-values indicated. In the boxplots, stars (*) indicate the average values of the datasets.

2.7. EC Interfaces Are Enriched in Negatively Charged Residues

We have also calculated the amino acid and amino acid group compositions of the trimmed protein segments (analyzed sequence ranges), all interfaces (all IFs), interfaces without any detectable EC pairs

(No-EC IFs), EC-carrying interfaces (EC IFs) and EC residues for the IDPs and partners separately. Comparing EC IFs and ECs to all IFs, we found no significant differences either for IDPs (Figure 4, Supplementary Figure S2) or their partners (Supplementary Figure S3), although the enrichment of the EC-carrying interfaces of IDPs in negative residues was nearly significant ($p = 0.0519$). We have to note though that we had only 28 IDP- and 30 partner EC residues, which is suboptimal for rigorous statistical analysis. Due to the small sample size, even seemingly very large differences were not statistically significant, for instance, IDP EC residues were not found to be enriched in positive amino acids compared to all IDP interface residues, despite having almost two times as high fraction of positives (28.6% vs. 14.5%; Figure 4). When comparing the compositions of EC interfaces to those of no-EC interfaces, we found that the EC-carrying interfaces of IDPs harbor significantly more negatively charged residues ($p = 0.0101$), in particular glutamates ($p = 0.0240$), than the IDP interfaces with no ECs (Figure 4, Supplementary Figure S2). For the partner interfaces no significant differences were found (Supplementary Figure S3).

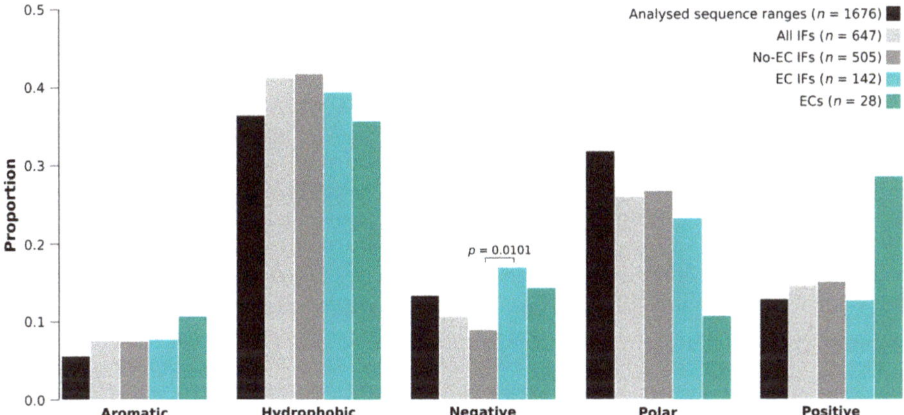

Figure 4. The amino acid group compositions of IDP EC-carrying interfaces and ECs. The amino acid group compositions of trimmed protein segments (analyzed sequence ranges), all interfaces (all IFs), interfaces with no ECs (no-EC IFs), EC-carrying interfaces (EC IFs) and EC residues of IDPs. Both EC-carrying interfaces and EC residues have been compared to all IFs. Also, EC interfaces were compared to no-EC interfaces. P-values are only indicated for amino acid group proportion differences that were found significant by the built-in test of equal proportions of R.

3. Discussion

Due to increased substitution rates [37], IDP homologs are difficult to identify and correctly align. Also, their extended, floppy conformational states manifest in less contacting residue pairs per unit length compared to folded proteins/domains, also leaving less room for residue co-evolution [39]. Furthermore, IDPs excel in moonlighting [57], meaning that they can interact with multiple partners (often through the very same sequence region) and thus their evolution might well be simultaneously constrained by many different partners. However, these multiple different interactions are only poorly represented in structural databases, and residue co-variation analysis can also only be performed for one pair of sequences at a time. Another important factor is that the ensembles and interactions of IDRs are frequently modulated by post-translational modifications (although these are largely limited to eukaryotic proteins), which currently cannot be taken into account by the available residue co-variation analysis methods. Therefore, we need to acknowledge that, at the moment, it is not feasible to comprehensively cover the full range of structural and interaction plasticity of disordered proteins by such analyses. Due to these reasons, IDPs and their interactions have hardly been investigated for the occurrence of residue co-variation [6,7,9,11,12,14,15]. Although a recent analysis suggested structured states of IDPs based on observable intrachain ECs [40], the co-evolution of IDPs with folded partners remained to be elucidated. To possibly demonstrate and understand IDP–partner co-evolution, we performed a comprehensive analysis of IDP–partner interprotein evolutionary couplings on the available bacterial complexes of the DIBS database using dedicated methods [14,15]. To ensure sufficient diversity and depth of sequence coverage by homologs, we restricted our analysis to bacterial IDPs and their cognate partners. Furthermore, to improve internal consistency of the predictions, we applied two different algorithms, Gremlin and EVcomplex. We could identify high-scoring interprotein ECs for seven interacting protein pairs, typically the ones with the best alignment coverage. Alignment coverage is, however, not strictly correlated with phylogenetic spread; it also depends on the length of interacting protein regions. For complexes where the IDP contacts the partner over a relatively long sequence bit, like, for instance, for the FlgM-FliA complex (Figure 1E), we had a good chance to identify some ECs. However, although SLiM-mediated interactions are highly preferred by IDPs and some of them are exceptionally widely conserved, they did not show high-scoring interprotein ECs, probably due to their short length, wherein the evolutionary information on a relatively small fraction of specificity-determining residues is diluted by many more surrounding other residues undergoing fast evolutionary turnover [31]. The absence of detected ECs for SLiM-mediated interactions could be a major limitation of performing residue co-variation analyses on IDPs, as SLiMs are crucial and widely applied interaction units of IDPs that mediate contacts with many of their partners.

Supporting the validity of the predicted ECs, i.e., that they represent true physical contacts between the IDP and its partner, or at least could be spatially restrained by each other, the majority of the visible ECs fall within 8 Å. Residue contact networks from PCA also showed that several EC pairs had multiple atomic contacts; moreover, ECs had more atomic contacts than other contacting IDP–partner residue pairs. Regarding their structural preferences, EC residues tend to be located in helices in both IDPs and folded partners. Although EC-carrying interfaces did not significantly differ in size or in the surface-normalized number of H-bonds and salt bridges, it is important to note that no ECs could be identified in interfaces that are small (<600 Å^2) or have no hydrogen bonds or salt bridges, no matter how widely conserved the interaction is. Since for the partners we did not find enrichment in any residue type, we do not claim that the enrichment of EC-carrying IDP interfaces in negative residues would imply the employment of increased electrostatic attraction compared to other IDP–partner interfaces. Therefore, we cannot ascertain the biological relevance of the noted preference for negatively charged residues.

Importantly, ECs could be identified for both permanent and transient IDP–partner interactions. This shows that not only obligatory, permanent associations, such as the subunits of the ribosome [15] or heteromeric enzymes, but also well-established signaling relationships, such as regulatory antisigma factors, have a trace of detectable co-evolution between the IDP and its interaction partner. Regarding the evolutionary history and genomic proximity of the genes encoding the protein pairs with high-scoring ECs, the CcdA-CcdB (PDB: 3HPW) and MazE-MazF (5CQX) toxin–antitoxin pairs, the alpha and delta subunits of the ATP synthase (2A7U) and the Sigma-E/Anti-Sigma-E regulatory factors RseB and RseA (3M4W), are each encoded on common operons. This suggests that these protein pairs are co-expressed and their co-occurrence is strongly preferred in evolution. To the contrary, the other two pairs, FliA with its anti-sigma factor FlgM (1SC5) and proteasome-associated ATPase (product of *mpa* gene) with prokaryotic ubiquitin-like protein Pup (3M91) are located in different operons and are thus substantially less strictly associated both in gene-regulatory and evolutionary terms. Therefore, based on our findings, a phylogenetically widely preserved protein–protein interaction that buries an interface larger than SLiM-mediated interactions typically do, might be enough for the reliable identification of interprotein evolutionary couplings. No other assumption regarding permanence of the interaction or co-regulation of the corresponding genes needs to be made prior to analysis, although such factors could increase the chance of finding true ECs.

In this first dedicated study of IDP–partner co-evolution, we also show that IDPs are difficult to investigate by methods addressing residue co-variation, due to their fast evolutionary changes, limited sequence representation, increased propensity for moonlighting functions, frequent use of short linear motifs for partner binding and relatively few co-evolving residue pairs both within their chains [39,40] and with their partners. Nevertheless, by demonstrating detectable footprints of IDP–partner co-evolution for interactions with largely different functional readouts, our results are also promising. They imply that the explosion in the number of sequenced genomes, the continuous improvement of techniques of sequence homology detection [58], and advances in sequence alignment approaches optimized for IDPs [40] could soon empower residue covariation analyses of IDPs to provide predictions and new insights into the structures and interactions of IDPs whose experimental investigation proved to be challenging.

4. Materials and Methods

4.1. Dataset Preparation

Bacterial protein complexes were obtained from the DIBS database [45] (Supplementary Table S1). We identified 4 of the 42 as redundant: DIBS: DI2210001–PDB: 3TCJ, DI1200012–3UF7, DI1210005–3C94, and DI1210010–5CW7 were excluded due to redundancy to DI2200001–3HPW, DI1200011–3UF7, DI1210004–3C94, and DI1210007–5CZF, respectively. The DI1200014–5F56 complex was also excluded due to the IDP peptide being <5 residues long. For the remaining 37 complexes, the constituent chains were trimmed or extended to make them optimal for co-evolution analysis. IDP chains shorter than 30 residues were extended based on UniProt [59] to reach the minimum of 30 residues length required for Gremlin analysis. Terminal segments could obviously only be extended to one direction. For the two short IDP chains representing internal segments, more extension was added to the end with residues forming an interface with the partner chain. Partner chains/very long IDP chains were trimmed to interacting domains/subdomains to ensure that the best possible coverage values are obtained (sequence coverage values are highly dependent on the total length of the analyzed sequences). The resulting trimmed UniProt regions used for further analysis are indicated in Table 1. Then for each complex, the IDP counterpart was checked in PFAM 31 [46] to get an idea of their phylogenetic spread. If, for the IDP region in the complex or at least for a neighboring protein domain/region of the protein there were no PFAM families available, or the number of sequences in the full alignments of the relevant PFAM families were <130, then the complex was excluded from correlated mutation analysis (for PFAM families see Supplementary Table S2).

Information on the complexes representing permanent or transient interactions was taken from the literature and UniProt subunit structure annotations. To assign if the genes are encoded within a single operon/gene neighborhood, we used information in the STRING database [60] and Ensembl Bacteria [61]. Assignments on the phylogenetic spread of the interactions were based on literature mining.

4.2. Co-Evolution Analysis

The trimmed regions of the 19 remaining complexes were run with the Gremlin [15] and EVcomplex [14] webservers for interprotein co-variation analysis. When using Gremlin, for regions >60 residues the e-value threshold was set to E-06 and the number of iterations with Jackhmmer to 4, while for regions ≤60 residues, we applied a less stringent e-value threshold of E-04 with 8 iterations. The reason for using a less stringent e-value threshold for shorter sequences is explained and supported by the Introduction/Updates section of the Gremlin webserver. Δgene was set to 1–∞ to identify the closest homologs in the analyzed proteomes regardless of their genomic location. In 6 cases, Gremlin stopped due to insufficient alignments for further analysis. It finished the analysis in 13 cases. We accepted interprotein residue pairs with a scaled score ≥1.30 and a probability >0.88 as co-varying pairs; evolutionary couplings (ECs). These thresholds were selected based on our observation that among the predicted possible EC pairs with a scaled score >1.3 the probability values are the main determinants of the residue pairs representing true positive predictions with residue–residue distances reflecting direct contacts or false positive predictions with large interresidue distances. Therefore, we have selected a strict probability value threshold (top 12%) to avoid false positive ECs compromising the dataset used for analysis. The 7 complexes with such high-scoring interface ECs identified by Gremlin were analyzed further and compared to the remaining 30 complexes with no high-scoring ECs.

For the EVcomplex analysis, we applied the default e-value (10^{-5}) in searching for homologues and chose the option of selecting the closest homologs of the query sequences from the analyzed proteomes. Here, interprotein ECs with an EVcomplex score >0.9 have been accepted as ECs. EVcomplex identified ECs for 7/19 complexes.

4.3. Interface Properties

For each complex, information on the smallest meaningful biological assembly (chains that can represent the biologically relevant interaction) was taken from DIBS and is indicated in Supplementary Table S1. These contain strictly one IDP chain, which interacts with one or more (but in this case identical) partner chains. An exception is 3O0E, where from the DIBS-indicated L and A chains, L was not present in PDBe PISA, so the equivalent M and B chains have been used. Interface areas, interfacing residues, and interchain physical interactions were derived from PDBe PISA [47] assignments. Total interface areas were defined as the sum of interchain interface areas between the IDP chain and the different interfacing partner chains. The total number of IDP–partner interface H-bonds and salt bridges were also obtained as the sum of such bonds between the IDP chain and the interfacing partner chains, then these were normalized for 1000 Å^2 interface area before statistical comparisons. Interface residues were assigned for each chain based on PISA.

4.4. Amino Acid Compositions

In the residue and residue group composition analyses, we have calculated the compositions of the trimmed protein segments (analyzed sequence ranges), all interfaces (all IFs), interfaces with no ECs (no-EC IFs), EC-carrying interfaces (EC IFs) and EC residues for the IDPs and partners separately. Identical partner chains have been only included once into these composition analyses. The interfaces have been defined based on PDBe PISA, so regions of the trimmed protein segments falling outside the segments with PDB coordinates are not represented by the interface composition values. However, they are represented in the analyzed sequence ranges and among ECs. Residue groups were defined as: hydrophobic (G, A, V, L, M, I), aromatic (F, Y, W), polar (S, T, C, P, N, Q), negative (D, E), and positive (K, R, H). Residue group and residue proportion significances were obtained by the built-in test of equal proportions of R [62].

4.5. Residue–Residue Contact Networks

Interchain residue–residue atomic contacts have been downloaded from the Protein Contacts Atlas [51] for all the X-ray structures (34/37 complexes) using the default 0.5 Å distance cut-off value in PCA. Only the contact lists between the IDP chain and the relevant partner chains have been used for analysis. The number of atomic contacts between each contacting residue pair has been obtained. Also, the total number of atomic contacts (with all contacting residues) has been calculated for all the residues with at least one such interchain atomic contact. If the same partner residue had atomic contacts with the IDP from more than one partner chain, then those different contact numbers were added up to get the total contact number. Also, for IDP residues, all atomic contacts with partner residues were added up regardless of the partner chain. The calculated totals have been compared between EC and non-EC residues for the IDPs and partners separately.

4.6. Secondary Structure Assignments

DSSP secondary structure assignments [63] have been obtained for all the residues present in the complex structures in a way that only one of the identical partner chains has been used. The eight DSSP secondary structure element type assignments have been simplified and grouped in the traditional way into three larger classes: helix (G, H, and I), strand (E and B) and loop (S, T, and C). The distribution of residues among different secondary structure element types have been compared between EC residues and non-EC residues for the IDPs and partners separately using the built-in test of equal proportions of R.

4.7. Residue–Residue Distances

For the ECs being in direct atomic contact according to the Protein Contacts Atlas, the shortest atomic contact distance (between heavy atoms) has been accepted as residue–residue distance. For the rest of the ECs, the shortest distance between the heavy atoms of the two residues has been measured using PyMOL (https://pymol.org/2/; The PyMOL Molecular Graphics System, Version 2.0 by Schrödinger, LLC, New York, USA). Then, for each EC we have randomly selected an interface IDP residue and an interface partner residue from the same interface, and measured their distance similarly. The distances of ECs and the thereby assembled equivalent random reference interface pairs have been compared by Mann–Whitney U test. To check if the observed difference remains consistent, we have obtained 100 additional samples of random residue pairs and compared their distances to those of ECs.

Supplementary Materials: Supplementary materials can be found at http://www.mdpi.com/1422-0067/19/11/3315/s1.

Author Contributions: Conceptualization, R.P.; Data curation, R.P. and F.Z.; Formal analysis, R.P. and F.Z.; Funding acquisition, P.T.; Methodology, R.P.; Resources, P.T.; Supervision, R.P. and P.T.; Visualization, R.P. and F.Z.; Writing—original draft, R.P.; Writing—review & editing, R.P., F.Z. and P.T.

Funding: This work was supported by the European Molecular Biology Organization (ALTF 702-2015) and the Hungarian Academy of Sciences Premium_2017-48 fellowships to R.P. and the Odysseus grant G.0029.12 from Research Foundation Flanders to P.T.

Conflicts of Interest: The authors declare no conflict of interest.

References

1. Stein, R.R.; Marks, D.S.; Sander, C. Inferring Pairwise Interactions from Biological Data Using Maximum-Entropy Probability Models. *PLoS Comput. Biol.* **2015**, *11*, e1004182. [CrossRef] [PubMed]
2. Marks, D.S.; Hopf, T.A.; Sander, C. Protein structure prediction from sequence variation. *Nat. Biotechnol.* **2012**, *30*, 1072–1080. [CrossRef] [PubMed]
3. Taylor, W.R.; Hamilton, R.S.; Sadowski, M.I. Prediction of contacts from correlated sequence substitutions. *Curr. Opin. Struct. Biol.* **2013**, *23*, 473–479. [CrossRef] [PubMed]
4. Adhikari, B.; Bhattacharya, D.; Cao, R.; Cheng, J. CONFOLD: Residue-residue contact-guided ab initio protein folding. *Proteins* **2015**, *83*, 1436–1449. [CrossRef] [PubMed]
5. Jones, D.T.; Singh, T.; Kosciolek, T.; Tetchner, S. MetaPSICOV: Combining coevolution methods for accurate prediction of contacts and long range hydrogen bonding in proteins. *Bioinformatics* **2015**, *31*, 999–1006. [CrossRef] [PubMed]
6. Kosciolek, T.; Jones, D.T. De novo structure prediction of globular proteins aided by sequence variation-derived contacts. *PLoS ONE* **2014**, *9*, e92197. [CrossRef] [PubMed]
7. Kosciolek, T.; Jones, D.T. Accurate contact predictions using covariation techniques and machine learning. *Proteins* **2015**. [CrossRef] [PubMed]
8. Ovchinnikov, S.; Kim, D.E.; Wang, R.Y.; Liu, Y.; DiMaio, F.; Baker, D. Improved de novo structure prediction in CASP11 by incorporating Co-evolution information into rosetta. *Proteins* **2015**. [CrossRef]
9. Ovchinnikov, S.; Kinch, L.; Park, H.; Liao, Y.; Pei, J.; Kim, D.E.; Kamisetty, H.; Grishin, N.V.; Baker, D. Large-scale determination of previously unsolved protein structures using evolutionary information. *eLife* **2015**, *4*, e09248. [CrossRef] [PubMed]
10. Wang, S.; Sun, S.; Li, Z.; Zhang, R.; Xu, J. Accurate De Novo Prediction of Protein Contact Map by Ultra-Deep Learning Model. *PLoS Computat. Biol.* **2017**, *13*, e1005324. [CrossRef] [PubMed]
11. Hayat, S.; Sander, C.; Marks, D.S.; Elofsson, A. All-atom 3D structure prediction of transmembrane beta-barrel proteins from sequences. *Proc. Natl. Acad. Sci. USA* **2015**, *112*, 5413–5418. [CrossRef] [PubMed]

12. Hopf, T.A.; Colwell, L.J.; Sheridan, R.; Rost, B.; Sander, C.; Marks, D.S. Three-dimensional structures of membrane proteins from genomic sequencing. *Cell* **2012**, *149*, 1607–1621. [CrossRef] [PubMed]

13. Halabi, N.; Rivoire, O.; Leibler, S.; Ranganathan, R. Protein sectors: Evolutionary units of three-dimensional structure. *Cell* **2009**, *138*, 774–786. [CrossRef] [PubMed]

14. Hopf, T.A.; Scharfe, C.P.; Rodrigues, J.P.; Green, A.G.; Kohlbacher, O.; Sander, C.; Bonvin, A.M.; Marks, D.S. Sequence co-evolution gives 3D contacts and structures of protein complexes. *eLife* **2014**, *3*. [CrossRef] [PubMed]

15. Ovchinnikov, S.; Kamisetty, H.; Baker, D. Robust and accurate prediction of residue-residue interactions across protein interfaces using evolutionary information. *eLife* **2014**, *3*, e02030. [CrossRef] [PubMed]

16. Kamisetty, H.; Ovchinnikov, S.; Baker, D. Assessing the utility of coevolution-based residue-residue contact predictions in a sequence- and structure-rich era. *Proc. Natl. Acad. Sci. USA* **2013**, *110*, 15674–15679. [CrossRef] [PubMed]

17. Dyson, H.J.; Wright, P.E. Intrinsically unstructured proteins and their functions. *Nat. Rev. Mol. Cell Biol.* **2005**, *6*, 197–208. [CrossRef] [PubMed]

18. Habchi, J.; Tompa, P.; Longhi, S.; Uversky, V.N. Introducing protein intrinsic disorder. *Chem. Rev.* **2014**, *114*, 6561–6588. [CrossRef] [PubMed]

19. Van der Lee, R.; Buljan, M.; Lang, B.; Weatheritt, R.J.; Daughdrill, G.W.; Dunker, A.K.; Fuxreiter, M.; Gough, J.; Gsponer, J.; Jones, D.T.; et al. Classification of intrinsically disordered regions and proteins. *Chem. Rev.* **2014**, *114*, 6589–6631. [CrossRef] [PubMed]

20. Wright, P.E.; Dyson, H.J. Intrinsically unstructured proteins: Re-assessing the protein structure-function paradigm. *J. Mol. Biol.* **1999**, *293*, 321–331. [CrossRef] [PubMed]

21. Pancsa, R.; Fuxreiter, M. Interactions via intrinsically disordered regions: What kind of motifs? *IUBMB Life* **2012**, *64*, 513–520. [CrossRef] [PubMed]

22. Tompa, P.; Davey, N.E.; Gibson, T.J.; Babu, M.M. A Million Peptide Motifs for the Molecular Biologist. *Mol. Cell* **2014**, *55*, 161–169. [CrossRef] [PubMed]

23. Iakoucheva, L.M.; Radivojac, P.; Brown, C.J.; O'Connor, T.R.; Sikes, J.G.; Obradovic, Z.; Dunker, A.K. The importance of intrinsic disorder for protein phosphorylation. *Nucleic Acids Res.* **2004**, *32*, 1037–1049. [CrossRef] [PubMed]

24. Wright, P.E.; Dyson, H.J. Intrinsically disordered proteins in cellular signalling and regulation. *Nat. Rev. Mol. Cell Biol.* **2014**, *16*, 18–29. [CrossRef] [PubMed]

25. Hegyi, H.; Schad, E.; Tompa, P. Structural disorder promotes assembly of protein complexes. *BMC Struct. Biol.* **2007**, *7*, 65. [CrossRef] [PubMed]

26. Balazs, A.; Csizmok, V.; Buday, L.; Rakacs, M.; Kiss, R.; Bokor, M.; Udupa, R.; Tompa, K.; Tompa, P. High levels of structural disorder in scaffold proteins as exemplified by a novel neuronal protein, CASK-interactive protein1. *FEBS J.* **2009**, *276*, 3744–3756. [CrossRef] [PubMed]

27. Mark, W.Y.; Liao, J.C.; Lu, Y.; Ayed, A.; Laister, R.; Szymczyna, B.; Chakrabartty, A.; Arrowsmith, C.H. Characterization of segments from the central region of BRCA1: An intrinsically disordered scaffold for multiple protein-protein and protein-DNA interactions? *J. Mol. Biol.* **2005**, *345*, 275–287. [CrossRef] [PubMed]

28. Dosztanyi, Z.; Chen, J.; Dunker, A.K.; Simon, I.; Tompa, P. Disorder and sequence repeats in hub proteins and their implications for network evolution. *J. Proteome Res.* **2006**, *5*, 2985–2995. [CrossRef] [PubMed]

29. Dunker, A.K.; Cortese, M.S.; Romero, P.; Iakoucheva, L.M.; Uversky, V.N. Flexible nets. The roles of intrinsic disorder in protein interaction networks. *FEBS J.* **2005**, *272*, 5129–5148. [CrossRef] [PubMed]

30. Uversky, V.N.; Dave, V.; Iakoucheva, L.M.; Malaney, P.; Metallo, S.J.; Pathak, R.R.; Joerger, A.C. Pathological unfoldomics of uncontrolled chaos: Intrinsically disordered proteins and human diseases. *Chem. Rev.* **2014**, *114*, 6844–6879. [CrossRef] [PubMed]

31. Davey, N.E.; Van Roey, K.; Weatheritt, R.J.; Toedt, G.; Uyar, B.; Altenberg, B.; Budd, A.; Diella, F.; Dinkel, H.; Gibson, T.J. Attributes of short linear motifs. *Mol. BioSyst.* **2012**, *8*, 268–281. [CrossRef] [PubMed]

32. Van Roey, K.; Uyar, B.; Weatheritt, R.J.; Dinkel, H.; Seiler, M.; Budd, A.; Gibson, T.J.; Davey, N.E. Short linear motifs: Ubiquitous and functionally diverse protein interaction modules directing cell regulation. *Chem. Rev.* **2014**, *114*, 6733–6778. [CrossRef] [PubMed]

33. Fuxreiter, M.; Tompa, P.; Simon, I. Local structural disorder imparts plasticity on linear motifs. *Bioinformatics* **2007**, *23*, 950–956. [CrossRef] [PubMed]

34. Buljan, M.; Chalancon, G.; Eustermann, S.; Wagner, G.P.; Fuxreiter, M.; Bateman, A.; Babu, M.M. Tissue-specific splicing of disordered segments that embed binding motifs rewires protein interaction networks. *Mol. Cell* **2012**, *46*, 871–883. [CrossRef] [PubMed]

35. Weatheritt, R.J.; Davey, N.E.; Gibson, T.J. Linear motifs confer functional diversity onto splice variants. *Nucleic Acids Res.* **2012**, *40*, 7123–7131. [CrossRef] [PubMed]

36. Weatheritt, R.J.; Gibson, T.J. Linear motifs: Lost in (pre)translation. *Trends Biochem. Sci.* **2012**, *37*, 333–341. [CrossRef] [PubMed]

37. Brown, C.J.; Takayama, S.; Campen, A.M.; Vise, P.; Marshall, T.W.; Oldfield, C.J.; Williams, C.J.; Dunker, A.K. Evolutionary rate heterogeneity in proteins with long disordered regions. *J. Mol. Evol.* **2002**, *55*, 104–110. [CrossRef] [PubMed]

38. Brown, C.J.; Johnson, A.K.; Daughdrill, G.W. Comparing models of evolution for ordered and disordered proteins. *Mol. Biol. Evol.* **2010**, *27*, 609–621. [CrossRef] [PubMed]

39. Jeong, C.S.; Kim, D. Coevolved residues and the functional association for intrinsically disordered proteins. In Proceedings of the Pacific Symposium on Biocomputing, Kohala Coast, HI, USA, 3–7 January 2012; pp. 140–151. [CrossRef]

40. Toth-Petroczy, A.; Palmedo, P.; Ingraham, J.; Hopf, T.A.; Berger, B.; Sander, C.; Marks, D.S. Structured States of Disordered Proteins from Genomic Sequences. *Cell* **2016**, *167*, 158–170. [CrossRef] [PubMed]

41. Tompa, P.; Schad, E.; Tantos, A.; Kalmar, L. Intrinsically disordered proteins: Emerging interaction specialists. *Curr. Opin. Struct. Biol.* **2015**, *35*, 49–59. [CrossRef] [PubMed]

42. Dunker, A.K.; Obradovic, Z.; Romero, P.; Garner, E.C.; Brown, C.J. Intrinsic protein disorder in complete genomes. *Genome Inform.* **2000**, *11*, 161–171.

43. Pancsa, R.; Tompa, P. Structural disorder in eukaryotes. *PLoS ONE* **2012**, *7*, e34687. [CrossRef] [PubMed]

44. Ward, J.J.; Sodhi, J.S.; McGuffin, L.J.; Buxton, B.F.; Jones, D.T. Prediction and functional analysis of native disorder in proteins from the three kingdoms of life. *J. Mol. Biol.* **2004**, *337*, 635–645. [CrossRef] [PubMed]

45. Schad, E.; Ficho, E.; Pancsa, R.; Simon, I.; Dosztanyi, Z.; Meszaros, B. DIBS: A repository of disordered binding sites mediating interactions with ordered proteins. *Bioinformatics* **2018**, *34*, 535–537. [CrossRef] [PubMed]

46. Finn, R.D.; Coggill, P.; Eberhardt, R.Y.; Eddy, S.R.; Mistry, J.; Mitchell, A.L.; Potter, S.C.; Punta, M.; Qureshi, M.; Sangrador-Vegas, A.; et al. The Pfam protein families database: Towards a more sustainable future. *Nucleic Acids Res.* **2016**, *44*, D279–D285. [CrossRef] [PubMed]

47. Krissinel, E.; Henrick, K. Inference of macromolecular assemblies from crystalline state. *J. Mol. Biol.* **2007**, *372*, 774–797. [CrossRef] [PubMed]

48. Zor, T.; Mayr, B.M.; Dyson, H.J.; Montminy, M.R.; Wright, P.E. Roles of phosphorylation and helix propensity in the binding of the KIX domain of CREB-binding protein by constitutive (c-Myb) and inducible (CREB) activators. *J. Biol. Chem.* **2002**, *277*, 42241–42248. [CrossRef] [PubMed]

49. Selenko, P.; Gregorovic, G.; Sprangers, R.; Stier, G.; Rhani, Z.; Kramer, A.; Sattler, M. Structural basis for the molecular recognition between human splicing factors U2AF65 and SF1/mBBP. *Mol. Cell* **2003**, *11*, 965–976. [CrossRef]

50. Tompa, P.; Fuxreiter, M. Fuzzy complexes: Polymorphism and structural disorder in protein-protein interactions. *Trends Biochem. Sci.* **2008**, *33*, 2–8. [CrossRef] [PubMed]

51. Kayikci, M.; Venkatakrishnan, A.J.; Scott-Brown, J.; Ravarani, C.N.J.; Flock, T.; Babu, M.M. Visualization and analysis of non-covalent contacts using the Protein Contacts Atlas. *Nat. Struct. Mol. Biol.* **2018**, *25*, 185–194. [CrossRef] [PubMed]

52. Anishchenko, I.; Ovchinnikov, S.; Kamisetty, H.; Baker, D. Origins of coevolution between residues distant in protein 3D structures. *Proc. Natl. Acad. Sci. USA* **2017**, *114*, 9122–9127. [CrossRef] [PubMed]

53. Ait-Bara, S.; Carpousis, A.J.; Quentin, Y. RNase E in the gamma-Proteobacteria: Conservation of intrinsically disordered noncatalytic region and molecular evolution of microdomains. *Mol. Genet. Genom.* **2015**, *290*, 847–862. [CrossRef] [PubMed]

54. Savvides, S.N.; Raghunathan, S.; Futterer, K.; Kozlov, A.G.; Lohman, T.M.; Waksman, G. The C-terminal domain of full-length E. coli SSB is disordered even when bound to DNA. *Protein Sci.* **2004**, *13*, 1942–1947. [CrossRef] [PubMed]

55. Shereda, R.D.; Kozlov, A.G.; Lohman, T.M.; Cox, M.M.; Keck, J.L. SSB as an organizer/mobilizer of genome maintenance complexes. *Crit. Rev. Biochem. Mol. Biol.* **2008**, *43*, 289–318. [CrossRef] [PubMed]

56. Lu, D.; Keck, J.L. Structural basis of Escherichia coli single-stranded DNA-binding protein stimulation of exonuclease I. *Proc. Natl. Acad. Sci. USA* **2008**, *105*, 9169–9174. [CrossRef] [PubMed]

57. Tompa, P.; Szasz, C.; Buday, L. Structural disorder throws new light on moonlighting. *Trends Biochem. Sci.* **2005**, *30*, 484–489. [CrossRef] [PubMed]

58. Finn, R.D.; Clements, J.; Arndt, W.; Miller, B.L.; Wheeler, T.J.; Schreiber, F.; Bateman, A.; Eddy, S.R. HMMER web server: 2015 update. *Nucleic Acids Res.* **2015**, *43*, W30–W38. [CrossRef] [PubMed]

59. Apweiler, R.; Bairoch, A.; Wu, C.H.; Barker, W.C.; Boeckmann, B.; Ferro, S.; Gasteiger, E.; Huang, H.; Lopez, R.; Magrane, M.; et al. UniProt: The Universal Protein knowledgebase. *Nucleic Acids Res.* **2004**, *32*, D115–D119. [CrossRef] [PubMed]

60. Szklarczyk, D.; Franceschini, A.; Wyder, S.; Forslund, K.; Heller, D.; Huerta-Cepas, J.; Simonovic, M.; Roth, A.; Santos, A.; Tsafou, K.P.; et al. STRING v10: Protein-protein interaction networks, integrated over the tree of life. *Nucleic Acids Res.* **2015**, *43*, D447–D452. [CrossRef] [PubMed]

61. Kersey, P.J.; Allen, J.E.; Armean, I.; Boddu, S.; Bolt, B.J.; Carvalho-Silva, D.; Christensen, M.; Davis, P.; Falin, L.J.; Grabmueller, C.; et al. Ensembl Genomes 2016: More genomes, more complexity. *Nucleic Acids Res.* **2016**, *44*, D574–D580. [CrossRef] [PubMed]

62. Newcombe, R.G. Interval estimation for the difference between independent proportions: Comparison of eleven methods. *Stat. Med.* **1998**, *17*, 873–890. [CrossRef]

63. Touw, W.G.; Baakman, C.; Black, J.; te Beek, T.A.; Krieger, E.; Joosten, R.P.; Vriend, G. A series of PDB-related databanks for everyday needs. *Nucleic Acids Res.* **2015**, *43*, D364–D368. [CrossRef] [PubMed]

© 2018 by the authors. Licensee MDPI, Basel, Switzerland. This article is an open access article distributed under the terms and conditions of the Creative Commons Attribution (CC BY) license (http://creativecommons.org/licenses/by/4.0/).

International Journal of
Molecular Sciences

Article

Physical Background of the Disordered Nature of "Mutual Synergetic Folding" Proteins

Csaba Magyar [1,†], Anikó Mentes [1,†], Erzsébet Fichó [1], Miklós Cserző [1,2] and István Simon [1,*]

[1] Institute of Enzymology, Research Centre for Natural Sciences, Hungarian Academy of Sciences, Magyar Tudósok krt. 2, H-1117 Budapest, Hungary; magyar.csaba@ttk.mta.hu (C.M.); mentes.aniko@ttk.mta.hu (A.M.); ficho.erzsebet@ttk.mta.hu (E.F.); cserzo.miklos@ttk.mta.hu (M.C.)

[2] Department of Physiology, Faculty of Medicine, Semmelweis University, Tűzoltó u. 37-47, H-1094 Budapest, Hungary

* Correspondence: simon.istvan@ttk.mta.hu; Tel.: +36-1-3826-710

† These authors contributed equally to the work.

Received: 28 September 2018; Accepted: 21 October 2018; Published: 26 October 2018

Abstract: Intrinsically disordered proteins (IDPs) lack a well-defined 3D structure. Their disordered nature enables them to interact with several other proteins and to fulfil their vital biological roles, in most cases after coupled folding and binding. In this paper, we analyze IDPs involved in a new mechanism, mutual synergistic folding (MSF). These proteins define a new subset of IDPs. Recently we collected information on these complexes and created the Mutual Folding Induced by Binding (MFIB) database. These protein complexes exhibit considerable structural variation, and almost half of them are homodimers, but there is a significant amount of heterodimers and various kinds of oligomers. In order to understand the basic background of the disordered character of the monomers found in MSF complexes, the simplest part of the MFIB database, the homodimers are analyzed here. We conclude that MFIB homodimeric proteins have a larger solvent-accessible main-chain surface area on the contact surface of the subunits, when compared to globular homodimeric proteins. The main driving force of the dimerization is the mutual shielding of the water-accessible backbones and the formation of extra intermolecular interactions.

Keywords: dehydron; homodimer; hydrogen bond; inter-subunit interaction; intrinsically disordered protein; ion pair; mutual synergistic folding; solvent-accessible surface area; stabilization center

---

1. Introduction

Since the millennium it has been clear that Anfinsen's long-standing paradigm that was alleged to be valid for all proteins: "Protein structure is uniquely determined by its amino acid sequences" [1,2] is only valid for a specific subclass of proteins, while the rest of the proteins, termed intrinsically disordered proteins (IDPs), have no permanent 3D structures [3–6]. In our earlier effort to identify the physical background of protein disorder, the lack of sufficient pairwise interaction energy between the residues to ensure a stable 3D structure was pinpointed. When this energy is not enough to compensate the entropy-related free energy loss in the course of the formation of a unique structure, intrinsically disordered proteins are witnessed [7]. It has been shown that this pairwise energy can be calculated from the amino acid sequences without any structural information. On this basis we developed a widely used method, IUPred, to predict disordered proteins or protein segments from local composition data [8]. Another application of the estimation of the pairwise interaction energies led us to recognize the physical properties of the binding regions of disordered proteins, which can bind to ordered proteins [9]. When certain segments of a disordered protein interact with an ordered protein structure, part of their interactions will be manifested through elements of this stable globular protein having enough pairwise energy to stabilize their structures, i.e., to be folded, on the surface of ordered proteins. The contribution of a single residue depends only on the composition of the surrounding residues. Since ordered proteins have different amino acid compositions to disordered proteins, the resulting interaction energies of the residues at the contact surface can stabilize the structure (coupled folding and binding).

On the basis of this phenomenon, a binding site prediction method, termed ANCHOR, was developed [10]. These interacting segments generally appeared as short motifs of polypeptide chains (ELMs) [9,10]. More recently, the upgraded version of IUPred and ANCHOR were combined into a new server called IUPred2A [11].

While this phenomenon appeared to be general, over the years the number of "exceptions" increased, suggesting that the insufficient pairwise energy calculated by the IUPred algorithms was only valid for certain intrinsically disordered proteins and protein segments (IDSs), and that another kind of IDP and IDS also existed. Even in the early age of IDP studies, there was sporadic information that some IDPs exhibit mutual folding and binding together with other IDPs, without the help of already stable proteins or other stable macromolecules [12,13]. For example, NCDB segments of CBP form a complex with the ACTR domain of p160, see: protein data bank (PDB) entry 1kbh [14] or region C of WASP is I complex with the GBD segment of WASP [15]. In these examples, the interacting parts of the disordered proteins were not ELM sized, but rather have structural domain sizes [16]. In many cases the interacting disordered protein segments were alike, forming homodimer or homo-oligomers. Here the coupled folding and binding should not appear due to the difference in residue composition, as in the case of ELMs stabilized on the surface of an ordered protein. Therefore there should be another mechanism for coupled folding and binding than the one we can recognize by ANCHOR. Since macromolecular interactions are part of almost all the activity of disordered proteins, a new mechanism for coupled folding and binding, where there is no stable template to use, define a new subset of IDPs. Despite the sporadic information about these interactions, not too many of this kind of complexes were reported in the literature [17,18]. Therefore we performed a detailed analysis of several databases and on the scientific literature and collected information on these complexes and created the Mutual Folding Induced by Binding (MFIB) database [19]. These complexes exhibit large structural variations (see Figure 1).

Almost half of the MSF-complexes are homodimers, but there is a significant amount of heterodimers and other oligomeric states, including homo- and heterotetramers, as well as trimers, pentamers, and hexamers. To explore the unique features of the entries in the MFIB database and pinpoint those characters that differ between these entries and those of those disordered segments that can participate in coupled folding and binding with already stable proteins, we created the Disordered Binding Site (DIBS) database of the latter complexes [20]. Currently, a publication of the comparison of the structural differences of proteins of the MFIB and DIBS databases is in progress [21].

The elements of the pairwise interaction matrices used in the IUPred and ANCHOR algorithms were derived from the structure data of folded globular proteins, therefore this data includes the free energy from the average hydration of the residues in these proteins. We showed that this is similar for most globular protein, therefore a fair free energy contribution of a particular residue can be calculated from the composition of the rather large polypeptide segment centered by the particular residue, using the pairwise energy interaction matrix [7]. In the IUPred algorithm, when a particular residue is processed, whether it belongs to an ordered segment or a disordered one, the interaction of this residue in question and all other residues in a large surrounding region are considered. Therefore this calculated energy value has to be the same for all permutations of the residues of the segments located at both sides of the center residue, until or unless the compositions of the segments are changed. The amino acid sequences of proteins that have stable folded structures evolved in such a way that the side chains together shield the backbone from water, which minimizes the energetically unfavored water-accessible area on the polypeptide backbone. In this work we show that this statement is not valid for the disordered proteins listed in MFIB.

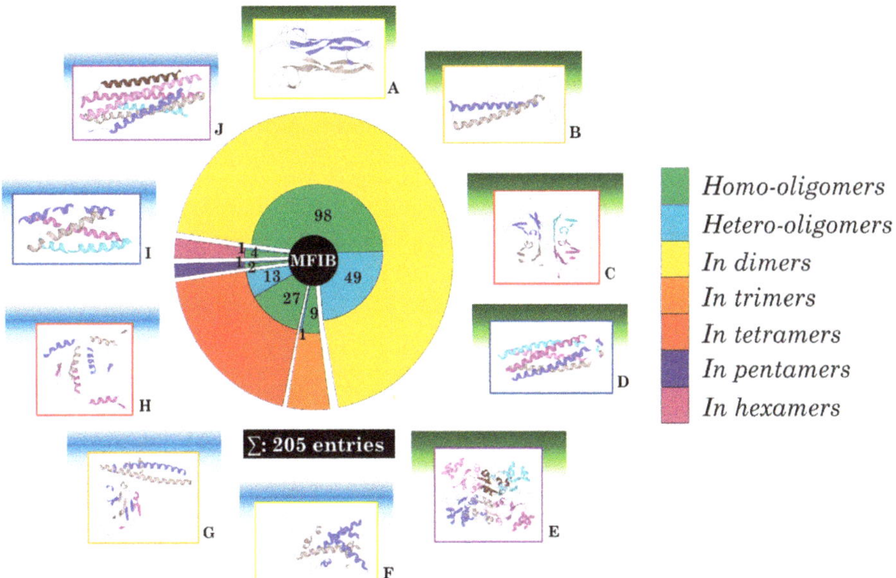

Figure 1. Oligomeric states in the MFIB database with example complexes. (**A**: 1BET, nerve growth factor (*Mus musculus*); **B**: 1AQ5, assembly domain of cartilage oligomeric matrix protein (*Gallus gallus*); **C**: 1GKE, Transthyretin (*Rattus norvegicus*); **D**: 1MZ9, assembly domain of cartilage oligomeric matrix protein (*Mus musculus*); **E**: 1NPK, nucleoside diphosphate kinase (*Dictyostelium discoideum*); **F**: 5GT0, H2A-H2B histone dimer, containing histone variants H2A type 1-A and H2B type 1-J (*Homo sapiens*); **G**: 2AZE, Rb C-terminal domain bound to an E2F1-DP1 heterodimer (*Homo sapiens*); **H**: 2NB1, p63/p73 hetero-tetramerization domain (*Homo sapiens*); **I**: 1VZJ, The synaptic acetylcholinesterase tetramer assembled around a polyproline-II helix (*Homo sapiens*); **J**: 1G2C, respiratory syncytial virus fusion protein core (*Homo sapiens*).

We investigated whether the interacting regions of these proteins can be identified based on their location in the whole polypeptide chain, on their biased amino acid composition or on specific physical properties. We discovered that their most unique characteristic is the high water accessibility of their peptide backbone, compared to the water accessibility of the folded proteins, which have similar amino acid compositions.

2. Results

2.1. Sequence-Based Analysis

In this study, homodimeric protein complexes from MFIB were analyzed regarding sequence and structural properties. First we checked the location of the MFIB homodimeric dataset (MFHD) PDB segments with a known 3D structure in the full UniProt protein sequences. In some cases, the MFHD PDB segments were located near the N-terminus, near the C-terminus, in the middle of the sequence or they were identical with the full sequence (Figure 2).

We examined the residue composition of the MFHD proteins (Figure 3, Table S2), which were compared with two reference datasets, the globular homodimeric dataset (GLHD) and the globular monomeric dataset (GLMD, see Section 4). To better understand the amino acid composition of the sequences, it was depicted by principal component analysis (PCA) (Figure 4, Table S3). PCA showed that the amino acid composition of the MFHD proteins did not differ significantly from the amino acid composition of the globular proteins (GLHD, GLMD). The PCA also demonstrated that MFHD formed a diverse group based on their amino acid composition.

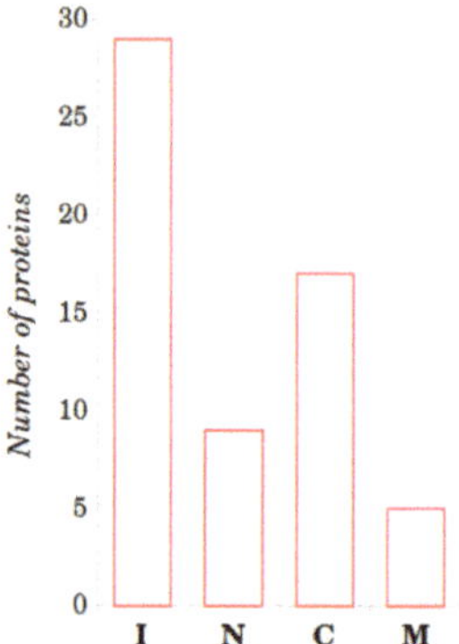

Figure 2. Distribution of MFHD PDB segments in the full UniProt sequences. (I: Amino acid sequence from UniProt is identical with amino acid sequences of MFHD PDB segment amino acid sequences; N: MFHD PDB segment is located in N-terminus of the amino acid sequences from UniProt; C: MFHD PDB segment is located in C-terminus of the amino acid sequences from UniProt; M: MFHD PDB segment is located in middle of the full amino acid sequence from UniProt).

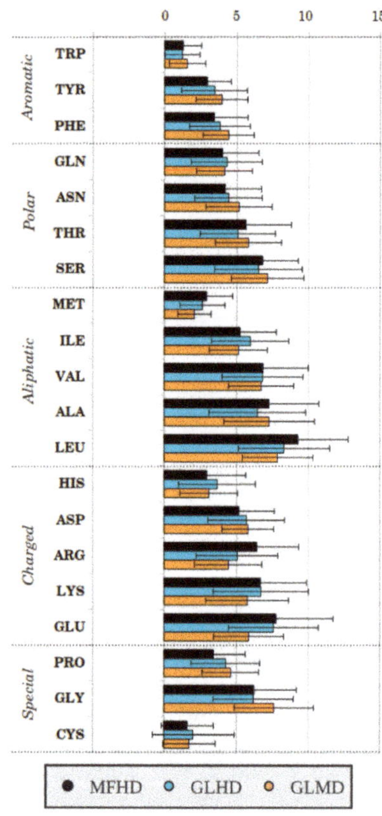

Figure 3. Sequence properties of MFHD, GLHD, and GLMD proteins (For values, see Table S2).

We investigated the MFHD with several protein disorder predictors (IUPred, ESpritz, GlobPlot, VSL2b, MobiDB Lite, MetaDisorder) [8,22–26], which worked well on the IDPs listed in DIBS, but did not recognize the polypeptide of MFHD complexes and other members of the MFIB database as disordered proteins. All methods predicted less than 30% of the protein residues as disordered, while the IUPred long/short methods, relying on a physical basis, predicted only 8 and 10% of the protein residues as disordered, respectively (for values, see Table S4). Other prediction methods based on amino acid composition bias also failed to detect MFHD PDB segments. Methods developed from the DAS and DAS-TMfilter [27,28] algorithms were tested on the dataset.

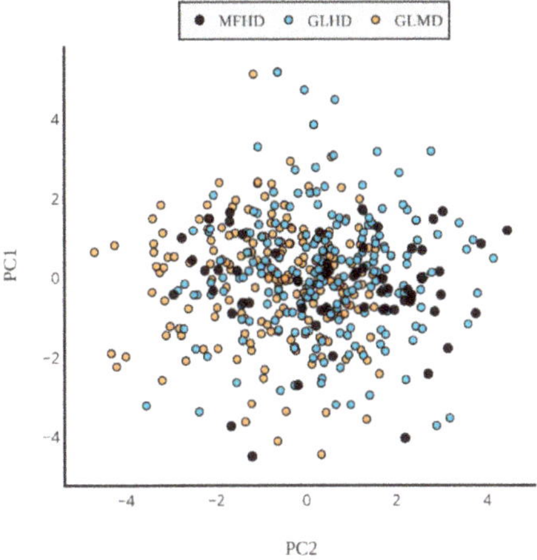

Figure 4. PCA ordination of the proteins from MFHD, GLHD, and GLMD based on their amino acid compositions (for values, see Table S3).

2.2. Structure-Based Analysis

We will use the term "interface" for the contact surface area of the two identical subunits in the dimeric structures. In cases where the term "monomeric structure" is used, calculations were carried out on structures from which the second chain was deleted since the PDB files contained dimer forms of the complexes. Residues belonging to the interface region were identified based on solvent accessible surface area (SASA) calculations. All-atom SASA values were calculated for the residues. Residues where the SASA value calculated from the dimer form were less than or equal to 20% of its counterpart from the monomeric structure defining the interface. We found that on average there were 26.4 interface residues per polypeptide-chain in the MFIB homodimeric dataset and 21.0 interface residues per polypeptide-chain in the reference globular homodimeric dataset. Considering the average size of the protein, this means that 27.13% of all residues in the MFHD and 22.34% of all residues in the GLHD belonged to the interface region. The higher value obtained for the MFIB homodimeric structures indicates that inter-subunit interactions may play an essential role in the stabilization of MFHD proteins.

We were looking for residues in the interface that have solvent-accessible spots in their main-chain in the monomeric structure, which become buried in the dimeric structures. We identified residues where the main-chain SASA in the dimeric form was less than 20% of the monomeric form value. Only residues with exposed main-chains, with a relative main-chain SASA larger than 0.2 in the monomeric structure, were taken into account. These residues with solvent-accessible main-chain patches (RSAMPs) were believed to be the main driving force of the dimerization of the disordered polypeptide chains collected in the MFIB database. We found a total of 183 such residues in the MFHD proteins; all structures contained at least one such residue. This was 3.14% of all residues. Considering that 27.13% of the residues were forming the interface, this means that 11.57% of the MFHD interface residues were RSAMPs. In the GLHD, 40.83% of the proteins did not contain such residues, on average 1.56% of all residues were RSAMPs. Since 22.34% of the residues form the interface, only 6.98% of the interface residues were RSAMPs. We calculated the average solvent-accessible surface area of the main chains. In the MFHD, the average solvent-accessible, main-chain area belonging to the interface region was 1154.56 Å^2 per polypeptide-chain, while in the GLHD this value was 790.54 Å^2. We can see that in the case of MFIB proteins a larger main-chain surface area is solvent accessible, which is energetically not favorable. The amino acid composition of the interface region and RSAMPs of the MFHD and GLHD complexes can be seen in Figure 5, Tables S5 and S6. Alanine and glycine were the most abundant residues under RSAMPs, which might be responsible for the higher solvent accessibility of the main chain in the MFHD. In the interface region, aliphatic residues are predominant. In the MFHD this was 50.6%, while in the GLHD 45.4% of the interface residues were aliphatic, making inter-subunit hydrophobic interactions even more prominent in MFIB proteins.

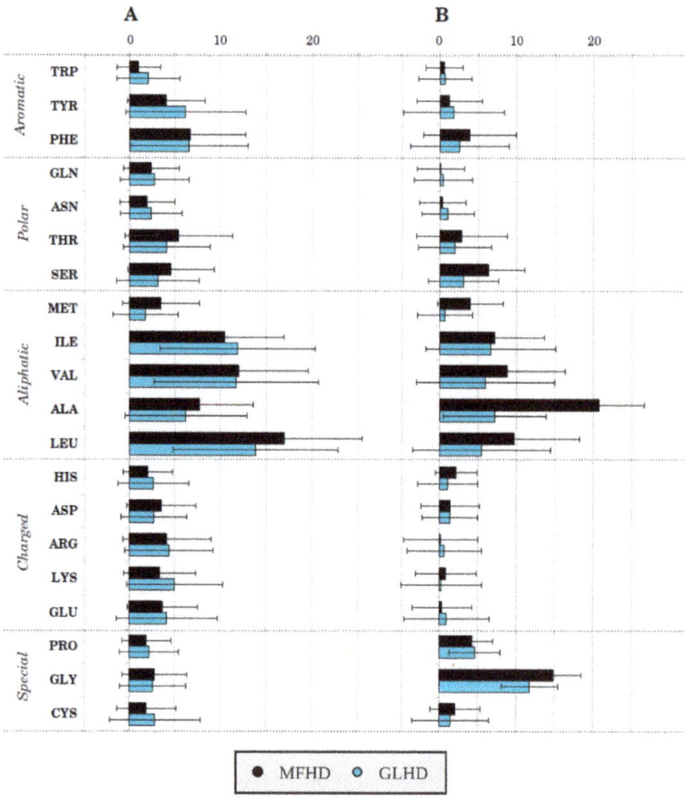

Figure 5. Amino acid composition of interface region (**A**) and RSAMPs (**B**) of the MFIB and globular homodimeric datasets (For values, see Tables S5 and S6).

We determined the secondary structural propensities in the MFHD, GLHD, and GLMD. We found that in the MFHD a significantly higher percentage of residues (39.4%) belonged to α-helices when compared to GLHD and GLMD (39.4% and 27.9%). In the MFHD, 21.2% of the residues belong to β-sheets, while in the GLHD and GLMD this value was 27.1% and 28.3%, respectively. The MFIB proteins show higher helical propensities than globular proteins.

We identified the hydrogen bonds formed between the two subunits. In the MFHD 6.97 inter-subunit H-bonds per structure were found, while in the GLHD this was only 4.58. Furthermore we identified underwrapped hydrogen bonds that are not well-enough shielded from the solvent, called dehydrons, in all structures [29]. In the MFHD we found 3.11 dehydrons per polypeptide chain under the inter-subunit H-bonds, while only 2.18 were found in the GLHD. Contrary to these results is the average wrapping of inter-subunit H-bonds, which was 16.0 for the MFHD and 13.6 for the GLHD. Although there were more dehydrons—i.e., underwrapped H-bonds—in the MFHD, the average wrapping value was still higher.

Due to the large difference found in the inter-subunit H-bonds, other inter-subunit interactions were also investigated. First we identified inter-subunit ion-pairs. We found that in the MFHD there were 1.17 inter-subunit ion-pairs on average, with only 0.66 in the GLHD. Charged residues tend to occur at the surface due to the desolvation of buried charges being energetically not favorable. Charged residues buried either in the interior of a protein or in the interface region of the dimeric structure should form ion pairs in order to compensate the desolvation penalty through favorable electrostatic interactions. Since the occurrence of charged residues is a bit higher in the interface region of the GLHD (16.2% vs. 14.7%), the lower number of inter-subunit ion-pairs was unexpected. We already noted in an earlier publication that inter-subunit ion pairs might contribute to the stabilization of proteins [30].

Stabilization centers (SCs) are pairs of residues involved in more than average long-range interactions [31]. These residue clusters are believed to contribute to the stabilization of protein structures through the cooperativity of the individual interactions [32,33]. The stabilization centers formed between different polypeptide chains can contribute to the stabilization of a protein complex [34]. We identified inter-subunit SCs in both the MFHD and GLHD. The two residues that form a stabilization center are called stabilization center elements (SCEs). We identified the SCEs belonging to the interface. In the MFHD, 3.86% of all residues form inter-subunit SCs, that is on average 14.22% of the interface residues form inter-subunit SCs. In the GLHD, only 1.83% of the residues belong to inter-subunit SCs. This means that only 8.19% of the interface residues form inter-subunit SCs. In MFIB dimers, the inter-subunit SCs were much more frequent than in the GLHD. We investigated whether SCEs overlap with RSAMPs or whether they are segregated. We found that there was a significant overlap, as 29.51% of the RSAMPs were SCEs in the MFHD. In the GLHD, we obtained a similar value of 29.19% for the overlap.

3. Discussion

In a recent study, we compared the residue composition of IDPs from the MFIB with complexes from the DIBS and other human protein databases [21] and we found that the composition of MFIB complexes was significantly different from that of the DIBS and only slightly different from that of human proteins. IDPs from the DIBS database are capable of coupled binding and folding on the surface of ordered proteins and can be predicted through bioinformatics methods like the ANCHOR algorithm, which is based on the different residue composition of the disordered monomer and the disordered–ordered protein complex. Therefore, in this work we studied MSF-homodimers to exclude this explanation for the case of mutual synergistic folding. We observed that in some cases the interacting segment of MFIB homodimers was the full polypeptide chain, while in other cases only a part of the chain was involved in the dimerization (Figure 2). We showed that they could be an order of magnitude longer than ELMs, which can be recognized by ANCHOR in other proteins.

In our current study the residue composition of the homodimeric complexes from the MFIB was determined and compared with that of homodimeric and monomeric globular proteins in similar amino acid sequence lengths (Figures 3 and 4). Our results showed that the IDPs listed in the MFIB had a similar amino acid composition to that of globular proteins. The PCA showed that the globular (GLHD, GLMD) and the MFHD proteins were not distinguishable. Although the points belonging to the complexes in the PCA figure were not certainly clustered, suggesting that MFHD is a distinct subgroup of IDPs. This was confirmed by the comparison of MFHD with the UniRef50 database, which showed that the main part of MFHD belongs to a distinct cluster and there is no significant similarity between their Pfam domains.

We investigated the MFHD with several protein disorder predictors (IUPred, ESpritz, GlobPlot, VSL2b, MobiDB Lite, MetaDisorder), which work well on the IDPs listed in DIBS. These methods did not recognize the full-length polypeptide chains of the MFHD complexes and other members of the MFIB database as disordered proteins. Since the disorder predictors IUPred and ANCHOR rely exclusively on solid physical principles, these methods were used to discover the physical principles behind the disordered character of the protein and the origin of the coupled folding and binding of the homodimers in the MFIB database. Our current study indicated that in the case of MFHD, the IUPred algorithm using its standard 20×20 pairwise free energy matrix overestimated the stabilizing energy because the energetically-unfavorable large solvent-accessible surface area of the peptide backbone in single protein chains resulted in less stabilizing energy. This can explain why these proteins were disordered in monomeric form. On the one hand, members of the MFIB dataset can be disordered for similar reason than other disordered proteins. That is, the sum of their pairwise interaction enthalpy did not compensate the free energy contribution of the entropy loss during folding. However, this is not the consequence of the amino acid composition of these polypeptides. Pairwise interactions of residue pairs, which have backbone parts not sufficiently shielded from the solvent, contribute less enthalpy to the stabilization than that found in globular proteins, from which the standard 20×20 pairwise free energy matrix was derived. Therefore, by using the free energy matrix in IUPred, we overestimated the stabilizing free energy of the proteins listed in the MFIB. This is why the IUPred algorithm predicted these monomers as structured proteins, while the experiments showed that they are disordered in their monomeric form [16].

We can conclude that the residue composition of MFHD is rather similar to that of the globular proteins (GLHD and GLMD), we were looking for structural differences among them. We found that the interface region had more residues in the MFHD than in the GLHD. MFIB homodimeric proteins had a larger solvent-accessible main-chain surface area in the interface when compared to globular homodimeric proteins. The polypeptide backbone of MFHD proteins was more accessible for water than in globular proteins. During dimerization, the solvent-accessible surface area of the backbone decreased and a high number of inter-subunit interactions (H-bonds, ion-pairs and stabilization centers) formed, leading to the stabilization of the of the disordered polypeptide-chains, enabling an ordered structure of MFIB proteins in the dimeric form. The driving force of the dimerization was the mutual shielding of the water-accessible backbones and the formation of extra intermolecular interactions.

4. Materials and Methods

Filters were applied to the homodimeric structures of the MFIB database. A reference dataset was created from homodimeric globular proteins, where the monomeric form was also globular in itself. Another reference dataset was created from monomeric globular proteins.

All homodimeric structures were collected from the MFIB database, and the modified PDB files were used. Entries belonging to the "coils and zippers" structure class were discarded since structures belonging to this class are both sequentially and structurally different from other homodimers. It is evident that a structure like a leucine-zipper cannot exist in monomeric form, thus no reference dataset can be created from "coils and zippers" where the monomer is not disordered in itself. A contact map matrix for all remaining structures was created. Entries with unusual contact maps were manually inspected. After inspection, the following entries were discarded: 2adl, 1r05, 4ath, 1aa0, 4w4k, 1ejp, resulting in a dataset of 60 homodimeric structures (Table S1). Heteroatoms were deleted from the structure. This dataset was referred to as MFIB homodimeric dataset (MFHD). We checked the secondary structure of the databases using the DSSP 2.0.4 program [35]. We found that 39.4% of the residues belonged to α-helices and 21.2% to β-sheets. The size distribution of the dataset was investigated. We counted the number of residues belonging to the N, N + 20 intervals. We found that the 140–240 interval was predominant, thus the reference datasets were created according to this size distribution.

A non-redundant reference dataset was created from homodimeric globular proteins. All homodimeric structures within the 140–240 amino acid size range were collected from the non-homologous PDB_Select database as of November, 2017 [36]. Structures containing coiled-coil structural elements identified with the Socket 3.0.3 program were excluded from the dataset [37]. The proper quaternary structure of the homodimers was created according to the BIOMT records of the PDB files. Entries with the following PDB ligand summary "ids" of large molecular sizes ligands and cofactors were discarded from the dataset because they could significantly alter the results of the solvent-accessible surface area calculations (017, 1BG, 1PE, 1PG, 5GP, C2E, FAD, HEC, KI1, MYA, MYR, MYS, NER, O8N, OLC, P33, P6G, PE5, UNL). Heteroatoms were deleted from the remaining structures. This procedure resulted in a list of 218 protein structures. This dataset was referred to as the globular homodimeric dataset (GLHD). For the PDB codes, see Table S1. According to DSSP, 27.9% of the residues belonged to α-helices and 27.1% to β-sheets.

An additional non-redundant reference dataset of the monomeric structures in the 140–240 amino acid size range containing only one structural domain was created from the PDB_SELECT database. The initial database was filtered by size and monomeric state criteria. All entries proved to be single domain according to the DDomain program using authors-trained parameters [38]. This dataset was referred to as the globular monomeric dataset (GLMD) and contained 191 entries (Table S1). According to DSSP, 24.9% of the residues belonged to α-helices and 28.3% to β-sheets.

Differences in the amino acid composition of the proteins sequences from the MFHD, GLHD, and GLMD datasets were revealed by principal component analysis (PCA) ordination using the plotly software according to Raska [39].

Hydrogen bonds were identified using the find_pairs command of PyMOL using 3.5 Å distance and 45 degree angle criteria between the donor and acceptor groups [40]. The calculation of the wrapping of hydrogen bonds and the identification of dehydrons was performed with the dehydron_ter.py program [41].

Stabilization centers (SCs) are pairs of residues, called stabilization center elements (SCEs), which are involved in several long-range interactions. These residues can be identified with our publicly available web server at http://scide.enzim.hu [42].

The solvent-accessible surface area (SASA) was calculated using the FreeSASA 2.03 program [43]. A residue was classified as buried when its relative SASA was below or equal to 0.2. Residues with a relative SASA value of over 0.2 were considered as exposed. A residue was classified as part of the interface region when its all-atom SASA calculated from the dimeric structure was less than 20% of the value calculated from the monomeric structure (created by deleting the second chain from the PDB file).

Ion-pairs were defined as pairs of negatively and positively charged residues, where the distance between the charged groups was equal to or less than 4 Å [44]. Ion pairs were identified using our own C++ program.

Supplementary Materials: Supplementary materials can be found at http://www.mdpi.com/1422-0067/19/11/3340/s1. Table S1. List of PDB entries in the MFHD, GLHD, GLMD datasets. Table S2. Average amino acid sequence composition of proteins from MFHD, GLMD, and GLHD. Table S3. Amino acid sequence composition of proteins from MFHD, GLMD, and GLHD for PCA. Table S4. Disorder content by various predictors. Table S5. Average amino acid composition of interface region of the proteins from MFHD and GLMD. Table S6. Average amino acid composition of RSAMPs of the proteins from MFHD and GLMD.

Author Contributions: Conceptualization, I.S., C.M., M.C.; methodology, A.M., E.F., C. M.; software, A.M., E.F., M.C., C.M.; validation, A.M., C.M; formal analysis, C.M.; investigation, A.M., E.F.; resources, A.M, E.F.; data curation, A.M., E.F., C.M.; writing—original draft preparation, A.M., C.M., I.S.; writing—review and editing, E.F., A.M., C.M; visualization, A.M.; supervision, I.S., M.C.; project administration, I.S.; funding acquisition, I.S.

Funding: This work was financially supported by the National Research, Development and Innovation Office (grant no. K115698). IS was supported by project no. FIEK_16-1-2016-0005 financed under the FIEK_16 funding scheme (National Research, Development and Innovation Fund of Hungary). The work of AM was supported by the ÚNKP-18-3 New National Excellence Program of the Ministry of Human Capacities (Hungary).

Conflicts of Interest: The authors declare no conflict of interest. The funders had no role in the design of the study; in the collection, analyses, or interpretation of data; in the writing of the manuscript, or in the decision to publish the results.

Abbreviations

IDPs	Intrinsically disordered proteins
MFIB	Mutual Folding Induced by Binding database
DIBS	Disordered Binding Site database
ELMs	Short motifs of polypeptide chains
MFHD	MFIB homodimeric dataset
MSF	Mutual synergistic folding
GLHD	Globular homodimeric dataset
GLMD	Globular monomeric dataset
PCA	Principal component analysis
PDB	Protein data bank
RSAMPs	Residues with solvent accessible main-chain patches
SC/SCE	Stabilization centers/stabilization center elements
SASA	Solvent accessible surface area

References

1. Anfinsen, C.B.; Haber, E.; Sela, M.; White, F.H., Jr. The kinetics of formation of native ribonuclease during oxidation of the reduced polypeptide chain. *Proc. Natl. Acad. Sci. USA* **1961**, *47*, 1309–1314. [CrossRef] [PubMed]
2. Anfinsen, C.B. The formation and stabilization of protein structure. *Biochem. J.* **1972**, *128*, 737–749. [CrossRef] [PubMed]
3. Wright, P.E.; Dyson, H.J. Intrinsically unstructured proteins: Re-assessing the protein structure-function paradigm. *J. Mol. Biol.* **1999**, *293*, 321–331. [CrossRef] [PubMed]
4. Uversky, V.N.; Dunker, A.K. Understanding protein non-folding. *Biochim. Biophys. Acta* **2010**, *1804*, 1231–1264. [CrossRef] [PubMed]
5. Tompa, P. Intrinsically unstructured proteins. *Trends Biochem. Sci.* **2002**, *27*, 527–533. [CrossRef]
6. Dyson, H.J.; Wright, P.E. Intrinsically unstructured proteins and their functions. *Nat. Rev. Mol. Cell Biol.* **2005**, *6*, 197–208. [CrossRef] [PubMed]
7. Dosztányi, Z.; Csizmók, V.; Tompa, P.; Simon, I. The pairwise energy content estimated from amino acid composition discriminates between folded and intrinsically unstructured proteins. *J. Mol. Biol.* **2005**, *347*, 827–839. [CrossRef] [PubMed]
8. Dosztányi, Z.; Csizmok, V.; Tompa, P.; Simon, I. IUPred: Web server for the prediction of intrinsically unstructured regions of proteins based on estimated energy content. *Bioinformatics* **2005**, *21*, 3433–3434. [CrossRef] [PubMed]
9. Mészáros, B.; Simon, I.; Dosztányi, Z. Prediction of Protein Binding Regions in Disordered Proteins. *PLoS Comput. Biol.* **2009**, *5*, e1000376. [CrossRef] [PubMed]
10. Dosztányi, Z.; Mészáros, B.; Simon, I. ANCHOR: Web server for predicting protein binding regions in disordered proteins. *Bioinformatics* **2009**, *25*, 2745–2746. [CrossRef] [PubMed]
11. Mészáros, B.; Erdos, G.; Dosztányi, Z. IUPred2A: Context-dependent prediction of protein disorder as a function of redox state and protein binding. *Nucleic Acids Res.* **2018**, *46*, W329–W337. [CrossRef] [PubMed]
12. Gunasekaran, K.; Tsai, C.-J.; Nussinov, R. Analysis of Ordered and Disordered Protein Complexes Reveals Structural Features Discriminating Between Stable and Unstable Monomers. *J. Mol. Biol.* **2004**, *341*, 1327–1341. [CrossRef] [PubMed]
13. Rumfeldt, J.A.O.; Galvagnion, C.; Vassall, K.A.; Meiering, E.M. Conformational stability and folding mechanisms of dimeric proteins. *Prog. Biophys. Mol. Biol.* **2008**, *98*, 61–84. [CrossRef] [PubMed]
14. Demarest, S.J.; Martinez-Yamout, M.; Chung, J.; Chen, H.; Xu, W.; Dyson, H.J.; Evans, R.M.; Wright, P.E. Mutual synergistic folding in recruitment of CBP/p300 by p160 nuclear receptor coactivators. *Nature* **2002**, *415*, 549–553. [CrossRef] [PubMed]

15. Garrard, S.M.; Capaldo, C.T.; Gao, L.; Rosen, M.K.; Macara, I.G.; Tomchick, D.R. Structure of Cdc42 in a complex with the GTPase-binding domain of the cell polarity protein, Par6. *EMBO J.* **2003**, *22*, 1125–1133. [CrossRef] [PubMed]

16. Tompa, P.; Fuxreiter, M.; Oldfield, C.J.; Simon, I.; Dunker, A.K.; Uversky, V.N. Close encounters of the third kind: Disordered domains and the interactions of proteins. *Bioessays* **2009**, *31*, 328–335. [CrossRef] [PubMed]

17. Zheng, Y.; Wu, Q.; Wang, C.; Xu, M.-Q.; Liu, Y. Mutual synergistic protein folding in split intein. *Biosci. Rep.* **2012**, *32*, 433–442. [CrossRef] [PubMed]

18. Ganguly, D.; Zhang, W.; Chen, J. Synergistic folding of two intrinsically disordered proteins: Searching for conformational selection. *Mol. Biosyst.* **2012**, *8*, 198–209. [CrossRef] [PubMed]

19. Fichó, E.; Reményi, I.; Simon, I.; Mészáros, B. MFIB: A repository of protein complexes with mutual folding induced by binding. *Bioinformatics* **2017**, *33*, 3682–3684. [CrossRef] [PubMed]

20. Schad, E.; Fichó, E.; Pancsa, R.; Simon, I.; Dosztányi, Z.; Mészáros, B. DIBS: A repository of disordered binding sites mediating interactions with ordered proteins. *Bioinformatics* **2017**, *34*, 535–537. [CrossRef] [PubMed]

21. Mészáros, B.; Dobson, L.; Fichó, E.; Tusnády, G.E.; Dosztányi, Z.; Simon, I. Interplay between folding and binding modulates protein sequences, structures, functions and regulation. *bioRxiv* **2017**, 211524. [CrossRef]

22. Walsh, I.; Martin, A.J.M.; Di Domenico, T.; Tosatto, S.C.E. ESpritz: Accurate and fast prediction of protein disorder. *Bioinformatics* **2012**, *28*, 503–509. [CrossRef] [PubMed]

23. Linding, R. GlobPlot: Exploring protein sequences for globularity and disorder. *Nucleic Acids Res.* **2003**, *31*, 3701–3708. [CrossRef] [PubMed]

24. Peng, K.; Radivojac, P.; Vucetic, S.; Dunker, A.K.; Obradovic, Z. Length-dependent prediction of protein intrinsic disorder. *BMC Bioinform.* **2006**, *7*, 208. [CrossRef] [PubMed]

25. Necci, M.; Piovesan, D.; Dosztányi, Z.; Tosatto, S.C.E. MobiDB-lite: Fast and highly specific consensus prediction of intrinsic disorder in proteins. *Bioinformatics* **2017**, *33*, 1402–1404. [CrossRef] [PubMed]

26. Kozlowski, L.P.; Bujnicki, J.M. MetaDisorder: A meta-server for the prediction of intrinsic disorder in proteins. *BMC Bioinform.* **2012**, *13*, 111. [CrossRef] [PubMed]

27. Cserzö, M.; Wallin, E.; Simon, I.; von Heijne, G.; Elofsson, A. Prediction of transmembrane alpha-helices in prokaryotic membrane proteins: The dense alignment surface method. *Protein Eng.* **1997**, *10*, 673–676. [CrossRef] [PubMed]

28. Cserzö, M.; Eisenhaber, F.; Eisenhaber, B.; Simon, I. On filtering false positive transmembrane protein predictions. *Protein Eng.* **2002**, *15*, 745–752. [CrossRef] [PubMed]

29. Fernández, A.; Scott, R. Dehydron: A structurally encoded signal for protein interaction. *Biophys. J.* **2003**, *85*, 1914–1928. [CrossRef]

30. Németh, A.; Svingor, A.; Pócsik, M.; Dobó, J.; Magyar, C.; Szilágyi, A.; Gál, P.; Závodszky, P. Mirror image mutations reveal the significance of an intersubunit ion cluster in the stability of 3-isopropylmalate dehydrogenase. *FEBS Lett.* **2000**, *468*, 48–52. [CrossRef]

31. Dosztányi, Z.; Fiser, A.; Simon, I. Stabilization centers in proteins: Identification, characterization and predictions. *J. Mol. Biol.* **1997**, *272*, 597–612. [CrossRef] [PubMed]

32. Magyar, C.; Gromiha, M.M.; Sávoly, Z.; Simon, I. The role of stabilization centers in protein thermal stability. *Biochem. Biophys. Res. Commun.* **2016**, *471*, 57–62. [CrossRef] [PubMed]

33. Simon, Á.; Dosztányi, Z.; Magyar, C.; Szirtes, G.; Rajnavölgyi, É.; Simon, I. Stabilization centers and protein stability. *Theor. Chem. Acc.* **2001**, *106*, 121–127. [CrossRef]

34. Simon, A.; Dosztányi, Z.; Rajnavölgyi, E.; Simon, I. Function-related regulation of the stability of MHC proteins. *Biophys. J.* **2000**, *79*, 2305–2313. [CrossRef]

35. Kabsch, W.; Sander, C. Dictionary of protein secondary structure: Pattern recognition of hydrogen-bonded and geometrical features. *Biopolymers* **1983**, *22*, 2577–2637. [CrossRef] [PubMed]

36. Griep, S.; Hobohm, U. PDBselect 1992–2009 and PDBfilter-select. *Nucleic Acids Res.* **2010**, *38*, D318–D319. [CrossRef] [PubMed]

37. Walshaw, J.; Woolfson, D.N. Socket: A program for identifying and analysing coiled-coil motifs within protein structures. *J. Mol. Biol.* **2001**, *307*, 1427–1450. [CrossRef] [PubMed]

38. Zhou, H.; Xue, B.; Zhou, Y. DDOMAIN: Dividing structures into domains using a normalized domain-domain interaction profile. *Protein Sci.* **2007**, *16*, 947–955. [CrossRef] [PubMed]

39. Raschka, S. *Python Machine Learning*; Packt Publishing Ltd.: Birmingham, UK, 2015; ISBN 9781783555147.

40. *The PyMOL Molecular Graphics System*; Version 1.6; Schrodinger, LLC: New York, NY, USA, 2011.

41. Martin, O.A. *Wrappy: A Dehydron Calculator Plugin for PyMOL*; IMASL-CONICET: San Luis, Argentina, 2012.
42. Dosztanyi, Z. Servers for sequence-structure relationship analysis and prediction. *Nucleic Acids Res.* **2003**, *31*, 3359–3363. [CrossRef] [PubMed]
43. Mitternacht, S. FreeSASA: An open source C library for solvent accessible surface area calculations. *F1000Research* **2016**, *5*, 189. [CrossRef] [PubMed]
44. Barlow, D.J.; Thornton, J.M. Ion-pairs in proteins. *J. Mol. Biol.* **1983**, *168*, 867–885. [CrossRef]

 © 2018 by the authors. Licensee MDPI, Basel, Switzerland. This article is an open access article distributed under the terms and conditions of the Creative Commons Attribution (CC BY) license (http://creativecommons.org/licenses/by/4.0/).

International Journal of *Molecular Sciences*

MDPI

Article

Disordered Regions of Mixed Lineage Leukemia 4 (MLL4) Protein Are Capable of RNA Binding

Beáta Szabó [1], Nikoletta Murvai [1], Rawan Abukhairan [1], Éva Schád [1], József Kardos [2], Bálint Szeder [1], László Buday [1] and Ágnes Tantos [1,*]

[1] Institute of Enzymology, Research Centre for Natural Sciences, Hungarian Academy of Sciences, H-1117 Budapest, Hungary; szabo.beata@ttk.mta.hu (B.S.); murvai.nikoletta@ttk.mta.hu (N.M.); rawan.abukhairan@ttk.mta.hu (R.A.); schad.eva@ttk.mta.hu (E.S.); szeder.balint@ttk.mta.hu (B.S.); buday.laszlo@ttk.mta.hu (L.B.)

[2] ELTE NAP Neuroimmunology Research Group, Department of Biochemistry, Eötvös Loránd University, H-1117 Budapest, Hungary; kardos@elte.hu

* Correspondence: tantos.agnes@ttk.mta.hu; Tel.: +36-1-382-6705

Received: 30 September 2018; Accepted: 2 November 2018; Published: 5 November 2018

Abstract: Long non-coding RNAs (lncRNAs) are emerging as important regulators of cellular processes and are extensively involved in the development of different cancers; including leukemias. As one of the accepted methods of lncRNA function is affecting chromatin structure; lncRNA binding has been shown for different chromatin modifiers. Histone lysine methyltransferases (HKMTs) are also subject of lncRNA regulation as demonstrated for example in the case of Polycomb Repressive Complex 2 (PRC2). Mixed Lineage Leukemia (MLL) proteins that catalyze the methylation of H3K4 have been implicated in several different cancers; yet many details of their regulation and targeting remain elusive. In this work we explored the RNA binding capability of two; so far uncharacterized regions of MLL4; with the aim of shedding light to the existence of possible regulatory lncRNA interactions of the protein. We demonstrated that both regions; one that contains a predicted RNA binding sequence and one that does not; are capable of binding to different RNA constructs in vitro. To our knowledge, these findings are the first to indicate that an MLL protein itself is capable of lncRNA binding.

Keywords: MLL proteins; MLL4; lncRNA; HOTAIR; MEG3; leukemia; histone lysine methyltransferase; RNA binding; intrinsically disordered protein

---

1. Introduction

Long non-coding RNAs (lncRNAs) are transcribed RNA molecules longer than 200 nucleotides that do not code for translated proteins. The human genome is estimated to code for about 58,000 lncRNAs [1], that are being more and more recognized as central players in a plethora of biological processes. They can act as flexible scaffolds providing binding platforms for different proteins, they can interfere with other endogenous RNAs acting as microRNA "sponges" and they can modify chromatin state [2], thus regulating the expression of various proteins. LncRNAs have also been shown to play a role in several layers of epigenetic regulation: they are involved in DNA methylation and demethylation, they can modify chromatin conformation through binding to remodelers [3] and many of them interact with histone modifier enzyme complexes such as PRC2, coREST or SMCX [4].

The physiological processes where lncRNA regulation have been suggested involve cell cycle regulation, epithelial mesenchymal transition (EMT) [5], cancer progression [6] and maintenance of cancer stem cells [5], hypoxia [7] and leukemia [8].

Various lncRNAs are shown to have altered expression levels in different leukemias, resulting in a crucial influence on cellular transformation [9], chromosomal translocation [10], apoptosis [11] and on drug resistance [12]. Accumulating evidence regarding the involvement of lncRNAs in leukemic processes prompted the suggestion to use them as prognostic and classification factors. It was found that lncRNA expression has prognostic value in AML patients [13] and multiple pathways were involved in lncRNA expression, including chromosome organization and trans-membrane receptor protein tyrosine kinase signalling pathway.

As lncRNAs are also considered valuable drug targets, it is essential that the molecular details of their functions are uncovered.

Polycomb repressive complex (PRC2) is the most studied histone modifier that relies on lncRNA binding in its function, being able to bind several lncRNAs including HOTAIR, Xist, RepA, Braveheart, MALAT1 and MEG3 [14]. In vitro experiments revealed that not only EZH2, but other PRC2 subunits are also capable of lncRNA binding [15], thus providing a pattern of binding regions distributed along the surface of the complex. Even though there remain open questions regarding the specificity of the RNA binding by PRC2 [16], it is widely accepted that lncRNA binding plays a defining role in PRC2 targeting and the ensuing gene silencing [14]. It is interesting to note that despite the numerous experimental results that show EZH2 to be an RNA binding protein, it cannot be found in databases that list RNA binding proteins, furthermore no RNA binding site is predicted to be located in the region that is shown to be responsible for the RNA-protein interaction [17].

Apart from PRC2, other histone lysine methyltransferases (HKMTs) or HKMT complex components also appear to bind lncRNAs with a relevant physiological outcome.

LncRNA EZR-AS1 enhances EZR expression through recruiting SMYD2 to the upstream region of its promoter region and elevating the activating H3K4 methylation [18].

G9a interacts with lncRNA PARTICLE to regulate MATA2 expression upon mild irradiation [19]. The interaction was shown using ChIP assay and apart from G9a, the PRC2 subunit Suz12 was also pulled down. In a later experiment, it was found that PARTICLE can also interact with DNA methylase DNMT1 and that it increases H3K27 methylation as well as EZH2 expression. It was suggested that PARTICLE may serve as a functional platform that enables the specific targeting of chromatin modifiers, such as PRC2 [20].

WDR5, a component of the MLL1-4 and SET1a/1b complexes was proven to interact with lncRNAs NeST and HOTTIP with an effect on microbial susceptibility through the enhancement of interferon-γ expression [21]. Further investigation of the WDR5-HOTTIP interaction led to the recognition that lncRNA binding by WDR5 is essential in maintaining embryonic stem cell pluripotency [22]. However, not this work nor any previous studies investigated the possibility that the enzymatic component of the methyltransferase complex may also be capable of lncRNA binding.

The family of mammalian MLL (Mixed Lineage Leukemia) proteins consist of Set1a, Set1b and four MLL proteins, MLL1, MLL2, MLL3 and MLL4. They work in COMPASS-like complexes and catalyze H3K4 mono-, di- or tri-methylation, each complex having different specificity and methylase activity [23]. MLL3 and MLL4 are responsible for the monomethylation of H3K4 at enhancer regions [24] and has been linked to a high number of different cancers. Properly functioning MLL3 and MLL4 act as tumor suppressors [23], therefore mutations affecting their activity or stability can result in cancer development. Despite their central role in several types of cancers, many open questions regarding the regulation of the activity and the targeting of the MLL complexes remain unanswered. The exact molecular details of how MLL3 and MLL4 are targeting enhancer regions [23] as well as the specific molecular effects of the interactions of their different regulatory domains [25] are largely unknown. It is also worth noting that the known structured domains represent only 15–21% of the sequences of MLL proteins, leaving the vast majority of these proteins uncharacterized both structurally and functionally.

In a previous work [26] we suggested that the disordered regions of HKMTs may harbor so far unrecognized interaction sites, adding more layers of the regulation of their activity. Based on the observation that many lncRNAs are involved in processes governed by HKMTs, we hypothesized that lncRNA binding might be one of the functions of these regions.

Since multiple evidence point in the direction that leukemic processes are fundamentally affected by lncRNAs and MLL complexes are involved in this regulation, we concentrated on MLL proteins. Taken the analogy of the PRC2 complex, where more than one complex subunits are capable of lncRNA binding, we aimed at testing the ability of MLL4 to bind different RNA molecules.

2. Results

2.1. In Silico Analysis of the RNA Binding Capacity of MLL Proteins

As a first step, we mapped the predicted RNA binding motifs on the sequence of four MLL proteins. We used DisoRDPbind, an RNA interaction prediction tool specifically designed to find RNA interaction sites in the disordered regions of proteins. Results shown in Table 1 indicate that all MLL proteins contain several putative RNA interaction motifs in their disordered regions. These regions are found at various positions in the proteins and vary in length from a couple of amino acids to almost a hundred residues, suggesting that RNA binding might be a common feature in MLL proteins.

Table 1. Predicted RNA binding regions in the disordered regions of Mixed Lineage Leukemia (MLL) proteins (aa positions).

MLL1	MLL2	MLL3	MLL4
296–327	84–107	1068–1079	1559–1567
348–408	184–234	1678–1695	3526–3581
415–418	241–244	1701–1709	3899–3983
1155–1194	536–560	1715–1737	4960–5014
1977–1992	783–806	2406–2409	5147–5165
3854–3861	820–828	3052–3073	5227–5251
	1753–1778	3246–3250	
	2600–2616	3394–3427	
	2685–2709	4330–4356	
		4514–4524	
		4586–4625	

A comparison with our earlier studies [26] revealed that two conserved disordered binding sites (residues 3537–3545 and 3560–3567) reside within one of the predicted RNA binding regions (residues 3526–3581, Figure 1A) of MLL4, underlining the reliability of the predictions. This region also harbors several cancer-related point mutations, two of them corresponding to a predicted binding site at positions 3560 (D-N) and 3561 (A-D). All these evidences point to the physiological importance of this protein region, making its structural and functional study worthwhile. ANCHOR prediction [27] shows that within the C-terminal border of the predicted RNA binding region there is a region with a strong tendency of the protein chain to form protein-protein interactions (residues 3597–3613, Figure 1A) that corresponds to a run of 14 glutamine residues. Since polyQ repeats in RNA binding proteins have been linked to protein-RNA droplet formation [28], this raises the intriguing possibility of granule formation potency of this segment. Therefore, we chose to test the RNA binding capacity of the MLL4 region between residues 3500–3630 (Figure 1A). As an internal control, another disordered region with no predicted RNA or protein binding sites was selected between residues 4210–4280 of MLL4 (Figure 1D).

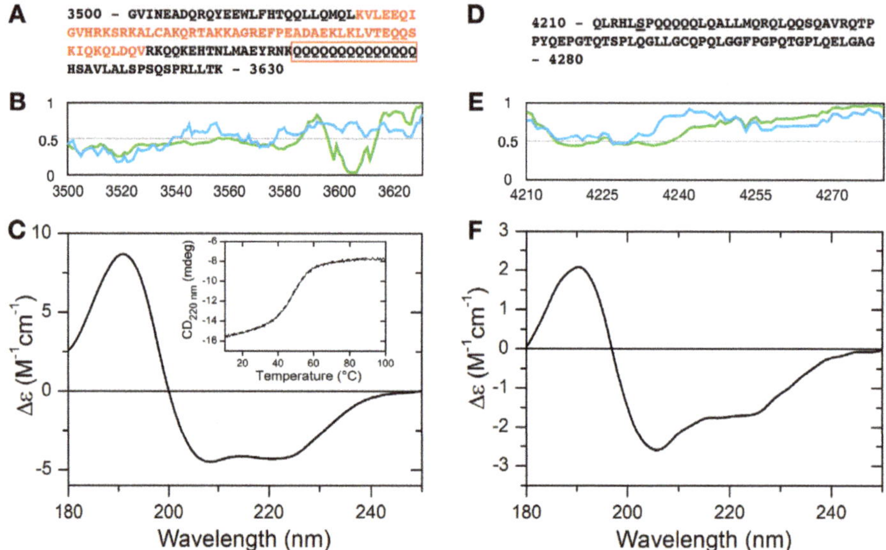

Figure 1. Structural characterization of the MLL4 regions. Sequences of MLL4$_{3500-3630}$ (**A**) and MLL4$_{4210-4280}$ (**D**). Predicted RNA binding region is indicated by red letters and the polyQ stretch is framed with red. IUPRed (blue) and Anchor (green) prediction of MLL4$_{3500-3630}$ (**B**) and MLL4$_{4210-4280}$ (**E**). Residues having an IUPred score above 0.5 are considered to be disordered, while residues with an Anchor score below 0.5 constitute predicted binding sites. Far-UV CD spectra of MLL4$_{3500-3630}$ (**C**) and MLL4$_{4210-4280}$ (**F**). Inset: temperature-dependent changes in the structure of MLL4$_{3500-3630}$ as observed by monitoring the changes in the absorbance at 220 nm.

As for binding RNAs, we opted to test two different lncRNA constructs, both having been reported to play a role in leukemias. The first is HOTAIR, that has the ability to bind EZH2 (PRC2). The 5′ 300 nucleotides of HOTAIR are thought to mediate its binding to PRC2 complex subunits, but the latest annotation in the NCBI database contains an additional 140 bases at the beginning of HOTAIR sequence, compared to the one reported earlier. Therefore, we prepared the longer version (HOTAIR$_{440}$) that encompasses the 300 nucleotides already known to be involved in protein-RNA interactions and also the nucleotides that has not been studied yet. Since there is no information available about the region of MEG3 that is able to bind proteins, we used the full length MEG3 for our experiments.

2.2. Secondary Structure of MLL4$_{3500-3630}$ and MLL4$_{4210-4280}$

Disorder prediction profiles (Figure 1B,E) indicated that both protein regions have a significant disorder tendency. Disorder profile of MLL4$_{3500-3630}$ indicates a rather ambiguous disorder state, with prediction scores fluctuating around the 0.5 limit between ordered and disordered states. This disorder prediction might indicate a disordered region that has an elevated tendency to fold or a relatively unstable folded segment as well. Far-UV CD measurements revealed that MLL4$_{3500-3630}$ has a helical structure in isolation (Figure 1C). The CD spectrum of this region of MLL4 showed a typical alpha helical conformation with a pronounced double minimum at 208 and 220 nm. Secondary structure content calculation using the BeStSel algorithm [29,30] gave an α-helix content of ~36.2%, while another ~36% of the secondary structure content was characterized as "Others", which mainly corresponds to the disordered structure. Thermal unfolding of the observed helical structure was followed by gradually heating the sample to 100 °C while recording the absorbance at 220 nm (Figure 1C inset). The melting curve indicated a cooperative unfolding of the structure with a melting point of 48 °C. The CD spectrum of the thermal denatured state is shown in Supplementary Figure S1, demonstrating a complete loss of structure at high temperatures.

MLL4$_{4210-4280}$ has a more pronounced disorder tendency, as demonstrated by the IUPred profile and is devoid of any predicted ANCHOR binding sites (Figure 1E). Its sequence contains a significant portion of glutamines (Figure 1D), but it does not contain Q stretches longer than 4 residues. Far-UV CD measurements confirmed the disorder predictions, indicating that the protein is mostly disordered in solution, with a considerable α-helical tendency. Secondary structure calculations gave a result of 16% α-helix and ~45% "Others" content, underlining that this segment of MLL4 is not fully disordered and contrary to interaction site predictions, might be involved in molecular recognition.

2.3. RNA Binding of MLL4$_{3500-3630}$ and MLL4$_{4210-4280}$

Microscale thermophoresis measurements were performed to characterize the RNA binding of the expressed protein regions. We used two lncRNA constructs, HOTAIR$_{440}$, a segment of HOTAIR that contains the region involved in binding to EZH2 [31], MEG3, a lncRNA involved in leukemias [32] and a 50 nt long RNA with random nucleotide sequence. Contradicting to the lack of predicted binding sites, MLL4$_{4210-4280}$ showed a relatively strong binding to HOTAIR$_{440}$ with an apparent Kd of 13.05 μM (Figure 2A), while the negative control Thymosin beta 4 (Tβ4) did not bind to the RNA, showing any sign of interaction at only the highest concentrations applied.

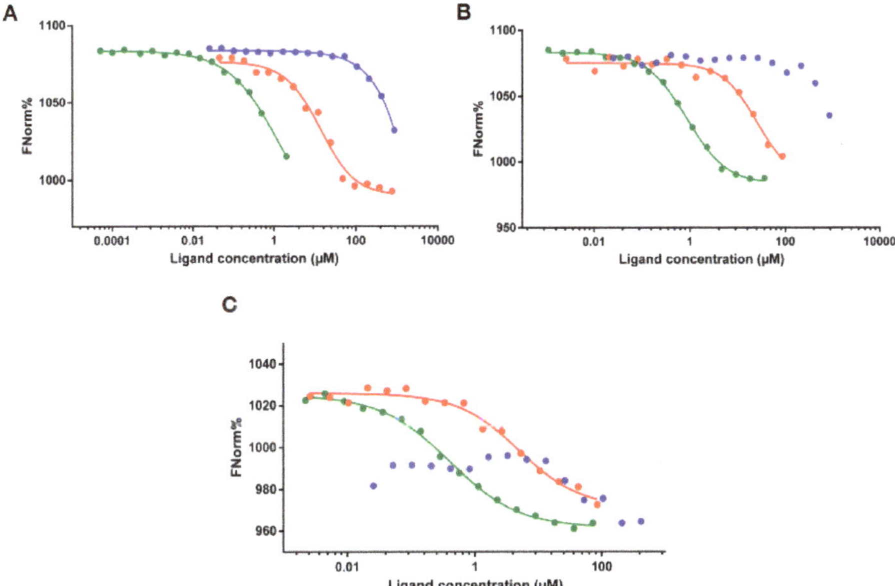

Figure 2. RNA binding detected by microscale thermophoresis. MST binding curves of MLL4$_{3500-3630}$ (green), MLL4$_{4210-4280}$ (red) and thymosin beta 4 (blue) to different RNAs: HOTAIR$_{440}$ (**A**), MEG3 (**B**) and 50 nt RNA (**C**).

In the case of MLL4$_{3500-3630}$, saturation of the reaction could not be reached because of marked aggregation above 1:20 RNA:protein ratio (Supplementary Figure S2) but using the T-jump values of the MST measurement (Supplementary Figure S3) an approximate binding constant of 0.1 μM could be determined. The appearance of large particles in the solution, generally considered to be aggregates, is indicated by a "wavy" MST curve and a randomly fluctuating normalized fluorescent percentage as shown on Supplementary Figures S2 and S5. The observed aggregation was dependent on the RNA species, since it was not seen with either of the other tested RNAs (Figure 3B,C), or with a shorter, 300 nt long HOTAIR construct (Supplementary Figure S4). The HOTAIR$_{300}$ construct overlaps with HOTAIR$_{440}$ in the 3' 300 nucleotides but lacks the first 140 nucleotides of the latter. This shorter HOTAIR construct bound to MLL4$_{3500-3630}$ with a Kd of 0.97 μM, with no sign of irregular behavior. Centrifugation (15 min at 13,000× *g*) of the samples resulted in the loss of fluorescent signal in a protein concentration-dependent manner (Supplementary Figure S5), indicating a formation of structures containing both RNA and protein. Such phenomenon was not observed with MLL4$_{4210-4280}$, or Tβ4 upon mixing them with HOTAIR$_{440}$, even at significantly higher protein concentrations than MLL4$_{3500-3630}$. Also, MLL4$_{3500-3630}$ did not show aggregation-prone behavior in the absence of RNA.

As we experienced no anomaly in the behavior of MLL4$_{3500-3630}$ when titrated to MEG3, determination of a binding constant was straightforward for this interaction. As shown in Figure 2B, affinity to MEG3 of this region of MLL4 was higher than that of MLL4$_{4210-4280}$. The Kd of MLL4$_{3500-3630}$ binding to MEG3 was calculated to be 0.722 μM, while Kd calculation for MLL4$_{4210-4280}$ was not reliable since saturation of the reaction could not be reached throughout the protein concentration range tested. Tβ4 did not show significant affinity to MEG3, resulting in a failure of binding curve fitting.

To check for any specificity of binding that the expressed MLL4 regions may possess, we also tested a physiologically non-relevant 50 nt RNA construct. Binding curves presented in Figure 2C indicate that both MLL4$_{3500-3630}$ and MLL4$_{4210-4280}$ are capable of binding to this RNA species, but with a remarkably lower affinity than to the lncRNA constructs, while Tβ4 could not bind to it at all. The extended shape of the binding curve and the absence of saturation in the case of both MLL4 constructs indicate weak binding that resulted in an inability to reliably determine the binding constants. Nevertheless, MLL4$_{3500-3630}$ still displayed a stronger affinity towards the RNA than MLL4$_{4210-4280}$.

Electrophoretic Mobility Shift Assay (EMSA) experiments confirmed the findings of the MST measurements (Figure 3) as both MLL4 regions caused a significant change in RNA mobility in the case of HOTAIR$_{440}$ and MEG3 (Figure 3A,B) RNAs. This shift was drastically less pronounced with the 50 nt RNA sample (Figure 3C), resulting only in a minor weakening of the RNA signal in the lane with the highest protein concentration. This observation corresponds to the outcome of the MST experiments, underlining the existence of a certain level of specificity in the RNA recognition by these two MLL4 regions. The negative control Tβ4 failed to cause any visible change in the RNA mobility, indicating a lack of interaction with any of the tested RNAs. Competitive RNA binding (Figure 3, compare the 3rd and 5th lanes) demonstrated that the observed shift in mobility was indeed a result of RNA-protein interaction, since the shift could be prevented at least to some extent by adding excess unlabeled RNA to the reaction mixtures.

The anomalous behavior of the MLL4$_{3500-3630}$:HOTAIR$_{440}$ interaction observed in MST was seen in the EMSA experiments as well, since at high protein:RNA ratios the samples obtained a highly viscous quality and completely remained in the wells during the electrophoretic run. Successful experiments could only be carried out by lowering the applied protein concentration, but the interaction was clearly observable even under these circumstances.

In all of the tested interactions, MLL4$_{3500-3630}$, which contains a predicted RNA binding region presented higher affinities to RNAs than the other MLL4 segment, indicating the validity of the prediction. On the other hand, binding of MLL4$_{4210-4280}$ could also be detected in all cases, raising the possibility of the existence of RNA binding sequences differing from the already described interaction motifs. EZH2, a known RNA binding HKTM also interacts with RNAs through a region [17] that has no recognizable RNA binding sequence, emphasizing our lack of complete knowledge of the sequential determinants of protein-RNA interactions.

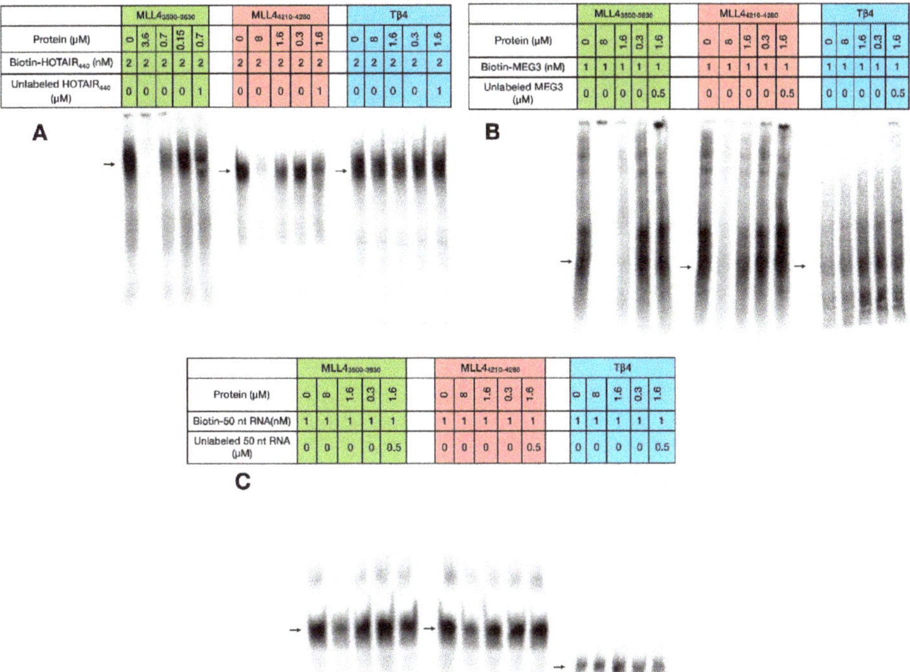

Figure 3. Electrophoretic Mobility Shift Assay. Interaction of MLL4$_{3500-3630}$, MLL4$_{4210-4280}$ and Tβ4 with HOTAIR$_{440}$ (**A**), MEG3 (**B**) and 50 nt RNA (**C**). For easier understanding, the coloring scheme of Figure 2 is followed (MLL4$_{3500-3630}$: green MLL4$_{4210-4280}$: red, Tβ4: blue). Free RNA is indicated by arrows.

3. Discussion

Histone methylation is one of the most studied and best-characterized histone modifications that drive the regulation of complete genetic programs in the cells. However, many details of the regulation and targeting of the enzyme complexes mediating histone methylation remain elusive and a subject of debate [23]. One possible regulatory pathway is represented by the ability of certain HKMT complexes to bind different lncRNAs that serve as a targeting platform, bridging transcription factors and HKMT complexes [20,33] at the promoter regions of target genes. PRC2 is one example where it was shown by multiple experiments that it's binding to different lncRNAs results in different physiological outcomes [34]. lncRNAs are involved in many other processes connected to histone modification and there are examples in the literature of direct interaction between lncRNAs and histone modifier complexes [4,22]. Experimental evidence supports the direct binding of WDR5, a canonical MLL complex subunit, to different lncRNAs in cells [22] indicating the involvement of lncRNAs in the regulation of MLL complexes. Taken the analogy of the PRC2, where multiple subunits are shown to be involved in lncRNA binding (Figure 4A) [15], we hypothesized that MLL proteins might also interact with lncRNAs. This hypothesis was supported by our earlier bioinformatics studies that suggested the existence of several interaction sites in the so far uncharacterized, mostly disordered regions of HKMTs [26] and our prediction presented here that the disordered segments of MLL proteins contain several putative RNA binding sequences. We chose to test the RNA binding capability of one such region of MLL4 that also contains a polyQ stretch and is affected by mutations in different cancers. As an internal control, we also tested a different region of MLL4 that contains no such predicted RNA interaction site.

Our expectation was that the isolated small regions of the MLL4 protein would bind RNAs in a nonspecific manner, such as was observed for the isolated PRC2 complex components [34]. Surprisingly, we found that $MLL4_{4210-4280}$ bound MEG3 stronger than $HOTAIR_{440}$ or the 50 nt random RNA, even though the determination of the exact Kd-s was not successful in all cases.

More interesting was the behavior of the $MLL4_{3500-3630}$ region that showed dramatically different behavior with the different RNAs. Binding to MEG3 gave a Kd of 0.722 µM, while the binding to the 50 nt random RNAs proved to be so weak that a Kd calculation was not successful. Binding to $HOTAIR_{440}$ seemed to be the strongest with an apparent Kd of 0.1 µM, but it led to the aggregation of the protein-RNA complex. The aggregation was dependent on protein-RNA ratio and could be detected through a wide protein concentration range. The same aggregation could not be observed with a shorter HOTAIR construct that consisted of 300 bases (Supplementary Figure S3). The fact that we could not induce such aggregation by the addition of MEG3, which is much longer than $HOTAIR_{440}$, points to specific recognition rather than a side-effect of RNA length. We also observed the aggregation at low protein concentrations, but only in the presence of an appropriate amount of $HOTAIR_{440}$, indicating that the process is not driven by the protein in itself and is not a derivative of sample preparation errors.

It has been recently revealed that many proteins can go through liquid-liquid phase separation when interacting with RNAs, leading to the formation of membraneless organelles that have a significant importance in cellular processes [35]. Experimental evidence supports the involvement of polyQ regions of proteins in the RNA mediated phase separation [28], sometimes in an RNA secondary structure-dependent manner [36]. Since $MLL4_{3500-3630}$ sequence contains 22.9% glutamine residues and a continuous run of 15 glutamines (Figure 1A), it is not unfounded to speculate that this specific region plays a role in the observed anomaly but the fact that it only occurs with one of the tested RNA constructs, indicates that the process is coordinated by the RNA itself. One possibility is that the longer HOTAIR construct contains more than one binding sites for $MLL4_{3500-3630}$, thus facilitating the formation of higher order protein-RNA structures. Alternatively, $HOTAIR_{440}$ may have the ability to form secondary structures not found in $HOTAIR_{300}$ or MEG3, which would also provide an explanation for the different behavior of the three systems. As MLL4 is the only HKMT that contains long polyglutamine repeat stretches [26], phase separation might be a regulatory step specific for this protein. Therefore, it is certainly promising to investigate this peculiar phenomenon in more detail.

Since both tested lncRNAs are implicated in different cancers [5,37,38] involving leukemias, our finding that MLL4 has a capacity to bind them raises the possibility that lncRNAs play a role in MLL/COMPASS complex targeting and regulation to a larger extent than currently recognized.

Although cellular experiments are necessary to prove the validity of the observed interactions, our findings provide the first insights into the structure and function of two regions of MLL4 that have been uncharacterized so far. We were able to show that these regions are capable of RNA binding and may be involved in the lncRNA mediated regulation of the MLL4 complexes. Based on our results, we suggest that and MLL4 complexes utilize different regions on their surface to bind lncRNAs (Figure 4B), similarly to the way PRC2 subunits take part in lncRNA binding. As it was shown that lncRNA binding to WDR5 increases the dwelling time of the protein on the chromatin surface [22], binding of the same RNA to MLL4 might facilitate and accelerate the assembly of a functional methyltransferase complex. Since lncRNAs are large molecules that can adopt various secondary structures and interact with many different partners simultaneously, it is plausible to speculate that a specific and high-affinity interaction can be achieved by the combination of different binding sites distributed along the large surfaces of multi-subunit complexes. Given the central role of histone modifications in gene regulation, it is essential to understand the mechanisms that regulate this process. Mounting evidence supports the involvement of lncRNAs in the coordination of histone modifying enzymes but the exact molecular details of their interactions with proteins are yet to be discovered. Recognizing the importance of the disordered/structurally uncharacterized regions of HKMTs in these interactions might be the first step towards a more complete picture regarding the regulation of histone methylation.

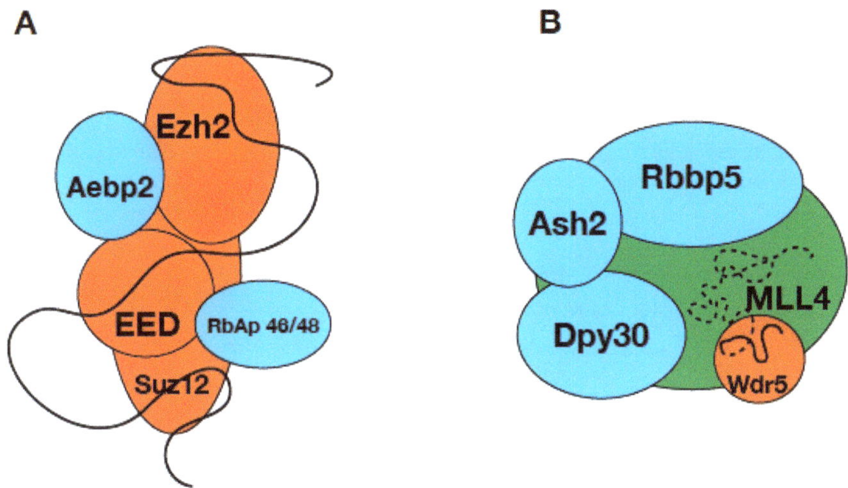

Figure 4. lncRNA binding of PRC2 and MLL4/COMPASS complex. Schematic representation of the PRC2 (**A**) and MLL4/COMPASS (**B**) complexes, where the known RNA binding subunits are shown in orange and the suggested lncRNA binding subunit MLL4 is green. Subunits currently not known to be involved in lncRNA binding are blue and the lncRNA is represented by a black line. Suggested lncRNA-MLL4 interaction is indicated by dashed line.

4. Materials and Methods

4.1. Bioinformatics Analysis

Disorder and disordered binding site predictions were performed with the IUPred2A online prediction tool (https://iupred2a.elte.hu/) [39] which incorporates the IUPred and Anchor predictors. RNA binding regions located in disordered regions were predicted using the DisoRDPbind tool (http://biomine.cs.vcu.edu/servers/DisoRDPbind/) [40]. Cancer-related single nucleotide polymorphisms in the long conserved IDR regions were collected from the BioMuta v2.0 [41] and COSMIC databases [42].

4.2. Accession Numbers

HOTAIR: Gene ID: 100124700
MEG3: Gene ID: 55384
MLL1: Uniprot: Q03164
MLL2: Uniprot: Q9UMN6
MLL3: Uniprot: Q8NEZ4
MLL4: Uniprot: O14686

4.3. Overexpression and Purification of MLL4 Protein Regions

The same methods of protein overexpression and purification were used for both protein constructs, MLL4$_{3500-3630}$ and MLL4$_{4210-4280}$. DNA sequences coding for each protein were cloned into pET22b cloning vector. Induction was done for 4 h at 28 °C by 0.1 M IPTG, cells were pelleted by centrifugation (4000 rpm, 20 min, 4 °C) then lysed by sonication in lysis buffer (50 mM Tris, 200 mM NaCl, 0.5% Triton X-100 pH 8.0 and EDTA-free SIGMAFAST Protease

Inhibitor Cocktail Tablets), cell debris was removed by centrifugation (12,100 rpm, 40 min, 4 °C). The supernatant was filtered through 0.2 μm nitrocellulose filter then purified over HisTrap HP column on an AKTA Explorer system using a gradient elution of two buffers (Buffer A: 20 mM imidazole, 200 mM NaCl, 20 mM Tris. pH 7.5. Buffer B: 1 M imidazole, 200 mM NaCl, 20 mM Tris, pH 7.5). Representative purification results are shown on Supplementary Figure S7. The mostly disordered nature of the $MLL4_{4210-4280}$ region was highlighted by its appearance at a larger size than its actual molecular weight (17 kDa vs. 7 kDa). Elution fractions containing sufficiently pure proteins were dialyzed against distilled water then lyophilized and stored at −20 °C. Lyophilized proteins were dissolved before use in ultrapure water or the appropriate assay buffer. The identity of the purified proteins was confirmed by mass spectrometry.

4.4. RNA Preparation

HOX transcript antisense RNA (HOTAIR):

$HOTAIR_{300}$ (140–440 nt) and $HOTAIR_{440}$ (1–440 nt) DNA sequences cloned into pEX-A128 vector were purchased from Eurofins Genomics (Ebersberg, Germany). After 2 h digestion with EcoRV restriction enzyme at 37 °C, the gel-purified, linearized DNA templates were used to synthesize RNA by in-vitro transcription.

Maternally Expressed 3 (MEG3) lncRNA:

pCI-MEG3 was a gift from Anne Klibanski (Addgene plasmid #44727, Watertown, MA, USA) [43]. Primers to obtain the DNA template for in vitro transcription were as follows:
T7 RNA promoter region followed by:
MEG3 forward primer:
TAATACGACTCACTATAGGGGCAGAGAGGGAGCGCGCCTTGG
MEG3 reverse primer:
GATATCTTTTTGTTAAGACAGGAAACACATTTATTGAGAGC

50 nt RNA:

50 nucleotide RNA was an artificial randomized RNA sequence.
DNA templates were T7 promoter region followed by:
50 nt forward oligo:
TAATACGACTCACTATAGAAGAATGGCCTCGCGGAGGCATGCGTCATGCTAGCGTGCGGG GTACTCTT and
50 nt reverse oligo:
AAGAGTACCCCGCACGCTAGCATGACGCATGCCTCCGCGAGGCCATTCTTCTATAGTGAG TCGTATTA
Transcribed RNA:
GAAGAAUGGCCUCGCGGAGGCAUGCGUCAUGCUAGCGUGCGGGGUACUCUU
All primers and oligonucleotides were purchased from Sigma-Aldrich Ltd. (St. Louis, MO, USA).
Tested RNAs were synthesized by in vitro transcription carried out with New England BioLabs HiScribe™ T7 Quick High Yield RNA Synthesis Kit (Ipswich, MA, USA). Fluorescein-labelled, single-stranded RNA probes were generated by using Roche (Basel, Switzerland) Fluorescein RNA Labeling Mix (11685619910) and NEB 10× T7 reaction buffer (#B2041A). After transcription, remaining DNA templates were eliminated with DNaseI treatment. RNA sample purification was carried out using Macherey-Nagel NucleoSpin® RNA Clean-up XS Kit (Düren, Germany). The quality and intactness of the purified transcription products were analysed by native and formaldehyde agarose gel electrophoresis.

Biotinylation of the RNAs was performed using Pierce™ RNA 3′ End Biotinylation Kit (Cat. Number 20160, Thermo Fisher Scientific, Waltham, UK) according to the instructions of the manufacturer. Overnight incubation at 16 °C was applied for the ligation of the biotin label. Final RNA concentrations were determined using NanoDrop™ 1000 Spectrophotometer (Thermo Fisher Scientific, Waltham, UK).

Purified RNAs were stored −80 °C until usage in the presence of RNAINH-RO Roche Protector RNase Inhibitor (20U).

4.5. Far-UV CD Measurements

CD measurements were performed in quartz cells of 0.1 mm pathlengths using a Jasco J-810 (Jasco, Tokyo, Japan) spectropolarimeter. Far-UV CD spectra were recorded in the range of 180–260 nm with a scanning speed of 20 nm/min, bandwidth of 1 nm and integration time of 4 s. 6 scans were accumulated. Thermal denaturation was recorded in a 1 mm cell at 220 nm from 10 to 100 °C with scanning rate of 120 °C/h. The temperature was controlled using a PTE Peltier unit. The thermal denaturation profile was fitted according to the Gibbs-Helmholtz equation assuming a two-state model, which is represented by a sigmoidal curve [44]. CD spectra were quantitatively analyzed by the BeStSel method [29,30] (http://bestsel.elte.hu).

4.6. Microscale Thermophoresis

RNA-protein binding assays were carried out on a Microscale Thermophoresis system (Monolith NT. 115 from NanoTemper Technologies, München, Germany). Standard treated capillaries (Cat. Number: MO-K002) were used for measurements. Instrument settings are presented in Table 2.

Table 2. Instrument settings for MST.

Title	LED Power (%)	MST Power (%)	Before MST (s)	MST on (s)	After MST (s)	Delay (s)
Round 1	10 or 40	20	5	30	5	25
Round 2	10 or 40	40	5	30	5	25

Normalized fluorescence values after 1.25 s after turning on the IR laser were used as T-jump values.

RNA concentrations were set to give an initial raw fluorescence between 300 and 1000 counts and varied between 30 and 100 nM. All experiments were done at room temperature. DEPC-treated PBS buffer containing 0.05% NP-40 was used as assay buffer.

4.7. Electrophoretic Mobility Shift Assay (EMSA)

LightShift® Chemiluminescent RNA EMSA Kit (Thermo Scientific, Cat. No. 20158, Thermo Fisher Scientific, Waltham, UK) was used for the EMSA experiments. Assay control was performed according to the instructions of the manufacturer with the control reaction provided with the kit. In short, 6.25 nM biotin-labeled IRE RNA was incubated with 2 µg of cytosolic liver extract with or without 1 µM of unlabeled IRE RNA. The result of the assay control is presented on Supplementary Figure S6. Binding, electrophoresis and detection of the tested RNAs with the proteins were carried out following the protocol of the kit. Briefly, proteins of varying concentrations were incubated with 1 or 2 nM of RNAs for 30 min at room temperature, then loaded on 4 or 6% native polyacrylamide gels. RNA was transferred to nitrocellulose membranes using Trans-Blot® Turbo™ Transfer System (Bio-Rad, Hercules, CA, USA) and crosslinked to the membrane by UV-light crosslinking. After proper washing and blocking, biotin labeled RNA was detected by chemiluminescence using Streptavidin-Horseradish Peroxidase Conjugate.

Supplementary Materials: The following are available online at http://www.mdpi.com/1422-0067/19/11/3478/s1.

Author Contributions: Conceptualization, A.T.; Data curation, B.S. (Beáta Szabó), E.S. and B.S. (Bálint Szeder); Formal analysis, J.K.; Funding acquisition, J.K., L.B. and A.T.; Investigation, B.S. (Beáta Szabó), N.M., R.A. and J.K.; Methodology, B.S. (Beáta Szabó) and A.T.; Project administration, B.S. (Beáta Szabó) and A.T.; Resources, B.S. (Beáta Szabó) and N.M.; Software, E.S.; Supervision, L.B. and A.T.; Validation, J.K. and L.B.;

Visualization, A.T.; Writing—original draft, A.T.; Writing—review & editing, B.S. (Beáta Szabó), N.M., R.A., E.S., J.K., B.S. (Bálint Szeder) and L.B.

Funding: This research was funded by the National Research, Development and Innovation Office, Hungary (grants: K-125340 (A.T.), K-120391 (J.K.), KH-125597 (J.K. and A.T.), TÉT_16-1-2016-0134 (J.K.), 2017-1.2.1-NKP-2017-00002 (J.K.), Medinprot synergy grant (A.T. and J.K.)) and a Korea-Pan-European Research Agreement (A.T.).

Conflicts of Interest: The authors declare no conflict of interest. The founding sponsors had no role in the design of the study; in the collection, analyses, or interpretation of data; in the writing of the manuscript, and in the decision to publish the results.

Abbreviations

EMSA	Electrophoretic mobility shift assay
EZH2	Enhancer of zeste homolog 2
HKMT	Histone lysin methyltransferase
HOTAIR	HOX transcript antisense RNA
lncRNA	Long non-coding RNA
MEG3	Maternally Expressed 3
MLL	Mixed lineage leukemia
MST	Microscale thermophoresis
PRC2	Polycomb repressive complex
WDR5	WD repeat-containing protein 5

References

1. Iyer, M.K.; Niknafs, Y.S.; Malik, R.; Singhal, U.; Sahu, A.; Hosono, Y.; Barrette, T.R.; Prensner, J.R.; Evans, J.R.; Zhao, S.; et al. The landscape of long noncoding RNAs in the human transcriptome. *Nat Genet.* **2015**, *47*, 199–208. [CrossRef] [PubMed]

2. Bartonicek, N.; Maag, J.L.V.; Dinger, M.E. Long noncoding RNAs in cancer: Mechanisms of action and technological advancements. *Mol. Cancer* **2016**, *15*, 43. [CrossRef] [PubMed]

3. Yang, F.; Deng, X.; Ma, W.; Berletch, J.B.; Rabaia, N.; Wei, G.; Moore, J.M.; Filippova, G.N.; Xu, J.; Liu, Y.; et al. The lncRNA Firre anchors the inactive X chromosome to the nucleolus by binding CTCF and maintains H3K27me3 methylation. *Genome Biol.* **2015**, *16*, 52. [CrossRef] [PubMed]

4. Khalil, A.M.; Guttman, M.; Huarte, M.; Garber, M.; Raj, A.; Morales, D.R.; Thomas, K.; Presser, A.; Bernstein, B.E.; Van Oudenaarden, A.; et al. Many human large intergenic noncoding RNAs associate with chromatin-modifying complexes and affect gene expression. *Proc. Natl. Acad. Sci. USA* **2009**, *106*, 11667–11672. [CrossRef] [PubMed]

5. Heery, R.; Finn, S.P.; Cuffe, S.; Gray, S.G. Long Non-Coding RNAs: Key Regulators of Epithelial-Mesenchymal Transition, Tumour Drug Resistance and Cancer Stem Cells. *Cancers* **2017**, *9*, 38. [CrossRef] [PubMed]

6. Liu, Z.; Chen, Z.; Fan, R.; Jiang, B.; Chen, X.; Chen, Q.; Nie, F.; Lu, K.; Sun, M. Over-expressed long noncoding RNA HOXA11-AS promotes cell cycle progression and metastasis in gastric cancer. *Mol. Cancer* **2017**, *16*, 82. [CrossRef] [PubMed]

7. Bhan, A.; Deb, P.; Shihabeddin, N.; Ansari, K.I.; Brotto, M.; Mandal, S.S. Histone methylase MLL1 coordinates with HIF and regulate lncRNA HOTAIR expression under hypoxia. *Gene* **2017**, *629*, 16–28. [CrossRef] [PubMed]

8. Chen, S.; Liang, H.; Yang, H.; Zhou, K.; Xu, L.; Liu, J.; Lai, B.; Song, L.; Luo, H.; Peng, J.; et al. Long non-coding RNAs: The novel diagnostic biomarkers for leukemia. *Environ. Toxicol. Pharmacol.* **2017**, *55*, 81–86. [CrossRef] [PubMed]

9. Guo, G.; Kang, Q.; Zhu, X.; Chen, Q.; Wang, X.; Chen, Y.; Ouyang, J.; Zhang, L.; Tan, H.; Chen, R.; et al. A long noncoding RNA critically regulates Bcr-Abl-mediated cellular transformation by acting as a competitive endogenous RNA. *Oncogene* **2015**, *34*, 1768–1779. [CrossRef] [PubMed]

10. Wang, P.; Ren, Z.; Sun, P. Overexpression of the long non-coding RNA MEG3 impairs in vitro glioma cell proliferation. *J. Cell. Biochem.* **2012**, *113*, 1868–1874. [CrossRef] [PubMed]

11. Zhang, L.; Xu, H.-G.; Lu, C. A novel long non-coding RNA T-ALL-R-LncR1 knockdown and Par-4 cooperate to induce cellular apoptosis in T-cell acute lymphoblastic leukemia cells. *Leuk. Lymphoma* **2014**, *55*, 1373–1382. [CrossRef] [PubMed]

12. Zhou, X.; Yuan, P.; Liu, Q.; Liu, Z. LncRNA MEG3 Regulates Imatinib Resistance in Chronic Myeloid Leukemia via Suppressing MicroRNA-21. *Biomol. Ther.* **2017**, *25*, 490–496. [CrossRef] [PubMed]

13. Mer, A.S.; Lindberg, J.; Nilsson, C.; Klevebring, D.; Wang, M.; Grönberg, H.; Lehmann, S.; Rantalainen, M. Expression levels of long non-coding RNAs are prognostic for AML outcome. *J. Hematol. Oncol.* **2018**, *11*, 52. [CrossRef] [PubMed]

14. Davidovich, C.; Cech, T.R. The recruitment of chromatin modifiers by long noncoding RNAs: Lessons from PRC2. *RNA* **2015**, *21*, 2007–2022. [CrossRef] [PubMed]

15. Cifuentes-Rojas, C.; Hernandez, A.J.; Sarma, K.; Lee, J.T. Regulatory interactions between RNA and polycomb repressive complex 2. *Mol. Cell* **2014**, *55*, 171–185. [CrossRef] [PubMed]

16. Davidovich, C.; Wang, X.; Cifuentes-Rojas, C.; Goodrich, K.J.; Gooding, A.R.; Lee, J.T.; Cech, T.R. Toward a Consensus on the Binding Specificity and Promiscuity of PRC2 for RNA. *Mol. Cell* **2015**, *57*, 552–558. [CrossRef] [PubMed]

17. Kaneko, S.; Li, G.; Son, J.; Xu, C.F.; Margueron, R.; Neubert, T.A.; Reinberg, D. Phosphorylation of the PRC2 component Ezh2 is cell cycle-regulated and up-regulates its binding to ncRNA. *Genes Dev.* **2010**, *24*, 2615–2620. [CrossRef] [PubMed]

18. Zhang, X.D.; Huang, G.W.; Xie, Y.H.; He, J.Z.; Guo, J.C.; Xu, X.E.; Liao, L.D.; Xie, Y.M.; Song, Y.M.; Li, E.M.; et al. The interaction of lncRNA EZR-AS1 with SMYD3 maintains overexpression of EZR in ESCC cells. *Nucleic Acids Res.* **2018**, *46*, 1793–1809. [CrossRef] [PubMed]

19. O'Leary, V.B.; Ovsepian, S.V.; Carrascosa, L.G.; Buske, F.A.; Radulovic, V.; Niyazi, M.; Moertl, S.; Trau, M.; Atkinson, M.J.; Anastasov, N. PARTICLE, a Triplex-Forming Long ncRNA, Regulates Locus-Specific Methylation in Response to Low-Dose Irradiation. *Cell Rep.* **2015**, *11*, 474–485. [CrossRef] [PubMed]

20. O'Leary, V.B.; Hain, S.; Maugg, D.; Smida, J.; Azimzadeh, O.; Tapio, S.; Ovsepian, S.V.; Atkinson, M.J. Long non-coding RNA PARTICLE bridges histone and DNA methylation. *Sci. Rep.* **2017**, *7*, 1790. [CrossRef] [PubMed]

21. Gomez, J.A.; Wapinski, O.L.; Yang, Y.W.; Bureau, J.F.; Gopinath, S.; Monack, D.M.; Chang, H.Y.; Brahic, M.; Kirkegaard, K. The NeST Long ncRNA Controls Microbial Susceptibility and Epigenetic Activation of the Interferon-γ Locus. *Cell* **2013**, *152*, 743–754. [CrossRef] [PubMed]

22. Yang, Y.W.; Flynn, R.A.; Chen, Y.; Qu, K.; Wan, B.; Wang, K.C.; Lei, M.; Chang, H.Y. Essential role of lncRNA binding for WDR5 maintenance of active chromatin and embryonic stem cell pluripotency. *Elife* **2014**, *3*, e02046. [CrossRef] [PubMed]

23. Sze, C.C.; Shilatifard, A. MLL3/MLL4/COMPASS Family on Epigenetic Regulation of Enhancer Function and Cancer. *Cold Spring Harb. Perspect. Med.* **2016**, *6*, a026427. [CrossRef] [PubMed]

24. Herz, H.M.; Mohan, M.; Garruss, A.S.; Liang, K.; Takahashi, Y.H.; Mickey, K.; Voets, O.; Verrijzer, C.P.; Shilatifard, A. Enhancer-associated H3K4 monomethylation by Trithorax-related, the Drosophila homolog of mammalian Mll3/Mll4. *Genes Dev.* **2012**, *26*, 2604–2620. [CrossRef] [PubMed]

25. Henikoff, S.; Shilatifard, A. Histone modification: Cause or cog? *Trends Genet.* **2011**, *27*, 389–396. [CrossRef] [PubMed]

26. Lazar, T.; Schad, E.; Szabo, B.; Horvath, T.; Meszaros, A.; Tompa, P.; Tantos, A. Intrinsic protein disorder in histone lysine methylation. *Biol. Dir.* **2016**, *11*, 30. [CrossRef] [PubMed]

27. Mészáros, B.; Simon, I.; Dosztányi, Z. Prediction of Protein Binding Regions in Disordered Proteins. *PLoS Comput. Biol.* **2009**, *5*, e1000376. [CrossRef] [PubMed]

28. Zhang, H.; Elbaum-Garfinkle, S.; Langdon, E.M.; Taylor, N.; Occhipinti, P.; Bridges, A.A.; Brangwynne, C.P.; Gladfelter, A.S. RNA Controls PolyQ Protein Phase Transitions. *Mol. Cell* **2015**, *60*, 220–230. [CrossRef] [PubMed]

29. Micsonai, A.; Wien, F.; Bulyáki, É.; Kun, J.; Moussong, É.; Lee, Y.H.; Goto, Y.; Réfrégiers, M.; Kardos, J. BeStSel: A web server for accurate protein secondary structure prediction and fold recognition from the circular dichroism spectra. *Nucleic Acids Res.* **2018**, *46*, W315–W322. [CrossRef] [PubMed]

30. Micsonai, A.; Wien, F.; Kernya, L.; Lee, Y.H.; Goto, Y.; Réfrégiers, M.; Kardos, J. Accurate secondary structure prediction and fold recognition for circular dichroism spectroscopy. *Proc. Natl. Acad. Sci. USA* **2015**, *112*, E3095–E3103. [CrossRef] [PubMed]

31. Tsai, M.C.; Manor, O.; Wan, Y.; Mosammaparast, N.; Wang, J.K.; Lan, F.; Shi, Y.; Segal, E.; Chang, H.Y. Long noncoding RNA as modular scaffold of histone modification complexes. *Science* **2010**, *329*, 689–693. [CrossRef] [PubMed]

32. Schwarzer, A.; Emmrich, S.; Schmidt, F.; Beck, D.; Ng, M.; Reimer, C.; Adams, F.F.; Grasedieck, S.; Witte, D.; Käbler, S.; et al. The non-coding RNA landscape of human hematopoiesis and leukemia. *Nat. Commun.* **2017**, *8*, 218. [CrossRef] [PubMed]

33. Battistelli, C.; Cicchini, C.; Santangelo, L.; Tramontano, A.; Grassi, L.; Gonzalez, F.J.; de Nonno, V.; Grassi, G.; Amicone, L.; Tripodi, M. The Snail repressor recruits EZH2 to specific genomic sites through the enrollment of the lncRNA HOTAIR in epithelial-to-mesenchymal transition. *Oncogene* **2016**, *36*, 942–955. [CrossRef] [PubMed]

34. Davidovich, C.; Zheng, L.; Goodrich, K.J.; Cech, T.R. Promiscuous RNA binding by Polycomb repressive complex 2. *Nat. Struct. Mol. Biol.* **2013**, *20*, 1250–1257. [CrossRef] [PubMed]

35. Banani, S.F.; Lee, H.O.; Hyman, A.A.; Rosen, M.K. Biomolecular condensates: Organizers of cellular biochemistry. *Nat. Rev. Mol. Cell Biol.* **2017**, *18*, 285–298. [CrossRef] [PubMed]

36. Langdon, E.M.; Qiu, Y.; Niaki, A.G.; McLaughlin, G.A.; Weidmann, C.; Gerbich, T.M.; Smith, J.A.; Crutchley, J.M.; Termini, C.M.; Weeks, K.M.; et al. mRNA structure determines specificity of a polyQ-driven phase separation. *Science* **2018**, *360*, 922–927. [CrossRef] [PubMed]

37. Li, Z.-Y.; Yang, L.; Liu, X.-J.; Wang, X.-Z.; Pan, Y.-X.; Luo, J.-M. The Long Noncoding RNA MEG3 and its Target miR-147 Regulate JAK/STAT Pathway in Advanced Chronic Myeloid Leukemia. *EBioMedicine* **2018**, *34*, 61–75. [CrossRef] [PubMed]

38. Zhang, Y.-Y.; Huang, S.-H.; Zhou, H.-R.; Chen, C.-J.; Tian, L.-H.; Shen, J.-Z. Role of HOTAIR in the diagnosis and prognosis of acute leukemia. *Oncol. Rep.* **2016**, *36*, 3113–3122. [CrossRef] [PubMed]

39. Mészáros, B.; Erdos, G.; Dosztányi, Z. IUPred2A: Context-dependent prediction of protein disorder as a function of redox state and protein binding. *Nucleic Acids Res.* **2018**, *46*, W329–W337. [CrossRef] [PubMed]

40. Peng, Z.; Kurgan, L. High-throughput prediction of RNA, DNA and protein binding regions mediated by intrinsic disorder. *Nucleic Acids Res.* **2015**, *43*, e121. [CrossRef] [PubMed]

41. Wu, T.J.; Shamsaddini, A.; Pan, Y.; Smith, K.; Crichton, D.J.; Simonyan, V.; Mazumder, R. A framework for organizing cancer-related variations from existing databases, publications and NGS data using a High-performance Integrated Virtual Environment (HIVE). *Database* **2014**, *2014*, bau022. [CrossRef] [PubMed]

42. Forbes, S.A.; Beare, D.; Gunasekaran, P.; Leung, K.; Bindal, N.; Boutselakis, H.; Ding, M.; Bamford, S.; Cole, C.; Ward, S.; et al. COSMIC: Exploring the world's knowledge of somatic mutations in human cancer. *Nucleic Acids Res.* **2014**, *43*, D805–D811. [CrossRef] [PubMed]

43. Zhou, Y.; Zhong, Y.; Wang, Y.; Zhang, X.; Batista, D.L.; Gejman, R.; Ansell, P.J.; Zhao, J.; Weng, C.; Klibanski, A. Activation of p53 by MEG3 non-coding RNA. *J. Biol. Chem.* **2007**, *282*, 24731–24742. [CrossRef] [PubMed]

44. Shih, P.; Holland, D.R.; Kirsch, J.F. Thermal stability determinants of chicken egg-white lysozyme core mutants: Hydrophobicity, packing volume, and conserved buried water molecules. *Protein Sci.* **1995**, *4*, 2050–2062. [CrossRef] [PubMed]

© 2018 by the authors. Licensee MDPI, Basel, Switzerland. This article is an open access article distributed under the terms and conditions of the Creative Commons Attribution (CC BY) license (http://creativecommons.org/licenses/by/4.0/).

International Journal of
Molecular Sciences

Article

The Melting Diagram of Protein Solutions and Its Thermodynamic Interpretation

Kálmán Tompa [1], Mónika Bokor [1] and Péter Tompa [2,3,*]

[1] Institute for Solid State Physics and Optics, Wigner RCP of the HAS, Konkoly-Thege út 29-33,
     H-1121 Budapest, Hungary; tompa.kalman@wigner.mta.hu (K.T.); bokor.monika@wigner.mta.hu (M.B.)
[2] Institute of Enzymology, Research Centre for Natural Sciences of the HAS, Magyar Tudósok körút. 27,
     H-1117 Budapest, Hungary
[3] VIB Center for Structural Biology, Vrije Universiteit Brussel, Building E, Pleinlaan 2, 1050 Brussel, Belgium
* Correspondence: peter.tompa@vub.be; Tel.: +32-2-629-19-62

Received: 25 August 2018; Accepted: 30 October 2018; Published: 12 November 2018

Abstract: Here we present a novel method for the characterization of the hydration of protein solutions based on measuring and evaluating two-component wide-line ^{1}H NMR signals. We also provide a description of key elements of the procedure conceived for the thermodynamic interpretation of such results. These interdependent experimental and theoretical treatments provide direct experimental insight into the potential energy surface of proteins. The utility of our approach is demonstrated through the examples of two proteins of distinct structural classes: the globular, structured ubiquitin; and the intrinsically disordered ERD10 (early response to dehydration 10). We provide a detailed analysis and interpretation of data recorded earlier by cooling and slowly warming the protein solutions through thermal equilibrium states. We introduce and use order parameters that can be thus derived to characterize the distribution of potential energy barriers inhibiting the movement of water molecules bound to the surface of the protein. Our results enable a quantitative description of the ratio of ordered and disordered parts of proteins, and of the energy relations of protein–water bonds in aqueous solutions of the proteins.

Keywords: protein; hydration; wide-line ^{1}H NMR

---

1. Introduction

Wide-line ^{1}H NMR is an accepted method to delineate the structures of hydrogen-containing molecules determined primarily by X-ray and, to a lesser extent, by neutron-scattering. This way, it can provide information on the location and structural environment of hydrogen atoms in proteins. It has a unique capability, on the other hand, in the direct observation of translational and rotational movements of molecules in the condensed phase.

NMR characteristics of aqueous solutions rapidly frozen and then slowly thawed through equilibrium thermal states provide direct information on the immobile and partially or fully mobile parts of the molecules. We have previously reviewed relevant features of this approach in our works "Hydrogen skeleton, mobility and protein architecture" [1] and "Studying molecular motions in solid states by NMR" [2].

Based on these studies, we state that molecular motions in the sample result in narrowing of the wide-line NMR spectrum. This phenomenon is known as motional narrowing in the literature [3]. Our goal is to advance from this observation to arrive at the thermodynamic characterization of protein systems.

In Figure 1a, we show the typical ^{1}H NMR free-induction decay (FID) signal of a set of spins containing proton–proton pairs of different mobilities. In Figure 1b, we present the deduced NMR spectrum. Similar FID signals and NMR spectra are observed at certain temperatures when studying the aqueous solution of a protein that contains hydrogen pairs.

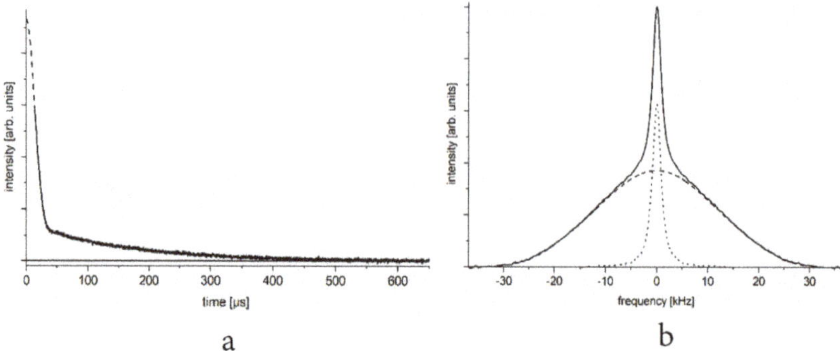

Figure 1. Free induction decay (FID, panel (**a**)) and spectrum (panel (**b**)) of a motionally two-state spin system. (We focus on the slow component of the FID, the initial part of which is lost in the dead time of the spectrometer, marked by dashed line, and can be disregarded).

Time domain (Figure 1a) and frequency or energy domain (Figure 1b) representation of the spectra are linked through Fourier transformation, yet it may be useful to consider both, as they provide information on different practical utilities. The amplitude of the FID signal (even considering its slow component) extrapolated to time zero gives the number of relevant nuclei (spins) through nuclear magnetization. The amplitude of response to the 90° radiofrequency pulse is proportional to the relevant x-y component of nuclear magnetization that is further proportional to $M_0 \approx (nB_0)/T$, in which B_0 is the constant magnetic induction, T is the absolute temperature, and n is the number of resonant nuclei (in our case, it equals the number of protons in water). On the other hand, the width of the spectrum gives direct information on the motional characteristics of proton spin pairs. In a system of two components (e.g., one that contains both mobile and immobile spin-pairs), it is important to have direct information on both parameters.

It is questionable whether such a simple approach can give significant novel information on the dynamics of a complex system, such as a protein and its environment in an aqueous solution. The independent measurement over a broad temperature range of the two parameters of the slow FID component (FID amplitude extrapolated to $t = 0$ and the spectral width) is debatable.

Therefore, here we address the behavior of the slow-FID, and the narrow-spectral component. Our working hypothesis (that we already partially proved) is that the narrow spectral component comes from water molecules bound to the protein, termed bound water-molecules [4]. One may ask a range of relevant questions about their number, their strength of binding to the protein vs. the neighboring water molecules, and about their potential field following molecular changes of the protein, etc. Similar questions can also be asked for the broad-spectrum component, which we have already addressed before [1,2].

In earlier studies [5–8], we have addressed in detail the behavior of globular and intrinsically disordered proteins (IDPs) in aqueous solutions and provided an initial and partial interpretation of experimental observations. As relevant examples, we refer to results with proteins, such as ubiquitin, bovine serum albumin, α-synuclein (and its point mutants), calpastatin, ERD10 (early response to dehydration 10), and lysozyme. Here, we demonstrate our point by focusing on two proteins, ubiquitin (Ubq) and ERD10, as they have been thoroughly studied earlier; one (Ubq) is a globular/structured protein and the other (ERD10) is intrinsically disordered, i.e., they are representatives of these distinct structural classes. We show the temperature dependence of the slowly-decaying component of the FID extrapolated to $t = 0$ (which gives directly the ratio of relevant mobile water protons). We show the observed behavior in the form of a melting diagram (*MD*). In Figure 2, we show the *MD* of three studied systems (bulk water and the aqueous solution of two proteins, ubiquitin and ERD10) in the usual °C scale.

The melting process of inhomogeneous systems (such as the protein solutions we study), basically differs from the first-order phase transition of homogeneous, single component material, such as the melting of ice at a given transition temperature.

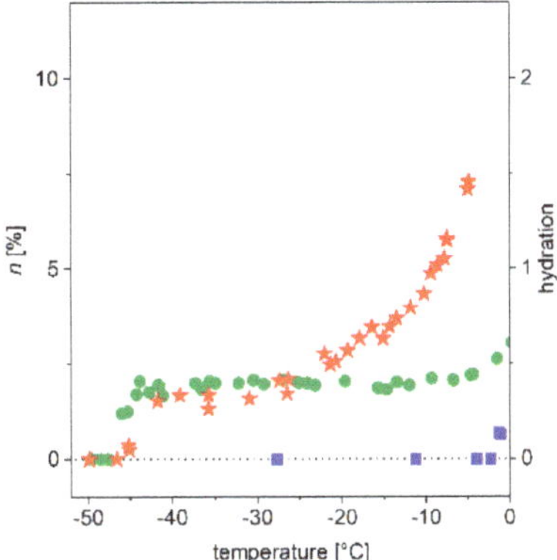

Figure 2. "Old fashioned" melting diagrams, i.e., the total number of mobile water molecules (through protons) normalized to the total number of water molecules, as a function of temperature (blue squares: bulk water, green circles: ubiquitin, red stars: ERD10 proteins in aqueous solutions). The data are given for 50 mg/mL protein concentration.

We consider melting as the process of the beginning of movement of a component of the mixture (such as a bound water-molecule, or a fragment of the protein of high symmetry, e.g., a methyl group or other terminal moiety), in which either translation or rotation begins. In our case, these (individual) events of initial movements show a temperature distribution characteristic of the given molecule, and the derived *MD*s link the well-defined, directly measurable NMR characteristics with atomic/molecular motions.

These characteristics can thus also give direct information on molecular interactions. The water molecules associated with the protein molecule constitute an integral part of the system. Thus, their nuclei, rather than large energy particles applied in scattering techniques (such as X-ray crystallography), monitor the potential energy surface of the protein as built in probes. In our previous works [5–8], however, we only drew qualitative conclusions from the *MD*s.

These were as follows. In aqueous solutions, melting (that is, beginning of molecular motions) of protein-bound water molecules begins at a much lower temperature than the melting of bulk ice. Each protein has a unique *MD* (individual profile or fingerprint) that results from its individual thermodynamic characteristics. The *MD* of globular and ID proteins vastly differ. They can be characterized by temperature-independent FID amplitudes—a plateau (globular protein)—or they can lack a plateau (IDP) or can have a plateau of small temperature extension (partly IDP).

1.1. Energetic Interpretation of Melting Diagrams

We have made significant advances in several respects of interpreting our results [9,10] since we last addressed these questions [5–8]. Key steps are detailed in chapters 4–6 of ref [9]; here, we add a new element and summarize these steps in more detail, following the logical order of the application.

As a reminder, we are following the beginning of the movement—probably the rotation—of water molecules bound to the surface of the protein, by observing motional narrowing in wide-line ^{1}H NMR spectroscopy. For the first time in the field—following the seminal work of Kittel and Kroemer [11]—we introduced the concept of fundamental temperature, T_f, and also introduced here the idea suggested by Waugh and Fedin [12] for connecting the thermal excitation energy, V_0, in which molecular motions begin with the temperature, T, as V_0 = constant $\times$ T.

In some detail, the key steps taken are as follows.

1.1.1. Fundamental Temperature

As a first step, we introduced the use of the scale of fundamental temperature, i.e., thermal excitation energy scale, T_f, and its version normalized to the melting point of ice, T_{fn}. By definition, $T_f = k_B T$, in which $k_B = 1.381 \cdot 10^{-16}$ erg/K ($k_B = 1.381 \cdot 10^{-23}$ J/K) is the Boltzmann constant, and T is the absolute temperature in K. We can also use the equation of $T_f = RT$, in which $R = 8.317$ J/mol·K, the universal gas constant. If we need a dimensionless scale, it is expedient to use the normalized fundamental temperature scale, T_{fn}, nor malized to the melting temperature of bulk water formally as $T_{fn} = k_B \cdot T / (k_B \times 273.15) = T/273.15$. This way, it becomes possible to characterize the events of the beginning of molecular motion on an energy scale.

1.1.2. Energy Scale and the Heterogeneity of the Protein Surface

As a next step, we invoked the formula of Waugh and Fedin, after the improvement of placing it on a fundamental temperature scale of the right dimension. The formula can then be used for aqueous solutions. The equation at atomic/molecular level is

$$E_{0a} \text{ [erg]} = c k_B T \text{ [erg]}, \tag{1a}$$

or applied to molar quantities it is

$$E_{0m} \text{ [kJ/mol]} = cRT \text{ [kJ/mol]}. \tag{1b}$$

In these equations, c is a dimensionless quantity, i.e., a number, the value of which was determined by applying Equation (1b) to the melting of bulk ice, considering the melting heat of ice (6.01 kJ/mol [13]). The fundamental temperature equivalent with 273.15 K is $T_f = RT = 2.272$ kJ/mol. In Equation (1b), the c proportionality constant is 2.65. When comparing Equation (1a) with the energy pertaining to one degree of freedom by the equation of equipartition ($1/2\,k_B T$), we may deduce the degree of freedom of a water molecule as 5.3, which seems to be in the right range for a rotating (and not translating) electric water dipole.

In addition, we introduced dynamic parameters for the quantitative characterization of the ordered/disordered state of protein molecules, which goes beyond their static structural description. Before formalizing the definitions, let us take a look at Figure 2 (and for details, Figures 3 and 4). There is a marked difference between the globular and intrinsically disordered proteins. On the melting diagram of the globular protein Ubq one can see a broad, temperature- (or excitation energy-) independent region (plateau). On the other hand, the plateau of the IDP ERD10 is significantly smaller. (A similar behavior was also seen for other proteins [5–10]). Significantly more information is provided by the initial (T_{fno}) and the ending temperature (T_{fne}) values of the plateau. The region between these two temperatures shows homogeneous bond (potential energy barrier) distribution, whereas the region above T_{fne} shows a heterogeneity in terms of protein–water-bond energy distribution. After this introduction, the following quantities can be defined.

Heterogeneity ratio, *HeR*. According to our observations [5–10] and the literature quoted therein, protein molecules can be characterized and categorized by the homogeneity/heterogeneity of the energy distribution of water binding. The basis of the classification is the measurement of the ratio, for which we suggest the relation

$$HeR = (1 - T_{fne})/(1 - T_{fno}), \tag{2}$$

in which $(1 - T_{\text{fne}})$ and $(1 - T_{\text{fno}})$ give the measured distances from the melting point of ice. These values can be easily read from the novel *MDs*. *HeR* is 1 (one) for systems showing heterogeneous water binding (lacking a plateau) and 0 (zero) for homogeneous binding systems (e.g., bulk water), and is between 0 and 1 for partially heterogeneous systems. *HeR* therefore gives the order parameter type specification for what extent of the surface of the protein molecule can be regarded as showing heterogeneous potential energy distribution (disordered) in terms of water binding. It must be emphasized that this correlation measures the heterogeneity ratio based on the comparison of the extent of the two possible regions and does not measure the number of actual protein–water bonds in them.

1.1.3. An Analytical Description of n

The introduction of fundamental temperature or energy scale makes it possible to describe *MD* by power series in the form

$$n = A + B(T_{\text{fn}} - T_{\text{fn1}}) + C(T_{\text{fn}} - T_{\text{fn2}})^2 + \ldots \tag{3}$$

That is, we can define the total number of water molecules, n, that are moving at a given thermal energy (temperature), as well as the change of *MD* on a normalized fundamental energy scale, i.e., the differential form of melting diagram, *DMD*

$$\Delta n / \Delta T_{\text{fn}} = B + 2C(T_{\text{fn}} - T_{\text{fn2}}) + \ldots , \tag{4}$$

which defines the number of water molecules that begin to move at the given excitation energy. T_{fnx} (with $x = 1, 2, \ldots , n$) is fitting parameter in Equations (3) and (4), in which x is equal to the exponent in each term (in the other terms too, with $n \geq 3$ not given here in detail). The present form of equations calls attention to the validity of any term in a given temperature range. It should be emphasized that all quantities and coefficients are dimensionless in these formulae.

Number of protein–water bonds, HeR_{n}. We can make a statement about the homogeneity/ heterogeneity of bonds (potential barriers) if we ask about the exact number of protein–water bonds in the given excitation energy range. Parameters that fit the power series provide the answer. In the simplest cases (including, in our experience, aqueous solutions with distilled water), in which there is only a wider heterogeneous range in *MD*, the number of water bonds in the heterogeneous region depends on the number of fitting members, $B/(1\ 2212\ T_{\text{fne}})$, and $2C/(1 - T_{\text{fne}})$; if both, then it depends on the sum of the two members. As simplification of the determination of the number of protein–water bonds (the degree of hydration), it can be directly read from the *DMDs*, i.e., the value or the sum of the areas colored in the figures enter (in principle, the definite integrals within the region T_{fne} to $T_{\text{fn}} \approx 1$). Let n_{ho} be the number of water molecules in the first hydrate shell and n_{he} the total number of water molecules in the entire heterogeneous region. In this case, the second relation suggested for the ratio of heterogeneity is

$$HeR_{\text{n}} = n_{\text{he}} / (n_{\text{he}} + n_{\text{ho}}). \tag{5}$$

The value of n_{ho} (approximately) is given by the area of the rectangle at the lowest excitation energy region, whereas n_{he} in our case is given by the areas of triangles (in general, those described by members of higher exponents; see Figures 3 and 4).

The numbers in the equation can be measured directly based on *MD*. n_{ho} can be determined with high accuracy as the average of all n points measured on the plateau, and $(n_{\text{he}} + n_{\text{ho}})$ as an approximate value by the n value reliably measured at a temperature close to the highest temperature, $T_{\text{fn}} \approx 1$. The process has a self-checking potential and thus improves the reliability of the data.

The measure of heterogeneity is *HeM*. We suggested [9] to introduce this as the parameter

$$HeM = (B + 2C)/(1 - T_{fne}).\tag{6}$$

This relationship is also correct in terms of dimensions, and the *HeM* value is generally a positive number. Its value is zero for proteins of almost equipotential molecular surface, so it can be considered as a quasi-order parameter. The denominator, $(1 - T_{fne})$ designates the energy range in which there are varying protein–water bonds (of heterogeneous distribution), and $B + 2C$ (going till the second term of non-zero exponent) is the number of bonds within this range. The fraction is thus a kind of slope of the *MD* function; its values cannot be limited to the range of 0 to 1, just like for the tangent function. Non-heterogeneously binding proteins, such as globular proteins by our experience, have a *HeM* value, by definition, which applies to the region above the plateau. It is not unfounded to suggest that there is a similar dynamic difference in the hydrogen mobility (*HM* [1]) and *HeM* parameters, i.e., in the mobility of all proton–proton pairs, and in the degree of heterogeneity of protein–water bonds.

In the power series of n Equation (3), we only went till the first two members of non-zero exponent (which is enough to interpret the results presented in most of our examples). Heterogeneity and homogeneity can be observed in both the nature and the magnitude of the respective potentials and their distance dependence. Variants of theoretical possibilities are found in the literature [14–17].

The determination of the *MD* function and its differential form that can also be described analytically allows for the unique and individual mapping of the energy distribution of the potential barriers that inhibit the motion of water molecules bound to the protein. Using the elements required for the interpretation of measured *MDs* we have introduced, the purpose of our present work can be easily formulated. Specifically, it is a deeper, thermodynamic interpretation of our results. The examples that illustrate this statement are presented through the analysis of the *MD* of the globular standard protein ubiquitin, and the intrinsically disordered ERD10.

2. Results and Discussion

In Figures 3 and 4, we show the *MDs* determined for the two proteins, dissolved in double-distilled water, with a panel (a) showing measurements on reference water too, and panel (b) the derived curves, *DMDs* (that is $\Delta n / \Delta T_{fn}$ the potential distribution of protein–water bonds). The information on the origin of the samples, the measuring equipment, and the details of the measurements is described in our above-mentioned articles and in book chapters [4,5].

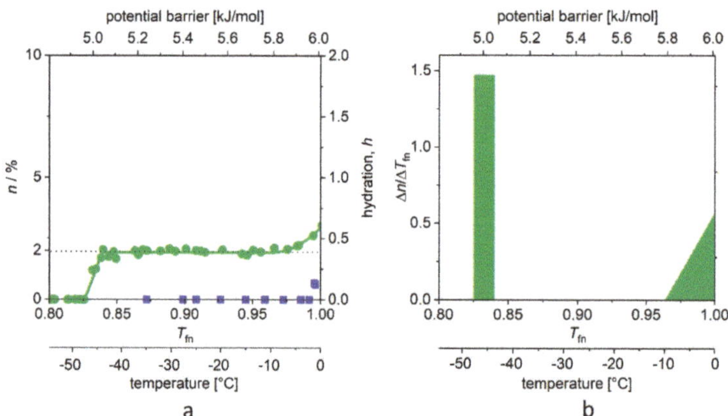

Figure 3. (a) Melting diagram (*MD*, green circles) of ubiquitin dissolved in double distilled water and that of frozen water under identical conditions (blue squares). (b) *DMD* curves (that is, the potential barrier distribution of protein–water bonds). There is no reliable measured data in the range -1–0 °C (0.995–1.00 T_{fn}). The data are given for 50 mg/mL protein concentration.

Perhaps it is not unnecessary to repeat that the amplitude value of the slow component of the measured FID signal extrapolated to $t = 0$ gives directly the number n of resonant protons (i.e., the protein-bound water molecules), whereas the temperature dependence of MD gives the dependence of n on thermal excitation energy.

The information can be read from Figures 3 and 4 as follows. Bulk water (blue squares) show the microscopic image of the ice-water phase transition. What would we expect of an absolute pure water sample of infinite size (in theory, one having a periodic boundary condition)? A single step of infinite slope at $T_{fn} = 1.00$ and $E_a = 6.01$ kJ/mol excitation energy (at 0.00 °C), in which all four bonds of the water molecule in the tetrahedral bond symmetry environment "melt" simultaneously. Instead moving water molecules are detected already below 0 °C. There are several reasons for this. The sample is not of infinite size, and the environment of the water molecules on the surface of the small sample is not the same as of those in the bulk environment. Secondly, the sample is not of absolute purity, so the environment of pollutions is not the same as in the clean environment. Third, the temperature of the sample in the measurement can be controlled and determined with limited accuracy only, especially at 0 °C.

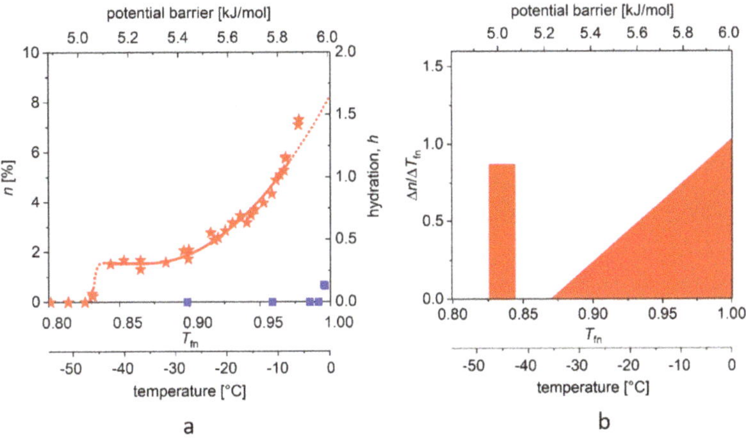

Figure 4. (**a**) The melting diagram (*MD*, red stars) of ERD10 dissolved in double-distilled water and the melting curve (blue squares) of the solvent (water). (**b**) *DMD* curves are shown (that is, the potential barrier distribution of protein–water bonds). There is no reliable measured data in the range -1–0 °C (0.995–$1.00\ T_{fn}$). The data are given for 50 mg/mL protein concentration.

In Figure 3a, the "melting point" (-46 (1) °C (for definition of error, see Table 1) of the aqueous solution of ubiquitin shows the thermal energy investment (ΔQ) that is required to start to move the water molecules that are bound to the protein. The steep step (with a narrow, $\approx$0.01 kJ/mol energy range) shows that there are water molecules in the first hydrate shell that are bound almost identically. It is a reasonable approximation to consider these energies nearly the same, and the relevant molecular surface equipotential. This potential field of nearly identical elements resembles the feature of the H-bridges [16–19] and is largely different from strongly distant dependent potentials (the variants can be find in the text-books [16–19]). The number of protein–water bonds in the actual region is given by the area of the rectangle. As a self-check, the same quantity can be more accurately determined from the average of all n points on the MD plateau.

The next wide region is the plateau. (This region begins at $T_{fno} = 0.832$ (4), in Figure 3b, in which the value of $\Delta n / \Delta T_{fn}$ is zero). The plateau carries very important information. No new water molecules begin to move in this excitation energy region, because there are no water molecules that are bound by corresponding energy to the protein. We can suggest that the H-bridges here, which link the bulk of the protein molecule to a globule. Thus, this can be an ancestral form of a higher order structure, which is represented not only by geometry but also by a bonding network of a certain energy. The heat invested within the plateau region does not start to move new water molecules; rather, it increases the specific heat, and the rotational speed of already rotating water molecules, as we have seen in a previous work on differential scanning calorimetry (DSC) measurements and data interpretation [7]. (Based on this interpretation, these statements can be made to be more accurate, which we intend to do in a short notice.)

T_{fne} is the end of the plateau, and here begins the energy region where there are binding energies close to the binding energy of water–water bonds, presumably on parts of the protein molecule that are better exposed to water. In principle, the temperature dependence of n can be described by the higher exponents of the power function; the quadratic member was sufficient in this case. All data are available; we summarize the values and the order parameters introduced by us in Table 1.

In Figure 4a, we show the "melting point" (approximately -42 (2) °C) for ERD10 (red stars). We also repeat the above procedure for ERD10 with different parameters. The steep step (with narrow, ≈ 0.01 kJ/mole energy range) shows the presence of water molecules of nearly identical binding energy in the first hydrate shell, but this region is followed by a plateau, which is significantly narrower than that observed in the case of globular proteins. We then observe a phase of continuous rise in MD, which can be well approximated by the quadratic (or even higher) component of the summation. A much larger part of the molecular surface is exposed to water than in the case of ubiquitin, i.e., about 69–77% of the protein molecule can be described as disordered. The range $(1 - T_{\mathrm{fne}})$ of energy barriers inhibiting water movements (which can be defined as disorder) is more than three times broader than for the selected globular protein.

Figure 3b and Figure 4b depict the changes (differential quotient) of the mobile water fractions by normalized functional temperature, i.e., they are the graphical representations of Equation (4). As outlined, the bars at low temperature (around -45 °C) correspond to the relatively high differential quotient values describing the first few data points greater than zero. The fraction of mobile hydration water increases here within a few degrees to the level of $n(E_{\mathrm{a,o}})$ or A while the first mobile hydration layer forms, which gives the high differential quotient values.

Table 1. Characteristic thermal quantities for two sample proteins. T_{fno} end T_{fne} give the start and the end points of the plateau in MDs, respectively, as normalized fundamental temperature. n_{ho} and n_{he} values are given as the mobile hydration water fraction and as the number of mobile hydration water per protein molecule. *HeR*, *HeR*$_{\mathrm{n}}$, and *HeM* are dynamic parameters describing heterogeneity from various aspects (see text).

Protein	T_{fno}	T_{fne}	*HeR* (4) *	n_{ho}	n_{he} **	*HeR*$_{\mathrm{n}}$ (6) *	*HeM*
UBQ	0.832 (4)	0.961 (5)	0.23 (2)	0.019 (1) 226 (3)	>0.009 (3) >102 (33)	0.3 (1)	241 (147)
ERD10	0.835 (3)	0.889(2)	0.73 (4)	0.0157 (4) 514 (13)	>0.098 (8) >3200 (275)	0.9 (1)	415 (60)

* The number in parentheses is the measurement error in the order of magnitude of the last number; the heterogeneity ratio is defined by the relation (4) or (6); ** Lower limit estimate due to the uncertainty of measured data is close to $T_{\mathrm{fn}} = 1$; at T_{fno} value given in Table 1 (-43 °C), the excitation energy is 5.06 (4) kJ/mol for both proteins; at T_{fne} for ubiquitin, the excitation energy is 5.798 (2) kJ/mol at -9.9 °C; and for ERD10, it is 5.31 (3) kJ/mol at -36 °C.

A comparison of *HeM* values (analogous with the tangent function) shows that in globular proteins the realization of the two extreme values, conditions in the first hydrate shell and water-water bonding, are very close. For ERD10, a much wider distribution of potential energy barriers is characteristic of structural disorder. The typical data are summarized in Table 1. The ordered/disordered state of the two protein molecules only approximates the ideal limiting values, *HeR* = 0 and *HeR* = 1.

The reality of the n_{ho} number of protein–water bonds in the homogeneous binding energy region (in other words, in the first hydrate shell) is better appreciated by reference to our knowledge of the hydration of protein-forming amino acids [18]. The sum of the numbers of the possible H-bridges of ubiquitin molecule gives 211. According to our measurements, the number of water molecules bound in the first hydration shell by similar binding energies is $n_{\text{ho}} = 226$ (3).

The summation of possible H-bridges within ERD10 yields 986. According to our measurements, the number of water molecules bound in the first hydration shell by similar binding energies is $n_{\text{ho}} = 514$ (13). The difference between measured and estimated values is unsurprising, especially in light of the good agreement found for ubiquitin. It is reasonable to ask the question whether approximately half of the H-bridges does not link with other water molecules, but realize some other type of bond.

Among the quantities given in Table 1, it is necessary to emphasize the determination of the relative number of bonds that fall into the heterogeneous region ($n_{\text{he}}/(n_{\text{he}} + n_{\text{ho}})$). The result is surprising if one is thinking in terms of a globular protein molecule, because for ubiquitin, the protein is in contact with an additional $n_{\text{he}} > 102$ (33) water molecules, which is approximately 36% of all bound water-molecules. The bonds of these water molecules are dominated by water–protein bonds, which are close in energy to the of water–water bonds. In the case of ERD10, the protein surface is in contact with an additional $n_{\text{he}} > 2200$ (220) water molecules, which is approximately 73% of all bound water-molecules. In the bonds of the latter water-molecules, water–protein bonds similar to water–water bonds dominate with a substantially wider energy distribution for this initially disordered protein.

It is maybe unnecessary to emphasize that the values we suggest are derived from direct measurements, i.e., they do not rely on assuming any hypothesis or model! They allow to determine the number of first-neighbor water molecules per amino acid (n_{ho}/amino acid), which is $226/76 \cong 3.0$ for UBQ and $514/260 \cong 2.0$ for ERD10. The round value within an error of 1% is surprising, as well as the close match of 2.0 with other values observed for other globular proteins (casein, lysozyme, and BSA, to be published).

Therefore, the measured number of bound water-molecules for ubiquitin is 328 (30). Molecular dynamics simulation estimation from the literature [19] gives a value of 379. For ERD10, the numbers per protein molecule is 2714 (263) (measured) and 881 (estimated by molecular dynamics simulation [19]). The difference between the two proteins and the reverse ratio raise many questions about the nature of protein–water bonds that are still difficult to answer.

3. Materials and Experimental Methods

3.1. Selection of Proteins

We have selected these proteins, as we and others have collected ample evidence for their function depending on their particular structural class, which is folded and intrinsically disordered. Ubiquitin (UBQ) is a small, 76-amino acid globular protein that is found ubiquitously in the cells of all eukaryotic organisms, carrying out basic and indispensable functions in regulating protein function [20]. That is, proteins targeted for degradation are covalently modified by a mono- or poly-ubiquitin chain and are directed for degradation by the 26S proteasome, whereas other proteins labeled with a mono-ubiquitin chain enter regulatory interactions in transcription regulation, for example.

ERD10 [21], on the other hand, is a plant dehydrin that has its cellular protection function strictly linked with structural disorder [22]. Its length is 260 amino acids; it is structurally disordered by a broad range of biophysical techniques, and it functions by protecting the structural integrity of client proteins under the conditions of dehydration and other stresses.

3.2. Expression and Purification of Proteins

Lyophilized ubiquitin (UBQ) was obtained from Sigma Chemical (St. Louis, MO, USA), whereas plant late embryogenesis abundant protein early response to dehydration 10 (ERD10, UniProt P42759) was produced via recombinant expression in *Escherichia coli* BL21(DE3) Star expression strain and purified as described previously [22]. In short, purification was carried out through three chromatographic steps: an ion exchange on HiTrap Q FF at pH 9.5 with gradient elution, followed by two gel-filtration steps on Superdex 200 and Superdex 75 columns, on an AKTA Avant (GE Healthcare, Little Chalfont, UK) FPLC system. The purity of the proteins was checked by SDS-PAGE and was found to be at least 98%.

3.3. Wide-Line NMR

We have reviewed the varieties of radiofrequency excitations applied in wide-line NMR in two book chapters [23,24], and here we only address the simplest excitation protocol that uses a 90° ($\pi/2$) radiofrequency pulse. NMR measurements and data acquisition were performed with a Bruker AVANCE III NMR pulse spectrometer at a frequency of 82.4 MHz with a stability of better than 10^{-6}. The $\pi/2$ pulse was 3–4 μs, and the dead time of the spectrometer was 6–8 μs. The inhomogeneity of the magnetic field was 2 ppm. The accumulated repeat number of the measurements was between 50 and 80.

The temperature was controlled by an open-cycle Janis cryostat with an uncertainty better than 0.5 K. The system was complemented by an adequate NMR head and by a closed sample holder.

4. Conclusions

By reinterpreting our previous results, we have determined the energy distribution of the potential barriers inhibiting the movement of water molecules bound to two protein molecules in aqueous solution. Based on our results, we could deduce quantitative conclusions about the ratios of the globular/ordered and more solvent exposed/disordered regions of the protein molecules and the extent of the latter, as well as the energy relations of the protein–water bonds. We suggest that short range forces (H-bonds) play a dominant role in the formation of the first hydrate shell.

The mapping of the water-binding characteristics of protein molecules is certainly not the only area of the application of wide-line NMR measurements and this novel interpretational procedure. The rapid, non-disruptive measurement and the data interpretation had already opened a novel avenue to study molecular interactions and to determine the moisture content of solid phase samples. In the outline of our previous work [9], we listed some additional possibilities. Three of these are also mentioned here: (i) the possibility to directly demonstrate the interaction between different molecules (e.g., protein and drug), (ii) the possibility of direct, non-destructive measurement of the different bonds between identical molecules, and (iii) the possibility to determine the effect of a standard (often NaCl containing) solvent on the structure and properties of protein molecules.

Author Contributions: Conceptualization, K.T., M.B., and P.T.; Methodology, M.B.; Validation, K.T., M.B. and P.T.; Formal Analysis, M.B.; Investigation K.T. and M.B.; Data Curation, M.B.; Writing-Original Draft, K.T.; Writing—Review & Editing K.T., M.B. and P.T.; Visualization M.B.; Supervision, P.T. Funding Acquisition P.T.

Funding: This research was funded by the Hungarian Scientific Research Fund (OTKA, grant no. K124670).

Acknowledgments: We would like to thank all colleagues, who contributed to previous studies that resulted in the results on which this work is based.

Conflicts of Interest: The authors declare no conflict of interest.

Abbreviations

FID	free induction decay
IDP	intrinsically disordered protein
MD	melting diagram
DMD	differential form of the melting diagram
DSC	differential scanning calorimetry

References

1. Tompa, K.; Bokor, M.; Han, K.H.; Tompa, P. Hydrogen skeleton mobility and protein architecture. *Intrinsically Disord. Proteins* **2013**, *1*, 77–86. [CrossRef] [PubMed]

2. Grüner, G.; Tompa, K. Molekuláris mozgások vizsgálata szilárdtestekben NMR módszerrel. *Kémiai Közlemények* **1968**, *30*, 315–356.

3. Slichter, C.P. *Principles of Magnetic Resonance*, 3rd ed.; Springer: Berlin/Heidelberg, Germany, 1990; ISBN 978-642080692.

4. Cooke, R.; Kuntz, I.D. The properties of water in biological systems. *Annu. Rev. Biophys. Bioeng.* **1974**, *3*, 95–124. [CrossRef] [PubMed]

5. Tompa, K.; Bánki, P.; Bokor, M.; Kamasa, P.; Lasanda, G.; Tompa, P. Interfacial water at protein sur-faces: Wide-line NMR and DSC characterization of hydration in ubiquitin solutions. *Biophys. J.* **2009**, *96*, 2789–2798. [CrossRef] [PubMed]

6. Csizmók, V.; Bokor, M.; Bánki, P.; Klement, E.; Medzihradszky, K.F.; Friedrich, P.; Tompa, K.; Tompa, P. Primary contact sites in intrinsically unstructured proteins: The case of calpastatin and microtubule-associated protein 2. *Biochemistry* **2005**, *44*, 3955–3964. [CrossRef] [PubMed]

7. Hazy, E.; Bokor, M.; Kalmar, L.; Gelencser, A.; Kamasa, P.; Han, K.H.; Tompa, K.; Tompa, P. Distinct hydration properties of wild-type and familial point mutant A53T of α-synuclein associated with parkinson's disease. *Biophys. J.* **2011**, *101*, 2260–2266. [CrossRef] [PubMed]

8. Tompa, P.; Bánki, P.; Bokor, M.; Kamasa, P.; Kovács, D.; Lasanda, G.; Tompa, K. Protein–water and protein-buffer interactions in the aqueous solution of an intrinsically unstructured plant dehydrin: NMR intensity and DSC aspects. *Biophys. J.* **2006**, *91*, 2243–2249. [CrossRef] [PubMed]

9. Tompa, K.; Bokor, M.; Verebélyi, T.; Tompa, P. Water rotation barriers on protein molecular sur-faces. *Chem. Phys.* **2015**, *448*, 15–25. [CrossRef]

10. Tompa, K.; Bokor, M.; Ágner, D.; Iván, D.; Kovács, D.; Verebélyi, T.; Tompa, P. Hydrogen mobility and protein–water interactions in proteins in the solid state. *ChemPhysChem* **2017**, *18*, 677–682. [CrossRef] [PubMed]

11. Kittel, C.; Kroemer, H. *Thermal Physics*, 2nd ed.; W. H. Freeman and Company: San Francisco, CA, USA, 1980; p. 445, ISBN 9780716710882.

12. Waugh, J.S.; Fedin, E.I. On the determination of hindered rotation barriers in solids. *Leningrad* **1962**, *4*, 2233–2237.

13. *CRC Handbook of Chemistry and Physics*, 97th ed.; Haynes, W.M. (Ed.) CRC Press, Taylor & Francis: Boca Raton, FL, USA, 2016; ISBN 9781498784542.

14. Atkins, P.; de Paula, J. *Physical Chemistry*, 9th ed.; Oxford University Press: Oxford, UK, 2010; ISBN 9781429218122.

15. Chaplin, M. Water's Hydrogen Bond Strength. Available online: https://arxiv.org/abs/0706.1355 (accessed on 19 October 2018).

16. Steiner, T. The hydrogen bond in the solid state. *Angew. Chem. Int. Ed. Engl.* **2002**, *41*, 49–76. [CrossRef]

17. Stone, A. *The Theory of Intermolecular Forces*, 2nd ed.; Oxford University Press: New York, NY, USA, 2013; ISBN 0198558848.

18. Kuntz, I.D. Hydration of macromolecules. III. Hydration of polypeptides. *J. Am. Chem. Soc.* **1971**, *93*, 514–516. [CrossRef]

19. Bizzarri, A.R.; Wang, C.X.; Chen, W.Z.; Cannistraro, S. Hydrogen bond analysis by MD simulation of copper plastocyanin at different hydration levels. *Chem. Phys.* **1995**, *201*, 463–472. [CrossRef]

20. Braten, O.; Livneh, I.; Ziv, T.; Admon, A.; Kehat, I.; Caspi, L.H.; Gonen, H.; Bercovich, B.; Godzik, A.; Jahandideh, S.; et al. Numerous proteins with unique characteristics are degraded by the 26S proteasome following monoubiquitination. *Proc. Natl. Acad. Sci. USA* **2016**, *113*, E4639–E4647. [CrossRef] [PubMed]

21. UniProtKB-P42759. Available online: https://www.uniprot.org/uniprot/P42759 (accessed on 19 October 2018).

22. Kovacs, D.; Kalmar, E.; Torok, Z.; Tompa, P. Chaperone activity of ERD10 and ERD14, two disordered stress-related plant proteins. *Plant Physiol.* **2008**, *147*, 381–390. [CrossRef] [PubMed]

23. Tompa, K.; Bokor, M.; Tompa, P. Hydration of Intrinsically Disordered Proteins from Wide-Line NMR. In *Instrumental Analysis of Intrinsically Disordered Proteins: Assessing Structures and Conformation*; Uversky, V., Longhi, S., Eds.; Wiley & Sons Inc.: Hoboken, NJ, USA, 2010; pp. 345–368, ISBN 9780470343418.
24. Tompa, K.; Bokor, M.; Tompa, P. Wide-Line NMR and Protein Hydration. In *Intrinsically Disordered Protein Analysis*; Uversky, V.N., Dunker, A.K., Eds.; Humana Press: New York, NY, USA, 2012; Volume 895, ISBN 10643745.

© 2018 by the authors. Licensee MDPI, Basel, Switzerland. This article is an open access article distributed under the terms and conditions of the Creative Commons Attribution (CC BY) license (http://creativecommons.org/licenses/by/4.0/).

 International Journal of
Molecular Sciences

Article

Both Intrinsically Disordered Regions and Structural Domains Evolve Rapidly in Immune-Related Mammalian Proteins

Keiichi Homma [1],*, Hiroto Anbo [1], Tamotsu Noguchi [2] and Satoshi Fukuchi [1]

[1]   Department of Life Science and Informatics, Maebashi Institute of Technology, 460-1 Kamisadori-machi,
    Maebashi-shi 371-0816, Japan; koume8@icloud.com (H.A.); sfukuchi@maebashi-it.ac.jp (S.F.)
[2]   Pharmaceutical Education Research Center, Meiji Pharmaceutical University, 2-522-1 Noshio, Kiyose-shi,
    Tokyo 204-8588, Japan; noguchit@my-pharm.ac.jp
*   Correspondence: khomma@maebashi-it.ac.jp; Tel.: +81-27-265-7360

Received: 30 September 2018; Accepted: 2 December 2018; Published: 4 December 2018

Abstract: Eukaryotic proteins consist of structural domains (SDs) and intrinsically disordered regions (IDRs), i.e., regions that by themselves do not assume unique three-dimensional structures. IDRs are generally subject to less constraint and evolve more rapidly than SDs. Proteins with a lower number of protein-to-protein interactions (PPIs) are also less constrained and tend to evolve fast. Extracellular proteins of mammals, especially immune-related extracellular proteins, on average have relatively high evolution rates. This article aims to examine if a high evolution rate in IDRs or that in SDs accounts for the rapid evolution of extracellular proteins. To this end, we classified eukaryotic proteins based on their cellular localizations and analyzed them. Moreover, we divided proteins into SDs and IDRs and calculated the respective evolution rate. Fractional IDR content is positively correlated with evolution rate. For their fractional IDR content, immune-related extracellular proteins show an aberrantly high evolution rate. IDRs evolve more rapidly than SDs in most subcellular localizations. In extracellular proteins, however, the difference is diminished. For immune-related proteins in mammals in particular, the evolution rates in SDs come close to those in IDRs. Thus high evolution rates in both IDRs and SDs account for the rapid evolution of immune-related proteins.

Keywords: secretion; immune; extracellular; protein-protein interaction; intrinsically disordered region; structural domain; evolution

---

1. Introduction

Mature eukaryotic proteins consist not only of structural domains (SDs), but also of intrinsically disordered regions (IDRs), i.e., regions that by themselves do not fold into unique three-dimensional structures [1]. Although some IDRs interact with proteins or other macromolecules, they are generally under less constraint than SDs and thus have higher evolution rates [2]. A positive correlation between fractional IDR contents of proteins and evolution rates is thus expected.

Proteins with more protein-to-protein interactions (PPIs) tend to be more evolutionarily constrained and have lower evolution rates [3,4]. Highly expressed proteins are also more constrained and evolve slowly [4–6]. These two factors partially account for the evolution rate of proteins.

Eukaryotic proteins have specific subcellular localizations in general, with different average fractional IDR contents in different cellular localizations [7]. For instance, IDR contents are generally high in nuclear proteins [7,8], while they tend to be low in mitochondrial proteins [9,10]. It is plausible that different fractional IDR contents in different subcellular localizations result in varied evolution rates.

Interestingly, extracellular proteins (synonymously called secreted proteins) in mammalian species were often found to evolve faster than intracellular proteins [11,12]. This finding is partly explainable by rapid evolution of immune-related extracellular proteins as many of the coding genes are subject to positive selection [13,14]. That is, the evolution rate, ω, defined by the nonsynonymous to synonymous substitution rate ratio, exceeds unity at sites under positive selection and the existence of many such sites result in high evolution rates of many immune-related genes. For instance, antimicrobial peptides, α- and β-defensins and cathelicidins, are reportedly subject to positive selection and evolve rapidly in mammals [15–17]. We consider it worthwhile to carry out research on evolutionary characteristics of immune-related secreted proteins, as they are involved in host defences [18], pathogen–host interactions [19,20], production of antibodies [21], colony-stimulating factors [22], haematopoiesis [23], and triggering proteolytic cascades [24,25], as well as enzyme replacement therapies [26]. The generally high evolution of immune-related proteins evinces their importance in evolution of mammalian species [27]. Further research may reveal how immune-related proteins function and may lead to pharmaceutical applications.

However, the difference in evolution rate with intracellular proteins remained significant even if analyses were limited to non-immune-related extracellular proteins. The generally low expression levels in secreted proteins partially explain the rapid evolution. Whether the substitution frequency in IDRs or SDs or both contributes to the increased evolution rate of extracellular proteins, however, has not been explored.

We examined the correlation of fractional IDR content and evolution rate and found it positive. We then analyzed the evolution rates of SDs and IDRs of proteins in different localizations. In most localizations, IDRs were found to evolve faster than SDs, as expected. Immune-related secreted proteins in mammals, however, exhibited extremely high evolution rates in SDs that approach those in IDRs. This surprising finding indicates that positive selection that is said to function on a number of immune-related genes operates strongly both on IDRs and SDs of the coded proteins.

2. Results and Discussion

2.1. Classification of Eukaryotic Proteins by Subcellular Localizations

For accurate analyses of evolution rates in different subcellular localizations, reliable localization annotations of most proteins are necessary. At present, only four species satisfy this criterion in UniProt: *Homo sapiens*, *Mus musculus* (mouse), *Arabidopsis thaliana* (thale cress), and *Saccharomyces cerevisiae* (budding yeast). We thus selected the human, mouse, thale cress, and budding yeast proteins with orthologs and classified the selected proteins by subcellular localization (Table 1). Proteins that are localized to both the nucleus and the cytosol were specifically grouped (abbreviated as NC), as the group reportedly contains many proteins with multiple PPIs [28]. We combined proteins residing in the endoplasmic reticulum and the Golgi apparatus (termed EG), since many proteins cycle between the two organells. Secreted proteins were divided into immune-related (SI) and non-immune-related (SN), because immune-related proteins generally evolve rapidly [11]. Thale cress had a limited number of immune-related proteins, while unsurprisingly budding yeast had none. Multiply localized proteins except for the aforementioned NC proteins were classified as one group (ML). Note that many proteins with orthologs were not classifiable due to the unavailability of pertinent information.

Table 1. Number of proteins in each subcellular localization.

Species	All	NU	NC	CY	MT	EG	PM	SN	SI	ML
H. sapiens	10,348	1639	632	455	377	400	1116	584	139	3023
M. musculus	10,068	1719	546	224	426	514	998	796	125	2787
A. thaliana	8910	1032	163	331	356	348	534	431	6	594
S. cerevisiae	5304	1532	232	241	639	458	281	69	0	416

NU: Nucleus; NC: Nucleus and cytosol; CY: Cytosol; MT: Mitochondria; EG: Endoplasmic reticulum or Golgi apparatus; PM: Plasma membrane; SN: Secreted, non-immune-related; SI: Secreted, immune-related; ML: Multiple localizations except NC (ML).

2.2. Evolution Rates and Other Properties of Proteins in Different Subcellular Localizations

For each pair of orthologs, we determined the evolution rate, ω, defined by the ratio of nonsynonymous to synonymous substitution rate, i.e., dN/dS. The median ω at each localization is shown (Figures 1 and 2). Note that for this and other data presented in the figure, different scales were used in different species. As the number of immune-related proteins (SI) in *A. thaliana* is small, no corresponding data were plotted in this species. Proteins of the four species showed similar patterns. For instance proteins in the cytosol (CY) and those that reside both in the nucleus and the cytosol (NC) had the median evolution rates lower than the overall median in all four species. In general the median evolution rates in intracellular proteins (NU, NC, CY, MT, and EG; shown in blue) were lower than those of secreted proteins (SN and SI; shown in red). Among the secreted proteins, immune-related proteins (SI) exhibited particularly high evolution rates, in agreement with the literature [13,14].

The fractional IDR content of each protein was predicted by DISOPRED [29], DICHOT [30], and POODLE-L [31] and the median in each localization was calculated (Figures 1 and 2). Although the medians of most localizations (Figures 1 and 2) were nearly always the lowest by DISOPRED, higher by DICHOT, and the highest by POODLE, we note that the overall averages by the three methods generally do not differ much. For instance, the fractions of IDRs in human proteins by DISOPRED, DICHOT, POODLE are 30.2%, 26.4%, and 30.1%, respectively. The differences in the medians are thus mostly attributable to differences in the distributions of fractional IDRs. Nevertheless the corresponding medians by the three prediction methods showed similar patterns. For instance, by all three methods in the four species, we got high fractional IDR contents in the nuclear proteins (NU) and low values in the mitochondrial proteins (MT), consistent with previous reports [7–10]. Intriguingly, the secreted non-immune proteins (SN) in budding yeast were revealed to have a high median IDR content, unlike the counterparts of the three multicellular eukaryotes. The difference may reflect the difference between unicellular and multicellular organisms. This issue needs to be addressed later with analyses of more eukaryotes.

We also calculated and graphed the median numbers of PPIs of proteins in the localizations (Figures 1 and 2). PPIs have been less studied in mouse and thale cress proteins than in human and budding yeast counterparts, as evidenced by the reduced numbers of PPIs in mouse and thale cress (Figures 1 and 2). The mouse and thale cress PPI data are therefore less reliable as those of the other two species. As reported [28], multiply localized proteins (NC and ML) generally showed more interactions with other proteins. Immune-related secreted proteins (SI), however, had fewer interacting partners on average.

Additionally, the median expression level in ppm of the proteins at each localization was determined and graphed as logarithms to the base of ten (Figures 1 and 2). Yeast proteins were generally expressed much more than mammalian proteins. The expression levels of the human immune-related proteins (SI) were generally high, but those of the mouse counterparts were indistinguishable from the average.

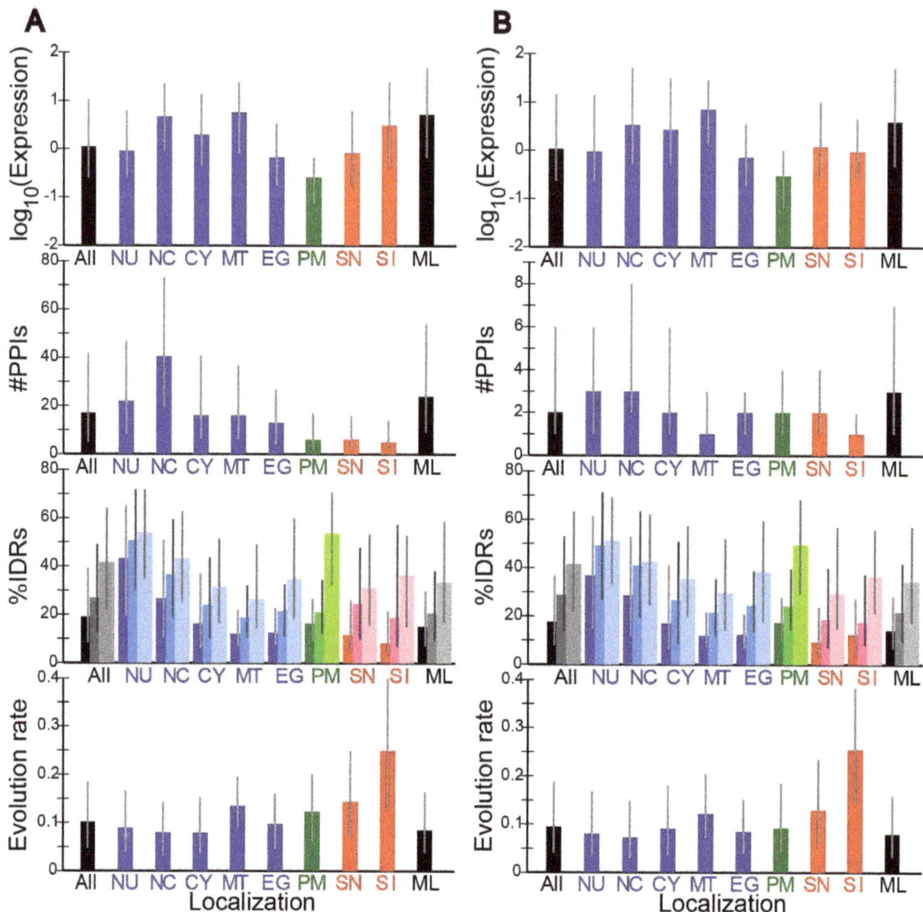

Figure 1. Medians and ranges of four quantities in different localizations in two mammals. (**A**) *H. sapiens*; (**B**) *M. musculus*; Rectangles in each panel from the bottom to the top represent the medians in evolution rate, fractional IDR content by DISOPRED (left), DICHOT (middle) and POODLE (right), the number of PPIs, and expression level. Grey vertical bars represent interquartile ranges, with their bottom and top corresponding to the 25th to the 75th percentile, respectively. The abbreviations for localizations are as in Table 1.

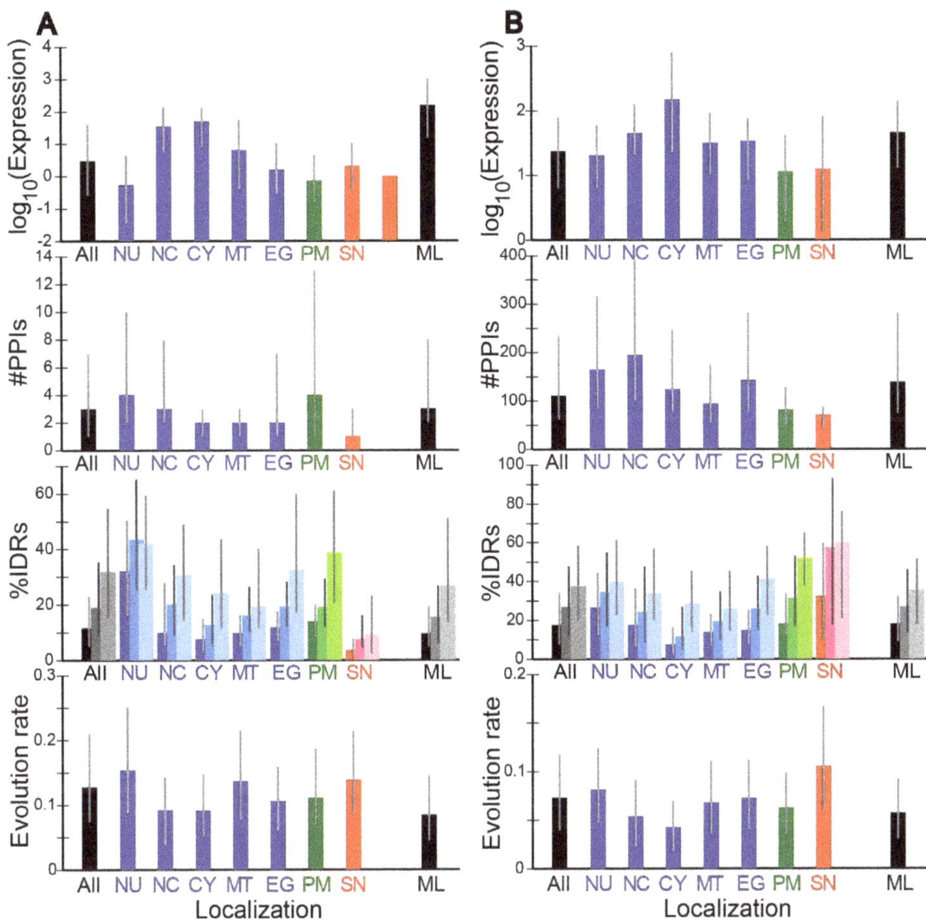

Figure 2. Medians of four quantities in different localizations in two non-mammalian eukaryotes. (**A**) *A. thaliana*; (**B**) *S. cerevisiae*; the data are presented as in Figure 1.

2.3. Correlation of Evolution Rates with Protein Properties

We computed Spearman's correlation coefficients (Rhos) of number of PPIs with evolution rate (ω) and found them to be weakly negative but significantly different from zero (all at $p < 0.01$) (Table 2). The negative correlation is consistent with previous results [3,4]. As the number of PPIs was generally low in extracellular proteins (SN and SI, Figures 1 and 2), the negative correlation partially explains their high evolution rates.

We also found small but significant (all at $p < 1 \times 10^{-113}$) negative correlations between expression level and ω (Table 2), corroborating previous findings [5,6]. The negative correlation was stronger in budding yeast. Since the expression levels of non-immune-related secreted proteins (SN) were not high (Figures 1 and 2), the negative correlation at least in part explains the high evolution rates of these proteins. By contrast the expression levels of immune-related secreted proteins (SI) were not significantly low (Figure 1) and do not contribute to the extremely high evolution rates.

As IDRs have a propensity to evolve faster than SDs, the more IDRs a protein has, the faster it is expected to evolve. To test this possibility, correlation coefficients of %IDR with ω were calculated. Fractional IDR content was positively correlated with evolution rate in all the four species (Table 2).

Although the correlation coefficients were generally small, they all significantly differed from zero (at $p < 1 \times 10^{-4}$). As the median fractional IDR contents in immune-related secreted proteins (SI) were lower than average, this factor does not make positive contribution to the evolution rates.

Table 2. Correlations between three properties and evolution rate ω.

Correlation with	*H. sapiens*	*M. musculus*	*A. thaliana*	*S. cerevisiae*
#PPI with ω	−0.293	−0.194	−0.054	−0.195
Expression level with ω	−0.264	−0.231	−0.337	−0.459
%IDR (DISOPRED) with ω	0.093	0.094	0.168	0.264
%IDR (DICHOT) with ω	0.113	0.146	0.052	0.303
%IDR (POODLE) with ω	0.096	0.097	0.113	0.179

Spearman's correlation coefficient (Rho) of each pair is shown.

2.4. Evolution Rates in SDs and IDRs in Different Subcellular Localizations

In order to see whether IDRs or SDs in immune-related proteins mostly account for the high evolution rates, we calculated the evolution rates in IDRs and SDs separately and compared the two. The median evolution rate in IDRs in all proteins was significantly higher than that in SDs, irrespective of species (Figures 3 and S1). We detected the same disparity at most localizations.

Upon closer examination of the mammalian rates, we noticed that the IDR/SD evolution rate ratio tended to be higher in intracellular localizations (NU, NC, CY, MT, and EG) than in extracellular ones (SN and SI). In the plant *A. thaliana* the inside–outside difference in evolution rate was detectable but was less pronounced (Figure S1A). In contrast, budding yeast failed to show this tendency (Figure S1B). In immune-relate secreted proteins (SI), the rates in IDRs and SDs were both higher than average, with the difference between them statistically insignificant in a majority of cases (Figure 3). SDs apparently evolve quite rapidly in immune-related proteins to approach the rates of IDRs to give rise to the anomalously high evolution rates. So far as we are aware, the phenomenon of the evolution rate in SDs that comes close that in IDRs in immune-related proteins is the first to be reported. The non-immune related extracellular proteins (SN) also tended to have higher than average evolution rate in SDs in *H. sapiens* and *M. musculus*, and *A. thaliana*, although the difference from the average was more conspicuous in the two mammals (Figure 3) than in the plant species (Figure S1A). In contrast SDs in non-immune related extracellular proteins (SN) did not show an above-average evolution rate in *S. cerevisiaie* (Figure S1B). In mammalian mitochondrial (MT) and plasma membrane (PM) proteins, the evolution rates of SDs and IDRs were close to each other (Figure 3), although the former was significantly higher than the latter in all cases. By contrast the counterparts in the two non-mammalian species failed to show the tendency (Figure S1). We need to investigate other species before attaching any significance to this possibly mammalian-specific phenomenon.

We recognize the need to analyze more animal species to check the generality of our finding on immune-related extracellular proteins. For accurate analyses by the same methodology, however, two closely related and entirely sequenced species must be available and at least one of them must have a majority of proteins annotated by UniProt to provide reliable subcellular localizations. Unfortunately no animal species other than *H. sapiens* and *M. musculus* currently meet the latter criterion. Since 3463 (~22% of the total) *Drosophila melanogaster* proteins have been annotated, however, we carried out preliminary analyses of this fly. Thirty-eight annotated immune-related extracellular proteins were identified in 13,957 orthologs. The results showed that the evolution rates in IDRs and SDs were both high in immune-related proteins but the former was much higher than the latter. The ratio of the median evolution rate in IDRs to that in SDs was 2.37, 1.60, and 2.99 by DISOPRED, DICHOT, and POODLE, respectively. As the corresponding ratios of all *Drosophila* proteins were 2.10, 2.45, and 1.79, the ratio was not necessarily diminished in immune-related proteins in fruit fly. Thus, the preliminary results indicate that the phenomenon of rapid evolution in both SD and IDRs in immune-related secreted proteins is possibly limited to vertebrates.

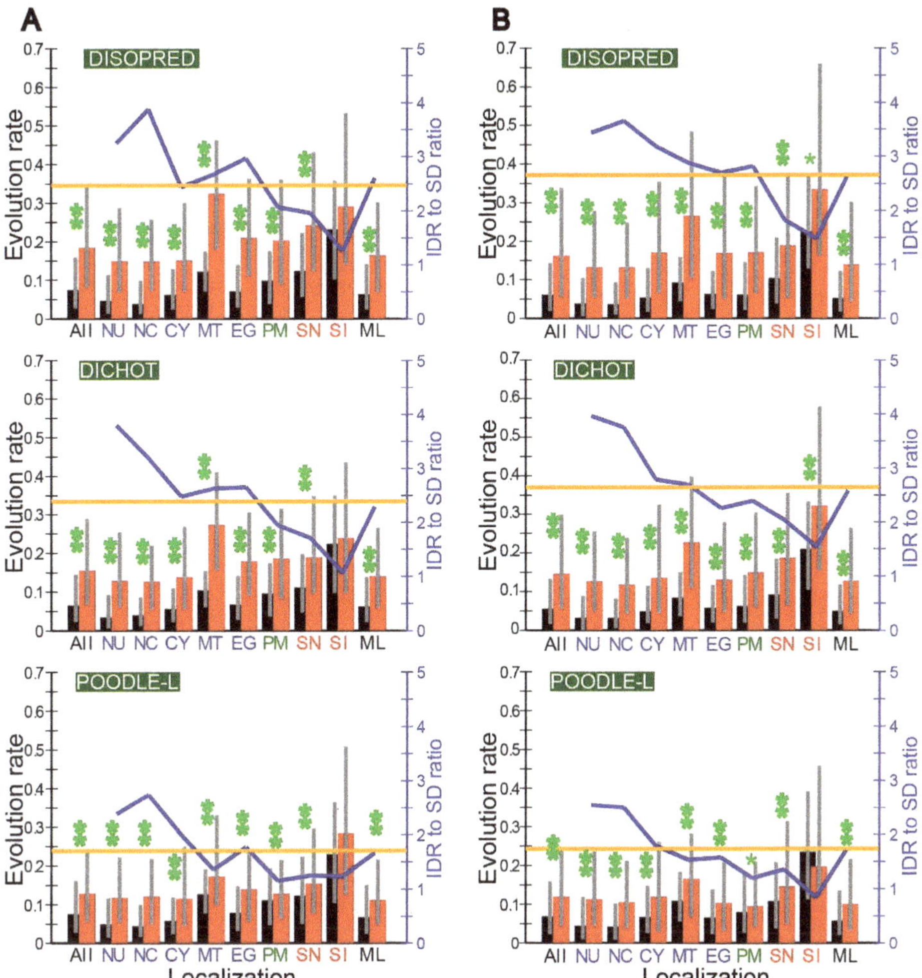

Figure 3. Evolution rates are higher in IDRs than in SDs except possibly for immune-related secreted proteins in mammals. (**A**) *H. sapiens*; (**B**) *M. musculus*; the diagrams (top to bottom) in each panel are based on DISOPRED, DICHOT and POODLE-L predictions. The median evolution rates in SDs are shown in black rectangles, while those in IDRs are depicted in red (left scale). Grey vertical lines show ranges from the 25th to the 75th percentile. Blue lines represent the median evolution rate ratios of IDRs to SDs at respective localizations, while horizontal orange lines show the ratio of all proteins (right scale). One asterisk signifies a statistically significant difference between the evolution rate distributions of IDRs and SDs at $p < 0.01$, while two asterisks denote a statistically significant difference at $p < 0.001$ (*U*-test). The same abbreviations for localizations as those in Table 1 are used.

In the cytosolic proteins (CY) of budding yeast, the median evolution rate in IDRs was only a little higher than that in SDs (Figure S1B). As noted before, budding yeast proteins generally interact with much more proteins than human proteins and did not exhibit intracellular-extracellular disparity in the IDR to SD evolution ratio.

2.5. Examles of Proteins with Nonsynonymous and Synonymous Substitutions

To give specific examples, we diagramed some human and mouse proteins with locations of nonsynonymous and synonymous substitutions (Figure 4). As we selected the proteins as they exhibit close-to-median ratios of nonsynonymous to synonymous substitution rates in SDs and IDRs, the frequencies of nonsynonymous to synonymous substitutions do not necessarily show median values. Although the three prediction methods gave different results, the major disparities were found in the boundaries of IDRs and did not affect main results. In immune-related secreted proteins (Figure 4A–D), nonsynonymous mutations (red bars) were almost as frequent as synonymous ones (black bars) both in IDRs (pink rectangles) and SDs (gray regions). In comparison, in proteins of other subcellular localizations, nonsynonymous substitutions occurred much less frequently than synonymous substitutions in SDs, while the difference was less pronounced in IDRs (Figure 4E–H).

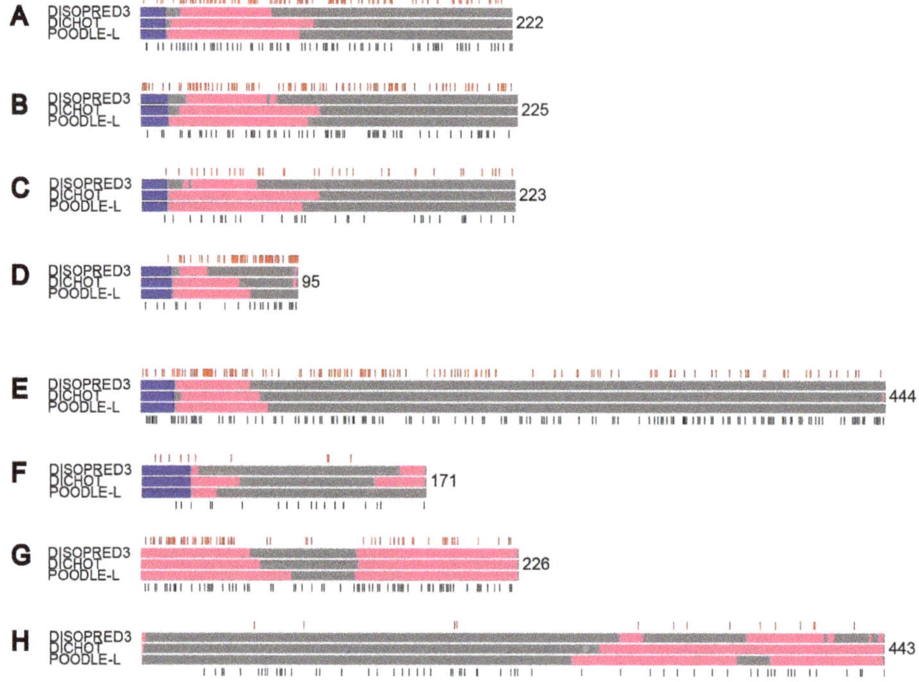

Figure 4. Examples of proteins with locations of nonsynonymous and synonymous substitutions. Each protein is represented by three rectangles with DISOPRED, DICHOT, and POODLE predictions (top to bottom) of IDRs (pink) and SDs (gray) as well as signal sequences (blue), if any, and the length shown on the right. The locations of nonsynonymous and synonymous substitutions are shown above (red lines) and below (black lines), respectively. (**A–D**): Immune-related secreted proteins, (**E,F**): non-immune-related secreted proteins, (**G,H**): nuclear proteins. (**A**) The human PRG2, (**B**) human PRG3, (**C**) mouse PRG2, (**D**) mouse DEFA20, (**E**) human SERPINA10, (**F**) mouse NENF, (**G**) human PROP1, (**H**) mouse NEK2 proteins.

2.6. Significance and Remaining Issues

The generally much lower frequency of nonsynonymous substitutions than synonymous substitutions in SDs reflects the fact that nonsynonymous changes very often destabilize the structures. By contrast, the difference between nonsynonymous and synonymous substitution rates is diminished in IDRs as nonsynonymous changes are frequently accommodated in IDRs. Consequently the ratio of nonsynonymous to synonymous substitution rate (ω) is expected to be much smaller in SDs than in IDRs. Although the actual results obtained in this research were mostly consistent with this expectation, those of immune-related secreted proteins of the two mammalian species were not; ω in SDs approaches that in IDRs.

As ω is larger than 1 at positively selected sites, the existence of numerous such sites in a region increases the average ω. Since many sites in immune-related secreted proteins were reported to be under positive selection [13–17], the heightened ω in SDs of such proteins indicates that many positively selected sites fall in SDs. The observation that IDRs of immune-related proteins exhibit higher ω than those of other proteins also implies that IDRs contain positively selected sites, too. The classification of positively selected sites in immune-related proteins into SDs and IDRs will probably lead to a better understanding of mechanisms of immunity. It is plausible that many nonsynonymous changes occur at the surface of SDs that interacts with other proteins.

It is also of interest to investigate known genes under positive selection that are associated with gamete recognition [32,33] and male reproduction [34,35] to find if SDs as well as IDRs of the encoded proteins evolve rapidly. We note that extracellular domains receive a number of posttranslational modifications such as phosphorylations, glycosylation, and lipidation. Investigation of evolution rates at posttranslational modification sites of immune-related proteins is another prospective area.

3. Materials and Methods

The nucleotide sequences of *H. sapience*, *M. musculus*, and *Rattus norvegicus* genes were downloaded from Ensembl (Release 91) [36]. The nucleotide sequences of *A. thaliana* (TAIR10), *Arabidopsis lyrata*, *Drosophila melanogaster* (BDGP6) genes were obtained from Ensembl, too. Ensembl also provided the orthologous relationships between *H. sapience* and *M. musculus* as well as those of *M. musculus* and *R. norvegicus*. The sequences of *S. cerevisiae* and *Saccharomyces paradoxus* were obtained from the Saccharomyces Genome Database [37], while those of *Drosophila pseudoobscura* genes were downloaded from FlyBase [38]. The orthologs of the two *Arabisopsis* species, the two yeast species, and the two *Drosophila* species were selected by bidirectional best hit analysis. The proteins were classified by subcellular localizations based on the Gene-Ontology (GO) annotations in UniProt (Release 2017_05) [39]. Specifically, the following GO IDs were used for subcellular classifications: nucleus: GO:0005634; cytoplasm: GO:0005829; mitochondria: GO:0005739; endoplasmic reticulum/Golgi apparatus: GO:0005783, GO:0005794, and GO:0005793; plasma membrane: GO:0005886; secreted: GO:0005576 and GO:0005615; immune-related: GO:0002376.

From the coding sequences, the signal peptides were removed based on UniProt annotations because they are unclassifiable as SDs or IDRs due to their absence in mature proteins. The remaining amino acid sequences of orthologs were aligned by MAFFT [40] and the corresponding nucleotide sequences were aligned according to the MAFFT results. dn/ds values were then computed using the codeml program (model M0) in PAML (version 4.9d) [41]. Statistical differences between two quantities were tested by Mann-Whitney's U-test, while statistical significance of correlations was evaluated by Spearman's rank correlation by means of in-house programs.

Number of PPIs and expression levels were taken from the BioGRID (version 3.4.158) [42] and the PaxDb (version 4.1) [43] databases, respectively. BioGRID is a curated database of interactions including protein-protein interactions obtained by two-hybrid studies, affinity purification coupled to mass spectrometry, and other methods, while PaxDB contains whole genome protein abundance information obtained by integrating numerous datasets using scores and ranks. Each protein was divided into SDs and IDRs by three methods: DISOPRED3 [29], DICHOT [30] and POODLE-L [31]. Briefly, DISOPRED3 assigns IDRs based on sequence profiles and other sequence-derived features, DICHOT classifies proteins into SDs and IDRs using sequence characteristics, alignments to existing protein structures, and sequence divergence, while POODLE-L is a prediction method for long IDRs that makes use of support-vector machine with 10 kinds of simple physico-chemical properties of amino acids. Based on the overall MAFFT alignments, the alignments of the corresponding sections were made. The evolution rate of each section was then determined as above.

4. Conclusions

In human and mouse, the SDs of immune-related proteins evolve at a high rate which comes close to that of the IDRs. This observation indicates that positive selection operates on both SDs and IDRs of the encoded proteins in many immune-related genes. Comparatively high evolution in SDs is also observed in non-immune-related secreted proteins in human and mouse, and to a lesser extent in thale cress, but not in budding yeast. Thus accelerated evolution in SDs as well as in IDRs contributes to rapid evolution of extracellular proteins in mammals.

Supplementary Materials: The following are available online at http://www.mdpi.com/1422-0067/19/12/3860/s1.

Author Contributions: Conceptualization, K.H.; Methodology, K.H.; Software, H.A., T.N., S.F.; Validation, K.H., H.A., T.N., S.F.; Formal Analysis, K.H.; Investigation, K.H., H.A.; Resources, K.H.; Data Curation, K.H.; Writing-Original Draft Preparation, K.H.; Writing-Review & Editing, K.H.; Visualization, K.H.; Supervision, K.H.; Project Administration, K.H.; Funding Acquisition, K.H.

Conflicts of Interest: The authors declare no conflict of interest.

Abbreviations

SD	Structural domain
IDR	Intrinsically disordered region
PPI	Protein-to-protein interaction
NU	Nucleus
NC	Nucleus and cytosol
CY	Cytosol
MT	Mitochondria
EG	Endoplasmic reticulum or Golgi apparatus
PM	Plasma membrane
SN	Secreted, non-immune-related
SI	Secreted, immune-related
ML	Multiple localizations except NC
dN	Nonsynonymous substitution rate
dS	Synonymous substitution rate
GO	Gene Ontology

References

1. Wright, P.E.; Dyson, H.J. Intrinsically unstructured proteins: Re-assessing the protein structure-function paradigm. *J. Mol. Biol.* **1999**, *293*, 321–331. [CrossRef] [PubMed]
2. Brown, C.J.; Takayama, S.; Campen, A.M.; Vise, P.; Marshall, T.W.; Oldfield, C.J.; Williams, C.J.; Dunker, A.K. Evolutionary rate heterogeneity in proteins with long disordered regions. *J. Mol. Evol.* **2002**, *55*, 104–110. [CrossRef] [PubMed]
3. Fraser, H.B.; Hirsh, A.E.; Steinmetz, L.M.; Scharfe, C.; Feldman, M.W. Evolutionary rate in the protein interaction network. *Science* **2002**, *296*, 750–752. [CrossRef] [PubMed]
4. Krylov, D.M.; Wolf, Y.I.; Rogozin, I.B.; Koonin, E.V. Gene loss, protein sequence divergence, gene dispensability, expression level, and interactivity are correlated in eukaryotic evolution. *Genome Res.* **2003**, *13*, 2229–2235. [CrossRef] [PubMed]
5. Pál, C.; Papp, B.; Hurst, L.D. Highly expressed genes in yeast evolve slowly. *Genetics* **2001**, *158*, 927–931. [PubMed]
6. Subramanian, S.; Kumar, S. Gene expression intensity shapes evolutionary rates of the proteins encoded by the vertebrate genome. *Genetics* **2004**, *168*, 373–381. [CrossRef]
7. Liu, J.; Perumal, N.E.; Oldfield, C.J.; Su, E.W.; Uversky, V.N.; Dunker, A.K. Intrinsic disorder in transcription factors. *Biochemistry* **2006**, *45*, 6873–6888. [CrossRef]

8. Minezaki, Y.; Homma, K.; Kinjo, A.R.; Nishikawa, K. Human transcription factors contain a high fraction of intrinsically disordered regions essential for transcriptional regulation. *J. Mol. Biol.* **2006**, *359*, 1137–1149. [CrossRef]

9. Homma, K.; Fukuchi, S.; Nishikawa, K.; Sakamoto, S.; Sugawara, H. Intrinsically disordered regions have specific functions in mitochondrial and nuclear proteins. *Mol. Biosyst.* **2012**, *8*, 247–255. [CrossRef]

10. Ito, M.; Tohsato, Y.; Sugisawa, H.; Kohara, S.; Fukuchi, S.; Nishikawa, I.; Nishikawa, K. Intrinsically disordered proteins in human mitochondria. *Genes Cells* **2012**, *17*, 817–825. [CrossRef]

11. Julenius, K.; Pedersen, A.G. Protein evolution is faster outside the cell. *Mol. Biol. Evol.* **2006**, *23*, 2039–2048. [CrossRef]

12. Liao, B.Y.; Weng, M.P.; Zhang, J. Impact of extracellularity on the evolutionary rate of mammalian proteins. *Genome Biol. Evol.* **2010**, *2*, 39–43. [CrossRef]

13. Hughes, A.L.; Nei, M. Pattern of nucleotide substitution at major histocompatibility complex class I loci reveals overdominant selection. *Nature* **1988**, *335*, 167–170. [CrossRef]

14. Hughes, A.L.; Nei, M. Nucleotide substitution at major histocompatibility complex class II loci: Evidence for overdominant selection. *Proc. Natl. Acad. Sci. USA* **1989**, *86*, 958–962. [CrossRef]

15. Patil, A.; Hughes, A.L.; Zhang, G. Rapid evolution and diversification of mammalian α-defensins as revealed by comparative analysis of rodent and primate genes. *Physiol. Genom.* **2004**, *20*, 1–11. [CrossRef]

16. Morrison, G.M.; Semple, C.A.M.; Kilanowski, F.M.; Hill, R.E.; Dorin, J.R. Signal sequence conservation and mature peptide divergence within subgroups of the murine β-defensin gene family. *Mol. Biol. Evol.* **2003**, *20*, 460–470. [CrossRef]

17. Zelezetsky, I.; Pontillo, A.; Puzzi, L.; Antcheva, N.; Segat, L.; Pacor, S.; Crovella, S.; Tossi, A. Evolution of the primate cathelicidin. Correlation between structural variations and antimicrobial activity. *J. Biol. Chem.* **2006**, *281*, 19861–19871. [CrossRef]

18. Baxt, L.A.; Garza-Mayers, A.C.; Goldberg, M.B. Bacterial subversion of host innate immune pathways. *Science* **2013**, *340*, 697–701. [CrossRef]

19. Sánchez, B.; Urdaci, M.C.; Margolles, A. Extracellular proteins secreted by probiotic bacteria as mediators of effects that promote mucosa-bacteria interactions. *Microbiology* **2010**, *156*, 3232–3242. [CrossRef]

20. Nobre, T.M.; Martynowicz, M.W.; Andreev, K.; Kuzmenko, I.; Nikaido, H.; Gidalevitz, D. Modification of Salmonella lipopolysaccharides prevents the outer membrane penetration of novobiocin. *Biophys. J.* **2015**, *109*, 2537–2545. [CrossRef]

21. Horlick, R.A.; Macomber, J.L.; Bowers, P.M.; Nebern, T.Y.; Tomlinson, G.L.; Krapf, I.P.; Dalton, J.L.; Verdino, P.; King, D.J. Simultaneous surface display and secretion of proteins from mammalian cells facilitate efficient in vitro selection and maturation of antibodies. *J. Biol. Chem.* **2013**, *288*, 19861–19869. [CrossRef]

22. Lieschke, G.J.; Burgess, A.W. Granulocyte colony-stimulating factor and granulocyte-macrophage colony-stimulating factor (1). *N. Engl. J. Med.* **1992**, *327*, 28–35. [CrossRef]

23. Jelkmann, W. Regulation of erythropoietin production. *J. Physiol.* **2011**, *589*, 1251–1258. [CrossRef]

24. Lucas, A.; McFadden, G. Secreted immunomodulatory viral proteins as novel biotherapeutics. *J. Immunol.* **2004**, *173*, 4765–4774. [CrossRef]

25. Lubbers, R.; van Essen, M.F.; van Kooten, C.; Trouw, L.A. Production of complement components by cells of the immune system. *Clin. Exp. Immunol.* **2017**, *188*, 183–194. [CrossRef]

26. Bonin-Debs, A.L.; Boche, I.; Gille, H.; Brinkmann, U. Development of secreted proteins as biotherapeutic agents. *Expert Opin. Biol. Ther.* **2004**, *4*, 551–558. [CrossRef]

27. Castillo-Davis, C.I.; Kondrashov, F.A.; Hartl, D.L.; Kulathinal, R.J. The functional genomic distribution of protein divergence in two animal phyla: Coevolution, genomic conflict, and constraint. *Genome Res.* **2004**, *14*, 802–811. [CrossRef]

28. Ota, M.; Gonja, H.; Koike, R.; Fukuchi, S. Multiple-localization and hub proteins. *PLoS ONE* **2016**, *11*, e0156455. [CrossRef]

29. Jones, D.T.; Cozetto, D. DISOPRED3: Precise disordered region predictions with annotated protein-binding activity. *Bioinformatics* **2015**, *31*, 857–863. [CrossRef]

30. Fukuchi, S.; Homma, K.; Minezaki, Y.; Gojobori, T.; Nishikawa, K. Development of an accurate classification system of proteins into structured and unstructured regions that uncovers novel structural domains: Its application to human transcription factors. *BMC Struct. Biol.* **2009**, *9*, 26. [CrossRef]

31. Hirose, S.; Shimizu, K.; Kanai, S.; Kuroda, Y.; Noguchi, T. POODLE-L: A two-level SVM prediction system for reliably predicting long disordered regions. *Bioinformatics* **2007**, *23*, 2046–2053. [CrossRef]

32. Lee, Y.H.; Ota, T.; Vacquier, V.D. Positive selection is a general phenomenon in the evolution of abalone sperm lysin. *Mol. Biol. Evol.* **1995**, *12*, 231–238. [CrossRef]

33. Swanson, W.J.; Vacquier, V.D. Extraordinary divergence and positive Darwinian selection in a fusagenic protein coating the acrosomal process of abalone spermatozoa. *Proc. Natl. Acad. Sci. USA* **1995**, *92*, 4957–4961. [CrossRef]

34. Tsauer, S.-C.; Wu, C.-I. Positive selection and the molecular evolution of a gene of male reproduction, *Acp26Aa*, of *Drosophila*. *Mol. Biol. Evol.* **1997**, *14*, 544–549. [CrossRef]

35. Wyckoff, G.J.; Wang, W.; Wu, C.-I. Rapid evolution of male reproductive genes in the descent of man. *Nature* **2000**, *403*, 304–309. [CrossRef]

36. Zerbino, D.R.; Achuthan, P.; Akanni, W.; Amode, M.R.; Barrell, D.; Bhai, J.; Billis, K.; Cummins, C.; Gall, A.; Girón, C.G.; et al. Ensembl 2018. *Nucleic Acids Res.* **2018**, *46*, D754–D761. [CrossRef]

37. Skrzypek, M.S.; Nash, R.S.; Wong, E.D.; MacPherson, K.A.; Hellerstedt, S.T.; Engel, S.R.; Karra, K.; Weng, S.; Sheppard, T.K.; Binkley, G.; et al. Saccharomyces genome database informs human biology. *Nucleic Acids Res.* **2018**, *46*, D736–D742. [CrossRef]

38. Gramates, L.S.; Marygold, S.J.; dos Santos, G.; Urbano, J.M.; Antonazzo, G.; Matthews, B.B.; Rey, A.J.; Tabone, C.J.; Crosby, M.A.; Emmert, D.B.; et al. FlyBase at 25: Looking to the future. *Nucleic Acids Res.* **2017**, *45*, D663–D671. [CrossRef]

39. The UniProt Consortium. UniProt: The universal protein knowledgebase. *Nucleic Acids Res.* **2017**, *45*, D158–D169. [CrossRef]

40. Kuraku, S.; Zmasek, C.M.; Nishimura, O.; Katoh, K. aLeaves facilitates on-demand exploration of metazoan gene family trees on MAFFT sequence alignment server with enhanced interactivity. *Nucleic Acids Res.* **2013**, *41*, W22–W28. [CrossRef]

41. Yang, Z. PAML4: Phylogenetic analysis by maximum likelihood. *Mol. Biol. Evol.* **2007**, *24*, 1586–1591. [CrossRef]

42. Chatr-Aryamontri, A.; Oughtred, R.; Boucher, L.; Rust, J.; Chang, C.; Kolas, N.K.; O'Donnell, L.; Oster, S.; Theesfeld, C.; Sellam, A.; et al. The BioGRID interaction database: 2017 update. *Nucleic Acids Res.* **2017**, *45*, D369–D379. [CrossRef]

43. Wang, M.; Herrmann, C.J.; Simonovic, M.; Szklarczyk, D.; von Mering, C. Version 4.0 of PaxDB: Protein abundance data, integrated across model organisms, tissues, and cell-lines. *Proteomics* **2015**, *15*, 3163–3168. [CrossRef]

© 2018 by the authors. Licensee MDPI, Basel, Switzerland. This article is an open access article distributed under the terms and conditions of the Creative Commons Attribution (CC BY) license (http://creativecommons.org/licenses/by/4.0/).

International Journal of
Molecular Sciences

Communication

Direct Single-Molecule Observation of Sequential DNA Bending Transitions by the Sox2 HMG Box

Mahdi Muhammad Moosa, Phoebe S. Tsoi, Kyoung-Jae Choi, Allan Chris M. Ferreon * and Josephine C. Ferreon *

Department of Pharmacology and Chemical Biology, Baylor College of Medicine, Houston, TX 77030, USA; Mahdi.Moosa@bcm.edu (M.M.M.); Phoebe.Tsoi@bcm.edu (P.S.T.); Kyoungjae.Choi@bcm.edu (K.-J.C.)
* Correspondence: Allan.Ferreon@bcm.edu (A.C.M.F.); Josephine.Ferreon@bcm.edu (J.C.F.);
  Tel.: +1-713-798-1754 (A.C.M.F.); +1-713-798-1756 (J.C.F.)

Received: 2 October 2018; Accepted: 30 November 2018; Published: 4 December 2018

Abstract: Sox2 is a pioneer transcription factor that initiates cell fate reprogramming through locus-specific differential regulation. Mechanistically, it was assumed that Sox2 achieves its regulatory diversity via heterodimerization with partner transcription factors. Here, utilizing single-molecule fluorescence spectroscopy, we show that Sox2 alone can modulate DNA structural landscape in a dosage-dependent manner. We propose that such stoichiometric tuning of regulatory DNAs is crucial to the diverse biological functions of Sox2, and represents a generic mechanism of conferring functional plasticity and multiplicity to transcription factors.

Keywords: transcription factors; DNA-protein interactions; Sox2 sequential DNA loading; smFRET; DNA conformational landscape; sequential DNA bending; transcription factor dosage

1. Introduction

Sox2 regulates a remarkable variety of genes differentially; it activates some and represses others [1–3]. This functional diversity is assumed to be mediated by Sox2 heterodimerization with other transcription factors (TFs) such as Oct4, Oct1, Pax6, and Nanog [4,5]. Recent reports, however, suggest that these canonical partners often remain spatiotemporally separated from Sox2 during genome engagement [6–10]. This raises an important question regarding the TF's mechanism of action as to how Sox2 alone can exert differential loci-specific regulatory effects.

Sox2 is a sequence-specific high-mobility group transcription factor (HMG-TF) [11]. These TFs have conserved DNA binding domains [12,13], also known as HMG box. These DNA binding domains are partly disordered and are assumed to undergo binding-induced functional disorder-to-order transitions [14]. HMG-TFs are known to cooperatively form heterodimers on DNA regulatory elements [13,15–17]; each heteromeric TF pair induces characteristic DNA bend and differentially regulates target gene transcription [18–20]. Interestingly, a number of recent studies suggested that Sox2 can also function as homodimers [21–23]. Whether and how such Sox2 assemblies alter DNA conformations remain largely unknown. Here, we utilize the strengths of single-molecule Förster/fluorescence resonance energy transfer (smFRET) measurements along with ensemble methods to understand the effects of Sox2 binding on regulatory DNA structural landscape in the context of the HMG box (Sox2^{HMG}). Our results suggest that Sox2^{HMG} induces stoichiometry-dependent alternate DNA bends and we propose that the resulting alternate DNA conformations may drive different transcriptional outcomes.

2. Results

2.1. Multiple Sox2^{HMG} Domains Cooperatively Interact with dsDNA^{NANOG}

In our initial ensemble experiments, we observe that Sox2^{HMG} cooperatively binds to the *NANOG* composite promoter (DNA^{NANOG}; Figure 1). We utilized fluorescence anisotropy to detect Sox2^{HMG} binding to dsDNA^{NANOG} (Supplementary Methods). Anisotropy reports on fluorophore rotational properties, dependent on both probe local and global environment perturbations; fluorescence anisotropy of labeled macromolecule usually increases upon ligand binding. To characterize Sox2^{HMG}-DNA binding, we singly-labeled dsDNA^{NANOG} with Alexa Fluor 647 (Supplementary Methods) and monitored changes in DNA fluorescence anisotropy with increasing Sox2^{HMG} concentrations (Figure 1a). Nonlinear least squares (NLS) fitting of the anisotropy data to a Hill equation yields an apparent dissociation constant (K_D) of 15.1 ($\pm$2.0) nM and Hill coefficient of 1.5 ($\pm$0.3). The estimated K_D is similar to that previously reported for specific DNA-Sox2 interactions [18]. A Hill coefficient greater than 1 indicates that multiple Sox2 HMG boxes bind to the DNA in a TF concentration-dependent fashion [24]. Anisotropy measurements also indicate that Sox2^{HMG} alone (i.e., the DNA-binding domain in the absence of dsDNA^{NANOG}) fails to dimerize/oligomerize (Figure S1). To verify the binding of multiple Sox2 molecules to DNA^{NANOG}, we carried out fluorescence electrophoretic mobility shift assay (fEMSA) of DNA with increasing [Sox2^{HMG}]. The fEMSA micrograph shows concentration-dependent appearance of multiple electrophoretic species (Figure 1b). This suggests a multistep Sox2^{HMG} interaction with the *NANOG* proximal promoter. The non-equilibrium nature of mobility shift assays, however, precludes precise estimation of binding affinities of individual Sox2-DNA assemblies on the basis the fEMSA micrograph [25].

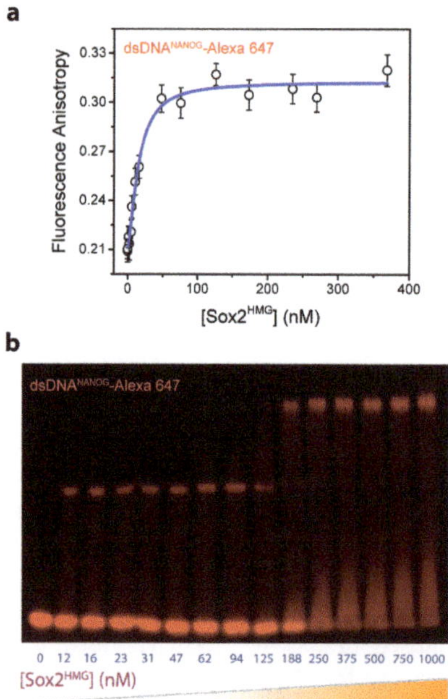

Figure 1. Sox2 cooperatively binds to the *NANOG* upstream promoter (DNA^{NANOG}). (**a**) DNA binding of Sox2^{HMG} was probed by monitoring changes in fluorescence anisotropy of Alexa Fluor 647-labeled dsDNA with increasing [Sox2^{HMG}]. The solid line represents nonlinear least squares (NLS) fit of the data to a Hill equation. NLS-derived parameters: K_D = 15.1 ($\pm$2.0) nM, Hill coefficient = 1.5 ($\pm$0.3). (**b**) Fluorescence electrophoretic mobility assay (fEMSA) of Sox2^{HMG}-DNA^{NANOG} binding suggests a multistep Sox2^{HMG} complex formation with dsDNA^{NANOG} involving multiple protein molecules that are able to bind the DNA partner. (See also Figure S2.)

2.2. Sox2^{HMG} Induces Sequential dsDNA^{NANOG} Bending Transitions

Next, we focused on understanding the mechanism of the TF-DNA complex formation. Although the mobility shift assay clearly demonstrates a multistep higher-order Sox2^{HMG} complex formation with the dsDNA (Figure 1b), our ensemble experiments (i.e., fluorescence anisotropy and fEMSA) were not sensitive enough to determine the stoichiometries of respective TF-DNA complexes. To directly observe Sox2^{HMG}-DNA^{NANOG} binding steps, we performed single-molecule fluorescence microscopy experiments that provide key advantages over conventional ensemble methods: (1) individual conformational sub-populations that are averaged out in ensemble measurements can be directly detected; and (2) experiments can be performed with extremely low concentrations of the labeled molecule (typically 50–100 pM). The ability to carry out experiments at low biomolecule concentration provides access and resolution for characterizing individual interaction steps in tightly interacting systems.

We utilized the distance-dependence of FRET to characterize Sox2^{HMG}-DNA^{NANOG} interaction at single-molecule resolution. smFRET is sensitive to distance changes in the 20–70 Å range [26], and provides the necessary spatial resolution to probe changes in dsDNA^{NANOG} conformations as induced by TF binding (estimated end-to-end distance of DNA^{NANOG} is 57.4 Å, assuming inter-base axial rise of 3.4 Å [27]). For the smFRET experiments, we labeled DNA^{NANOG} with Alexa Fluor 488 and 594 donor-accepter dye-pair (Supplementary Methods). Bursts of fluorescence from donor and acceptor dyes were recorded as dual-labeled *NANOG* promoter DNA passed through the sub-fL observation volume of our custom-built ISS Alba confocal laser microscopy system (described previously [28]). These fluorescence intensities were converted to FRET efficiency (E_{FRET}) histograms, providing a scheme for direct visualization of DNA conformational distributions. Without Sox2^{HMG}, the dual-labeled DNA showed a single-peak in its E_{FRET} histogram with histogram width typical of smFRET studies of freely diffusing dsDNA molecules [29,30] (Figure 2a; top panel). An NLS fit of the histogram to a Gaussian function yielded E_{FRET} value of 0.39 ($\pm$0.04). On the basis of this E_{FRET} value, we estimate the apparent distance between the two dyes to be approximately 64.6 Å (assuming a Förster distance of 60 Å between Alexa 488/594 dyes [31]). This is consistent with the estimated end-to-end distance of dsDNA^{NANOG}, where the slight increase in the apparent distance (compared to the estimated distance) can be attributed to the linkers present in Alexa dyes.

Often, histograms of data collected in diffusion-based smFRET experiments show an additional peak at zero E_{FRET} that arise from molecules with active donor(s) and either inactive or absent acceptor [29,30,32–35]. These zero E_{FRET} peaks tend to significantly overlap with low E_{FRET} peak populations and hamper direct estimation of the position of the non-zero peak(s) [36–39]. Interestingly, our smFRET histograms lack zero E_{FRET} peaks (Figure 2a). We attribute this to the absence of dual donor-labeled dsDNA molecules as ensured by sequential labeling of individual DNA strands (Supplementary Methods). Therefore, sequential labeling and purification of individual fluorophore-conjugated oligos prior to duplex formation can be utilized to minimize zero peaks.

DNA bending (also known as DNA looping) is critical for many eukaryotic TF function [40–44]. Accordingly, Sox2 was shown to induce binding-mediated *FGF* (fibroblast growth factor) enhancer bending [18]. We postulate that similar spatially precise bending is induced in Sox2-DNA^{NANOG} complexes during gene regulation. To characterize Sox2^{HMG} binding-induced *NANOG* promoter DNA bending, we carried out isothermal smFRET Sox2^{HMG} titration against approximately 100 pM dual-labeled DNA (Figure 2). Our smFRET experiments provide a direct way to distinguish between subtle conformational changes of DNA^{NANOG} induced upon Sox2 binding. In our smFRET experiments, we observed a multistep bending transition in the DNA structural landscape (Figure 2a). Initially, DNA^{NANOG} undergoes a cooperative bending to a 0.45 ($\pm$0.01) E_{FRET} state that corresponds to 32.1° ($\pm$1.4°) apparent bend angle at low Sox2^{HMG} concentrations ($\leq$4 nM) (see Supplementary Methods for the details of FRET-to-apparent-angle conversion). NLS fit of the data yields an estimated K_D of 305 ($\pm$39) pM (Figure 2c). Such a tight interaction is unlikely to be driven by higher order

Sox2^{HMG} assemblies and we therefore postulate that this dsDNA conformation (henceforth referred as B$_I$) is induced by binding to single Sox2^{HMG} molecules.

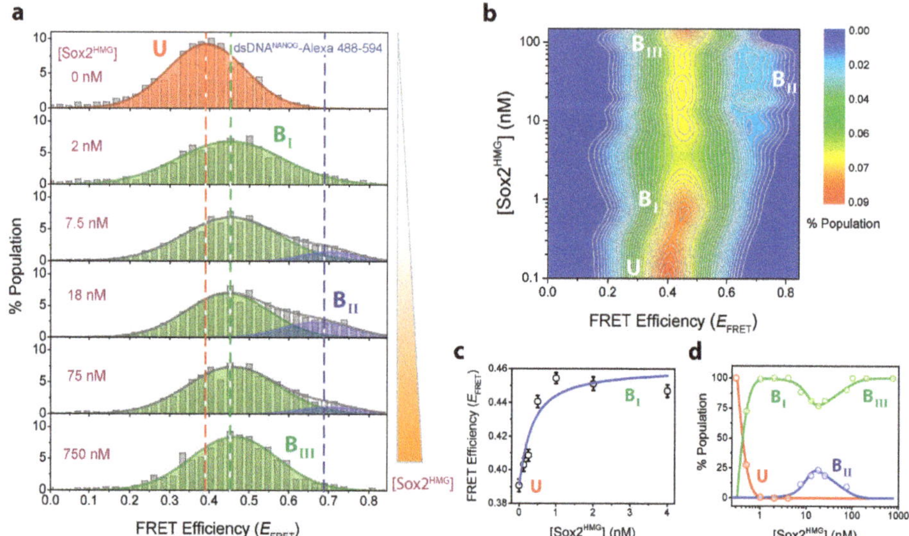

Figure 2. smFRET reveals Sox2^{HMG} concentration-dependent multistep bending of DNA^{NANOG}. (**a**) E_{FRET} histograms of DNA^{NANOG} with increasing [Sox2^{HMG}]. (**b**) [Sox2^{HMG}]-E_{FRET} contour map color coded based on fractional occupancy of individual DNA conformations. Corresponding DNA conformations are marked on the contour map. (**c**) Sox2 binding isotherm of the U $\rightleftharpoons$ B$_I$ transition as probed by detecting changes in E_{FRET}, linked to dsDNA bending transition. The NLS-derived apparent K_D for this binding step is 0.30 ($\pm$0.04) nM (binding equation with fixed Hill coefficient of 1). (**d**) dsDNA^{NANOG} conformational distributions as modulated by Sox2^{HMG} concentration, determined from NLS fitting of individual smFRET histograms to Gaussian functions.

Our ensemble results suggested that multiple Sox2^{HMG} can form higher order TF-DNA assemblies (Figure 1). To characterize the complex formation, we probed for changes in DNA^{NANOG} conformations upon further addition of Sox2 on preformed monomeric Sox2^{HMG}-DNA^{NANOG} complexes. With increasing [Sox2^{HMG}], we observe a progressive reduction of the B$_I$ population and the emergence of a new population exhibiting higher E_{FRET} (~0.68). This higher E_{FRET} population corresponds to a DNA^{NANOG} apparent bend angle of 70° ($\pm$2.4°; henceforth referred to as B$_{II}$ DNA conformation). We infer that this DNA conformation is induced by sequential binding of two individual Sox2 TFs on the dsDNA, where binding of each monomer induces an approximate 32° bend at respective binding sites. Our observed apparent bend angle in the ternary complex (two Sox2 monomers and DNA) is similar to the DNA bend angle previously resolved for heterodimeric HMG box TF-DNA complexes [11,17].

Interestingly, an additional transition is visible in our isothermal smFRET titration when additional Sox2^{HMG} is added (i.e., >75 nM [Sox2^{HMG}]). We observe progressive depopulation of the B$_{II}$ bent DNA conformation and coupled emergence of a population at E_{FRET} ~0.44 as [Sox2^{HMG}] increases further (henceforth referred as B$_{III}$; Figure 2a). We estimate the apparent bend angle for the B$_{III}$ population to be 30.4° ($\pm$4.5°) from the E_{FRET} data (Supplementary Methods). A longer fEMSA run also indicates higher-order oligomer formation that is consistent with the formation of B$_{III}$ population (Figure S2). Mechanistically, Sox family TFs induce DNA bends via FM dipeptide intercalation between two Thymine (T) bases at the minor groove interface [45,46]. Within the *NANOG* composite promoter, three TT pairs are present: two within the two HMG-TF binding sites (Oct/Sox motifs) identified by

Rodda et al. [47] and one in between. We hypothesize that the initial two DNA bends are induced by sequential Sox2 binding to the two high-affinity HMG-TF binding motifs, where each binding induces an apparent 32° bend at the sites of interactions (a net 70° DNA apparent bend angle in the ternary complex). As Sox2^{HMG} concentration further increases (>75 nM), an additional TF molecule interacts with the DNA at the remaining TT site and induces similar bend albeit at the opposite DNA face. This results in effective reversal of the second bend as evidenced by the increased inter-dye distance (i.e., reduced E_{FRET}) at higher [Sox2^{HMG}]. The final bend remains relatively unchanged upon further increase in Sox2 (up to 1 µM; Figure 2d). Overall, our smFRET data directly demonstrates multistep sequential DNA bending transitions dependent on Sox2 concentration.

3. Discussion

Sox2 is a tightly regulated transcription factor; both significant increases and decreases in Sox2 dosage can be detrimental to its biological function [48,49]. Alterations in Sox2 dosage result in multiple developmental and acquired disorders [50–54]. We show that the Sox2 HMG box can induce concentration-dependent alternate DNA bends (Figure 3). Alternate promoter bends are likely to regulate genes differentially and initiate downstream cascades crucial for Sox2's diverse functions. Our results provide a mechanism for Sox2's strict dosage dependence in its function-dysfunction dichotomy [50,55–58].

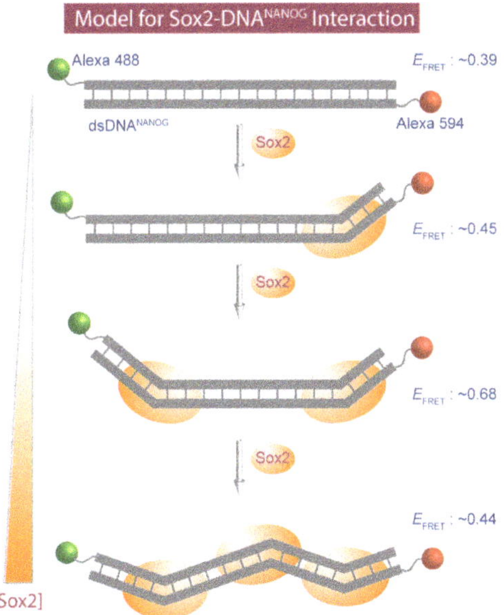

Figure 3. Schematic representation of the Sox2 stoichiometry-dependent dsDNA bending transitions.

In summary, our smFRET experiments clearly demonstrate the role of Sox2 dosage in modulating the conformational landscape of HMG box-binding DNA motifs. Previous studies on Sox family members suggested that heterodimeric homeodomain TFs can induce sequential bending as they interact with their DNA partners [59–62]. Here, we utilize the strengths of smFRET to demonstrate that a representative sequence-specific HMG-TF alone induces concentration-dependent multistep DNA bending transitions. We envision additional layers of tunability for heteromeric HMG-TFs in respective regulatory complexes where affinities of individual transcription factors for DNAs as well as inter-TF interactions can vary dramatically.

4. Materials and Methods

Experimental details are provided in the Supplementary Materials. Briefly, ensemble fluorescence anisotropy and fluorescence electrophoretic mobility assay (fEMSA) experiments were performed in Buffer E (20 mM Tris, 50 mM NaCl, 0.10 mg/mL BSA, 5% glycerol, 0.1 mM DTT/0.05 mM TCEP, pH 8) with Alexa Fluor-647 labeled dsDNA^{NANOG} (Forward: ACTTTTGCATTACAATG; 17 bp). smFRET experiments were performed in the same buffer using a custom-built confocal fluorescence microscopy set up as described previously [28].

Supplementary Materials: Supplementary materials can be found at http://www.mdpi.com/1422-0067/19/12/3865/s1.

Author Contributions: A.C.M.F. and J.C.F. conceived and designed the experiments; M.M.M., P.S.T., K.-J.C., A.C.M.F. and J.C.F. performed the experiments; M.M.M., P.S.T., A.C.M.F. and J.C.F. analyzed the data; M.M.M., A.C.M.F. and J.C.F. wrote the paper.

Funding: This work was supported by laboratory startup funds from the Baylor College of Medicine (A.C.M.F. and J.C.F.). J.C.F. is supported by R01 GM122763 from the NIGMS, NIH.

Conflicts of Interest: The authors declare no conflict of interest.

Abbreviations

fEMSA	fluorescence electrophoretic mobility shift assay
FGF	fibroblast growth factor
HMG	high mobility group
NLS	nonlinear least squares
smFRET	single-molecule Förster resonance energy transfer
TF	transcription factor

References

1. Chew, L.J.; Gallo, V. The Yin and Yang of Sox proteins: Activation and repression in development and disease. *J. Neurosci. Res.* **2009**, *87*, 3277–3287. [CrossRef] [PubMed]
2. Zhang, S.; Cui, W. Sox2, a key factor in the regulation of pluripotency and neural differentiation. *World J. Stem Cells* **2014**, *6*, 305–311. [CrossRef] [PubMed]
3. Liu, Y.R.; Laghari, Z.A.; Novoa, C.A.; Hughes, J.; Webster, J.R.; Goodwin, P.E.; Wheatley, S.P.; Scotting, P.J. Sox2 acts as a transcriptional repressor in neural stem cells. *BMC Neurosci.* **2014**, *15*, 95. [CrossRef] [PubMed]
4. Kondoh, H.; Kamachi, Y. SOX-partner code for cell specification: Regulatory target selection and underlying molecular mechanisms. *Int. J. Biochem. Cell Biol.* **2010**, *42*, 391–399. [CrossRef] [PubMed]
5. Kondoh, H.; Kamachi, Y. Chapter 8—SOX2–Partner Factor Interactions and Enhancer Regulation. In *Sox2*; Academic Press: Boston, MA, USA, 2016; pp. 131–144.
6. Thomson, M.; Liu, S.J.; Zou, L.N.; Smith, Z.; Meissner, A.; Ramanathan, S. Pluripotency factors in embryonic stem cells regulate differentiation into germ layers. *Cell* **2011**, *145*, 875–889. [CrossRef] [PubMed]
7. Soufi, A.; Donahue, G.; Zaret, K.S. Facilitators and impediments of the pluripotency reprogramming factors' initial engagement with the genome. *Cell* **2012**, *151*, 994–1004. [CrossRef] [PubMed]
8. Chen, J.; Zhang, Z.; Li, L.; Chen, B.C.; Revyakin, A.; Hajj, B.; Legant, W.; Dahan, M.; Lionnet, T.; Betzig, E.; Tjian, R. Single-molecule dynamics of enhanceosome assembly in embryonic stem cells. *Cell* **2014**, *156*, 1274–1285. [CrossRef]
9. Soufi, A.; Garcia, M.F.; Jaroszewicz, A.; Osman, N.; Pellegrini, M.; Zaret, K.S. Pioneer transcription factors target partial DNA motifs on nucleosomes to initiate reprogramming. *Cell* **2015**, *161*, 555–568. [CrossRef]
10. White, M.D.; Angiolini, J.F.; Alvarez, Y.D.; Kaur, G.; Zhao, Z.W.; Mocskos, E.; Bruno, L.; Bissiere, S.; Levi, V.; Plachta, N. Long-Lived Binding of Sox2 to DNA Predicts Cell Fate in the Four-Cell Mouse Embryo. *Cell* **2016**, *165*, 75–87. [CrossRef]
11. Hou, L.; Srivastava, Y.; Jauch, R. Molecular basis for the genome engagement by Sox proteins. *Semin. Cell Dev. Biol.* **2017**, *63*, 2–12. [CrossRef]

12. Soullier, S.; Jay, P.; Poulat, F.; Vanacker, J.M.; Berta, P.; Laudet, V. Diversification pattern of the HMG and SOX family members during evolution. *J. Mol. Evol.* **1999**, *48*, 517–527. [CrossRef] [PubMed]

13. Malarkey, C.S.; Churchill, M.E. The high mobility group box: The ultimate utility player of a cell. *Trends Biochem. Sci.* **2012**, *37*, 553–562. [CrossRef] [PubMed]

14. Weiss, M.A. Floppy SOX: Mutual induced fit in hmg (high-mobility group) box-DNA recognition. *Mol. Endocrinol.* **2001**, *15*, 353–362. [CrossRef] [PubMed]

15. Schlierf, B.; Ludwig, A.; Klenovsek, K.; Wegner, M. Cooperative binding of Sox10 to DNA: Requirements and consequences. *Nucleic Acids Res.* **2002**, *30*, 5509–5516. [CrossRef] [PubMed]

16. Ng, C.K.; Li, N.X.; Chee, S.; Prabhakar, S.; Kolatkar, P.R.; Jauch, R. Deciphering the Sox-Oct partner code by quantitative cooperativity measurements. *Nucleic Acids Res.* **2012**, *40*, 4933–4941. [CrossRef] [PubMed]

17. Clore, G.M. Chapter 3—Dynamics of SOX2 Interactions with DNA A2—Kondoh, Hisato. In *Sox2*; Lovell-Badge, R., Ed.; Academic Press: Boston, MA, USA, 2016; pp. 25–41.

18. Scaffidi, P.; Bianchi, M.E. Spatially precise DNA bending is an essential activity of the sox2 transcription factor. *J. Biol. Chem.* **2001**, *276*, 47296–47302. [CrossRef] [PubMed]

19. Dragan, A.I.; Read, C.M.; Makeyeva, E.N.; Milgotina, E.I.; Churchill, M.E.; Crane-Robinson, C.; Privalov, P.L. DNA binding and bending by HMG boxes: Energetic determinants of specificity. *J. Mol. Biol.* **2004**, *343*, 371–393. [CrossRef]

20. Slattery, M.; Riley, T.; Liu, P.; Abe, N.; Gomez-Alcala, P.; Dror, I.; Zhou, T.; Rohs, R.; Honig, B.; Bussemaker, H.J.; Mann, R.S. Cofactor binding evokes latent differences in DNA binding specificity between Hox proteins. *Cell* **2011**, *147*, 1270–1282. [CrossRef] [PubMed]

21. Li, J.; Pan, G.; Cui, K.; Liu, Y.; Xu, S.; Pei, D. A dominant-negative form of mouse SOX2 induces trophectoderm differentiation and progressive polyploidy in mouse embryonic stem cells. *J. Biol. Chem.* **2007**, *282*, 19481–19492. [CrossRef]

22. Cox, J.L.; Mallanna, S.K.; Luo, X.; Rizzino, A. Sox2 uses multiple domains to associate with proteins present in Sox2-protein complexes. *PLoS ONE* **2010**, *5*, e15486. [CrossRef]

23. Xia, P.; Wang, S.; Ye, B.; Du, Y.; Huang, G.; Zhu, P.; Fan, Z. Sox2 functions as a sequence-specific DNA sensor in neutrophils to initiate innate immunity against microbial infection. *Nat. Immunol.* **2015**, *16*, 366–375. [CrossRef] [PubMed]

24. Weiss, J.N. The Hill equation revisited: Uses and misuses. *FASEB J.* **1997**, *11*, 835–841. [CrossRef] [PubMed]

25. Hellman, L.M.; Fried, M.G. Electrophoretic mobility shift assay (EMSA) for detecting protein-nucleic acid interactions. *Nat. Protoc.* **2007**, *2*, 1849–1861. [CrossRef] [PubMed]

26. Ferreon, A.C.; Deniz, A.A. Protein folding at single-molecule resolution. *Biochim. Biophys. Acta* **2011**, *1814*, 1021–1029. [CrossRef]

27. Watson, J.D.; Crick, F.H.C. Molecular Structure of Nucleic Acids: A Structure for Deoxyribose Nucleic Acid. *Nature* **1953**, *171*, 737. [CrossRef] [PubMed]

28. Tsoi, P.S.; Choi, K.J.; Leonard, P.G.; Sizovs, A.; Moosa, M.M.; MacKenzie, K.R.; Ferreon, J.C.; Ferreon, A.C. The N-Terminal Domain of ALS-Linked TDP-43 Assembles without Misfolding. *Angew. Chem. Int. Ed. Engl.* **2017**, *56*, 12590–12593. [CrossRef] [PubMed]

29. Deniz, A.A.; Dahan, M.; Grunwell, J.R.; Ha, T.; Faulhaber, A.E.; Chemla, D.S.; Weiss, S.; Schultz, P.G. Single-pair fluorescence resonance energy transfer on freely diffusing molecules: Observation of Forster distance dependence and subpopulations. *Proc. Natl. Acad. Sci. USA* **1999**, *96*, 3670–3675. [CrossRef]

30. Dey, S.K.; Pettersson, J.R.; Topacio, A.Z.; Das, S.R.; Peteanu, L.A. Eliminating Spurious Zero-Efficiency FRET States in Diffusion-Based Single-Molecule Confocal Microscopy. *J. Phys. Chem. Lett.* **2018**, *9*, 2259–2265. [CrossRef]

31. Johnson, I.D.; Spence, M.T.Z. *The Molecular Probes Handbook: A Guide to Fluorescent Probes and Labeling Technologies*; Molecular Probes: Eugene, OR, USA, 2010.

32. Pljevaljcic, G.; Millar, D.P.; Deniz, A.A. Freely diffusing single hairpin ribozymes provide insights into the role of secondary structure and partially folded states in RNA folding. *Biophys. J.* **2004**, *87*, 457–467. [CrossRef]

33. Morgan, M.A.; Okamoto, K.; Kahn, J.D.; English, D.S. Single-molecule spectroscopic determination of lac repressor-DNA loop conformation. *Biophys. J.* **2005**, *89*, 2588–2596. [CrossRef]

34. Schuler, B. Single-molecule FRET of protein structure and dynamics—A primer. *J. Nanobiotechnol.* **2013**, *11* (Suppl. 1), S2. [CrossRef]

35. Tyagi, S.; VanDelinder, V.; Banterle, N.; Fuertes, G.; Milles, S.; Agez, M.; Lemke, E.A. Continuous throughput and long-term observation of single-molecule FRET without immobilization. *Nat. Methods* **2014**, *11*, 297–300. [CrossRef] [PubMed]

36. Ferreon, A.C.; Gambin, Y.; Lemke, E.A.; Deniz, A.A. Interplay of α-synuclein binding and conformational switching probed by single-molecule fluorescence. *Proc. Natl. Acad. Sci. USA* **2009**, *106*, 5645–5650. [CrossRef] [PubMed]

37. Ferreon, A.C.; Moran, C.R.; Ferreon, J.C.; Deniz, A.A. Alteration of the α-synuclein folding landscape by a mutation related to Parkinson's disease. *Angew. Chem. Int. Ed. Engl.* **2010**, *49*, 3469–3472. [CrossRef]

38. Gambin, Y.; VanDelinder, V.; Ferreon, A.C.; Lemke, E.A.; Groisman, A.; Deniz, A.A. Visualizing a one-way protein encounter complex by ultrafast single-molecule mixing. *Nat. Methods* **2011**, *8*, 239–241. [CrossRef] [PubMed]

39. Moosa, M.M.; Ferreon, A.C.; Deniz, A.A. Forced folding of a disordered protein accesses an alternative folding landscape. *Chemphyschem* **2015**, *16*, 90–94. [CrossRef]

40. Su, W.; Jackson, S.; Tjian, R.; Echols, H. DNA looping between sites for transcriptional activation: Self-association of DNA-bound Sp1. *Genes Dev.* **1991**, *5*, 820–826. [CrossRef]

41. Lim, F.L.; Hayes, A.; West, A.G.; Pic-Taylor, A.; Darieva, Z.; Morgan, B.A.; Oliver, S.G.; Sharrocks, A.D. Mcm1p-induced DNA bending regulates the formation of ternary transcription factor complexes. *Mol. Cell Biol.* **2003**, *23*, 450–461. [CrossRef]

42. Petrascheck, M.; Escher, D.; Mahmoudi, T.; Verrijzer, C.P.; Schaffner, W.; Barberis, A. DNA looping induced by a transcriptional enhancer in vivo. *Nucleic Acids Res.* **2005**, *33*, 3743–3750. [CrossRef]

43. Whittington, J.E.; Delgadillo, R.F.; Attebury, T.J.; Parkhurst, L.K.; Daugherty, M.A.; Parkhurst, L.J. TATA-binding protein recognition and bending of a consensus promoter are protein species dependent. *Biochemistry* **2008**, *47*, 7264–7273. [CrossRef]

44. Gietl, A.; Grohmann, D. Modern biophysical approaches probe transcription-factor-induced DNA bending and looping. *Biochem. Soc. Trans.* **2013**, *41*, 368–373. [CrossRef] [PubMed]

45. Williams, D.C.; Cai, M., Jr.; Clore, G.M. Molecular basis for synergistic transcriptional activation by Oct1 and Sox2 revealed from the solution structure of the 42-kDa Oct1.Sox2.Hoxb1-DNA ternary transcription factor complex. *J. Biol. Chem.* **2004**, *279*, 1449–1457. [CrossRef]

46. Palasingam, P.; Jauch, R.; Ng, C.K.; Kolatkar, P.R. The structure of Sox17 bound to DNA reveals a conserved bending topology but selective protein interaction platforms. *J. Mol. Biol.* **2009**, *388*, 619–630. [CrossRef] [PubMed]

47. Rodda, D.J.; Chew, J.L.; Lim, L.H.; Loh, Y.H.; Wang, B.; Ng, H.H.; Robson, P. Transcriptional regulation of nanog by OCT4 and SOX2. *J. Biol. Chem.* **2005**, *280*, 24731–24737. [CrossRef] [PubMed]

48. Yamaguchi, S.; Hirano, K.; Nagata, S.; Tada, T. Sox2 expression effects on direct reprogramming efficiency as determined by alternative somatic cell fate. *Stem Cell Res.* **2011**, *6*, 177–186. [CrossRef] [PubMed]

49. Prakash, N. Chapter 4—Posttranscriptional Modulation of Sox2 Activity by miRNAs A2—Kondoh, Hisato. In *Sox2*; Lovell-Badge, R., Ed.; Academic Press: Boston, MA, USA, 2016; pp. 43–71.

50. Bertolini, J.; Mercurio, S.; Favaro, R.; Mariani, J.; Ottolenghi, S.; Nicolis, S.K. Chapter 11—Sox2-Dependent Regulation of Neural Stem Cells and CNS Development A2—Kondoh, Hisato. In *Sox2*; Lovell-Badge, R., Ed.; Academic Press: Boston, MA, USA, 2016; pp. 187–216.

51. Van Heyningen, V. Chapter 13—Congenital Abnormalities and SOX2 Mutations A2—Kondoh, Hisato. In *Sox2*; Lovell-Badge, R., Ed.; Academic Press: Boston, MA, USA, 2016; pp. 235–242.

52. Rizzoti, K.; Lovell-Badge, R. Chapter 14—Role of SOX2 in the Hypothalamo–Pituitary Axis. In *Sox2*; Academic Press: Boston, MA, USA, 2016; pp. 243–262.

53. Iwafuchi-Doi, M.; Zaret, K.S. Cell fate control by pioneer transcription factors. *Development* **2016**, *143*, 1833–1837. [CrossRef] [PubMed]

54. Wuebben, E.L.; Rizzino, A. The dark side of SOX2: Cancer—A comprehensive overview. *Oncotarget* **2017**, *8*, 44917–44943. [CrossRef] [PubMed]

55. Sarkar, A.; Hochedlinger, K. The sox family of transcription factors: Versatile regulators of stem and progenitor cell fate. *Cell Stem Cell* **2013**, *12*, 15–30. [CrossRef] [PubMed]

56. Liu, K.; Lin, B.; Zhao, M.; Yang, X.; Chen, M.; Gao, A.; Liu, F.; Que, J.; Lan, X. The multiple roles for Sox2 in stem cell maintenance and tumorigenesis. *Cell Signal.* **2013**, *25*, 1264–1271. [CrossRef] [PubMed]

57. Kamachi, Y.; Kondoh, H. Sox proteins: Regulators of cell fate specification and differentiation. *Development* **2013**, *140*, 4129–4144. [CrossRef]

58. Hagey, D.W.; Muhr, J. Sox2 acts in a dose-dependent fashion to regulate proliferation of cortical progenitors. *Cell Rep.* **2014**, *9*, 1908–1920. [CrossRef] [PubMed]

59. Peirano, R.I.; Wegner, M. The glial transcription factor Sox10 binds to DNA both as monomer and dimer with different functional consequences. *Nucleic Acids Res.* **2000**, *28*, 3047–3055. [CrossRef] [PubMed]

60. Takayama, Y.; Clore, G.M. Impact of protein/protein interactions on global intermolecular translocation rates of the transcription factors Sox2 and Oct1 between DNA cognate sites analyzed by z-exchange NMR spectroscopy. *J. Biol. Chem.* **2012**, *287*, 26962–26970. [CrossRef] [PubMed]

61. Morimura, H.; Tanaka, S.I.; Ishitobi, H.; Mikami, T.; Kamachi, Y.; Kondoh, H.; Inouye, Y. Nano-analysis of DNA conformation changes induced by transcription factor complex binding using plasmonic nanodimers. *ACS Nano* **2013**, *7*, 10733–10740. [CrossRef] [PubMed]

62. Yamamoto, S.; De, D.; Hidaka, K.; Kim, K.K.; Endo, M.; Sugiyama, H. Single molecule visualization and characterization of Sox2-Pax6 complex formation on a regulatory DNA element using a DNA origami frame. *Nano Lett.* **2014**, *14*, 2286–2292. [CrossRef] [PubMed]

© 2018 by the authors. Licensee MDPI, Basel, Switzerland. This article is an open access article distributed under the terms and conditions of the Creative Commons Attribution (CC BY) license (http://creativecommons.org/licenses/by/4.0/).

International Journal of
Molecular Sciences

Article

The C Terminus of the Ribosomal-Associated Protein LrtA Is an Intrinsically Disordered Oligomer

José L. Neira [1,2,*,†], A. Marcela Giudici [1,†], Felipe Hornos [1], Arantxa Arbe [3] and Bruno Rizzuti [4,*]

[1] Instituto de Biología Molecular y Celular, Edificio Torregaitán, Universidad Miguel Hernández, Avda. del Ferrocarril s/n, 03202 Elche (Alicante), Spain; marcela@umh.es (A.M.G.); fhornos@umh.es (F.H.)

[2] Instituto de Biocomputación y Física de Sistemas Complejos, Joint Units IQFR-CSIC-BIFI, and GBsC-CSIC-BIFI, Universidad de Zaragoza, 50009 Zaragoza, Spain

[3] Centro de Física de Materiales (CFM) (CSIC-UPV/EHU)—Materials Physics Center (MPC), 20018 San Sebastián, Spain; mariaaranzazu.arbe@ehu.eus

[4] CNR-NANOTEC, Licryl-UOS Cosenza and CEMIF.Cal, Department of Physics, University of Calabria, Ponte P. Bucci, 87036 Rende, Italy

* Correspondence: jlneira@umh.es (J.L.N.); bruno.rizzuti@cnr.it (B.R.); Tel.: +34-96-6658475 (J.L.N.); +39-0984-49-6078 (B.R.)

† These two authors contributed equally to this work.

Received: 14 November 2018; Accepted: 2 December 2018; Published: 5 December 2018

Abstract: The 191-residue-long LrtA protein of *Synechocystis* sp. PCC 6803 is involved in post-stress survival and in stabilizing 70S ribosomal particles. It belongs to the hibernating promoting factor (HPF) family, intervening in protein synthesis. The protein consists of two domains: The N-terminal region (N-LrtA, residues 1–101), which is common to all the members of the HPF, and seems to be well-folded; and the C-terminal region (C-LrtA, residues 102–191), which is hypothesized to be disordered. In this work, we studied the conformational preferences of isolated C-LrtA in solution. The protein was disordered, as shown by computational modelling, 1D-^{1}H NMR, steady-state far-UV circular dichroism (CD) and chemical and thermal denaturations followed by fluorescence and far-UV CD. Moreover, at physiological conditions, as indicated by several biochemical and hydrodynamic techniques, isolated C-LrtA intervened in a self-association equilibrium, involving several oligomerization reactions. Thus, C-LrtA was an oligomeric disordered protein.

Keywords: disordered protein; folding; oligomer; ribosomal protein; protein stability

1. Introduction

The *lrtA* gene from *Synechococcus* sp. PCC702 is known to express a light-repressed protein [1,2]. Further investigations have shown that LrtA is involved in the stabilization the 70S ribosomal particles [3], as well as in cell survival during stress circumstances. LrtA is related with other proteins that take part in ribosome activity. Under environmental stress conditions, protein synthesis is stopped in a down-regulation process. Reduction of protein production involves: (i) Formation of the inactive 100S disome through dimerization of 70S particles [4], implicating the action of some proteins; or (ii) protein-ribosome interactions which involve the canonical ribosomal proteins [5,6]. The family of ribosomal proteins in *E. coli* includes YfiA (also known as PY or RaiA, ribosome associated inhibitor A); and YhbH (also known as HPF, hibernation promoting factor). YfiA likely inhibits translation indirectly, involving 70S particles [7,8]. Alternatively, HPF stops translation by stabilizing 100S dimers [8–10]. Most bacteria have one or more homologues related to HPF or YfiA [10]. These homologues can be classified in long HPF, short HPF and YFiA, on the basis of the length of their sequences and the presence of a specific domain. The conserved domain in all of them has the β-α-β-β-β-α fold [5,11], with a β-sheet formed by four strands and two α-helices packed against it. According to its sequence,

LrtA from *Synechocystis* sp. PCC 6803 belongs to the long HPF sub-family. We have previously shown that LrtA is involved in self-association equilibria [12], and has chameleonic structural properties. In particular, molecular dynamics (MD) simulations and experimental analyses suggest that the whole LrtA has a solvent-dependent conformation, where the N terminus adopts the β-α-β-β-β-α fold and the C terminus is disordered and compact [12].

In this work, we have studied the conformational preferences of the isolated C-terminal region of LrtA (residues 102-191), C-LrtA. We aimed to test whether: (i) C-LrtA was disordered and collapsed, as suggested by previous MD simulations of the whole LrtA; and (ii) isolated C-LrtA was oligomeric in solution. Characterizing the degree of disorder in proteins or protein domains, and whether this contributes to attaining a quaternary structure, is important to explain their functions; in fact, most of the intrinsically disordered proteins (IDPs) characterized so far are involved in protein-protein contacts [13], and it is essential to establish how specificity is achieved in those interactions. We show here that C-LrtA was disordered and with a strong self-association tendency, as shown by several biochemical, biophysical and hydrodynamic techniques: Blue-native gels, glutaraldehyde cross-linking, iodide quenching, small-angle X-ray scattering (SAXS), size exclusion chromatography (SEC) and isothermal titration calorimetry (ITC). MD simulations of isolated C-LrtA also predicted a disordered conformation, in reasonable agreement with the experiments. Therefore, we proved that: (i) former MD predictions on C-LrtA, based on the whole parental LrtA, were correct; (ii) the isolated domain has a tendency to self-associate; and (iii) the presence of quaternary interactions in C-LrtA did not induce any stable secondary nor tertiary structure, and therefore, C-LrtA was an oligomeric IDP.

2. Results

2.1. Isolated C-LrtA Was Intrinsically Disordered in Solution

To map the conformational features of C-LrtA in solution, we used NMR, fluorescence, far-UV (circular dichroism) CD and MD simulations. Fluorescence gives us information about the overall environment around the fluorescent residues (C-LrtA has 4 tyrosine residues). Far-UV CD provides information about the percentages of secondary structure. NMR gives further information about the presence of secondary and tertiary structures. Finally, MD simulations give indications on the conformation of isolated C-LrtA in solution and on the local propensity for secondary structure formation along the backbone.

(a) NMR: In the methyl (Figure 1A) and the amide (Figure 1B) regions of the NMR spectrum of C-LrtA, there was no dispersion, i.e., all the amide signals appeared clustered between 8.2 and 8.6 ppm and most of the alkyl chains appeared between 0.8 and 1.0 ppm. Only a small shoulder appeared up-field shifted at 0.7 ppm, indicative of a local conformation around the methyl group of a valine, leucine or isoleucine probably close to an aromatic residue; we hypothesize that this signal could correspond to the polypeptide patch VIYI (residues 173–176, in the numbering of the whole LrtA).

The NMR spectrum showed significant broadening in all the signals. On the basis of the results of the other techniques (see below in this section, and later in Section 2.2), this could be due to the presence of conformational exchange (equilibria) among protein species with different self-associated order. However, given the mobility of the protein (see MD simulations results, below in this section), we cannot exclude that the broadening observed could be due to conformational exchange in a single protein species.

Therefore, we can conclude from the NMR spectrum that C-LrtA was disordered.

(b) Far-UV CD: The far-UV CD spectrum of C-LrtA showed a minimum of around 200 nm and a wide shoulder at 222 nm (Figure 2A), which could be due to the presence of helix or turn-like structures, although the absorbance of aromatic residues (4 tyrosine and 3 phenylalanine ones) at the latter wavelength cannot be ruled out [14,15]. Although we cannot exclude protein adsorption to the cell at the lowest protein concentration used, the spectrum intensity was protein-concentration dependent (in the range of 10 to 20 µM of protein concentration, Figure S1) and its shape did not change. The fact that the other techniques used (see Section 2.2) indicate the presence of oligomeric species suggests that the variations observed in the far-UV CD spectra are due to the existence of self-associated species.

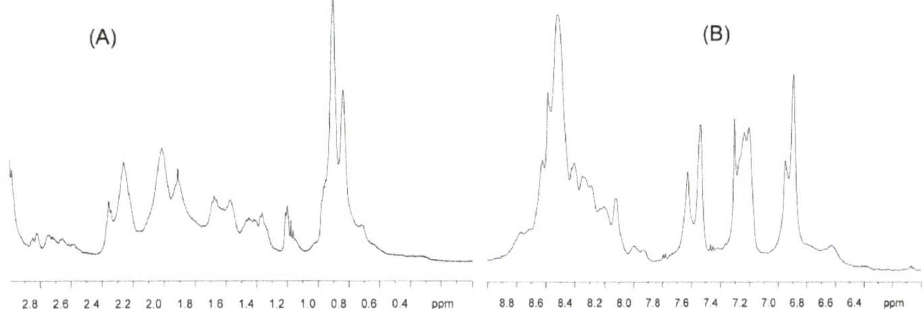

Figure 1. NMR characterization of C-LrtA: (**A**) Methyl and (**B**) amide regions of the 1D ^{1}H NMR spectrum of C-LrtA. Spectrum was acquired at 20 °C, and pH 7.2 (50 mM, Tris buffer).

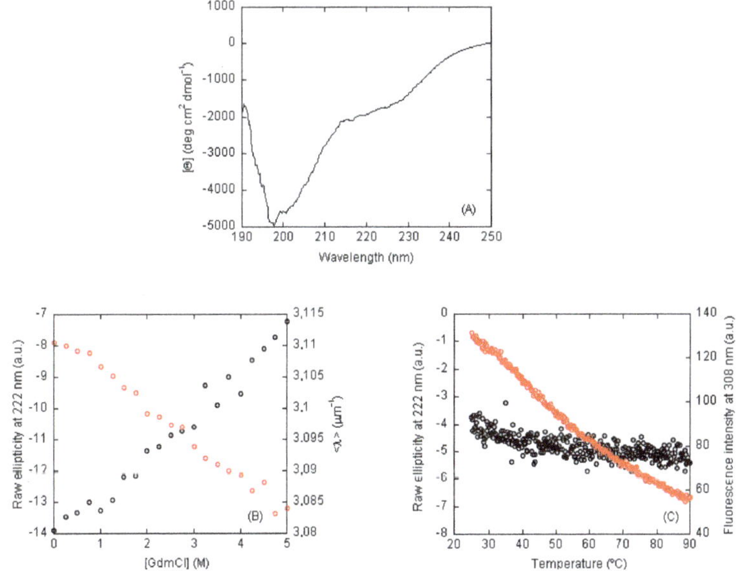

Figure 2. Spectroscopic characterization of C-LrtA: (**A**) Far-UV circular dichroism (CD) spectrum of C-LrtA. Spectrum was acquired at 20 °C, and pH 7.2 (50 mM, Tris buffer) with 20 µM (in protomer units) of protein concentration; (**B**) GdmCl-denaturations of C-LrtA followed by fluorescence (right axis, red circles) and CD (left axis, black circles), at 10 µM of protein concentration (in protomer units) and 20 °C; (**C**) Thermal denaturations of C-LrtA followed by fluorescence (right axis, red circles) and CD (left axis, black circles). Experiments were acquired at pH 7.2 (50 mM, Tris buffer) and 10 µM of protein concentration (in protomer units). The ellipticity (far-UV CD) units of thermal denaturations are arbitrary, because values are scaled up.

The shape of the spectrum of C-LrtA was characteristic of IDPs [16]. Its deconvolution, by using the algorithms available at the DICHROWEB site [17,18], yielded percentages of 7–8% for α-helix structure, 15–20% for β-turn, 28–44% for β-sheet and 45–48% for random-coil.

In the presence of increasing GdmCl concentrations, the shoulder at 222 nm of the far-UV CD spectrum decreased (Figure 2B, left axis, black circles) at the two concentrations explored (10 and 20 μM). These results suggest that the shoulder was not due to the presence of any well-fixed structure, but rather to flickering helix- or turn-like motifs, or even local conformations of the aromatic residues [14,15] Attempts to fit these data to the linear extrapolation model failed, as they led to thermodynamic parameters with non-physical meaning (i.e., negative values of m or large values of $[GdmCl]_{1/2}$). We further tested the disordered nature of C-LrtA by performing thermal denaturations. We observed a decrease in the ellipticity as the temperature was increased (Figure 2C, left axis, black circles) and therefore we did not observe a sigmoidal co-operative behaviour, as it should be expected for a well-folded globular domain [16,19].

Therefore, we can conclude from the far-UV CD data that C-LrtA was disordered.

(c) Fluorescence: Fluorescence spectra of C-LrtA showed a maximum at 307 nm, corresponding to its 4 tyrosine residues [20,21]. We carried out GdmCl denaturations by following the <λ> (at two different C-LrtA concentrations) after excitation at 280 nm. At both protein concentrations, we observed a linear decrease in the <λ> as the concentration of chemical denaturant was increased (Figure 2B, right axis, red circles). We could not fit these data to the linear extrapolation model, as fitting led to thermodynamic parameters (m- or $[GdmCl]_{1/2}$-values) with non-physical meaning (i.e., either negative values or values higher than the protein concentration explored). A similar linear tendency was observed in thermal denaturations (Figure 2C, right axis, red circles).

Therefore, we can conclude from the fluorescence data that C-LrtA was disordered.

(d) MD simulations: C-LrtA (the sequence present in the wild-type protein, i.e., without the His-tag) was simulated starting from an extended conformation with a radius of gyration R_g = 64 Å. The protein structure spontaneously collapsed in 15 ns, and for the subsequent 10 ns maintained a size of R_g = 24 ± 2 Å, in excellent agreement with the value (R_g = 25 Å) predicted for a tag-free IDP of 90 residues with the sequence features of C-LrtA [22]. As reported in Figure 3, the secondary structure propensity of the protein in the time interval considered was reasonably consistent. In contrast, the conformations sampled at sufficiently large intervals (e.g., every 0.1 ns) were relatively different in terms of their three-dimensional arrangement, due to the dynamics of C-LrtA. Although not being long enough to obtain a complete statistical ensemble of conformations, the simulation was not prolonged to prevent well-known artifacts, due to over-compaction of the protein [23,24], because a small drift was observed in its size (decrease of R_g was ~ 0.3 Å/ns on average, Figure 3).

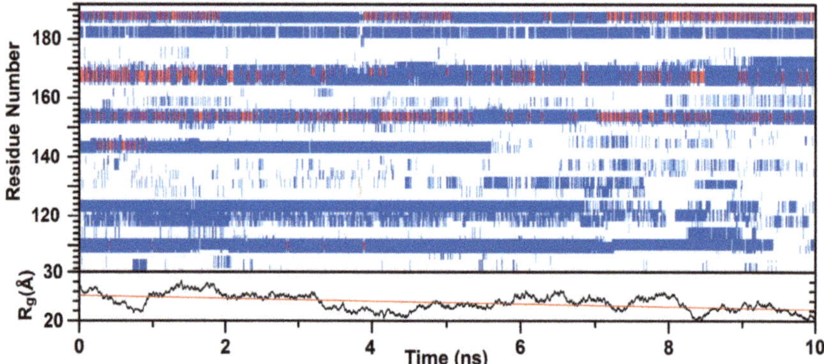

Figure 3. Simulated secondary structure and radius of gyration of C-LrtA: Backbone properties of the protein without the His-tag are calculated in a 10 ns time interval, following 15 ns of equilibration after starting from an elongated conformation. (**Up**) Secondary structure propensities calculated with VMD [25]: (Blue) β-structure, (red) helical structure, and (white) random coil; (**Down**) Radius of gyration of C-LrtA; the drift leading to a small decrease of R_g is also shown (red line).

The simulation results concurred to indicate that C-LrtA in solution was a very flexible protein with little secondary structure. The percentages of helical/β-structure were in good agreement with the range of those obtained from the deconvolution of far-UV CD spectra, but the corresponding backbone conformations were in all cases local and did not extend for more than a few residues. Interestingly, among the four tyrosine residues of C-LrtA, only Tyr182 (according to the numbering in intact LrtA) was in a region with β-structure propensity, whereas the other three were in random-coil regions and showed a large conformational freedom in the isolated domain.

In summary, taking into account all the data, as concluded from fluorescence, far- UV CD, NMR and MD simulations, C-LrtA appeared disordered in solution.

2.2. Isolated C-LrtA Was Involved in Self-Association Equilibria in Solution

To map the hydrodynamic properties of LrtA we used several biochemical, biophysical and hydrodynamic techniques: Blue-native gels; glutaraldehyde cross-linking; iodide quenching; DOSY-NMR (diffusion ordered spectroscopy NMR); SAXS (small-angle X-ray scattering); SEC (size exclusion chromatography) and ITC (isothermal titration calorimetry). We used such a plethora of different techniques to provide an unambiguous evidence of the presence of oligomerization in disordered C-LrtA. It is important to pinpoint, however, that with NMR we shall obtain information about the low-molecular weight species whose overall rotational tumbling is very fast, and then, we shall be able to obtain information only on the monomer and/or dimer species.

(a) DOSY-NMR: The DOSY-NMR measurements of C-LrtA yielded a translational diffusion coefficient, D, of $(5.0 \pm 0.2) \times 10^{-7}$ cm^2 s^{-1} (Figure 4A). By taking into account the hydrodynamic radius, R_S, of dioxane (2.12 Å), and its D under our conditions (($8.53 \pm 0.02) \times 10^{-6}$ cm^2 s^{-1}), the estimated R_S for C-LrtA was 36 ± 4 Å. We can compare this value with that theoretically determined for a polypeptide with the sequence length of C-LrtA (including the N-terminal His-tag). The R value for an unsolvated, ideal, spherical molecule can be estimated from [21]: $R = \sqrt[3]{3M\overline{V}/4N_A\pi}$, where N_A is Avogadro's number, M is the molecular weight of the C-LrtA construct (12,449.89 Da), and $\overline{V}$ the specific volume of C-LrtA construct (0.721 mL/g). The calculated radius for C-LrtA is 15.3 Å, but taking into account the water shell [21,26], the hydration radius is 18.5 Å; this value is different from that obtained from experimental DOSY measurements. The R_s for a spherical, folded protein is given by [27]: $R_S = (4.75 \pm 1.11)N^{0.29}$, where N is the number of residues; in a 109-residue-long protein, such as C-LrtA (the His-tag and the 90-residue-long domain), this expression yields 18 ± 4 Å, in good agreement with the other theoretically calculated value. On the other hand, for an unfolded polypeptide chain, the R_S could be estimated from [27]: $R_S = (2.21 \pm 1.07)N^{0.57}$; for C-LrtA, the value is $32 + 15$ Å, which is closer to the values measured in the DOSY-NMR experiments; however, the use of that expression yields a value slightly higher (43 ± 15 Å) for a dimeric species. Therefore, by the DOSY-NMR experiments we are only detecting low-molecular weight species, which seemed to be unfolded.

(b) SEC: Different amounts of C-LrtA were loaded in an analytical Superose 12 10/300 GL column at pH 8.0 (50 mM Tris) and 0.250 M NaCl. The chromatograms did not show a sole peak (Figure S2), and the smaller the concentration of the protein the more were the peaks that appeared. This finding, given protein purity (Figure S3A), could be attributed to protein-column interactions of some species, which eluted at larger volumes than expected from their size. Similar delayed peaks, due to protein-column interactions, have been observed in the intact LrtA [12]. It is important to note that a small peak appearing at 16.48 mL was also present in the chromatogram of the most concentrated sample (500 μM) (Figure S2), as well as at any other protein concentration; we interpreted this peak as due to a monomeric species interacting with the column.

The elution volumes of one of the peaks showed a hyperbolic dependence as the concentration of protein was changed (Figure 4B). At very high concentrations (500 μM), the protein had elution volumes of 11.98 mL (obtained as the mean of three different measurements, although the elution peak was very broad). This value would correspond to a molecular weight of 100 kDa (for a comparison, a protein, such as ferritin, with a molecular weight of 400 kDa, elutes in the column at 10.11 mL, Figure S2); these results suggest that C-LrtA was probably an octamer under these conditions. On the other hand, at 30 µM, C-LrtA eluted at 13.88 mL, which would correspond to a molecular weight of 31.6 kDa, close to the expected molecular weight of a dimeric species. Therefore, in the column matrix the protein behaved as a self-associated species with several oligomerization orders, depending on the concentration used.

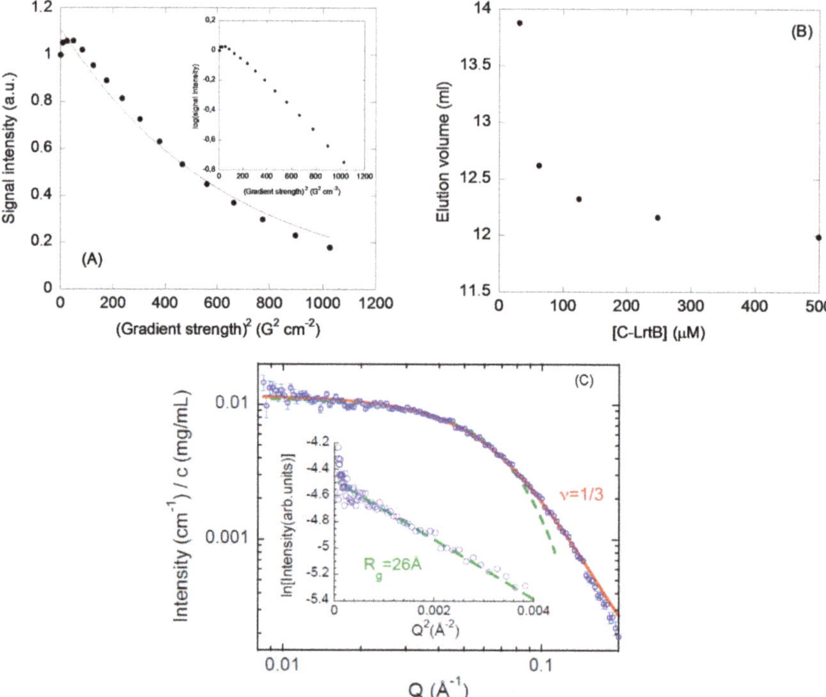

Figure 4. Hydrodynamic and biophysical measurements of C-LrtA: (**A**) DOSY measurements: Intensity decay (arbitrary units) of the methyl signals as the pulse field gradient strength was increased (x-axis). The line is the fitting to equation, as described in Section 4.6. The inset shows the linear relationship between the logarithm of the intensity and the square of the gradient strength; (**B**) size exclusion chromatography (SEC) measurements: Elution volumes of one of the peaks observed for C-LrtA in a Superose 12 10/300 GL at different protein concentrations in buffer pH 8.0 (50 mM Tris) and 0.250 M NaCl; the data have an error of 0.1 mL as obtained from three independent measurements at each particular C-LrtA concentration; (**C**) small-angle X-ray scattering (SAXS) results of C-LrtA are shown with the solid line representing a fit with a generalized Gaussian coil with a scaling exponent value of 1/3 (Section 4.11), and the dashed line is the Guinier description of the low-Q limit (see Guinier plot in the inset, and Section 4.11).

(c) BN-PAGE (blue native polyacrylamide gel electrophoresis): C-LrtA exhibited two species in these experiments, which corresponded to different self-associated species (Figure S3B). The protein species in the fastest migrating band corresponded to an apparent molecular weight of 33 kDa (close to the molecular weight of a dimer, and similar to that observed at the most diluted protein concentration in the SEC experiments). On the other hand, the other band corresponded to an apparent molecular weight of 66 kDa, denoting a pentamer. Our results also suggest that increasing the amount of SDS (well-below the concentration used in denaturing SDS-PAGE gels: 33 mM [28]) had significant effects on the population of self-associated C-LrtA species: the larger the proportion of SDS, the higher the amount of self-associated species detected (the critic micellar concentration of SDS is 1.33 mM). It is important to indicate, at this stage, that the BN-PAGE technique can lead to overestimation of the molecular weights, as some proteins can bind Coomassie dye [29].

(d) Glutaraldehyde cross-linking: To detect the presence of oligomeric species in C-LrtA, we also used the glutaraldehyde agent. We observed dimers (close to the band of the protein marker at 32 kDa) (Figure S3C) at shorter times after addition of the cross-linking agent, and other high-molecular-weight species at the top of the SDS-PAGE lanes. The population of these high-molecular weight self-associated species increased at the largest incubation times (Figure S3C).

(e) KI quenching: It is reasonable to assume that if the self-associated species form at low C-LrtA concentrations, and the tyrosine residues were involved in the association interfaces, then we should be able to follow protein self-association by KI quenching, and we should expect a decrease in the K_{sv} constant as the C-LrtA concentration was increased. We observed the following K_{sv} values in C-LrtA: 1.5 ± 0.3 M^{-1} (at 5 µM of protein); 1.1 ± 0.2 M^{-1} (at 20 µM of protein); and 1.04 ± 0.05 M^{-1} (at 40 µM of protein, all of them in protomer units). Then, there was a protein-concentration behaviour in the 5–40 µM concentration range for the K_{sv} and C-LrtA self-associates.

(f) ITC experiments: We also tried to test whether C-LrtA dissociated upon dilution, using the heat evolved in the reaction monitored by ITC. For experiments performed at a high protein concentration stock, the heat released upon dilution of the protein into the calorimetric cell was consistent with a dissociation reaction for all injections (Figure S4). We tried to fit the heat released to a simple dimer-monomer equilibrium, but the results of the fitting indicated that this assumption was not good enough, suggesting the presence of higher-order equilibria, as indicated by SEC results (see above in this section).

(g) SAXS experiments: The experiments with C-LrtA indicate a $R_g \approx 26$ Å with a $\upsilon \approx 0.33$, close to a compact species value, but with a value of R_g larger than that of a well-folded protein (Section 4, Figure 4C), which is within the range observed for unfolded polypeptide chains [27]. We obtained a good agreement with the expected Guinier regime at low Q values, indicating that, although the protein was self-associated, the size of the C-LrtA species was relatively small.

3. Discussion

The structural propensities and association features of IDPs and disordered protein domains are still poorly understood compared to those of well-folded protein regions. The difference is especially important for proteins, such as LrtA, which is formed by two distinct domains (i.e., N-LrtA and C-LrtA) with roughly the same sequence length, but with completely distinct conformational features. In a previous work [12], we have hypothesized that N-LrtA has a distinct folding topology that is in common with other members of the HPF protein family [5,11], whereas C-LrtA was predicted to be unfolded. In the present work, we have tested that hypothesis and we have found evidence that isolated C-LrtA is an IDP. The intrinsically disordered nature of C-LrtA was suggested by several pieces of evidence: (i) the lack of dispersion in NMR spectra (Figure 1); (ii) the shape (Figure 2A) and the deconvolution of far-UV CD spectrum; (iii) the absence of all-or-none co-operative transitions in the thermal and chemical denaturations (Figure 2B,C); and (iv) a model of the structure without the His-tag obtained with MD simulation (Figure 3).

It could be thought that the observed C-LrtA self-association is non-specific, that is, the protein has solvent-exposed hydrophobic patches which induce highly unspecific, self-association. However, there are two pieces of evidence that suggest that association was not the result of random solvent-exposed hydrophobic residues. First, the fact that Tyr residues were implicated in the oligomerization indicates that only regions containing those residues were involved. Second, if self-association was to be unspecific very large high-molecular species should be observed in some of the techniques used; in contrast, the highest molecular-weight species is observed to be an octamer (in SEC experiments) and glutaraldehyde cross-linking showed the presence of higher molecular-weight species only at very long incubation times (Figure S3C). Finally, it is interesting to note that we have shown that the intact protein, which also self-associates, did not have any large amount of close solvent-exposed hydrophobic patch at physiological pH [12].

Although it is devoid of secondary and tertiary structures, C-LrtA is involved in quaternary contacts, involving different orders of self-associated species, as indicated by several of the techniques used. In general, IDPs intervene in protein-protein contacts [13], but only a few are reported as self-associated species in solution, while keeping their disordered nature. In C-LrtA, some of the four tyrosine residues were involved in its self-association, as judged by the changes in the quenching parameters as the concentration of protein was increased. In the parental, whole protein, tyrosine residues were already found to participate in the oligomerization interface [12]. With the new results in hand, we suggest that some of the four tyrosine residues in C-LrtB were responsible for the self-oligomerization of the whole protein. Then, whereas tyrosine residues in N-LrtA (residues 1–101) seem to contribute to the rigidity of the β-sheet scaffold, tyrosine amino acids in C-LrtA (residues 102–191) seem to be responsible for the quaternary structure of the intact protein. It is important to note that the detected order of the self-associated species in C-LrtA varied among the different biochemical and hydrodynamic techniques used (as it happens for the whole LrtA [12]) indicating the presence of different oligomerization equilibria. That is, the self-association did not involve the simple dimer-monomer equilibrium, but rather a sequence of different order equilibria.

The exact biological significance of self-association in LrtA remains unknown, and it is still a matter of debate. In fact, in spite of similar structural organization and high sequence homology among the members of the HPF family, their functions during stress responses are very different in the organisms to which they belong to [4,7–10,30]. There are some examples of oligomeric HPFs reported in the literature: for instance, the short *Vibrio choleras* HPF is a dimer, whose dimerization occurs through Zn ions at one of the β-strands of the β-α-β-β-β-α fold; however, it is not known if such dimerization is due to the crystallization process [11]. On the other hand, the HPF of *Staphylococcus aureus* is also a member of the long HPF family, but its C-terminal region is shorter (60-residue long) and it is folded [5]. The protein is a dimer, and its C-terminal region is responsible for this dimerization; furthermore, interactions of the dimeric C-terminal region with ribosomes are responsible for ribosome dimerization. Thus, we hypothesize that in LrtA the oligomeric, disordered C-LrtA domain might be responsible for the dimerization of 70S particles (whose abundance in the cell has been associated with the presence of LrtA [3]) during stress conditions. The fact that C-LrtA is disordered (in contrast to that of *Staphylococcus aureus* HPF [5]) could provide the advantage that LrtA may be bound not only to the 70S particles, but also to other macromolecules, regulating several cyanobacterial processes triggered by stress conditions.

Evidence of the possibility of self-association and supra-molecular order in IDP are starting to be mounting. One of the first reported examples of oligomeric IDPs was the "fuzzy" dimer formed by the cytoplasmic domain of the T-cell receptor zeta subunit [31,32], although this putative dimer was later shown to be a monomer under a wide range of conditions, and its oligomerization was attributed to non-ideal protein-column-resin interactions [33]. However, other studies have used a plethora of biophysical and biochemical techniques to unambiguously show the existence of intact self-associated IDPs. For instance, there are reports describing oligomeric plant IDPs [34]; dimeric proteins in the disordered *umuD* gene products [35]; oligomeric mitochondrial IDPs [36]; oligomeric IDPs which bind to the Polycomb complex [37]; disordered, oligomeric acid-rich proteins of rod photoreceptors [38]; and disordered oligomeric oncogen products [39]. Many of these homoligomeric interactions are deposited in the MFIB database (http://mfib.enzim.ttk.mta.hu) [40], together with others involving hetero-oligomer assemblies, where in all cases mutual folding of the IDP chains occurs. The presence of such a quaternary (homo or hetero-oligomeric) structural organization for an IDP has been explained as due to multiple transient interactions or long-range contacts, which yield a fuzzy self-associated species [41–43], involving different degrees of organization in the association process [44].

4. Materials and Methods

4.1. Materials

Deuterium oxide, d_{11}-Tris acid and isopropyl-β-D-1-tiogalactopyranoside (IPTG) were purchased from Apollo Scientific (Stockport, UK). DNase, kanamycin, the Trizma base and its acid, sodium trimethylsilyl (2,2,3,3-$^2H_4$) propionate (TSP), imidazole, and the His-Select HF nickel resin were from Sigma-Aldrich (Madrid, Spain). The β-mercaptoethanol (β-ME) was provided by BioRad (Madrid, Spain). The protein marker, PAGEmark Tricolor, and Triton X-100 were from VWR (Barcelona, Spain). The protein gel-filtration-column calibration markers were from GE Healthcare (Barcelona, Spain). Dialysis tubing was from Spectrapor (Spectrum Laboratories, Shiga, Japan). Amicon centrifugal devices with a cut-off molecular weight of 3000 Da were from Millipore (Barcelona, Spain). The rest of the materials used were of analytical grade. Water was deionized and purified on a Millipore system.

4.2. Protein Expression and Purification

The C-LrtA region (residues 102-191 of LrtA) was cloned by NZytech (Lisbon, Portugal) in a pHTP1 *E. coli* expression vector (between XhoI and NcoI sites), with kanamycin resistance. The final construct contained an N-terminal His-tag to allow for purification (MGSSHHHHHHSSGPQQGLR), and had the overall sequence: MGSSHHHHHHSSGPQQGLRQHGNVKTSEIVEDKPVEENLIGDRA PELPSEVLRMKYFAMPPMAIEDALEQLQLVDHDFYMFRNKDTDEINVIYIRNHGGYGVIQPHQAS.

Expression of C-LrtA was carried out in the *E. coli* BL21 (DE3) strain (Novagen, VWR, Barcelona; Spain) strains with a final kanamycin concentration of 50 mg/mL. We used 1 L flasks to culture the cells. The expression of the protein was induced with a final concentration of 1.0 mM IPTG, when the absorbance observed for the cell culture was 0.4–0.9 at 600 nm, and the growth of the cells continued for 15–16 h at 25 °C 15–16 h at 25 °C (this temperature was chosen to decrease the possibility of aggregates in C-LrtA, as the parental LrtA had a tendency to form inclusion bodies). We harvested the cells in a JA-10 rotor (Beckman Coulter) for 15 min at 8000 rpm. We re-suspended the cellular pellet from 5 L of culture in 50 mL buffer A (500 mM NaCl, 5 mM imidazole, 20 mM Tris buffer (pH 8), 0.1% Triton X-100 and 1 mM β-ME) and adding a tablet of Sigma Protease Cocktail EDTA-free. In the first attempts to get the protocol of purification, we added 2 mg of DNase (per 1 L of culture). We incubated the mixture for 10 min, with gentle agitation in the fridge (4 °C). Next, we sonicated the mixture (by using a Branson sonicator, 750 W), with 10 cycles of 45 s each at 55% of maximal power output, with intervals of 15 s between the cycles, and always keeping the cells in ice. We separated the supernatant by centrifugation at 18,000 rpm for 40 min at 4 °C in a Beckman JSI30 centrifuge with a JA-20 rotor. C-LrtA was present in the supernatant, and we purified it by immobilized affinity chromatography (IMAC), by adding 5 mL of Ni-resin previously equilibrated in buffer A. The mixture was incubated for 20 min at 4 °C, and afterwards, the lysate was separated from the resin by gravity. The washing step was carried out with 20 mL of buffer B (20 mM Tris buffer (pH 8.0), 500 mM NaCl, 1 mM β-ME, and 20 mM imidazole); the protein was eluted by gravity from the column with buffer C (20 mM Tris buffer (pH 8.0), 500 mM NaCl, 1 mM β-ME, and 500 mM imidazole). The solution was dialyzed against 50 mM Tris buffer (pH 8.0). The protein was further purified in a Hi-Trap Mono Q (GE Healthcare) column by using a gradient step from 0 to 1 M NaCl (50 mM Tris buffer, pH 8.0) in 60 min and a flow of 1 mL/min, in an AKTA Basic system (GE Healthcare), while monitoring the absorbance at 280 nm. This column purification step was used based on the theoretical isoelectric point of C-LrtA (pI = 5.72).

In the initial purification attempts (carried out in the presence of DNase), we observed that the main peak coming from the Hi-Trap Mono Q column did not show an emission fluorescence spectrum expected for a protein containing only 4 tyrosine residues (i.e., with a maximum at 308 nm), but rather the spectrum of a protein containing tryptophan (a maximum at ~330 nm). Therefore, we thought of the possibility of contamination of the protein with the DNase used in the first steps of the purification, which also has a similar pI (5.2), and it has a tendency to self-associate. Then, in subsequent purifications we did not add DNase to the cell lysate. As a consequence, the protein coming from the Hi-Trap Mono Q showed absorbance at 260 nm, probably due to the presence of oligonucleotides resulting from the sonication step; these oligonucleotides must be present in a small amount, since no evidence of sharp peaks (contaminants of low-molecular weight) were observed in the 1D ^{1}H-NMR spectrum (Figure 1). The presence of traces of oligonucleotides is not infrequent in proteins of the same family as it has been also observed in the recombinant HPF from *S. aureus* after its purification [5]. After elution from the Hi-Trap Mono Q column, the sample was extensively dialyzed against water, and no precipitation was observed in the dialysis tubing. The concentration of the protein Pc (mg/mL) was calculated by [45]: $P_c = 1.55\ A_{280} - 0.75\ A_{260}$, where A_{280} and A_{260} are the absorbances of the dialyzed solution of the protein at 280 and 260 nm, respectively.

4.3. Fluorescence

Fluorescence spectra were collected on a Cary Varian spectrofluorimeter (Agilent, Santa Clara, CA, USA), interfaced with a Peltier, at 25 °C. The C-LrtA concentrations used were 5 and 10 μM of protein (in protomer units) in the GdmCl denaturations carried out at pH 7.5 (50 mM Tris buffer). A 1-cm-pathlength quartz cell (Hellma; Sigma Aldrich, Madrid, Spain) was used. In the denaturation experiments with GdmCl, we prepared the samples the day before starting from a 7 M GdmCl concentrated stock that equilibrated overnight; samples were left at 25 °C for 1 h before performing the measurements.

The emission intensity weighted average of the inverse wavelengths (also called the spectrum mass centre, or the spectral average energy of emission), <λ>, was calculated as described [46].

(a) Steady-state spectra: The experimental set-up used in the case of the denaturation experiments with GdmCl was as previously described [46]. In brief, excitation of the protein samples was at 278 nm, and excitation and emission slits were 5 nm in all cases. The experiments were recorded between 300 and 400 nm. The signal was acquired for 1 s and the increment of wavelength was set to 1 nm.

(b) Thermal denaturations: Thermal denaturations of isolated C-LrtA were carried out with the same experimental set-up previously described [46]. These experiments were performed at constant heating rates of 60 °C/h, with an average time of 1 s. Thermal scans were collected at 308 nm after excitation at 278 nm from 25 to 90 °C and acquired every 0.2 °C. Protein concentration was 10 μM (in protomer units).

(c) Fluorescence quenching: Quenching by iodide was examined with concentrations ranging from 5 to 40 μM (in protomer units) at pH 7.0 (phosphate buffer, 50 mM). The experimental set-up for KI was the same described above for the intrinsic fluorescence experiments. The data were fitted to [47]: $F_0/F = 1 + K_{sv}[KI]$, where K_{sv} is the Stern-Volmer constant for collisional quenching; F_0 is the fluorescence intensity in the absence of KI; and F is that at any KI concentration. The range of KI concentrations explored was 0–0.7 M. Fittings to the above equation were carried out by using Kaleidagraph (Synergy software).

4.4. CD

The far-UV CD spectra were recorded at 25 °C on a Jasco J815 spectropolarimeter (Jasco, Easton, MD, USA City, Japan) equipped with a thermostated cell holder, and interfaced with a Peltier unit. A periodical calibration was performed with (+)-10-camphorsulphonic acid. We used two concentration values for C-LrtA (10 and 20 μM, in protomer units) to check for concentration-dependence of the shape and intensity of spectra, performed at pH 7.0 (50 mM, phosphate buffer). Molar ellipticity was determined as previously indicated [46].

(a) *Steady-state spectra*: Experiments were performed using the experimental set-up described previously [46]. Spectra were corrected by subtracting the baseline in all cases. Protein concentration was 10 and 20 μM (in protomer units) for GdmCl-denaturation experiments.

(b) Thermal denaturations: Experiments were carried out with the same experimental set-up described previously [46] and a protein concentration of 10 μM (in protomer units). Briefly, thermal denaturations were performed at a constant heating rate of 60 °C/h from 25 to 85 °C, a response time of 8 s, a band width of 1 nm, acquired every 0.2 °C, and following the ellipticity at 222 nm.

4.5. NMR Spectroscopy

The NMR experiments were acquired at 20 °C on a Bruker Avance DRX-500 spectrometer equipped with a triple resonance probe and z-pulse field gradients. All spectra were processed and analysed by using TopSpin 2.1 (Bruker GmbH, Karlsruhe, Germany). We used TSP as the external chemical shift reference [26].

(a) 1D ^{1}H-NMR experiments: The 1D-^{1}H-NMR spectrum were acquired with a C-LrtA concentration of 120 μM (in protomer units) in 0.5 mL, 50 mM d_{11}-Tris buffer (pH 7.2) in H_2O/D_2O (90%/10%, v/v), without any correction for deuterium isotope effects. The spectrum was acquired with 16 K data points. We acquired 2 K scans with a 6000 Hz spectral width (12 ppm), by using the WATERGATE sequence [48]. Before processing the data, baseline correction and zero-filling were applied.

(b) Translational diffusion measurements: The DOSY experiments of C-LrtA at pH 7.2 were performed with the pulse-field gradient (PFG) spin-echo sequence, as described previously [12,46], with sixteen gradient strengths ranging linearly from 2% to 95% of the total power of the gradient unit. The intensity of the methyl signals, I, was fit to: $\frac{I}{I_0} = -\exp\left(D\gamma_H^2\delta^2G^2\left(\Delta - \frac{\delta}{3} - \frac{\tau}{2}\right)\right)$, where I_0 is the maximum peak intensity of the methyl resonances at the smallest gradient strength; δ is the duration (in s) of the gradient (2.7 ms); G is the gradient strength (in T cm^{-1}); Δ is the time (in s) between the gradients (150 ms); γ_H is the gyromagnetic constant of the proton; and, τ is the recovery delay between the bipolar gradients (100 μs). Samples were exchanged in D_2O buffer (50 mM d_{11}-Tris buffer, pH 7.2, not corrected for isotope effects) by using Amicon centrifugal devices for 4 to 6 h.

4.6. Blue-Native PAGE (BN-PAGE)

BN-PAGE was performed in linear 4% to 16% (w/v) polyacrylamide-gradient gels [28,49,50]. Before running the BN-PAGE, sample aliquots (10 μL) containing 20 μg of C-LrtA were mixed with 1 μL of 5% Coomassie Brilliant blue G stock solution in 750 mM aminocaproic acid. Electrophoresis was initiated at 85 V for 30 min, and then continued at 200 V for 2.5 h, at 4 °C. After electrophoresis, the gels were stained overnight with colloidal Coomassie Blue G 250 [51]. In these gels, it is the negative charge from the Coomassie Blue, capable of binding to the hydrophobic protein surfaces, what determines the electrophoretic mobility. The technique is a method of choice to study the organization of protein complexes in their native state [28,49–51].

4.7. Glutaraldehyde Cross-Linking

Glutaraldehyde cross-linking was carried out at pH 7.2 (50 mM, Tris buffer). The volume of the sample was 200 μL. The C-LrtA concentration was 120 μM (in protomer units). The protocol used has been described previously [12], and experiments were repeated twice.

4.8. SEC

SEC experiments were performed at pH 8.0 (50 mM Tris) and 0.250 M NaCl in a Superose 12 10/300 GL column, which was connected to an AKTA FPLC (GE Healthcare, Barcelona, Spain); absorbance at 280 nm was monitored during the elution. Flow rate was 0.8 mL/min, and C-LrtA was loaded in a volume of 100 μL at several protein concentrations, in the range of 30–500 μM (in protomer concentration). The markers used to calibrate the column were ferritin, catalase, aldolase, albumin, bovine RNase A and blue dextran (GE Healthcare, Barcelona, Spain); the isolated protein markers were loaded in the column in the above described buffer, at the same flow rate, and three independent measurements were performed with each marker.

4.9. ITC

The protocol used during the ITC experiments was that previously described [46]. Briefly, the ITC experiments were carried out by using a VP-ITC instrument (Microcal, Northampton, MA, USA). Before the experiments, C-LrtA was dialysed at 4 °C against water at physiological pH. Dilution ITC experiments involved sequential injections of microliter amounts (10 µL) of a concentrated protein solution (498 µM, in protomer units) into the calorimetric cell (1.4 mL), which initially contained water alone.

4.10. SAXS

SAXS experiments were performed on a Rigaku 3-pinhole PSAXS-L instrument, at 45 kV and 0.88 mA. The MicroMax-002+ X-ray Generator Systems includes a module with a microfocus sealed tube source, and an X-ray generator unit producing Cu-Kα transition photons, with λ = 1.54 Å wavelength. Vacuum was maintained both in the flight path and sample chamber. A two-dimensional multiwire X-ray detector (Gabriel design, 2D-2000X) was used as a detector of the scattered X-rays. We obtained the azimuthally averaged scattered intensities as a function of the scattering vector Q (where $Q = 4\pi(\lambda)^{-1}\sin\theta$, and θ represents half the scattering angle). Silver behenate was used as standard for calibration (reciprocal space). The solutions were filling Boron-rich capillaries with an outside diameter of 2 mm and wall thickness of 0.01 mm. The contribution from the corresponding buffer (measured on the same capillary) was subtracted by applying the proper factors obtained from transmission measurements. The sample-detector distance was 2 m, allowing covering a Q-range from 0.008 to 0.2 Å^{-1}.

From the intensity scattered at low-Q values –in the so-called Guinier regime– we can determine the average gyration radius, R_g, of the protein under several conditions, by using the Guinier law: $I(Q) = A\exp(\frac{-R_g^2 Q^2}{3})$. The pre-exponential factor, A, is determined by the molecule concentration, the scattering contrast and the mass of the macromolecules dispersed in the solution. On the other hand, we can estimate the compaction grade through the scaling exponent υ, which relates to Q as $I(Q) \approx Q^{-1/\upsilon}$, in the high Q-range here explored. The values of υ are 1/3 for a polymeric chain collapsed into a globule; 0.5 for a random-coil polymer (which is the conformation of a linear polymer chain in Θ-conditions); and 0.6 for a swollen chain in a good solvent (self-avoiding-walk conformation). The form factor of a coil, with scaling exponent υ, is described in terms of the so-called generalized Gaussian coil function, given by the expression [52]: $I(Q) \approx \frac{1}{\upsilon U^{1/2\upsilon}}\gamma\left(\frac{1}{2\upsilon}, U\right) - \frac{1}{\upsilon U^{1/\upsilon}}\gamma\left(\frac{1}{\upsilon}, U\right)$, where $U = (2\upsilon + 1)(2\upsilon + 2)Q^2 R_g^2/6$ and $\gamma(a, x) = \int_0^x t^{a-1}\exp(-t)dt$. From the fits of this function to the experimental data, the value of the radius of gyration can also be obtained.

4.11. Molecular Modelling

A model of C-LrtA (without the His-tag) was obtained in MD simulations by using a protocol previously adopted for IDPs [23,24]. In brief, simulations were performed with the GROMACS package [53] starting from a protein model built by using VMD [25], and collapsing it in a brute-force run carried out in the isobaric-isothermal ensemble. C-LrtA was initially in an extended conformation, except for backbone turns in correspondence with proline residues. The protein was centered in a rhombic dodecahedron box with a minimum distance of 1 nm from the edge of the simulation box, and surrounded with explicit water molecules. Amino acid residues were adjusted to mimic neutral pH and Na$^+$ counterions were added to obtain an overall neutral molecular system. The AMBER ff99SB-ILDN force field [54] was used for the protein, and the TIP3P [55] model for water. Other simulation conditions, including modelling of the electrostatics and van der Waals interactions, and reference values and coupling times for the thermostat and barostat, were as previously described [56,57].

5. Conclusions

The C-terminal region of the cyanobacterial protein LrtA, a member of the HPF family, was a disordered domain that self-associated through a mechanism that involved some of its tyrosine residues. These findings clarify a number of important features of intact LrtA, including demonstrating, in an unambiguous way, the presence of two distinct and structurally different domains in this protein. Moreover, in the absence of studies of the N-terminal region (residues 1-101) our results suggest that the unstructured C-terminal one is the region driving the supramolecular self-organization of the whole protein. This study contributes to clarify the structural and stability features of proteins that interact and modulate the ribosome activity, and especially those belonging to the almost unexplored subfamily of long HPF proteins.

Supplementary Materials: Supplementary materials can be found at http://www.mdpi.com/1422-0067/19/12/3902/s1. There are four figures, showing: Figure S1: the far-UV CD spectra of C-LrtA at two concentrations; Figure S2: the size exclusion chromatograms of C-LrtA at different concentrations; Figure S3: details of the purification and self-association of C-LrtA (Figure S3), which include the SDS-PAGE gels of purified C-LrtB (Figure S3A); the BN-PAGE (Figure S3B); and the cross-linked protein (Figure S3C) at different reaction times; Figure S4: the ITC thermogram.

Author Contributions: Conceptualization, J.L.N. and B.R.; Methodology, J.L.N., A.M.G., F.H., A.A. and B.R.; Software, B.R.; Investigation, J.L.N., A.M.G., F.H., A.A. and B.R.; Writing-Original Draft Preparation, J.L.N., A.M.G., F.H., A.A. and B.R.; Writing-Review and Editing, J.L.N. and B.R.

Funding: This research was funded by Spanish Ministry of Economy and Competitiveness [CTQ2015-64445-R (to J.L.N.) and MAT2015-63704-P (to A.A.), with Fondo Social Europeo (ESF)], and by the Basque Government [IT-654-13 (to A.A.)].

Acknowledgments: B.R. acknowledges kind hospitality and use of computational resources in the European Magnetic Resonance Center (CERM), Sesto Fiorentino (Florence), Italy. J.L.N. thanks F. J. Florencio and M. I. Muro-Pastor for introducing to him the LrtA system. J.L.N. and F.H. thank Javier Gómez for helpful discussions. All the authors thank the three anonymous reviewers for their encouragement to formulate some hypotheses on the biological significance of the oligomerization in LrtA.

Conflicts of Interest: The authors declare no conflict of interest.

Abbreviations

β-ME	β-mercaptoethanol
BN	Blue native
CD	Circular dichroism
C-LrtA	C-terminal half of LrtA protein (comprising residues 102–191)
DOSY	Diffusion ordered spectroscopy
GdmCl	Guanidine hydrochloride
HPF	Hibernating promoting factor
IDP	Intrinsically disordered protein
IMAC	Immobilized affinity chromatography
IPTG	Isopropyl-β-D-1-tiogalactopyranoside
ITC	Isothermal titration calorimetry
MD	Molecular dynamics
N-LrtA	N-terminal half (residues 1–101) of LrtA protein
NMR	Nuclear Magnetic Resonance
PAGE	Polyacrylamide gel electrophoresis
PFG	Pulse field gradient
RaiA	Ribosome associate inhibitor A
RNase	Ribonuclease
SAXS	Small-angle X-ray scattering
SDS	Sodium dodecyl sulphate
SEC	Size exclusion chromatography
UV	Ultraviolet

References

1. Tan, X.; Varughese, M.; Widger, W.R. A light-repressed transcript found in *Synechococcus* sp. PCC 7002 is similar to a chloroplast-specific small subunit ribosomal protein and to a transcription modulator protein associated with sigma 54. *J. Biol. Chem.* **1994**, *269*, 20905–20912. [PubMed]
2. Samartzidou, H.; Widger, W.R. Transcriptional and post-transcriptional control of mRNA from lrtA, a light-repressed transcript in *Synechococcus* sp. PC 7002. *Plant Physiol.* **1998**, *117*, 225–234. [CrossRef] [PubMed]
3. Galmozzi, C.V.; Florencio, F.J.; Muro-Pastor, M.I. The cyanobacterial ribosomal-associated protein LrtA is involved in post-stress survival in *Synechocystis* sp. PCC 6803. *PLoS ONE* **2016**. [CrossRef] [PubMed]
4. Yoshida, H.; Wada, A. The 100S ribosome: Ribosomal hibernation induced by stress. *Wiley Interdiscipl. Rev. RNA* **2014**, *5*, 723–732. [CrossRef] [PubMed]
5. Khusainov, I.; Vicens, Q.; Ayupov, R.; Usachev, K.; Myasnikov, A.; Simonetti, A.; Validov, S.; Kieffer, B.; Yuspova, G.; Yusupov, M.; et al. Structures and dynamics of hibernating ribosomes from *Staphylococcus aureus* mediated by intermolecular interactions of HPF. *EMBO J.* **2017**, *36*, 2073–2087. [CrossRef] [PubMed]
6. Starosta, A.L.; Lasak, J.; Jung, K.; Wilson, D.N. The bacterial translation stress response. *FEMS Microbiol. Rev.* **2014**, *38*, 1172–1201. [CrossRef] [PubMed]
7. Agafonov, D.E.; Spirin, A.S. The ribosome-associated inhibitor A reduces translation errors. *Biochem. Biophys. Res. Commun.* **2004**, *320*, 354–358. [CrossRef]
8. Polikanov, Y.S.; Blaha, G.M.; Steitz, T.A. How hibernation factors RMF, HPF and YfiA turn off protein synthesis. *Science* **2012**, *336*, 915–918. [CrossRef]
9. Ueta, M.; Yoshida, H.; Wada, C.; Baba, T.; Mori, H.; Wada, A. Ribosome binding proteins YHbH and YfiA have opposite functions during 100S formation in the stationary phase of *Escherichia coli. Genes Cells* **2005**, *10*, 1103–1112. [CrossRef]
10. Ueta, M.; Ohniwa, R.L.; Yoshida, H.; Maki, Y.; Wada, C.; Wada, A. Role of HPF (hibernation promoting factor) in translational activity in *Escherichia coli. J. Biochem.* **2008**, *143*, 425–433. [CrossRef]
11. De Bari, H.; Berry, E.A. Structure of *Vibrio cholerae* ribosome hibernation factor. *Acta Cryst. Sect. F* **2013**, *69*, 228–236. [CrossRef] [PubMed]
12. Contreras, L.M.; Sevilla, P.; Cámara-Artigas, A.; Hernández-Cifre, J.G.; Rizzuti, B.; Florencio, F.J.; Muro-Pastor, M.I.; García de la Torre, J.; Neira, J.L. The Cyanobacterial ribosomal-associated protein LrtA from *Synechocystis* sp. PCC 6803 is an oligomeric protein in solution with chameleonic sequence properties. *Int. J. Mol. Sci.* **2018**. [CrossRef] [PubMed]
13. Sickmeier, M.; Hamilton, J.A.; LeGall, T.; Vacic, V.; Cortese, M.S.; Tantos, A.; Szabo, B.; Tompa, P.; Chen, J.; Uversky, V.N.; et al. DisProt: The database of disordered proteins. *Nucleic Acids Res.* **2007**, *35*, D786–D793. [CrossRef] [PubMed]
14. Woody, R.W. Circular dichroism. *Methods Enzymol.* **1995**, *246*, 34–71. [PubMed]
15. Kelly, S.M.; Jess, T.J.; Price, N.C. How to study proteins by circular dichroism. *Biochim. Biophys. Acta* **2005**, *1751*, 119–139. [CrossRef] [PubMed]
16. Chemes, L.B.; Alonso, L.G.; Noval, M.G.; de Prat-Gay, G. Circular dichroism techniques for the analysis of intrinsically disordered proteins and domains. *Methods Mol. Biol.* **2012**, *895*, 387–404. [PubMed]
17. Whitmore, L.; Wallace, B.A. Protein secondary structure analysis from circular dichroism spectroscopy: Methods and reference databases. *Biopolymers* **2008**, *89*, 392–400. [CrossRef]
18. Whitmore, L.; Wallace, B.A. DICHROWEB, an online server for protein secondary structure analyses from circular dichroism spectroscopic data. *Nucleic Acids Res.* **2004**, *32*, W668–w673. [CrossRef]
19. Receveur-Bréchot, V.; Bourhis, J.M.; Uversky, V.N.; Canard, B.; Longhi, S. Assessing protein disorder and induced folding. *Proteins* **2006**, *62*, 24–45.
20. Neira, J.L. Fluorescence, circular dichroism and mass spectrometry as tools to study virus structure. *Subcell. Biochem.* **2013**, *68*, 177–202.
21. Cantor, C.R.; Schimmel, P.R. *Biophysical Chemistry*; W. H. Freeman: New York, NY, USA, 1980.
22. Marsh, J.A.; Forman-Kay, J.D. Sequence determinants of compaction in intrinsically disordered proteins. *Biophys. J.* **2010**, *98*, 2383–2390. [CrossRef] [PubMed]
23. Neira, J.L.; Rizzuti, B.; Iovanna, J.L. Determinants of the pK_a values of ionizable residues in an intrinsically disordered protein. *Arch. Biochem. Biophys.* **2016**, *598*, 18–27. [CrossRef] [PubMed]

24. Cozza, C.; Neira, J.L.; Florencio, F.J.; Muro-Pastor, M.I.; Rizzuti, B. Intrinsically disordered inhibitor of glutamine synthetase is a functional protein with random-coil-like pK_a values. *Protein Sci.* **2017**, *26*, 1105–1115. [CrossRef] [PubMed]

25. Humphrey, W.; Dalke, A.; Schulten, K. VMD: Visual molecular dynamics. *J. Mol. Graph. Model* **1996**, *14*, 33–38. [CrossRef]

26. Cavanagh, J.F.; Wayne, J.; Palmer, A.G., III; Skelton, N.J. *Protein NMR Spectroscopy: Principles and Practice*, 1st ed.; Academic Press: San Diego, CA, USA, 1996.

27. Wilkins, D.K.; Grimshaw, S.B.; Receveur, V.; Dobson, C.M.; Jones, J.A.; Smith, L.J. Hydrodynamic radii of native and denatured proteins measured by pulse field gradient NMR techniques. *Biochemistry* **1999**, *38*, 16424–16431. [CrossRef] [PubMed]

28. Giudici, A.M.; Molina, M.L.; Ayala, J.L.; Montoya, E.; Renart, M.L.; Fernandez, A.M.; Encinar, J.A.; Ferrer-Montiel, A.V.; Poveda, J.A.; Gonzalez-Ros, J.M. Detergent-labile, supramolecular assemblies of KcsA: Relative abundance and interactions involved. *Biochim. Biophys. Acta* **2013**, *1828*, 193–200. [CrossRef] [PubMed]

29. Wittig, I.; Beckhaus, T.; Wumaier, Z.; Karas, M.; Schägger, H. Mass estimation of native proteins by blue native electrophoresis: Principles and practical hints. *Mol. Cell Proteom.* **2010**, *9*, 2149–2161. [CrossRef] [PubMed]

30. Basu, A.; Yap, M.F.N. Ribosome hibernation factor promotes Staphylococcal survival and diferentially represses translation. *Nucleic Acids Res.* **2016**, *44*, 4881–4893. [CrossRef]

31. Sigalov, A.B.; Aivazian, D.; Stern, L. Homo-oligomerization of the cytoplasmic domain of the T-cell receptor zeta chain and of other proteins containing the immunoreceptor tyrosine-based activation motif. *Biochemistry* **2004**, *43*, 2049–2061. [CrossRef]

32. Sigalov, A.B.; Zhurauleva, A.V.; Orekhov, V.Y. Binding of intrinsically disordered proteins is not necessarily accompanied by a structural transition to a folded form. *Biochimie* **2007**, *89*, 419–421. [CrossRef]

33. Nourse, A.; Mittag, T. The cytoplasmic domain of the T-cell receptor zeta subunit does not form disordered dimers. *J. Mol. Biol.* **2014**, *426*, 62–70. [CrossRef]

34. Rivera-Nájera, L.Y.; Saab-Rincón, G.; Battaglia, M.; Amero, C.; Pulido, N.O.; García-Hernández, E.; Solórzano, R.M.; Reyes, J.L.; Covarrubias, A.A. A group 6 late embryogenesis abundant protein from common bean is a disordered protein with extended helical structure and oligomer-forming properties. *J. Biol. Chem.* **2014**, *289*, 31995–32009. [CrossRef] [PubMed]

35. Simon, S.M.; Sousa, F.J.R.; Mohana-Borges, R.; Walker, G.C. Regulation of *Escherichia coli* SOS mutagenesis by dimeric intrinsically disordered *umu*D gene products. *Proc. Natl. Acad. Sci USA* **2008**, *105*, 1152–1157. [CrossRef] [PubMed]

36. Neira, J.L.; Martínez-Rodríguez, S.; Hernández-Cifre, J.G.; Cámara-Artigas, A.; Clemente, P.; Peralta, S.; Fernández-Moreno, M.Á.; Garesse, R.; García de la Torre, J.; Rizzuti, B. Human COA3 is an oligomeric highly flexible protein in solution. *Biochemistry* **2016**, *55*, 6209–6220. [CrossRef] [PubMed]

37. Neira, J.L.; Román-Trufero, M.; Contreras, L.M.; Prieto, J.; Singh, G.; Barrera, F.N.; Renart, M.L.; Vidal, M. The transcriptional repressor RYBP is a natively unfolded protein which folds upon binding to DNA. *Biochemistry* **2009**, *48*, 1348–1360. [CrossRef] [PubMed]

38. Batra-Safferling, R.; Abarca-Heidermann, K.; Körscehn, H.G.; Tziatzios, C.; Stoldt, M.; Budyak, I.; Willbold, D.; Schwalbe, H.; Klein-Seetharaman, J.; Kaupp, U.B. Glutamic acid-rich proteins of rod photoreceptors are natively unfolded. *J. Biol. Chem.* **2006**, *281*, 1449–1460. [CrossRef] [PubMed]

39. Neira, J.L.; López, M.B.; Sevilla, P.; Rizzuti, B.; Cámara-Artigas, A.; Vidal, M.; Iovanna, J.L. The chromatin nuclear protein NUPR1L is intrinsically disordered and binds to the same proteins as its paralogue. *Biochem. J.* **2018**, *475*, 2271–2291. [CrossRef] [PubMed]

40. Fichó, E.; Reményi, I.; Simon, I.; Mészáros, B. MFIB: A repository of protein complexes with mutual folding induced by binding. *Bioinformatics* **2017**, *33*, 3682–3684. [CrossRef] [PubMed]

41. Uversky, V.N. Intrinsically disordered proteins in overcrowded milieu: Membrane-less organelles, phase separation and intrinsic disorder. *Cur. Opin. Struct. Biol.* **2017**, *44*, 17–30. [CrossRef] [PubMed]

42. Tompa, P.; Fuxreiter, M. Fuzzy complexes: Polymorphism and structural disorder in protein-protein interactions. *Trends. Biochem. Sci.* **2008**, *33*, 2–8. [CrossRef] [PubMed]

43. Fuxreiter, M. Fuzziness in protein interactions: A historical perspective. *J. Mol. Biol.* **2018**, *430*, 2278–2287. [CrossRef] [PubMed]

44. Borgia, A.; Borgia, M.B.; Bugge, K.; Kissling, V.M.; Heidarsson, P.O.; Fernandes, C.B.; Sottini, A.; Soranno, A.; Buholzer, K.J.; Nettels, D.; et al. Extreme disorder in an ultrahigh-affinity protein complex. *Nature* **2018**, *555*, 61–66. [CrossRef] [PubMed]

45. Dunn, M.J. Initial planning: Determination of total protein concentration. In *Protein Purification Methods*; Harris, E.L.V., Angal, S., Eds.; Oxford University Press: Oxford, UK, 1995; pp. 10–20.

46. Neira, J.L.; Hornos, F.; Bacarizo, J.; Cámara-Artigas, A.; Gómez, J. The monomeric species of the regulatory domain of Tyrosine Hydroxylase has a low conformational stability. *Biochemistry* **2016**, *55*, 6209–6220. [CrossRef] [PubMed]

47. Lakowicz, J.R. *Principles of Fluorescence Spectroscopy*, 2nd ed.; Plenum Press: New York, NY, USA, 1999.

48. Piotto, M.; Saudek, V.; Sklenar, V. Gradient-tailored excitation for single-quantum NMR spectroscopy of aqueous solutions. *J. Biomol. NMR* **1993**, *2*, 661–665. [CrossRef]

49. Schagger, H.; von Jagow, G. Blue native electrophoresis for isolation of membrane protein complexes in enzymatically active form. *Anal. Biochem.* **1991**, *199*, 223–231. [CrossRef]

50. Schagger, H.; Cramer, W.A.; von Jagow, G. Analysis of molecular masses and oligomeric states of protein complexes by blue native electrophoresis and isolation of membrane protein complexes by two-dimensional native electrophoresis. *Anal. Biochem.* **1994**, *217*, 220–230. [CrossRef] [PubMed]

51. Neuhoff, V.; Stamm, R.; Pardowitz, I.; Arold, N.; Ehrhardt, W.; Taube, D. Essential problems in quantification of proteins following colloidal staining with coomassie brilliant blue dyes in polyacrylamide gels and their solution. *Electrophoresis* **1990**, *11*, 101–117. [CrossRef] [PubMed]

52. Hammouda, B. Small angle scattering from branched polymers. *Macromol. Theory Simul.* **2012**, *21*, 372–381. [CrossRef]

53. Abraham, M.J.; Murtola, T.; Schulz, R.; Páll, S.; Smith, J.C.; Hess, B.; Lindahl, E. GROMACS: High performance molecular simulations through multi-level parallelism from laptops to supercomputers. *SoftwareX* **2015**, *1–2*, 19–25. [CrossRef]

54. Lindorff-Larsen, K.; Piana, S.; Palmo, K.; Maragakis, P.; Klepeis, J.L.; Dror, R.O.; Shaw, D.E. Improved side-chain torsion potentials for the Amber ff99SB protein force field. *Proteins* **2010**, *78*, 1950–1958. [CrossRef] [PubMed]

55. Jorgensen, W.L.; Chandrasekhar, J.; Madura, J.D.; Impey, R.W.; Klein, M.L. Comparison of simple potential functions for simulating liquid water. *J. Chem. Phys.* **1983**, *79*, 926–935. [CrossRef]

56. Rizzuti, B.; Bartucci, R.; Sportelli, L.; Guzzi, R. Fatty acid binding into the highest affinity site of human serum albumin observed in molecular dynamics simulation. *Arch. Biochem. Biophys.* **2015**, *579*, 18–25. [CrossRef] [PubMed]

57. Evoli, S.; Mobley, D.L.; Guzzi, R.; Rizzuti, B. Multiple binding modes of ibuprofen in human serum albumin identified by absolute binding free energy calculations. *Phys. Chem. Chem. Phys.* **2016**, *18*, 32358–32368. [CrossRef] [PubMed]

© 2018 by the authors. Licensee MDPI, Basel, Switzerland. This article is an open access article distributed under the terms and conditions of the Creative Commons Attribution (CC BY) license (http://creativecommons.org/licenses/by/4.0/).

International Journal of
Molecular Sciences

Article

The Effect of FG-Nup Phosphorylation on NPC Selectivity: A One-Bead-Per-Amino-Acid Molecular Dynamics Study

Ankur Mishra [1], Wouter Sipma [1], Liesbeth M. Veenhoff [2], Erik Van der Giessen [1] and Patrick R. Onck [1,*]

[1] Zernike Institute for Advanced Materials, University of Groningen, Groningen, 9747 AG, The Netherlands; a.mishra@rug.nl (A.M.); woutersipma@gmail.com (W.S.); e.van.der.giessen@rug.nl (E.V.G.)

[2] European Research Institute for the Biology of Ageing, University of Groningen, University Medical Centre, Groningen, 9713 AV, The Netherlands; l.m.veenhoff@rug.nl

* Correspondence: p.r.onck@rug.nl; Tel.: +31-503-638-039

Received: 4 December 2018; Accepted: 23 January 2019; Published: 30 January 2019

Abstract: Nuclear pore complexes (NPCs) are large protein complexes embedded in the nuclear envelope separating the cytoplasm from the nucleoplasm in eukaryotic cells. They function as selective gates for the transport of molecules in and out of the nucleus. The inner wall of the NPC is coated with intrinsically disordered proteins rich in phenylalanine-glycine repeats (FG-repeats), which are responsible for the intriguing selectivity of NPCs. The phosphorylation state of the FG-Nups is controlled by kinases and phosphatases. In the current study, we extended our one-bead-per-amino-acid (1BPA) model for intrinsically disordered proteins to account for phosphorylation. With this, we performed molecular dynamics simulations to probe the effect of phosphorylation on the Stokes radius of isolated FG-Nups, and on the structure and transport properties of the NPC. Our results indicate that phosphorylation causes a reduced attraction between the residues, leading to an extension of the FG-Nups and the formation of a significantly less dense FG-network inside the NPC. Furthermore, our simulations show that upon phosphorylation, the transport rate of inert molecules increases, while that of nuclear transport receptors decreases, which can be rationalized in terms of modified hydrophobic, electrostatic, and steric interactions. Altogether, our models provide a molecular framework to explain how extensive phosphorylation of FG-Nups decreases the selectivity of the NPC.

Keywords: Nuclear pore complex; FG-Nups; phosphorylation

1. Introduction

Eukaryotic cells are characterized by the presence of the nuclear envelope (NE), a lipid bilayer membrane that separates the cells into two compartments, i.e., the nucleus and the cytoplasm. The NE contains many nuclear pore complexes (NPCs), which are the sole gateway for the exchange of essential biomolecules between the two compartments. NPCs are large protein complexes, with a molecular mass of ~55–66 MDa [1,2] in yeast and ~ 125 MDa in vertebrates [3]. The NPC is composed of 30 different types of proteins called nucleoporins (Nups) [4,5]. One third of these Nups are intrinsically disordered proteins (IDPs), which are anchored to the inner wall of the NPC and are rich in phenylalanine-glycine (FG) repeats. Inside the NPC, these FG-Nups form a central meshwork that provides a permeability barrier for translocating molecules. Various studies have revealed that the NPC allows rapid transport of small molecules (30 kDa or ~5 nm in diameter), but drastically slows down the translocation of larger molecules from one compartment to the other [6–8]. It also has been found that FG Nups bind to nuclear transport receptors (NTRs) [9,10] by means of hydrophobic interactions, which facilitates the translocation of NTRs by lowering the permeability barrier [11]. Cargoes of diameter up to 40 nm are known to translocate by this facilitated transport mechanism [12,13]. Therefore, the FG-Nups are considered to be crucial in establishing the selective permeability barrier of the NPC.

Nucleocytoplasmic trafficking can be altered by a change in the surface properties of translocating molecules [14], by the deletion of FG-Nups [6,8], and by the change in cohesiveness of the FG-Nups [15]. For example, mutation of the hydrophobic F residues of Nsp1, a representative yeast FG-Nup, into the hydrophilic Serine S reduces the propensity of Nsp1 to form a hydrogel [16]. These experiments revealed that the hydrogels exclude inert molecules, but allow hydrophobic NTRs to enter, which is explained in terms of a local disruption of the cohesive gel network [16,17]. In a separate study [15], the Nsp1 molecules tethered onto the inner surface of solid state NPC mimics formed a dense phase (over 100 mg/mL) and enabled transport selectivity. Kap95 (a yeast NTR) traversed the pore whereas the translocation of tCherry (an inert molecule of similar size) was inhibited. The F, I, L, V to S mutation of Nsp1 resulted in a remarkably less dense FG-Nup network inside the pore, which led to a loss of selectivity, as both tCherry and Kap95 were able to translocate. Taken together, these studies show that the transition from a dense, hydrophobic phase to a dispersed, hydrophilic phase results in the nanopores losing their selective barrier function.

The hydrophobicity of FG-Nups can also be altered through phosphorylation, one of the most abundant protein modifications inside the cell [18,19]. Phosphorylation is catalyzed by kinases and can be reversed by phosphatases. It has been shown that extracellular signal-regulated kinase (ERK), a phosphorylating agent, can directly interact with FG-Nups [20,21], causing FG-Nups to phosphorylate [22–24]. Several in vitro studies have revealed that specifically Nup62, Nup98, Nup153, Nup214, and Nup358 can undergo phosphorylation [25,26]. Furthermore, there is evidence which confirms that FG-Nups undergo phosphorylation in vivo as well [21,23,27,28]. Transport studies demonstrated that the phosphorylation of nucleoporins results in decreased kinetics of active transport of Kap95 [25,27,29] and Kap-cargo complexes [30,31], and increased kinetics of passive transport [32]. These studies indicate that phosphorylation can modulate the selective permeability of the NPCs. However, the molecular mechanism behind the alteration in nucleocytoplasmic transport due to phosphorylation is not well understood.

Molecular dynamics (MD) simulations have proved to be a powerful tool to study the disordered protein structure inside the NPC and the transport through native and biomimetic nanopores [8,15,33–35]. Therefore, in order to understand the molecular mechanism behind the phosphorylation-induced alteration in transport kinetics, we carried out MD studies using our earlier developed one-bead-per-amino-acid (1BPA) coarse grained (CG) model for FG-Nups [35], extended here for phosphorylated FG-Nups. This 1BPA model has been successfully applied to probe the (doughnut-like) density distribution of the disordered domain of yeast NPCs [35], the facilitated transport of NTRs through yeast [33] and biomimetic [15] NPCs, and the size selectivity for passive transport [6], in good agreement with experiments. Although the transport experiments on phosphorylated NPCs cited above were carried out on mammalian NPCs, we here used the yeast NPC model, which has structural and functional similarities to the vertebrate NPC [36].

In the current study, we extended our 1BPA model to phosphorylated FG-Nups and carried out MD simulations of FG-Nups in isolation, as well as within the NPC. We studied the impact of phosphorylation on the structure of the disordered phase and the transport across the NPC in two scenarios. In the first (referred to as the Phos_N scenario), we used the NetPhosYeast 1.0 server [37] to obtain the phosphorylated residues of the yeast FG-Nups (yielding phosphorylated serine (S) and threonine (T) residues only), and in the second (referred to as the Phos_Max scenario), we assumed that all phosphorylatable residues (serine (S), histidine (H), threonine (T), and tyrosine (Y)) were phosphorylated. We investigated the changes in conformation of phosphorylated FG-Nups compared to FG-Nups in their native state by using the Stokes radius (R_S) as a measure for their size (see Section 2.1). We found that phosphorylation causes FG-Nups to extend by an amount that depends on the fraction of phosphorylatable residues and positively charged residues. In Section 2.2, we present a study on the collective interaction of phosphorylated FG-Nups inside the confined environment of the NPC. We found that phosphorylation drastically reduces the FG-Nup density inside the NPC. Finally, in Section 2.3 we report on simulation results of the phosphorylation-affected transport of inert particles and Kap95, and discuss these results in light of the various contributions to the interaction energy inside the NPC. Our transport simulations are in qualitative agreement with the experimentally-observed increase and decrease in transport rate of the passive and active transport pathways, respectively. Note that the Phos_N scenario predicts more phosphorylation sites than other phosphorylation databases, such as the fungi phosphorylation database (FPD), which provides a comprehensive list of experimentally validated phosphorylation sites [38]. The prediction from the FPD database is incorporated in the Supplementary Materials (see the section "Sensitivity analysis") to provide a scenario for experimentally validated phosphorylation sites. It is important to note that it is unclear which phosphosites predicted in either scenario are phosphorylated simultaneously in vivo, and hence the predictions provided in this study are not meant to mimic specific biological conditions, but rather to shed light on the fundamental mechanisms underlying the changes in transport kinetics of phosphorylated NPCs.

2. Results

In order to study the effect of phosphorylation on FG-Nups, we started with our previously developed MD model for intrinsically disordered proteins (IDPs) in their native state, coarse-grained at a resolution of one bead per amino acid (1BPA) [35]. This 1BPA model accounts for non-bonded hydrophobic and electrostatic interactions between the amino acids, including the effect of solvent polarity and ionic screening to mimic the solvent conditions inside the NPC. The model is accurate (within 20% error) in predicting the Stokes radius R_S [35] for a range of FG-Nups and FG-Nup segments [39]. In the current study, we extended the model for phosphorylation by accounting for the change in hydrophobicity and charge of four amino acids: S, H, T, and Y. We used a weighted average scheme of five predictor programs KOWWIN, ClogP, ChemAxon, ALOGPS, and miLogP [40–43] to predict the change in hydrophobicity due to the change in the chemical structure. For details on the model development for phosphorylated FG-Nups we refer to the Materials and Methods section (Section 4.2). The new parameters for phosphorylated amino acids are summarized in Table 1.

Table 1. Parameters in the 1BPA forcefield for phosphorylated amino acids. Note that: ε_{1BPA} and $\varepsilon_{weighted}$ are the normalized hydrophobicity values (between 0 and 1) from the 1BPA model [35] and the weighted average scheme (see Section 4) for the amino acids in their native state, respectively; ε_p is the hydrophobicity of the phosphorylated amino acid; and q and q_p denote the charge of the amino acids in their native and phosphorylated conditions, respectively.

AA	ε_{1BPA}	$\varepsilon_{weighted}$	ε_p	q	q_p
Ser (S)	0.45	0.41	0.07	0	$-2e$
His (H)	0.53	0.44	0.06	0	$-2e$
Thr (T)	0.51	0.52	0.23	0	$-2e$
Tyr (Y)	0.82	0.83	0.67	0	$-2e$

2.1. Effect of Phosphorylation on Isolated FG Nups

We used our newly developed parametrization for phosphorylation and performed MD simulations to study the effect of phosphorylation on the conformation of isolated FG-Nup segments [35]. The simulated trajectories were analyzed to determine the time averaged R_S using the Hydro program [44,45]. The predicted R_S values for the phosphorylated FG-Nups are compared with that of FG-Nups in their native state (from experiments [39] and simulations [35]) in Figure 1. The error bars for the simulation data represent the standard deviation in time for R_S (See Table S1 in the supplementary data for the source data).

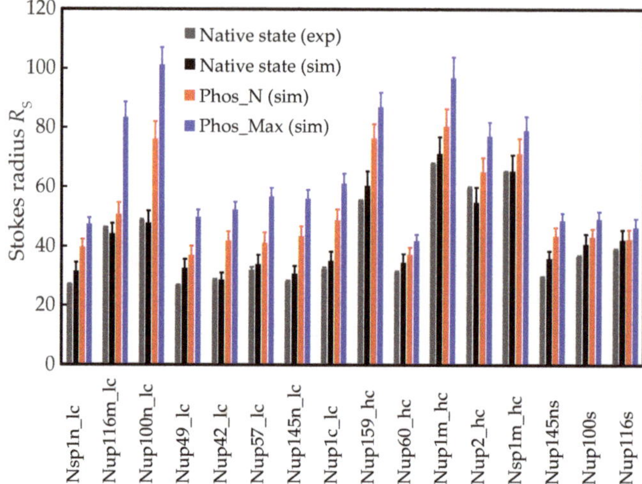

Figure 1. Phosphorylation-induced extension of FG-Nup segments. The Stokes radius R_S (in Angstrom) is depicted for a range of FG-Nup segments in their native and phosphorylated states. The suffix *lc* denotes low charge, *hc* high charge and *s* refers to the stalk region of the Nup. The grey and black bars represent the data in the native state from experiments [39] and simulations (results reproduced from [35]), respectively, and the prediction for the phosphorylated states are plotted in red for Phos_N and blue for Phos_Max. For the simulation data, the error bars represent the standard deviation in time (see Table S1 for the source data).

As a result of phosphorylation, the amino acids become more hydrophilic and negatively charged (see Table 1). Thus, compared to the native state, the phosphorylated FG-Nups exhibit enhanced electrostatic repulsion and reduced hydrophobic attraction leading to an overall decrease in intra-molecular cohesion and thus a more extended configuration (see Figure 1). In Table S2, we have summarized the number of amino acids that can be phosphorylated in each FG-Nup segment. The FG-segments are grouped as low charged (*lc*), high charged (*hc*), and stalk (*s*) domains, following the definition of Yamada et al. [39]. We found that the relative abundance of phosphorylatable residues in all FG-Nup segments ranges from ~15% (for Nup116s) to ~33% (for Nsp1n_lc) for the maximally phosphorylated (Phos_Max) condition, whereas for the Phos_N scenario the range is from ~4% (Nup116s) to ~17% (Nup159_hc). In order to quantify the change in Stokes radius in terms of the number of residues undergoing phosphorylation, we plot the normalized change in R_S as a function of the percentage of phosphorylatable residues (*n*) for the low charged, high charged, and stalk groups in blue, red, and green data points, respectively, for the Phos_Max (Figure 2a) and Phos_N (Figure 2b) scenarios. The change ΔR_S is normalized as $\Delta R_S/(N-1)b$, where N is the total number of residues of the FG-Nup segment and b is the coarse-grained bond length (3.8 Angstrom). We fitted the data points for individual groups to a straight line passing through the origin, represented as colored lines in Figure 2a,b. We note that for both Phos_Max and Phos_N, the FG-segments from the *lc* group show the highest normalized change in R_S (blue line in Figure 2a,b), whereas phosphorylation has a smaller effect on size for the *hc* and *s* groups (red and green lines in Figure 2a,b).

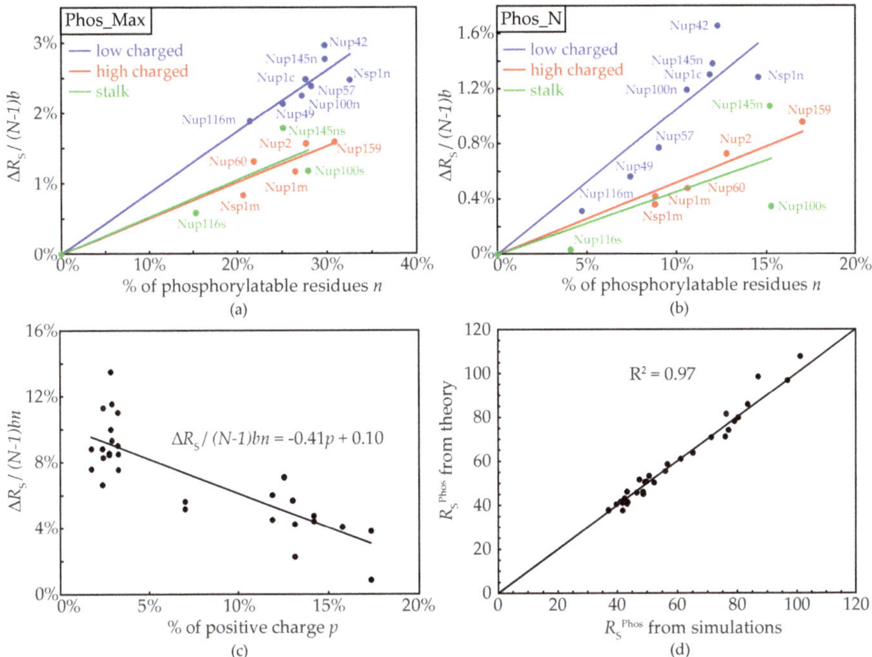

Figure 2. The normalized change in R_S (i.e., $\Delta R_S / (N-1)b$) due to phosphorylation as a function of the fraction of phosphorylatable residues (n) for (**a**) the Phos_Max and (**b**) the Phos_N scenarios [37]. The expression for the change in R_S (i.e., $\Delta R_S = R_S^{phos} - R_S^{native}$), with R_S^{phos} and R_S^{native} being the Stokes radii of the FG-Nup segments in the phosphorylated and native states, respectively, is normalized with $(N-1)b$ where N is the total number of residues of the FG-Nup segment, and b (= 3.8 Angstrom) is the coarse-grained bond length between neighboring amino acids [35,46]. The data for the FG-Nups from the high charged (*hc*), low charged (*lc*), and stalk (*s*) segments [35,39] are represented in red, blue, and green data points, respectively. The data points of each group are fitted to a straight line passing through the origin revealing different slopes for different groups. For Phos_Max we observe slopes of 0.087 for *lc* (R^2 = 0.94), 0.051 for *hc* (R^2 = 0.92), and 0.053 for *s* (R^2 = 0.80), respectively, whereas for Phos_N the slopes are 0.1 for *lc* (R^2 = 0.86), 0.052 for *hc* (R^2 = 0.95) and 0.045 for *s* (R^2 = 0.61). (**c**) The ratio of normalized change in R_S to the fraction of phosphorylatable residues (n) is plotted as a function of the fraction of positively charged residues (p) for all data points (black) from the Phos_Max and Phos_N scenarios. These data points are fitted to a linear equation, as shown in the figure (giving R^2 = 0.68). (**d**) The R_S^{phos} predicted from the theory in Equation (1) compared to R_S^{phos} computed from the MD simulations, both in Angstrom, show a good agreement with a fitness measure of R^2 = 0.97.

In order to investigate the varying response for the three groups, as shown in Figure 2a,b, we analyzed the change in hydrophobicity upon phosphorylation and found that it is roughly similar for the three groups, i.e., for FG-Nups from the *lc*, *hc*, and *s* groups, the hydrophobicity drops by 13-20%, 16–22%, and 12–21%, respectively, for Phos_Max (see Table S2). Similarly, for Phos_N, the reduction in hydrophobicity amounts to 3–10%, 7–10%, and 4–12% for the *lc*, *hc*, and *s* groups, respectively, showing no major difference across the three groups. Clearly, the effect of phosphorylation on hydrophobicity alone cannot account for the different R_S of the groups. It has been argued that the net proline content in IDPs plays an important role in determining the effective Stokes radius [47], as proline provides additional stiffness to the peptide chain because of its ring structure. This effect of proline is included in our 1BPA model in the form of the bonded potentials [46]. However, all the FG-Nup segments analyzed in this study (Figure 1) have a similar 3–8% proline content, and therefore cannot explain the different

Stokes radii across the three groups (see Table S2). Next, we analyzed the effect of charge. Since the net charge of the three families is quite similar (i.e., 2–3%, 0–3%, and 0–1% for *lc*, *hc*, and *s*, respectively), we investigated the occurrence of positively charged residues R and K in the FG-segments and found that the *lc* group contains only 2–3% of positively charged residues in contrast to the *hc* and *s* groups, which have more positively charged residues (7–16% for *hc* and 13–17% for *s*, see Table S2). Thus, it seems that the larger amount of positive charge in *hc* and *s* is more efficient in screening the effect of the negative charge increase induced from phosphorylation than the small amount of positive charges in the *lc* group (see Table S2). In order to confirm this, we plotted the ratio of $\Delta R_S/(N-1)b$ normalized by the fraction of phosphorylatable residues (n) as a function of the percentage of positive charge content (p) in the FG-segments (see Figure 2c). The data points in Figure 2c can be fitted to a straight line with a slope of -0.41 and y-intercept of 0.1, with an R^2 value of 0.68. Using this observation, the Stokes radius for a phosphorylated FG-segment can be predicted using the following expression:

$$R_S^{\text{phos}} = R_S^{\text{native}} + bn\,(N-1)(-0.41p + 0.1) \tag{1}$$

We show the predictive power of this formula in Figure 2d, where the R_S^{phos} predicted using Equation (1) is plotted against the computed R_S^{phos} from the MD simulations, showing a very good correlation (with $R^2 = 0.97$).

2.2. Effect of Phosphorylation on NPC Structure

For the next step, we analyzed the disordered protein distribution inside the yeast NPC upon phosphorylation. Due to phosphorylation according to the Phos_N scheme, 6432 (7.4%) of all residues (86520) are phosphorylated inside the NPC, whereas for Phos_Max, 20992 (24.3%) of the NPC residues are phosphorylated (see Table S5). We tethered the FG-Nups inside the yeast scaffold at the same anchoring points as in the wild type yeast NPC [35,48], and then switched on phosphorylation. In Figure 3, we show snapshots of the wild type and phosphorylated NPCs (for both the Phos_N and Phos_Max conditions). As expected from the conformational analysis of isolated FG-Nups (Section 2.1), the FG-Nups assumed extended conformations in the phosphorylated NPC compared to the wild type NPC. For the maximally phosphorylated (Phos_Max) state, the FG-Nups spilled out of the NPC over a distance of almost 100 nm from the NPC surface, which is considerably larger than that for the Phos_N state. In the case of isolated FG-Nups, only intra-molecular interactions are present, whereas inside the NPC, the residues of each FG-Nup also interact with residues from other FG-Nups (i.e., both intra- and inter-molecular interactions are present). Therefore, the enhanced electrostatic repulsion and reduced hydrophobic attraction due to phosphorylation is much more pronounced inside a confined space like the NPC, resulting in a strong reduction of protein density.

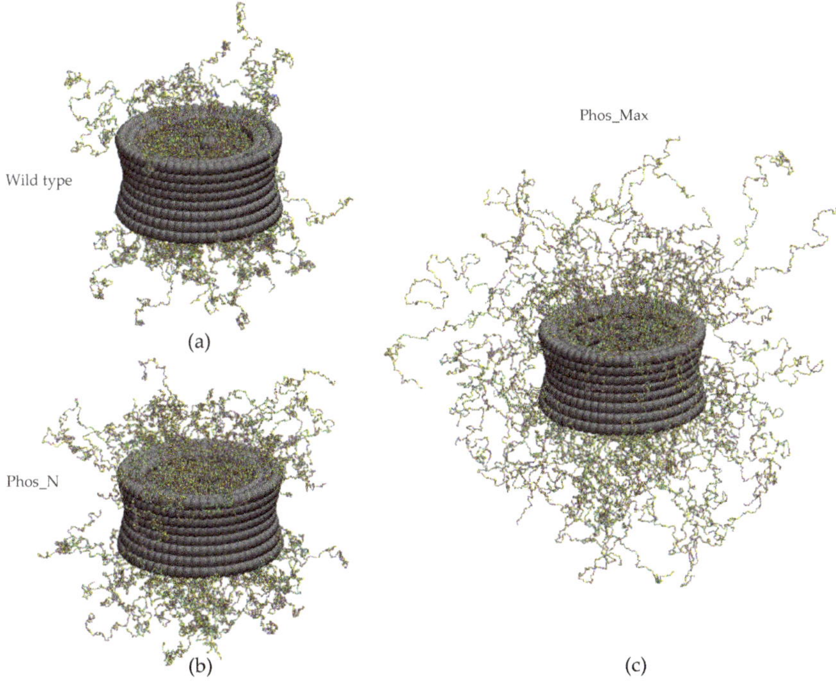

Figure 3. Snapshots of the coarse-grained MD models of (**a**) the wild type NPC, (**b**) the Phos_N NPC, and (**c**) the Phos_Max NPC. The FG-Nups (different color beads denote different amino acids) are attached at the same anchor positions, as in [35,48]. Phosphorylation-induced changes in the interaction leads to spilling of FG-Nups out of the NPC.

We analyzed the collective distribution of the FG Nups inside the NPC by calculating the time averaged radial density distribution (averaged over the axial and circumferential direction) inside the pore (i.e., for $|z| < 15.5$ nm) (note that the origin of the coordinate system coincides with the center of the NPC). The radial density profile for all residues and hydrophobic residues are plotted in Figure 4a,b, respectively, for the wild type and phosphorylated NPCs. In addition, the 2-dimensional (rz) density distribution (averaged over the circumferential direction) is plotted in Figure 4c for the wild type and phosphorylated NPCs. For technical details on calculating the density distributions, the reader is referred to the Materials and Methods section. Figure 4 clearly shows that phosphorylation significantly alters the density distribution of the FG-Nups inside the NPC. For a wild type NPC, the mean density at the center (0 nm < r < 5 nm) is ~ 80 mg/mL (see Figure 4a), which gradually increases with the radial distance r from the center and attains a peak value (~180 mg/mL) at r ~ 15 nm, after which the density gradually decreases. This is consistent with the doughnut-like structure in Figure 4c (left panel). However, for the phosphorylated NPCs, the density drops drastically inside the pore, amounting to only ~50 mg/mL and ~20 mg/mL at the center (see Figure 4a) for the Phos_N and Phos_Max scenarios, respectively. In the case of Phos_N, the radial density profile follows a similar trend (but lower in magnitude) as the wild type. This results in a less dense doughnut-like structure, as shown in Figure 4c (middle panel). For Phos_Max, the density remains lower than 20 mg/mL throughout the full range of r-values, as shown in Figure 4a,c (right panel). The density distribution of hydrophobic residues (see Figure 4b) is highly correlated with the density distribution of the total amount of residues (see Figure 4a).

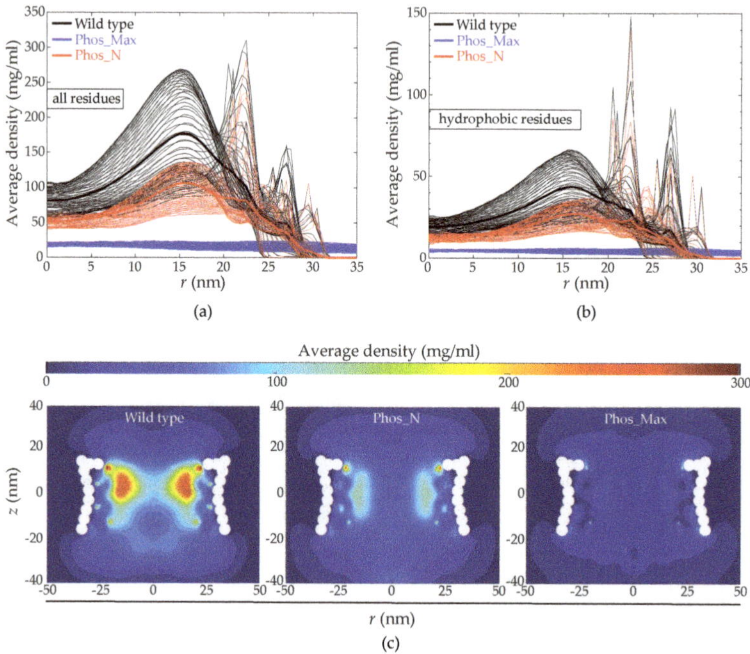

Figure 4. Disordered protein structure inside the wild-type and phosphorylated NPCs. (**a**,**b**) Time-averaged radial density distribution inside wild type (black), Phos_N (red) and Phos_Max (blue) NPCs for (**a**) all residues, and (**b**) hydrophobic residues. The thin lines represent the density at different positions along the z-axis separated by 1 nm, in the range of $|z| < 15.5$ nm (height of the NPC), with the mean density plotted as thick lines. (**c**) The rz-density map for the wild type (left panel), Phos_N (middle panel), and Phos_Max (right panel) NPCs. The wild type shows the characteristic highly dense doughnut-like structure [35]. The Phos_N NPC shows a density-depleted doughnut-like structure, whereas the Phos_Max NPC shows a significantly less dense and rather uniform density distribution.

To explore the main reason for the reduction in protein density for both phosphorylated NPCs, we computed the relative contribution of the hydrophobic and electrostatic energy to the total interaction energy inside the wild-type and phosphorylated NPCs [15,34,35], as shown in Figure 5a,b, respectively. In the wild type NPC, the time-averaged hydrophobic interaction energy amounts to approximately −76,300 kJ/mol, whereas for the Phos_N and Phos_Max NPCs, these values are about −49,000 kJ/mol and −6100 kJ/mol, respectively. Here, by far the largest reduction is in the Phos_Max NPC, with almost a twelve-fold decrease in hydrophobic interaction energy relative to the wild type. Note that this twelve-fold decrease cannot be explained by the reduction in net hydrophobicity alone (a reduction of 16%, see Table S5); also, the distance between the hydrophobic amino acids plays an important role, being much larger for Phos_Max than for the wildtype and Phos_N NPCs (see Figure 4). On the other hand, the Coulomb energy was measured to be two orders of magnitude smaller than the hydrophobic energy for the wild type and Phos_N NPCs (around −750 kJ/mol for wild type and −800 kJ/mol for Phos_N), while for Phos_Max, the total repulsive Coulomb energy (around 31,200 kJ/mol) is much larger than the hydrophobic energy. In summary, the wild type and Phos_N NPCs are highly hydrophobic, with only a small contribution from electrostatics. In sharp contrast to this, the energy in the Phos_Max NPC has a much more dominant (repulsive) Coulombic contribution corresponding to the large net negative charge, while the hydrophobic energy is much lower.

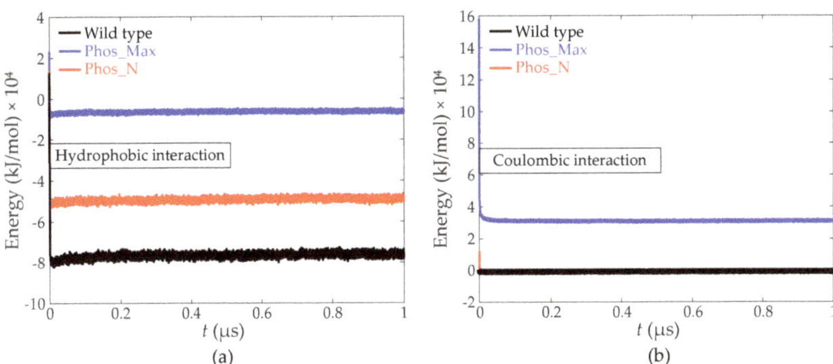

Figure 5. Time evolution of (**a**) the hydrophobic interaction energy, and (**b**) the coulombic interaction energy of the FG-Nups for the wild type (black), Phos_Max (blue), and Phos_N (red) NPCs.

2.3. Effect of Phosphorylation on Active and Passive Transport

In this section, we focus on the selective permeability barrier of the NPC and how phosphorylation affects this. The nuclear transport receptor Kap95 is known to interact with the FG-Nups via its hydrophobic binding sites and to translocate through the NPC by facilitated transport in the presence of RanGTP at the nucleoplasmic side, which dissociates the Kaps from the NPC [49,50]. Beside being hydrophobic, the Kaps are also negatively charged [51]. It is therefore expected that the interaction between the Kaps and FG-Nups is strongly affected by the phosphorylation-induced charge modification and reduction in hydrophobicity of the FG-Nups. To investigate this, we modelled a yeast NPC in the presence of ten Kap95 particles (with a diameter of 8.5 nm, 10 hydrophobic binding sites, and a uniformly-distributed surface charge of $-43e$, as used previously [15]) that are released at the cytoplasmic side to probe facilitated transport. After equilibrating the system (see Materials and Methods for details), the simulations were carried out for the same initial positions of the Kaps for the wild type and phosphorylated pores. Snapshots of the final state (at $t = 2$ μs) are shown in Figure 6, illustrating the inhibition of facilitated transport in phosphorylated NPCs. The bottom panels of Figure 6a–c depict a reduced binding affinity of the Kaps with the phosphorylated FG-Nups. To assess the propensity for translocation, we plot the initial ($t = 0$ μs) and final ($t = 2$ μs) z coordinate of the center of mass of the Kap95 particles in Figure 7a. We observed that for the wild type NPC, the Kap95-FG-Nup affinity is larger compared to the phosphorylated NPCs (see Figure 6), so that the Kap95 particles are able to enter the pore and translocate (Figure 7a). The large affinity is due to the fact that the pore is hydrophobic (see Figure 5 and Table S5) and has a weak positive charge [35]. In the course of 2 μs, a total of 9 Kaps translocated through the pore and the remaining Kap ended up inside the pore (Figures 6a and 7a). In the phosphorylated NPC, however, despite the lower FG-Nup density (see Figure 4), the Kaps are excluded from the pore (Figure 6b,c and Figure 7a). The results from Figure 5, Figure 6, and Figure 7a point towards a phosphorylation-induced decrease in hydrophobicity (for Phos_Max and Phos_N) and increase in coulombic repulsion (for the Phos_Max case) resulting in lowering of Kap95-FG-Nup binding affinity, which is instrumental for active transport.

Next, we study the effect of phosphorylation on passive transport by probing the transport of inert particles of the same size as the Kap95 particles (i.e., 8.5 nm in diameter) but without charge and hydrophobic binding spots. We used the same initial positions for the inert particles as for the Kap95 particles used in the case of active transport (Figures 6 and 7a). In Figure 7b, we plot the initial and final z location of the center of mass of the inert particles for the wild type and phosphorylated NPCs. For the wild type NPCs, it can be clearly observed that the inert particles stay at the cytoplasmic side and do not enter the NPC. On the other hand, in the Phos_Max NPC, two inert particles managed to translocate through the pore within 2 μs, showing that the permeability barrier of the Phos_Max NPC is jeopardized. For the Phos_N NPC, we did not observe any translocation. To further test the size-dependent permeability barrier of the Phos_N and wild type NPCs, we performed two additional transport simulations for ten spherical inert particles of diameter 4 nm. The results are shown in Figure S1 (see the Supplementary Information), revealing that in the wild type, the total number of translocations is 129, whereas in the Phos_N pore, there were 328 translocation events, indicating a 2.5-fold increase in passive transport rate compared to wild type. For the wild type, this is consistent with our previous work [33], where we found the energy barrier for inert particles of size 4 nm to be lower than $\kappa_B T$ (and thus likely to go through), while that for inert particles of 7 nm (and up) was found to be larger than 2 $\kappa_B T$ (and thus likely to not pass through). Our results for the wild type can be understood by recourse to the scaling relation of Timney et al. [8], which states that the characteristic time constant of passive transport scales with the third power of the molecular mass. Since the two inert particles used here are of size 4 and 8.5 nm, the mass dependence rule predicts that the translocation of the 8.5 nm particle should be 883 times slower compared to the smaller 4 nm particle. This is in qualitative agreement with our simulations, where for the 4 nm particles we see 129 translocation events within 2 μs of simulation time, whereas no translocation is observed for the 8.5 nm inert particle. Since Phos_Max NPCs have a lower protein density compared to Phos_N NPCs, we can expect the transport rates to be even higher for a 4 nm particle. In the case of passive transport, the inert particles interact with the FG-nups by means of steric repulsion only, and therefore the translocation events (Figure 7b and Figure S1) of inert molecules can be understood in terms of the density distribution of FG-Nups in the wild type and phosphorylated NPCs. Figure 4 shows that the FG-Nup density inside the NPC is significantly higher for the wild type NPC compared to the phosphorylated NPCs, which explains why in the wild type NPC no passive transport is observed (Figure 7b), whereas in the phosphorylated NPCs, passive transport occurs due to the lower permeability barrier.

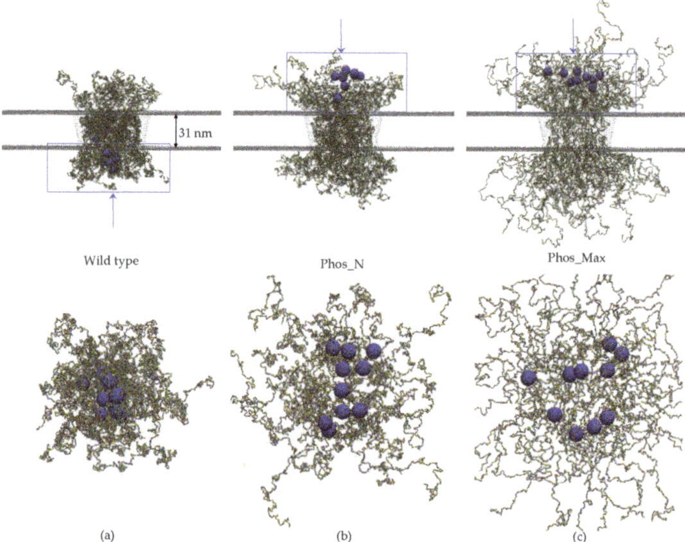

Figure 6. Snapshots at $t = 2$ μs of the FG-Nups and model Kap95 particles inside a wild type NPC (**a**), a phosphorylated NPC according to the Phos_N scheme (**b**), and a phosphorylated NPC according to the Phos_Max scheme (**c**). The Kap95 particles are shown in blue with the red hydrophobic binding spots on its surface. The 20 different amino acids of the FG-Nups are represented by different colors. The size of the scaffold beads (grey) is scaled down to make the Kap particles better visible. In the bottom panel we provide a magnified view from the bottom for (**a**) and from the top for (**b**) and (**c**), focusing on the region indicated by the blue boxes in the top panels. The Kaps are shown to traverse the wild type NPC, whereas the Kaps are not able to strongly partition into the FG-Nup meshwork for the phosphorylated NPCs.

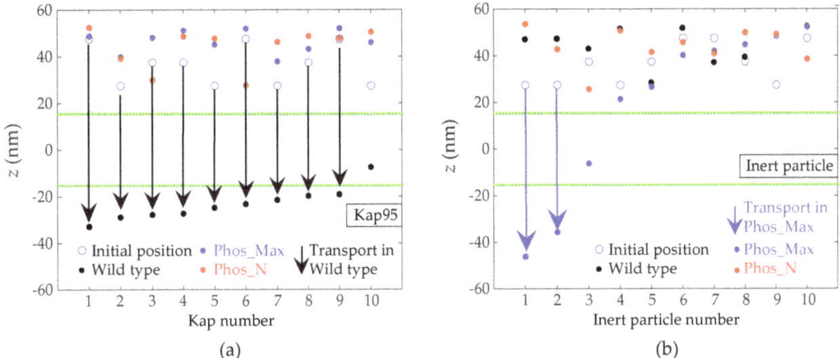

Figure 7. Effect of phosphorylation on active (**a**) and passive (**b**) transport. In both cases, the particles are released from the same position at the cytoplasmic side and are ordered from left to right based on the end position at $t = 2$ μs. (**a**) Initial ($t = 0$ μs) and final ($t = 2$ μs) axial (z) position of ten Kap95 particles, and (**b**) Initial ($t = 0$ μs) and final ($t = 2$ μs) axial (z) position of ten inert particles of the same size as the Kap95 particles (diameter = 8.5 nm, no charge and no hydrophobic binding spots). In both (**a**) and (**b**), the boundaries of the NPCs ($|z| = 15.5$ nm) are represented by green lines and the arrows represent translocations from the cytoplasm to the nucleoplasm.

Finally, we summarize our findings on active (from Figure 7a) and passive transport (Figure 7b and Figure S1) for the wild type, Phos_N, and Phos_Max NPCs in Figure 8. The wild type NPC is seen to have a selective permeability barrier, as it allows Kaps and small inert particles (diameter = 4 nm) to pass through, whereas larger inert particles (diameter = 8.5 nm) are excluded. The phosphorylated Phos_N NPC loses its selectivity, as transport of Kap95 is not observed, but its permeability barrier is still intact (8.5 nm particles are excluded). This indicates that the reduced steric hindrance due to the reduced amino acid density in the center (Figure 4a) is still sufficient to exclude large particles, but that the reduced hydrophobicity is no longer able to attract Kap95 particles into the FG-Nup mesh-work. Finally, the heavily phosphorylated Phos_Max NPC is observed to lose both its ability to facilitate active transport (due to the reduced hydrophobic attraction and increased electrostatic repulsion with respect to Kap95 particles), as well as its permeability barrier (due to the drastically reduced amino acid density allowing for transport of both inert particles). This, of course, is subject to the constraint of the limited time frame of our simulations (i.e., 2 μs).

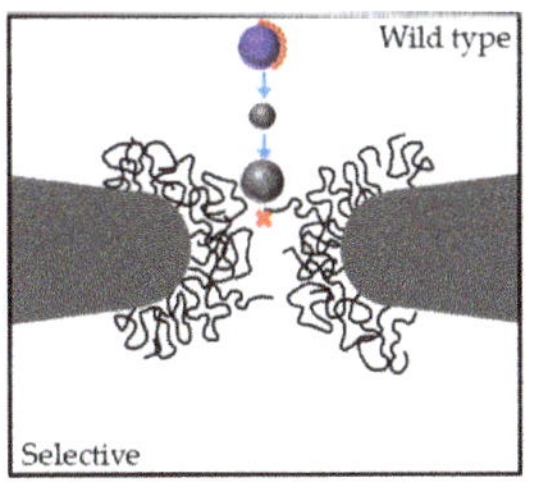

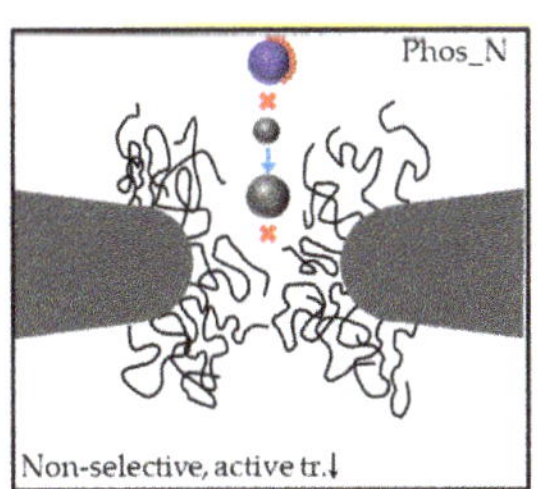

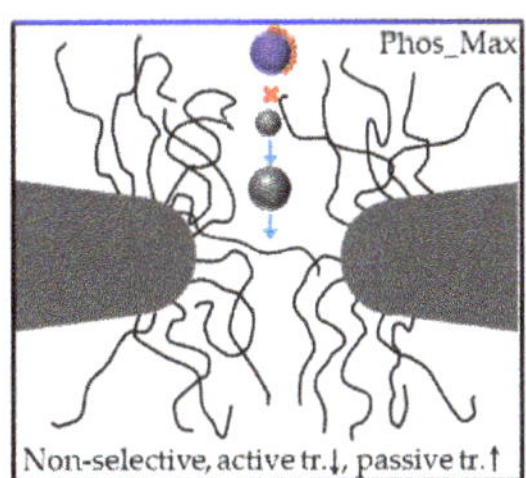

Figure 8. Summary of simulation results on wildtype and phosphorylated yeast NPCs, showing active transport of Kap95 (blue sphere with red dots representing the hydrophobic binding sites) and passive transport of inert particles (grey spheres of diameter 4 nm and 8.5 nm). The FG-Nups are represented by black filaments. The blue arrows indicate observed transport and the red crosses indicate prohibited transport within the 2 μs simulation time.

3. Discussion

In the current study, we incorporated phosphorylation-induced modifications of the hydrophobicity and charge of the four amino acids S, H, Y, and T (see Table 1) into our 1BPA model for IDPs. We addressed the effect of phosphorylation on the conformational changes of 16 different isolated FG-Nup segments of varying length and with varying numbers of charged and hydrophobic residues. We compared the predicted Stokes radius R_S of these FG-Nup segments in their phosphorylated state with the values in their native state (see Figure 1), and observed an increase in size due to phosphorylation. We found that R_S increases linearly with the fraction of phosphorylatable residues and decreases linearly with the percentage of positively-charged amino acids (see Equation (1) and Figure 2d. While the former dependence is straight-forward, the latter is subtler and points to the important role of positive charge in screening the effect of the phosphorylation-induced increase in negative charge.

Next, we investigated how the FG-Nups interact when confined inside the NPC and how phosphorylation alters these interactions and the resulting protein distribution. The density distributions demonstrate a considerable difference between the wild type and phosphorylated NPCs, with the total amino acid density and hydrophobic density dropping by almost a factor two and four for the Phos_N and Phos_Max NPCs, respectively, in comparison to wild type. Whereas the hydrophobicity changed both in terms of density (Figure 4b) and energy (Figure 5a) for the two phosphorylated NPCs, only the Phos_Max NPC showed a large increase in (repulsive) electrostatic energy, while for the Phos_N and wild type NPC, the electrostatic energy remained negligible compared to the hydrophobic energy (Figure 5a,b). All considered, we can conclude that the phosphorylated FG-nups resulted in a higher negative charge and lower hydrophobicity, resulting in a strong depletion of amino acid density in phosphorylated NPCs, with the effects (especially the electrostatic) much more pronounced in Phos_Max NPCs.

For those molecules that translocate through the NPC by means of active transport (for example Kap95 in this study) [15,33–35], the molecular interactions can be divided into three components: (i) steric repulsion by means of excluded volume; (ii) hydrophobic interactions; and (iii) Coulombic interactions. Firstly, as the density inside the phosphorylated pores (both Phos_N and Phos_Max NPCs) is significantly lower than in the wild type (see Figure 4), the steric repulsion component is lower. Secondly, as illustrated in Figure 5a, phosphorylation results in a serious reduction of the hydrophobic interaction energy, as the residues become more hydrophilic upon phosphorylation. Finally, as Kap95 carries negative charge, it will face electrostatic attraction when the pore is positively charged (wild type) and electrostatic repulsion when the pore is negatively charged (phosphorylated). All the energy components taken together indicate that the negatively-charged and hydrophobic Kap95 experiences a much more repulsive environment inside the phosphorylated NPCs due to the increased negative charge and reduced hydrophobicity compared to wild type. Therefore, the overall energy barrier for the translocation of Kap95 particles through a phosphorylated pore is much higher compared to the wild type pore. Our transport simulations (Figures 6 and 7a) for Kaps indeed reveal inhibition of facilitated transport upon phosphorylation, while wild type NPCs facilitate Kap translocation. We do not account for the presence of RanGTP in our model, which is known to play an important role in dissociating the Kaps from the FG-Nups. As a result, the model Kaps remain in the bound state towards the nuclear side of the pore thanks to a slightly higher affinity of the nuclear FG-Nups with the Kap95 [52]. In contrast to this, the phosphorylated pore inhibits the Kaps to enter the pore. Of course, we cannot completely rule out the fact that some of the Kaps might translocate through phosphorylated NPCs at longer simulation times. Nevertheless, the trend of a reduced probability for active transport through phosphorylated pores and an increased probability for passive transport upon phosphorylation is in accordance with experimental observations [27,29–32].

The results for wild type and Phos_N are comparable to our previous studies on biomimetic nanopores coated with Nsp1 and a more hydrophilic mutant, Nsp1-S, which illustrated that a lack of cohesion in the hydrophilic Nsp1-S pore can result in a depleted density (~twofold decrease) compared to the hydrophobic Nsp1 pore [15,53]. Despite the different nature of the modification (mutation versus phosphorylation), both cases result in a reduced protein density due to a depleted hydrophobicity, while the electrostatic interactions remain approximately unaffected (see Figure 5). However, the effect on the selective permeability was found to be different. Whereas the Nsp1-S pore lost its selectivity due to the fact that the permeability barrier was jeopardized (large inert particles were found to go through), the Phos_N NPC retained its permeability barrier, but lost its ability to actively transport Kap95 particles. This difference is most likely related to the different unfolded protein composition of the yeast NPC and the biomimetic nanopore, with the former consisting of 10 different FG-Nups, while the latter has only one.

It should be noted that in this study, we compared the wild type results with an NPC in which all S, H, T, and Y residues are maximally phosphorylated (Phos_Max), and a second variant (Phos_N) in which the phosphorylation sites are extracted from the NetPhosYeast 1.0 server [37], accounting for the phosphorylation of a subset of all S and T residues. Clearly, the Phos_Max scenario is not very relevant from a biology point of view, as phosphorylation of all S, T, H, and Y does not occur simultaneously in reality. The results of the Phos_Max scenario therefore serve as a theoretical limiting case of phosphorylated NPCs that feature a maximal phosphorylation-induced modification of charge and hydrophobicity. The Phos_N scenario predicts a higher number of phosphorylation sites compared to other phosphorylation databases, such as the fungi phosphorylation database (FPD) (see Tables S6 and S7 for FPD phosphosites). Despite this difference, for both scenarios we observe loss of selectivity (see Figure 7a and Figure S2), while the permeability barrier is retained (see the section "Sensitivity analysis" in the Supplementary Materials).

It is still not known what fraction of the phosphorylatable residues predicted in these databases actually undergo simultaneous phosphorylation inside the NPC in vivo. This will be an interesting aspect to be explored further, since our study indicates that the degree of phosphorylation can have a large impact on the structure of the NPC and the rate of transportation in passive and active pathways. Our work should therefore not be seen as an exact mimic of specific biological conditions, but as a qualitative study, in which NPC phosphorylation is explored in order to shed light on the fundamental mechanisms underlying in vitro experiments on the decreased kinetics for active import [25,27,29] and the increased kinetics of passive import [32] in phosphorylated NPCs.

4. Materials and Methods

4.1. Coarse-Grained Molecular Dynamics Simulations

The 1BPA Molecular Dynamics model used in this study accounts for the exact amino acid sequence of the FG-Nups, in which each bead is located at the C_α positions of the polypeptide chain [35,46]. We set the mass of each bead to the average amino acid mass (120 Da), and the distance between neighboring beads to ~0.38 nm through a stiff harmonic spring potential. The bending and torsion potentials are extracted from the Ramachandran data of the coiled regions of protein structures [46]. The solvent molecules are treated in an implicit manner. A distance-dependent dielectric constant is used to account for the solvent polarity, and ionic screening is incorporated through Debye screening with a screening constant $k = 1$ nm^{-1} corresponding to the physiological salt concentration inside the NPC [54]. The hydrophobic interactions between the amino acids are incorporated through a modified Lennard-Jones potential, which accounts for hydrophobicity scales of all 20 amino acids derived from normalized experimental partition energy data renormalized in a range from 0 to 1. For details of the method, the reader is referred to [35].

All MD simulations were carried out with a time step of 0.02 ps [35]. The simulations for the isolated disordered FG-Nup segments were carried out for 2.5×10^7 steps [35], which was found to be sufficiently long to reach convergence. For the NPC simulations with particles (Figure 6, Figure 7 and Figure S1) and without particles (Figure 3, Figure 4, and Figure 5), the systems were first energy minimized to remove any overlap of the amino acid beads. Then, all long-range forces were gradually switched on, and for the NPC with particle systems, the inert/Kap95 particles were kept at a fixed position on the cytoplasmic side. In the final production run for the NPC without particles, the simulations were carried out for 5×10^7 steps (with the first 5×10^6 steps ignored so that only the statistically meaningful results are extracted), which was found to be long enough to have converged results for the density distribution inside the pores. For the NPC with particles, we included one additional step before the production runs, in which we equilibrated the system for 5×10^6 steps with all long-range forces switched on while keeping the inert/Kap95 particles fixed at their position. In the final production runs for the NPC with transporting particles, the inert/Kap95 particles were allowed to move and the simulations were carried out for 10^8 steps. For the Kap95 simulations we modelled the hydrophobic binding sites on the Kaps as F beads [15,33].

The time-averaged density calculations presented in the main text (see Figure 4) were derived by using the "gmx densmap" tool in GROMACS. The nanopore is centered inside a box of size 100 nm $\times$ 100 nm $\times$ 200 nm, which was divided into discrete cells of size 0.5 nm $\times$ 0.5 nm $\times$ 0.5 nm. The trajectory files from the simulations were analyzed to compute the number density in each cell as a function of simulation time. A time averaged 3D mass density profile was obtained by multiplying the number density with the mass of each bead and then averaging over the simulation time. The 3D density was averaged in the circumferential direction to obtain two-dimensional (2D) rz density plots (as shown in Figure 4c). Finally, the radial density distribution was obtained by averaging these 2D density maps in the vertical direction (as shown in Figure 4a,b). To compute the Coulombic and hydrophobic interaction inside the NPC (see Figure 5), we used the "gmx energy" tool from GROMACS.

4.2. Parametrization of Phosphorylated Amino Acids

We used five different hydrophobicity-predictor programs to estimate the hydrophobicity of phosphorylated residues. These programs calculate the logarithmic value of the equilibrium partition coefficient P, i.e., the ratio of concentrations in a mixture of two immiscible phases, water, and 1-octanol, as a measure for hydrophobicity. They use experimental log P values of fragments (small groups of atoms) to calculate the log P for bigger molecules by adding the individual contributions from the constituting fragments, based on the structure additivity principle for hydrophobicity [55]. There are several challenges in incorporating the log P estimates for the complete molecules directly in our 1 BPA model, which are: (i) the error from the estimate of the log P values for the fragments accumulate while calculating the log P for the entire molecule; and (ii) each of the hydrophobicity-predictor programs are trained with different experimental data, and therefore generate different estimates of log P for a given molecular structure. In order to have the hydrophobicity values comparable to our 1BPA model, first we rescaled and then normalized the log P estimates for all amino acids obtained from a hydrophobicity-predictor program k, so that hydrophobicity of any amino acid i (i.e., $\varepsilon_{k,i}$) falls in the range from 0 to 1. Here, 0 and 1 corresponds to the hydrophobicity of the most hydrophilic and most hydrophobic amino acids, according to hydrophobicity-predictor program k. Next, to minimize the error we decided to incorporate the change in $\varepsilon_{k,i}$ values of the phosphorylatable residues (obtained from the five hydrophobicity-predictor programs) compared to the 1BPA model, for which the hydrophobicity values are extracted from three different partition coefficient measurements [35]. To account for the variation in the prediction of $\varepsilon_{k,i}$ by the different hydrophobicity-predictor programs, a weighted average approach is considered. The weights are assigned to individual predictor programs based on their accuracy in predicting the $\varepsilon_{k,i}$ values of the amino acids in their native state, as used in [35]. Thus, the assigned weight for hydrophobicity-predictor program k for amino acid i can be written as,

$$w_{k,i} = \frac{(1/\Delta\varepsilon_{k,i})^2}{\sum_{k=1}^{5}(1/\Delta\varepsilon_{k,i})^2},$$

(2)

where $\Delta\varepsilon_{k,i} = \varepsilon_{k,i} - \varepsilon_{1\text{BPA},i}$ represents the difference between the hydrophobicity for an amino acid in its native state used in our 1BPA model [35] and the hydrophobicity-predictor programs (see Tables S3 and S4 for the source data). Next, the change in hydrophobicity upon phosphorylation is calculated as

$\Delta\varepsilon_{k,i\text{-phos}} = \varepsilon_{k,i\text{-phos}} - \varepsilon_{1BPA,i}$, where "*i*-phos" represents the amino acid *i* in its phosphorylated state. Finally, using the weights for the hydrophobicity-predictor programs (see Table S4) we computed the hydrophobicity for the phosphorylated amino acid as $\varepsilon_{p,i} = \varepsilon_{1BPA,i} + \sum_{k=1}^{5} w_{k,i}\,\Delta\varepsilon_{k,i\text{-phos}}$. The amino acids Serine (S), Histidine (H), Tyrosine (Y), and Threonine (T) undergo phosphorylation [18,19], and the introduction of a phosphate group results in the introduction of a $-2e$ charge, as shown in Table 1. The phosphorylation of these amino acids results in a more hydrophilic atomic composition, which can be seen in Table 1. As a reference, the prediction of the hydrophobicity of amino acid *i* in the native state, $\varepsilon_{\text{weighted},i} = \varepsilon_{1BPA,i} + \sum_{k=1}^{5} w_{k,i}\,\Delta\varepsilon_{k,i}$, is also shown in Table 1. Note that the subscript *i* is dropped from $\varepsilon_{p,i}$ and $\varepsilon_{\text{weighted},i}$ in Table 1 for clarity.

Supplementary Materials: Supplementary materials can be found at http://www.mdpi.com/1422-0067/20/3/596/s1.

Author Contributions: Conceptualization, A.M. and P.R.O.; methodology, A.M., W.S.; software, A.M.; validation, A.M.; formal analysis, A.M.; investigation, A.M..; writing—original draft preparation, A.M.; writing—review and editing, A.M., W.S., L.M.V., E.V.G., and P.R.O.; visualization, A.M..; supervision, P.R.O. and E.V.G.; project administration, P.R.O.; funding acquisition, P.R.O.

Funding: This research was funded by the Zernike Institute for Advanced Materials (University of Groningen), the UMCG, and NWO ECHO (grant number: 711.013.008 to A.M., P.R.O., and L.M.V.).

Acknowledgments: We acknowledge the use of the Peregrine cluster (University of Groningen) and the Cartesius cluster (SURFsara, funding grant by NWO) for the large scale simulations carried out during this project.

Conflicts of Interest: The authors declare no conflict of interest.

Abbreviations

NPC	Nuclear pore complex
FG	phenylalanine-glycine
lc	Low charged
hc	High charged
s	Stalk
Kap	Karyopherin protein

References

1. Rout, M.P.; Blobel, G. Isolation of the yeast nuclear pore complex. *J. Cell Biol.* **1993**, *123*, 771–783. [CrossRef] [PubMed]
2. Yang, Q.; Rout, M.P.; Akey, C.W. Three-dimensional architecture of the isolated yeast nuclear pore complex: Functional and evolutionary implications. *Mol. Cell* **1998**, *1*, 223–234. [CrossRef]
3. Reichelt, R.; Holzenburg, A.; Buhle, E.L.; Jarnik, M.; Engel, A.; Aebi, U. Correlation between structure and mass distribution of the nuclear pore complex and of distinct pore complex components. *J. Cell Biol.* **1990**, *110*, 883–894. [CrossRef] [PubMed]
4. Hurt, E.; Beck, M. Towards understanding nuclear pore complex architecture and dynamics in the age of integrative structural analysis. *Curr. Opin. Cell Biol.* **2015**, *34*, 31–38. [CrossRef] [PubMed]
5. Hoelz, A.; Glavy, J.S.; Beck, M. Toward the atomic structure of the nuclear pore complex: When top down meets bottom up. *Nat. Struct. Mol. Biol.* **2016**, *23*, 624–630. [CrossRef] [PubMed]
6. Popken, P.; Ghavami, A.; Onck, P.R.; Poolman, B.; Veenhoff, L.M. Size-Dependent Leak of Soluble and Membrane Proteins Through the Yeast Nuclear Pore Complex. *Mol. Biol. Cell* **2015**, *26*, 1386–1394. [CrossRef] [PubMed]
7. Schmidt, H.B.; Görlich, D. Transport Selectivity of Nuclear Pores, Phase Separation, and Membraneless Organelles. *Trends Biochem. Sci.* **2016**, *41*, 46–61. [CrossRef]
8. Timney, B.L.; Raveh, B.; Mironska, R.; Trivedi, J.M.; Kim, S.J.; Russel, D.; Wente, S.R.; Sali, A.; Rout, M.P. Simple rules for passive diffusion through the nuclear pore complex. *J. Cell Biol.* **2016**, *215*. [CrossRef]
9. Iovine, M.K.; Watkins, J.L.; Wente, S.R. The GLFG repetitive region of the nucleoporin Nup116p interacts with Kap95p, an essential yeast nuclear import factor. *J. Cell Biol.* **1995**, *131*, 1699–1713. [CrossRef]

10. Bayliss, R.; Ribbeck, K.; Akin, D.; Kent, H.M.; Feldherr, C.M.; Görlich, D.; Stewart, M. Interaction between NTF2 and xFxFG-containing nucleoporins is required to mediate nuclear import of RanGDP. *J. Mol. Biol.* **1999**, *293*, 579–593. [CrossRef]

11. Rout, M.P.; Aitchison, J.D.; Magnasco, M.O.; Chait, B.T. Virtual gating and nuclear transport: The hole picture. *Trends Cell Biol.* **2003**, *13*, 622–628. [CrossRef] [PubMed]

12. Pante, N.; Kann, M. Nuclear Pore Complex Is Able to Transport Macromolecules with Diameters of 39 nm. *Mol. Biol. Cell* **2002**, *13*, 425–434. [CrossRef] [PubMed]

13. Lowe, A.R.; Siegel, J.J.; Kalab, P.; Siu, M.; Weis, K.; Liphardt, J.T. Selectivity mechanism of the nuclear pore complex characterized by single cargo tracking. *Nature* **2010**, *467*, 600–603. [CrossRef] [PubMed]

14. Frey, S.; Rees, R.; Schünemann, J.; Ng, S.C.; Fünfgeld, K.; Huyton, T.; Görlich, D. Surface Properties Determining Passage Rates of Proteins through Nuclear Pores. *Cell* **2018**, *174*, 202–217.e9. [CrossRef] [PubMed]

15. Ananth, A.N.; Mishra, A.; Frey, S.; Dwarkasing, A.; Versloot, R.; van der Giessen, E.; Görlich, D.; Onck, P.; Dekker, C. Spatial structure of disordered proteins dictates conductance and selectivity in nuclear pore complex mimics. *Elife* **2018**, *7*. [CrossRef] [PubMed]

16. Frey, S.; Richter, R.P.; Görlich, D. FG-rich repeats of nuclear pore proteins form a three-dimensional meshwork with hydrogel-like properties. *Science* **2006**, *314*, 815–817. [CrossRef] [PubMed]

17. Frey, S.; Görlich, D. A saturated FG-repeat hydrogel can reproduce the permeability properties of nuclear pore complexes. *Cell* **2007**, *130*, 512–523. [CrossRef]

18. Polyansky, A.A.; Zagrovic, B. Protein electrostatic properties predefining the level of surface hydrophobicity change upon phosphorylation. *J. Phys. Chem. Lett.* **2012**, *3*, 973–976. [CrossRef]

19. Petrov, D.; Margreitter, C.; Grandits, M.; Oostenbrink, C.; Zagrovic, B. A Systematic Framework for Molecular Dynamics Simulations of Protein Post-Translational Modifications. *PLoS Comput. Biol.* **2013**, *9*. [CrossRef]

20. Lee, T.; Hoofnagle, A.N.; Kabuyama, Y.; Stroud, J.; Min, X.; Goldsmith, E.J.; Chen, L.; Resing, K.A.; Ahn, N.G. Docking motif interactions in Map kinases revealed by hydrogen exchange mass spectrometry. *Mol. Cell* **2004**, *14*, 43–55. [CrossRef]

21. Vomastek, T.; Iwanicki, M.P.; Burack, W.R.; Tiwari, D.; Kumar, D.; Parsons, J.T.; Weber, M.J.; Nandicoori, V.K. Extracellular Signal-Regulated Kinase 2 (ERK2) Phosphorylation Sites and Docking Domain on the Nuclear Pore Complex Protein Tpr Cooperatively Regulate ERK2-Tpr Interaction. *Mol. Cell. Biol.* **2008**, *28*, 6954–6966. [CrossRef] [PubMed]

22. Kosako, H.; Imamoto, N. Phosphorylation of nucleoporins: Signal transduction-mediated regulation of their interaction with nuclear transport receptors. *Nucleus* **2010**. [CrossRef] [PubMed]

23. Lusk, C.P.; Waller, D.D.; Makhnevych, T.; Dienemann, A.; Whiteway, M.; Thomas, D.Y.; Wozniak, R.W. Nup53p is a target of two mitotic kinases, Cdk1p and Hrr25p. *Traffic* **2007**. [CrossRef] [PubMed]

24. Ficarro, S.B.; McCleland, M.L.; Stukenberg, P.T.; Burke, D.J.; Ross, M.M.; Shabanowitz, J.; Hunt, D.F.; White, F.M. Phosphoproteome analysis by mass spectrometry and its application to Saccharomyces cerevisiae. *Nat. Biotechnol.* **2002**. [CrossRef] [PubMed]

25. Ciomperlik, J.J.; Basta, H.A.; Palmenberg, A.C. Three cardiovirus Leader proteins equivalently inhibit four different nucleocytoplasmic trafficking pathways. *Virology* **2015**, *484*, 194–202. [CrossRef] [PubMed]

26. Porter, F.W.; Palmenberg, A.C. Leader-induced phosphorylation of nucleoporins correlates with nuclear trafficking inhibition by cardioviruses. *J. Virol.* **2009**. [CrossRef]

27. Kosako, H.; Yamaguchi, N.; Aranami, C.; Ushiyama, M.; Kose, S.; Imamoto, N.; Taniguchi, H.; Nishida, E.; Hattori, S. Phosphoproteomics reveals new ERK MAP kinase targets and links ERK to nucleoporin-mediated nuclear transport. *Nat. Struct. Mol. Biol.* **2009**, *16*, 1026–1035. [CrossRef]

28. Carlson, S.M.; Chouinard, C.R.; Labadorf, A.; Lam, C.J.; Schmelzle, K.; Fraenkel, E.; White, F.M. Large-scale discovery of ERK2 substrates identifies ERK-mediated transcriptional regulation by ETV3. *Sci. Signal.* **2011**, *4*. [CrossRef]

29. Kehlenbach, R.H.; Gerace, L. Phosphorylation of the nuclear transport machinery down-regulates nuclear protein import in vitro. *J. Biol. Chem.* **2000**, *275*, 17848–17856. [CrossRef]

30. Hazawa, M.; Lin, D.; Kobayashi, A.; Jiang, Y.; Xu, L.; Dewi, F.R.P.; Mohamed, M.S.; Hartono; Nakada, M.; Meguro-Horike, M.; Horike, S.; et al. ROCK-dependent phosphorylation of NUP62 regulates p63 nuclear transport and squamous cell carcinoma proliferation. *EMBO Rep.* **2018**. [CrossRef]

31. Borlido, J.; D'Angelo, M.A. Nup62-mediated nuclear import of p63 in squamous cell carcinoma. *EMBO Rep.* **2018**, *19*, 3–4. [CrossRef]

32. Shindo, Y.; Iwamoto, K.; Mouri, K.; Hibino, K.; Tomita, M.; Kosako, H.; Sako, Y.; Takahashi, K. Conversion of graded phosphorylation into switch-like nuclear translocation via autoregulatory mechanisms in ERK signalling. *Nat. Commun.* **2016**, *7*. [CrossRef]

33. Ghavami, A.; Van Der Giessen, E.; Onck, P.R. Energetics of transport through the nuclear pore complex. *PLoS ONE* **2016**, *11*. [CrossRef]

34. Tagliazucchi, M.; Peleg, O.; Kröger, M.; Rabin, Y.; Szleifer, I. Effect of charge, hydrophobicity, and sequence of nucleoporins on the translocation of model particles through the nuclear pore complex. *Proc. Natl. Acad. Sci. USA* **2013**, *110*, 3363–3368. [CrossRef]

35. Ghavami, A.; Veenhoff, L.M.; Van Der Giessen, E.; Onck, P.R. Probing the disordered domain of the nuclear pore complex through coarse-grained molecular dynamics simulations. *Biophys. J.* **2014**, *107*, 1393–1402. [CrossRef]

36. Hayama, R.; Rout, M.P.; Fernandez-Martinez, J. The nuclear pore complex core scaffold and permeability barrier: Variations of a common theme. *Curr. Opin. Cell Biol.* **2017**, *46*, 110–118. [CrossRef]

37. Ingrell, C.R.; Miller, M.L.; Jensen, O.N.; Blom, N. NetPhosYeast: Prediction of protein phosphorylation sites in yeast. *Bioinformatics* **2007**. [CrossRef]

38. Bai, Y.; Chen, B.; Li, M.; Zhou, Y.; Ren, S.; Xu, Q.; Chen, M.; Wang, S. FPD: A comprehensive phosphorylation database in fungi. *Fungal Biol.* **2017**. [CrossRef]

39. Yamada, J.; Phillips, J.L.; Patel, S.; Goldfien, G.; Calestagne-Morelli, A.; Huang, H.; Reza, R.; Acheson, J.; Krishnan, V.V.; Newsam, S.; et al. A bimodal distribution of two distinct categories of intrinsically disordered structures with separate functions in FG nucleoporins. *Mol. Cell. Proteomics* **2010**, *9*, 2205–2224. [CrossRef]

40. Leo, A.J. Calculating log Poct from Structures. *Chem. Rev.* **1993**, *93*, 1281–1306. [CrossRef]

41. Viswanadhan, V.N.; Ghose, A.K.; Revankar, G.R.; Robins, R.K. Atomic Physicochemical Parameters for Three Dimensional Structure Directed Quantitative Structure-Activity Relationships. 4. Additional Parameters for Hydrophobic and Dispersive Interactions and Their Application for an Automated Superposition of Certain. *J. Chem. Inf. Comput. Sci.* **1989**, *29*, 163–172. [CrossRef]

42. Meylan, W.M.; Howard, P.H. Atom/fragment contribution method for estimating octanol-water partition coefficients. *J. Pharm. Sci.* **1995**, *84*, 83–92. [CrossRef]

43. Tetko, I.V.; Gasteiger, J.; Todeschini, R.; Mauri, A.; Livingstone, D.; Ertl, P.; Palyulin, V.A.; Radchenko, E.V.; Zefirov, N.S.; Makarenko, A.S.; Tanchuk, V.Y.; Prokopenko, V.V. Virtual computational chemistry laboratory—Design and description. *J. Comput. Aided Mol. Des.* **2005**, *19*, 453–463. [CrossRef]

44. Carrasco, B.; De La Torre, J.G. Hydrodynamic properties of rigid particles: Comparison of different modeling and computational procedures. *Biophys. J.* **1999**. [CrossRef]

45. Garcia de la Torre, J.; Navarro, S.; Lopez Martinez, M.C.; Diaz, F.G.; Lopez Cascales, J.J. HYDRO: A computer program for the prediction of hydrodynamic properties of macromolecules. *Biophys. J.* **1994**. [CrossRef]

46. Ghavami, A.; van der Giessen, E.; Onck, P.R. Coarse-Grained Potentials for Local Interactions in Unfolded Proteins. *J. Chem. Theory Comput.* **2013**, *9*, 432–440. [CrossRef]

47. Marsh, J.A.; Forman-Kay, J.D. Sequence determinants of compaction in intrinsically disordered proteins. *Biophys. J.* **2010**, *98*, 2383–2390. [CrossRef]

48. Alber, F.; Dokudovskaya, S.; Veenhoff, L.M.; Zhang, W.; Kipper, J.; Devos, D.; Suprapto, A.; Karni-Schmidt, O.; Williams, R.; Chait, B.T.; et al. The molecular architecture of the nuclear pore complex. *Nature* **2007**, *450*, 695–701. [CrossRef]

49. Rexach, M.; Blobel, G. Protein import into nuclei: Association and dissociation reactions involving transport substrate, transport factors, and nucleoporins. *Cell* **1995**, *83*, 683–692. [CrossRef]

50. Görlich, D.; Panté, N.; Kutay, U.; Aebi, U.; Bischoff, F.R. Identification of different roles for RanGDP and RanGTP in nuclear protein import. *EMBO J.* **1996**, *15*, 5584–5594. [CrossRef]

51. Kersey, P.J.; Staines, D.M.; Lawson, D.; Kulesha, E.; Derwent, P.; Humphrey, J.C.; Hughes, D.S.T.; Keenan, S.; Kerhornou, A.; Koscielny, G.; et al. Ensembl Genomes: An integrative resource for genome-scale data from non-vertebrate species. *Nucleic Acids Res.* **2012**, *40*. [CrossRef]

52. Pyhtila, B.; Rexach, M. A Gradient of Affinity for the Karyopherin Kap95p along the Yeast Nuclear Pore Complex. *J. Biol. Chem.* **2003**, *278*, 42699–42709. [CrossRef]

53. Ketterer, P.; Ananth, A.N.; Laman Trip, D.S.; Mishra, A.; Bertosin, E.; Ganji, M.; Van Der Torre, J.; Onck, P.; Dietz, H.; Dekker, C. DNA origami scaffold for studying intrinsically disordered proteins of the nuclear pore complex. *Nat. Commun.* **2018**, *9*. [CrossRef]
54. Colwell, L.; Brenner, M.; Ribbeck, K. Charge as a selection criterion for translocation through the nuclear pore complex. *PLoS Comput. Biol.* **2010**. [CrossRef]
55. Fujita, T.; Iwasa, J.; Hansch, C. A New Substituent Constant, π, Derived from Partition Coefficients. *J. Am. Chem. Soc.* **1964**, *86*, 5175–5180. [CrossRef]

© 2019 by the authors. Licensee MDPI, Basel, Switzerland. This article is an open access article distributed under the terms and conditions of the Creative Commons Attribution (CC BY) license (http://creativecommons.org/licenses/by/4.0/).

International Journal of
Molecular Sciences

Article

Hydrodynamic Behavior of the Intrinsically Disordered Potyvirus Protein VPg, of the Translation Initiation Factor eIF4E and of their Binary Complex

Jocelyne Walter *, Amandine Barra, Bénédicte Doublet, Nicolas Céré, Justine Charon [†] and Thierry Michon *

UMR 1332 Biologie du Fruit et Pathologie, INRA, Université de Bordeaux, CS 20032, 33140 Villenave d'Ornon, France; amandine.barra@inra.fr (A.B.); benedicte.doublet@inra.fr (B.D.); nicolas.cere@hotmail.fr (N.C.); justine.charon@sydney.edu.au (J.C.)
* Correspondence: Jocelyne.walter@inra.fr (J.W.); thierry.michon@inra.fr (T.M.)
† Present address: School of Life & Environmental Sciences, University of Sydney, Sydney, NSW 2006, Australia.

Received: 13 March 2019; Accepted: 5 April 2019; Published: 11 April 2019

Abstract: Protein intrinsic disorder is involved in many biological processes and good experimental models are valuable to investigate its functions. The potyvirus genome-linked protein, VPg, displays many features of an intrinsically disordered protein. The virus cycle requires the formation of a complex between VPg and eIF4E, one of the host translation initiation factors. An in-depth characterization of the hydrodynamic properties of VPg, eIF4E, and of their binary complex VPg-eIF4E was carried out. Two complementary experimental approaches, size-exclusion chromatography and fluorescence anisotropy, which is more resolving and revealed especially suitable when protein concentration is the limiting factor, allowed to estimate monomers compaction upon complex formation. VPg possesses a high degree of hydration which is in agreement with its classification as a partially folded protein in between a molten and pre-molten globule. The natively disordered first 46 amino acids of eIF4E contribute to modulate the protein hydrodynamic properties. The addition of an N-ter His tag decreased the conformational entropy of this intrinsically disordered region. A comparative study between the two tagged and untagged proteins revealed the His tag contribution to proteins hydrodynamic behavior.

Keywords: intrinsically disordered protein; plant virus; eIF4E; VPg; potyvirus; molten globule; protein-protein interaction; fluorescence anisotropy; protein hydrodynamics

1. Introduction

Many biologically functional protein regions do not fold spontaneously. This class of proteins, termed intrinsically disordered proteins (IDP), contains intrinsically disordered regions (IDR) which are devoid of stable secondary and tertiary structures under physiological conditions and rather, exist as dynamic ensembles of inter-converting conformers [1]. Many of these proteins gain a stable 3D structure only when they interact with their target molecules [2]. The ability to exert specific biological functions and to interact with various partners in spite of the lack of a precise 3D scaffold, challenges the classic paradigm according to which specificity can only be achieved through surface complementation between structured and conserved domains. It is now well accepted that intrinsic disorder is involved in a large spectrum of functional properties modulated through multi-partnership interactions with proteins and nucleic acids. With between 7.3% and 77% of residues being disordered, the proteome of viruses on the whole presents the highest variability of intrinsic disorder in the living world [3–5]. The genus *Potyvirus* represents one of the largest and most economically damaging genus of plant-infecting viruses [6]. These viruses possess a single-stranded, polyadenylated, positive-sense genomic RNA which is covalently linked at its 5′ end to a viral protein, the viral protein genome-linked (VPg) [7,8]. The VPgs from several potyviruses, namely lettuce mosaic virus (LMV) [9], potato virus Y (PVY) [10], and Potato virus A (PVA) [11] have been experimentally characterized as intrinsically disordered. The potyviral VPg has been shown to interact with several viral and host factors. It is assumed to be a multifunctional protein involved in essential steps of the virus infectious cycle, translation, replication, and movement [12,13]. The VPg recruits the host eukaryotic translation initiation factor 4E (eIF4E), or its isoform eIF(iso)4E, in an interaction that is crucial for virus infection [14–16]. Mutations in the central region of VPg (residues 80–125) are associated with host resistance breakdown, [17–19]. In the LMV VPg, this central region interacts with eIF4E [20]. This region has been predicted to be an IDR for VPg from twelve potyviral species [9,21]. The study reported here analyzes the contribution of VPg and eIF4E flexible regions to the hydrodynamic properties of the two proteins, either monomeric, or associated in a binary complex.

2. Results and Discussion

2.1. Hydrodynamic Behavior of the Histidine Tagged Forms Assessed by Size Exclusion Chromatography

The secondary and tertiary structures of a protein involve non-covalent interactions. The more the amino-acid residues will interact, the more compact (globular) the protein will be. IDRs will involve fewer interactions between residues and lower compaction. Consequently, the hydration sphere of an intrinsically disordered protein is often larger than what would be expected for globular proteins of a similar molecular weight. A simple way to assess the degree of compactness of a protein in solution is to measure its hydrodynamic radius (R_h). A commonly used method for measuring R_h is size exclusion chromatography, SEC (Figure 1).

When submitted to SEC (Figure 1A), the hydrodynamic behavior of His$_6$ eIF4E, M 28,550 Da (MALDI-TOF spectrometry of the purified recombinant protein), was in agreement with that of a 30 kDa globular protein. R_h values were deduced from experimentally determined apparent molecular weights (M_{app}) (see Section 4). The elution profile of purified recombinant His$_6$ VPg, M 26,250 Da (mass spectrometry), featured two populations, with the major one (69%) suggesting that of a 40 kDa globular protein, and the minor species (27%) averaging 90 kDa. Clearly, His$_6$ VPg did not behave like a globular protein. The trypsin hydrolysis kinetics of His$_6$ VPg shows a moderate proteolytic resistance profile comparable to that of α-casein, a disordered protein (Figure S1). A method was developed to classify IDPs according to the relationship between their apparent molecular density (ρ) and their true molecular weight (M) [22]. From Figure 1C,D it can be deduced that His$_6$ VPg shares hydrodynamic features of molten and pre-molten globules. The binary complex His$_6$ VPg-His$_6$ eIF4E showed a more complex elution profile (Figure 1A). Its major component (peak 2) displayed a M_{app} of 56kDa, which is close to the weight (M 54,772 Da) calculated by adding the two partners molecular weights. SDS-PAGE analysis showed that this elution fraction contained equal amounts of His$_6$ VPg and His$_6$ eIF4E indicating a possible compaction of VPg upon its association with eIF4E. The major component eluted under peak 1, M_{app} = 110 kDa, corresponds to oligomeric forms of His$_6$ VPg (Figure 1B). The VPg propensity to aggregate was previously described [10].

Using PONDR-VLXT, intrinsic disorder was predicted for the tagged and untagged proteins. The His$_6$ tag potentially brings disorder to the N-terminus of all proteins. After the His$_6$ tag removal, the first 46 amino acids segment of the native eIF4E was still predicted as unstructured. This prediction was validated by previously reported structural data showing that the first 40 amino acids of eIF4E were intrinsically disordered and fold upon binding with eIF4G [23]. In addition, upon His$_6$ tag removal, the first 25 amino acids of VPg were predicted as disordered (Figure S2). This region was recently reported as being a conformational switch [24]. Therefore, the contribution of these regions, predicted as disordered, to the hydrodynamic properties of the proteins was assessed.

The hydrodynamic behavior of VPg, eIF4E and their binary complex was deduced from fluorescence anisotropy measurements.

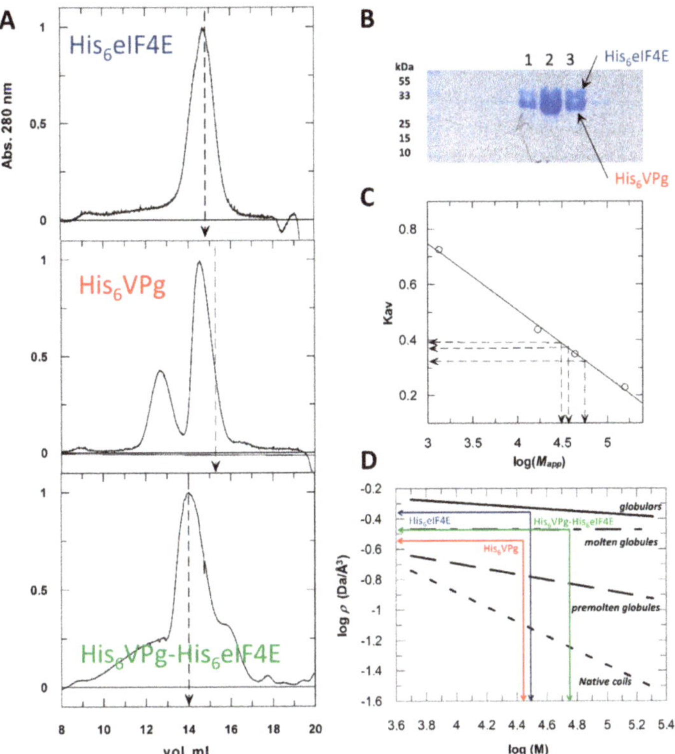

Figure 1. Size exclusion chromatography of His$_6$ tagged VPg, His$_6$ tagged 4E (eIF4E) and His$_6$ VPg-His$_6$ eIF4E, their binary complex. (**A**) Separated runs of the purified monomers (1 mL of 0.7–1 mg/mL) were performed. Vertical dashed lines refer to the elution volumes expected for globular proteins of 28,549 Da (upper panel) and 26,137 Da (middle panel), the molecular weights of recombinant His$_6$ eIF4E and His$_6$ VPg respectively (MALDI-TOF spectrometry determinations). The His$_6$ VPg-His$_6$ eIF4E binary complex was pre-formed by mixing His$_6$ eIF4E (0.7 mg/mL with an excess of His$_6$ VPg (1.2 mg/mL) and loaded up to the column (lower panel). A mass of 54,772 Da was estimated for the binary complex (summing His$_6$ VPg and His$_6$ eIF4E molecular weights) vertical dashed line. For comparison, absorbance values were standardized to the maximum value of each peak. (**B**) Distribution of the various molecular species through the size exclusion chromatography (SEC) separation of a His$_6$ VPg-His$_6$ eIF4E mix. Upon elution, fractions 1. 2 and 3 were recovered and submitted to SDS-PAGE analysis. (**C**) Determination of the three molecular species apparent molecular weights (M_{app}) deduced from standard calibration with a set of known globular proteins. K_{av} is a mean value determined from at least three independent SEC runs. (**D**) Apparent molecular densities (ρ) of the three molecular species were deduced from their experimentally determined hydrodynamic radius (R_h) values (see material and methods). For each protein, the intersection between logρ and log(M), M being the true molecular mass, allows to deduce the conformational families to which they belong to.

Because SEC leads to proteins diluting and to complexes partly dissociating during the chromatography process, it required substantial amount of proteins at concentration above 0.5 mg/mL. We experienced difficulties to obtain isolated untagged VPg and eIF4E at the concentrations suited for SEC experiments. Indeed, in vitro enzymatic tag cleavage resulted in a mixture of molecular species, which after separation (0.1–0.2 mg/mL), were not concentrated enough for SEC. Consequently, hydrodynamic parameters were deduced from fluorescence anisotropy measurements, which return a more detailed analysis of hydrodynamic behavior and can be operated at much lower concentrations. Anisotropy measurements give access to the rotational correlation time (θ) of the proteins. This parameter is strongly related to their hydrodynamic properties, as it depends on the protein shape and it is linked to the effective solvent shell accompanying the protein rotational diffusion. For that purpose, a fluorescent probe, N-acetyl-N'-(5-sulfo-1-naphtyl)ethylenediamine (AEDANS) was linked to the VPg single cysteine with an efficiency of 0.9 AEDANS moiety per VPg. In another set of experiments, AEDANS was also coupled to His_6 eIF4E, eIF4E and eIF4E$^{\Delta 1-46}$ to evaluate the first 1–46 disordered residues contribution to eIF4E compaction and also, more generally, the effect of the His_6 tag on the compaction of monomeric forms. There are four cysteine residues within lettuce eIF4E, among which two are strictly conserved in plant orthologues. The modification resulted in a mean of 1.7 AEDANS moieties per molecule. The addition of the fluorophore did not alter the binding properties of His_6 eIF4E, eIF4E, and eIF4E$^{\Delta 1-46}$ to VPg (Figure S3). This result was not surprising as wheat eIF4E, either reduced, oxidized, or with a cysteine-to-serine mutation do not undergo structural changes and are functional, all binding m7GTP in a similar and labile manner [25]. In addition, in the structure of pea eIF4E, the two sulfur atoms are in close proximity but are clearly not bridged [26].

2.2. His Tagging Modulates Proteins Hydrodynamic Parameters

As expected, upon addition of His_6 eIF4E, the fluorescence anisotropy of His_6 VPg*, increased proportionally to the amount of His_6 VPg*-His_6 eIF4E complex formed. It reached a plateau value indicating a saturation, Figure 2A inset.

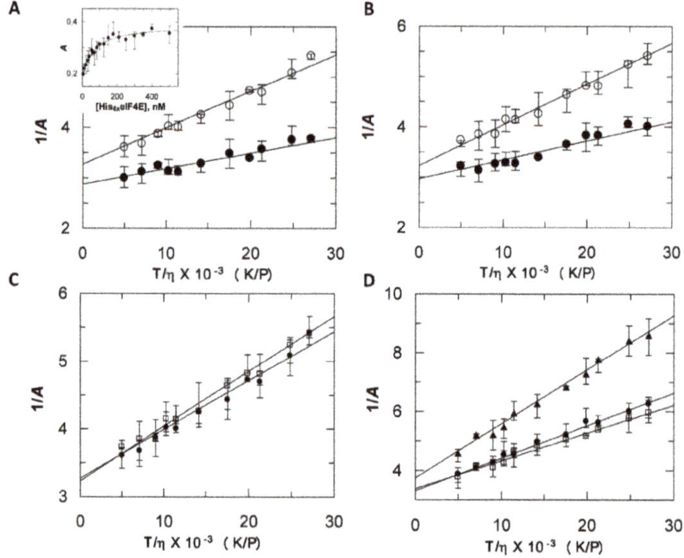

Figure 2. Fluorescence anisotropy of the various molecular forms of eIF4E and VPg. (**A**) The fluorescence anisotropy of His_6 VPg* (300 nM, open circles) and His_6 VPg*-His_6 eIF4E complex (mix of 300 nM His_6 VPg* and 2 µM His_6 eIF4E filled circles) was recorded as a function of the viscosity increase at 25 °C. The reciprocal of the emitted light anisotropy (Perrin's plot), $1/A$, is plotted as a function of T/η. V, the apparent molecular volume of the proteins and their complexes and A_0, the fundamental anisotropy were obtained respectively from the slope and intercept at infinite viscosity. Inset, the fluorescence anisotropy increase upon association of His_6 eIF4E with His_6 VPg* (300 nM). (**B**) Perrin's plots of untagged VPg* (open circles) and VPg*-eIF4E complex (filled circles) in the same conditions as (**A**). (**C**) Perrin's plots of His_6 VPg* (open squares), VPg* (filled circles) and D. His_6 eIF4E* (filled circles), eIF4E* (open squared) and eIF4E$^{\Delta 1-46}$* (filled triangles). Experimental conditions were the same as in B. All measurements were obtained using VPg from lettuce mosaic virus (LMV) AF199 strain.

The modification by the probe did not significantly change the binding strength as comparable dissociation constants (K_D) values were found for probed and unprobed proteins. A K_D value of 63 nM could be extracted from the data, in agreement with intrinsic fluorescence measurements (Figure S3). The hydrodynamic molar volume (V) and the rotational correlation time (θ_{exp}) of the various molecular species were determined from steady state fluorescence anisotropy measurements as described in the experimental procedure section.

2.3. Discrepancies between SEC and Fluorescence Anisotropy Suggest a Contribution of Tags in Proteins Hydrodynamic Behavior

The parameters V and θ_{exp} were derived for the various monomeric species (Figure 2C,D, and Table 1). The more the ratio $\theta_{exp}/\theta_{calc}$ differs from unit, the more asymmetric the protein is. The $\theta_{exp}/\theta_{calc}$ ratio of His_6 VPg (2.41) and His_6 eIF4E (1.62) indicated that their shape differed from a sphere. In addition, although the His_6 VPg molecular weight was 4295 Da less than His_6 eIF4E, its V value was 1.37 times larger, indicating that it was significantly less compact, a pre-molten globule feature. The untagged form of VPg displayed an expected decrease of its hydrodynamic molar volume with respect to its tagged form. Interestingly, the eIF4E* untagged form showed a higher V value (46.5 L/mol) and hence, a significant decrease in compaction compared to His_6 eIF4E* (42.9 L/mol). This could be due to interactions between the His_6 tag and the eIF4E N-ter IDR [23], an effect previously discussed [27]. Because of the intrinsic conformational entropy and the positive charges cluster of the disordered His_6 tag, its interactions with other parts of the protein are more likely to occur. This could account for the more compact hydrodynamic behavior of His_6 eIF4E. For most 3D structures solved from tagged proteins, the peptide tag is too disordered to be resolved. However, it is worth mentioning that, in most of the cases, comparisons of tagged with corresponding untagged structures determined from X ray diffraction data revealed only minor structural differences of the type that might be observed when comparing two identical sequences solved in different space groups [28]. This tends to show that these purification tags generally had no significant effect on the structure of the native protein. However, the importance of the proteins hydrodynamics properties cannot be understated as it accounts for the polypeptide chain dynamics, which drives most of biological functions. This is exemplified through the richness of the functional interpretations provided by NMR data related to disordered segments [29].

A comparison of V values between tagged and untagged species allows for the discussing of tag contribution to compaction. A decrease of the Perrin's plot slope was observed for His_6 VPg*-His_6 eIF4E binary complex when compared to the slope of His_6 VPg* (Figure 2A). Upon analysis of this data, it was shown that the hydrodynamic molar volume V linked to the labeled VPg enlarged from 58 L/mol to 121 L/mol, (Table 1). The later, attributed to the complex, was 20% larger than could be expected by adding the V value of His_6 VPg* and His_6 eIF4E monomers.

Hence, the anisotropy approach revealed a possible contribution of the two flexible His tags conformational entropy to the proteins hydrodynamic behavior. By contrast, the V value experimentally determined for the native untagged binary complex was close to that of the sum of the monomer values (Table 1). Compaction can be estimated by ρ, the molecular density value. Interestingly, the compaction within the binary complex suggested by SEC was not observed by anisotropy. In the SEC experiments, elution is concomitant to dilution, and hence to a modification of the species distribution. This could contribute to average the observed apparent molecular weight of each species. As stated in the experimental procedures section, anisotropy measurements on binary complexes as a function of various viscosity values were performed after the molecular species in presence have reached an equilibrium. The ρ value is directly obtained from V the experimental molecular volume, and thus, the anisotropy approach provides a likely more accurate way to estimate compaction than SEC does through the use of M_{app} in the empirical linear Equation (4). More generally, ρ values provided by anisotropy were 1.5 time higher than those deduced from the SEC (Table 1).

Table 1. Physical parameters of the various molecular forms of VPg, eIF4E and of their binary complexes.

Parameter	Units	HV	V	HE	E	EΔ(1–46)	HVHE	VE
MW	g/mol	26,222 [†]	21,781 [†]	28,550 [†]	26,076 [†]	21,246 [†]	54,772 [‡]	47,857 [‡]
SEC								
V_R	mL	14.4	nd	14.8	nd	nd	13.7	nd
M_{app}	kg/mol	38.0 ± 1.0	nd	30.2 ± 0.8	nd	nd	56.2 ± 1.5	nd
R_h	Å	28.1 ± 0.7	nd	26.4 ± 1.1	nd	nd	33.0 ± 0.9	nd
ρ	g/mol·Å^3	0.28	nd	0.40	nd	nd	0.38	nd
Conformation		MG/PMG		F/MG			MG	
[1] Fluorescence Anisotropy								
V_{app}	L/mol	58.7 ± 1.8	51.6 ± 0.2	42.9 ± 0.8	46.5 ± 1.0	24.3 ± 0.5	121.3 ± 3.2	102.4 ± 0.9
θ_{exp}	s (×10^9)	23.7	20.9	17.3	18.8	9.8	49 [3]	41.4
θ_{calc}	s (×10^9)	9.8	8.9	10.7	9.8	7.9	20.6	18.0
$(\theta_{exp})/(\theta_{calc})$			2.41	2.55	1.62	1.92	1.23	2.38	2.30
h	g water/g	1.44	1.57	0.72	1	0.38 (0.53) [2]	1.41	1.35
R_h	Å	24	23	22	22	18	31	29
ρ	g/mol·Å^3	0.45	0.42	0.67	0.56	0.88	0.45	0.47

HV, His$_6$ VPg; HE, His$_6$ eIF4E; V, His$_6$ VPg; E, eIF4E; EΔ(1–46), 1–46 residues deleted eIF4E; HVHE, His$_6$ VPg-His$_6$ eIF4E binary complex; VE, VPg-eIF4E binary complex. MW (molecular weight), V_R Retention volume), M_{app} (apparent molecular mass), R_h (hydrodynamic radius), ρ (apparent molecular density), V_{app} (apparent molecular volume), θ_{exp} (experimental rotational correlation time), θ_{calc} (calculated rotational correlation time), h (estimated hydration degree). [1] Determined for η = 0.01 P (water) at 298 K; τ, AEDANS fluorescence lifetime: 15.6 ns; [2] Calculated from pea eIF4E structure (2WMC) using the Hydropro software (http://leonardo.inf.um. es/macromol/programs/hydropro/hydropro.htm); [3] value obtained from His$_6$ VPg*-His$_6$ eIF4E binary complex; [†] Mass spectrometry (MALDI-TOF); [‡] Calculated from amino acid sequence.

2.4. His$_6$ VPg* and His$_6$ VPg*-His$_6$ eIF4E Binary Complex Shapes Differ from the Globular State

The deduced experimental rotational correlation time, θ_{exp}, increased from 23.7 ns for His$_6$ VPg* to 49 ns for the His$_6$ VPg*-His$_6$ eIF4E binary complex. The θ_{calc} values for hydrated rigid spheres of 26 kDa and 54.7 kDa were 9.8 ns and 20.6 ns (His$_6$ VPg* and His$_6$ VPg*-His$_6$ eIF4E respectively), indicating that His$_6$ VPg* and His$_6$ VPg*-His$_6$ eIF4E shapes differ from the compact globular state. On an indicative basis, the compact EΔ(1–46) [26] displayed an θ_{exp} value (7.9 ns) which was close to the θ_{calc} value (9.8 ns) calculated for a globular protein of comparable molecular weight. The $\theta_{exp}/\theta_{calc}$ ratio for the His$_6$ VPg*-His$_6$ eIF4E complex and His$_6$ VPg* were close, suggesting that the complex formation has no effect on the probe segmental motion within the VPg [30].

2.5. Compaction and Hydration are Experimentally Correlated

Because of their non-compact structures and more solvent-accessible surface area, disordered regions tend to display a higher hydration water density as compared to more ordered regions [31]. The degree of hydration (h) of each species was estimated from θ_{exp} (see experimental section). For compact globular proteins, this value is usually in between 0.2 and 0.4 g water/g protein. Our data suggests that His$_6$ VPg and VPg forms are more solvated than standard folded proteins (1.44 and 1.57 g/g respectively). This is in agreement with previous hydrodynamic experiments on IDPs. For instance, in the absence of calcium, the adenylate cyclase toxin calcium binding domain is intrinsically disordered and displays a high hydration propensity [32]. Moreover, molecular dynamics simulations show that partially disordered proteins like VPg have a higher capacity to bind hydration water as compared to globular proteins [33]. Interestingly, His$_6$ eIF4E and eIF4E, which both include a long disordered N-terminal region, have a rather high hydration degree (0.72 and 1 g/g respectively) although the presence of the His$_6$ tag seems to slightly decrease the hydration degree in accordance to its compacting effect discussed above. Conversely, eIF4E$^{\Delta1-46}$, which corresponds to the globular part of eIF4E, possesses a hydration value expected for compact proteins (0.38 g/g). Using the PDB coordinates of eIF4E$^{\Delta1-46}$ from pea (PDB file 2WMC), we calculated the protein rotational diffusion coefficient (D) [34] from Equation (12). From this value (1.49 × 10^7 s^{-1}) the rotational coefficient time was deduced and, in turn a h value of 0.53 g/g, which is in fairly good agreement with our estimation

(Table 1). Finally, the binary complex formation is not associated with a significant modification of hydration.

3. Conclusions

As opposed to fully disordered proteins, which display a random coil state, VPg possesses a significant content in secondary structure. These more compact intermediates have led to the concept of molten globule (MG) and the somewhat less compact pre-molten globule (PMG) states. A MG is characterized by a large internal flexibility of its side chains and backbone, with a R_h 1.5–2.0 times larger than that of globular proteins. As they are usually distributed in solutions between a limited number of conformers, these proteins prove to be more complex to analyze than fully disordered polypeptides. Modern NMR approaches provide an excellent way to study such proteins [35,36]. However the concentrations required [37], usually from 2 to 10 mg/mL, are far beyond what can be stabilized in solution in the present case. Because it enables the development of low-resolution structural models, taking into account the contribution of intrinsically disordered regions, a hydrodynamic analysis can provide useful data. Hydrodynamic parameters are usually assessed using SEC, AUC (analytical ultracentrifugation), and DLS (dynamic light scattering). However, these techniques also require protein concentrations above 1 mg/mL. We propose an elegant way to deal with especially difficult proteins. Although less used, fluorescence anisotropy can prove quite resolutive and especially suitable when protein concentration is the limiting factor. One can argue that this method gives access to the rotational mobility of the reporter fluorophore and not of the macromolecule itself. However, if, as it is the case here, the probe motion displays a low degree of freedom within the macromolecule, the extracted parameters reflect the hydrodynamic properties of the macromolecule well. The data obtained by SEC and fluorescence spectroscopy were in good agreement. Because analytical ultracentrifugation (AUC) is based on equilibrium and non-equilibrium thermodynamics, it is referred as a gold standard for characterizing the hydrodynamic properties. The expected sedimentation coefficients ($s_{20,w}$) for the LMV VPg and lettuce eIF4E were derived from their experimentally measured diffusion coefficients (see Equation (13) in the experimental section). They were in agreement with values reported for their homologous counterparts PVY VPg (46.7% identity, 76.4% similar with LMV VPg) and human eIF4E (43.2% identity, 67.6% similarity with lettuce eIF4E), (Table 2). A fine analysis of potyviral VPg conformers distribution on the basis of their sedimentation properties was recently reported [24]. A $s_{20,w}$ value of 3.2 Svedberg was determined for the major VPg molecular species (70%). [10]. This suggests that most of the VPg was present as a dimer. Indeed, this value is in accordance with the value obtained for non-reduced VPg from PVY [10], Table 2.

Table 2. VPgs and eIF4E sedimentation coefficients in water at 20 °C.

Molecular Species	MM, kDa *	$s_{20,w}$, s ($\times 10^{13}$)	Ref.	Method
VPg (LMV)-R	21.8	1.8	This study	Fluorescence
VPg (PVY)-R	22	1.7	[10]	UAC
VPg (PVY)	22	3.0	[10]	UAC
VPg (PVBV)	21.8	3.2	[24]	UAC
eIF4E (lettuce)	26.1	2.2	This study	Fluorescence
eIF4E (human)	24.3	2.03	[38]	UAC

LMV (lettuce mosaic virus), PVY (potato virus Y), PVBV (Pepper vein banding virus). * theoretical molecular weight. VPg (LMV)-R and VPg (PVY)-R, VPg under reduced conditions.

4. Materials and Methods

4.1. Protein Preparation

The gene coding for eIF4E initiation factor from the lettuce (Lactuca sativa cultivar Salinas GenBank AF 530162) and its derived molecular species were cloned into the vector pENTR/D-TOPO®

(Invitrogen, Carlsbad, CA, USA). They were transferred into pDEST^{TM}17 using the Gateway$^{®}$ recombinant Technology to allow production of N-terminal fusions with an hexahistidine tag (Invitrogen). In addition, full length eIF4E and eIF4E $^{(\Delta 1-46)}$ were cloned into the vector pENTR/SD/D-TOPO$^{®}$ (Invitrogen) and transferred into the Gateway pDEST^{TM}14 expression vector according to manufacturer's instructions to obtain untagged full length eIF4E and eIF4E$^{(\Delta 1-46)}$. The constructs were introduced into *E. coli* (BL21-AI strain), expression and purification of the His-tagged proteins were performed on ion metal affinity chromatography followed by m7GTP-Sepharose 4B (GE Healthcare, Amersham, UK) as previously described [39]. Untagged proteins were obtained by one step affinity purification on m7GTP sepharose 4B. The Lettuce mosaic virus VPg (isolate AF199GenBank AJ2 78854) coding sequence was cloned into the pTrcHis C expression vector downstream from an hexahistidine tag (Invitrogen). The vector contains a specific enterokinase cleavage site in frame with the protein for proteolytic tag removal. The protein was produced and purified as previously reported [39] except that the final monoQ chromatographic step was replaced by a size exclusion chromatography on a Superdex 75 HR 10/30 column (GE Healthcare) in 20 mM Hepes pH 8, 300 mM NaCl, and 2 mM DTT. For His-tag removal, this last step was omitted and the protein was diluted twice in the same buffer with reduced ionic strength (150 mM NaCl) containing 1 mM CaCl$_2$, 0.1% Tween-20. His tagged enterokinase (0.02 mg for 0.2 mg VPg) was added and the protein mix was dialyzed overnight at 4 °C against the same buffer. The protease and uncleaved tagged VPg were subsequently trapped on a NiNTA resin and the pure free VPg form was recovered in the flow through (30–40% yield).

Protein labelling with *N*-(iodoacetyl)-*N'*-(5-sulfo-1-naphtyl)ethylenediamine, (IAEDANS).

Cysteine residues were reduced before labelling. The VPg or eIF4E solutions (0.2 mg/mL) were dialyzed overnight in 25 mM HEPES, pH 7.5, 6 mM βMe, 2 mM AcNa, 5 mM EDTA, 25 mM DTT, NaCl 0.3 M at 4 °C. The protein solutions (1 mL) were loaded on a 5 mL G25 column equilibrated in the coupling buffer (25 mM HEPES, pH 7.5, NaCl 0.3 M). A 10 times molar excess of IAEDANS was added in the same buffer and the mix was incubated in the dark at 25 °C for 2 h. The reaction was stopped by addition of DTT (25 mM final concentration). A buffer exchange was performed over a G25 equilibrated in 25 mM HEPES, pH 7.5, 150 mM NaCl. About 1 and 2 AEDANS molecules were bound per VPg and eIF4E respectively.

4.2. Size Exclusion Chromatography

A superose 12 HR 10/30 column (GE Healthcare) was calibrated with separate 500 µL injections of the following native globular proteins: γ-globulin (bovine, 158 kDa), Ovalbumin (Chiken) 44 kDa, myoglobin (horse, 17 kDa and vitamin B$_{12}$ (1.35 kDa). Excluded volume (V_0, 8.74 mL) was determined with blue dextran. The volume of the column, V_t was 24 mL. Chromatography conditions were 20 mM HEPES/KOH, 0.4M KCl and 1.4 mM 2-Mercaptoethanol, flowrate 0.5 mL/min.

4.3. Hydrodynamic Radius Measure

The Stokes radius, also termed R_h, is the radius of a hard sphere that diffuses at the same rate as the protein. A commonly used method for measuring R_h is SEC. The size exclusion column is calibrated using the elution volume (V_e) of standard folded proteins (i.e., Globular) of known molecular weight. The apparent molecular weight of the protein of interest is then deduced from V_e. R_h is determined as the R_h expected for a globular protein of that apparent molecular weight, for which simple relations exist [22,40].

The retention factor K_{av} of the proteins was determined as follows:

$$K_{av} = \frac{V_e - V_0}{V_t - V_0} \tag{1}$$

There was a linear relationship between K_{av} and $\log M_{app}$ (Figure 1C):

$$K_{av} = -0.24 \log M_{app} + 1.47 \tag{2}$$

$$M_{app} = 10^{\frac{1.47-K_{av}}{0.24}} \tag{3}$$

It was assumed that the protein considered has the hydrodynamic behavior of its equivalent rigid sphere of R_h and an M_{app}. A linear relationship exists between $\log M_{app}$ and $\log R_h$ which, knowing the SEC derived apparent molecular weight of a protein, allows the determination of its R_h [22]. For a globular protein, the expression is:

$$\log(R_h) = -0.2 + 0.36 \log(M_{app}) \tag{4}$$

4.4. Structural Feature Estimation

The apparent molecular density or compaction (ρ) of a globular protein is:

$$\rho = \frac{M}{4/3\pi R_h^3} \tag{5}$$

with M, the molecular weight calculated from the protein amino acid composition. A plot of $\log(\rho)$ vs. $\log(M)$ allowed us to estimate the structural family; ordered globular, molten globule, pre-molten globule, or native coil-like protein (Figure 1D). Straight lines that define the different groups of conformational states were calculated from [41].

Fluorescence measurements. All steady state fluorescence acquisitions were obtained at 25 °C in 20 mM HEPES pH 7.5, 0.25 mM NaCl and 1 mM DTT using a SAFAS Xenius spectrofluorimeter (Monaco) equipped with a Peltier temperature controller. For optical characteristics of the instrument, see [39]. Affinity constants were deduced from steady state eIF4E tryptophan intrinsic fluorescence decrease upon titration by VPg as previously described [39]. In order to ensure that the system reached an equilibrium before measurements, a thorough mixing (gentle back and forth syringe flushing) of the various molecular species, was followed by a 10 min incubation. Then, an average value was collected during another 10 min, both for the acquisition of eIF4E steady state intrinsic fluorescence and anisotropy measurements.

Fluorescence anisotropy measurement. The fluorescent probe bound to the VPg was chosen so that τ its fluorescent lifetime be of the order of magnitude of θ, the rotational correlation time of the VPg in solution. A, the measured anisotropy is defined as:

$$\frac{A_0}{A} = 1 + \frac{\tau}{\theta} \tag{6}$$

A_0 is the fundamental anisotropy observed in the absence of other depolarizing processes such as rotational diffusion or energy transfer. If $\theta \gg \tau$ then the measured anisotropy is equal to A_0 (infinite viscosity, no motion of the macromolecule). If $\theta \ll \tau$ then the anisotropy is zero. For the AEDANS group, a fluorescence lifetime of 15.6 ns was determined by phase-modulation fluorimetry on a SLM 4800 fluorimeter. This value is in the range of θ values for proteins (15–70 kDa).

The anisotropy of the AEDANS labelled VPg either free or associated with eIF4E was measured at 25 °C in solutions of various viscosity (η). The dependency of the anisotropy on viscosity is given by the Perrin equation:

$$\frac{1}{A} = \frac{1}{A_0} + \alpha \frac{T}{\eta} \tag{7}$$

with

$$\alpha = \frac{\tau R}{A_0 V} \tag{8}$$

Plotting $1/A$ versus T/η gives usually a straight line. The viscosity was experimentally increased by addition of sucrose in the buffer. V the hydrodynamic molar volume was determined from α, the slope value:

$$V = \frac{\tau R}{A_0 \alpha} \tag{9}$$

The experimental rotational correlation time θ_{exp} was deduced from V:

$$\theta_{exp} = \frac{\eta V}{RT} \tag{10}$$

The rotational correlation time of an equivalent rigid sphere of the same molecular weight M was calculated as follows:

$$\theta_{calc} = \frac{\eta M}{RT}(\bar{v} + h) \tag{11}$$

where $\bar{v}$ is the protein partial specific volume (usually 0.73 cm^3/g) and h is the degree of hydration (g H$_2$O per g of protein; usually $0.2 < h < 0.4$); $R = 8.31 \times 10^7$ erg mol$^{-1}\cdot$K^{-1}. From Equation (13), replacing the calculated rotational time by rotational correlation times experimentally determined for each molecular species, leads to an estimation of h their degree of hydration (Table 1). Alternatively, knowing D, the rotational diffusion coefficient of the protein, its rotational correlation time can be obtained:

$$\theta = \frac{1}{6D} \tag{12}$$

The expected sedimentation coefficient can be deduced as follows:

$$s = \frac{MD(1 - \rho_{20}\bar{v})}{RT} \tag{13}$$

with M, molecular weight, ρ_{20} solvent density (water).

Supplementary Materials: Supplementary materials can be found at http://www.mdpi.com/1422-0067/20/7/1794/s1.

Author Contributions: T.M. and J.W. conceived and designed the experiments. A.B., N.C., B.D., J.C., J.W., and T.M. performed the experiments. T.M. and J.W. analyzed the data. T.M. wrote the manuscript. T.M. and J.W. discussed the results and commented on the manuscript.

Funding: This work was partially supported by Le Ministère Français de l'enseignement supérieur et de la Recherche (JC Fellowship).

Acknowledgments: We thank Sonia Longhi (CNRS-Marseille, France) for fruitful discussions. We are indebted to Stephane Claverol (centre de genomique fonctionnelle, Bordeaux) for mass spectrometry analysis.

Conflicts of Interest: The authors declare that they have no conflicts of interest with the contents of this article.

Abbreviations

LMV	Lettuce mosaic virus
PVY	Potato virus Y
TEV	Tobacco etch virus
TuMV	Turnip mosaic virus
IAEDANS	N-(iodoacetyl)-N'-(5-sulfo-1-naphtyl)ethylenediamine
AEDANS	N-(acetyl)-N'-(5-sulfo-1-naphtyl)ethylenediamine
His$_6$ eIF4E	N-ter hexahistidine tagged lettuce eIF4E
His$_6$ VPg	N-ter hexahistidine tagged LMV VPg
eIF4E$^{\Delta 1-46}$	untagged lettuce eIF4E deleted from its first 46 N-ter amino acid
His$_6$ VPg*	AEDANS labelled N-ter hexahistidine tagged LMV VPg
His$_6$ eIF4E	AEDANS labelled N-ter hexahistidine tagged lettuce eIF4E
eIF4E*	AEDANS labelled untagged lettuce eIF4E
eIF4E$^{\Delta 1-46}$*	AEDANS labelled untagged lettuce eIF4E deleted from its first 46 N-ter amino acids
[His$_6$ VPg*-His$_6$ eIF4E]	AEDANS labelled binary complex

References

1. Boehr, D.D.; Wright, P.E. Biochemistry. How do proteins interact? *Science* **2018**, *320*, 1429–1430. [CrossRef]
2. Habchi, J.; Tompa, P.; Longhi, S.; Uversky, V.N. Introducing Protein Intrinsic Disorder. *Chem. Rev.* **2014**, *114*, 6561–6588. [CrossRef]
3. Xue, B.; Blocquel, D.; Habchi, J.; Uversky, A.V.; Kurgan, L.; Uversky, V.N.; Longhi, S. Structural Disorder in Viral Proteins. *Chem. Rev.* **2014**, *114*, 6880–6911. [CrossRef]
4. Pushker, R.; Mooney, C.; Davey, N.E.; Jacqué, J.-M.; Shields, D.C. Marked Variability in the Extent of Protein Disorder within and between Viral Families. *PLoS ONE* **2013**, *8*, e60724. [CrossRef]
5. Xue, B.; Dunker, A.K.; Uversky, V.N. Orderly order in protein intrinsic disorder distribution: Disorder in 3500 proteomes from viruses and the three domains of life. *J. Biomol. Struct. Dyn.* **2012**, *30*, 137–149. [CrossRef]
6. Adams, M.; Zerbini, F.; French, R.; Rabenstein, F.; Stenger, D.; Valkonen, J. *Potyviridae. Virus Taxonomy, 9th Report of the International Committee for Taxonomy of Viruses*; Elsevier Academic Press: San Diego, CA, USA, 2011.
7. Murphy, J.F.; Klein, P.G.; Hunt, A.G.; Shaw, J.G. Replacement of the tyrosine residue that links a potyviral VPg to the viral RNA is lethal. *Virology* **1996**, *220*, 535–538. [CrossRef]
8. Murphy, J.F.; Rychlik, W.; Rhoads, R.E.; Hunt, A.G.; Shaw, J.G. A tyrosine residue in the small nuclear inclusion protein of tobacco vein mottling virus links the VPg to the viral RNA. *J. Virol.* **1991**, *65*, 511–513.
9. Hebrard, E.; Bessin, Y.; Michon, T.; Longhi, S.; Uversky, V.N.; Delalande, F.; Van Dorsselaer, A.; Romero, P.; Walter, J.; Declerck, N.; et al. Intrinsic disorder in Viral Proteins Genome-Linked: Experimental and predictive analyses. *Virol. J.* **2009**, *6*, 23–36.
10. Grzela, R.; Szolajska, E.; Ebel, C.; Madern, D.; Favier, A.; Wojtal, I.; Zagorski, W.; Chroboczek, J. Virulence Factor of Potato Virus Y, Genome-attached Terminal Protein VPg, Is a Highly Disordered Protein. *J. Biol. Chem.* **2008**, *283*, 213–221. [CrossRef] [PubMed]
11. Rantalainen, K.I.; Uversky, V.N.; Permi, P.; Kalkkinen, N.; Dunker, A.K.; Makinen, K. Potato virus A genome-linked protein VPg is an intrinsically disordered molten globule-like protein with a hydrophobic core. *Virology* **2008**, *377*, 280–288. [CrossRef]
12. Jiang, J.; Laliberte, J.F. The genome-linked protein VPg of plant viruses-a protein with many partners. *Curr. Opin. Virol.* **2011**, *1*, 347–354. [CrossRef] [PubMed]
13. Martínez, F.; Rodrigo, G.; Aragonés, V.; Ruiz, M.; Lodewijk, I.; Fernández, U.; Elena, S.F.; Daròs, J.-A. Interaction network of tobacco etch potyvirus NIa protein with the host proteome during infection. *BMC Genom.* **2016**, *17*, 87. [CrossRef]
14. Charron, C.; Nicolaï, M.; Gallois, J.L.; Robaglia, C.; Moury, B.; Palloix, A.; Caranta, C. Natural variation and functional analyses provide evidence for co-evolution between plant eIF4E and potyviral VPg. *Plant J.* **2008**, *54*, 56–68. [CrossRef] [PubMed]
15. Leonard, S.; Plante, D.; Wittmann, S.; Daigneault, N.; Fortin, M.G.; Laliberte, J.F. Complex formation between potyvirus VPg and translation eukaryotic initiation factor 4E correlates with virus infectivity. *J. Virol.* **2000**, *74*, 7730–7737. [CrossRef]
16. Robaglia, C.; Caranta, C. Translation initiation factors: A weak link in plant RNA virus infection. *Trends Plant Sci.* **2006**, *11*, 40–45. [CrossRef]
17. Moury, B.; Charron, C.; Janzac, B.; Simon, V.; Gallois, J.L.; Palloix, A.; Caranta, C. Evolution of plant eukaryotic initiation factor 4E (eIF4E) and potyvirus genome-linked protein (VPg): A game of mirrors impacting resistance spectrum and durability. *Infect. Genet. Evol.* **2014**, *27*, 472–480. [CrossRef] [PubMed]
18. Ayme, V.; Souche, S.; Caranta, C.; Jacquemond, M.; Chadoeuf, J.; Palloix, A.; Moury, B. Different mutations in the genome-linked protein VPg of potato virus Y confer virulence on the pvr2(3) resistance in pepper. *Mol. Plant Microbe Interact.* **2006**, *19*, 557–563. [CrossRef]
19. Ayme, V.; Petit-Pierre, J.; Souche, S.; Palloix, A.; Moury, B. Molecular dissection of the potato virus Y VPg virulence factor reveals complex adaptations to the pvr2 resistance allelic series in pepper. *J. Gen. Virol.* **2007**, *88*, 1594–1601. [CrossRef]
20. Roudet-Tavert, G.; Michon, T.; Walter, J.; Delaunay, T.; Redondo, E.; Le Gall, O. Central domain of a potyvirus VPg is involved in the interaction with the host translation initiation factor eIF4E and the viral protein HcPro. *J. Gen. Virol.* **2007**, *88*, 1029–1033. [CrossRef]

21. Charon, J.; Theil, S.; Nicaise, V.; Michon, T. Protein intrinsic disorder within the Potyvirus genus: From proteome-wide analysis to functional annotation. *Mol. BioSyst.* **2016**, *12*, 634–652. [CrossRef]

22. Uversky, V.N. What does it mean to be natively unfolded? *Eur. J. Biochem.* **2002**, *269*, 2–12. [CrossRef]

23. von der Haar, T.; Oku, Y.; Ptushkina, M.; Moerke, N.; Wagner, G.; Gross, J.D.; McCarthy, J.E. Folding transitions during assembly of the eukaryotic mRNA cap-binding complex. *J. Mol. Biol.* **2006**, *356*, 982–992. [CrossRef]

24. Sabharwal, P.; Srinivas, S.; Savithri, H.S. Mapping the domain of interaction of PVBV VPg with NIa-Pro: Role of N-terminal disordered region of VPg in the modulation of structure and function. *Virology* **2018**, *524*, 18–31. [CrossRef]

25. Monzingo, A.F.; Dhaliwal, S.; Dutt-Chaudhuri, A.; Lyon, A.; Sadow, J.H.; Hoffman, D.W.; Robertus, J.D.; Browning, K.S. The structure of eukaryotic translation initiation factor-4E from wheat reveals a novel disulfide bond. *Plant Physiol.* **2007**, *143*, 1504–1518. [CrossRef]

26. Ashby, J.A.; Stevenson, C.E.M.; Jarvis, G.E.; Lawson, D.M.; Maule, A.J. Structure-based mutational analysis of eIF4E in relation to sbm1 resistance to Pea seed-borne mosaic virus in Pea. *PLoS ONE* **2011**, *6*, e15873. [CrossRef]

27. Marsh, J.A.; Forman-Kay, J.D. Sequence determinants of compaction in intrinsically disordered proteins. *Biophys. J.* **2010**, *98*, 2374–2382. [CrossRef]

28. Carson, M.; Johnson, D.H.; McDonald, H.; Brouillette, C.; Delucas, L.J. His-tag impact on structure. *Acta Crystallogr. D Biol. Crystallogr.* **2007**, *63*, 295–301. [CrossRef]

29. Mollica, L.; Bessa, L.M.; Hanoulle, X.; Jensen, M.R.; Blackledge, M.; Schneider, R. Binding Mechanisms of Intrinsically Disordered Proteins: Theory, Simulation, and Experiment. *Front. Mol. Biosci.* **2016**. [CrossRef]

30. Granon, S.; Kerfelec, B.; Chapus, C. Spectrofluorimetric investigation of the interactions between the subunits of bovine pancreatic procarboxypeptidase A-S6. *J. Biol. Chem.* **1990**, *265*, 10383–10388.

31. Aggarwal, L.; Biswas, P. Hydration Water Distribution around Intrinsically Disordered Proteins. *J. Phys. Chem. B.* **2018**, *122*, 4206–4218. [CrossRef]

32. Chenal, A.; Guijarro, J.I.; Raynal, B.; Delepierre, M.; Ladant, D. RTX calcium binding motifs are intrinsically disordered in the absence of calcium: Implication for protein secretion. *J. Biol. Chem.* **2009**, *284*, 1781–1789. [CrossRef]

33. Rani, P.; Biswas, P. Local Structure and Dynamics of Hydration Water in Intrinsically Disordered Proteins. *J. Phys. Chem. B.* **2015**, *119*, 10858–10867. [CrossRef]

34. Ortega, A.; Amoros, D.; Garcia de la Torre, J. Prediction of hydrodynamic and other solution properties of rigid proteins from atomic- and residue-level models. *Biophys. J.* **2011**, *101*, 892–898. [CrossRef]

35. Jensen, M.R.; Ruigrok, R.W.H.; Blackledge, M. Describing intrinsically disordered proteins at atomic resolution by NMR. *Curr. Opin. Struct. Biol.* **2013**. [CrossRef]

36. Wells, M.; Tidow, H.; Rutherford, T.J.; Markwick, P.; Jensen, M.R.; Mylonas, E.; Svergun, D.I.; Blackledge, M.; Fersht, A.R. Structure of tumor suppressor p53 and its intrinsically disordered N-terminal transactivation domain. *Proc. Natl. Acad. Sci. USA* **2008**. [CrossRef]

37. Kosol, S.; Contreras-Martos, S.; Cedeño, C.; Tompa, P. Structural characterization of intrinsically disordered proteins by NMR spectroscopy. *Molecules* **2013**, *18*, 10802–10828. [CrossRef]

38. Modrak-Wojcik, A.; Gorka, M.; Niedzwiecka, K.; Zdanowski, K.; Zuberek, J.; Niedzwiecka, A.; Stolarski, R. Eukaryotic translation initiation is controlled by cooperativity effects within ternary complexes of 4E-BP1, eIF4E, and the mRNA 5′ cap. *FEBS Lett.* **2013**. [CrossRef]

39. Michon, T.; Estevez, Y.; Walter, J.; German-Retana, S.; Le Gall, O. The potyviral virus genome-linked protein VPg forms a ternary complex with the eukaryotic initiation factors eIF4E and eIF4G and reduces eIF4E affinity for a mRNA cap analogue. *FEBS J.* **2006**, *273*, 1312–1322. [CrossRef]

40. Uversky, V.N. Natively unfolded proteins: A point where biology waits for physics. *Protein Sci.* **2002**, *11*, 739–756. [CrossRef]

41. Uversky, V.N.; Santambrogio, C.; Brocca, S.; Grandori, R. Length-dependent compaction of intrinsically disordered proteins. *FEBS Lett.* **2012**, *586*, 70–73. [CrossRef]

© 2019 by the authors. Licensee MDPI, Basel, Switzerland. This article is an open access article distributed under the terms and conditions of the Creative Commons Attribution (CC BY) license (http://creativecommons.org/licenses/by/4.0/).

International Journal of
Molecular Sciences

Article

Intrinsically Disordered Linkers Impart Processivity on Enzymes by Spatial Confinement of Binding Domains

Beata Szabo [1], Tamas Horvath [1], Eva Schad [1], Nikoletta Murvai [1], Agnes Tantos [1], Lajos Kalmar [1], Lucía Beatriz Chemes [2], Kyou-Hoon Han [3,4] and Peter Tompa [1,5,*]

[1] Institute of Enzymology, Center of Natural Sciences, Hungarian Academy of Sciences, 1117 Budapest, Hungary; szabo.beata@ttk.mta.hu (B.S.); hotafin@gmail.com (T.H.); schad.eva@ttk.mta.hu (E.S.); murvai.nikoletta@ttk.mta.hu (N.M.); tantos.agnes@ttk.mta.hu (A.T.); lk397@cam.ac.uk (L.K.)

[2] Instituto de Investigaciones Biotecnológicas IIB-INTECH, Consejo Nacional de Investigaciones Científicas y Técnicas (CONICET), Universidad Nacional de San Martín, Buenos Aires 1650, Argentina; lchemes@iibintech.com.ar

[3] Genome Editing Research Center, Division of Biomedical Science, Korea Research Institute of Bioscience and Biotechnology (KRIBB), Daejeon 34113, Korea; khhan600@kribb.re.kr

[4] Department of Nano and Bioinformatics, University of Science and Technology (UST), Daejeon 34113, Korea

[5] VIB Center for Structural Biology, Vrije Univresiteit Brussel, 1050 Belgium, Brussel

[*] Correspondence: peter.tompa@vub.be; Tel.: +32-2-629-19-62

Received: 3 April 2019; Accepted: 26 April 2019; Published: 29 April 2019

Abstract: (1) Background: Processivity is common among enzymes and mechanochemical motors that synthesize, degrade, modify or move along polymeric substrates, such as DNA, RNA, polysaccharides or proteins. Processive enzymes can make multiple rounds of modification without releasing the substrate/partner, making their operation extremely effective and economical. The molecular mechanism of processivity is rather well understood in cases when the enzyme structurally confines the substrate, such as the DNA replication factor PCNA, and also when ATP energy is used to confine the succession of molecular events, such as with mechanochemical motors. Processivity may also result from the kinetic bias of binding imposed by spatial confinement of two binding elements connected by an intrinsically disordered (ID) linker. (2) Method: By statistical physical modeling, we show that this arrangement results in processive systems, in which the linker ensures an optimized effective concentration around novel binding site(s), favoring rebinding over full release of the polymeric partner. (3) Results: By analyzing 12 such proteins, such as cellulase, and RNAse-H, we illustrate that in these proteins linker length and flexibility, and the kinetic parameters of binding elements, are fine-tuned for optimizing processivity. We also report a conservation of structural disorder, special amino acid composition of linkers, and the correlation of their length with step size. (4) Conclusion: These observations suggest a unique type of entropic chain function of ID proteins, that may impart functional advantages on diverse enzymes in a variety of biological contexts.

Keywords: enzyme efficiency; polymeric substrate; processive enzyme; disordered linker; binding motif; binding domain; spatial search; local effective concentration

---

1. Introduction

Processivity is a kinetic phenomenon widespread among enzymes that act on polymeric substrates, such as DNA, RNA, polysaccharides, and proteins [1]. Once committed, processive enzymes engage in multiple rounds of modification instead of releasing their substrate after modifying it once. Served by different sliding mechanism(s), very effective enzymatic modifiers arose in evolution that can carry out hundreds or thousands of elementary steps upon a single engagement with the substrate [1].

Processivity occurs in: (i) synthesis (e.g., DNA by DNA polymerase [2], RNA by RNA polymerase, and protein by the ribosome [3]); (ii) degradation (e.g., DNA by DNAse [4], RNA by RNAse [5], polysaccharides by glycohydrolases [6] or proteins by the proteasome [7,8]); (iii) structural modification (e.g., DNA by helicase [9]); (iv) chemical modification (e.g., ubiquitination of proteins by ubiquitin ligases [10,11]); or (v) cargo transport (e.g., movement by mechanochemical motors kinesin, dynein and myosin [12–15] along actin and tubulin tracks).

A compilation of domain-linker-domain (DLD)-type monomeric processive enzymes is taken from the comprehensive list given in Supplementary Table S1. Important parameters including the length of predicted disordered linker, mean linker length of orthologous proteins (see Table S2 for species), κ value describing charge distribution, and the level of processivity (such as the length of processive move, the number of steps taken or the number of elementary substrate units covered, if determined at all), are given.

Given the extreme diversity of substrates upon which these processive enzymes act and also the variability of the chemical/mechanochemical changes they make, it is of little surprise that the molecular details of processivity are rather diverse, yet they are based on combinations of two basic designs principles. The classic and amply studied mechanism relies on structural confinement by circular/cylindrical or asymmetric binding domains or subunits of the enzymes. The former occurs, for example, when the PCNA subunit of DNA polymerase encircles the template DNA (Figure 1A) to ensure that the enzyme adds a practically unlimited number of nucleotides [16,17] to the growing DNA polymer. A closely related solution is used by HIV reverse transcriptase [18], which has an asymmetric binding domain that strongly favors sliding along the RNA substrate over dissociating from it (Figure 1B). A completely different mechanism has evolved in mechanochemical motors, such as kinesin and dynein, which move along polymeric protein tracks of tubulin [15]. These dimeric proteins have long coiled-coil stalks and ATPase binding domains, which undergo conformational changes that result in a strong preference for rebinding following dissociation due to a proximity effect, i.e., spatial confinement (Figure 1C). The region connecting the dimerization domain with the binding domain may even undergo transitions between ordered and disordered states [19]. The latter class of processive motors suggests that the presence of two binding elements (motifs or domains) connected by long, conformationally adaptable/flexible linker region(s) appears to be a key element of processivity, which combines deterministic and probabilistic elements of binding [20].

Here we generalize this concept by observing and analyzing that proteins in which binding domains are connected by a disordered linker may show probabilistic bias for re-binding over dissociation from their substrate, due to which they possess processive capacity. As structural disorder of proteins (intrinsically disordered protein/region, IDP/IDR) is widespread in eukaryotic proteomes [21,22], this may be a frequently applied mechanism. IDPs/IDRs often engage in protein-protein interactions [23,24] but their function may also directly stem from the disordered state, termed entropic-chain functions [25]. Binding and entropic-chain functions can actually be combined because often part of the IDP remains disordered even in the bound state, a phenomenon termed fuzziness [26]. Of particular relevance to the observed processivity is that binding motifs embedded in disordered regions, due to the arising "proximity effect" or "optimal effective concentration" around binding sites, may feature facilitated binding, which is central to the concepts of: (i) acceleration of binding by "fly casting" [27], (ii) reduction of binding dimensionality by the "monkey-bar" mechanism [28], and (iii) "ultrasensitive" binding by repetitive binding motifs in signaling proteins [29,30].

By statistical-physical modeling and bioinformatics analysis we show that this kinetic proximity effect is also a widespread inherent property of many monomeric processive enzymes that are capable of multiple rounds of modification of their polymeric substrate. These enzymes, such as a variety of glycohydrolases (e.g., cellulases) [6,31,32], Ribonuclease H1 (RNAse-H1) [5] and matrix metalloproteinase-9 (MMP-9) [33], need no ATP energy for processivity, which makes it a robust and widespread mechanism in the proteome. Here we have selected 12 such monomeric (ATP-independent) processive enzymes from the literature and provide a comprehensive analysis of their physical and structural properties. We show that once engaged with their substrate, their structural organization kinetically biases binding of their free binding domain over dissociation of both its domains, resulting in multiple successive binding events without ever fully releasing the polymeric partner (Figure 1D). We suggest that this type of processivity represents a unique type of "entropic chain" function enabled by the structural disorder of their linker region [25,34], which may be a general mechanism that arises in a broad range of biological contexts.

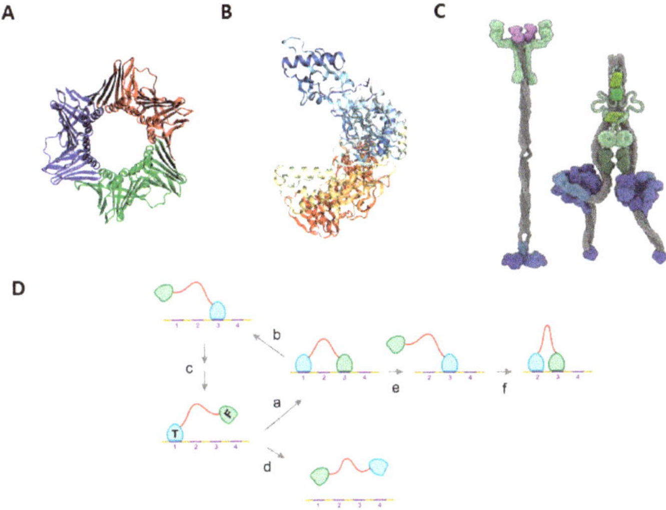

Figure 1. Basic mechanisms of processivity. The figure illustrates the two basic types (and four subtypes) of the mechanism of processivity. The classical mechanism based on structural confinement are represented by folded proteins that either (**A**) completely surround their partner by an oligomeric structure of toroidal shape, such as PCNA (PDB: 1AXC) [16,17], or (**B**) use an asymmetric binding domain to restrict its dissociation, such as in HIV reverse transcriptase (PDB: 1REV) [18]. Basically, different mechanisms are based on spatial confinement allowed by two binding motifs connected by a long, adaptable or flexible linker, as appears in (**C**) the ATP-dependent dimeric mechanochemical motors kinesin-1 and dynein (adapted from [20]), or (**D**) monomeric processive enzymes of domain-disordered linker-domain arrangement. These types of enzymes analyzed here in detail (for cases, see Table 1) bind their polymeric substrate via two binding domains, termed "bound" or "tethered" (T) for the one that anchors the enzyme to the substrate and "unbound" or "free" (F) for the one that is in search for substrate "target" binding sites), connected by a structurally disordered linker. We show by statistical-kinetic modeling that binding via the tethering domain kinetically favors binding via the free domain (a) over full dissociation of the protein (d), which may then result in processive diffusional moves (c) or directed movements driven by energy-dependent binding and/or modification of the substrate (e,f).

Table 1. ATP-independent monomeric domain-linker-domain (DLD)-type processive enzymes.

	Protein Name	UniProt ID	ATP	Partner	Linker Length	Kappa Value (Plot Region)	Processivity
1	*H. sapiens* **RNAse H1**	O60930	-	RNA	50 aa (78–127)	0.254 (2)	
2	*H. sapiens* **XPF**	Q92889	-	DNA	22 aa (821–842)	0.187 (1)	60 nucleotides
3	*T. reesei* **Cel7A**	P62694	-	cellulose	33 aa (445–477)	0.503 (1)	21 catalytic steps
4	*H. insolens* **Cel6A**	Q9C1S9	-	cellulose	46 aa (68–113)	0.288 (1)	
5	*C. cellulolyticum* **Cel48F** *	P37698	-	cellulose	28 aa (106–133)	0.069 (2)	
6	*C. thermocellum* **1,4-beta-glucanase** *	Q5TIQ4	-	cellulose	103 aa (688–790)	0.238 (1)	
7	*H. sapiens* **Telomerase**	O14746	-	DNA	94 aa (231–324)	0.252 (1)	
8	*X. laevis* **XMAP215**	Q9PT63	-	tubulin	121 aa (1079–1199)	0.189 (1)	25 tubulin dimers
9	*H. sapiens* **Chitotriosidase-1**	Q13231	-	chitooligosaccharides	31 aa (387–417)	0.263 (1)	8.6 cleavage steps
10	*B. circulans* **Chitinase A1**	P20533	-	crystalline-chitin	23 aa (444–466)	0.353 (1)	
11	*O. sativa subsp. Japonica* **Chitinase 2**	Q7DNA1	-	chitin	17 aa (74–90)	0.848 (1)	
12	*H. sapiens* **MMP-9**	P14780	-	gelatine	76 aa (434–509)	0.112 (1)	

* no sufficient number of orthologous proteins.

2. Results

2.1. The Classical Mechanisms of Processivity

For rationalizing the diverse mechanisms of processivity, we suggest that they fall into two broad mechanistic categories (cf. Table S1). The structural underpinning of the mechanism is straightforward when the enzyme uses structural confinement to make dissociation from the substrate highly unfavorable [1]. Complete confinement may result from ring-shaped oligomeric structures (e.g., PCNA [16,17] (Figure 1A)), whereas asymmetric structures of a single polypeptide chain can also either fully (e.g., exonuclease I [1]) or partially (e.g., HIV reverse transcriptase [18] (Figure 1B)) enclose the substrate. These mechanisms can be interpreted in terms of a preferred 1D sliding of the substrate (template) within the well-defined structural element of the enzyme.

Processivity of a completely different structural rationale can be observed in motor enzymes that use chemical energy for unidirectional movement along cytoskeletal tracks [12,13]. These motors usually have a dimeric structure, with their dimerization region and ATPase domains connected to their substrate-binding domains by long and extended structures (stalk) (Figure 1C). Large-scale conformational changes elicited by ATP hydrolysis in the ATPase domain(s) propagate to these binding domains, which result in a preference for the re-binding to the substrate track vs. full dissociation [14,15]. In these mechanisms, passive diffusional moves and energy-driven directional steps are combined, i.e., they represent a combination of confining the sequence of events by structural and spatial means. As outlined in the next paragraph, confinement by the limitation of search space by a disordered linker connecting binding domains (Figure 1D) can also account for processivity of enzymes, which appears to be widely applied in biology.

2.2. Statistical Physical Modelling of Domain-Linker-Domain Enzymes

In order to determine how the disordered linker influences (re)binding kinetics of binding domains within a DLD-type enzyme, we used a statistical-kinetic approximation of their binding/unbinding behavior. As the effect of linker length will depend on distances between binding sites and on/off rates of binding domains, we used as a representative example the cellulose/cellulase (Cel7A in Table 1) system. To describe the kinetic behavior of the system, we used a Gaussian approximation of the exact Freely Jointed Chain (FJC) model (see Supplementary Methods and Figure S1). Figure 2 shows the results of varying parameters of a sample case where the tethering domain (cf. Figure 1D) is bound at a substrate site, and we calculate the average binding time (the time it takes for half the free domains to bind a target binding site on the substrate; cf. Supplementary Methods, Equations (S9) and (S10)). By considering the distribution of concentration of the free domain around the bound tethering domain (Figure S1) and integrating binding events (kinetics) based on the binding rate of cellulases (Table S3) over all binding sites within the reach of the free domain, it appears (Figure 2A) that the average time required for re-binding (Supplementary Equation (S10)) increases with increasing linker length. By assuming a threshold set by the kinetics of the dissociation of the tethering domain (for illustration, dissociation half-time (i.e., the time taken for half the bound domains to dissociate) taken as 3×10^{-3} s), the system is processive below a certain linker length (re-binding will be preferred over dissociation), and becomes non-processive for longer linkers (e.g., the threshold linker length is 50 residues in Figure 2A). It should not be forgotten here that the domains in this modelling are dimensionless, due to which there is no minimum on the curve (although there appears to be a minimum imposed by the separation between binding sites, setting a minimum to Kuhn segments).

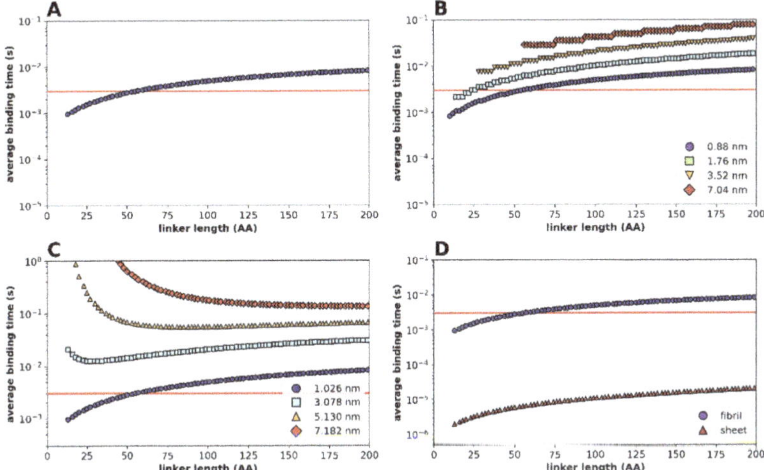

Figure 2. Modelling linker length in processive enzymes. Average binding times (tb) of a free domain linked to the tethering domain already bound to the substrate by a disordered linker of the given length (cf. Figure 1D, and Supplementary Equations (S9) and (S10)). The substrate is modelled based on cellulose geometry: it is assumed to contain binding sites spaced equidistantly every 1.026 nm (1 cellobiose unit) in the X dimension for a thread, and every 2 nm in the Y dimension in case of a sheet. (**A**) Average binding time of the free domain with a random-coil linker (length of Kuhn segment (lk) = 0.88 nm) and binding domains with no physical dimensions. (**B**) Lengthening the Kuhn segment length from 0.88 nm (random-coil) to 7.04 nm (PPII helix) significantly slows binding and reduces processivity. (**C**) "Diluting" binding sites on the substrate (by lengthening the distance between binding sites from 1 cellobiose unit to 7) has a dramatic effect on binding time. (**D**) Binding to a 2D substrate (sheet) is much faster than binding to a 1D substrate (fibril), making the enzyme more processive. On all the panels, if we assume a dissociation half-time of 3×10^{-3} s (limited by catalysis), the enzyme is typically processive at shorter, but not at longer, linker lengths (see text for details).

Therefore, spatially confined diffusional search by the free domain can result in processivity under certain circumstances, when (re)binding by the free domain is kinetically favored over dissociation of the tethering domain. Next, we asked how the flexibility of the linker affects binding time by the free domain. To this end, we ran the statistical kinetic model by varying the length of Kuhn segments (and therefore the persistence length of the chain, see Supplementary Methods) from 0.88 nm (characteristic of random coil chains) to 7.04 nm (characteristic of a polyproline II (PPII) helix), and found a marked effect (Figure 2B), with a more rigid linker providing longer binding times, making the enzyme less processive (e.g., at a length of 30 residues, the enzyme is processive with a linker of 0.88 nm, but not of 3.52 nm, Kuhn-segment length), which may be a prime factor in determining the amino acid composition and sequence conservation of processive linkers, as shown later.

As the calculated binding time is an aggregate value (integrating binding events over all substrate binding sites that can be reached by the free domain, see Supplementary Equation (S10)), we intuitively expect that processivity is increased when possible binding sites are closer to each other, i.e., there are more sites within the reach of the free domain. This is formally demonstrated by varying the spacing of sites (Figure 2C), showing that a processive enzyme can be made non-processive by moving the target sites farther away (this will depend on linker length and could actually be a tuned feature of each system). Along a similar logic, one might expect that the level of processivity is higher when target sites are spread on a two-dimensional surface, by making more sites available for binding. This is formally shown in Figure 2D, where clearly the enzyme is much more processive with a two-dimensional substrate.

Another caveat to the model calculations is if, besides qualitatively assessing whether an enzyme is processive or not, we can draw quantitative conclusions on the level of processivity (average number of steps taken before releasing the substrate). For this, one has to note that the extent of processivity (average number of elementary steps upon engagement with the substrate) is straightforward to define, but not trivial—and is probably not unequivocal—to measure. Furthermore, being a kinetic phenomenon, it may show high stochastic fluctuations and may be very sensitive to experimental conditions.

Nevertheless, one can infer the typical linker-length range where a particular enzyme may behave processive (say, 10–100 residues, cf. intersection of red and blue traces in Figure 2A). This inference may also suggest that linker length and the distance between substrate binding sites must have co-evolved. As an additional note, whereas preferential binding (over dissociation) follows from the kinetic setup of the system, its capacity for unidirectionality does not. As a diffusive move can equally well occur in the backward direction (Figure 1D), directionality may stem from additional mechanistic elements, such as the use of energy and/or post-translational modifications of the substrate. This may even include its degradation, such as that of extracellular matrix proteins in the case of MMP-9 [33,35] or cellulose in the case of cellulases [31,32,36]. This may hinder backward movement and result in rapid unidirectional, forward translocation (Figure 1D).

2.3. Multiple Examples of DLD-Type Processive Enzymes

The foregoing modelling studies show the potential for processivity encoded in the DLD arrangement of enzymes. Next, we demonstrate that there are many such enzymes in biology. Out of 47 processive enzymes of various mechanisms (Table S1), a simple literature search identified 12 processive systems that appear to rely on the DLD domain arrangement, such as MMP-9 [33,37], RNAse H1 [5], or a variety of glycohydrolases [6,31,32]. These ATP-independent enzymes enlisted in Table 1, are analyzed further.

2.3.1. Structural Disorder of Linkers in Monomeric Processive Enzymes

A critical element of processivity in these DLD-type of processive enzymes is the structural disorder of the linker region connecting the binding domains, which has been experimentally demonstrated in only a few cases. For example, the cellulose-binding domain can be effectively separated from the catalytic domain of cellobiohydrolase I by limited proteolysis [38], in agreement with the extreme proteolytic sensitivity of IDPs [34]. Structural disorder was directly observed in cellulase Cel6A and

Cel6B by small-angle X-ray scattering (SAXS) [39], in xylanase 10C by X-ray crystallography [40], and in MMP-9 by atomic-force microscopy (AFM) [33]. Besides these few examples, however, structural disorder has not yet been systematically analyzed in monomeric processive enzymes.

To this end, we applied bioinformatic predictions for the local structural disorder of the linker regions of DLD enzymes in Table 1 (Figure 3). Prediction of structural disorder of three processive enzymes MMP-9, Cel6A and RNAse H1 by IUPred [41] shows a distinctive pattern of a very sharp transition from local order in the binding domains to structural disorder within the linker region. Given the reliability of disorder prediction [42], we may conclude that the linker region in processive enzymes is always disordered, as confirmed for all the cases collected from literature (cf. Table 1, predicted disorder values). Interestingly, the length of the linkers in these processive enzymes always falls within the critical range suggested by model calculations above (cf. Figure 2).

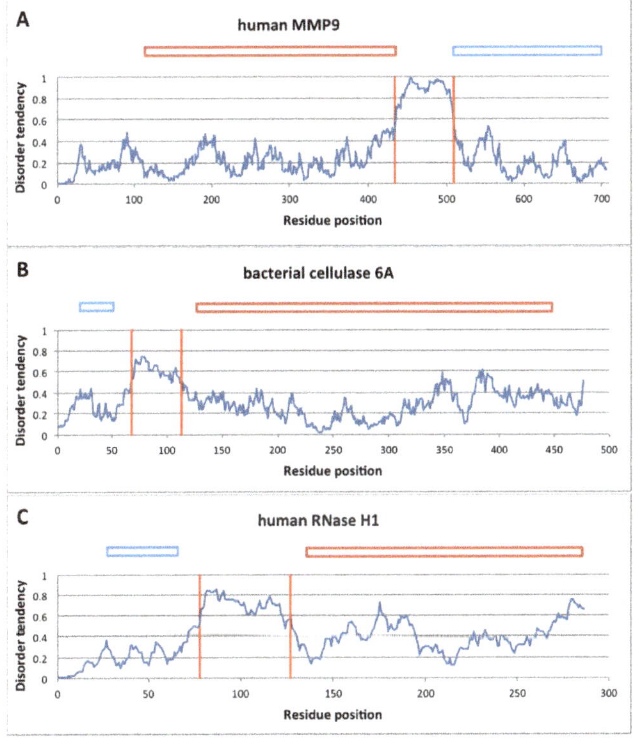

Figure 3. Structural disorder of linker regions in processive enzymes. The linker region in monomeric processive enzymes tends to be highly disordered, as shown here for three illustrative examples by the IUPred algorithm [41]. Traces of disorder score are given for the human and matrix metalloproteinase-9 (MMP-9) sequence (**A**), bacterial cellulase 6A (**B**) and Ribonuclease H1 (RNAseH1) (**C**). In each case, the sharp transition from order to disorder (IUPred score > 0.5) and again to order clearly delimits the linker as a disordered element connecting two globular domains. Globular domains are visualized on top of the diagrams, with blue rectangles representing binding domains and red ones representing catalytic domains.

2.3.2. Conservation of Sequence, Length and Dynamics of Linkers

Modelling (Figure 2) suggests that the length, structural disorder and rigidity of the linker are key elements of processive behavior, which may be in (co)evolutionary link with the typical distance between binding sites (step size) of the given system. This inference also suggests evolutionary constraints on the length and physical properties of the linker regions in these enzymes. We address this issue next.

Regarding evolutionary conservation, IDPs/IDRs have been roughly classified into three classes [43], constrained (where both sequence and structural disorder are conserved), flexible, where sequence varies but structural disorder is conserved, and non-conserved where both lack evolutionary conservation. The underlying assumption in this classification is that disordered regions that function by molecular recognition tend to have conserved sequences, whereas those having linker function are free to evolve, as long as they preserve their structural disorder. As shown in our modelling studies (Figure 2), however, spatial confinement does limit the acceptable length and flexibility of the linker. We assessed these features of the linkers for the 12 DLD-type processive enzymes in Table 1.

In agreement with this expectation, their length shows notably narrower distribution than that of all disordered regions and all disordered linker regions in the DisProt database [44]. Processive enzymes have no short (<30 residues) or long (>150 residues) linkers, although there are many such examples of IDRs in general (Figure 4A). Furthermore, there are characteristic differences between the different DLD enzyme families (Figure S2), which also suggests a co-evolutionary relationship with the typical step size the enzyme takes. When the mean of the linker length of different families is plotted as a function of unit size of different substrates (Table S2), we can see an increase in linker length with the lengthening of processive steps (Figure 5).

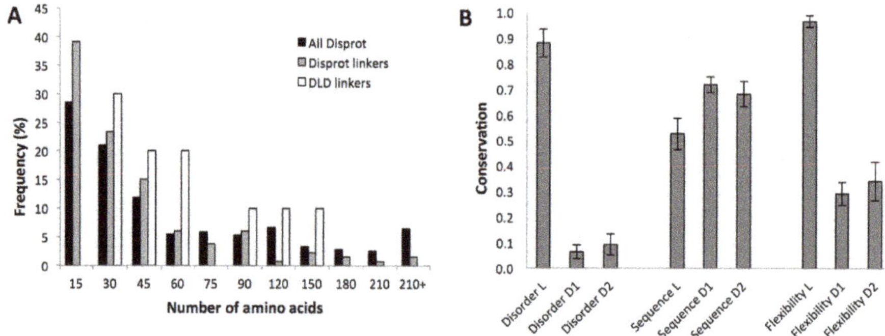

Figure 4. Length distribution and conservation of linker regions in DLD type processive enzymes. (**A**) Length distribution of linkers in DLD enzymes (Table 1), in comparison with that of all disordered regions and disordered linkers in the DisProt database [44]. (**B**) Comparison of the variance (mean values of the data ± SD) of structural disorder (predicted by IUPred [41]) flexibility (as approximated by the ratio of flexible residues predicted by DynaMine [45]) and sequence (assessed by DisCons [22]) of the linkers (L) and their flanking domains (D1 and D2) of the processive DLD type of enzymes (from Table 1) calculated for sequences in species given in (Table S2). Sequence conservation is defined in Section 4 Data and Methods.

This suggests an adaptation of linker length to the geometry of the actual substrate, which also explains: (i) very similar linker length of different processive enzymes functioning on the same substrate, and (ii) the lack of very short and very long linkers in this functional class (Figures 4A and 5).

Their particular function also suggests that selection pressure may also act on their flexibility. As suggested by the above classification [43], classical entropic-chain linker functions are manifested in flexible disorder, where the sequence of the disordered region is rather free to vary, but structural disorder itself is conserved; this is what is expected for the linkers of DLD-type processive enzymes. Therefore, we analyzed the evolution of these features next (Figure 4B). First, we have shown that structural disorder of DLD linkers is highly conserved (as defined in Section 4 Data and Methods), i.e., it shows very little variation. This does not necessarily entail conservation of the sequence (as suggested by flexible disorder [43]), in fact we observe that linker sequences are rather free to vary. Even though structural disorder of the linkers is conserved, it may not necessarily mean that their level of flexibility is maintained at the same level, although this is a critical feature of linkers for the level of processivity (cf. Figure 2). Actually, it was experimentally shown for a similar linker by NMR that despite extreme sequence variation, the flexibility of a linker is maintained [46]. To formally address this issue in DLD linkers, we applied the DynaMine tool developed for assessing local dynamics of IDP backbones [45]. As expected, the overall flexibility of the linker is very high and hardly varies in any of the processive enzymes (Figure 4B).

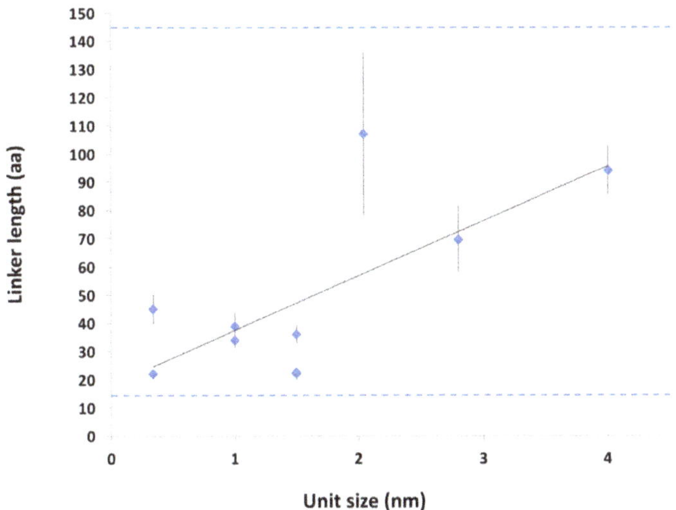

Figure 5. Linker length in DLD enzymes correlates with step size. Linker length in amino acids of the DLD-type processive enzymes (Table 1) is plotted as a function of the unit (step) size in the given substrate. The unit size is the size of the elementary unit (e.g., cellobiose in cellulose, nucleotides in RNA and DNA cf. Table S4) derived from the geometry of the substrate, which is the first approximation of the size of elementary steps the enzyme may take along the given substrate. The linear fit shows the correlation between the two (R^2 = 0.4998), whereas horizontal dashed lines show the shortest and longest linker that occurs in DLD processive enzymes (Figure 4A).

Another characteristic closely linked with flexibility of linkers is their charge state, i.e., net charge and charge distribution, because they are among the primary determinants of the chain dimensions and conformational classes of IDPs [47], and even in the lack of hydrophobic groups, polar IDPs/IDRs may favor collapsed ensembles in water. To evaluate sequence polarity, usually the net charge per residue (NCPR), total fraction of charged residues (FCR) and the linear distribution of opposite charges (characterized by κ value) [48] are considered. Interestingly, for all the DLD linkers, their NCPR is low and their FCR is below the threshold of 0.2 (Figure S3), suggesting that they tend to have very similar behavior (they are weak polyampholytes), preferentially populate collapsed states [48]. Their low κ value (Table 1), however, suggests that they tend to have coil-like conformations. It is of note that high proline content may make the structure more extended than simply suggested by charge distribution suggests. In our case, eight out of 12 proteins have high proline content, with the exception of the two proteins in the boundary region (1: Human RNAse H1 and 5: *Clostridium cellulolyticum* Cel48F, cf. Table 1), which do not have high proline content.

2.3.3. Specific Sequence Features of Processive Linkers

Disordered linkers can also be classified by their amino acid composition [49]. Processive linkers in DLD enzymes may also be under special pressure in this regard, because their potential to interact with the flanking domains and/or with other protein partners, or to undergo regulatory post-translational modifications (PTMs), may be of paramount importance. To assess these features, we analyzed the amino acid composition of disordered linkers in DLD enzymes and compared them to that of DisProt linkers and all disordered regions and annotated disordered linkers in the DisProt database [44] (Figure 6). Our results show that processive linkers have significantly less hydrophobic residues than other linkers and disordered proteins in general, which suggests they have to avoid hydrophobic collapse (cf. restraints on κ value stated above) and/or interactions with partners, which most often is mediated by motifs of hydrophobic character [50]. On the other hand, they are enriched in Pro and Gly (denoted as special residues, Figure 5A only shows P under 'special'), which entails that they have to remain extended and flexible and have a balance in oppositely-charged residues (D + E vs. R + K). Probably also for the same reason, they are, on average, more polar.

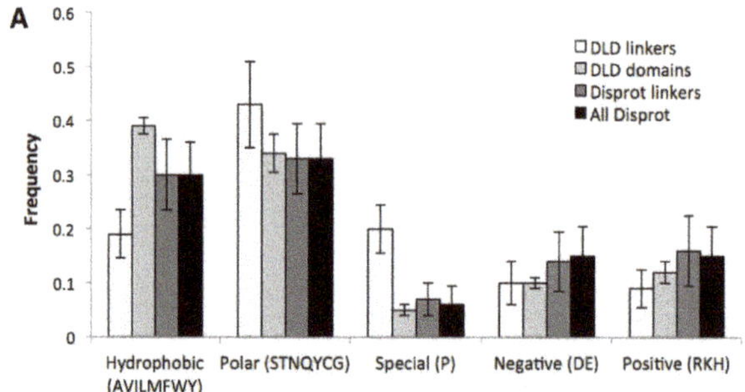

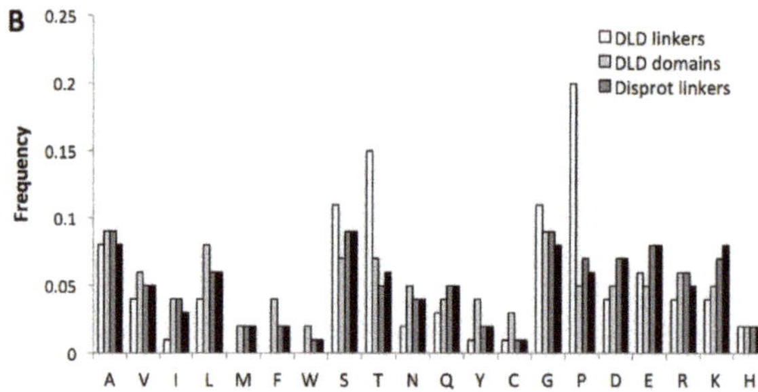

Figure 6. Special features of amino acid composition of linkers. Amino acid composition of linkers in DLD processive enzymes was analyzed and depicted with reference to similar measures of other data. (**A**) Amino acids of linkers were grouped into five categories and compared to the composition of non-linker (binding domain) regions of DLD enzymes (in Table 1) and also of all disordered linkers and assigned disordered linkers in the DisProt database [44]. (**B**) The abundance of amino acids in linkers and non-linker regions in DLD processive enzymes and in all disordered regions and assigned linker regions in the DisProt database.

A further notable feature of DLD linkers is their enrichment in Ser and Thr, which may be indicative of frequent O-linked glycosylation and/or regulatory phosphorylation. A search in UniProt [51] for post-translational modifications (PTMs) of the DLD linkers shows several such modifications in these enzymes (Table 2).

These modifications may impact their kinetic and structural parameters and may tune their interaction with one of the domains of the flanking domains or with external partners. For example, the linker of cellulase emerges from a point not proximal to the cellulose substrate, rather from a point behind, i.e., the kinetic behavior of the enzyme is fine-tuned by the binding of the linker to the surface of the catalytic domain (see next section). Regulated linker-domain interactions are also instrumental in MMP-9, in which the linker has two short binding motifs, that bind the catalytic domain of the enzyme [35].

Table 2. Additional functions of linkers in DLD processive enzymes. Cases where the linker was shown to bind to its adjacent domain are marked with "+".

Enzyme	UniProt ID	PTMs	Domain Binding	Ref.
H. sapiens RNASEH1	O60930	Phosphorylation: S74, S76		[52]
T. reesei Cel7A	P62694	Glycosylation: T461, T462, T463, T462, T469, T470, T471, S473, S474	+	[53]
H. sapiens Telomerase	O14746	Phosphorylation: S227		[54–56]
H. sapiens Nedd4-1	P46934	Phosphorylation: S670, S742, S743, S747, Y785, S884, S888. Ubiqutination: K882		
H. sapiens MMP-9	P14780		+	

The primary function of linkers in DLD processive enzymes is to ensure relatively unrestricted spatial search of domains for binding sites along a multivalent (polymeric) substrate partner. They, however, are also often involved in the regulation of the functioning of the enzyme, as witnessed by additional binding functions and/or PTM events within the linkers themselves (for PTMs, data are either taken from UniProt or from the reference given).

2.3.4. Modelling Cellulase, a Processive Enzyme

Based on all the foregoing analyses, it appears compelling that the DLD arrangement makes enzymes processive. This seems a general phenomenon, which can be demonstrated by low-resolution statistical-kinetic modelling (Figure 2). Here we proceed to show that by incorporating structural details, i.e., atomistic structural models of the domains, into the model and considering domain-linker interactions (Figure 7), we can quantitatively describe the mechanistic and kinetic behavior of one of the most-studied DLD processive enzymes, that of bacterial cellulase (*Trichoderma. reesei* Cel7A, cf. Table 1). Cel7A has two domains of different size, a larger catalytic domain (CD) that confines the linear cellulose substrate, i.e., in itself tends to be processive, and a smaller cellulose binding domain (also termed motif, CBM) attached with a disordered linker of 33 amino acids in length (Figure 7A). The enzyme is processive, typically carrying out about 20–100 cleavage events before dissociating form its substrate. By modeling all parameters of: (i) linker length and flexibility, (ii) catalytic parameters of the enzymatic domain (for the range of kinetic parameters within the Cel7A family, cf. Table S3) and binding parameters of the free (binding) domain, (iii) structural hindrance arising from the actual structures of the domains and domain-linker interaction, and (iv) distance of cellulose binding sites, we show that average binding time of the CBM domain (Figure 7B) undergoes a minimum at a linker length range that is very close to the observed linker lengths in cellulases (Table 1). Furthermore, binding of the linker to the CD has an effect on the behavior of the system (Figure 7B, cf. blue region in color scheme) as it restricts the freedom of movement of the domains, making it less processive. Since all the known cellulase linkers are highly flexible and contain little or no secondary structural elements, changing the Kuhn-segment length is not applicable in this system. The level of processivity that can be approximated as the ratio of the time of binding of CBM to the time of the catalytic reaction (for the CD of cellulase, Table S3, measured with rather artificial substrates) is on the order of 10–100, which agrees with the values reported (Table 1).

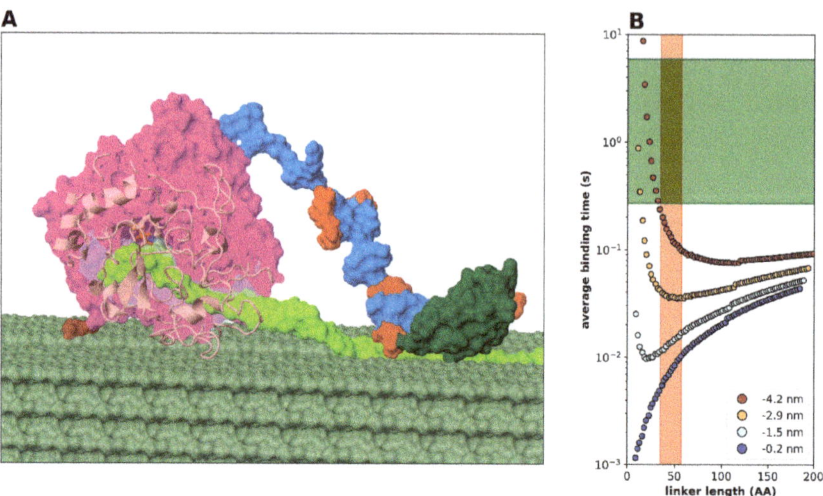

Figure 7. Cellulase: a model processive enzyme. (**A**) Model of the Cel7A cellulase based on the structure PDB 8cel for the catalytic domain (CD) and PDB 2mwk for the cellulose-binding domain (CBM). The CD is purple with the cellulose tunnel shown in transparent blue. One glycosylation of the CD is visible in dark red. Further elements marked are the two catalytic amino acids (red and blue stick-and-ball), the linker region (blue with orange mannose glycosylation), the CBM (dark green), and the cellulose sheet (pale green) of which one fibril (yellow-green) is being processed. The sequence and glycosylation is based on UniProt P62694. (**B**) Statistical kinetic modelling considering geometry (size) and binding of the linker to CD shows binding times characteristic of this system. The green area represents typical catalytic times for Cel7A cellulase family (Table S3), whereas the red area marks typical linker region lengths (Figure S2). The four curves correspond to various values of the linker region's partial binding to the CD, which results in it emerging from the CD at different points (see color mark). If we consider the beginning of the CD domain as the origin of the coordinate system and the cellulose filament moves along the X axis, and assume no binding between the linker and the CD, then free end of the linker region reaches −4.2 nm (red in color scale). When the largest portion of the linker is bound to the CD, the starting point of the free linker end is at zero (blue in color scale). Yellow and light blue colors represent intermediate back-binding cases, with −2.5 and −1.5 nm starting points, respectively.

3. Discussion

Processivity is a basic device of enzymes working on (generating, modifying or moving along) polymeric substrates [1]. By its very molecular logic, it increases cellular economy by limiting the production of metabolic by-products and the dissipation of energy, and it enables large-scale molecular changes to occur, thus it is at the heart of many key cellular processes. Due to the all-or-none character of the operation of processive enzymes, however, there have to be very precise and highly controlled cellular mechanisms for turning them on.

As outlined, there are diverse molecular mechanisms underlying processivity, falling into two general categories, structural confinement by well-folded binding elements and spatial confinement by independent binding elements connected through a linker region. This latter mechanism is apparent in dimeric mechanochemical motors and also in monomeric enzymes. The importance of the general kinetic consequence of processivity can be deduced from its convergent appearance in many independent systems. Whereas its mechanistic underpinning is rather well understood in the case of enzymes that rely on structural confinement and is also analyzed rather extensively in the case of mechanochemical motors, it has so far been largely overlooked in the case of monomeric enzymes.

The typical design of such enzymes is embodied by certain bacterial cellulases, which have a modular structure that combines a large CD linked to a smaller CBM by an intrinsically disordered linker [39] that enables a continuum of conformations. A similar feature has been suggested for the matrix metalloproteinase MMP-9 [33,37], which progressively degrades polymeric components of the extracellular matrix, such as collagen. This enzyme also has a modular structure, with an N-terminal unit of a catalytic domain and three fibronectin type II exosite modules, connected by a 54-residues long linker to a C-terminal hemopexin C domain. SAXS and AFM demonstrated that it can assume multiple conformations and that it can crawl in an inchworm-like manner along its substrate [57]. A similar architecture has been suggested and/or theoretically modelled in the case of glycohydrolases, such as Cel7A [58], cellobiohydrolase I [59] and chitinases [60]. The importance of this arrangement is underscored by cellobiohydrolase I, in which the deletion of the linker dramatically reduces the rate of crystalline cellulose degradation [32] and also other glycoside hydrolases, in which the removal of the carbohydrate-binding module results in a significant decrease in their activity [6], without directly affecting their catalytic domain. Apparently, the unifying feature of all these examples is the structural disorder of their linkers, which ensures a high local concentration and relatively restricted conformational search of binding domains around their binding sites.

Here, we used statistical-kinetic modelling of such systems that this structural arrangement can endow such an enzyme with the capacity of processive movements along a polymeric substrate of spatially repeating binding sites. We characterized these enzymes by the time of (re)binding as a function of linker length, and found that within a certain length range, they have a preference for binding over dissociation, i.e., they show processive kinetic behavior. Geometric features of the domains, direct binding of the linker with the domains themselves and PTMs of the linkers all influence binding kinetics and may thus serve as points of regulatory input. This might be of no negligible importance, as the processive chain of events past the point of activation appears uncontrolled, which may have dire consequences. A proper regulatory input halting the reaction may be a remedy under some circumstances, as suggested by frequent PTMs of processive linkers (Table 2) and their regulated binding to the flanking domains, as shown for MMP-9, for example [33].

These theoretical observations have general relevance and are supported by a collection of 12 such enzymes that all have highly disordered linkers. Notably, despite rapid evolution and sequence variability of IDPs/IDRs in general, and disordered linker regions in particular, the length and flexibility of linkers in the processive enzymes is conserved. Quantitative modelling of the cellulase enzymes is in general agreement with the observed level of processivity and suggests that this functional-kinetic property is manifest in a relatively limited range of linker lengths, which appear to be in co-evolutionary link with the particular step size along their typical substrate. This has been also suggested by the behavior of the related mechanochemical motors kinesin-1 and kinesin-2, the degree of processivity of which sharply changes by changing the length of their linker regions [15]. This feature is also underlined by the observation that short and long linkers are entirely missing in DLD-type processive enzymes.

In a broader functional context, we suggest that this observed behavior is a special case of the entropic chain functions of IDPs/IDRs and appears as a conceptual extension of mechanisms, such as fly casting [27] and monkey-bar mechanism [28]. Processivity appears to draw on all these mechanisms and may represent one of the primary benefits of the flexibility emanating from structural disorder [25,61]. This type of function cannot be supported by a structured protein; thus it is an appealing addition to the functional arsenal of structural disorder, understanding of which may even enable the design and generation of enzymes of improved capacity for the needs of biotechnology.

4. Data and Methods

4.1. Collection of Processive Enzymes and Intrinsically Disordered Proteins

Processive enzymes were collected from the literature by searching for keywords "processive" or "processivity." We aimed for a full coverage of all types of processive enzymes, which resulted in

47 illustrative examples (Table S1), many of which were covered previously [1]. From this collection we selected 12 monomeric enzymes, for further analysis (Table 1). Due to their dominant modular arrangement, we term these monomeric processive enzymes domain-linker-domain (DLD) type. For comparative purposes, we also downloaded 1274 IDP/IDR sequences from the DisProt database (version 7.0) and selected 133 of the IDRs annotated as "linkers" [44].

4.2. Statistical Kinetic Modelling of Linker Regions

To assess the statistical kinetic behavior of DLD proteins we chose the Freely Jointed Chain (FJC) model and simulated it with a Gaussian approximation [36,62]. As shown by details of the model (Supplementary Methods and Figure S1), this only causes minor deviations from the analytical solution at extreme linker lengths.

An important parameter in modelling is the stiffness of the chain that characterizes its nature of spatial distribution. In the FJC model, this is described by Kuhn segments (l_k), whose measure is two times the persistence length. In a freely moving random-coil polypeptide chain this persistence length is 0.44 nm [62], whereas in a stiff polyproline helix it is roughly an order of magnitude longer. To get the number of Kuhn segments, an amino acid chain can be simulated by calculating the contour length of the chain, l_c, divided by l_k.

It is to be noted that the approximation of a kinetic phenomenon of binding and/or dissociation is only tenable if reaching the equilibrium in spatial distribution is much faster than the event of binding and unbinding, i.e., binding/unbinding is not rate-limiting. As diffusion rates of small proteins in water are on the order of 10^{-6} cm^2 s^{-1} [63], which is equivalent to 102 nm$^2 \cdot$s^{-1}, the typical µs time of the unbound ("free," for domain definitions, cf. Figure 1D) domain equilibrating within the boundaries of the model is well below the time scale of processivity steps.

4.3. Assessing Structural Disorder of Linkers

Structural disorder of processive enzymes was predicted by the IUPred algorithm [41], which is based on estimating the total pairwise inter-residue interaction energy gained upon folding of a polypeptide chain. The predictor returns a position-specific disorder score in the range 0.0–1.0, and a residue with score ≥0.5 is considered as locally disordered. To characterize the disorder tendency of domains and linkers, we calculated the ratio of disordered residues within the given region.

4.4. Flexibility of Linker Regions

To quantify the flexibility of linkers, we used DynaMine [45], a backbone dynamics predictor that has been trained on proteins for which NMR-based chemical shifts and experimental amide bond order parameters (S2) were available. Its score falls between 0.0 and 1.0, with a threshold 0.78 separating flexible (below) and rigid (above) regions. Residue-level DynaMine values were averaged for the entire sequence of linkers to calculate an overall measure of flexibility.

4.5. Charge State and Kappa Value Calculation of Linkers

The charge state of linkers was characterized by three parameters [47,48]. The net charge per residue value (NCPR) is defined as $|f+ - f-|$, where $f+$ and $f-$ are the fractions of positively- and negatively-charged residues within the linker region, respectively. The total fraction of charged residues (FCR) is defined as $(f+) + (f-)$. The linear distribution of opposite charges is described by the kappa (κ) parameter [48], which is the mean-square deviation of local charge asymmetry from the overall sequence charge asymmetry weighted on the maximal asymmetry allowed for a given amino-acid composition. Kappa can range from 0 (when opposite charges are evenly distributed) to 1 (when opposite charges are segregated into two clusters). Kappa has a basic influence on IDP/IDR conformation, as there appears to be an inverse correlation between the kappa value and the radius of gyration of the polypeptide chain.

4.6. Amino-Acid Composition and Length Distribution of Linkers

The length and amino acid composition of each processive linker (Table 1) and all IDPs/IDRs in DisProt [44] were calculated. For classification purposes, we also determined composition in terms of a reduced set of amino acid types (positive/basic: Arg, Lys; negative/acidic: Asp, Glu; polar: Ser, Thr, Cys, Gln, His, Tyr, Asn; hydrophobic: Ala, Val, Met, Trp, Phe, Leu, Ile; and special: Pro, Gly).

4.7. Variability and Conservation of Linker Regions

The DLD-type processive enzymes studied here contain two globular domains connected by a disordered linker. To analyze their evolutionary relatedness, we applied the MAFFT (Multiple Alignment using Fast Fourier Transform) program to generate multiple alignments [64] of the sequences from several species, anchored by the flanking ordered binding domain(s), which are highly conserved. Evolutionary conservation of a given region (either disordered or folded) was calculated by an algorithm that computes the average of genetic distances between each pair of sequences in the alignment. The details of the applied method are given in [65]. The species used for alignments and conservation analysis are listed for each protein in Table S2.

Supplementary Materials: Supplementary materials can be found at http://www.mdpi.com/1422-0067/20/9/2119/s1.

Author Contributions: Conceptualization, P.T., A.T., L.B.C., K.H.H.; methodology, T.H., E.S., B.S.; software, L.K., T.H.; formal analysis, B.S., E.S., T.H., L.B.C.; data curation, B.S., E.S., N.M., writing—original draft preparation, A.T., L.B.C.; writing—review and editing, P.T., L.B.C.; A.T., P.T.; funding acquisition, P.T., L.B.C., K.H.H.

Funding: This work was supported by the Odysseus grant G.0029.12 from Research Foundation Flanders (FWO), a "Korea-Hungary & Pan EU consortium for investigation of IDP structure and function" from National Research Council of Science and Technology (NST) of Korea (NTM2231611, to K.H. and P.T.) and grants K124670 (to P.T.) and K125340 (to A.T.) from National Research, Development and Innovation Office (NRDIO). LBC is a career investigator from Consejo Nacional de Investigaciones Científicas y Técnicas (CONICET, Argentina).

Conflicts of Interest: The authors declare no conflict of interest.

Abbreviations

AFM	atomic-force microscopy
DLD	domain-linker-domain
FJC	freely jointed chain
ID	intrinsically disordered
IDP	intrinsically disordered protein
IDR	intrinsically disordered region
MMP-9	matrix metalloproteinase-9
PTM	post-translational modification
RNAse-H1	ribonuclease H1
SAXS	small-angle X-ray scattering

References

1. Breyer, W.A.; Matthews, B.W. A structural basis for processivity. *Protein Sci.* **2001**, *10*, 1699–1711. [CrossRef]
2. Bambara, R.A.; Uyemura, D.; Choi, T. On the processive mechanism of *Escherichia coli* DNA polymerase I. Quantitative assessment of processivity. *J. Biol. Chem.* **1978**, *253*, 413–423.
3. Bonderoff, J.M.; Lloyd, R.E. Time-dependent increase in ribosome processivity. *Nucleic Acids Res.* **2010**, *38*, 7054–7067. [CrossRef] [PubMed]
4. Breyer, W.A.; Matthews, B.W. Structure of *Escherichia coli* exonuclease I suggests how processivity is achieved. *Nat. Struct. Biol.* **2000**, *7*, 1125–1128. [PubMed]
5. Gaidamakov, S.A.; Gorshkova, I.I.; Schuck, P.; Steinbach, P.J.; Yamada, H.; Crouch, R.J.; Cerritelli, S.M. Eukaryotic RNases H1 act processively by interactions through the duplex RNA-binding domain. *Nucleic Acids Res.* **2005**, *33*, 2166–2175. [CrossRef]

6. Boraston, A.B.; Bolam, D.N.; Gilbert, H.J.; Davies, G.J. Carbohydrate-binding modules: fine-tuning polysaccharide recognition Carbohydrate-binding modules: Fine-tuning polysaccharide recognition. *Biochem. J.* **2004**, *382*, 769–781. [CrossRef]

7. Akopian, T.N.; Kisselev, A.F.; Goldberg, A.L. Processive degradation of proteins and other catalytic properties of the proteasome from Thermoplasma acidophilum. *J. Biol. Chem.* **1997**, *272*, 1791–1798. [CrossRef]

8. Schrader, E.K.; Harstad, K.G.; Matouschek, A. Targeting proteins for degradation. *Nat. Chem. Biol.* **2009**, *5*, 815–822. [CrossRef]

9. Gyimesi, M.; Sarlos, K.; Kovacs, M. Processive translocation mechanism of the human Bloom's syndrome helicase along single-stranded DNA. *Nucleic Acids Res.* **2010**, *38*, 4404–4414. [CrossRef] [PubMed]

10. Hochstrasser, M. Lingering mysteries of ubiquitin-chain assembly. *Cell* **2006**, *124*, 27–34. [CrossRef] [PubMed]

11. Sowa, M.E.; Harper, J.W. From loops to chains: Unraveling the mysteries of polyubiquitin chain specificity and processivity. *ACS Chem. Biol.* **2006**, *1*, 20–24. [CrossRef] [PubMed]

12. Gyimesi, M.; Sarlós, K.; Derényi, I.; Kovács, M. Streamlined determination of processive run length and mechanochemical coupling of nucleic acid motor activities. *Nucleic Acids Res.* **2010**, *38*, e102. [CrossRef] [PubMed]

13. Kolomeisky, A.B.; Fisher, M.E. Molecular motors: A theorist's perspective. *Annu. Rev. Phys. Chem.* **2007**, *58*, 675–695. [CrossRef] [PubMed]

14. Rock, R.S.; Ramamurthy, B.; Dunn, A.R.; Beccafico, S.; Rami, B.R.; Morris, C.; Spink, B.J.; Franzini-Armstrong, C.; Spudich, J.A.; Sweeney, H. A flexible domain is essential for the large step size and processivity of myosin VI. *Mol. Cell* **2005**, *17*, 603–609. [CrossRef] [PubMed]

15. Shastry, S.; Hancock, W.O. Neck linker length determines the degree of processivity in kinesin-1 and kinesin-2 motors. *Curr. Biol.* **2010**, *20*, 939–943. [CrossRef] [PubMed]

16. Krishna, T.S.; Fenyö, D.; Kong, X.-P.; Gary, S.; Chait, B.T.; Burgers, P.; Kuriyan, J. Crystallization of proliferating cell nuclear antigen (PCNA) from Saccharomyces cerevisiae. *J. Mol. Biol.* **1994**, *241*, 265–268. [CrossRef]

17. Krishna, T.S.; Kong, X.-P.; Gary, S.; Burgers, P.M.; Kuriyan, J. Crystal structure of the eukaryotic DNA polymerase processivity factor PCNA. *Cell* **1994**, *79*, 1233–1243. [CrossRef]

18. Huang, H.; Chopra, R.; Verdine, G.L.; Harrison, S.C. Structure of a covalently trapped catalytic complex of HIV-1 reverse transcriptase: Implications for drug resistance. *Science* **1998**, *282*, 1669–1675. [CrossRef]

19. Asenjo, A.B.; Weinberg, Y.; Sosa, H. Nucleotide binding and hydrolysis induces a disorder-order transition in the kinesin neck-linker region. *Nat. Struct. Mol. Biol.* **2006**, *13*, 648–654. [CrossRef] [PubMed]

20. Carter, A.P. Crystal clear insights into how the dynein motor moves. *J. Cell Sci.* **2013**, *126*, 705–713. [CrossRef]

21. Tompa, P. Unstructural biology coming of age. *Curr. Opin. Struct. Biol.* **2011**, *21*, 419–425. [CrossRef]

22. Varadi, M.; Guharoy, M.; Zsolyomi, F.; Tompa, P. DisCons: A novel tool to quantify and classify evolutionary conservation of intrinsic protein disorder. *BMC Bioinform.* **2015**, *16*, 153. [CrossRef]

23. Tompa, P.; Fuxreiter, M.; Oldfield, C.J.; Simon, I.; Dunker, A.K.; Uversky, V.N. Close encounters of the third kind: Disordered domains and the interactions of proteins. *Bioessays* **2009**, *31*, 328–335. [CrossRef] [PubMed]

24. Wright, P.E.; Dyson, H.J. Linking folding and binding. *Curr. Opin. Struct. Biol.* **2009**, *19*, 31–38. [CrossRef] [PubMed]

25. Tompa, P. The interplay between structure and function in intrinsically unstructured proteins. *FEBS Lett.* **2005**, *579*, 3346–3354. [CrossRef]

26. Tompa, P.; Fuxreiter, M. Fuzzy complexes: Polymorphism and structural disorder in protein-protein interactions. *Trends Biochem. Sci.* **2008**, *33*, 2–8. [CrossRef] [PubMed]

27. Shoemaker, B.A.; Portman, J.J.; Wolynes, P.G. Speeding molecular recognition by using the folding funnel: The fly-casting mechanism. *Proc. Natl. Acad. Sci. USA* **2000**, *97*, 8868–8873. [CrossRef]

28. Vuzman, D.; Azia, A.; Levy, Y. Searching DNA via a "Monkey Bar" mechanism: The significance of disordered tails. *J. Mol. Biol.* **2010**, *396*, 674–684. [CrossRef]

29. Mittag, T.; Orlicky, S.; Choy, W.-Y.; Tang, X.; Lin, H.; Sicheri, F.; Kay, L.E.; Tyers, M.; Forman-Kay, J.D. Dynamic equilibrium engagement of a polyvalent ligand with a single-site receptor. *Proc. Natl. Acad. Sci. USA* **2008**, *105*, 17772–17777. [CrossRef]

30. Song, J.; Ng, S.C.; Tompa, P.; Lee, K.A.; Chan, H.S. Polycation-pi interactions are a driving force for molecular recognition by an intrinsically disordered oncoprotein family. *PLoS Comput. Biol.* **2013**, *9*, e1003239. [CrossRef]

31. Carrard, G.; Koivula, A.; Söderlund, H.; Béguin, P. Cellulose-binding domains promote hydrolysis of different sites on crystalline cellulose. *Proc. Natl. Acad. Sci. USA* **2000**, *97*, 10342–10347. [CrossRef] [PubMed]

32. Srisodsuk, M.; Reinikainen, T.; Penttilä, M.; Teeri, T.T. Role of the interdomain linker peptide of Trichoderma reesei cellobiohydrolase I in its interaction with crystalline cellulose. *J. Biol. Chem.* **1993**, *268*, 20756–20761.

33. Rosenblum, G.; Steen, P.E.V.D.; Cohen, S.R.; Grossmann, J.G.; Frenkel, J.; Sertchook, R.; Slack, N.; Strange, R.W.; Opdenakker, G.; Sagi, I. Insights into the structure and domain flexibility of full-length pro-matrix metalloproteinase-9/gelatinase B. *Structure* **2007**, *15*, 1227–1236. [CrossRef]

34. Tompa, P. Intrinsically unstructured proteins. *Trends Biochem. Sci.* **2002**, *27*, 527–533. [CrossRef]

35. Chen, Y.; Jiang, T.; Mao, A.; Xu, J. Esophageal cancer stem cells express PLGF to increase cancer invasion through MMP9 activation. *Tumour Biol.* **2014**, *35*, 12749–12755. [CrossRef]

36. Gao, D.; Chundawat, S.P.S.; Sethi, A.; Balan, V.; Gnanakaran, S.; Dale, B.E. Increased enzyme binding to substrate is not necessary for more efficient cellulose hydrolysis. *Proc. Natl. Acad. Sci. USA* **2013**, *110*, 10922–10927. [CrossRef]

37. Rosenblum, G.; Meroueh, S.; Toth, M.; Fisher, J.F.; Fridman, R.; Mobashery, S.; Sagi, I. Molecular structures and dynamics of the stepwise activation mechanism of a matrix metalloproteinase zymogen: Challenging the cysteine switch dogma. *J. Am. Chem. Soc.* **2007**, *129*, 13566–13574. [CrossRef]

38. Tilbeurgh, H.V.; Tomme, P.; Claeyssens, M.; Bhikhabhai, R.; Pettersson, G. Limited proteolysis of the cellobiohydrolase I from Trichoderma reesei. Separation of functional domains. *FEBS Lett.* **1986**, *204*, 223–227. [CrossRef]

39. Von Ossowski, I.; Eaton, J.T.; Czjzek, M.; Perkins, S.J.; Frandsen, T.P.; Schülein, M.; Panine, P.; Henrissat, B.; Receveur-Bréchot, V. Protein disorder: Conformational distribution of the flexible linker in a chimeric double cellulase. *Biophys. J.* **2005**, *88*, 2823–2832. [CrossRef]

40. Pell, G.; Szabo, L.; Charnock, S.J.; Xie, H.; Gloster, T.M.; Davies, G.J.; Gilbert, H.J. Structural and biochemical analysis of Cellvibrio japonicus xylanase 10C: How variation in substrate-binding cleft influences the catalytic profile of family GH-10 xylanases. *J. Biol. Chem.* **2004**, *279*, 11777–11788. [CrossRef]

41. Dosztanyi, Z.; Csizmok, V.; Tompa, P.; Simon, I. IUPred: Web server for the prediction of intrinsically unstructured regions of proteins based on estimated energy content. *Bioinformatics* **2005**, *21*, 3433–3434. [CrossRef]

42. Noivirt-Brik, O.; Prilusky, J.; Sussman, J.L. Assessment of disorder predictions in CASP8. *Proteins* **2009**, *77* (Suppl. S9), 210–216. [CrossRef]

43. Bellay, J.; Han, S.; Michaut, M.; Kim, T.; Costanzo, M.; Andrews, B.J.; Boone, C.; Bader, G.D.; Myers, C.L.; Kim, P.M. Bringing order to protein disorder through comparative genomics and genetic interactions. *Genome Biol.* **2011**, *12*, R14. [CrossRef] [PubMed]

44. Piovesan, D.; Tabaro, F.; Mičetić, I.; Necci, M.; Quaglia, F.; Oldfield, C.J.; Aspromonte, M.C.; Davey, N.E.; Davidović, R.; Dosztányi, Z.; et al. DisProt 7.0: A major update of the database of disordered proteins. *Nucleic Acids Res.* **2017**, *45*, D219–D227. [CrossRef] [PubMed]

45. Cilia, E.; Pancsa, R.; Tompa, P.; Lenaerts, T. From protein sequence to dynamics and disorder with DynaMine. *Nat. Commun.* **2013**, *4*, 2741. [CrossRef] [PubMed]

46. Daughdrill, G.W.; Narayanaswami, P.; Gilmore, S.H.; Belczyk, A.; Brown, C.J. Dynamic behavior of an intrinsically unstructured linker domain is conserved in the face of negligible amino acid sequence conservation. *J. Mol. Evol.* **2007**, *65*, 277–288. [CrossRef] [PubMed]

47. Mao, A.H.; Crick, S.L.; Vitalis, A.; Chicoine, C.L.; Pappu, R.V. Net charge per residue modulates conformational ensembles of intrinsically disordered proteins. *Proc. Natl. Acad. Sci. USA* **2010**, *107*, 8183–8188. [CrossRef]

48. Das, R.K.; Pappu, R.V. Conformations of intrinsically disordered proteins are influenced by linear sequence distributions of oppositely charged residues. *Proc. Natl. Acad. Sci. USA* **2013**, *110*, 13392–13397. [CrossRef] [PubMed]

49. George, R.A.; Heringa, J. An analysis of protein domain linkers: Their classification and role in protein folding. *Protein Eng.* **2002**, *15*, 871–879. [CrossRef]

50. Fuxreiter, M.; Tompa, P.; Simon, I. Local structural disorder imparts plasticity on linear motifs. *Bioinformatics* **2007**, *23*, 950–956. [CrossRef] [PubMed]

51. The UniProt Consortium. UniProt: The universal protein knowledgebase. *Nucleic Acids Res.* **2017**, *45*, D158–D169. [CrossRef] [PubMed]

52. Pan, C.; Olsen, J.V.; Daub, H.; Mann, M. Global effects of kinase inhibitors on signaling networks revealed by quantitative phosphoproteomics. *Mol. Cell. Proteom.* **2009**, *8*, 2796–2808. [CrossRef] [PubMed]

53. Harrison, M.J.; Nouwens, A.S.; Jardine, D.R.; Zachara, N.E.; Gooley, A.A.; Nevalainen, H.; Packer, N.H. Modified glycosylation of cellobiohydrolase I from a high cellulase-producing mutant strain of Trichoderma reesei. *Eur. J. Biochem.* **1998**, *256*, 119–127. [CrossRef]

54. Chung, J.; Khadka, P.; Chung, I.K. Nuclear import of hTERT requires a bipartite nuclear localization signal and Akt-mediated phosphorylation. *J. Cell Sci.* **2012**, *125*, 2684–2697. [CrossRef] [PubMed]

55. Jeong, S.A.; Kim, K.; Lee, J.H.; Cha, J.S.; Khadka, P.; Cho, H.S.; Chung, I.K. Akt-mediated phosphorylation increases the binding affinity of hTERT for importin alpha to promote nuclear translocation. *J. Cell Sci.* **2015**, *128*, 2287–2301. [CrossRef]

56. Kang, S.S.; Kwon, T.; Kwon, D.Y.; Do, S.I. Akt protein kinase enhances human telomerase activity through phosphorylation of telomerase reverse transcriptase subunit. *J. Biol. Chem.* **1999**, *274*, 13085–13090. [CrossRef]

57. Overall, C.M.; Butler, G.S. Protease yoga: Extreme flexibility of a matrix metalloproteinase. *Structure* **2007**, *15*, 1159–1161. [CrossRef]

58. Zhao, Y.; Wang, Y.; Zhu, J.; Ragauskas, A.; Deng, Y.; Ragauskas, A. Enhanced enzymatic hydrolysis of spruce by alkaline pretreatment at low temperature. *Biotechnol. Bioeng.* **2008**, *99*, 1320–1328. [CrossRef]

59. Igarashi, K.; Koivula, A.; Wada, M.; Kimura, S.; Penttilä, M.; Samejima, M. High speed atomic force microscopy visualizes processive movement of Trichoderma reesei cellobiohydrolase I on crystalline cellulose. *J. Biol. Chem.* **2009**, *284*, 36186–36190. [CrossRef]

60. Seidl, V. Chitinases of filamentous fungi: A large group of diverse proteins with multiple physiological functions. *Fungal Biol. Rev.* **2008**, *22*, 36–42. [CrossRef]

61. Van der Lee, R.; Buljan, M.; Lang, B.; Weatheritt, R.J.; Daughdrill, G.W.; Dunker, A.K.; Fuxreiter, M.; Gough, J.; Gsponer, J.; Jones, D.T.; et al. Classification of intrinsically disordered regions and proteins. *Chem. Rev.* **2014**, *114*, 6589–6631. [CrossRef] [PubMed]

62. Czovek, A.; Szollosi, G.J.; Derenyi, I. The relevance of neck linker docking in the motility of kinesin. *Biosystems* **2008**, *93*, 29–33. [CrossRef] [PubMed]

63. Czovek, A.; Szollosi, G.J.; Derenyi, I. Neck-linker docking coordinates the kinetics of kinesin's heads. *Biophys. J.* **2011**, *100*, 1729–1736. [CrossRef] [PubMed]

64. Katoh, K. MAFFT: A novel method for rapid multiple sequence alignment based on fast Fourier transform. *Nucleic Acids Res.* **2002**, *30*, 3059–3066. [CrossRef] [PubMed]

65. Capra, J.A.; Singh, M. Predicting functionally important residues from sequence conservation. *Bioinformatics* **2007**, *23*, 1875–1882. [CrossRef] [PubMed]

© 2019 by the authors. Licensee MDPI, Basel, Switzerland. This article is an open access article distributed under the terms and conditions of the Creative Commons Attribution (CC BY) license (http://creativecommons.org/licenses/by/4.0/).

International Journal of
Molecular Sciences

Article

Repeats in S1 Proteins: Flexibility and Tendency for Intrinsic Disorder

Andrey Machulin [1], Evgenia Deryusheva [2], Mikhail Lobanov [3] and Oxana Galzitskaya [3,*

[1]   Skryabin Institute of Biochemistry and Physiology of Microorganisms, Russian Academy of Sciences, Federal Research Center "Pushchino Scientific Center for Biological Research of the Russian Academy of Sciences, 142290 Pushchino, Russia; and.machul@gmail.com
[2]   Institute for Biological Instrumentation, Federal Research Center "Pushchino Scientific Center for Biological Research of the Russian Academy of Sciences, 142290 Pushchino, Russia; evgenia.deryusheva@gmail.com
[3]   Institute of Protein Research, Russian Academy of Sciences, 142290 Pushchino, Russia; mlobanov@phys.protres.ru
*   Correspondence: ogalzit@vega.protres.ru; Tel.: +7-903-675-0156

Received: 9 April 2019; Accepted: 10 May 2019; Published: 14 May 2019

Abstract: An important feature of ribosomal S1 proteins is multiple copies of structural domains in bacteria, the number of which changes in a strictly limited range from one to six. For S1 proteins, little is known about the contribution of flexible regions to protein domain function. We exhaustively studied a tendency for intrinsic disorder and flexibility within and between structural domains for all available UniProt S1 sequences. Using charge–hydrophobicity plot cumulative distribution function (CH-CDF) analysis we classified 53% of S1 proteins as ordered proteins; the remaining proteins were related to molten globule state. S1 proteins are characterized by an equal ratio of regions connecting the secondary structure within and between structural domains, which indicates a similar organization of separate S1 domains and multi-domain S1 proteins. According to the FoldUnfold and IsUnstruct programs, in the multi-domain proteins, relatively short flexible or disordered regions are predominant. The lowest percentage of flexibility is in the central parts of multi-domain proteins. Our results suggest that the ratio of flexibility in the separate domains is related to their roles in the activity and functionality of S1: a more stable and compact central part in the multi-domain proteins is vital for RNA interaction, terminals domains are important for other functions.

Keywords: ribosomal proteins S1; structural domains; intrinsically flexibility; FoldUnfold program; IsUnstruct program

---

1. Introduction

It is known that multi-domain proteins are frequently characterized by the occurrence of domain repeats in proteomes across the three domains of life: Bacteria, Archaea, and Eukaryotes [1,2]. Proteins with repeats participate in nearly every cellular process from transcriptional regulation in the nucleus to cell adhesion at the plasma membrane [3]. In addition, due to their flexibility, domain repeats can be found in cytoskeleton proteins, proteins responsible for transport and cell cycle control [4]. Proteins with structural repeats are believed to be ancient folds.

One such unique protein family is a family of bacterial ribosomal proteins S1 in which structural domain S1 (one of the oligonucleotide/oligosaccharide-binding fold (OB-fold) options) repeats and changes in a strictly limited range from one to six [5]. As demonstrated in our recent paper [5], the family of polyfunctional ribosomal proteins S1 contains about 20% of all bacterial proteins, including the S1 domain. This fold also could be found in different eukaryotic protein families and protein complexes in different number variations. Such multiple copies of the structure increase the affinity and/or specificity of the protein binding to nucleic acid molecules.

Recently we have shown that the sequence alignments of S1 proteins between separate domains in each group reveal a rather low percentage of identity. In addition, the verification of the equivalence of the domain characteristics showed that for long S1 proteins (five- and six-domain containing S1 proteins) the central part of the proteins (the third domain) is more conservative than the terminal domains and apparently is vital for the activity and functionality of S1. Data obtained indicated that for general functioning of these proteins, the structure scaffold (OB-fold) is obviously more important than the amino acid sequence [6]. This statement is in good agreement with the fact that there is a high degree of conservatism and topology position of the binding site on the OB-fold surface in others proteins, as well as "fold resistance" to mutations and the ability to adapt to a wide range of ligands, which allows us to consider this fold as one of the ancient protein folds. For example, the author of article [7] proposed considering this core structure of inorganic pyrophosphatase as the evolutionary precursor of all other superfamilies.

At present, the structure of S1 from *Escherichia coli* was obtained only with a very low resolution of 11.5 Å using cryo-electron microscopy [8]. In the Protein Data Bank, there are only 3D structures of separate domains of ribosomal S1 from *E. coli* obtained by NMR [9,10]. Recently, protein S1 on the 70S ribosome was visualized by ensemble cryo-electron microscopy [11]. It was shown that S1 cooperates with other ribosomal proteins (S2, S3, S6, and S18) to form a dynamic mesh near the mRNA exit and entrance channels to modulate the binding, folding and movement of mRNA. The cryo-electron microscopy was also used to obtain the structure of the inactive conformation of the S1 protein as part of a hibernating 100S ribosome [12].

A separate S1 domain from the ribosomal proteins S1 [9] and other bacterial proteins containing an S1 domain [13–16] represents a β-barrel with an additional α-helix between the third and fourth β-sheets. As shown in the articles [13–16], the S1 domain as a part of different bacterial proteins (as well as in eukaryotic proteins) itself is quite compact, therefore it crystallizes and is visualized very well.

At the same time, there are currently no determined structures for full-length, intact ribosomal S1 proteins containing a different number of structural domains (six in *E. coli*, five in *Thermus thermophilus*, etc.). This may be due to the increased flexibility of multi-domain proteins as was noted in [17]. In addition, some biochemical studies suggest that in solution and on the ribosome, S1 can have an elongated shape stretching over 200 Å long [17–20].

Moreover, recently it was shown that the prediction of intrinsic disorder within proteins with the tandem repeats supports the conclusion that the level of repetition correlates with their tendency to be unstructured and the chance to find natural structured proteins in the Protein Data Bank (PDB) increases with a decrease in the level of repeat perfection. Also, the authors suggested that in general, the repeat perfection is a sign of recent evolutionary events rather than of exceptional structural and/or functional importance of the repeat residues [21].

Despite all these observations, the flexibility of S1 proteins, their tendency for intrinsic disorder, and the structural characteristics of this family have not been studied as of yet. To fill this gap, we have analyzed here the flexibility of the bacterial S1 proteins within and between structural domains, as well as the tendency for intrinsic disorder of the S1 protein family.

2. Results and Discussion

2.1. Analysis of Tendency for Intrinsic Disorder of the Bacterial S1 Proteins

Binary disorder analysis using the charge–hydrophobicity plot cumulative distribution function (CH-CDF) plot [22] showed that most of the bacterial S1 proteins (1374 sequences) (53%) are expected to be mostly ordered (or folded, 'F') (Figure 1a).

Mixed or molten globular ('MG') forms comprised the remaining 47% of the bacterial S1 proteins. Major protein states for separate groups of the S1 proteins (different number of structural domains) according to the CH-CDF analysis are shown in Figure 1b. In the case of S1 proteins containing one, two or six structural domains (1S1, 2S1, 6S1) the ordered state prevailed (83%, 78% and 67%, respectively). S1 proteins containing three, four and five domains were classified as molten globule state according to the CH-CDF analysis in 69%, 74% and 56% cases, respectively. It was seen that with an increase in the number of structural domains (starting from the three-domain containing proteins), the MG state prevailed, but for six-domain proteins only 34% of the records belonged to this area. Despite the fact that one-domain and two-domain containing proteins were the least represented in our dataset, the data obtained for these groups results are in good agreement with the fact that the separate S1 domain is stable and has rather rigid structure [13–16]. Note that for other structural variants of the OB-fold (for example, CSD domain [23], inorganic pyrophosphatase [24], *MOP-like* [25], etc.) there are available structures that also have only one or two (repeated) domains [5].

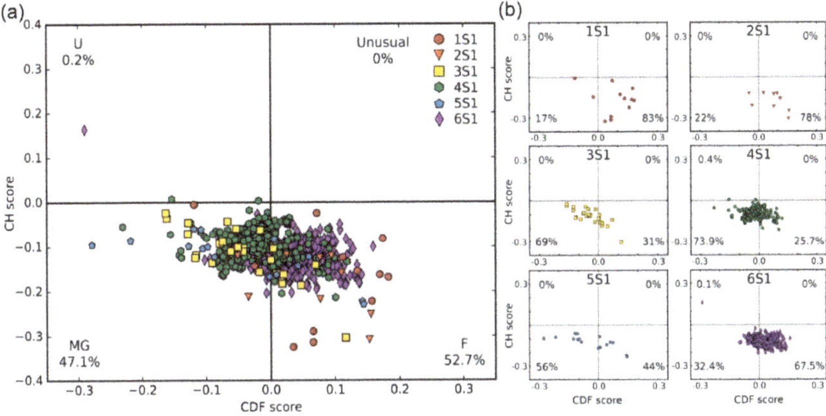

Figure 1. (**a**) Binary disorder analysis (charge–hydrophobicity plot cumulative distribution function (CH-CDF) plots [22]) of 1374 S1 proteins; (**b**) separate S1 proteins groups containing different numbers of structural domains.

2.2. Analysis of Intrinsic Flexibility and Disorder of the Bacterial S1 Proteins and Its Domains.

For analysis of intrinsic flexibility and disorder of the full length bacterial S1 proteins and its separate structural domains we used the FoldUnfold (average window 11 aa and 5 aa) and IsUnstruct programs; their possibilities and accuracy were described in [26–29]. The obtained results are given in Table 1.

Analysis of the percentage of disorder in the full length S1 proteins and in their separate domains by the FoldUnfold (average window 11 aa and 5 aa) and IsUnstruct programs revealed their close similarity (Table 1).

For full-length proteins, the highest percentage of disorder was detected for four- (30%) and five-domain (30%) containing proteins using the FoldUnfold program (average window 5 aa). The smallest percentage was in the six-domain proteins (13%) when using the FoldUnfold program (average window 11 aa). This indicates the predominance of relatively short flexible or unstructured regions in the considered sequences of the proteins of this group, consistent with the fact that the binary predictor of the CH-CDF plot revealed the ordered states for 67% of proteins in this group.

Most of the separate S1 domains exhibited disorder values around 20%. The lowest percentage of disorder (except the third domain in three-domain containing proteins and the separate domains in the one-domain containing proteins) predicted by the FoldUnfold program (average window 5 aa) was the third domain in six-domain containing proteins (13%). Using the FoldUnfold program (average window 11 aa) and IsUnstruct for this domain also revealed a relatively low percentage of intrinsically disorder compared with other domains in this group and other groups (by the number of domains), 19% and 21%, respectively. The largest percentage of disorder predicted by the IsUnstruct program belonged to the sixth domain in the six-domain containing proteins (45%). Using the FoldUnfold program for six-domain containing proteins, a propensity for a more disordered state in the terminal domains was also identified. Note that, earlier, we have shown that for long S1 proteins (six-domain S1 proteins) the central part of the proteins (the third domain) is more conservative (as a percent of identity between separate domains) than the terminal domains, and apparently is vital for the activity and functionality of S1 proteins [6].

The concept of order and disorder in protein segments has often been investigated in correlation with the presence or absence of protein repeats at the sequence level. It is noticed that intrinsically disordered proteins often correspond to regions of low compositional complexity (low sequence entropy) and sometimes to repetitive sub-sequences, for example, in fibrillar proteins [30]. Also in some special cases, protein repeats (for example, in the PEVK ((Pro-Glu-Val-Lys) domain) regions of human titin, the prion proteins, or the CTD domain of RNA polymerase) are discussed in detail [31]. However, these findings on specific instances are hard to generalize. A general property observed is that a higher level of repeat perfection correlates positively with the disordered state of protein sub-chains [21].

S1 proteins, having a low degree of conservatism (not perfect repeats) [6], in addition to the found low degree of disorder within and between the domains, demonstrate the unique structural organization of proteins of this family. Apparently, the organization is closer to the formation of the quaternary structure of globular proteins, with the same structural organization of individual structural domains.

Table 1. Intrinsic flexibility and disorder of S1 protein family and its structural domains. The largest and smallest values are highlighted in bold.

Number of Structural S1 Domains		FoldUnfold (11 aa)		FoldUnfold (5 aa)		IsUnstruct	
		% Disorder for Each Domain	Full Length Proteins	% Disorder for Each Domain	Full Length Proteins	% Disorder for Each Domain	Full Length Proteins
1S1		20 ± 3	25 ± 10	17 ± 11	22 ± 13	17 ± 11	24 ± 17
2S1	1	16 ± 1	20 ± 11	13 ± 6	20 ± 5	18 ± 5	19 ± 10
	2	24 ± 10		20 ± 10		28 ± 11	
3S1	1	17 ± 1	15 ± 9	20 ± 6	26 ± 7	36 ± 13	26 ± 9
	2	21 ± 7		21 ± 7		36 ± 16	
	3	0		13 ± 6		20 ± 4	
4S1	1	21 ± 5	18 ± 5	25 ± 7	**30 ± 4**	24 ± 9	22 ± 5
	2	18 ± 1		13 ± 5		24 ± 5	
	3	21 ± 6		17 ± 8		28 ± 10	
	4	18 ± 3		16 ± 7		23 ± 7	
5S1	1	21 ± 3	17 ± 13	22 ± 12	**30 ± 11**	28 ± 16	21 ± 15
	2	21 ± 5		15 ± 8		23 ± 13	
	3	20 ± 3		22 ± 12		28 ± 13	
	4	24 ± 1		22 ± 8		35 ± 16	
	5	18 ± 2		22 ± 5		28 ± 10	
6S1	1	24 ± 9	**13 ± 4**	22 ± 8	27 ± 3	27 ± 12	16 ± 4
	2	18 ± 3		14 ± 8		22 ± 7	
	3	18 ± 4		**12 ± 6**		21 ± 3	
	4	19 ± 3		19 ± 6		24 ± 4	
	5	20 ± 5		27 ± 7		25 ± 5	
	6	22 ± 7		32 ± 9		**45 ± 19**	

2.3. Flexibility of S1 Domain in the Bacterial Proteins

Besides the ribosomal proteins, S1 domains are identified in different quantities in different archaeal, bacterial and eukaryotic proteins [5]. As we recently showed, archaeal proteins contain one copy of the S1 domain, while the number of repeats in the eukaryotic proteins varies between 1 and 15 and correlates with the protein size. In the bacterial proteins, the number of repeats is no more than 6, regardless of the protein size. To compare the obtained data on the flexibility of ribosomal proteins S1, S1 domains from some bacterial proteins [5] were investigated using the approaches described above (Table 2).

Table 2. Intrinsic flexibility and disorder of S1 domains in some bacterial proteins.

Protein Name	Source Organism	UniProt Code	Percent of Flexibility/Disorder		
			FoldUnfold (11 aa)	FoldUnfold (5 aa)	IsUnstruct
S1 domain PNPase	*E. coli*	P05055	0	17	17
Protein YhgF	*E. coli*	P46837	0	0	11
Antitermination protein NusA	*E. coli*	P0AFF6	0	36	26
Ribonuclease R	*E. coli*	P21499	13	6	27
Ribonuclease E	*E. coli*	P21513	0	20	26
Tex-like protein N-terminal domain protein	*Kingella denitrificans*	F0F1S0	0	0	13

In all proteins (Table 2, Figure 2), one S1 domain was identified and had a low degree of disorder (about 20%). It can be seen that when the size of average window of the FoldUnfold program decreases, this percentage increases, indicating the presence of flexible sections of short length in the considered proteins. This is consistent with the fact that S1 domains in these proteins are well determined by various methods (Figure 2).

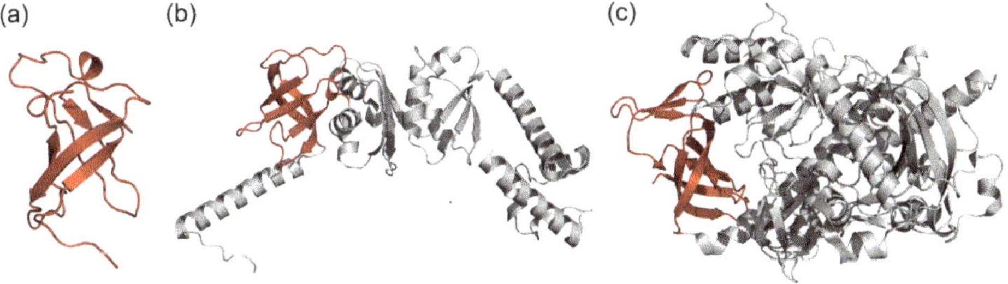

Figure 2. Protein structures with the S1 domain from different bacterial proteins. The S1 domain in each structure is highlighted with red color. (**a**) S1 domain PNPase, PDB code: 1sro; (**b**) antitermination protein NusA, PDB code: 5ml9; (**c**) Ribonuclease R, PDB code: 5xgu.

However, structures of proteins containing three or more S1 domains have not been determined yet. In the eukaryotic proteins containing more than two S1 domain (from 7 to 15) determined structures also are not available. Note that in these proteins, functions of separate S1 domains are not defined, for example, Rrp5p [32], Prp22p [33].

2.4. Analysis of the Ratio of Secondary Structures in the Bacterial S1 Proteins and Its Domains

Obtained ratios of regions connecting secondary structure according to the JPred predictions are shown in Table 3.

Table 3. Ratio of regions connecting elements of the secondary structure according to the JPred predictions.

Number of Structural S1 Domains	JPred			
	% aa in the Regions Connecting Secondary Structures (Separate Domains)		% aa in the Regions Connecting Secondary Structures (Full Length Proteins)	% aa of the Linkers between Structural Domains (Full Length Proteins)
1S1	41 ± 6		49 ± 7	45 ± 13
2S1	1	50 ± 6	47 ± 3	33 ± 13
	2	41 ± 4		
3S1	1	48 ± 3	46 ± 4	38 ± 7
	2	44 ± 10		
	3	43 ± 3		
4S1	1	47 ± 2	51 ± 2	38 ± 7
	2	47 ± 6		
	3	51 ± 4		
	4	44 ± 2		
5S1	1	49 ± 3	53 ± 4	33 ± 8
	2	51 ± 4		
	3	51 ± 5		
	4	48 ± 3		
	5	48 ± 6		
6S1	1	47 ± 2	52 ± 2	27 ± 3
	2	52 ± 5		
	3	49 ± 3		
	4	50 ± 4		
	5	51 ± 4		
	6	47 ± 3		

It can be seen that the ratio of regions connecting the secondary structure in separate domains was approximately the same and equal to about 50%, which in addition to conservative secondary structure indicates about the same organization of separate S1 domains. For full length proteins this ratio (linkers and regions connecting secondary structures within domains) was also about 50%, indicating about the same organization of multi-domains containing S1 proteins. The average percent of linkers between structural domains was about 30–40%. The obtained results are in a good agreement with the predictions of the FoldUnfold and IsUnstruct programs and CH-CDF plots, and characterized the family of S1 proteins as proteins with relatively short flexible regions within domains and between them that apparently prefer to be in the folded or MG state. In addition to the aforementioned lower conservatism between separate domains in each group, it can be argued that the unique S1 protein family is different in the classical sense from a protein with tandem repeats, such as the ANK family, leucine-rich-repeat proteins, etc. [4]. This family having repeats (separate structural domains) with 70 residues is close to a "beads-on-a-string" organization with each repeat being folded into a globular domain, for example, Zn-finger domains [34], Ig-domains [35] and the human matrix metalloproteinase [36]. Thus, one of the reasons for the absence of allowed three-dimensional structures of multi-domain S1 proteins may be the mobility of domains relative to each other due to the flexibility of interdomain linkers.

In fact, the biochemical experimental study of various fragments allowed establishing the functions of individual protein domains and parts only for the well-studied 30S ribosomal protein S1 with six S1 domain repeats from *E. coli*. For example, it has been shown that cutting one S1 domain from the C-terminus or two S1 domains from the N-terminus of the protein reduces only the effectiveness of protein functions but not its functional abilities; the sixth domain is bound with the process of autoregulation of synthesis, thus cutting off the fifth and sixth domain leads to effective participation of the remaining part of protein only in synthetic mRNA translation [37,38]. Our results indicated about the same organization of separate S1 domains and full-length proteins (conservative secondary structure, ratio of linkers and regions connecting secondary structures within domains). In addition, the percent of intrinsic flexibility is less for the central domains in the multi-domain proteins. These facts allowed us to assume that for all multi-domain S1 proteins more stable and compact domain are located in the central part and are vital for RNA interaction, while more flexible terminals domains are for other functions. The obtained results will be used as a base for investigation of the proposed theories on the evolutionary development of proteins with structural repeats: From the multi-repeat assemblies to single repeat or vice versa.

3. Materials and Methods

3.1. Construction of Ribosomal Proteins S1 Dataset

To make a representative dataset of records for the family of ribosomal proteins S1 from the UniProt database, all records for the bacteria containing any one of the keywords «30s ribosomal protein s1», «ribosomal protein s1», «30s ribosomal protein s1 (ec 1.17.1.2)», «30s ribosomal protein s1 (ribosomal protein s1)», «ribosomal protein s1 domain protein», «rna binding protein s1», «rna binding s1 domain protein», «s1 rna binding domain protein» in the protein name were selected (UniProt release 2018_04). Then the obtained array of data was used to choose only proteins encoded by the rpsA gene or its analog; for example, rpsA_1, rpsA_2, rpsA_3, etc. Only this gene, coding the ribosomal protein S1, in the European nucleotide archive (ENA, http://www.ebi.ac.uk/ena) is affiliated to the STD class, that is, the class of standard annotated sequences. From the obtained dataset records, those with six-digital identification numbers (annotated records in the UniProt database) were selected. All data were collected in one file that was the basis for further analysis, namely for collection of data on the number of structural domains and for phylogenetic grouping in the main bacterial phyla (http://bioinfo.protres.ru/other/uniprot_S1.xlsx). Records characterized by the presence of the word "candidate" were removed from our dataset. The automated advanced exhaustive analysis allowed us to choose 1374 records corresponding to these search parameters.

3.2. Number and Identification of Structural Domains in Protein Sequences

The values of the number of S1 domains corresponding to the SMART database (about 1200 domains), were selected for each analyzed record. If no data on the number of domains in one of the analyzed bases was available (None), this number was taken to be zero (these records were removed from investigated dataset). Accurate borders for each S1 domain for each record were taken from the UniProt database (position, domain and repeats field).

3.3. Prediction of Disordered Regions and Tendency for Intrinsic Disorder

3.3.1. FoldUnfold and IsUnstruct Programs

The FoldUnfold program is accessible at http://bioinfo.protres.ru/ogu/. The principle of its operation is described elsewhere [26,27]. Such a property of residues as the observed average number of contacts in a globular state, closed at a given distance, was used. To predict IDRs (intrinsically disordered regions) in the protein chain using the amino acid sequence, every residue was given an expected number of contacts in the globular state. Then averaging was done by the residue equal to the window width. The obtained average value of expected contacts was ascribed to the central residue in the chosen window. After that the window was shifted by one residue, and the procedure was repeated. On the profile of expected contacts, a boundary was marked that separated structured and unstructured residues. The mean expected number of closed residues, estimated from the sequence, was equal to the sum of expected contact residues divided by the number of amino acid residues in the protein. According to the algorithm of the program, the size of disordered (flexible) regions in such a protein must be equal to or greater than the size of the averaged window. Therefore, the number of predicted regions depended on the window size. The window size in 11 amino acid residues was optimal for the search for relatively short disordered regions in the polypeptide chain. In the case of searching for long disordered regions in partially disordered proteins, the window size must be increased to several tens of amino acid resides. At the same time, for searching for short loops one should use the averaged window size of five amino acid residues, which is optimal for this task.

The IsUnstruct program (v.2.02) is accessible at http://bioinfo.protres.ru/IsUnstruct/. The algorithm of the IsUnstruct program is based on the Ising model. For estimation the energy of any state, the energy of the border between ordered and disordered residues and the energies of initiation of disordered state at the ends were used [39]. After the optimization procedure [28], 20 energetic potentials for residues were obtained which were considered to be in a disordered state, the energy of border, and the energies of initiation of disordered state at the ends. The energy of the completely ordered state was taken to be zero.

3.3.2. CH-CDF Analysis

The charge–hydrophobicity plots (CH-plots) [40] and the cumulative distribution function (CDF) analysis [41] were used for binary prediction of protein stability based of its amino acid sequence.

The Y-coordinate in the CH-CDF plot corresponded to the distance from the obtained ordinate value to the correlation line separating the structured and unstructured conformational state of the protein on the CH (charge-hydrophobicity) plot. The X-coordinate on the CH-CDF plot corresponded to the distance from the obtained ordinate value to the correlation line separating the structured and unstructured conformational state of the protein in the CDF. Thus, in the coordinates of CH-CDF plot it was possible to assign the sequence to one of four quadrants (four conformational states). I quadrant ($CH > 0$, $CDF > 0$) were rare proteins for which it was impossible to determine accurately the state (unusual/rare); II quadrant ($CH > 0$, $CDF < 0$) were unfolded proteins (U), III quadrant ($CH < 0$, $CDF < 0$) was the state of the molten globule (MG), IV quadrant ($CH < 0$, $CDF > 0$) were structured proteins (F) [22]. Calculation of the Y-coordinate (CH-coordinate) was performed automatically. The CH coordinate values were calculated as a distance between the CH values calculated using PONDR® online service (http://www.pondr.com/) and the linear border between IDPs and structured proteins ($y = 2.743 \times x - 1.109$) [41]. Values of the X-coordinate (CDF) were the average of the vertical distances from the CDF curve to the seven boundary points. To obtain CDF-values, the version VSL2 PONDR was used [42].

3.4. Prediction of Secondary Structure

Jpred4 (http://www.compbio.dundee.ac.uk/jpred/) was used for prediction of secondary structure for each sequence in our dataset [43].

3.5. Analysis and Visualization

Algorithms of search, collection, representation and analysis by the described methods of the data were realized using the freely available programming language Python 3 (https://www.python.org/). The result of the obtained two-dimensional array of data (for CH-CDF plots) was visualized using the Matplotlib library.

4. Conclusions

In this work, we show that S1 proteins belong to a unique family, which differs in the classical sense from proteins with tandem repeats. We found that the one-domain and two-domain containing S1 proteins apparently have more stable and rigid structure. An increase in the number of structural domains contributes to the possible transition of a portion of proteins from the folded state to the MG state. For example, for three- and four-domain containing proteins, the ratio of predicted MG state is about 70%. A relatively small percentage of internal flexibility/disorder within individual structural domains could be seen as an indicator of the stability of the S1 domain as one of the OB-fold in this family. At the same time the ratio of flexibility in the separate domains apparently is related to their roles in the activity and functionality of S1. A more stable, compact and conservative central part in the multi-domain proteins is vital for RNA interaction, while terminals domains are for other functions. At the same time, an equal ratio of regions connecting the secondary structure in separate domains and between structural domains indicates about the same organization of multi-domains containing S1 proteins, as well as position and ratio of the secondary structures within separate domains. Reasons for the lack of intact 3D structure of full-length ribosomal protein S1 is not well-understood Perhaps this is due to the high mobility of domains relative to each other in the multi-domain proteins. Further investigation of the flexibility of the available 3D structures for separate S1 domains and the full length S1 domain from *E. coli* in complex with 70S ribosomal subunit will allow finding an accurate explanation.

Author Contributions: Conceptualization and experiment design, O.G., A.M., E.D.; software, A.M., M.L.; formal analysis, A.M., M.L., E.D., O.G.; data analysis, A.M., E.D., O.G.; visualization, A.M.; writing—original draft preparation, E.D. and O.G.; writing—review & editing, O.G.; supervision, O.G.

Funding: This research was funded by the RUSSIAN SCIENCE FOUNDATION, grant number 18-14-00321.

Conflicts of Interest: The authors declare no conflict of interest.

Abbreviations

OB-fold	Oligonucleotide/oligosaccharide-binding fold
CH-plot	Charge–hydrophobicity plot
CDF	Cumulative distribution function
U	Unfolded
MG	Molten globule
F	Folded
SMART	Simple Modular Architecture Research Tool
ANK	Ankyrin
Ig	Immunoglobulin

References

1. Björklund, A.K.; Ekman, D.; Elofsson, A. Expansion of protein domain repeats. *PLoS Comput. Biol.* **2006**, *2*, e114. [CrossRef] [PubMed]
2. Jernigan, K.K.; Bordenstein, S.R. Tandem-repeat protein domains across the tree of life. *PeerJ* **2015**, *3*, e732. [CrossRef] [PubMed]
3. Andrade, M.A.; Petosa, C.; O'Donoghue, S.I.; Müller, C.W.; Bork, P. Comparison of ARM and HEAT protein repeats. *J. Mol. Biol.* **2001**, *309*, 1–18. [CrossRef] [PubMed]
4. Andrade, M.A.; Perez-Iratxeta, C.; Ponting, C.P. Protein Repeats: Structures, Functions, and Evolution. *J. Struct. Biol.* **2001**, *134*, 117–131. [CrossRef] [PubMed]
5. Deryusheva, E.I.; Machulin, A.V.; Selivanova, O.M.; Galzitskaya, O. V Taxonomic distribution, repeats, and functions of the S1 domain-containing proteins as members of the OB-fold family. *Proteins* **2017**, *85*, 602–613. [CrossRef]
6. Machulin, A.; Deryusheva, E.; Selivanova, O.; Galzitskaya, O. Phylogenetic bacterial grouping by numbers of structural domains in the family of ribosomal proteins S1. *Sci. Rep.* under review.
7. Arcus, V. OB-fold domains: A snapshot of the evolution of sequence, structure and function. *Curr. Opin. Struct. Biol.* **2002**, *12*, 794–801. [CrossRef]
8. Sengupta, J.; Agrawal, R.K.; Frank, J. Visualization of protein S1 within the 30S ribosomal subunit and its interaction with messenger RNA. *Proc. Natl. Acad. Sci. USA* **2001**, *98*, 11991–11996. [CrossRef]
9. Salah, P.; Bisaglia, M.; Aliprandi, P.; Uzan, M.; Sizun, C.; Bontems, F. Probing the relationship between gram-negative and gram-positive S1 proteins by sequence analysis. *Nucleic Acids Res.* **2009**, *37*, 5578–5588. [CrossRef]
10. Giraud, P.; Créchet, J.-B.; Uzan, M.; Bontems, F.; Sizun, C. Resonance assignment of the ribosome binding domain of E. coli ribosomal protein S1. *Biomol. NMR Assign.* **2015**, *9*, 107–111. [CrossRef]

11. Loveland, A.B.; Korostelev, A.A. Structural dynamics of protein S1 on the 70S ribosome visualized by ensemble cryo-EM. *Methods* **2018**, *137*, 55–66. [CrossRef]

12. Beckert, B.; Turk, M.; Czech, A.; Berninghausen, O.; Beckmann, R.; Ignatova, Z.; Plitzko, J.M.; Wilson, D.N. Structure of a hibernating 100S ribosome reveals an inactive conformation of the ribosomal protein S1. *Nat. Microbiol.* **2018**, *3*, 1115–1121. [CrossRef]

13. Bycroft, M.; Hubbard, T.J.; Proctor, M.; Freund, S.M.; Murzin, A.G. The solution structure of the S1 RNA binding domain: a member of an ancient nucleic acid-binding fold. *Cell* **1997**, *88*, 235–242. [CrossRef]

14. Schubert, M.; Edge, R.E.; Lario, P.; Cook, M.A.; Strynadka, N.C.J.; Mackie, G.A.; McIntosh, L.P. Structural characterization of the RNase E S1 domain and identification of its oligonucleotide-binding and dimerization interfaces. *J. Mol. Biol.* **2004**, *341*, 37–54. [CrossRef]

15. Beuth, B.; Pennell, S.; Arnvig, K.B.; Martin, S.R.; Taylor, I.A. Structure of a Mycobacterium tuberculosis NusA-RNA complex. *EMBO J.* **2005**, *24*, 3576–3587. [CrossRef]

16. Battiste, J.L.; Pestova, T.V.; Hellen, C.U.; Wagner, G. The eIF1A solution structure reveals a large RNA-binding surface important for scanning function. *Mol. Cell* **2000**, *5*, 109–119. [CrossRef]

17. Giri, L.; Subramanian, A.R. Hydrodynamic properties of protein S1 from Escherichia coli ribosome. *FEBS Lett.* **1977**, *81*, 199–203. [CrossRef]

18. Laughrea, M.; Moore, P.B. Physical properties of ribosomal protein S1 and its interaction with the 30 S ribosomal subunit of Escherichia coli. *J. Mol. Biol.* **1977**, *112*, 399–421. [CrossRef]

19. Labischinski, H.; Subramanian, A.R. Protein S1 from Escherichia coli ribosomes: an improved isolation procedure and shape determination by small-angle X-ray scattering. *Eur. J. Biochem.* **1979**, *95*, 359–366. [CrossRef]

20. Sillers, I.Y.; Moore, P.B. Position of protein S1 in the 30 S ribosomal subunit of Escherichia coli. *J. Mol. Biol.* **1981**, *153*, 761–780. [CrossRef]

21. Jorda, J.; Xue, B.; Uversky, V.N.; Kajava, A. V Protein tandem repeats - the more perfect, the less structured. *FEBS J.* **2010**, *277*, 2673–2682. [CrossRef]

22. Huang, F.; Oldfield, C.; Meng, J.; Hsu, W.L.; Xue, B.; Uversky, V.N.; Romero, P.; Dunker, A.K. Subclassifying disordered proteins by the CH-CDF plot method. *Pac. Symp. Biocomput.* **2012**, 128–139. [CrossRef]

23. Schindelin, H.; Jiang, W.; Inouye, M.; Heinemann, U. Crystal structure of CspA, the major cold shock protein of Escherichia coli. *Proc. Natl. Acad. Sci. USA* **1994**, *91*, 5119–5123. [CrossRef]

24. Heikinheimo, P.; Tuominen, V.; Ahonen, A.K.; Teplyakov, A.; Cooperman, B.S.; Baykov, A.A.; Lahti, R.; Goldman, A. Toward a quantum-mechanical description of metal-assisted phosphoryl transfer in pyrophosphatase. *Proc. Natl. Acad. Sci. USA* **2001**, *98*, 3121–3126. [CrossRef] [PubMed]

25. Delarbre, L.; Stevenson, C.E.; White, D.J.; Mitchenall, L.A.; Pau, R.N.; Lawson, D.M. Two crystal structures of the cytoplasmic molybdate-binding protein ModG suggest a novel cooperative binding mechanism and provide insights into ligand-binding specificity. *J. Mol. Biol.* **2001**, *308*, 1063–1079. [CrossRef] [PubMed]

26. Galzitskaya, O.V.; Garbuzynskiy, S.O.; Lobanov, M.Y. Prediction of amyloidogenic and disordered regions in protein chains. *PLoS Comput. Biol.* **2006**, *2*, 10. [CrossRef]

27. Galzitskaya, O.V.; Garbuzynskiy, S.O.; Lobanov, M.Y. FoldUnfold: Web server for the prediction of disordered regions in protein chain. *Bioinformatics* **2006**, *22*, 2948–2949. [CrossRef] [PubMed]

28. Lobanov, M.Y.; Galzitskaya, O.V. The Ising model for prediction of disordered residues from protein sequence alone. *Phys. Biol.* **2011**, *8*, 035004. [CrossRef]

29. Deryusheva, E.; Machulin, A.; Nemashkalova, E.; Glyakina, A.; Galzitskaya, O. Search for functional flexible regions in the G-protein family: new reading of the FoldUnfold program. *Protein Pept. Lett.* **2018**, *25*, 589–598. [CrossRef]

30. Dunker, A.K.; Lawson, J.D.; Brown, C.J.; Williams, R.M.; Romero, P.; Oh, J.S.; Oldfield, C.J.; Campen, A.M.; Ratliff, C.M.; Hipps, K.W.; et al. Intrinsically disordered proteins. *J. Mol. Graph. Model.* **2001**, *19*, 26–59. [CrossRef]

31. Tompa, P.; Fersht, A. *Structure and Function of Intrinsically Disordered Proteins*; Chapman and Hall/CRC: New York, NY, USA, 2010.

32. Hierlmeier, T.; Merl, J.; Sauert, M.; Perez-Fernandez, J.; Schultz, P.; Bruckmann, A.; Hamperl, S.; Ohmayer, U.; Rachel, R.; Jacob, A.; et al. Rrp5p, Noc1p and Noc2p form a protein module which is part of early large ribosomal subunit precursors in S. cerevisiae. *Nucleic Acids Res.* **2013**, *41*, 1191–1210. [CrossRef] [PubMed]

33. Mayas, R.M.; Maita, H.; Staley, J.P. Exon ligation is proofread by the DExD/H-box ATPase Prp22p. *Nat. Struct. Mol. Biol.* **2006**, *13*, 482–490. [CrossRef]

34. Lee, M.S.; Gippert, G.P.; Soman, K.V.; Case, D.A.; Wright, P.E. Three-dimensional solution structure of a single zinc finger DNA-binding domain. *Science* **1989**, *245*, 635–637. [CrossRef]

35. Sawaya, M.R.; Wojtowicz, W.M.; Andre, I.; Qian, B.; Wu, W.; Baker, D.; Eisenberg, D.; Zipursky, S.L. A double S shape provides the structural basis for the extraordinary binding specificity of Dscam isoforms. *Cell* **2008**, *134*, 1007–1018. [CrossRef]

36. Elkins, P.A.; Ho, Y.S.; Smith, W.W.; Janson, C.A.; D'Alessio, K.J.; McQueney, M.S.; Cummings, M.D.; Romanic, A.M. Structure of the C-terminally truncated human ProMMP9, a gelatin-binding matrix metalloproteinase. *Acta Crystallogr. D Biol. Crystallogr.* **2002**, *58*, 1182–1192. [CrossRef] [PubMed]

37. Amblar, M.; Barbas, A.; Gomez-Puertas, P.; Arraiano, C.M. The role of the S1 domain in exoribonucleolytic activity: substrate specificity and multimerization. *Rna* **2007**, *13*, 317–327. [CrossRef]

38. Boni, I.V.; Artamonova, V.S.; Dreyfus, M. The last RNA-binding repeat of the Escherichia coli ribosomal protein S1 is specifically involved in autogenous control. *J. Bacteriol.* **2000**, *182*, 5872–5879. [CrossRef] [PubMed]

39. Lobanov, M.Y.; Sokolovskiy, I.V.; Galzitskaya, O.V. IsUnstruct: prediction of the residue status to be ordered or disordered in the protein chain by a method based on the Ising model. *J. Biomol. Struct. Dyn.* **2013**, *31*, 1034–1043. [CrossRef]

40. Kyte, J.; Doolittle, R.F. A simple method for displaying the hydropathic character of a protein. *J. Mol. Biol.* **1982**, *157*, 105–132. [CrossRef]

41. Xue, B.; Oldfield, C.J.; Dunker, A.K.; Uversky, V.N. CDF it all: Consensus prediction of intrinsically disordered proteins based on various cumulative distribution functions. *FEBS Lett.* **2009**, *583*, 1469–1474. [CrossRef]

42. Pace, C.N.; Vajdos, F.; Fee, L.; Grimsley, G.; Gray, T. How to measure and predict the molar absorption coefficient of a protein. *Protein Sci.* **1995**, *4*, 2411–2423. [CrossRef] [PubMed]

43. Drozdetskiy, A.; Cole, C.; Procter, J.; Barton, G.J. JPred4: a protein secondary structure prediction server. *Nucleic Acids Res.* **2015**, *43*, W389–394. [CrossRef] [PubMed]

© 2019 by the authors. Licensee MDPI, Basel, Switzerland. This article is an open access article distributed under the terms and conditions of the Creative Commons Attribution (CC BY) license (http://creativecommons.org/licenses/by/4.0/).

International Journal of
Molecular Sciences

Article

Raman Evidence of p53-DBD Disorder Decrease upon Interaction with the Anticancer Protein Azurin

Sara Signorelli, Salvatore Cannistraro * and Anna Rita Bizzarri

Biophysics & Nanoscience Centre, DEB, Università della Tuscia, 01100 Viterbo, Italy; signorellis@unitus.it (S.S.); bizzarri@unitus.it (A.R.B.)
* Correspondence: cannistr@unitus.it; Tel.: +39-0761-357136

Received: 15 May 2019; Accepted: 20 June 2019; Published: 24 June 2019

Abstract: Raman spectroscopy, which is a suitable tool to elucidate the structural properties of intrinsically disordered proteins, was applied to investigate the changes in both the structure and the conformational heterogeneity of the DNA-binding domain (DBD) belonging to the intrinsically disordered protein p53 upon its binding to Azurin, an electron-transfer anticancer protein from *Pseudomonas aeruginosa*. The Raman spectra of the DBD and Azurin, isolated in solution or forming a complex, were analyzed by a combined analysis based on peak inspection, band convolution, and principal component analysis (PCA). In particular, our attention was focused on the Raman peaks of Tyrosine and Tryptophan residues, which are diagnostic markers of protein side chain environment, and on the Amide I band, of which the deconvolution allows us to extract information about α-helix, β-sheet, and random coil contents. The results show an increase of the secondary structure content of DBD concomitantly with a decrease of its conformational heterogeneity upon its binding to Azurin. These findings suggest an Azurin-induced conformational change of DBD structure with possible implications for p53 functionality.

Keywords: Raman spectroscopy; p53; intrinsically disordered protein; blue copper protein Azurin; protein–protein interaction; Amide I band deconvolution; principal component analysis

---

1. Introduction

p53 is an important tumor suppressor protein working as a central hub in a complex interaction network in which it regulates numerous cellular processes, including cell cycle progression, apoptosis induction, and DNA repair [1,2]. p53 is a member of the important class of intrinsically disordered proteins (IDPs), possessing both structured and disordered domains under physiological conditions and different conformations coexisting in solution [3]. Such a structural plasticity confers to IDP an extremely high conformational adaptability, allowing them to act according to functional modes not achievable by ordered proteins, with these properties having been recently exploited to develop engineered protein and peptide drugs [4–6].

p53 is a tetrameric protein composed of four identical subunits and acts as a transcription factor. Each monomer of p53 consists of an N-terminal transactivation domain (NTD), a C-terminal domain (CTD), and a core DNA-binding domain (DBD) [7–10] The presence of unstructured portions allows p53 to adopt widely different conformations, which are at the basis of a vast repertoire of available interactions to different biological partners [11]. Among them, Azurin (AZ), a copper-containing electron-transfer anticancer protein secreted by *Pseudomonas aeruginosa* bacteria, has demonstrated the ability to specifically bind to p53, leading both to its stabilization and to an intracellular level increase both in vitro and in vivo [12–17]. Therefore, the formation of the p53-AZ complex has opened new perspectives in cancer treatment, such as the development of an AZ-derived anticancer peptide [18].

Keeping in mind the crucial role of AZ in assisting the oncosuppressive function of p53, in our group, we investigated the interaction between p53 and AZ at the single molecule level by Atomic Force

Microscopy (AFM) and Atomic Force Spectroscopy (AFS) and by computational approaches [12,19–21]. These studies have provided information about the interaction kinetics between p53 or its DBD and AZ, obtaining also some relevant insights on the possible binding sites [21]. However, no experimental evidences on possible structural alterations of p53 upon its binding to AZ are so far available [3]. In this respect, Raman spectroscopy represents a suitable approach to extract information about the secondary structure of proteins as well as to probe their conformational heterogeneity, including IDPs [22]. Indeed, we have previously applied such a technique to investigate the structure and the conformational heterogeneity of wild-type and mutants p53 and, also, of the AZ-derived anticancer p28 peptide, even in different environmental conditions [18,23,24].

In the present work, we have employed a Raman-based approach to investigate if and how the native conformation of DBD is modified by its interaction with AZ. To such an aim, we have focused on an accurate inspection of the Fermi doublets relative to Tyrosine (830 and 850 cm^{-1}; Tyr) and of Tryptophan peaks (1340 and 1360 cm^{-1}), with these Raman signals having been recognized as suitable diagnostic markers of protein side chain environment [25,26]. Additionally, we have investigated the Amide I Raman band (1600–1700 cm^{-1}), of which the deconvolution has demonstrated to be particularly effective in both extracting conformational information (α-helix, β-sheet, and random coil motifs) and which is a reliable reporter on the structural heterogeneity of proteins [22,27–32]. The Raman spectra have also been analyzed by applying principal component analysis (PCA), which performs a dimensionality reduction of the spectra, allowing a revelation of the differences between the complex Raman spectra of the samples and helping to understand the principal factors affecting the spectral variation [33].

The combination of these approaches has put into evidence the occurrence of structural changes within p53DBD upon its interaction with AZ. In particular, passing from isolated DBD to DBD bound to AZ, we found a variation in Tyrosine (Tyr) and Tryptophan (Trp) residues hydrophobicity and an increase of the DBD secondary structure concomitantly with a significant reduction of the conformational heterogeneity. The observed changes in both the structure and conformational heterogeneity of DBD strongly support the ability of AZ to modulate the DBD structure, and this, in turn, may result in a stabilization of the oncosuppressive function of p53.

2. Results and Discussion

2.1. Raman Analysis of AZ and DBD

Figure 1 shows the Raman spectra of AZ and DBD in the 600–1725 cm^{-1} frequency range. The spectra display a complex set of bands arising from the modes of the aromatic amino acids (Tyr, Trp, and Phenylalanine (Phe) and of the peptide backbone, consistent with the typical Raman spectra of proteins [27,34]. The assignments of the main peaks are summarized in Table 1 [27].

Table 1. Typical proteins' Raman vibrational modes (Raman cm^{-1}) and related assignments.

Raman (cm^{-1})	Assignment
643	Tyr
805	Tyr
830,850	Tyr
870	Trp
902	ν_{CC}
930,980	ν_{CCN}
1001	Phe
1103	$\nu_{CC}, \nu_{CN}, \nu_{CO}$
1127	ν_{CC}

Table 1. *Cont.*

Raman (cm^{-1})	Assignment
1174	Tyr
1180	Phe
1210	Tyr
1230–1240	Amide III (α-helices)
1250–1255	Amide III (β-sheets)
1270–1300	Amide III (Random coils)
1320	CH_2 deformation
1340,1360	Trp
1403	Symmetric $\nu_{co2}{}^{-}$
1424	CH_2, CH_3 deformation
1451	CH_2, CH_3 deformation
1552	Trp
1604	Phe
1615	Tyr
1650–1680	Amide I

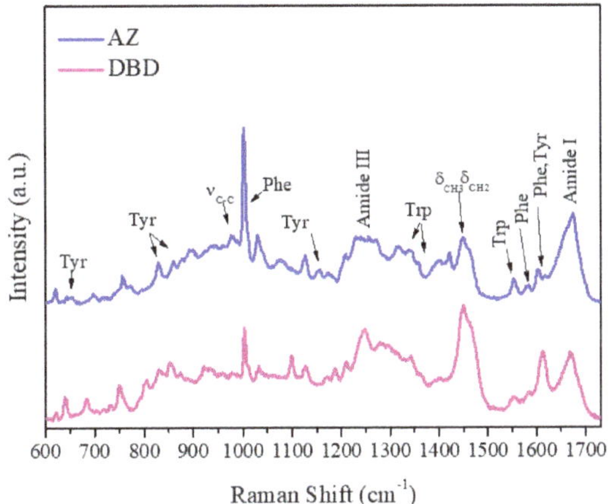

Figure 1. Raman spectra (600–1730 cm^{-1}) with excitation at 532 nm of Azurin (AZ; blue) and DNA-binding domain (DBD; magenta) in Phosphate Buffer Solution (PBS): The principal proteins' vibrational modes are marked. Spectra were normalized in the all spectrum frequency region and baseline corrected for a better visualization.

Among the main Raman markers, we focused our attention on the Raman peaks of Tyr and Trp residues, which allow the extraction of information on protein side-chain local environment and on the Raman band of Amide I, which provides a diagnostic of the protein secondary structure.

Concerning the Tyr residues, the ratio $I_Y = I_{850}/I_{830}$ between the intensity of doublet peaks at 850 and 830 cm^{-1} is related to the donor or acceptor role of the Tyr phenoxyl group. Specifically, a low I_Y value (around 0.3) indicates the phenolic hydroxyl (OH) group acting as a strong hydrogen bond donor, as occurring for buried tyrosine residues. As the I_Y value increases (until 2.5), the phenolic oxygen becomes a stronger hydrogen bond acceptor, while a largely enhanced value ($I_Y > 6.7$) represents a non-hydrogen-bonded state [25,26]. Experimental results on isolated AZ reveal an I_Y value of 0.38 ± 0.07, representative of a buried environment for its two Tyr residues (Tyr^{72} and Tyr^{108}) in agreement with the X-ray structure of AZ, in which Tyr^{72} belongs to the peripheral α-helix region with a moderate solvent accessibility and Tyr^{108} is practically inaccessible to solvent (see Figure 2A) [35].

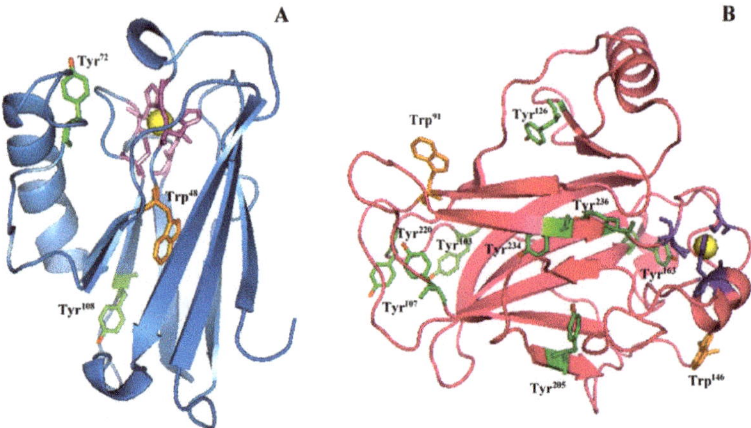

Figure 2. Three-dimensional structures of (**A**) AZ (PDB code: 4AZU) and (**B**) DBD (PDB code: 2XWR): The active site of AZ and the zinc-finger of the DBD are shown as yellow ball and stick models. The aromatic residues are Tyr (green) and Trp (orange). The OH groups in Tyr residues are marked in red.

Isolated DBD exhibits an I_Y ratio of 1.37 ± 0.16, indicating a predominant exposition to the solvent surfaces of the eight Tyr residues. From X-ray structure, the phenolic OH groups of Tyr[103] and Tyr[107] are highly oriented towards the solvent (see Figure 2B) [7]. Moreover, Tyr[126] and Tyr[205], located at a crucial protein region interfacing with the DNA, show moderate accessibility, similar to that of Tyr[220] located on the surface of the protein [7,8]. Finally, the remaining Tyr[163], Tyr[234], and Tyr[236] are almost inaccessible to the solvent [8]. Therefore, our results are consistent with the X-ray data endorsing the DBD–Tyr high solvent exposition.

Further information about the side chains can be achieved by analyzing the Fermi doublet bands of Trp residues at 1340 and 1360 cm^{-1}, which are reporters of the hydrophobicity/hydrophilicity neighboring the Trp indole ring [32]. In particular, an intensity ratio $I_W = I_{1360}/I_{1340}$ smaller than 1.0 reflects a hydrophilic environment, while a ratio greater than 1.0 indicates a hydrophobic one [9,32].

For AZ, we found an I_W ratio of 1.54 ± 0.10, indicative of a buried and solvent inaccessible environment for the lone Trp residue. This is in accordance with the AZ X-ray data, showing that the Trp[48] is deeply embedded in a highly hydrophobic core and surrounded by a closely packed β barrel structure (Figure 2A) [35].

We found for DBD an I_W ratio of 0.68 ± 0.10, which implies, on average, a moderate hydrophilic environment for its Trp[91] and Trp[146]. The latter is positioned in a hydrophobic side chain and oriented towards the solvent, while the former is located at the N-terminus of DBD and displays a high solvent accessibility, as it comes out from the X-ray data (Figure 2B) [36]. However, Trp[91] has been shown to be crucially involved in the packing process of DBD through interaction with the Arg[174] residue, which reduces its solvent exposure [36]. Therefore, our data suggest that both Trp residues in DBD globally experience a hydrophilic environment.

Information on protein secondary structure can be extracted by the Amide I band (1600–1700 cm^{-1}), mainly arising from C=O stretching and the combination of the C–N stretching, the Cα-C–N bending, and the N–H in-plane bending modes of peptide group. Such a band is usually used as a marker for secondary structure components. In particular, when Amide I band is centered at 1655 cm^{-1}, it indicates a prevailing α-helix conformational arrangement, while a shift of this band peak toward 1670 cm^{-1} is indicative of β-sheet conformation [22]. On the other hand, an analysis of the Amide I shape means an appropriate deconvolution strategy allows for the quantification of the percentage content of secondary structure components present in the protein [22,28].

Specifically, the Amide I band of AZ emerges at about 1670 cm^{-1} (Figure 3A), suggesting a predominant β-sheet conformation [37]. The curve-fitting procedure points out β-sheet conformations predominant for 60%, while the α-helices and random coils account for about 22% and 18%, respectively. The obtained AZ secondary structure is agreement with that determined by X-ray diffraction for the crystals of AZ. Indeed, the major AZ components are β strands and turns (≈69%), which form two sheets arranged in a Greek key motif and with a minor contribution from a rigid α-helix (about 31%), conferring to AZ a low level of flexibility and structural disorder (see Figure 3A) [21,35,38].

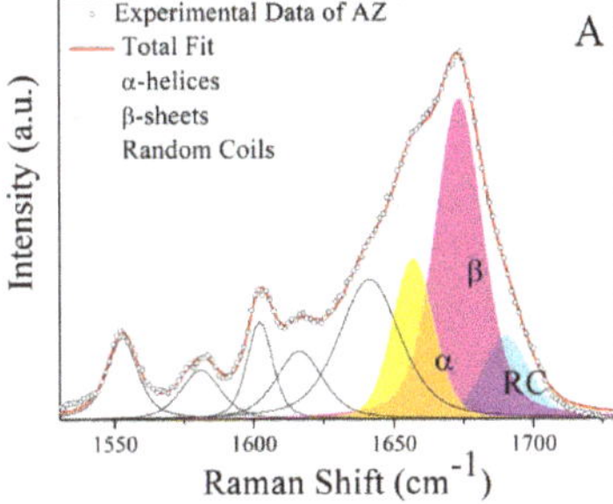

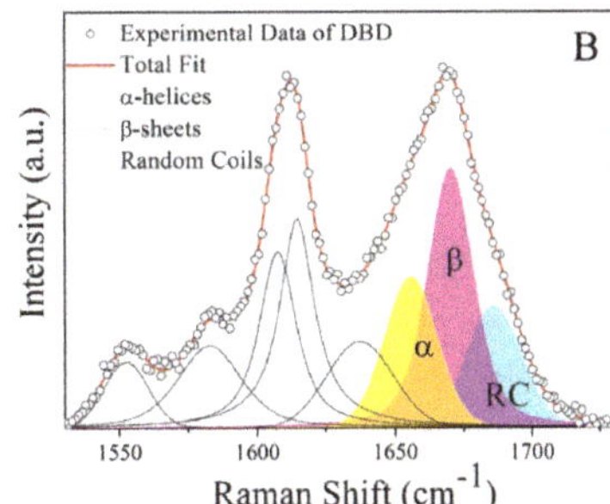

Figure 3. The Amide I band of AZ (**A**) and DBD (**B**) in PBS: The percentage of secondary structure for these proteins has been estimated from the relative area of deconvoluted bands of this spectral region of which the fitting parameters are reported in Table 2.

Table 2. Assignments, relative central frequency (Raman cm^{-1}), and integrated intensities (Area %) of the main Amide I band components (α-helix, β-sheet, or random coil) for AZ, DBD, and DBD:AZ complex obtained by a fitting procedure. $\chi2 = 0.002$ for all curve fitting analysis.

Sample In PBS	Secondary Structure	Raman Shift cm^{-1}	Area (%)
	α-helix	1659	22
AZ	β-sheet	1674	60
	random coil	1688	18
	α-helix	1655	25
DBD	β-sheet	1670	46
	random coil	1686	29
	α-helix	1655	26
DBD:AZ	β-sheet	1669	51
	random coil	1687	23

Concerning DBD (Figure 3B), the band corresponding to the β-structures provides 46% of the total, while those related to α-helix and to random coils have 25% and 29%, respectively. These results indicate that DBD is characterized by a partially ordered structure, combined with the presence of significant disordered regions. Additionally, the results confirm those reported in our recent study on different sample batches of DBD (aminoacids 81–300), from which a content of 27% and 50% for α-helical and β conformations, respectively, have been estimated [23]. Moreover, these data are in agreement with X-ray data indicating a 30% of β-arrangement with an 18% of α-structures (see Figure 3B) [36]. The DBD propensity to adopt a predominant β-conformation is actually related to the large presence in its sequence of hydrophobic residues, such as Cysteine (Cys), Trp, and Leucine (Leu), generally promoting an ordered structure [39].

2.2. Raman Analysis of the DBD:AZ Complex

The previous analysis on the Raman spectra of AZ and DBD proteins, isolated in solution, has provided information on their structural properties paving the way to investigate possible structural change when they are involved in the formation of a complex. The spectrum of DBD:AZ solution, obtained by mixing equimolar amounts of DBD and AZ in the 600–1725 cm^{-1} frequency region is shown in Figure 4. We note almost the same general features displayed by the isolated protein spectra with no significant shifts in frequency for the main vibrational modes. From the Tyr peaks visible at 828 and 854 cm^{-1}, the Fermi doublet ratio I_Y is 0.58 ± 0.08, which is indicative of a predominant hydrophobic environment. Such a value is closer to that of AZ (I_Y = 0.38 ± 0.07) with respect to that of DBD (I_Y = 1.37 ± 0.16), suggesting some changes in the Tyr microenvironment resulting from the interaction between the two biomolecules. To further support such a hypothesis, we have analyzed the Raman spectrum obtained by directly summing the spectra of isolated DBD and AZ molecules acquired at the same concentration used to form the complex (Figure 4), with the resulting spectrum being called added spectrum (AS) in the following. The analysis of the Tyr peaks in the AS spectrum reveals an I_Y of 1.11 ± 0.18, which is indicative of an average hydrophilic environment for all the Tyr residues in the system, as expected for isolated proteins. Therefore, the marked differences between the I_Y values from the complex and AS spectra can be ascribed to changes due to the interaction between the molecules. Although the spectroscopic results alone cannot allow us to identify the Tyrs that are involved in the structural changes, they support literature data that point out the involvement of the S_7–S_8 loops, comprising Tyr^{220}, Tyr^{234}, and Tyr^{236} and also Tyr^{126} at DBD binding sites with AZ, which, in turn, is engaged through its a.a 50–77 fragment, including Tyr^{72} [40].

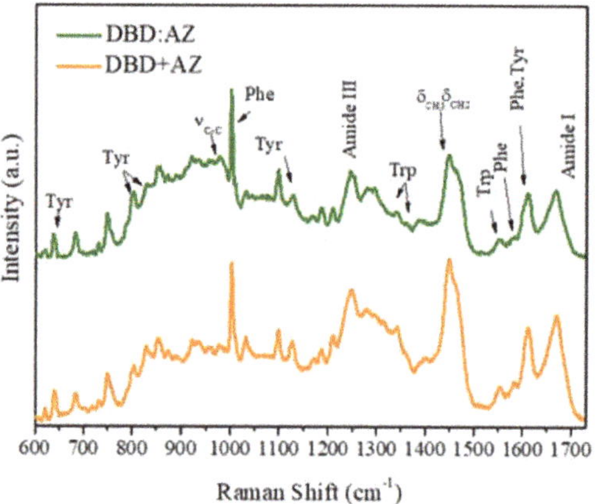

Figure 4. Comparison among the 532-nm-excited Raman spectra (600–1730 cm^{-1}) of DBD:AZ complex (green) and of added spectrum AS (orange) in PBS: The principal proteins' vibrational modes are marked. Spectra were normalized in the all spectrum frequency region and baseline corrected for a better visualization.

The spectrum of DBD:AZ shows that the Trp Raman peaks are located at the same frequencies as in the isolated proteins, with an I_W ratio of 1.15 ± 0.17, indicating a high hydrophobicity for the three Trps residues (AZ–Trp^{48} and DBD–Trp^{91}/Trp^{146}). The found value of I_W for DBD:AZ, slightly lower than that for AZ (I_W = 1.54 ± 0.10) and higher than both isolated DBD (I_W = 0.68 ± 0.10) and AS (I_W = 0.53 ± 0.10), suggests some modifications in the environment experienced by these residues upon complex formation. The AZ–Trp^{48} is well-known to be strongly buried in the central hydrophobic core of the AZ; therefore, these changes can be due to variations in the DBD–Trp neighboring. Additionally, since the DBD–Trp^{91} has been shown to be engaged with the Arg^{174} [36], we suggest that the observed modifications of DBD as due to AZ interactions occurring within the DBD–Trp^{146} environment, with this being in agreement with Docking and Molecular Dynamics (MD) data showing the involvement of Trp^{146} in AZ-binding site [40].

Figure 5A,B shows the fitted curves of the Amide I band for the DBD:AZ and AS spectra, respectively. The results of the best fit for both experimental and AS Amide I bands, obtained by applying the same method used for isolated molecules, are reported in Table 2. DBD:AZ shows a predominant contribution from β-sheet structures (51%) and an α-helix amount of about 26%, while the random coil conformations contributes to 23% of the total Amide I band area. Best fit of AS reveals a predominant β structure (41%) with α-helices and random coils percentages of 31% and 28%, respectively. The observed changes in the secondary structure composition in DBD:AZ with respect to those of DBD and AZ can be attributed to the interaction between these proteins. Furthermore, since AZ is characterized by a highly structured conformation, the decrease of random coil structures can be mainly attributed to DBD. Such a result is supported by previously reported molecular dynamics simulations and docking studies showing that the binding of AZ at the peripheral, unstable, L_1 and S_7–S_8 loops of DBD can enhance their stability upon restraining their flexibility [21], with this being in agreement with a reduction of the DBD disordered regions upon binding to AZ. Accordingly, it could be hypothesized that the increase of structural stability of DBD could be at the basis of the anticancer effect exerted by AZ. Since the structural dynamics and the interactions between proteins are strictly connected, a deeper characterization of the structural–functional relations is of fundamental interest for developing AZ-based drugs, of which effective action in vivo requires, however, further validations [41].

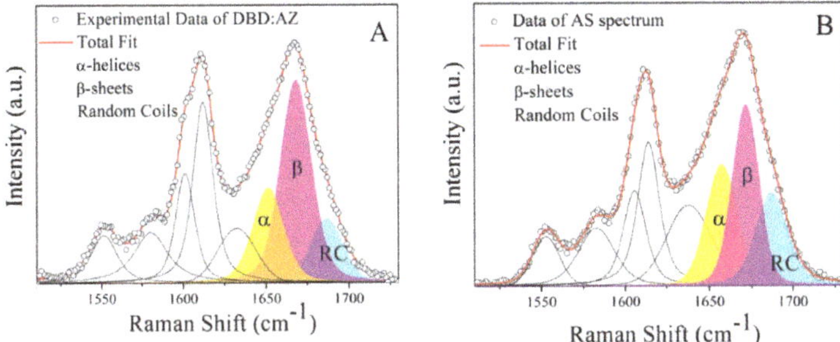

Figure 5. Amide I Raman band (open circles) excited at 532 nm of (**A**) DBD:AZ and (**B**) AS spectra, fitted through the Levenberg–Marquardt minimization algorithm (LMA; red line) in which the AZ total fit has been imposed as a constraint: Solid curves indicate the main structural conformations (α-helices, β-sheets, and random coils). Fitting results are summarized in Table 2.

2.3. Principal Component Analysis of DBD, AZ, and the DBD:AZ Complex

Different combinations of scores for the first three principal components (PC1, PC2, and PC3) have been used to build two-dimensional plots; in the following, the components releasing the highest structural information will be shown. Figure 6A shows the PCA scores of PC1 vs. PC2 components (providing about the 90% of the total variance) for the Fermi Doublet region relative to Tyr residues (790–870 cm^{-1}) for the AZ and DBD isolated molecules and for the DBD:AZ complex. In the scatter plot, two distinct groupings along the PC1 axis can be identified (see the ellipses drawn as a guide). Indeed, the AZ scores (blue symbols) are located in the positive portion of the plot along PC1 with a low spread along both the axes (10 and 5 along PC1 and PC2, respectively), while the DBD and DBD:AZ scores (magenta and green symbols, respectively) occupy the negative range of PC1 values. Along PC1, a larger variability is detected for DBD with respect to AZ and DBD:AZ. Additionally, along PC2, for AZ and DBD, negative values of PC2 are obtained while positive values are detected for DBD. To correlate the position of the scores in the plot with the samples' spectral features, we have analyzed the loadings with the variables mostly contributing to the PCA scores. As shown in Figure 6B,C, high levels of variance are detected in correspondence with the peaks at 829 cm^{-1} and 851 cm^{-1} related to the Fermi Doublet of Tyr modes and with a weaker peak at 805 cm^{-1}, associated with Tyr [25,32,42];

the latter provides the largest variance for PC2 loadings (see Figure 6B) [42]. These results show that Tyr vibrational modes are responsible for sample differentiation in PCA, consistent with our previous study supporting the important role played by Tyr modes as structural markers [25,42].

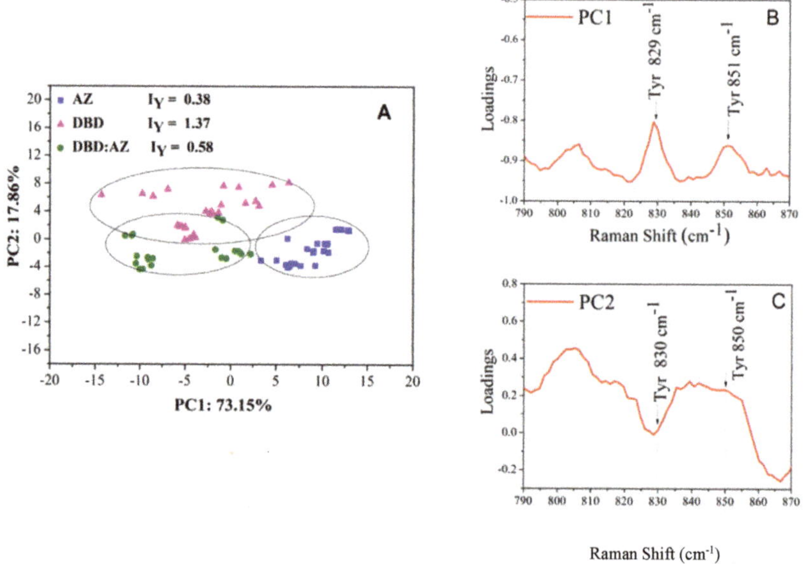

Figure 6. (**A**) Two-dimensional scores plot PC1 versus PC2 of the Raman spectra for AZ (blue squares), DBD (magenta triangles), and DBD:AZ complex (green circles) in the PBS performed on Fermi Doublet of Tyr region (790–870 cm^{-1}): The three groupings are indicated by ellipses. The Fermi Doublet ratio for Tyr residues are also reported. (**B**) PC1 (73% of total variance) and (**C**) PC2 (18% of total variance) one-dimensional loadings plot versus frequency. The Raman markers are indicated.

We then applied PCA to the Fermi Doublet region relative to Trp residues (1310–1380 cm^{-1}), with the PC1 and PC2 components providing about 91% of the total variance (see Figure 7A). AZ clusters at the upper side, DBD clusters at the middle, while DBD:AZ clusters at the lower region in correspondence to negative values of PC2 axis. Along PC1, DBD, and DBD:AZ, scores are mixed within an overlapped cloud, while AZ are well-clustered in a well-separated group. Concerning the loading plots, shown in Figure 8B,C, PC2 presents a broad band with a loading positive value of about 0.2. At 1360 cm^{-1}, a single evident peak emerges as ascribed to one of the Fermi Doublet of the Trp modes. This suggests that the separation among the three groups along PC2 depends on Trp vibrational modes. Since such a frequency changes according to the different Trp side-chain environment taken into consideration [28], a different spatial arrangement of this residue should be envisaged in the DBD isolated molecule and in DBD:AZ complex.

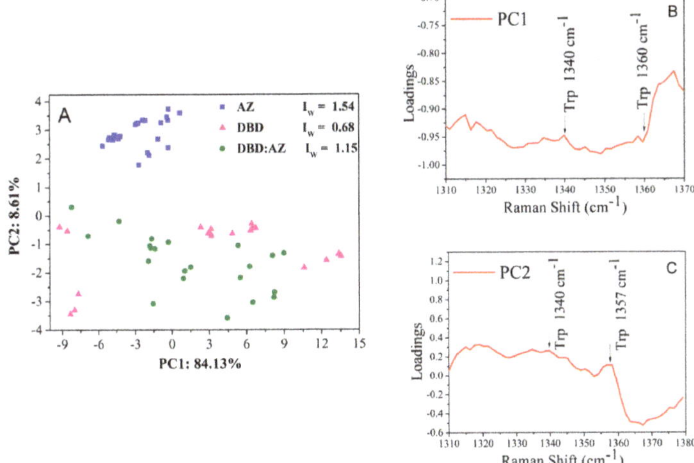

Figure 7. (**A**) Two-dimensional scores plot PC1 versus PC2 of the Raman spectra for AZ (blue squares), DBD (magenta triangles), and DBD:AZ complex (green circles) in the PBS performed on Fermi Doublet of Trp region (1310–1380 cm^{-1}): The Fermi Doublet ratio for Tyr residues are also reported. (**B**) PC1 (84% of total variance) and (**C**) PC2 (8% of total variance) one-dimensional loadings plot versus frequency. The Raman markers are indicated.

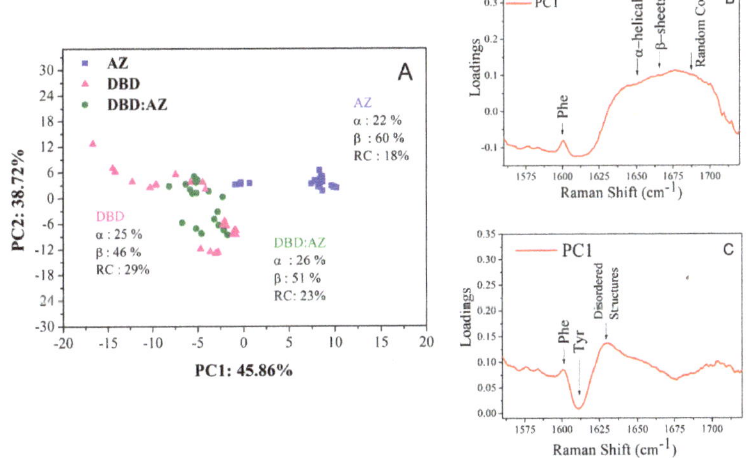

Figure 8. (**A**) Two-dimensional scores plot PC1 versus PC2 of the Raman spectra for AZ (blue squares), DBD (magenta triangles), and DBD:AZ complex (green circles) in the PBS performed on the Amide I band (1560–1720 cm^{-1}): The secondary structure percentages as obtained by the curve fitting analysis are also reported. (**B**) PC1 (46% of total variance) and (**C**) PC2 (39% of total variance) one-dimensional loadings plot versus frequency. The Raman markers are indicated.

Finally, the PCA was performed on the Amide I Raman band of AZ, DBD, and the DBD:AZ complex. From the scatter plot of PC1 versus PC2 (see Figure 8A), AZ data (blue squares) cluster at the positive side of the PC1 axis, with a low variance along both of the components, while DBD (magenta triangles) and DBD:AZ data (green circles) are characterized by negative values of PC1, with some overlap between them. A significantly larger variability is detected in DBD with respect to that of DBD:AZ.

The PC1 loading curve, accounting for 46% of the variance (Figure 8B), is characterized by a very broad band from 1625 to 1730 cm^{-1}, including all the frequencies related to the secondary structure of a protein. Such a band shows the lowest value for the 1650 cm^{-1} frequency and the highest for the 1680 cm^{-1} one, which are associated to the α-helices and disordered structures, respectively [19,24]. In PC2, accounting for a 39% of the variation in the spectra, the major source of variance comes from peak at 1627 cm^{-1}, consistent with the disordered structures component (Figure 8C) [19]. This indicates that PC1 discriminates the data based on different amounts of secondary structure of the sample, while PC2 reflects the amount of conformational disorder. Indeed, the AZ scores are very close to each other, indicating a very low variability in the secondary structure within different batches of samples, with this being consistent with the AZ ordered secondary structure [32]. Additionally, the superposition of DBD and DBD:AZ data along PC1 can be explained by assuming that the intrinsic disordered nature of DBD is able to populate an ensemble of different conformations. On the other hand, DBD:AZ distribution on the plot is narrower than that of DBD, reflecting a lower degree of disorder in the complex. These results confirm that PCA is a good reporter of the different structural differences among AZ, DBD, and DBD:AZ. Moreover, PCA is sensitive to changes in the conformational heterogeneity of DBD in the presence of AZ.

3. Materials and Methods

3.1. Sample Preparation

AZ (purity > 80%; MW = 14.6 kDa) was purchased from Sigma–Aldrich (St. Louis, MO, USA). The effective purity of the sample was checked by determining the ratio of spectral absorption at 630 nm and at 280 nm. AZ batches with a ratio value higher than 0.48 were used, with this indicating a good degree of purity [43]. AZ was dissolved in MilliQ water at a concentration of 200 μM. DBD (a.a 89–293; MW= 23 kDa) was purchased from GenScript (Piscataway, NJ, USA). DBD were dissolved in Phosphate Buffer Solution (PBS; 95.3% H_2O, 3.8% NaCl, 0.1% di KCl, 0.7% Na_2HPO_4, 0.1% KH_2PO_4; pH = 7.4), reaching a final concentration of 40 μM. The DBD:AZ complex in PBS solution were prepared by mixing equimolar amounts of the components.

3.2. Raman Spectroscopy

Raman measurements were carried out using a Super Labram confocal spectrometer (Horiba, France), equipped with several objectives, a diode-pumped solid-state laser (532 nm) and a spectrograph, with an 1800 g/mm grating allowing a resolution of 5 cm^{-1}. Raman spectra were collected by means of a liquid nitrogen-cooled charged coupled device (CCD) (back illuminated; pixel format: 1024×128 detector) and in the back-scattering geometry in which a notch filter was used to reject the elastic contribution. All the experiments were performed using a laser power of 10 mW (4.4 mW on the sample) and a 50× objective with a numerical aperture NA = 0.6 (laser spot diameter reaching the sample was about 1 μm). A large confocal diaphragm (400 μm) and a slit of 200 μm were used to obtain a good Raman signal.

Protein drops (10–15 μL) were deposited onto an optical glass, and spectra were acquired on partially dried samples. Indeed, it was demonstrated that there are no significant differences between the Raman spectra of protein in solution and the corresponding drop coating deposition, in which the protein remains substantially hydrated and the secondary structure is largely preserved [44].

Each Raman spectrum was acquired at room temperature by averaging 10 scans of 10 s integration time. For each sample, twenty-five Raman spectra were collected from different regions of the drops. Raman data processing and analysis were performed with OPUS software version 6.5 (Bruker Optics, Ettlingen, Germany). All the spectra were normalized with respect to the phenylalanine (Phe) ring breathing band at 1002 cm^{-1} due to its insensitivity to conformation or microenvironment [45], and the fluorescence background was removed by applying a rubber band baseline correction [46]. Finally, the spectra used for the structural analysis were obtained by averaging five measurements to improve the spectral signal/noise ratio.

3.3. Analysis of the Raman Spectra

The secondary structure content of isolated DBD and AZ was quantified through a deconvolution procedure of the Amide I Raman bands by using three pseudo-Voigt profiles. The model parameters were optimized with the Levenberg–Marquardt minimization algorithm (LMA), and the goodness of the fit was assessed by the reduced chi-square value. The AZ curves as extracted from the fit were used in the DBD:AZ complex analysis, under the hypothesis that the AZ secondary structure does not change upon the interaction [38]. The three pseudo-Voigt profiles were centered at 1650–1656, 1664–1670, and 1680 cm^{-1} and assigned to α-helix, β-strand, and random coil conformations, respectively, as validated on other IDPs [22,23,27]. In each fitting analysis, additional peaks had to be included in the band-fitting protocol to account for aromatic residue modes (1550, 1580, 1604, and 1615 cm^{-1}) and for disordered structure and/or vibronic coupling (1637 cm^{-1}) not baseline separated from Amide I features [22]. The errors relative to secondary structure percentages were evaluated by repeating the curve-fitting procedure on five different spectra and the accuracy associated with the determined secondary structure content was about 10% for each sample.

In order to improve the performance of deconvolution analysis, we performed a dimensionality reduction of the Raman spectra based on principal component analysis (PCA) [33]. The PCA transforms the original data set into a new data set with transformed variables (principal components) that are linear combinations of the original variables. The principal components were arranged in a swat that the variability of the original data set was contained in descending order in the first principal components. PCA was applied to the isolated DBD and AZ molecules and to the DBD:AZ complex (number of spectra n = 75) in three different spectral regions: (i) Fermi doublets relative to Tyr (830 and 850 cm^{-1}), satisfactorily described by a number of components N= 79; (ii) Fermi doublets relative to Trp (1340 and 1360 cm^{-1}) described by N = 79; and (iii) Amide I Raman band (1600–1700 cm^{-1}) described by N = 77. The number of components of the correlation matrix to be considered was defined as the number required to explain at least 80% of the total variance. STATISTICA 7.0 software (StatSoft Inc., Tulsa, OK, USA, 2004) was used for all the analyses.

4. Conclusions

The structural and conformational changes in the DBD region of the intrinsically disordered protein p53 upon interacting with the anticancer blue copper protein AZ were investigated by applying Raman spectroscopy. A careful inspection of the Raman spectra combined with a PCA analysis on the Fermi doublets of the Raman markers corresponding to the tyrosine and tryptophan residues allowed us to monitor the changes in their microenvironment as induced by the formation of a complex between DBD and AZ. Interestingly, we found a direct involvement of DBDTrp^{146} in the complex formation, as suggested by other experimental investigations. Additionally, a deconvolution of the Amide I band, remarkably sensitive to the α-helix, β-sheets, and random coil structures, allowed us to quantify the main secondary structural motifs of the DBD and its changes as induced upon binding to AZ. We found that DBD undergoes a slight increase of the β-conformation, with a concomitant lowering of its disordered portions as well as of its conformational heterogeneity. These findings are in agreement with our previous computational results and suggest that the binding of AZ to some unstructured motifs of DBD can restrain their flexibility. Collectively, the observed modulation the DBD structure when bound to AZ may represent a ground for understanding the molecular mechanisms of the AZ anticancer activity and could provide some hints for designing other molecules for p53-targeted therapies. Finally, we would remark that our Raman-based approach can be applied to investigate the structural changes of other biomolecules undergoing specific complex formation in order also to elucidate the molecular mechanisms which regulate their biological functions.

Author Contributions: S.S. data curation, formal analysis, investigation, and writing—original draft preparation; S.C. and A.R.B. writing—review and editing; S.C. and A.R.B. conceptualization supervision.

Funding: This research was funded by the Italian Association for Cancer Research (AIRC) Grant IG15866 to S.C.

Conflicts of Interest: The authors declare no conflict of interest.

References

1. Vousden, K.H.; Lane, D.P. p53 in health and disease. *Nat. Rev. Mol. Cell Biol.* **2007**, *8*, 275–283. [CrossRef] [PubMed]

2. Kruiswijk, F.; Labuschagne, C.F.; Vousden, K.H. p53 in survival, death and metabolic health: A lifeguard with a licence to kill. *Nat. Rev. Mol. Cell Biol.* **2015**, *16*, 393–405. [CrossRef] [PubMed]

3. Uversky, V.N. Unusual biophysics of intrinsically disordered proteins. *Biochim. Biophys. Acta (BBA)-Proteins Proteom.* **2013**, *1834*, 932–951. [CrossRef] [PubMed]

4. Tompa, P. Intrinsically unstructured proteins. *Trends Biochem. Sci.* **2002**, *27*, 527–533. [CrossRef]

5. Habchi, J.; Tompa, P.; Longhi, S.; Uversky, V.N. Introducing Protein Intrinsic Disorder. *Chem. Rev.* **2014**, *114*, 6561–6588. [CrossRef] [PubMed]

6. Minde, D.P.; Halff, E.F.; Tans, S. Designing disorder. *Intrinsically Disord. Proteins* **2013**, *1*, e26790. [CrossRef] [PubMed]

7. Cañadillas, J.M.P.; Tidow, H.; Freund, S.M.V.; Rutherford, T.J.; Ang, H.C.; Fersht, A.R. Solution structure of p53 core domain: Structural basis for its instability. *Proc. Natl. Acad. Sci. USA* **2006**, *103*, 2109–2114. [CrossRef]

8. Cho, Y.; Gorina, S.; Jeffrey, P.D.; Pavletich, N.P. Crystal structure of a p53 tumor suppressor-DNA complex: Understanding tumorigenic mutations. *Science* **1994**, *265*, 346–355. [CrossRef]

9. Pagano, B.; Jama, A.; Martinez, P.; Akanho, E.; Bui, T.T.T.; Drake, A.F.; Fraternali, F.; Nikolova, P.V. Structure and stability insights into tumour suppressor p53 evolutionary related proteins. *PLoS ONE* **2013**, *8*, e76014. [CrossRef]

10. Bell, S.; Klein, C.; Müller, L.; Hansen, S.; Buchner, J. P53 Contains Large Unstructured Regions in Its Native State. *J. Mol. Biol.* **2002**, *322*, 917–927. [CrossRef]

11. Berlow, R.B.; Dyson, H.J.; Wright, P.E. Functional advantages of dynamic protein disorder. *FEBS Lett.* **2015**, *589*, 2433–2440. [CrossRef]

12. Punj, V.; Das Gupta, T.K.; Chakrabarty, A.M. Bacterial cupredoxin azurin and its interactions with the tumor suppressor protein p53. *Biochem. Biophys. Res. Commun.* **2003**, *312*, 109–114. [CrossRef]

13. Yamada, T.; Goto, M.; Punj, V.; Zaborina, O.; Chen, M.L.; Kimbara, K.; Majumdar, D.; Cunningham, E.; Das Gupta, T.K.; Chakrabarty, A.M. Bacterial redox protein azurin, tumor suppressor protein p53, and regression of cancer. *Proc. Natl. Acad. Sci. USA* **2002**, *99*, 14098–14103. [CrossRef]

14. Goto, M.; Yamada, T.; Kimbara, K.; Horner, J.; Newcomb, M.; Gupta, T.K.; Chakrabarty, A.M. Induction of apoptosis in macrophages by Pseudomonas aeruginosa azurin: Tumour-suppressor protein p53 and reactive oxygen species, but not redox activity, as critical elements in cytotoxicity. *Mol. Microbiol.* **2003**, *47*, 549–559. [CrossRef]

15. Yamada, T.; Hiraoka, Y.; Ikehata, M.; Kimbara, K.; Avner, B.S.; Das Gupta, T.K.; Chakrabarty, A.M. Apoptosis or growth arrest: Modulation of tumor suppressor p53's specificity by bacterial redox protein azurin. *Proc. Natl. Acad. Sci. USA* **2004**, *101*, 4770–4775. [CrossRef]

16. Yamada, T.; Fialho, A.M.; Punj, V.; Bratescu, L.; Gupta, T.K.; Chakrabarty, A.M. Internalization of bacterial redox protein azurin in mammalian cells: Entry domain and specificity. *Cell. Microbiol.* **2005**, *7*, 1418–1431. [CrossRef]

17. Yamada, T.; Gupta, E.; Beattie, C.W. p28-Mediated Activation of p53 in G2–M Phase of the Cell Cycle Enhances the Efficacy of DNA Damaging and Antimitotic Chemotherapy. *Cancer Res.* **2016**, *76*, 2354–2365. [CrossRef]

18. Signorelli, S.; Santini, S.; Yamada, T.; Bizzarri, A.R.; Beattie, C.W.; Cannistraro, S. Binding of Amphipathic Cell Penetrating Peptide p28 to Wild Type and Mutated p53 as studied by Raman, Atomic Force and Surface Plasmon Resonance spectroscopies. *Biochim. Biophys Acta Gen. Subj.* **2017**, *1861*, 910–921. [CrossRef]

19. Domenici, F.; Frasconi, M.; Mazzei, F.; D'Orazi, G.; Bizzarri, A.R.; Cannistraro, S. Azurin modulates the association of Mdm2 with p53: SPR evidence from interaction of the full-length proteins. *J. Mol. Recognit.* **2011**, *24*, 707–714. [CrossRef]

20. Funari, G.; Domenici, F.; Nardinocchi, L.; Puca, R.; D'Orazi, G.; Bizzarri, A.R.; Cannistraro, S. Interaction of p53 with Mdm2 and azurin as studied by atomic force spectroscopy. *J. Mol. Recognit.* **2010**, *23*, 343–351. [CrossRef]

21. De Grandis, V.; Bizzarri, A.R.; Cannistraro, S. Docking study and free energy simulation of the complex between p53 DNA-binding domain and azurin. *J. Mol. Recognit.* **2007**, *20*, 215–226. [CrossRef]

22. Maiti, N.C.; Apetri, M.M.; Zagorski, M.G.; Carey, P.R.; Anderson, V.E. Raman spectroscopic characterization of secondary structure in natively unfolded proteins: Alpha-synuclein. *J. Am. Chem. Soc.* **2004**, *126*, 2399–2408. [CrossRef]

23. Signorelli, S.; Cannistraro, S.; Bizzarri, A.R. Structural Characterization of the Intrinsically Disordered Protein p53 Using Raman Spectroscopy. *Appl. Spectrosc.* **2016**, *71*, 823–832. [CrossRef]

24. Yamada, T.; Signorelli, S.; Cannistraro, S.; Beattie, C.W.; Bizzarri, A.R. Chirality switching within an anionic cell-penetrating peptide inhibits translocation without affecting preferential entry. *Mol. Pharm.* **2015**, *12*, 140–149. [CrossRef]

25. Siamwiza, M.N.; Lord, R.C.; Chen, M.C.; Takamatsu, T.; Harada, I.; Matsuura, H.; Shimanouchi, T. Interpretation of the doublet at 850 and 830 cm-1 in the Raman spectra of tyrosyl residues in proteins and certain model compounds. *Biochemistry* **1975**, *14*, 4870–4876. [CrossRef]

26. Arp, Z.; Autrey, D.; Laane, J.; Overman, S.A.; Thomas, G.J. Tyrosine Raman signatures of the filamentous virus Ff are diagnostic of non-hydrogen-bonded phenoxyls: Demonstration by Raman and infrared spectroscopy of p-cresol vapor. *Biochemistry* **2001**, *40*, 2522–2529. [CrossRef]

27. Tuma, R. Raman spectroscopy of proteins: From peptides to large assemblies. *J. Raman Spectrosc.* **2005**, *36*, 307–319. [CrossRef]

28. Torreggiani, A.; Fini, G. Raman spectroscopic studies of ligand-protein interactions: The binding of biotin analogues by avidin. *J. Raman Spectrosc.* **1998**, *29*, 229–236. [CrossRef]

29. Krimm, S.; Bandekar, J. Vibrational spectroscopy and conformation of peptides, polypeptides, and proteins. *Adv. Protein Chem.* **1986**, *38*, 181–364. [CrossRef]

30. Altose, M.D.; Zheng, Y.; Dong, J.; Palfey, B.A.; Carey, P.R. Comparing protein-ligand interactions in solution and single crystals by Raman spectroscopy. *Proc. Natl. Acad. Sci. USA* **2001**, *98*, 3006–3011. [CrossRef]

31. Carey, P.R. *Biochemical Applications of Raman and Resonance Raman Spectroscopies*; Academic Press: Cambridge, UK, 1982; ISBN 9780121596507.

32. Harada, I.; Miura, T.; Takeuchi, H. Origin of the doublet at 1360 and 1340 cm−1 in the Raman spectra of tryptophan and related compounds. *Spectrochim. Acta. Part A Mol. Spectrosc.* **1986**, *42*, 307–312. [CrossRef]

33. David, C.C.; Jacobs, D.J. Principal component analysis: A method for determining the essential dynamics of proteins. *Methods Mol. Biol.* **2014**, *1084*, 193–226. [CrossRef]

34. Thomas, G.J. Raman spectroscopy of protein and nucleic acid assemblies. *Annu. Rev. Biophys. Biomol. Struct.* **1999**, *28*, 1–27. [CrossRef]

35. Nar, H.; Messerschmidt, A.; Huber, R.; Van De Kamp, M.; Canters, G.W. Crystal structure of Pseudomonas aeruginosa apo-azurin at 1.85 A resolution. *FEBS Lett.* **1992**, *306*, 119–124. [CrossRef]

36. Natan, E.; Baloglu, C.; Pagel, K.; Freund, S.M.V.; Morgner, N.; Robinson, C.V.; Fersht, A.R.; Joerger, A.C. Interaction of the p53 DNA-binding domain with its n-terminal extension modulates the stability of the p53 tetramer. *J. Mol. Biol.* **2011**, *409*, 358–368. [CrossRef]

37. Wen, Z.Q. Raman spectroscopy of protein pharmaceuticals. *J. Pharm. Sci.* **2007**, *96*, 2861–2878. [CrossRef]

38. Apiyo, D.; Wittung-Stafshede, P. Unique complex between bacterial azurin and tumor-suppressor protein p53. *Biochem. Biophys. Res. Commun.* **2005**, *332*, 965–968. [CrossRef]

39. Uversky, V.N.; Oldfield, C.J.; Midic, U.; Xie, H.; Xue, B.; Vucetic, S.; Iakoucheva, L.M.; Obradovic, Z.; Dunker, A.K. Unfoldomics of human diseases: Linking protein intrinsic disorder with diseases. *BMC Genom.* **2009**, *10*, S7. [CrossRef]

40. Yamada, T.; Christov, K.; Shilkaitis, A.; Bratescu, L.; Green, A.; Santini, S.; Bizzarri, A.R.; Cannistraro, S.; Gupta, T.K.; Beattie, C.W. p28, A first in class peptide inhibitor of cop1 binding to p53. *Br. J. Cancer* **2013**, *108*, 2495–2504. [CrossRef]

41. Minde, D.P.; Dunker, A.K.; Lilley, K.S. Time, space, and disorder in the expanding proteome universe. *Proteomics* **2017**, *17*, 1600399. [CrossRef]

42. Kengne-Momo, R.P.; Daniel, P.; Lagarde, F.; Jeyachandran, Y.L.; Pilard, J.F.; Durand-Thouand, M.J.; Thouand, G. Protein Interactions Investigated by the Raman Spectroscopy for Biosensor Applications. *Int. J. Spectrosc.* **2012**, *2012*, 1–7. [CrossRef]

43. Domenici, F.; Bizzarri, A.R.; Cannistraro, S. Surface-enhanced Raman scattering detection of wild-type and mutant p53 proteins at very low concentration in human serum. *Anal. Biochem.* **2012**, *421*, 9–15. [CrossRef] [PubMed]

44. Ortiz, C.; Zhang, D.; Xie, Y.; Ribbe, A.E.; Ben-Amotz, D. Validation of the drop coating deposition Raman method for protein analysis. *Anal. Biochem.* **2006**, *353*, 157–166. [CrossRef] [PubMed]

45. Krafft, C.; Hinrichs, W.; Orth, P.; Saenger, W.; Welfle, H. Interaction of Tet repressor with operator DNA and with tetracycline studied by infrared and Raman spectroscopy. *Biophys. J.* **1998**, *74*, 63–71. [CrossRef]

46. Yang, H.; Yang, S.; Kong, J.; Dong, A.; Yu, S. Obtaining information about protein secondary structures in aqueous solution using Fourier transform IR spectroscopy. *Nat. Protoc.* **2015**, *10*, 382–396. [CrossRef] [PubMed]

© 2019 by the authors. Licensee MDPI, Basel, Switzerland. This article is an open access article distributed under the terms and conditions of the Creative Commons Attribution (CC BY) license (http://creativecommons.org/licenses/by/4.0/).

International Journal of
Molecular Sciences

Article

Structural and Functional Properties of the Capsid Protein of Dengue and Related *Flavivirus*

André F. Faustino [1,†], Ana S. Martins [1], Nina Karguth [1], Vanessa Artilheiro [1], Francisco J. Enguita [1], Joana C. Ricardo [2,‡], Nuno C. Santos [1,*] and Ivo C. Martins [1,*]

[1] Instituto de Medicina Molecular, Faculdade de Medicina, Universidade de Lisboa, Av. Prof. Egas Moniz, 1649-028 Lisbon, Portugal
[2] Centro de Química-Física Molecular, Instituto Superior Técnico, Universidade de Lisboa, 1049-001 Lisbon, Portugal
* Correspondence: nsantos@fm.ul.pt (N.C.S.); ivomartins@fm.ul.pt (I.C.M.); Tel.: +351-217-999-480 (N.C.S.); +351-217-999-476 (I.C.M.)
† Present address: Instituto de Biologia Experimental e Tecnológica (iBET), Apartado 12, 2780-901 Oeiras, Portugal.
‡ Present address: Department of Biophysical Chemistry, J. Heyrovský Institute of Physical Chemistry, Czech Academy of Sciences, Dolejškova 3, 182 23 Prague 8, Czech Republic.

Received: 21 June 2019; Accepted: 6 August 2019; Published: 8 August 2019

Abstract: Dengue, West Nile and Zika, closely related viruses of the Flaviviridae family, are an increasing global threat, due to the expansion of their mosquito vectors. They present a very similar viral particle with an outer lipid bilayer containing two viral proteins and, within it, the nucleocapsid core. This core is composed by the viral RNA complexed with multiple copies of the capsid protein, a crucial structural protein that mediates not only viral assembly, but also encapsidation, by interacting with host lipid systems. The capsid is a homodimeric protein that contains a disordered N-terminal region, an intermediate flexible fold section and a very stable conserved fold region. Since a better understanding of its structure can give light into its biological activity, here, first, we compared and analyzed relevant mosquito-borne *Flavivirus* capsid protein sequences and their predicted structures. Then, we studied the alternative conformations enabled by the N-terminal region. Finally, using dengue virus capsid protein as main model, we correlated the protein size, thermal stability and function with its structure/dynamics features. The findings suggest that the capsid protein interaction with host lipid systems leads to minor allosteric changes that may modulate the specific binding of the protein to the viral RNA. Such mechanism can be targeted in future drug development strategies, namely by using improved versions of pep14-23, a dengue virus capsid protein peptide inhibitor, previously developed by us. Such knowledge can yield promising advances against Zika, dengue and closely related *Flavivirus*.

Keywords: Dengue virus (DENV); capsid protein (C protein); *Flavivirus*; intrinsically disordered protein (IDP); protein–RNA interactions; protein–host lipid systems interaction; circular dichroism; time-resolved fluorescence anisotropy

---

1. Introduction

Viral hemorrhagic fever is a global problem, with most cases due to dengue virus (DENV), which originates over 390 million infections per year worldwide, being a major socio-economic burden, mainly for tropical and subtropical developing countries [1]. A working vaccine was registered in Mexico in December 2015, approved for official use in some endemic regions of Latin America and Asia and, as of October 2018, also in Europe [2–4]. However, this vaccine is not 100% effective against all DENV serotypes. Thus, research into new prophylactics is still ongoing, with a new vaccine proposed recently being now in phase 3 clinical trials [5]. In spite of these recent developments, fully effective prophylactics approaches are lacking and there are no effective therapies. This is in part, due to a poor understanding of key steps of the viral life cycle.

There are four dengue serotypes occurring: DENV-1, DENV-2, DENV-3 and DENV-4 [6]. Here, if not otherwise indicated, DENV refers to DENV-2. DENV is a member of the *Flavivirus* genus, part of the Flaviviridae family, a genus which comprises 53 viral species [6]. Many of these are important human pathogens as well, such as hepatitis C (HCV), tick-borne encephalitis (TBEV), yellow fever (YFV), West Nile (WNV) and Zika (ZIKV) viruses [6–9]. Flaviviridae are single-stranded positive-sense RNA viruses with approximately 11 kb, containing a single open reading frame [10]. Using the host cell translation machinery, the *Flavivirus* RNA genome is translated into a polyprotein that is co- and post-translationally cleaved by cellular and viral proteases into three structural proteins and seven non-structural proteins [10]. Structural proteins are named as such since they are present in the mature virion structure [11]. Nevertheless, they may also have non-structural roles, such as the capsid (C) protein. This is a structural protein that also mediates viral assembly and encapsidation, crucial steps of the viral life cycle. Given the C protein key roles, it is the focus of this work and will be described in detail below.

DENV C contains 100 amino acid residues, which form an homodimer with an intrinsically disordered protein (IDP) region in the N-terminal followed by four α-helices, α1 to α4, per monomer [12]. Overall, the main structural/dynamics regions consist of the disordered N-terminal, a short flexible intermediate fold and, finally, a large conserved fold region, which greatly stabilizes the protein homodimer structure [12–16]. The C protein has an asymmetric charge distribution: one side of the dimer contains a hydrophobic pocket (α2–α2′ interface), responsible for, alongside the disordered N-terminal, the binding to host lipid droplets (LDs) [12–16]. The other is the positively charged C-terminal side (α4–α4′ interface), proposed to mediate the C protein binding to the viral RNA [12]. It is noteworthy that several transient conformations for DENV C N-terminal were proposed, which may help modulate DENV C interaction with host lipid systems, via an autoinhibition mechanism [15].

DENV infection affects the host lipid metabolism, increasing host intracellular LDs and unbalancing plasma lipoprotein levels and composition [17–19]. Importantly, DENV C binds LDs, an interaction essential for viral replication [18,20]. DENV C-LDs binding requires potassium ions, the LDs surface protein perilipin 3 (PLIN3) and involves specific amino acid residues of DENV C α2–α2′ helical hydrophobic core and of the N-terminal [14,20]. This knowledge led us to design pep14-23, a patented peptide, based on a *Flavivirus* C protein conserved N-terminal motif. We then established that pep14-23 inhibits DENV C-LDs binding [14], acquiring α-helical structure in the presence of anionic phospholipids [15]. Moreover, we also found that DENV C binds specifically to very low-density lipoproteins (VLDL), requiring K$^+$ ions and a specific VLDL surface protein, apolipoprotein E (APOE), being also inhibited by pep14-23 [21]. This is analogous to DENV C-LDs interaction. The similarities between APOE and PLIN3 further reinforce this, suggesting a common mechanism [22]. The role of LDs in *Flavivirus* infection is well known and has been recently reviewed [14,18,20,23–25]. Given that, pep14-23 is an excellent drug development lead. Further developments require a better understanding of the function of the C protein of dengue and of *Flavivirus* in general.

Therefore, here, we seek to contribute to understand the C proteins biological activity, with a special focus on DENV C. Briefly, we studied DENV C structure-activity relationship in the context of similar and highly homologous mosquito-borne *Flavivirus* C proteins. Our findings shed light into the structure-function relationship behind the C protein biological roles, which may contribute to future therapeutic approaches against DENV and closely related *Flavivirus*.

2. Results

2.1. Analysis of Amino Acid Sequence Conservation Among Flavivirus C proteins

A phylogenetic analysis of the *Flavivirus* C protein and the polyprotein amino acid residue sequences reveals if the C protein is an indicator of phylogenetic similarity (Figure 1). C proteins of Spondweni group viruses, i.e., ZIKV, Spondweni virus (SPOV) and Kedougou virus (KEDV), cluster together, being the most similar to DENV (Figure 1a). Another cluster corresponds to mosquito-borne encephalitis-causing *Flavivirus*: Saint Louis encephalitis (SLEV), WNV, WNV serotype Kunjin (WNV-K), Alfuy (ALFV), Murray Valley encephalitis (MVEV), Usutu (USUV) and Japanese encephalitis (JEV) viruses. The *Flavivirus* polyproteins sequences show similar clusters (Figure 1b). As such, the C protein is a good indicator of viral genetic similarity. Thus, we investigated the C protein amino acid sequences, seeking common patterns relevant to biological activity.

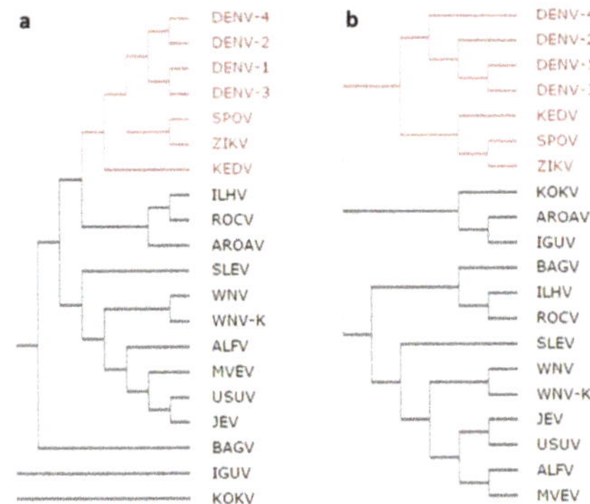

Figure 1. *Flavivirus* phylogenetic trees. Phylogenetic trees of (**a**) *Flavivirus* C proteins, highlighting in red the viruses with the C protein most similar to dengue virus (DENV) C (Spondweni group viruses (ZIKV), Spondweni virus (SPOV) and Kedougou virus (KEDV)) and of the (**b**) entire viral polyproteins of the same *Flavivirus*. Overall, despite some differences, the same general clusters are seen regardless of the clustering being based on the polyprotein or the capsid protein.

The amino acid residues sequences of the *Flavivirus* C proteins identified above were analyzed in the context of the three main regions identified in DENV C sequence, *i.e.*, the conserved fold region, the flexible fold region and the N-terminal IDP region (Figure 2). This was done for all mosquito-borne *Flavivirus* relevant for human diseases (Figure 2a), as well as for the four main DENV C serotypes (Figure 2b). For this, the 16 mosquito-borne *Flavivirus* and the 4 DENV serotypes amino acid sequence of the C protein are jointly aligned. In agreement with previous work [12,14], five conserved motifs are found in the mosquito-borne *Flavivirus* C proteins and deserve attention, namely: the N-terminal conserved [13]hNML+R[18]; [40]GXGP[43] in loop L1-2; [44]h+hhLAhhAFF+F[56] in α2 helix; [68]RW[69] of α3 helix; and, finally, the [84]F++−h[88] motif from α4 (with 'h', '+' and '−' representing hydrophobic, positively charged and negatively charged residues, respectively). Between residues 70–100, other motifs, not previously reported and containing hydrophobic and positively charged residues, are visible. Moreover, amino acid residues G and P, that can break the continuity of α-helices, are conserved in specific positions of the protein, especially in the disordered N-terminal and the flexible fold regions (Figure 2c). Charged residues are also conserved in specific locations. They are mostly in the conserved fold region, especially after position 95 (Figure 2d). Overall, the disordered N-terminal and the flexible fold regions, when compared with the conserved fold region, have an average of, respectively, 10 versus 4 G and P residues (Figure 2c), green, 10 versus 15 K and R residues (Figure 2d), blue, and 1 versus 2 D and E residues (Figure 2d), magenta.

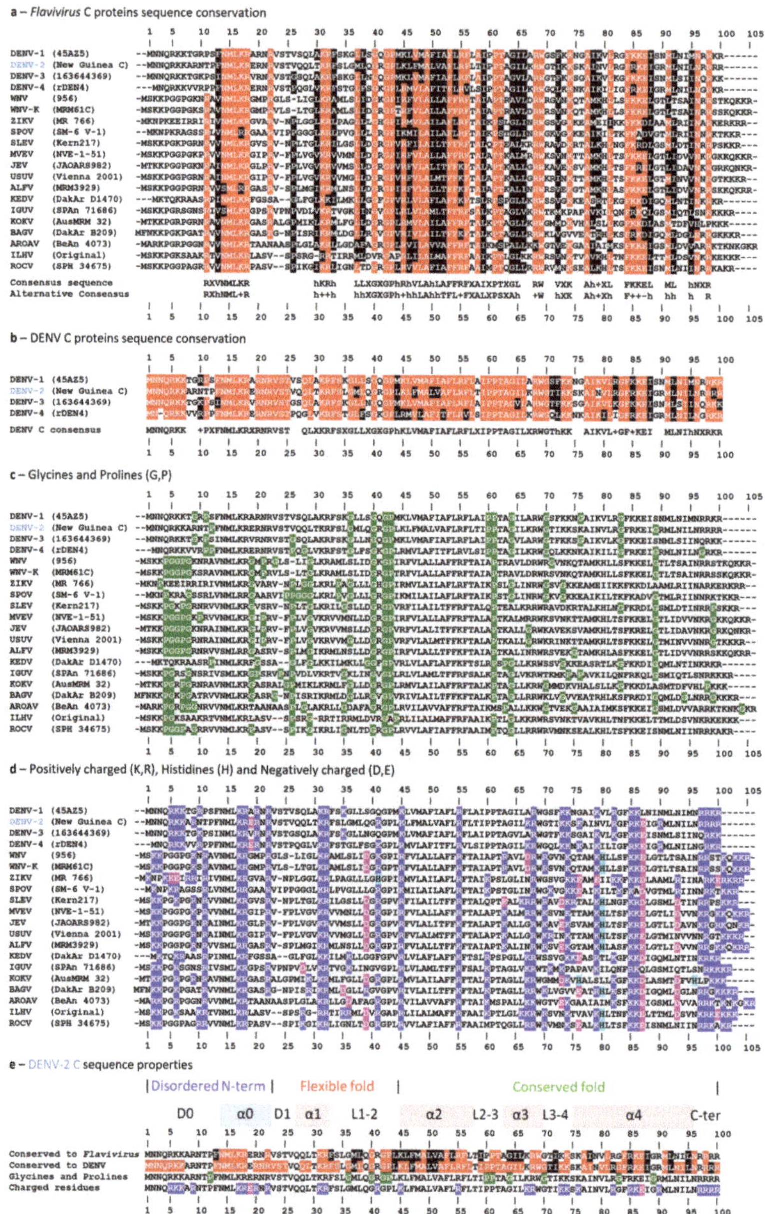

Figure 2. *Flavivirus* C proteins amino acid residues sequence conservation. (**a**) Mosquito-borne *Flavivirus* C protein are 55% conserved, with residues being considered conserved if, in a given position, more than 15 are equal (red) or stereochemically similar (black). (**b**) Conservation between DENV serotypes is 80%, with the same criteria as in (**a**). (**c**) Structure-breaking residues G and P (green). (**d**) Charged residues: dark blue for positively charged residues (K and R), light blue for H, and magenta for negatively charged residues (D or E). (**e**) Overall conserved regions of *Flavivirus* C proteins: the disordered N-terminal and the conserved fold are clearly conserved in terms of charged and G/P amino acids. In contrast, the flexible fold region allows higher variability. Thus, its main role seems to be to connect the disordered N-terminal and the conserved fold regions, and to enable alternative conformations. DENV C serotype 2 is highlighted in blue, with amino acid residues numbered according to its sequence. Amino acid residues are numbered according to the consensus, coinciding with DENV-2 residues numbers. The viruses' full designation is found in the abbreviations section.

Several motifs in the *Flavivirus* C protein sequences can be identified. These represent the main sections of the protein, conserved during evolution as these must be crucial to protein function (Figure 2e). The N-terminal region, although disordered, is highly conserved, in terms of charged amino acid and G/P residues. The flexible fold section allows greater variability, in line with previous reports by us and others, suggesting that it can adopt several conformations [15].

2.2. Analysis of the Flavirus C Protein Sequences Hydrophobicity and Secondary Structure Propensity

Hydrophobicity and α-helical propensity predictions were performed as previously reported [15], using the Kite-Doolittle [26] and the Deleage-Roux [27] scales on ProtScale server, respectively, for the 16 mosquito-borne *Flavivirus* C proteins analyzed (Figure 3). The hydrophobicity scale ranges from −4.5, for highly polar amino acids (hydrophilic), to 4.5, for highly hydrophobic amino acid residues [26]. Therefore, when plotting the average values for each amino acid residue of the *Flavivirus* C sequences, negative local minima and positive local maxima indicate, respectively, hydrophilic and hydrophobic regions (Figure 3a,b). All proteins display a similar profile even in the N-terminal and flexible fold regions despite the slightly higher amino acid residues variability (Figure 2). The α0 domain, homologous to pep14-23, is amphipathic, with average values near 0. In the flexible fold region, which is mostly amphipathic too, there is a peak of hydrophobicity between residues 30 and 40, possibly explaining its intermediate structure/dynamics behavior [13,14]. Some peaks of hydrophobicity are observed in the α3 and α4 domains, with the most hydrophobic domain being α2, as expected from the sequence analysis (Figure 2) and from the literature [12,14,18].

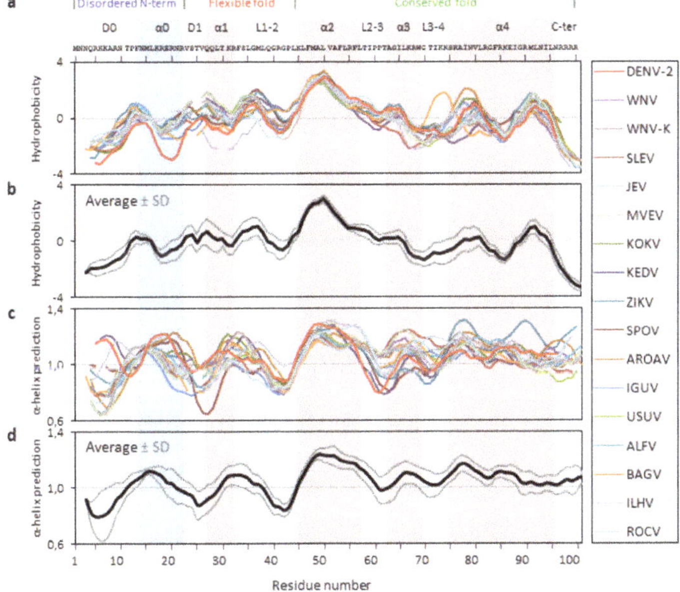

Figure 3. *Flavivirus* C proteins hydrophobicity and secondary structure predictions. (**a**) Hydrophobicity predictions and (**b**) respective average (black line) ± standard deviation, SD (gray lines). (**c**) α-helical secondary structure predictions and (**d**) respective average (black line) ± SD (gray lines). Amino acid residues are numbered according to the consensus, coinciding with DENV 2 residues numbers.

For α-helical predictions secondary structure is highly probable above a threshold of 1.0 [27]. *Flavivirus* C proteins secondary structure predictions correlate well with the known secondary structure of DENV C (Figure 2e) [12]. Such agreement supports the concept of a transient α0 occurring for these proteins, as hypothesized earlier [15]. Roughly, between positions 12 to 20, occurs a disordered region with high tendency to acquire α-helical secondary structure. Importantly, the values of the predictions are similar and the same tendencies are found in all proteins, with peaks and valleys co-localizing (Figure 3). Along with data from the last subsection, these results strengthen the idea that *Flavivirus* C proteins have similar structure and dynamics properties.

2.3. Analysis of the Flavivirus C Protein Tertiary Structure Propensity

Flavivirus C proteins tertiary structure was then investigated, complementing the α-helical predictions, to help understanding the disordered N-terminal region role(s). Following previous work [15], I-TASSER [28–30] was used to predict tertiary structures for the 16 closely related mosquito-borne *Flavivirus* C proteins (Figure 4). Eighty monomer conformations were obtained (several for each sequence) and superimposed with the DENV C homodimer partial structure deposited at the Protein Data Bank (PDB) and obtained via nuclear magnetic resonance (NMR) spectroscopy (PDB ID: 1R6R). Noteworthy, DENV [12,16], WNV [31] and ZIKV [25] C proteins form homodimers, stabilized by hydrophobic and electrostatic interactions involving their conserved fold region [12–14,25,31–33]. Since this is the most conserved region of *Flavivirus* C proteins sequences (Figure 2), a homodimer is thus not only a stable conformational arrangement, but also likely to occur. Thus, as 28 conformers had more than 5 backbone clashes with the other monomer when superimposed in a homodimer structure (not allowing a viable homodimer), those conformers were discarded Table 1. The remaining 52 *Flavivirus* C proteins conformational models were analyzed, while superimposed with DENV C homodimer (PDB ID: 1R6R, model 21 [12]). These were then grouped into four clusters by visual inspection of their similarity (Figure 4).

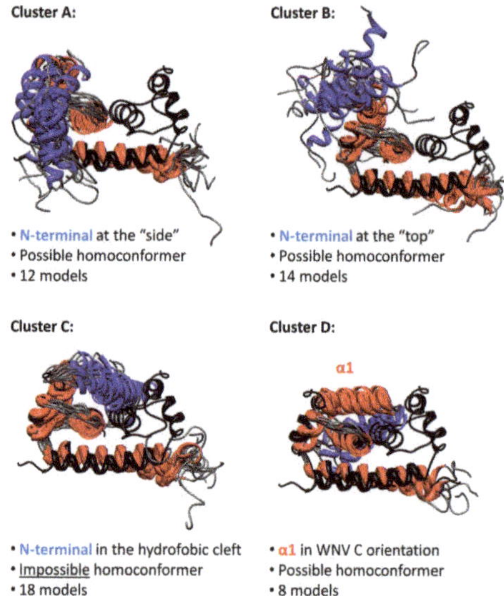

Figure 4. *Flavivirus* C proteins tertiary structure predictions, organized into four conformational clusters. The *Flavivirus* C proteins conformations predicted by I-TASSER are superimposed with DENV C experimental homodimer structure (black). Amino acid residues of the N-terminal region in α-helix conformation are in blue, the other α-helices in red and the loops in gray. From the 80 conformers, 52 can be clustered by similarity of conformations, from cluster A to D. Clusters A, B and C have the α1 helix in the DENV C experimentally determined conformation (Protein Data Bank (PDB) ID: 1R6R [12]). In cluster D the α1 is in West Nile Virus (WNV) C and ZIKV C conformation (PDB IDs: 1SFK [31] and 5YGH [25], respectively). The closed autoinhibitory conformation of cluster C seems the most probable, having the highest number of models. Although unlikely given their transient unstable nature, N-terminal IDP regions may interact with each other. Table 1 specifies each cluster composition.

Most sequences have a conformer in each cluster (Figure 1 and Table 1). In cluster A, some N-terminal amino acid residues are close to $\alpha4$–$\alpha4'$ and may interact with RNA, namely the positively charged residues. Cluster B has the most scattered conformers, with the N-terminal region at the "top", not interacting with other protein regions, resembling a transition between more ordered states. In cluster C, the N-terminal region is in an autoinhibitory conformation, blocking the access to the $\alpha1$–$\alpha2$–$\alpha2'$–$\alpha1'$ region, as previously suggested by us for DENV C [15]. 18 conformer models are predicted in this closed conformation with, at least, one model from most of the C proteins tested (except JEV C and ZIKV C; see Table 1). Therefore, it can occur in most *Flavivirus* C proteins. As for cluster D conformation, the $\alpha1$ helix is in the conformation of WNV [14,31] and ZIKV [25] C experimental structures, an arrangement not previously reported for DENV C [15]. This closed conformation also involves the N-terminal region and $\alpha1$ domain, and partially blocks the $\alpha2$–$\alpha2'$ hydrophobic cleft (or totally blocks it, when both monomers are in the same conformation). Importantly, both cluster C and D are closed conformations, supporting the autoinhibition hypothesis.

Table 1. Distribution of the I-TASSER predicted models through the four clusters.

Protein	Cluster A	Cluster B	Cluster C	Cluster D	Excluded
ALFV C	1	0	1	0	3
AROAV C	1	1	1	0	2
BAGV C	1	0	2	1	1
DENV C	1	2	1	0	1
IGUV C	1	2	1	0	1
ILHV C	0	2	2	0	1
JEV C	1	1	0	1	2
KEDV C	0	1	1	1	2
KOKV C	1	0	1	1	2
MVEV C	0	2	1	0	2
ROCV C	1	0	1	2	1
SLEV C	1	0	2	0	2
SPOV C	1	2	1	0	1
USUV C	1	0	2	0	2
WNV C	1	0	1	1	2
ZIKV C	0	1	0	1	3
Total	**12**	**14**	**18**	**8**	**28**

Dimers with A or B conformers in one monomer enable the simultaneous co-existence of all other conformers (A to D) on the other monomer. The C conformer neither permits the existence of C-C' homoconformers (i.e., both monomers in the same conformation) nor the heteroconformers of C-D' and D-C'. Despite that, D-D' homoconformers are allowed, similarly to the conformation that WNV C adopts in the crystal form [31]. Moreover, to go from cluster A to cluster C or D, the N-terminal region should pass by cluster B. These constraints suggest a path for transitions between conformations, discussed ahead. Overall, the autoinhibition hypothesis proposed for DENV C [15] is supported and such conformation can occur in other *Flavivirus* C proteins.

2.4. Analysis of Dengue Virus (DENV) C Protein Rotational Correlation Time

Given the close similarities between *Flavivirus* C proteins (Figures 1–4), DENV C can be used as a general model for them. Hence, we proceeded to determine DENV C overall rotational correlation time (τ_c), taking advantage of the tryptophan residue in position 69 (W69) intrinsic fluorescence. Our computational data support three main structure/dynamics regions, including a disordered N-terminal region, which would increase its expected apparent size (as it would not be globular and folded), a property detectable by such an approach. Upon testing molecules in aqueous solution and at room temperature, fluorescence lifetimes are usually in the ns timescale, and the fluorescence decays are sensitive to the anisotropy of the fluorophore, which depends on its τ_c (vd. Equations (1)–(8), describing these relations, in the Methods section [34,35]). Thus, the time-resolved fluorescence decay of DENV C W69 and the corresponding anisotropy decay were determined, both at pH 6.0 and 7.5 (Figure 5).

Table 2. Fitting parameters of DENV C time-resolved fluorescence anisotropy data analysis. Parameters obtained from fitting Equations (5) and (8) to the data of Figure 5. Values are average (±% standard error, SE). * Statistically significant differences ($p < 0.05$) between the values obtained for the two pH values tested.

Parameter	pH 6.0	pH 7.5
τ_1 (ns) *	0.209 (± 3.9%)	0.520 (± 4.0%)
τ_2 (ns)	3.106 (± 0.4%)	3.108 (± 0.9%)
τ_3 (ns) *	6.328 (± 0.4%)	6.506 (± 0.4%)
α_1 *	0.275 (± 0.7%)	0.178 (± 3.4%)
α_2 *	0.315 (± 0.9%)	0.385 (± 0.4%)
α_3 *	0.410 (± 0.4%)	0.437 (± 0.4%)
τ_c (ns) *	16.46 (± 2.9%)	16.41 (± 3.4%)
r_0	0.130 (± 0.8%)	0.131 (± 1.1%)

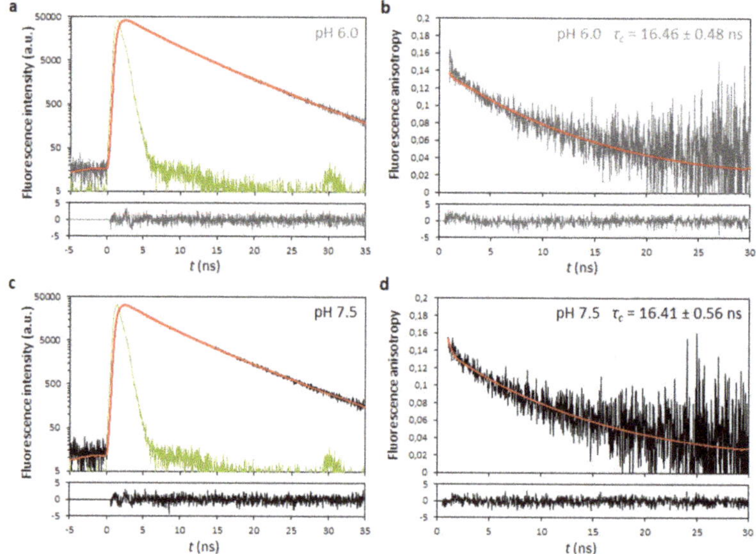

Figure 5. DENV C time-resolved fluorescence anisotropy. Time-resolved fluorescence decay at pH (**a**) 6.0 and (**c**) 7.5, with the corresponding anisotropy decays at pH (**b**) 6.0 and (**d**) 7.5. Fluorescence and anisotropy decays at both pH values are similar (gray and black decays, respectively). Fitting of experimental data (red) took into account the instrument response function (IRF; in green) and the corresponding residuals distribution, displayed below each graph. The equations used for fitting are presented on the Methods Equations (5) and (8). The parameters obtained are shown in Table 2.

Time-resolved fluorescence anisotropy decays at both pH values are similar (Figure 5b,d). Fluorescence lifetime components (τ_1, τ_2 and τ_3) were obtained from the intensity decays Equations (2)–(6) [34,35], with a triple-exponential retrieving the best fit (Figure 5a,c). Fitting the data retrieves similar values Table 2 for τ_1, τ_2 and τ_3, and corresponding weights (α_1, α_2 and α_3 pre-exponential factors, respectively). For accurate calculation of τ_c, the condition $\tau_c < 3 \times \tau_3$ must occur [34,35]. Since τ_3 values were ~6.4 ns (with a significant weight α_3 of ~0.42), this means that, at both pH values, we could measure τ_c values up to a limit of ~19 ns. In both pH conditions, the τ_c measured was

16.4 ± 0.5 ns at 22 °C, within the limit and higher than expected for a purely globular protein of DENV C size, as predicted [13].

Rossi et al. [36] correlated the τ_c of 16 globular proteins at 20 °C with their molecular weight (MW in kDa), based on NMR data, leading to the relation: $\tau_c \approx 0.6$ MW. Assuming DENV C as a 23.5 kDa fully globular homodimer and correcting for the temperature (T) and viscosity (η) [37], the τ_c predicted is 12.0 ns. However, the correlational time must be slightly higher, as the protein will be partially unfolded and disordered (in the N-terminal). Jones et al. [16] measured a τ_c of 13 ns at 27 °C, by NMR, which with the corrections from Equation (10) [37], corresponds to 13.4 ns at 25 °C. Given DENV C size, this implies that the protein is not globular, in line with current knowledge of DENV C structure and dynamics [12–16]. Fluorescence anisotropy supports an even more open and partially disordered DENV C structure, given the τ_c value of 15.2 ± 0.5 ns at 25 °C Table 3, in line with in silico data (Figures 1–4).

Table 3. Comparing DENV C τ_c values (τ_c at 25 °C in H_2O were calculated using Equation (10)).

τ_c (ns) at T	T (°C)	τ_c (ns) at 25 °C in H_2O	Method	Source
16.4 ± 0.5	22	15.2 ± 0.5	Time-resolved fluorescence anisotropy	This work
13.0	27	13.4	Overall NMR relaxation analysis	Jones et al., 2003 [16]
14.1	20	12.0	τ_c (ns) $\approx 0.6\times$MW (kDa)	Rossi et al., 2010 [36]

2.5. Analysis of DENV C Conformational Stability

Circular dichroism (CD) spectroscopy was used to study DENV C secondary structure, via its thermal denaturation in solution from 0 to 96 °C, at pH 6.0 and 7.5 (2 °C steps, Figure 6). At both pH values, the α-helical structure is partially lost upon increasing temperature (Figure 6a,b). However, even at 96 °C, the protein does no become completely random coil, as seen from the spectrum shape and its high ellipticity at 222 nm (Figure 6c). Plotting the mean residue molar ellipticity at 222 nm, [θ], as a function of temperature, T, reveals a transition at ~70 °C at both pH (Figure 6c).

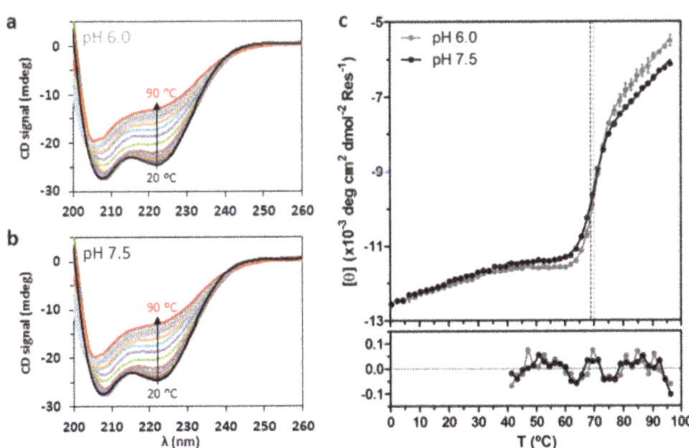

Figure 6. DENV C temperature denaturation followed via circular dichroism (CD) spectroscopy. CD spectra of DENV C, between 20 and 90 °C, at pH (**a**) 6.0 and (**b**) 7.5. For the sake of simplicity, the spectra from 0 to 18 °C and from 92 to 96 °C are not displayed, as they are similar to the 20 °C and the 90 °C spectra, respectively. (**c**) Mean residue molar ellipticity at 222 nm, [θ], as a function of temperature (dots) for pH 6.0 (gray) and 7.5 (black), between 0 and 96 °C. Lines correspond to the fitting of Equation (21) (combined with Equations (20), (22), (24) and (28)). Vertical dashed lines represent experimentally observed T_m, colored according to pH. Error bars represent SD, from three independent experiments. Residuals are shown below the graph, being lower than SD.

DENV C does not display a typical unfolding profile, as the denaturation curves do not reach a flat plateau. Still, ellipticity data were successfully fitted to a denaturation curve (Figure 6c), assuming a homodimer with one-step denaturation [32]. Briefly, Equation (21) was combined with Equations (20), (22) and (24) and fitted to the data. This allows to obtain the thermodynamic parameters of DENV C unfolding Table 4, namely the melting temperature (T_m°), the enthalpy variation at T_m° ($\Delta H^\circ{}_{T_m^\circ}$) and the entropy variation at T_m° ($\Delta S^\circ{}_{T_m^\circ}$), with all parameters at standard thermodynamics conditions (symbolized by '$^\circ$'). Equation (28) was then used to calculate the melting temperature (T_m) at the actual $[P_m]$ (instead of the value at $[P] = 1$ M, details in the Methods). Despite small differences, the parameters obtained are not significantly different between pH values Table 4. A small but consistent variation of the CD spectra between 0 and 40 °C is observable, implying: (i) a conformational equilibrium with temperature and/or (ii) some flexibility of the structure and/or (iii) a transition between alternative conformations. This temperature range covers the physiological conditions of both mosquitoes (20 to 40 °C, depending on the environment) and humans (36 to 40 °C). DENV C can continuously transition between conformations as temperature varies, in line with the previously hypothesized conformational equilibrium [15]. As temperature increases, the disordered conformations become more abundant but only a partial loss of structure is seen. This indicates that the C protein conserved region is thermodynamically stable. Similar observations are expected for other *Flavivirus* C proteins.

Table 4. Fitting parameters of DENV C temperature denaturation CD data. Parameters were estimated by fitting Equation (21) (combined with Equations (20), (22), (24) and (28)) to the data. T_m is the experimentally observed melting temperature (represented by the vertical lines in Figure 6c). Estimations are average ± SE. There were no significant variations between the two pH values tested ($p < 0.05$).

Parameter	pH 6.0	pH 7.5
T_m (°C)	70.02 ± 0.63	69.03 ± 0.65
T_m° (°C)	88.26 ± 0.80	88.80 ± 0.83
$\Delta H^\circ{}_{T_m^\circ}$ (kJ mol^{-1})	612 ± 26	564 ± 23
$\Delta S^\circ{}_{T_m^\circ}$ (kJ mol^{-1} K^{-1})	1.693 ± 0.073	1.557 ± 0.065

3. Discussion

Flavivirus C proteins are known to have similar sequences and structure [12–16,25,31]. Here, we go further by examining common features at different structural levels, complemented with data on DENV C size and thermodynamic stability. The phylogenetic analysis of the C proteins and the polyproteins (Figure 1) shows that the former is a marker of *Flavivirus* evolution. There are several conserved motifs, highlighted in previous studies with 16 *Flavivirus* [12,14]. The work is now expanded to include the four DENV serotypes (Figure 2). When these 20 *Flavivirus* C amino acid sequences, with between 96 and 107 amino acid residues each, are jointly analyzed, it is clear that 55% of the residues are conserved or stereochemically similar (Figure 2a). About 80% of amino acid residues are equal or similar and, thus, conserved among the four DENV C serotypes (Figure 2b). From the five major conserved motifs, four are known to be involved in dimer stabilization [14]: the ^{40}GXGP^{43} motif at loop L1-2, that marks the transition from the flexible to the conserved fold region [14]; the ^{68}RW^{69} at α3 forms an hydrophobic pocket that accommodates the W69 side chain involving residues from α2, α3 and α4 [12,32]; and, the ^{44}h+hhLAhhAFF+F^{56} and ^{84}F++–h^{88} motifs, respectively from α2 and α4 helices, maintain the homodimer structure both via the α2–α2′ hydrophobic interaction and via the salt bridges of residues [RK]^{45} and [RK]$^{55'}$ with [ED]^{87} [12,14,32]. *Flavivirus* C proteins must have similarly sized secondary structure domains, since G/P are in the same positions and these amino acid residues tend to break the secondary structure (Figure 2c). Charged residues are also conserved (Figure 2d), which makes sense as charges would promote the interaction of the C protein with the negatively charged host lipid systems [12,14,20–22] and the viral RNA [12]. C proteins have a common homodimer conserved fold region (roughly, residues 45–100), as observed for DENV, WNV and ZIKV C structures [12,14,25,31]. Conserved motifs are summarized in (Figure 2e).

The above explains the C proteins similar hydrophobic and α-helix propensities (Figure 3). The conserved motif ^{13}hNML+R^{18}, at the N-terminal region, and the $\alpha2$–$\alpha2'$ hydrophobic cleft are of particular importance for DENV C interaction with LDs and VLDL [14,20–22,38]. Mutations in specific residues of DENV C $\alpha2$–$\alpha2'$ and $\alpha4$–$\alpha4'$ also impair RNA binding. Likewise, ZIKV C also accumulates on LDs surface, with specific mutations on this protein disrupting the association [25]. ZIKV C also binds single-stranded and double-stranded RNAs [25], with, as for DENV C, the high positively charged residues density prompting the binding to LDs and RNA [12,39,40]. Given the match at the level of N-terminal α-helical propensity and $\alpha2$–$\alpha2'$ hydrophobicity (Figure 3), the C proteins may all be self-regulated by an autoinhibition mechanism, as proposed for DENV C [15].

The autoinhibition hypothesis is corroborated by the quaternary structure analysis (Figure 4); Table 1. Two clusters, C and D, are autoinhibited conformations. Importantly, cluster D $\alpha1$ aligns with WNV C [14,31] and ZIKV C [25]. Moreover, if two monomers are in a D conformation (D–D' homoconformer), the dimer $\alpha2$–$\alpha2'$ region is totally inaccessible. Cluster C does not allow a C–C' homoconformer nor a C–D heteroconformer, imposing restrictions to the simultaneous transitions that are possible between A, B, C and D, as homodimer. The interaction between N-terminal regions within a dimer may be considered. Nonetheless, the disordered nature and high density of positively charged amino acid residues will mostly favor the repulsion between these IDP regions.

It is important to look at the clusters (Figure 4), while considering the number of positively charged residues (Figure 2) in the disordered N-terminal and flexible fold (10 K and R residues) versus those in the conserved fold (15 K and R). The charge distribution in some arrangements implies that the disordered N-terminal is at least in theory able to bind the viral RNA [39,40]. Such binding would be governed by the N-terminal region cationic amino acid residues [41,42]. Here, the structure predictions reveal that, indeed, the first 12 N-terminal residues can locate near $\alpha4$–$\alpha4'$ Cluster A (Figure 4), the most likely RNA binding site [12,39,40]. Furthermore, binding to RNA via the C-terminal $\alpha4$–$\alpha4'$ interface may be favored by a previous or simultaneous interaction of the protein with host LDs via the N-terminal region and $\alpha2$–$\alpha2'$ interface. Access to $\alpha2$–$\alpha2'$ (controlled by the N-terminal region) would modulate the interaction (Figure 4) and, thus, viral assembly. In agreement, the binding of the related hepatitis C virus core protein (homologous to DENV C) to host LDs is what enables efficient viral assembly [43]. Thus, the C protein disordered N-terminal would be critical to protein function, enabling crucial structural and functional roles.

To evaluate this, we used DENV C as a model system, measuring its τ_c value by time-resolved fluorescence anisotropy (Figure 5) and its thermal stability by CD spectroscopy (Figure 6), at pH 6.0 and 7.5 (within the usual pH range of its biological microenvironment). A similar τ_c, 15.2 ± 0.5 ns, is obtained at both pH values (Figure 5; Tables 2 and 3), in line with previous work [13]. DENV C maintains its homodimer structure and dynamics behavior between pH 6.0 and 7.5. The τ_c value and respective size are higher than expected, due to the N-terminal disordered nature.

Regarding DENV C thermodynamic stability (Figure 6, Table 4), the protein T_m is ~70 °C at both pH values. These denaturation parameters are in line with other authors, as a chemically synthesized DENV C 21–100 fragment (without most of the disordered N-terminal region) displays a T_m = 71.6 °C [32]. DENV C high thermal stability in physiological conditions is likely due to the large hydrophobic area that is shared by the two monomers [12], but also to the W69 stabilizing interactions and, as experimentally observed [32], the formation of salt bridges (residues K45 and R55' with E87). As structure/dynamics properties are conserved among *Flavivirus* C proteins (Figures 2–4), these observations can probably be generalized for all these proteins.

These findings must also be considered in light of DENV C biologically relevant interactions with LDs [22] and RNA (Figure 7). DENV C experimental structure [12] contains three distinct structural regions [13]: a disordered N-terminal region (from the N-terminal up to residue R22), a flexible fold (residues V23 to L44, where α-helix 1 is located) and a conserved fold with helices $\alpha2$, $\alpha3$ and α4, containing the R68 and W69 amino acid residues, highly conserved among *Flavivirus* [12]. R68 terminates α3 helix, with its side chain pointing to the protein interior [12]. W69 locates at DENV C α4–α4′ interface, having a crucial role in the dimer structural stabilization [12]. Along with dimer structural stability, these interactions enable allosteric communication and movements between DENV C more hydrophobic section (α2–α2′dimer interface) and its remaining sections, namely the α4–α4′ region. Figure 7 displays this, in the context of the C protein biologically relevant interactions, as they are understood on the basis of recent studies [12–15,18,20–24].

Looking further, it is important to consider that the binding of DENV C to host LDs is mediated by both the N-terminal IDP region and the α2–α2′ interface [14]. V51 of α2 is affected by the interaction with LDs and stabilizes the dimer by contacting with α3 (I65). Another interaction via salt bridges, between α2 (K45 and R55′) and α4 (E87), stabilizes the homodimer (Figure 7a). The C protein binding to host LDs, which affects the α2–α2′, can lead to changes in the α4–α4′ structural arrangement (Figure 7b). To investigate this we searched for similar proteins. An RNA-binding protein with a two-helix domain similar to DENV C α4–α4′ was identified (Figure 7c), influenza A non-structural protein 1 (NS1, PDB ID: 2ZKO [44]). Influenza NS1 has interesting features: it accumulates in the nuclei of host cells after being translocated by importin α and β and works as a viral immuno-suppressor by weakening the host cell gene expression [45]. DENV C was also reported to have an importin α-like motif in the N-terminal [15,46]. Regarding the targets that may interact with importin α and be transported to the nucleus, they normally contain a nuclear localization sequence (NLS), consisting of a motif of at least 2 consecutive positively charged residues [47–51]. Some of these proteins contain 2 NLS motifs, with at least 8 (up to 40 or even more) residues in between, designated as a bipartite NLS motif [49–51]. Strikingly, *Flavivirus* C proteins have three motifs of two consecutive cationic residues in the N-terminal region and α1 domain, which could form a bipartite NLS. A bipartite NLS formed by the cationic residues before position 10 and at positions 17 and 18, with a spacer of 7 to 13 residues can occur. The other bipartite NLS possibility may be formed by residues at positions 17 and 18, and at positions 31 and 32, with 9 to 12 spacer residues. Possible bipartite NLS are also seen in the conserved fold region but its static nature precludes activity as NLS. If DENV C binds to importin α, it may act as a cargo protein to be transported to the nucleus. This could explain why has DENV C been found in the nucleus of DENV infected cells [46,52,53]. DENV C may directly bind importin β, given the similarities between the N-terminal region of DENV C and importin α [49]. This may allow it to disrupt the normal nuclear import/export system in DENV-infected cells. The conformational plasticity of the N-terminal and flexible fold regions is certainly compatible with interactions with importin(s). As the hypothesized bipartite NLS are conserved among *Flavivirus* C proteins, this may occur in other *Flavivirus*.

The C protein may act as an immuno-suppressor, similarly to influenza NS1, by interacting with importins α and/or importin β. Ivermectin, a specific inhibitor of importin α/β-mediated nuclear import, is able to inhibit HIV-1 and DENV replication [54]. The mechanism of DENV C inhibition might involve the C protein, specifically the intrinsically disordered N-terminal IDP region, which is similar to importin α disordered N-terminal region [15]. Moreover, influenza NS1 can counteract the RNA-activated protein kinase (PKR)-mediated antiviral response through a direct interaction with PKR [55]. Besides, influenza NS1 blocks interferon (IFN) regulatory factor 3 activation, which in turn prevents the induction of IFN-related genes [56]. DENV inhibits the IFN signaling pathway in a similar manner [57]. By its N-terminal region dsRNA-binding ability, influenza NS1 inhibits the nuclear export of mRNAs and modulates pre-mRNA splicing, suppressing antiviral response [44]. Similarities between DENV C and influenza NS1 also extend to the later ability to bind RNA (Figure 7c). Recognition of dsRNA is made by the influenza NS1 RNA-binding domain, which forms a homodimer [44]. Afterwards, a slight change in R38-R38′ orientation leads to anchoring the dsRNA to the protein by a hydrogen bond network to the protein [44]. One of the main functions of influenza NS1 binding to RNA is sequestering dsRNA from the 2′–5′ oligo(A) synthetase [58]. We propose that, as with influenza NS1, a small conformational change in DENV C α4–α4′ interface occurs after the contact of its α2–α2′ interface with LDs, modulated by transitions between alternative N-terminal "open" and "closed" conformations. Binding to LDs requires an open conformation (Figure 7d), decreasing the conformational variability and entropy of the C protein, which trigger the allosteric movements affecting the C-terminal α4–α4′. As with influenza NS1, the *Flavivirus* C protein would remain in the same overall fold, but a small opening of α4–α4′ would facilitate its binding to RNA.

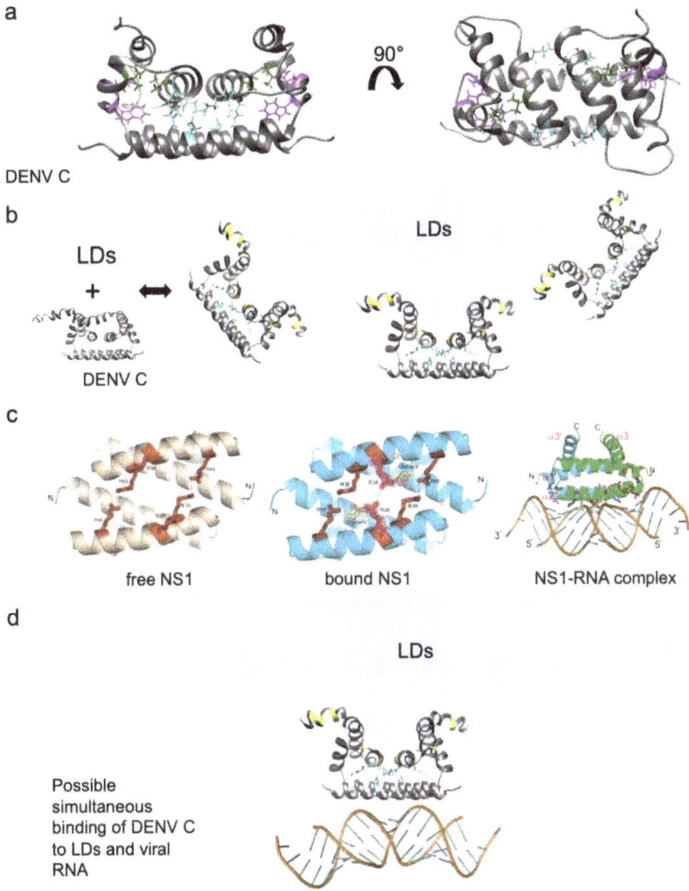

Figure 7. Protein structures of DENV C and influenza NS1. (**a**) DENV C structure from two different angles with the conserved residues R68 and W69 (purple) and the interface stabilizing residues V51 and I65 (green), as well as E87, R55 and K45, forming the salt bridge (cyan). (**b**) DENV C structure in a N-terminal region closed conformation and, next, in an open conformation with schematic binding of lipid droplets (LDs) and the affected amino acid residues (yellow). (**c**) The RNA-binding domain of NS1 protein from influenza A in a RNA-free (left) and RNA-bound state (middle and right), showing an organization similar to DENV C α4–α4′ region (adapted from Cheng et al., 2009 [44]). (**d**) DENV C with schematically bound to a LD and to RNA. DENV C amino acid residues affected by the binding to LDs are colored yellow, while a key internal salt bridge is shown in cyan. DENV C binding to host LDs may enable allosteric rearrangements (eventually involving the salt bridge), allowing a small conformational change in α4 side chains, namely the positively charged residues, prompting stable RNA-C protein binding.

The C-terminal is likely to be the crucial section for RNA binding given its similarity with influenza NS1 (Figure 7). Nevertheless, the N-terminal conformers must also be considered in the context of RNA binding (Figure 4). The A and D conformers allow RNA to be bound to the α4–α4′ interface and, simultaneously, to the N-terminal cationic amino acid residues. A–A′ and D–D′ conformations result in the possible binding of a single continuous portion of RNA to both the C-terminal α4–α4′ and the N-terminal IDP region, making the RNA more tightly bound. Moreover, the A–B′, B–B′ and B–C′ conformations would enable the protein to bind two distinct sections of the RNA, one bound to α4–α4′ and another to the N-terminal regions. That arrangement may allow to further compact the viral RNA. The N-terminal IDP region putative binding to RNA should not be disregarded given its positive net charge (+7). It compares very well with the C-terminal α-helical region net charge (+8 for a monomer, +16 for α4–α4′ dimer interface). Both may thus bind RNA due to, mostly, electrostatic forces. This IDP region can thus provide multi-functionality by several modes of binding and different ligands, enabled by alternative conformations. It must be stressed that this is not unlikely. Viral proteins tend to have IDP regions that increase their biological activity [59–61]. In a proteome as small as that of flaviviruses (10 proteins), IDP regions augment the number of ligands with which it can interact. Less structure often means more function. This is an increasingly hot topic of recent research, leading to design of algorithms to identify these regions [62,63]. Further analysis will help understand the interaction between DENV C and its ligands.

To conclude, the data imply a common structure and functions for mosquito-borne *Flavivirus* C proteins. Moreover, studying DENV C rotational diffusion and thermodynamics reveals a stable protein due to the conserved fold maintaining the homodimer structure. These findings apply to other *Flavivirus* C proteins, supporting a common mechanism for their biological activity. Such understanding of this key protein structure and dynamics properties may contribute to the future development of C protein-targeted drugs to impair dengue virus and other *Flavivirus* infections.

4. Materials and Methods

4.1. Materials

Chromatography columns HiTrap Heparin (1 and 5 mL), Sephadex S200 and the chromatography equipment AKTA-explorer were from GE Healthcare (Little Chalfont, UK). Sodium dodecyl sulphate-polyacrylamide gel electrophoresis (SDS-PAGE) reagents were from BioRad (Hercules, CA, USA). Unless otherwise stated, other chemicals were purchased from Sigma-Aldrich (St. Louis, MO, USA).

4.2. Flavivirus C Proteins Primary, Secondary and Tertiary Structural Predictions

For primary structure alignments we used the 16 non-DENV *Flavivirus* polyprotein sequences identified in reference [14], plus the four DENV reference sequences from NCBI, namely: DENV serotype 1, strain 45AZ5, NCBI ID NP_059433.1; DENV serotype 2, strain New Guinea C, NCBI ID NP_056776.2; DENV serotype 3, strain D3/H/IMTSSA-SRI/2000/1266, NCBI ID YP_001621843.1; and, DENV serotype 4, strain rDEN4, NCBI ID NP_073286.1. For the phylogenetic trees, both the entire polyproteins and the C protein regions were used. For the alignments and subsequent data analysis, the residues next to the NS2B-NS3 protease cleavage site [64,65] were excluded, leaving only the C protein sequences. Alignments and the derived phylogenetic trees were performed via Clustal Omega web tool (http://www.ebi.ac.uk/Tools/msa/clustalo/) [66,67].

Statistical comparison of the disordered N-terminal plus flexible fold regions with the conserved fold region of *Flavivirus* C proteins, for G and P content, as well as charged amino acid residues, was performed via a paired *t*-test, using GraphPad Prism v5 software. *p*-values were always lower than 0.001.

Predictions of hydrophobicity and α-helix propensity were done using ProtScale server (http://web.expasy.org/protscale/) [26,27], tertiary structure predictions were performed via I-TASSER server (http://zhanglab.ccmb.med.umich.edu/I-TASSER/) [28–30], following previous approaches [15]. Briefly, *Flavivirus* C protein sequences from our previous work were employed [14]. DENV and WNV (serotype Kunjin) C structures were excluded, not serving as templates for the tertiary structure prediction. ZIKV C protein structure was also not included, as it was not yet determined when the modeling was conducted. This avoids a bias towards known homologous protein structures. Five I-TASSER models were obtained for each C protein sequence. These were superimposed with DENV C experimental structure (PDB ID 1R6R, model 21) [12] after root-mean-square deviation (RMSD) minimization in UCSF Chimera v1.9 software [68]. Clusters were formed based on the visual similarity between predictions. The number of N-terminal amino acid residues with backbone clashes with the other monomer backbone was calculated for each model. In our previous work [15], a DENV C predicted structure was excluded from further analysis if it had 6 clashes or more, as it would not be viable as an homodimer [15]. Here we excluded models with more than 5 clashes (28 models rejected). These would preclude homodimer formation and, thus, were not considered in the clusters analysis (Table 1 excluded models column).

4.3. Structure Comparison Between DENV C and Influenza NS1

Protein structures coordinates were extracted from the Protein Data Bank (PDB, www.pdb.org). PDB identification codes are specified ahead after each protein name. The protein structures were superimposed through UCSF Chimera 1.13.1 software MatchMaker tool. After that, we carefully analyzed the superposition visually. Then, using the Match-Align tool of UCSF Chimera, which returns a sequence alignment based on the regions and taking into account the structure superimposition, we identified the residues simultaneously similar in structure and sequence. Protein structure figures were obtained using UCSF Chimera 1.13.1 version [68].

4.4. DENV C Recombinant Protein Production and Purification

Recombinant DENV C protein expression and purification was conducted based on previous approaches [13]. We used a pET-21a plasmid containing DENV serotype 2 strain New Guinea C capsid protein gene (encoding amino acid residues 1–100) [69]. The protein was expressed in *Escherichia coli* C41 and C43 bacteria grown in lysogeny broth (LB) medium. The only differences in the purification protocol are the abolition of the ammonium sulfate precipitation step and the addition of a size exclusion chromatography step (with Sephadex S200) after the heparin affinity column chromatography, using an AKTA chromatography equipment. The C protein was purified in a 55 mM KH_2PO_4, pH 6.0, 550 mM KCl. DENV C protein purified fractions were concentrated with Amicon Ultra-4 Centrifugal Filters of 3 or 10 kDa nominal cut-off, from Millipore (Billerica, MA, USA). Concentrated protein samples were stored at −80 °C. Protein samples quality was assessed by SDS-PAGE and matrix-assisted laser desorption/ionization, time-of-flight mass spectrometry (MALDI-TOF MS) analysis. Very low degradation and the highest peak consistent with the expected mass of the protein monomer (11765 Da).

4.5. Time-Resolved Fluorescence Anisotropy

Time-resolved fluorescence spectroscopy measurements were performed in a Life Spec II equipment with an EPLED-280 pulsed excitation light-emitting diode (LED) of 275 nm (Edinburgh Instruments, Livingston, UK), acquiring the emission at 350 nm. DENV C (monomer) concentration was 20 μM in 50 mM KH_2PO_4, 200 mM KCl, pH 6.0 or pH 7.5, with 550 μL total volume, in 0.5 cm × 0.5 cm quartz cuvettes. The instrument response function, IRF(t), was obtained with the same settings, except emission, which was at 280 nm, with a solution of polylatex beads of 60 nm diameter diluted in Mili-Q water. Measurements were performed at 22 °C. Time-resolved fluorescence intensity measurements with picosecond-resolution were obtained by the time-correlated single-photon timing (TCSPT) methodology [35]. Measurements were performed at constant time, with 15 min per decay, acquiring 2048 time points in a 50 ns window. Four intensity decays, $I(t)$, were acquired in each condition, with excitation/emission polarizers, respectively at vertical/vertical positions, $I_{VV}(t)$, vertical/horizontal positions, $I_{VH}(t)$, horizontal/vertical positions, $I_{HV}(t)$, and horizontal/horizontal positions, $I_{HH}(t)$. The instrumental G-factor was calculated as [35]:

$$G = \frac{\int_0^{50} I_{HV}(t)dt}{\int_0^{50} I_{HH}(t)dt} \tag{1}$$

The G-factor value obtained was 1.61. The intensity decay with emission polarizer at the magic angle ($\sim$54.7°, with respect to the vertical excitation polarizer), $I_m(t)$, avoids the effects of anisotropy. It can be calculated easily [35]:

$$I_m(t) = I_{VV}(t) + 2GI_{VH}(t) \tag{2}$$

with $I_{VV}(t)$ and $I_{VH}(t)$ depending on the time-resolved fluorescence anisotropy, $r(t)$, as:

$$I_{VV}(t) = \frac{I_m(t)}{3}(1 + 2r(t)) \tag{3}$$

$$I_{VH}(t) = \frac{I_m(t)}{3G}(1 - r(t)) \tag{4}$$

Thus, $I_m(t)$ was used to obtain the fluorescence lifetime components, τ_i, and the respective amplitudes, α_i, for the DENV C W69. $I_m(t)$ was described by a sum of three exponential terms:

$$I_m(t) = \sum_{i=1}^{3} \alpha_i e^{\left(-\frac{t}{\tau_i}\right)} \tag{5}$$

where the index i represents each component of the fluorescence decay. For the fitting to the data, α_i and τ_i values were obtained by iteratively convoluting $I_m(t)$ with the IRF(t):

$$I_m^{calc}(t) = I_m(t) \otimes \text{IRF}(t) \tag{6}$$

and fitting $I_m^{calc}(t)$ to the experimental data, $I_m^{exp}(t)$, using a non-linear least squares regression method. The usual statistical criteria, namely a reduced χ^2 value bellow 1.3 and a random distribution of weighted residuals, were used to evaluate the goodness of the fits [35]. Data analysis was performed using the TRFA Data Processing Package v1.4 (Scientific Software Technologies Centre, Belarusian State University, Minsk, Belarus) which allows calculating automatically the standard error (SE) for each fitted parameter [35].

The time-resolved fluorescence anisotropy, $r(t)$, is calculated via $I_{VV}(t)$, $I_{VH}(t)$ and G via_ENREF_52:

$$r(t) = \frac{I_{VV}(t) - GI_{VH}(t)}{I_{VV}(t) + 2GI_{VH}(t)} \tag{7}$$

In this case, the obtained $r(t)$ can be fitted to a single exponential decay [35]:

$$r(t) = r_0 e^{\left(-\frac{t}{\tau_c}\right)} \tag{8}$$

where r_0 is the anisotropy when $t\rightarrow 0$ and τ_c is the rotational correlation time. The $r(t)$ decays were globally analyzed in TRFA Data Processing Package v1.4 maintaining the previously obtained α_i and τ_i values constant, and convoluting Equations (3) and (4) with the respective IRF(t), analogously to the analysis of $I_m(t)$, using Equation (8) to fit $r(t)$. Values obtained for both pH conditions were considered statistically different if their 95% confidence intervals ($\sim$1.96 × SE) do not overlap (corresponding to $p < 0.05$).

4.6. Rotational Correlation Time Corrections

The τ_c of a molecule in solution is related with the solution viscosity, η, the molecular hydrodynamic volume, V, the Boltzmann constant, k_B, and the absolute temperature, T, as [35,70]:

$$\tau_c = \frac{\eta V}{k_B T} \tag{9}$$

Based on Equation (9), τ_c can be corrected for different temperatures, considering that the molecular volume does not change significantly in a small temperature interval ($\pm 5\,^\circ$C; i.e., V and k_B are constants), using [70]:

$$\frac{T_a \tau_{c,a}}{\eta_a} = \frac{T_b \tau_{c,b}}{\eta_b} \Leftrightarrow \tau_{c,b} = \tau_{c,a} \frac{\eta_b T_a}{\eta_a T_b} \tag{10}$$

where the indexes 'a' and 'b' represent a different condition of T and η, taking into account the variation of η with T [37]. The η values were assumed to be those of pure H_2O or 10% D_2O in the case of the corrections for the NMR-based values (those from the literature). In this way, Table 5 below shows the values employed on the calculations [37]:

Table 5. Values for η employed in this work, derived from the references and Equations above.

T (°C)	η in H_2O (cP)	η in 10% D_2O (cP)	$\frac{\eta_b T_a}{\eta_a T_b}$ in H_2O	$\frac{\eta_b T_a}{\eta_a T_b}$ in 10% D_2O
20	1.002	1.027	0.8736	0.8523
22	0.955	0.978	0.9231	0.9012
25	0.890	0.911	1	0.9770
27	0.851	0.871	1.0530	1.0293

4.7. Temperature Denaturation Measurements via Circular Dichroism (CD) Spectroscopy

Circular dichroism spectroscopy measurements were carried out in a JASCO J-815 (Tokyo, Japan), using 0.1 cm path length quartz cuvettes, data pitch of 0.5 nm, velocity of 200 nm/min, data integration time (DIT) of 1 s and performing 3 accumulations. Spectra were acquired in the far UV region, between 200 and 260 nm, with 1 nm bandwidth. The temperature was controlled by a JASCO PTC-423S/15 Peltier equipment. It was varied between 0 and 96 °C, in steps of 2 °C, increasing at a rate of 8 °C/min and waiting 100 s after crossing 5 times the target temperature, T. Then, the system was allowed, at least, 120 s to equilibrate (sufficient time for a stable CD signal). Before and after denaturation, spectra were acquired at 25 °C, to determine the reversibility of thermal denaturation. DENV C monomer concentration was 20 μM in 50 mM KH_2PO_4, 200 mM KCl, pH 6.0 or pH 7.5, with 220 μL of total volume. Spectra were smoothed through the means-movement method (using 7 points) and normalized to mean residue molar ellipticity, $[\theta]$ (in deg cm^2 dmol^{-1} Res^{-1}).

For the CD temperature denaturation data treatment, we assumed a dimer to monomer denaturation model [71–73] in which the folded dimer, F_2, separates into unfolded monomers, U, in a single step described by reaction R1:

$$F_2 \Leftrightarrow 2U \tag{R1}$$

In this system, the total protein concentration, $[P_m]$, in monomer equivalents, is described as:

$$[P_m] = 2[F_2] + [U] \tag{11}$$

Hereafter, concentrations are treated as dimensionless, being divided by the standard concentration of 1 M, in order to be at standard thermodynamic conditions. The fractions of monomer in the folded, f_F, and unfolded, f_U, states are calculated by [71,72]:

$$f_F = \frac{2[F_2]}{[P_m]} \tag{12}$$

$$f_U = \frac{[U]}{[P_m]} \tag{13}$$

$$f_F + f_U = 1 \tag{14}$$

and the concentrations of folded dimer and unfolded monomer can be written in terms of f_U:

$$[U] = f_U[P_m] \tag{15}$$

$$[F_2] = \frac{f_F[P_m]}{2} = \frac{(1 - f_U)[P_m]}{2} \tag{16}$$

Then, the equilibrium constant, K_{eq}, of R1 is defined in terms of $[U]$ and $[F_2]$, or f_U and $[P_m]$:

$$K_{eq} = \frac{[U]^2}{[F_2]} = \frac{(f_U[P_m])^2}{(1 - f_U)[P_m]/2} = \frac{2[P_m] \times f_U^2}{(1 - f_U)} \tag{17}$$

which can be solved in order to f_U, with the only solution in which $f_U \in [0; 1]$ being:

$$f_U = \frac{\sqrt{8[P_m]K_{eq} + K_{eq}^2} - K_{eq}}{4[P_m]} \tag{18}$$

The $[\theta]$ signal as a function of temperature [71,72,74], $[\theta]_T$, can be described as a linear combination of the signal of the folded, $[\theta]_{T,F}$, and unfolded states, $[\theta]_{T,U}$, weighted by f_U:

$$[\theta]_T = [\theta]_{T,F}(1 - f_U) + [\theta]_{T,U}f_U \tag{19}$$

where $[\theta]_{T,F}$ and $[\theta]_{T,U}$ have a variation with T described here by a straight line (*i* can be F or U) [72,74]:

$$[\theta]_{T,i} = m_i \times T + [\theta]_{0,i} \tag{20}$$

Equation (19) can be re-written to evidence f_U and then substitute it by Equation (18) [71,72]:

$$[\theta]_T = [\theta]_{T,F} + \left([\theta]_{T,U} - [\theta]_{T,F}\right)\frac{\sqrt{8[P_m]K_{eq} + K_{eq}^2} - K_{eq}}{4[P_m]} \tag{21}$$

K_{eq} can also be described by the standard Gibbs free-energy, ΔG°, of the reaction R1:

$$K_{eq} = e^{-\frac{\Delta G^\circ}{RT}} \tag{22}$$

where R is the rare gas constant and T is the absolute temperature. The ΔG° function used to fit the data contains both the enthalpic, ΔH°, and entropic, ΔS°, variations with temperature, which take into account $\Delta H^\circ_{T_m^\circ}$, the specific heat capacity at constant pressure, ΔC_p°, and the standard conditions' denaturation temperature, T_m°, according to [74]:

$$\Delta G^\circ = \Delta H^\circ_{T_m^\circ}\left(1 - \frac{T}{T_m^\circ}\right) - \Delta C_p^\circ\left(T_m^\circ - T + T \ln\left(\frac{T}{T_m^\circ}\right)\right) \tag{23}$$

In our data, ΔC_p° was statistically equal to 0 and, thus, Equation (23) can be simplified to:

$$\Delta G^{\circ} = \Delta H^{\circ}_{T_m^{\circ}}\left(1 - \frac{T}{T_m^{\circ}}\right) \tag{24}$$

Then, Equation (21) was combined with Equations (20), (22) and (24), and fitted to the data using GraphPad Prism v5 software, via the non-linear least squares method, to extract both the $\Delta H^{\circ}_{T_m^{\circ}}$ and T_m°, along with the respective SE values. Afterwards, $\Delta S^{\circ}_{T_m^{\circ}}$ can be obtained, since $\Delta G^{\circ} = 0$ kJ mol^{-1} at T_m°, via the following Equation:

$$\Delta H^{\circ}_{T_m^{\circ}} - T_m^{\circ}\Delta S^{\circ}_{T_m^{\circ}} = 0 \Rightarrow \Delta S^{\circ}_{T_m^{\circ}} = \frac{\Delta H^{\circ}_{T_m^{\circ}}}{T_m^{\circ}} \tag{25}$$

The SE of $\Delta S^{\circ}_{T_m^{\circ}}$ was calculated based on $\Delta H^{\circ}_{T_m^{\circ}}$, T_m°, and the respective SE values:

$$SE_{\Delta S^{\circ}_{T_m^{\circ}}} = \left|\frac{\Delta H^{\circ}_{T_m^{\circ}}}{T_m^{\circ}}\right| \times \sqrt{\left(\frac{SE_{\Delta H^{\circ}_{T_m^{\circ}}}}{\Delta H^{\circ}_{T_m^{\circ}}}\right)^2 + \left(\frac{SE_{T_m^{\circ}}}{T_m^{\circ}}\right)^2} \tag{26}$$

Interestingly, for a dimer to monomer denaturation, K_{eq} depends on $[P_m]$ and, consequently, ΔG° also depends on $[P_m]$. This implies that $\Delta G^{\circ} = 0$ at T_m° (T_m value estimated if $[P_m] = 1M$), which is considerably higher than the observed T_m (that occurs when $f_U = 0.5$). The dependence of T_m with $[P_m]$ is [72]:

$$\Delta G^{\circ}_{f_U=0.5} = -RT_m \ln([P_m]) \Rightarrow T_m = \frac{\Delta G^{\circ}_{f_U=0.5}}{-R\ln([P_m])} \tag{27}$$

$$T_m = \frac{\Delta H^{\circ}_{T_m^{\circ}}}{\Delta S^{\circ}_{T_m^{\circ}} - R\ln([P_m])} \tag{28}$$

The SE of T_m was based on the percentual SE value of T_m°.

Values obtained for both pH conditions were statistically evaluated via F-tests to compare two possible fits, one assuming a given parameter as being different for the distinct data sets, and another assuming that parameter to be equal between data sets (while maintaining the other parameters different). No statistically significant difference ($p < 0.05$) was observed.

Author Contributions: Conceptualization, A.F.F., N.C.S. and I.C.M.; In silico studies, A.F.F., V.A., A.S.M., N.K. and I.C.M.; Recombinant protein production, A.F.F., A.S.M., F.J.E. and I.C.M.; Time-resolved fluorescence anisotropy studies, A.F.F. and J.C.R.; Circular dichroism studies, A.F.F. and I.C.M.; Formal analysis, A.F.F., J.C.R. and I.C.M.; Resources, I.C.M., F.J.E., N.C.S.; Writing-original draft preparation, A.F.F., N.K., and I.C.M.; Writing-review and editing, A.F.F., A.S.M., N.K., N.C.S. and I.C.M.; Supervision, N.C.S. and I.C.M.; Project administration, N.C.S. and I.C.M.; Funding acquisition, N.C.S. and I.C.M.

Funding: This work was supported by "Fundação para a Ciência e a Tecnologia–Ministério da Ciência, Tecnologia e Ensino Superior" (FCT-MCTES, Portugal) project PTDC/SAU-ENB/117013/2010, Calouste Gulbenkian Foundation (FCG, Portugal) project Science Frontiers Research Prize 2010. A.F.F., A.S.M. and J.C.R. also acknowledge FCT-MCTES fellowships SFRH/BD/77609/2011, PD/BD/113698/2015 and SFRH/BD/95856/2013, respectively. I.C.M. acknowledges FCT-MCTES Programs "Investigador FCT" (IF/00772/2013) and "Concurso de Estímulo ao Emprego Científico" (CEECIND/01670/2017). This work was also supported by UID/BIM/50005/2019, project funded by Fundação para a Ciência e a Tecnologia (FCT)/ Ministério da Ciência, Tecnologia e Ensino Superior (MCTES) through Fundos do Orçamento de Estado.

Conflicts of Interest: The authors declare no conflict of interest. The funders had no role in the design of the study; in the collection, analyses, or interpretation of data; in the writing of the manuscript, and in the decision to publish the results.

Abbreviations

ALFV	Alfuy virus
APOE	Apolipoprotein E
AROAV	Aroa virus
BAGV	Bagaza virus
C protein	Capsid protein
CD	Circular dichroism
DENV	Dengue virus
ICTV	International Committee on Taxonomy of Viruses
IDP	Intrinsically disordered protein
IFN	Interferon
IGUV	Iguape virus
ILHV	Ilheus virus
JEV	Japanese encephalitis virus
KEDV	Kedougou virus
KOKV	Kokobera virus
LDs	Lipid droplets
MVEV	Murray Valley encephalitis virus
NS1	Non-structural protein 1 from influenza virus A
PDB	Protein Data Bank
pep14-23	Inhibitor peptide pep14-23 (amino acid sequence NMLKRARNRV)
PLIN3	Perilipin 3
ROCV	Rocio virus
SLEV	Saint Louis encephalitis virus
SPOV	Spondweni virus
USUV	Usutu virus
VLDL	Very low-density lipoproteins
WNV	West Nile virus
WNV-K	WNV serotype Kunjin
YFV	Yellow fever virus
ZIKV	Zika virus

References

1. Bhatt, S.; Gething, P.W.; Brady, O.J.; Messina, J.P.; Farlow, A.W.; Moyes, C.L.; Drake, J.M.; Brownstein, J.S.; Hoen, A.G.; Sankoh, O.; et al. The global distribution and burden of dengue. *Nature* **2013**, *496*, 504–507. [CrossRef] [PubMed]

2. Sanofi Pasteur. Available online: https://www.sanofipasteur.com/en/media-room/press-releases/dengvaxia-vaccine-approved-for-prevention-of-dengue-in-europe (accessed on 30 January 2019).

3. Durbin, A.P. A dengue vaccine. *Cell* **2016**, *166*, 1. [CrossRef] [PubMed]

4. Villar, L.; Dayan, G.H.; Arredondo-Garcia, J.L.; Rivera, D.M.; Cunha, R.; Deseda, C.; Reynales, H.; Costa, M.S.; Morales-Ramirez, J.O.; Carrasquilla, G.; et al. Efficacy of a tetravalent dengue vaccine in children in Latin America. *N. Engl. J. Med.* **2015**, *372*, 113–123. [CrossRef] [PubMed]

5. Takeda. Available online: https://www.takeda.com/newsroom/newsreleases/2019/takedas-dengue-vaccine-candidate-meets-primary-endpoint-in-pivotal-phase-3-efficacy-trial/ (accessed on 4 February 2019).

6. ICTV Taxonomy. Available online: https://talk.ictvonline.org/taxonomy/ (accessed on 17 April 2019).

7. Grard, G.; Moureau, G.; Charrel, R.N.; Holmes, E.C.; Gould, E.A.; de Lamballerie, X. Genomics and evolution of Aedes-borne flaviviruses. *J. Gen. Virol.* **2019**, *91*, 87–94. [CrossRef] [PubMed]

8. Schubert, A.M.; Putonti, C. Infection, genetics and evolution of the sequence composition of flaviviruses. *Infect. Genet. Evol.* **2010**, *10*, 129–136. [CrossRef] [PubMed]

9. Calisher, C.H.; Gould, E.A. Taxonomy of the virus family *Flaviviridae*. *Adv. Virus Res.* **2003**, *59*, 1–19. [PubMed]

10. Mukhopadhyay, S.; Kuhn, R.J.; Rossmann, M.G. A structural perspective of the flavivirus life cycle. *Nat. Rev. Microbiol.* **2005**, *3*, 13–22. [CrossRef]

11. Kuhn, R.J.; Zhang, W.; Rossmann, M.G.; Pletnev, S.V.; Corver, J.; Lenches, E.; Jones, C.T.; Mukhopadhyay, S.; Chipman, P.R.; Strauss, E.G.; et al. Structure of dengue virus: Implications for flavivirus organization, maturation, and fusion. *Cell* **2002**, *108*, 717–725. [CrossRef]

12. Ma, L.; Jones, C.T.; Groesch, T.D.; Kuhn, R.J.; Post, C.B. Solution structure of dengue virus capsid protein reveals another fold. *Proc. Natl. Acad. Sci. USA* **2004**, *101*, 3414–3419. [CrossRef]

13. Faustino, A.F.; Barbosa, G.M.; Silva, M.; Castanho, M.A.R.B.; da Poian, A.T.; Cabrita, E.J.; Santos, N.C.; Almeida, F.C.L.; Martins, I.C. Fast NMR method to probe solvent accessibility and disordered regions in proteins. *Sci. Rep.* **2019**, *9*, 1647. [CrossRef]

14. Martins, I.C.; Gomes-Neto, F.; Faustino, A.F.; Carvalho, F.A.; Carneiro, F.A.; Bozza, P.T.; Mohana-Borges, R.; Castanho, M.A.R.B.; Almeida, F.C.L.; Santos, N.C.; et al. The disordered N-terminal region of dengue virus capsid protein contains a lipid-droplet-binding motif. *Biochem. J.* **2012**, *444*, 405–415. [CrossRef]

15. Faustino, A.F.; Guerra, G.M.; Huber, R.G.; Hollmann, A.; Domingues, M.M.; Barbosa, G.M.; Enguita, F.J.; Bond, P.J.; Castanho, M.A.R.B.; da Poian, A.T.; et al. Understanding Dengue virus capsid protein disordered N-terminus and pep14-23-based inhibition. *ACS Chem. Biol.* **2015**, *10*, 517–526. [CrossRef]

16. Jones, C.T.; Ma, L.; Burgner, J.W.; Groesch, T.D.; Post, C.B.; Kuhn, R.J. Flavivirus capsid is a dimeric alpha-helical protein. *J. Virol.* **2003**, *77*, 7143–7149. [CrossRef] [PubMed]

17. Van Gorp, E.C.M.; Suharti, C.; Mairuhu, A.T.A.; Dolmans, W.M.V.; van der Ven, J.; Demacker, P.N.M.; van der Meer, J.W.M. Changes in the plasma lipid profile as a potential predictor of clinical outcome in dengue hemorrhagic fever. *Clin. Infect. Dis.* **2002**, *34*, 1150–1153. [CrossRef]

18. Samsa, M.M.; Mondotte, J.A.; Iglesias, N.G.; Assuncao-Miranda, I.; Barbosa-Lima, G.; da Poian, A.T.; Bozza, P.T.; Gamarnik, A. V Dengue virus capsid protein usurps lipid droplets for viral particle formation. *PLoS Pathog.* **2009**, *5*, e1000632. [CrossRef]

19. Suvarna, J.C.; Rane, P.P. Serum lipid profile: A predictor of clinical outcome in dengue infection. *Trop. Med. Int. Heal.* **2009**, *14*, 576–585. [CrossRef] [PubMed]

20. Carvalho, F.A.; Carneiro, F.A.; Martins, I.C.; Assunção-Miranda, I.; Faustino, A.F.; Pereira, R.M.; Bozza, P.T.; Castanho, M.A.R.B.; Mohana-Borges, R.; da Poian, A.T.; et al. Dengue virus capsid protein binding to hepatic lipid droplets (LD) is potassium ion dependent and is mediated by LD surface proteins. *J. Virol.* **2012**, *86*, 2096–2108. [CrossRef]

21. Faustino, A.F.; Carvalho, F.A.; Martins, I.C.; Castanho, M.A.R.B.; Mohana-Borges, R.; Almeida, F.C.L.; da Poian, A.T.; Santos, N.C. Dengue virus capsid protein interacts specifically with very low-density lipoproteins. *Nanomed. Nanotechnol. Biol. Med.* **2014**, *10*, 247–255. [CrossRef]

22. Faustino, A.F.; Martins, I.C.; Carvalho, F.A.; Castanho, M.A.R.B.; Maurer-Stroh, S.; Santos, N.C. Understanding dengue virus capsid protein interaction with key biological targets. *Sci. Rep.* **2015**, *5*, 10592. [CrossRef]

23. Martins, A.S.; Carvalho, F.A.; Faustino, A.F.; Martins, I.C.; Santos, N.C. West Nile virus capsid protein interacts with biologically relevant host lipid systems. *Front. Cell. Infect. Microbiol.* **2019**, *9*, 8. [CrossRef]

24. Martins, A.S.; Martins, I.C.; Santos, N.C. Methods for lipid droplet biophysical characterization in *Flaviviridae* infections. *Front. Microbiol.* **2018**, *9*, 1951. [CrossRef] [PubMed]

25. Shang, Z.; Song, H.; Shi, Y.; Qi, J.; Gao, G.F. Crystal structure of the capsid protein from Zika virus. *J. Mol. Biol.* **2018**, *430*, 948–962. [CrossRef] [PubMed]

26. Kyte, J.; Doolittle, R.F. A Simple method for displaying the hydropathic character of a protein. *J. Mol. Biol.* **1982**, *157*, 105–132. [CrossRef]

27. Deléage, G.; Roux, B. An algorithm for protein secondary structure prediction based on class prediction. *Protein Eng.* **1987**, *1*, 289–294. [CrossRef] [PubMed]

28. Zhang, Y. I-TASSER server for protein 3D structure prediction. *BMC Bioinform.* **2008**, *8*, 1–8. [CrossRef]

29. Yang, J.; Yan, R.; Roy, A.; Xu, D.; Poisson, J.; Arbor, A.; Arbor, A. The I-TASSER suite: Protein structure and function prediction. *Nat. Methods* **2015**, *12*, 7–8. [CrossRef] [PubMed]

30. Roy, A.; Kucukural, A.; Zhang, Y. I-TASSER: A unified platform for automated protein structure and function prediction. *Nat. Protoc.* **2011**, *5*, 725–738. [CrossRef]

31. Dokland, T.; Walsh, M.; Mackenzie, J.M.; Khromykh, A.A.; Ee, K.-H.; Wang, S. West Nile virus core protein; tetramer structure and ribbon formation. *Structure* **2004**, *12*, 1157–1163. [CrossRef]

32. Zhan, C.; Zhao, L.; Chen, X.; Lu, W.; Lu, W. Total chemical synthesis of dengue 2 virus capsid protein via native chemical ligation: Role of the conserved salt-bridge. *Bioorg. Med. Chem.* **2013**, *21*, 3443–3449. [CrossRef]

33. Morando, M.A.; Barbosa, G.M.; Cruz-Oliveira, C.; da Poian, A.T.; Almeida, F.C.L. Dynamics of Zika virus capsid protein in solution: The properties and exposure of the hydrophobic cleft are controlled by the α-helix 1 sequence. *Biochemistry* **2019**, *58*, 2488–2498. [CrossRef]

34. Kumar, S.; Ravi, V.K.; Swaminathan, R. How do surfactants and DTT affect the size, dynamics, activity and growth of soluble lysozyme aggregates? *Biochem. J.* **2008**, *415*, 275–288. [CrossRef]

35. Lakowicz, J. *Principles of Fluorescence Spectroscopy*, 3rd ed.; Springer Science, LLC: Berlin/Heidelberg, Germany, 2006; ISBN 9780387312781.

36. Rossi, P.; Yuanpeng, G.V.T.S.; James, J.H.; Anklin, C.; Conover, K.; Hamilton, K.; Xiao, R. A microscale protein NMR sample screening pipeline. *J. Biomol. NMR* **2010**, *46*, 11–22. [CrossRef] [PubMed]

37. Cho, C.H.; Urquidi, J.; Singh, S.; Robinson, G.W. Thermal offset viscosities of liquid H_2O, D_2O, and T_2O. *J. Phys. Chem. B* **1999**, *103*, 1991–1994. [CrossRef]

38. Martins, I.C.; Almeida, F.C.L.; Santos, N.C.; da Poian, A.T. DENV–Derived Peptides and Methods for the Inhibition of Flavivirus Replication. International Patent Publication Nr WO/2012/159187, 26 May 2011.

39. Ivanyi-Nagy, R.; Lavergne, J.; Gabus, C.; Ficheux, D.; Darlix, J.; Inserm, L.; Supe, E.N. RNA chaperoning and intrinsic disorder in the core proteins of *Flaviviridae*. *Nucleic Acids Res.* **2008**, *36*, 712–725. [CrossRef]

40. Ivanyi-Nagy, R.; Darlix, J. Core protein-mediated 5–3 annealing of the West Nile virus genomic RNA in vitro. *Virus Res.* **2012**, *167*, 226–235. [CrossRef]

41. Kumar, M.; Gromiha, M.M.; Raghava, G.P.S. SVM based prediction of RNA-binding proteins using binding residues and evolutionary information. *J. Mol. Recognit.* **2011**, *24*, 303–313. [CrossRef] [PubMed]

42. Järvelin, A.I.; Noerenberg, M.; Davis, I.; Castello, A. The new (dis)order in RNA regulation. *Cell Commun. Signal.* **2016**, *14*, 9. [CrossRef]

43. Shavinskaya, A.; Boulant, S.; Penin, F.; McLauchlan, J.; Bartenschlager, R. The lipid droplet binding domain of hepatitis C virus core protein is a major determinant for efficient virus assembly. *J. Biol. Chem.* **2007**, *282*, 37158–37169. [CrossRef]

44. Cheng, A.; Wong, S.M.; Yuan, Y.A. Structural basis for dsRNA recognition by NS1 protein of influenza A virus. *Cell Res.* **2009**, *19*, 187–195. [CrossRef]

45. Fernandez-Sesma, A.; Marukian, S.; Ebersole, B.J.; Kaminski, D.; Park, M.S.; Yuen, T.; Sealfon, S.C.; Garcia-Sastre, A.; Moran, T.M. Influenza virus evades innate and adaptive immunity via the NS1 protein. *J. Virol.* **2006**, *80*, 6295–6304. [CrossRef]

46. Wang, S.H.; Syu, W.J.; Huang, K.J.; Lei, H.Y.; Yao, C.W.; King, C.C.; Hu, S.T. Intracellular localization and determination of a nuclear localization signal of the core protein of dengue virus. *J. Gen. Virol.* **2002**, *83*, 3093–3102. [CrossRef] [PubMed]

47. Kobe, B. Autoinhibition by an internal nuclear localization signal revealed by the crystal structure of mammalian importin α. *Nat. Struct. Biol.* **1999**, *6*, 388–397. [CrossRef] [PubMed]

48. Catimel, B.; Teh, T.; Fontes, M.R.M.; Jennings, I.G.; Jans, D.A.; Howlett, G.J.; Nice, E.C.; Kobe, B. Biophysical characterization of interactions involving importin-α during nuclear import. *J. Biol. Chem.* **2001**, *276*, 34189–34198. [CrossRef] [PubMed]

49. Marfori, M.; Mynott, A.; Ellis, J.J.; Mehdi, A.M.; Saunders, N.F.W.; Curmi, P.M.; Forwood, J.K.; Boden, M.; Kobe, B. Molecular basis for specificity of nuclear import and prediction of nuclear localization. *Biochim. Biophys. Acta* **2011**, *1813*, 1562–1577. [CrossRef] [PubMed]

50. Fontes, M.R.M.; Teh, T.; Kobe, B. Structural basis of recognition of monopartite and bipartite nuclear localization sequences by mammalian importin-α. *J. Mol. Biol.* **2000**, *297*, 1183–1194. [CrossRef] [PubMed]

51. Marfori, M.; Lonhienne, T.G.; Forwood, J.K.; Kobe, B. Structural basis of high-affinity nuclear localization signal interactions with importin-alpha. *Traffic* **2012**, *13*, 532–548. [CrossRef] [PubMed]

52. Tadano, M.; Makino, Y.; Fukunaga, T.; Okuno, Y.; Fukai, K. Detection of dengue 4 virus core protein in the nucleus I. A monoclonal antibody to dengue 4 virus reacts with the antigen in the nucleus and cytoplasm. *J. Gen. Virol.* **1989**, *70*, 1409–1415. [CrossRef] [PubMed]

53. Makino, Y.; Tadano, M.; Anzai, T.; Ma, S.P.; Yasuda, S.; Žagar, E. Detection of dengue 4 virus core protein in the nucleus II. Antibody against dengue 4 core protein produced by a recombinant baculovirus reacts with the antigen in the nucleus. *J. Gen. Virol.* **1989**, *70*, 1417–1425. [CrossRef] [PubMed]

54. Wagstaff, K.M.; Sivakumaran, H.; Heaton, S.M.; Harrich, D.; Jans, D.A. Ivermectin is a specific inhibitor of importin α/β-mediated nuclear import able to inhibit replication of HIV-1 and dengue virus. *Biochem. J.* **2012**, *443*, 851–856. [CrossRef]

55. Bergmann, M.; Garcia-Sastre, A.; Carnero, E.; Pehamberger, H.; Wolff, K.; Palese, P.; Muster, T. Influenza virus NS1 protein counteracts PKR-mediated inhibition of replication. *J. Virol.* **2000**, *74*, 6203–6206. [CrossRef]

56. Kochs, G.; Garcia-Sastre, A.; Martinez-Sobrido, L. Multiple anti-interferon actions of the influenza A virus NS1 protein. *J. Virol.* **2007**, *81*, 7011–7021. [CrossRef] [PubMed]

57. Rodriguez-Madoz, J.R.; Bernal-Rubio, D.; Kaminski, D.; Boyd, K.; Fernandez-Sesma, A. Dengue virus inhibits the production of type I interferon in primary human dendritic cells. *J. Virol.* **2010**, *84*, 4845–4850. [CrossRef] [PubMed]

58. Min, J.Y.; Krug, R.M. The primary function of RNA binding by the influenza A virus NS1 protein in infected cells: Inhibiting the 2′–5′ oligo (A) synthetase/RNase L pathway. *Proc. Natl. Acad. Sci. USA* **2006**, *103*, 7100–7105. [CrossRef]

59. Uversky, V.N. Intrinsically disordered proteins and their "mysterious" (meta)physics. *Front. Phys.* **2019**, *7*, 10. [CrossRef]

60. Na, J.H.; Lee, W.K.; Yu, Y.G. How do we study the dynamic structure of unstructured proteins: A case study on nopp140 as an example of a large, intrinsically disordered protein. *Int. J. Mol. Sci.* **2018**, *19*, 381. [CrossRef]

61. Uversky, V.N. Introduction to intrinsically disordered proteins (IDPs). *Chem. Rev.* **2014**, *114*, 6557–6560. [CrossRef] [PubMed]

62. Minde, D.P.; Halff, E.F.; Tans, S. Designing disorder: Tales of the unexpected tails. *Intrinsically Disord. Proteins* **2013**, *1*, e26790. [CrossRef]

63. Krystkowiak, I.; Manguy, J.; Davey, N.E. PSSMSearch: A server for modeling, visualization, proteome-wide discovery and annotation of protein motif specificity determinants. *Nucleic Acids Res.* **2018**, *46*, W235–W241. [CrossRef]

64. Bera, A.K.; Kuhn, R.J.; Smith, J.L. Functional characterization of cis and trans activity of the Flavivirus NS2B-NS3 protease. *J. Biol. Chem.* **2007**, *282*, 12883–12892. [CrossRef]

65. Niyomrattanakit, P.; Yahorava, S.; Mutule, I.; Mutulis, F.; Petrovska, R.; Prusis, P.; Katzenmeier, G.; Wikberg, J.E. Probing the substrate specificity of the dengue virus type 2 NS3 serine protease by using internally quenched fluorescent peptides. *Biochem. J.* **2006**, *397*, 203–211. [CrossRef]

66. Sievers, F.; Higgins, D.G. Clustal omega. *Curr. Protoc. Bioinform.* **2014**, *13*, 1–16.

67. Sievers, F.; Wilm, A.; Dineen, D.; Gibson, T.J.; Karplus, K.; Li, W.; Lopez, R.; Thompson, J.D.; Higgins, D.G.; Mcwilliam, H.; et al. Fast, scalable generation of high-quality protein multiple sequence alignments using clustal omega. *Mol. Syst. Biol.* **2011**, *7*, 539. [CrossRef]

68. Pettersen, E.F.; Goddard, T.D.; Huang, C.C.; Couch, G.S.; Greenblatt, D.M.; Meng, E.C.; Ferrin, T.E. UCSF chimera—A visualization system for exploratory research and analysis. *J. Comput. Chem.* **2004**, *25*, 1605–1612. [CrossRef]

69. Irie, K.; Mohan, P.; Sasaguri, Y.; Putnak, R.; Padmanabhan, R. Sequence analysis of cloned dengue virus type 2 genome (New Guinea-C strain). *Gene* **1989**, *75*, 197–211. [CrossRef]

70. Smith, P.; van Gunsteren, W. Translational and rotational diffusion of proteins. *J. Mol. Biol.* **1994**, *236*, 629–636. [CrossRef] [PubMed]

71. Mok, Y.; de Prat Gay, G.; Butler, P.; Bycroft, M. Equilibrium dissociation and unfolding. *Protein Sci.* **1996**, *5*, 310–319. [CrossRef]

72. Rumfeldt, J.; Galvagnion, C.; Vassall, K.; Meiering, E. Conformational stability and folding mechanisms of dimeric proteins. *Prog. Biophys. Mol. Biol.* **2008**, *98*, 61–84. [CrossRef]

73. Neet, K.E.; Timm, D.E. Conformational stability of dimeric proteins: Quantitative studies by equilibrium denaturation. *Protein Sci.* **1994**, *3*, 2167–2174. [CrossRef]

74. Allen, D.L.; Pielak, G.J. Baseline length and automated fitting of denaturation data. *Protein Sci.* **1998**, *7*, 1262–1263. [CrossRef]

© 2019 by the authors. Licensee MDPI, Basel, Switzerland. This article is an open access article distributed under the terms and conditions of the Creative Commons Attribution (CC BY) license (http://creativecommons.org/licenses/by/4.0/).

International Journal of
Molecular Sciences

Article

Investigation into Early Steps of Actin Recognition by the Intrinsically Disordered N-WASP Domain V

Maud Chan-Yao-Chong [1,2], Dominique Durand [2] and Tâp Ha-Duong [1,*]

[1] BioCIS, University Paris-Sud, CNRS UMR 8076, University Paris-Saclay, 92290 Châtenay-Malabry, France;
   maud.chan-yao-chong@u-psud.fr
[2] Institute for Integrative Biology of the Cell (I2BC), CEA, CNRS, University Paris-Sud,
   University Paris-Saclay, 91190 Gif-sur-Yvette, France; dominique.durand@i2bc.paris-saclay.fr
* Correspondence: tap.ha-duong@u-psud.fr; Tel.: +33-1-46-83-57-38

Received: 31 August 2019; Accepted: 8 September 2019; Published: 11 September 2019

Abstract: Cellular regulation or signaling processes are mediated by many proteins which often have one or several intrinsically disordered regions (IDRs). These IDRs generally serve as binders to different proteins with high specificity. In many cases, IDRs undergo a disorder-to-order transition upon binding, following a mechanism between two possible pathways, the induced fit or the conformational selection. Since these mechanisms contribute differently to the kinetics of IDR associations, it is important to investigate them in order to gain insight into the physical factors that determine the biomolecular recognition process. The verprolin homology domain (V) of the Neural Wiskott–Aldrich Syndrome Protein (N-WASP), involved in the regulation of actin polymerization, is a typical example of IDR. It is composed of two WH2 motifs, each being able to bind one actin molecule. In this study, we investigated the early steps of the recognition process of actin by the WH2 motifs of N-WASP domain V. Using docking calculations and molecular dynamics simulations, our study shows that actin is first recognized by the N-WASP domain V regions which have the highest propensity to form transient α-helices. The WH2 motif consensus sequences "LKKV" subsequently bind to actin through large conformational changes of the disordered domain V.

Keywords: intrinsically disordered protein; protein–protein interaction; molecular docking; molecular dynamics

---

1. Introduction

Intrinsically disordered proteins (IDPs) play important roles in the regulation of many biological processes, such as cell growth, cell signaling, and cell survival. To exert these functions, their intrinsically disordered regions (IDRs) often bind to different proteins with high specificity and low affinity [1–4]. In many cases, it is observed that IDRs adopt well structured conformations when bound to their partners [5]. Segments that undergo such a disorder-to-order transition upon binding are frequently called Molecular Recognition Features (MoRFs) in the literature [4,6–10].

A typical IDR with a MoRF is the WASP-homology 2 (WH2) motif, which is found in about 50 proteins [11]. WH2 motifs are actin-binding modules of about 30–50 residues that are key players in regulation of the cytoskeleton actin polymerization, dynamics, and organization [11–13]. Proteins of the WH2 family can contain one to four WH2 motifs, each being able to bind one G-actin monomer (Table S1). In unbound state, WH2 motifs are intrinsically disordered, and, in complex with actin, they all share a similar binding mode: their N-terminal part folds into an α-helix which interacts with the barbed face of actin, between subdomains 1 and 3, while their central consensus sequence "LKKV" has an extended conformation which lies on the actin's surface, between subdomains 1 and 2 [11,14,15] (see Figure 2B). Although these actin–WH2 motif structures were determined by X-ray diffraction, the common folding of different WH2 motifs upon binding to actin indicates, with reasonable confidence, that it is probably similar to the one adopted in solution.

It should be noted that, when a WH2 motif or a peptide construct encompassing a WH2 motif is co-crystallized with actin, only the coordinates of about 20 residues, generally from the beginning of the helical segment to the consensus sequence "LKKV", were resolved in most crystallographic complexes (Table S1). Only the crystallographic structures 2A41, 2D1K, and 5YPU contain almost all residue coordinates of the co-crystallized WH2 motifs. The absence in most crystallographic structures of atomic coordinates for regions after the consensus sequence "LKKV" indicates that they probably keep a highly flexible and disordered conformation upon binding to actin, forming so-called fuzzy complexes. Questions that could be raised here are: What is the conformational dynamics of these invisible regions? Are they interacting with actin, and, if so, with which residues?

A more general and still debated question regarding IDRs concerns the mechanism of their specific binding to their partners. The formation of IDP–protein complexes can indeed follow a pathway between two possible mechanisms [16]: the "induced fit" pathway, in which the disordered region binds to its partner and folds into an ordered structure on its surface, and the "conformational selection" mechanism, in which the folded structure preexists among the ensemble of conformations of the unbound IDP and is recognized by the protein partner. However, the observation of preexisting structured segments in IDRs does not necessarily prove that the binding proceeds by a direct conformational selection [17]. For example, an alternative mechanism could be that the protein partner first binds to any IDR region and slides to the specific binding site which has the correct complementary conformation [18]. Thus, closer investigations are required to gain insight into the early events and pathways of the IDP–protein recognition mechanism.

In this report, we address these issues in the case of the verprolin homology domain (V) of the Neural Wiskott–Aldrich Syndrome Protein (N-WASP), which has two WH2 motifs. With the Arp2/3 complex, N-WASP stimulates actin filament branching and the formation of dendritic networks of filaments that shape or deform cell membranes in several cellular processes, such as cell motility or endocytosis [19,20]. The 505-residue sequence of the human N-WASP can be decomposed into seven domains: a primary WASP homology domain WH1 (segment 1–150), a basic domain B (186–200), a GTPase-binding domain GBD (203–274), a proline-rich domain PRD (277–392), a verprolin homology domain V (405–450), a cofilin homology domain C (451–485), and an acidic domain A (486–505) [21,22]. N-WASP domain V binds and recruits G-actin monomers, while domains CA are attached to the Arp2/3 complex. These associations allow the nucleation of new branch filaments [19,23,24]. N-WASP domain V is composed of two WH2 motifs (Table S1), each being able to bind one G-actin [25–27]. Interestingly, the presence of two WH2 motifs in N-WASP domain V induces more rapid actin polymerization than the other proteins of the WASP family which have only one WH2 motif [28]. However, the structural mechanism by which a tandem of WH2 motifs binds two actin monomers and accelerates polymerization and branching is not completely elucidated.

Two crystallographic structures of the N-WASP WH2 tandem in complex with actin are available in the Protein Data Bank: a 1:1 actin–domain VC (2VCP [27]) and a 2:1 actin–WH2 tandem (3M3N [26]). Nevertheless, in both 2VCP and 3M3N structures, we emphasize again that only about 20 residues of each WH2 motif, from the helical N-terminal part to the consensus sequence "LKKV", could be resolved by X-ray experiments (Table S1). It should be noted that the actin dimer in 3M3N complex has an overall longitudinal arrangement similar to that one in actin filament [26]. This suggests that N-WASP domain V might favor the formation of actin dimers in a longitudinal filament-like conformation, which might accelerate actin polymerization. However, to confirm this scenario, a detailed description of the formation of the 1:1 and 2:1 actin–domain V complexes in solution is required.

Previously, we structurally characterized the unbound state of a construct encompassing N-WASP domain V (Figure S1) by combining various biophysical techniques [29]. Multiple molecular dynamics (MD) simulations allowed generating a conformational ensemble of this construct (which we continue to call "N-WASP domain V" for simplicity) in very good agreement with both NMR chemical shifts and SAXS intensity measurements. In this ensemble, several conformations were identified with transient α-helices in the WH2 motifs, suggesting that these secondary structures might be selected by actin during the recognition process. We query here the validity of this hypothesis and, more generally, investigate the early events of actin recognition by these α-MoRFs, using protein–protein docking calculations and multiple MD simulations. In addition, since N-WASP has a tandem of WH2 motifs, we examine the possible molecular pathways leading to the ternary complex of domain V with two actins.

2. Results

NMR experiments and MD simulations previously showed that unbound N-WASP domain V has two transient α-helical structures (one per WH2 domain) at regions 10–15 and 37–43 corresponding to residues 407–412 and 434–440 in the whole protein sequence (Figure S1) [29].

2.1. Monomeric Actin–Domain V Encounter Complexes Generated by Docking Calculations

To examine whether these two helical MoRFs are preferential recognition sites for actin, we blindly docked the 527 most populated clusters of N-WASP domain V conformational ensemble (derived from MD simulations with the A03ws force field [29]) onto the actin chain B extracted from the PDB structure 2VCP [27]. Each docking generated about 1300 different poses of domain V on actin, yielding a total number of 702,920 encounter complexes. The likeliness of these complexes was evaluated with the scoring function 2/3B^{best} InterEvScore [30]. We delineated the 1% of complexes (i.e., 7030 conformers) having the highest 2/3B^{best} score as the most probable actin–domain V structures. It could be noted that, when compared to the 527 cluster representative structures, the domain V conformations that are retrieved in the 7030 most probable complexes are sightly more compact, as indicated by the radius of gyration distributions (Figure S2), indicating that extended conformations of domain V did not particularly favor their binding to actin. At the local level, the difference in probability for residues to be in α-helix, between the two ensembles of 527 clusters and of 7030 ligands, appears quite small and may not be significant (Figure S2).

We first analyzed the residues at the protein–protein interface in the 7030 most probable complex structures. The probability of N-WASP domain V residues to be in contact with actin was computed, as plotted in Figure 1. Clearly, it can be observed that actin preferentially recognizes two regions of domain V which can be delimited by residues 8–18 and 37–50. The first binding site is shorter than the second one, which might be related to the difference in propensity of the two WH2 motifs to form α-helical structures (Figure S2). Nevertheless, when the two regions with high probability to be contacted by actin are compared, a consensus sequence can be identified as the most probable recognition site for actin: ^{9}KAALLDQIRE^{18} and ^{37}RDALLDQIRQ^{46} in the first and second WH2 motif, respectively. It is worth noting that both recognition segments exhibit a similar pattern in which a positively charged residue (K9 or R37) precedes two moderate probability residues (A10/A11 or D38/A39), followed by two high probability hydrophobic residues (L12/L13 or L40/L41) and again two moderate probability ones (D14/Q15 or D42/Q43), before two other high probability residues (I16/R17 or I44/R45). This pattern suggests that the domain V recognized regions are rather α-helical structures than short linear motifs (SLiMs) in coil or extended conformations. The chemical nature of the mentioned residues also indicates that the central parts of the recognized segments are amphiphilic helices with their hydrophobic faces in contact with actin.

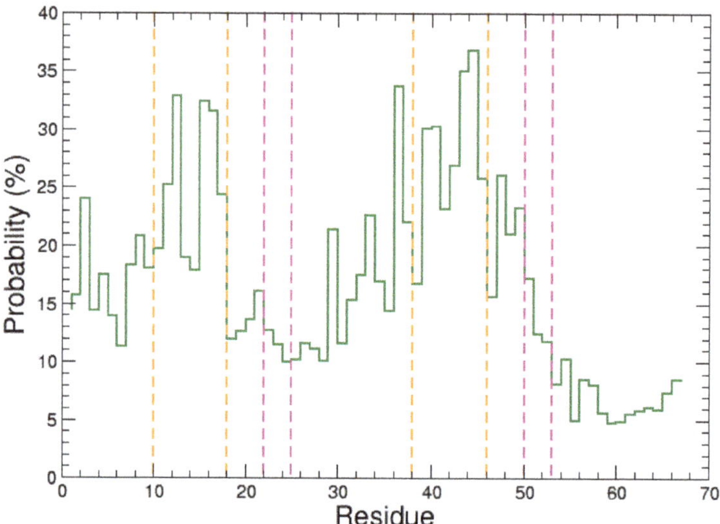

Figure 1. Probability of the N-WASP domain V residues to be distant by less than 4 Å from actin. Orange and magenta dashed lines indicate the protein regions in α-helix (as revealed by the X-ray structure 2VCP [27]) and the consensus sequences "LKKV" [14,31], respectively.

Besides, it could be noted that, among the most probable complexes, the conserved residues [22]LKKV[25] and [50]LKSV[53] have significantly lower probability to be in contact with actin than the two previous binding sites (Figure 1). This suggests that, after the recognition of regions 9–18 or 37–46 by actin, the N-WASP consensus sequences "LKKV" should move and anchor to the actin's surface in a second step. This scenario was further examined using MD simulations, as presented in the next section.

Before that, we investigated the preferential location of the two N-WASP regions 9–18 and 37–46 on actin's surface. To that end, the probability that actin residues are contacted by one of these two segments was computed over the 7030 most probable complexes predicted by docking, as plotted in Figure 2A. Among the actin residues which are frequently contacted by regions 9–18 and 37–46, we retrieved those (Y143, G146, T148, G168, Y169, L349, T351, M355, and F375) which make contacts with the N-WASP segment 37–46 in structure 2VCP [27]. However, we also observed that segments 9–18 or 37–46 can bind to other patches of the actin's surface with high probability, notably residues 171–173 and 283–290, which are not close to the cognate binding site (Figure 2). These observations could arise from various factors, including limitations of the rigid-body docking procedure and imperfections of the coarse-grained scoring function. This could be also related to the fact that, in most selected conformations of N-WASP domain V used in docking calculations, segments 9–18 and 37–46 were not fully helical, unlike in the crystallographic complex (Figure S2). This might favor the binding to pockets of the actin's surface with no particular shape, to the detriment of the groove that is expected to accommodate the WH2 motif helices. In these cases, the conformational transition of these N-WASP regions toward full α-helices might not lead to stable complexes. Besides, it could be noted that these non-specific binding sites on actin monomer also extend over the actin–actin interface in longitudinal dimers and, therefore, might be less observed in such actin assemblies.

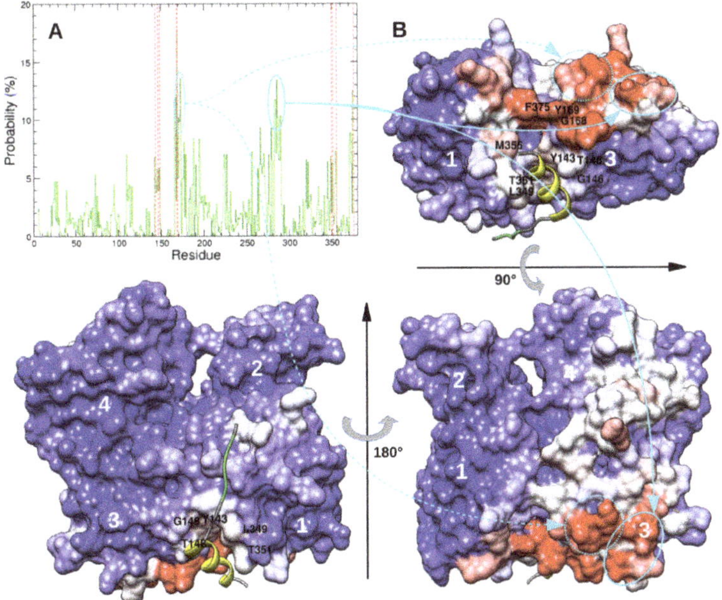

Figure 2. (**A**) Probability of actin residues to be distant by less than 4 Å from domain V regions 9–18 or 37–46. Red dashed lines indicate actin residues in contact with N-WASP helical segment in structure 2VCP [27]. (**B**) Views of actin's surface colored proportionally to previous probabilities. Blue, white, and red colors indicate actin residues with low, intermediate, and high probabilities to be contacted by domain V, respectively. As a reference, yellow and green ribbons represent the second WH2 motif helical region and conserved sequence LKSV as observed in 2VCP [27].

Overall, docking calculations of representative conformations of free domain V on actin monomer yielded many encounter complexes in which N-WASP segments 9–18 and 37–46 are preferentially bound to actin, but to both specific and non-specific sites. In these encounter complexes, consensus sequences "LKKV" have low probability to be in contact with actin, whereas they are found attached to actin in all available crystallographic complex structures. This suggests a two-step association mechanism involving large conformational rearrangements of domain V after the formation of a productive encounter complex with either segment 9–18 or 37–46 in cognate binding site of actin.

2.2. Identification and MD Simulations of Productive Actin–Domain V Encounter Complexes

The binding mechanism of N-WASP domain V to actin was further investigated using MD simulations of productive encounter complexes selected on the basis of the position and orientation of regions 9–18 or 37–46 in the cognate actin binding groove. More specifically, among the 7030 most probable complexes generated by docking, we identified those with residues 9–18 or 37–46 contacting at least six actin residues over the nine observed in contact with the N-WASP region 37–46 in the X-ray structure (Y143, G146, T148, G168, Y169, L349, T351, M355, and F375). We found a total of 194 complexes which have one of the two recognized segments in contact with at least six of the nine actin hot-spot residues. However, in a large number of these complexes, the segment 9–18 or 37–46 is oriented in the opposite direction of the crystallographic helix, so that the consensus sequence "LKKV" would not be able to reach its cognate binding site. Thus, we further filtered the 194 complexes based on the angle between the principal axis of segment 9–18 or 37–46 and that one of the helical region 37–46 in crystal. We obtained 16 and 18 complexes in which this angle is lower than 30° for N-WASP regions 9–18 and 37–46, respectively (Tables S2 and S3).

In these 34 productive actin–domain V encounter complexes, the recognized regions 9–18 and 37–46 are surprisingly not completely folded in α-helix, but can have various local conformations with 0–6 over 10 residues in helical structures. Nevertheless, it should be noted that the lack of helical residues is often balanced by several residues with a turn motif. This is notably the case for four over the five complexes which have region 9–18 or 37–46 RMSD lower than 5 Å relative to the crystallographic structure (Tables S2 and S3). In the 34 actin–domain V complexes, the consensus segments "LKKV" are variously far off from their cognate binding site on actin, as indicated by their RMSD values ranging from 8.7 to 37.7 Å. To study the complete association process of N-WASP WH2 motifs, we performed MD simulations of actin–bound domain V conformational changes starting from the two structures which have region 9–18 or 37–46 with the lowest RMSD relative the structure 2VCP (Figure 3). These selected productive encounter complexes are hereafter denoted CplxA and CplxB.

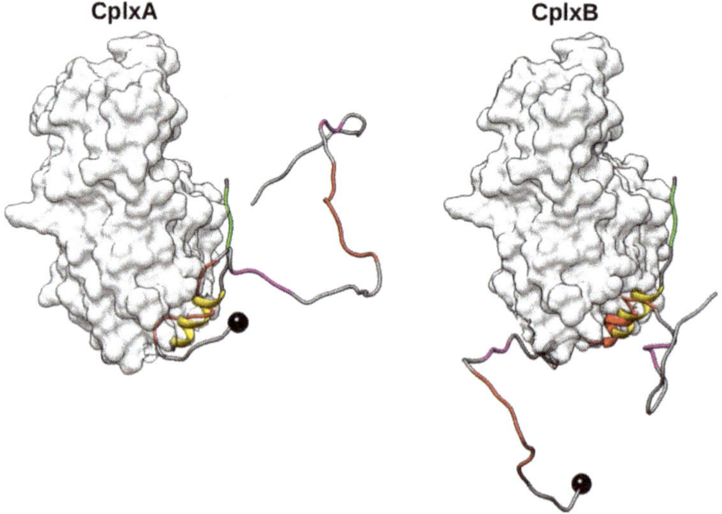

Figure 3. Side view of the two best 1:1 actin–domain V encounter complexes with N-WASP segment 9–18 (**left**) or 37–46 (**right**) located and oriented as in structure 2VCP. Black balls are N-terminal $C\alpha$-atoms of domain V. Red and magenta ribbons represent its regions 9–18 or 37–46 and consensus sequences "LKKV", respectively. As a reference, yellow and green ribbons indicate the helical and ^{50}LKSV^{53} regions of domain VC in 2VCP.

For each selected encounter complex, two MD simulations of about 350 ns were performed from the same coordinates but with different initial velocities. These four simulations will be referred to as CplxA_MD1, CplxA_MD2, CplxB_MD1, and CplxB_MD2. In all complex trajectories, the actin tertiary structure remains stable, with RSMD relative to structure 2VCP fluctuating below 5.2 Å (Figure 4). Regarding the N-WASP regions 9–18 and 37–46 (which are bound to actin in CplxA and CplxB, respectively), their position and orientation are maintained in the actin binding site in three over four simulations (CplxA_MD1, CplxA_MD2, and CplxB_MD1), as indicated by their average RMSD values relative to the complex 2VCP (4.4, 4.4, and 2.7 Å, respectively). A visual inspection of the CplxB_MD2 trajectory showed that segment 37–46 slid toward the bottom of actin, explaining its higher RMSD (8.2 Å on average). For the three other simulations, the N-WASP regions 9–18 and 37–46 remain attached to their binding site after the formation of productive encounter complexes.

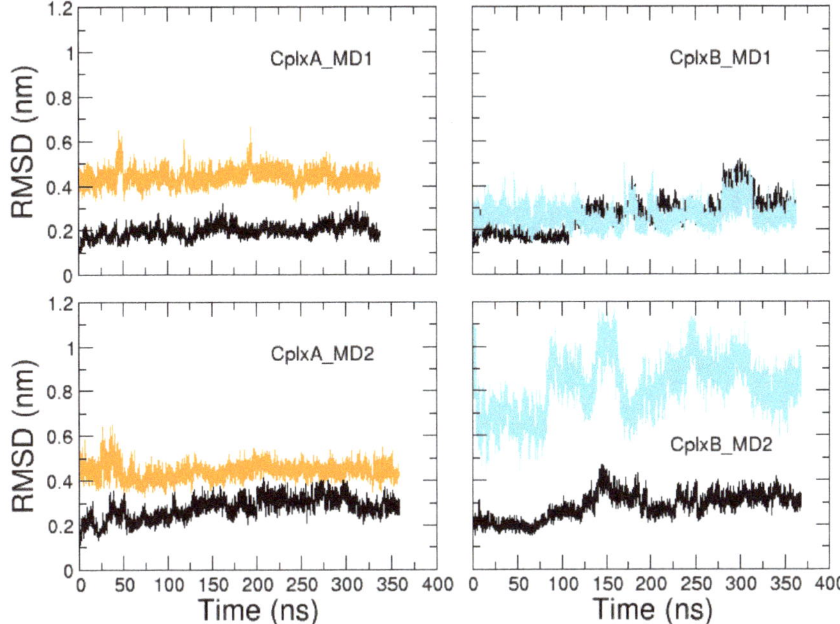

Figure 4. Time evolutions of RMSD relative to structure 2VCP, after fitting MD trajectories on crystallographic actin, for actin (black) and segments 9–18 (orange) and 37–46 (cyan) of N-WASP domain V.

Next, we monitored the dynamics of residues ^{22}LKKV^{25} and ^{50}LKSV^{53} relative to their cognate binding site on actin. As shown in Figure 5, segments ^{22}LKKV^{25} and ^{50}LKSV^{53} had large amplitude motions in all four simulations, without reaching stable bound positions on actin. Strikingly, the minimal distance to actin of these residues and their RMSD relative to structure 2VCP seem to be highly correlated, which can be explained as follows: Once N-WASP domain V helical region 9–18 or 37–46 is correctly positioned and oriented in its cognate binding site, if segment ^{22}LKKV^{25} or ^{50}LKSV^{53} is detached from actin's surface, it is largely free to move in solvent, accounting for large RMSD values. However, when it is bound to actin, its accessible space is narrowed down to a region close to the cognate site on actin, decreasing the RMSD relative to X-ray structure. However, in none of simulations, these segments were observed to persistently bind to their cognate binding site: In simulations CplxB_MD1 and CplxB_MD2, RMSD of residues ^{50}LKSV^{53} relative to the crystallographic structure never decreased below 13.8 Å. The observed large RMSD values are mainly due to the fact that segment ^{50}LKSV^{53} is, most of the time, detached from actin's surface in simulations of CplxB. In simulations of CplxA, segment ^{22}LKKV^{25} was able to reach its cognate site, with minimal RMSD of 2.4 and 4.3 Å in CplxA_MD1 and CplxA_MD2, respectively, but these associations were only transient (Figure 5). Overall, in three over four simulations, residues ^{22}LKKV^{25} or ^{50}LKSV^{53} were observed to bind the actin's surface during quite long periods, but not necessarily at their cognate locations, confirming that these N-WASP segments are not primary recognition sites for actin. Finally, we should point out that the auto-correlation functions of minimal distances to actin of residues ^{22}LKKV^{25} or ^{50}LKSV^{53} are characterized by relaxation times of 102, 126, 164, and 133 ns for simulations CplxA_MD1, CplxA_MD2, CplxB_MD1, and CplxB_MD2, respectively. This notably indicates that the two short simulations of CplxA still provide reliable information about the dynamics of segment ^{22}LKKV^{25}.

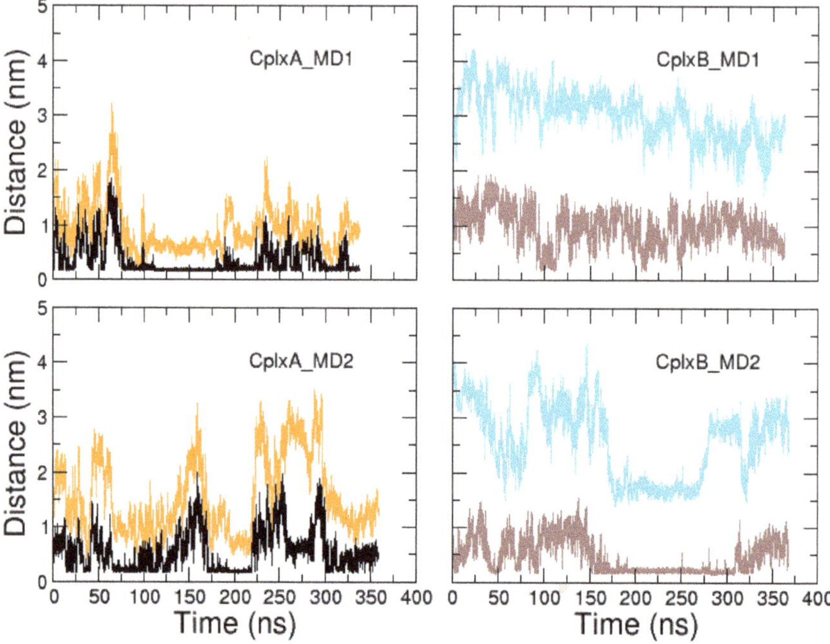

Figure 5. Time evolutions of minimal distance to actin of segments [22]LKKV[25] (black) and [50]LKSV[53] (brown) of N-WASP domain V. RMSD relative to structure 2VCP, after fitting trajectories on actin, are also displayed as a function of time for segments [22]LKKV[25] (orange) and [50]LKSV[53] (cyan).

The actin residues that have high probabilities to be contacted by these segments are shown in Figure 6. In both simulations of CplxA, segment [22]LKKV[25] was found in contact with several actin residues close to the cognate binding site. In contrast, due to the sliding of region 37–46 toward the bottom of actin in simulation CplxB_MD2, the segment [50]LKSV[53] is too far to reach and bind its cognate site on actin. All together, despite their limited number and length, our simulations suggest that CplxA (which has the N-WASP helical region 9–18 recognized by actin) is likely a productive encounter complex that can lead to a subsequent binding of segment [22]LKKV[25] to its specific site on actin. In contrast, simulations of CplxB suggest that the complete binding of N-WASP second WH2 motif is less favorable than for the first WH2 motif. Beyond the limited statistics, this could result from the fact that segment [50]LKSV[53] is less positively charged than [22]LKKV[25], whereas their cognate binding site on actin has two negatively charged residues (D24 and D25). Another possible explanation is that N-WASP region 37–46 has a higher propensity to form α-helices than segment 9–18. This would increase the stiffness of the second WH2 motif that might restrict the motion of residues [50]LKSV[53] and their ability to reach their cognate binding site on actin.

Finally, we studied the dynamics of domain V regions [28]NSRPVS[33] and [56]GQESTP[61] following the conserved sequences [22]LKKV[25] and [50]LKSV[53], respectively. Indeed, as mentioned in the introduction, most crystallographic structures of actin–WH2 motif lack atomic coordinates for regions after the consensus sequence "LKKV", indicating that they are highly flexible in their bound state. We thus characterized the preferential location of these two regions on actin's surface in our MD simulations. Figure 7 plots the minimal distance of regions [28]NSRPVS[33] and [56]GQESTP[61] to actin as a function of time in CplxA and CplxB simulations, respectively. It can be observed that these two regions mostly contact the actin's surface when the preceding conserved sequences [22]LKKV[25] or [50]LKSV[53] are already attached to actin, except in CplxB_MD1. In the latter, residues [56]GQESTP[61] make frequent contacts with actin when segment [50]LKSV[53] is not bound to actin.

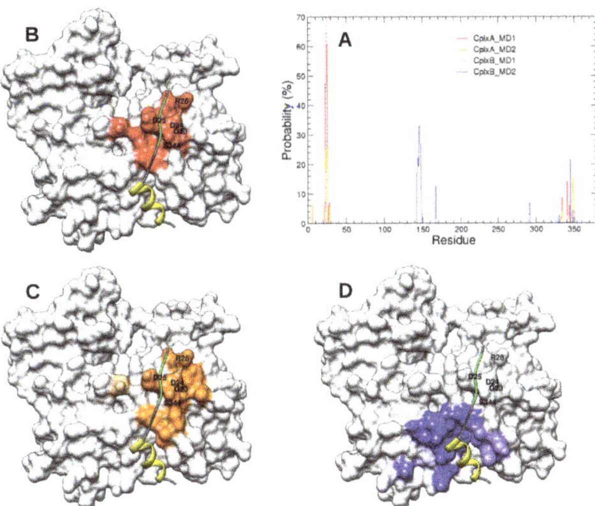

Figure 6. (**A**) Probability of actin residues to be distant by less than 4 Å from N-WASP segments [22]LKKV[25] or [50]LKSV[53] in CplxA_MD1 (red), CplxA_MD2 (orange), CplxB_MD1 (cyan), and CplxB_MD2 (blue). Brown dashed lines indicate the actin residues (G23, D24, D25, R28, and S344) in contact with N-WASP segment [50]LKSV[53] in structure 2VCP [27]). (**B–D**) Front views of the actin's surface colored proportionally to the previous probabilities. Red, orange, and blue colors indicate actin residues with high probabilities to be contacted by N-WASP segments [22]LKKV[25] or [50]LKSV[53] in simulations CplxA_MD1 (**B**), CplxA_MD2 (**C**), and CplxB_MD2 (**D**), respectively. As a reference, yellow and green ribbons represent the helical region and the conserved sequence LKSV of the second WH2 motif observed in structure 2VCP [27].

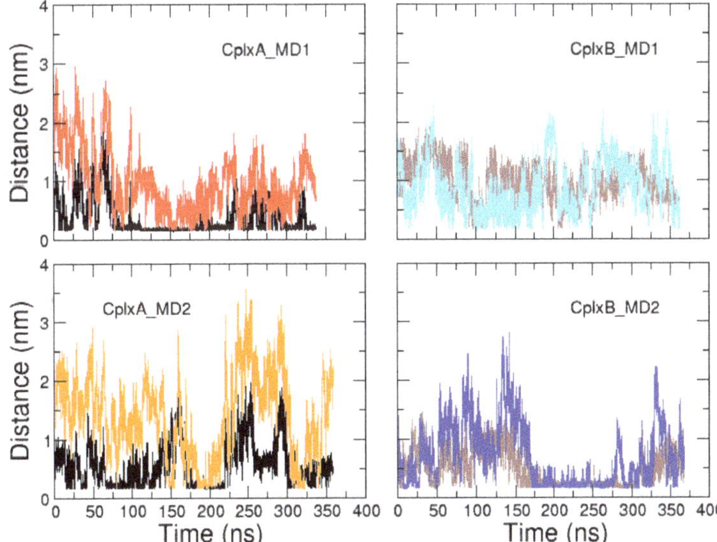

Figure 7. Time evolutions of minimal distances between actin and segment [28]NSRPVS[33] in simulations of CplxA (red and orange lines) and segment [56]GQESTP[61] in simulations of CplxB (cyan and blue lines). For comparison, time evolutions of minimal distances between actin and segments [22]LKKV[25] and [50]LKSV[53] are displayed with black and brown lines, respectively.

The actin residues that have high probabilities to be contacted by regions [28]NSRPVS[33] and [56]GQESTP[61] are displayed in Figure 8. In both simulations of CplxB, segment [56]GQESTP[61] was mostly found in contact with residues of the actin subdomain 3. In CplxB_MD1, this might be the reason the conserved segment [50]LKSV[53] cannot reach its cognate binding site on actin. In CplxB_MD2, this is probably because the helix 37–46 slid toward the bottom of actin and that segment [50]LKSV[53] is improperly located between actin subdomains 1 and 3 (Figure 6). Strikingly, in simulations of CplxA in which the helical segment 9–18 and conserved sequence [22]LKKV[25] are both satisfactorily positioned on actin's surface, the region [28]NSRPVS[33] is observed to contact several separated patches on actin's surface, mainly located on subdomains 2 and 4. This might explain why these disordered regions cannot crystallize in one homogeneous conformation and, therefore, are not visible in most crystallographic actin–WH2 complexes.

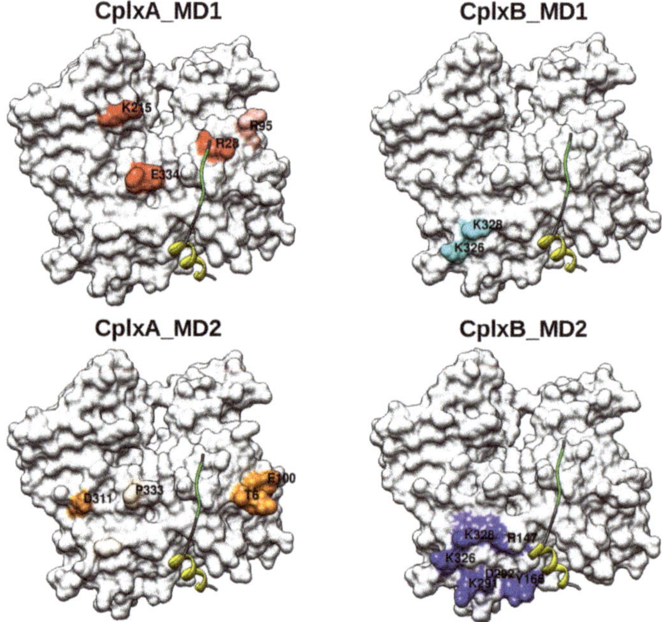

Figure 8. Actin residues distant by less than 4 Å from N-WASP segments [28]NSRPVS[33] or [56]GQESTP[61] in CplxA_MD1 (red), CplxA_MD2 (orange), CplxB_MD1 (cyan), and CplxB_MD2 (blue). As a reference, yellow and green ribbons represent the helical region and the conserved sequence LKSV of the second WH2 motif observed in structure 2VCP [27].

2.3. Dimeric Actin–Domain V Encounter Complexes Generated by Docking Calculations

As reported in the literature, a tandem of WH2 motifs, such as N-WASP domain V, can form a ternary complex with two actin molecules [26,32]. Rebowski et al. notably reported a 2:1 actin–domain V complex, in which two actins are assembled into a longitudinal filament-like dimer (PDB structure 3M3N) [26]. In this section, we investigate the early steps of formation of these ternary encounter complexes. As for actin monomer, we blindly docked the 527 most populated clusters of the MD-derived N-WASP domain V conformational ensemble [29], but here, onto the longitudinal actin dimer structure extracted from the PDB file 3M3N [26]. It should be noted that each chain of the 3M3N dimer is structurally very similar to actin in 2VCP (RMSD over Cα atoms being equal to 0.99 and 0.66 Å for chain A and B, respectively). Moreover, unlike in 2VCP structure, both chains of actin dimer 3M3N lack the coordinates of their last residue F375. A total number of 754,118 complex structures were generated. The likeliness of these complexes was evaluated with the scoring function 2/3B*best*

InterEvScore [30]. We delineated the 1% complexes (that is 7540 conformers) having the highest $2/3B^{best}$ score as the most probable actin dimer-domain V structures. As for actin monomer, when compared to the 527 cluster representative structures, the domain V conformations that are retrieved in the most probable complexes with actin dimer are in average more compact as indicated by the radius of gyration distributions (Figure S3). The dimeric state of actin did not favor the binding of extended conformations of domain V.

We then analyzed the probability of domain V residues to be in contact with each chain of actin dimer. We observed again that actin preferentially recognizes the domain V regions ^{9}KAALLDQIRE^{18} and ^{37}RDALLDQIRQ^{46}, with a similar pattern as for actin monomer (compare Figure 9 with Figure 1), indicating that the N-WASP recognized regions are rather in (partial) α-helical structures. It is also confirmed that the conserved sequences ^{22}LKKV^{25} and ^{50}LKSV^{53} have low probability to be contacted by actin dimer in the encounter complexes, suggesting again that they should move and anchor to the actin's surface after the recognition of the previously mentioned regions 9–18 and 37–46.

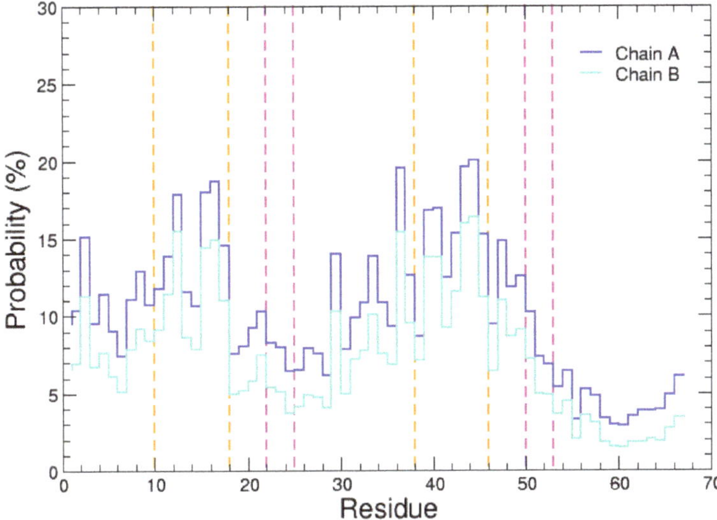

Figure 9. Probability of the N-WASP domain V residues to be distant by less than 4 Å from actin dimer. Orange and magenta dashed lines indicate the N-WASP regions in α-helix (as revealed by the X-ray structure 2VCP [27]) and the consensus sequences "LKKV" [14,31], respectively.

Finally, we determined the preferential location of the domain V regions 9–18 and 37–46 on actin dimer surface by computing over the 7540 most probable complexes the probability that actin residues are contacted by one of these segments (Figure 10). The N-WASP regions 9–18 and 37–46 can be retrieved in the cognate binding site of actin chain A but not of chain B. The presence of chain A at the bottom of chain B probably hinders the approach and accommodation of domain V in the binding site of chain B. As for actin monomer, we also observed that N-WASP segments 9–18 and 37–46 can bind to other patches of the actin's surface with high probability, notably at residues K191, E195, R256 and F266 which are located at the top of the back of actin dimer (Figure 10). It is not clear for us if these non-productive associations are artifacts or not. Nevertheless, since the consensus sequences "LKKV" have low probabilities to contact actin, large conformational changes of domain V are likely to occur after the formation of the encounter complexes. Only a productive encounter complex in which the cognate binding site of actin accommodates N-WASP segment 9–18 or 37–46 will be able to form the correct quaternary structure.

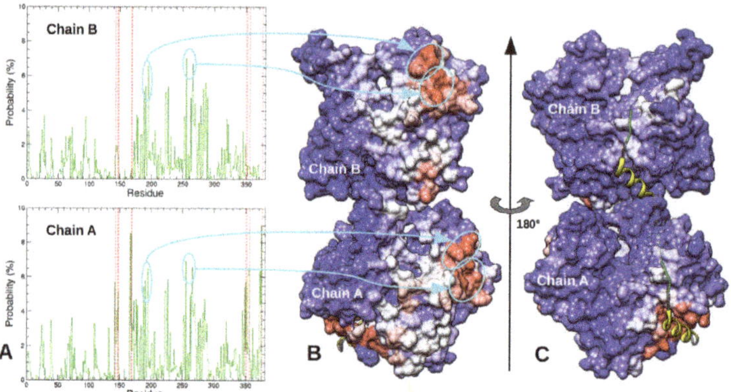

Figure 10. (**A**) Residue-specific probability of actin dimer chain A (bottom) and chain B (top) to be distant by less than 4 Å from N-WASP domain V regions 9–18 or 37–46 in the ensemble of 7540 ternary complexes generated by docking. Red dashed lines indicate the actin residues in contact with the N-WASP helical segment in the X-ray structure 2VCP [27]. (**B,C**) Back and front views of the actin dimer surface colored proportionally to the previous probabilities. Blue, white, and red colors indicate actin residues with low, intermediate, and high probabilities to be contacted by N-WASP domain V regions 9–18 or 37–46, respectively. As a reference, yellow and green ribbons represent helical regions and consensus sequences "LKKV" of the two WH2 motifs observed in the X-ray structure 3M3N [26].

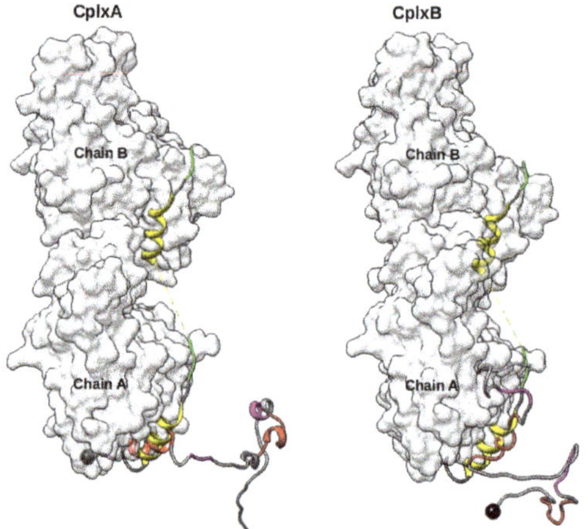

Figure 11. Side view of the two best 2:1 actin–domain V encounter complexes with N-WASP segment 9–18 (**left**) or 37–46 (**right**) located and oriented as in structure 3M3N. Black balls are N-terminal Cα-atoms of domain V. Red and magenta ribbons represent its regions 9–18 or 37–46 and consensus sequences "LKKV", respectively. As a reference, yellow and green ribbons indicate the helical and LKKV regions of domain V in 3M3N.

These productive actin–domain V encounter complexes were identified among the 7540 most probable complexes as those with segment 9–18 or 37–46 making contacts to at least 6 over the 8 hot-spot residues of 3M3N actin chain A (Y143, G146, T148, G168, Y169, L349, T351, and M355), and correctly oriented so that the conserved sequence ^{22}LKKV^{25} or ^{50}LKSV^{53} can reach their cognate binding site. We found 10 and 13 productive encounter complexes in which N-WASP segments 9–18 and 37–46 are bound to actin chain A, respectively (Tables S4 and S5). The two complexes for which the regions 9–18 or 37–46 have the lowest RMSD relative to structure 3M3N are displayed in Figure 11.

In all found productive encounter complexes, regions ^{22}LKKV^{25} or ^{50}LKSV^{53} are detached from actin, and actin chain B is not contacted by other parts of N-WASP domain V. The presence of chain B in the actin dimer does not seem to influence the recognition of N-WASP segments 9–18 or 37–46 by actin chain A. Besides, several representative structures of domain V conformational ensemble (clusters 105, 145, 230, 333, 407, and 411) were retrieved in the most probable encounter complexes on both the monomeric (2VCP) and dimeric (3M3N) states of actin. Nevertheless, as previously seen, the subsequent binding of residues ^{22}LKKV^{25} or ^{50}LKSV^{53} to actin was not persistent in our MD simulation of complexes with actin monomer, but this association might be stabilized by the presence of a second chain in complexes with actin dimer. This hypothesis can be assessed using extensive MD simulations. Unfortunately, our limited computational resources for this project did not allow us to perform these calculations.

3. Discussion

The characterization of the early events of protein–protein recognitions involving intrinsically disordered proteins is important for better understanding the molecular bases of regulation and signaling processes occurring in cells. This task is very challenging using current experimental techniques and can be fruitfully complemented by molecular modeling. However, MD simulations of encounter complexes starting from separated proteins are computationally very demanding and require extremely long trajectories in cases of IDPs. In this study, we propose a less expensive approach consisting, first, in discretizing the IDP large conformational ensemble into representative structures of the most populated clusters; secondly, in generating the protein–protein encounter complexes by rigid coarse-grained protein–protein docking; and, finally, in performing MD calculations of few selected productive complex conformations.

This approach was used to study the recognition by actin of the two WH2 motifs of N-WASP domain V, which is largely disordered in free state. Several crystallographic structures of actin–WH2 motif complexes show that the WH2 motif N-terminal part is folded into an amphiphilic α-helix located in a cleft at the bottom of actin, and that its consensus sequence "LKKV" has a rather extended conformation lying on the actin front surface (Figures 2 and 6). The pathway leading to these bound states remains largely unknown, especially in the case of tandems of WH2 motifs which bind two actins.

Previously, we identified several structures with transient α-helices at regions 9–18 and 37–46 in the unbound domain V conformational ensemble [29]. Our present docking calculations showed that these two regions are effectively preferential binding sites for actin (Figure 1). Our results also suggest that conformations with regions 9–18 or 37–46 completely structured in α-helix are not preferably recognized, but less folded conformations can be equally accommodated in the cognate binding site on actin (Tables S2–S5). Knowing the binding location on actin's surface of the conserved segments ^{22}LKKV^{25} or ^{50}LKSV^{53}, it is apparent that non-specific association and orientation of regions 9–18 and 37–46 on actin's surface cannot produce the observed quaternary structure of actin–WH2 motif complexes. Our MD simulations of a productive encounter complex even showed that, when the recognized helical region 37–46 of N-WASP is initially correctly located and oriented in the actin cognate binding site, a slight displacement of this region toward the bottom of actin prevents the segment ^{50}LKSV^{53} to reach and bind its specific site on actin (simulation CplxB_MD2).

In our modeling procedure, it could be noted that only the 7030 encounter complexes with the highest 2/3B^{best} score among the 702,920 generated by docking were deemed as probable and subsequently analyzed. Although this limited number could lead to possible missed relevant structures, it is much larger than the number of docking solutions that are usually analyzed to find near-native protein–protein interfaces (up to 1000) [30]. This provides reasonable confidence that our modeling generated relevant quaternary structures. Besides, the 7030 analyzed structures can be considered as representative of both the productive and non-productive encounter complexes (Figure 2), as they probably appear in vitro or in vivo. Strikingly, in all productive encounter complexes, the consensus sequence "LKKV" of WH2 motifs is found distant from actin's surface (Figure 3). This indicates that large amplitude motions of these segments are likely to occur in a second step to enable the formation of the final quaternary structure, as illustrated in our MD simulations of CplxA (Figures 5 and 7). Thus, we think that our modeling study has allowed going beyond the prediction of the actin–N-WASP complex quaternary structure and has also gained insight into its mechanism of formation. To sum up, our study of actin monomer recognition by N-WASP domain V indicates that actin first binds domain V regions 9–18 or 37–46 which are partially folded into amphiphilic helical structures, mainly through hydrophobic interactions. Then, the charged segments ^{22}LKKV^{25} or ^{50}LKSV^{53}, driven by electrostatic forces, move and attach to their cognate site on actin's surface.

When the binding of domain V to a longitudinal actin dimer was considered, our docking calculations showed that N-WASP helical regions 9–18 and 37–46 can bind their cognate binding sites, but preferentially on actin chain A, the access of the specific binding site on chain B being more restricted (Figure 10). Nevertheless, this result might depend on the quaternary structure of the actin dimer, particularly on the actin–actin interface, which can significantly vary, as observed in various crystallographic structures of actin oligomers (3M3N [26], 4JHD [32], and 6FHL [33]). All together, our results allow us to propose the following model for the early events of association of N-WASP domain V to two actins and the formation of a ternary complex with a longitudinal filament-like actin dimer, as observed in structure 3M3N (Figure 12): From isolated actin chains and N-WASP domain V, three possible binary complexes can be formed (States II-a, II-b, and II-c). In State II-a, the second WH2 motif attached to actin chain B prevents the approach and binding of chain A [11,15,34] and thus disfavors the formation of intermediate State III-a. When the actin dimer is already formed, our docking calculations indicate that the binding of N-WASP second WH2 motif to actin chain B is not favorable. Thus, the direct formation of the ternary State III-a from a preformed actin dimer or the evolution of intermediate State III-b toward the final complex are very unlikely. These considerations imply that the final state is likely formed through an intermediate ternary complex in which the two WH2 motifs are bound to two loosely interacting actin chains (State III-c). Then, this highly flexible assembly evolves toward the final state through the association of the two actin chains into a longitudinal dimer. This model suggests that the binding of N-WASP domain V to an actin dimer would not be a cooperative process, in line with fluorescence titration experiments reported by Gaucher et al. [27].

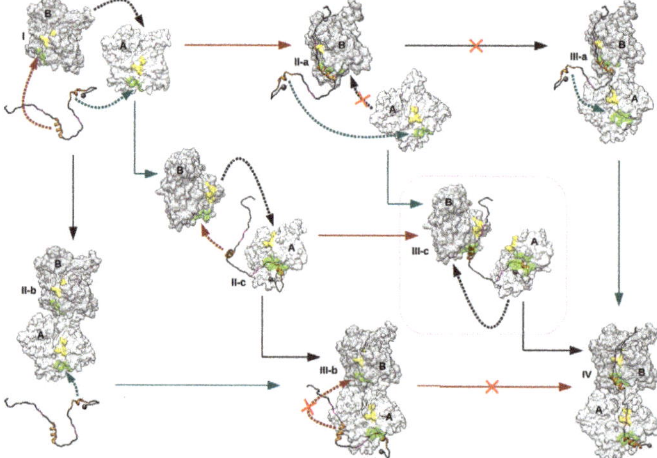

Figure 12. Possible pathways toward the formation of a 2:1 actin–domain V complex with a longitudinal actin dimer as observed in 3M3N. Starting from two actin chains and one N-WASP domain V (State I), three possible binary encounter complexes can be formed (States II-a, II-b, and II-c), leading to three possible intermediate ternary complexes (States III-a, III-b, and III-c) just before the final structure (State IV). Cyan and red arrows indicate the binding of the N-WASP first and second WH2 motif to actin chain A and B, respectively. Dark grey arrows represent the binding of two actins into a longitudinal dimer.

During this process, it is not clear whether the binding of the conserved sequences ^{22}LKKV^{25} and ^{50}LKSV^{53} to their cognate sites occurs before the formation of the longitudinal dimer. In crystallographic structure 2VCP, the four residues ^{50}LKSV^{53} are found attached to the actin's surface, but our MD simulations in explicit water indicate that this binding is rather transient in 1:1 actin–domain V complexes. We speculate that the interactions between the consensus sequences and actin might guide the dynamics of dimerization into longitudinal assemblies. All together, our model for the early events of domain V association to two actins might explain how the two WH2 motifs of N-WASP favor the formation of longitudinal filament-like conformation of actin dimer and why they induce more rapid actin polymerization than proteins of the WASP family with only one WH2 motif [28].

4. Methods

4.1. Conformational Clustering

The conformational ensemble of the studied construct encompassing N-WASP domain V and previously generated by MD simulations with the Amber-03ws force field [29] was clustered with the GROMACS tool *gmx cluster* using the *gromos* method [35] and a RMSD cutoff of 0.5 nm (computed over the mainchain atoms). We obtained 2467 clusters and decided, for subsequent protein–protein docking calculations, to keep only the 527 most populated ones, which represent 50% of the 40,000 conformations sampled by MD simulations. To verify that the 527 clusters are representative of the overall conformational ensemble, we compared the residue probabilities to be in α-helix and the distributions of gyration averaged over the 40,000 conformations or the 527 cluster structures. As shown in Figure S2, the probabilities to form α-helices of the 527 clusters and of the whole conformational ensemble are almost identical, and the protein radius of gyration has similar distributions when computed over the sub-ensemble of representative structures or over the 40,000 conformations. This indicates that the selected 527 conformers are locally and globally representative of the whole conformational ensemble of N-WASP domain V.

4.2. Protein–Protein Docking

The 527 representative conformations of N-WASP domain V were docked into two crystallographic structures of actin (PDB ID: 2VCP [27] and 3M3N [26]), using the molecular modeling library PTools [36]. This toolbox performs rigid-body docking of coarse-grained proteins by multiple energy minimizations, starting from regularly distributed initial positions and orientations of the ligand around the receptor surface. It should be emphasized that no conformational change was allowed during these docking calculations for both protein partners, notably the intrinsically disordered domain V. The energy function minimized here is the physics-based pairwise protein–protein interaction energy SCORPION [37,38]. Then, to better discriminate the near-native interface between actin and domain V, the complexes previously generated with PTools were rescored using a knowledge-based scoring function which additionally takes into account three-body interactions. We used in this study the 2/3B^{best} InterEvScore, without any evolutionary information from the actin or N-WASP domain V sequences [30].

The performance of 2/3B^{best} InterEvScore was positively evaluated on an ensemble of 131 protein–protein complexes which, as far as we know, did not include IDP case [30]. Thus, to assess the validity of our approach to study the actin–domain V recognition, we performed the redocking of the folded segment 433–451 of N-WASP domain V into actin structure 2VCP [27] and checked if the X-ray structure of the complex can be retrieved. The results of this test are reported in Figure S4, which displays the actin–ligand interaction 2/3B^{best} score as a function of the RMSD relative to the peptide conformation in the crystallographic structure. It can be seen that the coarse-grained protein–protein redocking is able to retrieve the experimental structure with a RMSD calculated over the Cα atoms of only 0.5 Å. In this particular case, the modeled complex structure, which is the closest to the experimental one is ranked first (the higher is the score, the more native-like is the interface). This benchmark led us to adopt this two-step approach consisting in generating complex structures with PTools and rescoring them with InterEvScore.

4.3. MD Simulations

From the docking results, several probable structures of the actin–domain V complex were selected and submitted to extensive MD simulations performed with the GROMACS software (versions 5.0.2 and 2016.1) [39]. Each selected complex initial conformation was put and solvated in a dodecahedral rhombic box of 14.0 nm edge, then neutralized by adding 175 sodium and 176 chloride ions to reach the salt concentration of 150 mM. The non-bonded interactions were treated using the smooth PME method [40] for the electrostatic terms and a cutoff distance of 1.2 nm for the van der Waals potentials. The solute and water covalent bond lengths were kept constant using the LINCS [41] and SETTLE [42] algorithms, respectively, allowing to integrate the equations of motion with a 2 fs time step. All simulations were run in the NPT ensemble, at T = 310 K and P = 1 bar, using the Nose–Hoover and Parrinello–Rahman algorithms [43–45] with the time coupling constants $\tau_T = 0.5$ ps and $\tau_P = 2.5$ ps.

In our previous study of the free state N-WASP domain V, short preliminary MD simulations indicated that the force field AMBER-03w [46] combined with the modified water model TIP4P/2005s [47] (a combination referred to as A03ws) allowed correctly exploring the protein conformational space. For consistency, we kept this force field for the study of its complex with actin. Each selected complex was submitted to about 350 ns MD simulations within the general conditions previously described. Data collected every 20 ps were kept for subsequent analyses. The latter were made using mostly the GROMACS tools, such as *gmx mindist* or *gmx cluster* for computing specific distances or structural clusters, respectively. The program STRIDE [48] was used to assign secondary structures to the protein residues.

Supplementary Materials: The following are available online at http://www.mdpi.com/1422-0067/20/18/4493/s1. Table S1: List of proteins with WH2 motifs which were co-crystallized with actin; Figure S1: Alignment of the studied construct sequence with those of N-WASP in 2VCP and 3M3N structures; Figure S2: N-WASP domain V residue probabilities to be in α-helix and distributions of gyration of the 7030 conformations in the most probable 1:1 actin–domain V complexes; Table S2: Most probable 1:1 actin–domain V encounter complexes in which domain V segment 9–18 is in contact with at least six over nine actin hot-spot residues; Table S3: Most probable 1:1 actin–domain V encounter complexes in which domain V segment 37–46 is in contact with at least six over nine actin hot-spot residues; Figure S3: N-WASP domain V residue probabilities to be in α-helix and distributions of gyration of the 7540 conformations in the most probable 2:1 actin–domain V complexes; Table S4: Most probable 2:1 actin–domain V encounter complexes in which segment 9–18 is in contact with at least six over eight actin hot-spot residues; Table S5: Most probable 2:1 actin–domain V encounter complexes in which segment 37–46 is in contact with at least six over eight actin hot-spot residues; Figure S4: $2/3B^{best}$ score of N-WASP segment 433–451 redocked into actin as a function of the ligand RMSD relative to the conformation found in structure 2VCP.

Author Contributions: Conceptualization, M.C.-Y.-C., D.D. and T.H.-D.; Formal analysis, M.C.-Y.-C. and T.H.-D.; Funding acquisition, D.D. and T.H.-D.; Investigation, M.C.-Y.-C. and T.H.-D.; Methodology, M.C.-Y.-C., D.D. and T.H.-D.; Project administration, D.D. and T.H.-D.; Resources, D.D. and T.H.-D.; Supervision, D.D. and T.H.-D.; Validation, M.C.-Y.-C. and T.H.-D.; Visualization, M.C.-Y.-C., D.D. and T.H.-D.; Writing—original draft, M.C.-Y.-C., D.D. and T.H.-D.; and Writing—review and editing, M.C.-Y.-C., D.D. and T.H.-D.

Funding: This research supported by the "IDI 2016" project funded by the IDEX Paris-Saclay (grant number ANR-11-IDEX-0003-02). MD simulations were performed using HPC resources from GENCI-CINES (grant number A0040710415).

Acknowledgments: We are grateful to L. Renault for fruitful discussions about actin and WH2 motifs.

Conflicts of Interest: The authors declare no conflict of interest. The funders had no role in the design of the study; in the collection, analyses, or interpretation of data; in the writing of the manuscript, or in the decision to publish the results.

Abbreviations

The following abbreviations are used in this manuscript:

IDP	Intrinsically Disordered Protein
IDR	Intrinsically Disordered Region
PDB	Protein Data Bank
N-WASP	Neural Wiskott–Aldrich Syndrome Protein
MoRF	Molecular Recognition Feature
NMR	Nuclear Magnetic Resonance
SAXS	Small-Angle X-ray Scattering
MD	Molecular Dynamics
RMSD	Root Mean Square Deviation

References

1. Wright, P.E.; Dyson, H.J. Intrinsically unstructured proteins: Re-assessing the protein structure-function paradigm. *J. Mol. Biol.* **1999**, *293*, 321–331. [CrossRef]
2. Dunker, A.K.; Lawson, J.D.; Brown, C.J.; Williams, R.M.; Romero, P.; Oh, J.S.; Oldfield, C.J.; Campen, A.M.; Ratliff, C.M.; Hipps, K.W.; et al. Intrinsically disordered protein. *J. Mol. Graph. Model.* **2001**, *19*, 26–59. [CrossRef]
3. Dyson, H.J.; Wright, P.E. Intrinsically unstructured proteins and their functions. *Nat. Rev. Mol. Cell Biol.* **2005**, *6*, 197–208. [CrossRef] [PubMed]
4. Dunker, A.K.; Silman, I.; Uversky, V.N.; Sussman, J.L. Function and structure of inherently disordered proteins. *Curr. Opin. Struct. Biol.* **2008**, *18*, 756–764. [CrossRef]
5. Zea, D.J.; Monzon, A.M.; Gonzalez, C.; Fornasari, M.S.; Tosatto, S.C.E.; Parisi, G. Disorder transitions and conformational diversity cooperatively modulate biological function in proteins. *Protein Sci.* **2016**, *25*, 1138–1146. [CrossRef]
6. Oldfield, C.J.; Cheng, Y.; Cortese, M.S.; Romero, P.; Uversky, V.N.; Dunker, A.K. Coupled Folding and Binding with α-Helix-Forming Molecular Recognition Elements. *Biochemistry* **2005**, *44*, 12454–12470. [CrossRef] [PubMed]
7. Mohan, A.; Oldfield, C.J.; Radivojac, P.; Vacic, V.; Cortese, M.S.; Dunker, A.K.; Uversky, V.N. Analysis of Molecular Recognition Features (MoRFs). *J. Mol. Biol.* **2006**, *362*, 1043–1059. [CrossRef]
8. Vacic, V.; Oldfield, C.J.; Mohan, A.; Radivojac, P.; Cortese, M.S.; Uversky, V.N.; Dunker, A.K. Characterization of Molecular Recognition Features, MoRFs, and Their Binding Partners. *J. Proteome Res.* **2007**, *6*, 2351–2366. [CrossRef] [PubMed]
9. Cheng, Y.; Oldfield, C.J.; Meng, J.; Romero, P.; Uversky, V.N.; Dunker, A.K. Mining α-helix-forming molecular recognition features with cross species sequence alignments. *Biochemistry* **2007**, *46*, 13468–13477. [CrossRef]
10. Lee, C.; Kalmar, L.; Xue, B.; Tompa, P.; Daughdrill, G.W.; Uversky, V.N.; Han, K.H. Contribution of proline to the pre-structuring tendency of transient helical secondary structure elements in intrinsically disordered proteins. *Biochim. Biophys. Acta Gen. Subj.* **2014**, *1840*, 993–1003. [CrossRef] [PubMed]
11. Carlier, M.F.; Husson, C.; Renault, L.; Didry, D. Chapter Two–Control of Actin Assembly by the WH2 Domains and Their Multifunctional Tandem Repeats in Spire and Cordon-Bleu. In *International Review of Cell and Molecular Biology*; Jeon, K.W., Ed.; Academic Press: Cambridge, MA, USA, 2011; Volume 290, pp. 55–85.
12. Derry, J.M.J.; Ochs, H.D.; Francke, U. Isolation of a novel gene mutated in Wiskott-Aldrich syndrome. *Cell* **1994**, *78*, 635–644. [CrossRef]
13. Palma, A.; Ortega, C.; Romero, P.; Garcia-V, A.; Roman, C.; Molina, I.; Santamaria, M. Wiskott-Aldrich syndrome protein (WASp) and relatives: A many-sided family. *Immunologia* **2004**, *23*, 217–230.
14. Chereau, D.; Kerff, F.; Graceffa, P.; Grabarek, Z.; Langsetmo, K.; Dominguez, R. Actin-bound structures of Wiskott-Aldrich syndrome protein (WASP)-homology domain 2 and the implications for filament assembly. *Proc. Natl. Acad. Sci. USA* **2005**, *102*, 16644–16649. [CrossRef] [PubMed]
15. Renault, L.; Deville, C.; van Heijenoort, C. Structural features and interfacial properties of WH2, β-thymosin domains and other intrinsically disordered domains in the regulation of actin cytoskeleton dynamics. *Cytoskeleton* **2013**, *70*, 686–705. [CrossRef]

16. Kiefhaber, T.; Bachmann, A.; Jensen, K.S. Dynamics and mechanisms of coupled protein folding and binding reactions. *Curr. Opin. Struct. Biol.* **2012**, *22*, 21–29. [CrossRef] [PubMed]

17. Liu, X.; Chen, J.; Chen, J. Residual Structure Accelerates Binding of Intrinsically Disordered ACTR by Promoting Efficient Folding upon Encounter. *J. Mol. Biol.* **2019**, *431*, 422–432. [CrossRef] [PubMed]

18. Kozakov, D.; Li, K.; Hall, D.R.; Beglov, D.; Zheng, J.; Vakili, P.; Schueler-Furman, O.; Paschalidis, I.C.; Clore, G.M.; Vajda, S. Encounter complexes and dimensionality reduction in protein–protein association. *eLife* **2014**, *3*, e01370. [CrossRef] [PubMed]

19. Pollard, T.D.; Borisy, G.G. Cellular Motility Driven by Assembly and Disassembly of Actin Filaments. *Cell* **2003**, *112*, 453–465. [CrossRef]

20. Takenawa, T.; Suetsugu, S. The WASP–WAVE protein network: connecting the membrane to the cytoskeleton. *Nat. Rev. Mol. Cell Biol.* **2007**, *8*, 37–48. [CrossRef]

21. Miki, H.; Miura, K.; Takenawa, T. N-WASP, a novel actin-depolymerizing protein, regulates the cortical cytoskeletal rearrangement in a PIP2-dependent manner downstream of tyrosine kinases. *Embo J.* **1996**, *15*, 5326–5335. [CrossRef]

22. Prehoda, K.E.; Scott, J.A.; Mullins, R.D.; Lim, W.A. Integration of Multiple Signals Through Cooperative Regulation of the N-WASP-Arp2/3 Complex. *Science* **2000**, *290*, 801–806. [CrossRef] [PubMed]

23. Fawcett, J.; Pawson, T. N-WASP Regulation—The Sting in the Tail. *Science* **2000**, *290*, 725–726. [CrossRef] [PubMed]

24. Luan, Q.; Zelter, A.; MacCoss, M.J.; Davis, T.N.; Nolen, B.J. Identification of Wiskott-Aldrich syndrome protein (WASP) binding sites on the branched actin filament nucleator Arp2/3 complex. *Proc. Natl. Acad. Sci. USA* **2018**, *115*, E1409–E1418. [CrossRef] [PubMed]

25. Dominguez, R. Actin filament nucleation and elongation factors— Structure–function relationships. *Crit. Rev. Biochem. Mol. Biol.* **2009**, *44*, 351–366. [CrossRef] [PubMed]

26. Rebowski, G.; Namgoong, S.; Boczkowska, M.; Leavis, P.C.; Navaza, J.; Dominguez, R. Structure of a Longitudinal Actin Dimer Assembled by Tandem W Domains: Implications for Actin Filament Nucleation. *J. Mol. Biol.* **2010**, *403*, 11–23. [CrossRef] [PubMed]

27. Gaucher, J.F.; Maugé, C.; Didry, D.; Guichard, B.; Renault, L.; Carlier, M.F. Interactions of Isolated C-terminal Fragments of Neural Wiskott-Aldrich Syndrome Protein (N-WASP) with Actin and Arp2/3 Complex. *J. Biol. Chem.* **2012**, *287*, 34646–34659. [CrossRef] [PubMed]

28. Yamaguchi, H.; Miki, H.; Suetsugu, S.; Ma, L.; Kirschner, M.W.; Takenawa, T. Two tandem verprolin homology domains are necessary for a strong activation of Arp2/3 complex-induced actin polymerization and induction of microspike formation by N-WASP. *Proc. Natl. Acad. Sci. USA* **2000**, *97*, 12631–12636. [CrossRef]

29. Chan-Yao-Chong, M.; Deville, C.; Pinet, L.; van Heijenoort, C.; Durand, D.; Ha-Duong, T. Structural Characterization of N-WASP Domain V Using MD Simulations with NMR and SAXS Data. *Biophys. J.* **2019**, *116*, 1216–1227. [CrossRef]

30. Andreani, J.; Faure, G.; Guerois, R. InterEvScore: a novel coarse-grained interface scoring function using a multi-body statistical potential coupled to evolution. *Bioinformatics* **2013**, *29*, 1742–1749. [CrossRef]

31. Kollmar, M.; Lbik, D.; Enge, S. Evolution of the eukaryotic ARP2/3 activators of the WASP family: WASP, WAVE, WASH, and WHAMM, and the proposed new family members WAWH and WAML. *BMC Res. Notes* **2012**, *5*, 88. [CrossRef]

32. Chen, X.; Ni, F.; Tian, X.; Kondrashkina, E.; Wang, Q.; Ma, J. Structural Basis of Actin Filament Nucleation by Tandem W Domains. *Cell Rep.* **2013**, *3*, 1910–1920. [CrossRef] [PubMed]

33. Merino, F.; Pospich, S.; Funk, J.; Wagner, T.; Küllmer, F.; Arndt, H.D.; Bieling, P.; Raunser, S. Structural transitions of F-actin upon ATP hydrolysis at near-atomic resolution revealed by cryo-EM. *Nat. Struct. Mol. Biol.* **2018**, *25*, 528–537. [CrossRef] [PubMed]

34. Hertzog, M.; van Heijenoort, C.; Didry, D.; Gaudier, M.; Coutant, J.; Gigant, B.; Didelot, G.; Préat, T.; Knossow, M.; Guittet, E.; et al. The β-Thymosin/WH2 Domain: Structural Basis for the Switch from Inhibition to Promotion of Actin Assembly. *Cell* **2004**, *117*, 611–623. [CrossRef]

35. Daura, X.; Gademann, K.; Jaun, B.; Seebach, D.; van Gunsteren, W.F.; Mark, A.E. Peptide Folding: When Simulation Meets Experiment. *Angew. Chem. Int. Ed.* **1999**, *38*, 236–240. [CrossRef]

36. Saladin, A.; Fiorucci, S.; Poulain, P.; Prévost, C.; Zacharias, M. PTools: An opensource molecular docking library. *BMC Struct. Biol.* **2009**, *9*, 27. [CrossRef] [PubMed]

37. Basdevant, N.; Borgis, D.; Ha-Duong, T. A Coarse-Grained Protein–Protein Potential Derived from an All-Atom Force Field. *J. Phys. Chem. B* **2007**, *111*, 9390–9399. [CrossRef] [PubMed]

38. Basdevant, N.; Borgis, D.; Ha-Duong, T. Modeling Protein–Protein Recognition in Solution Using the Coarse-Grained Force Field SCORPION. *J. Chem. Theory Comput.* **2013**, *9*, 803–813. [CrossRef] [PubMed]

39. Abraham, M.J.; Murtola, T.; Schulz, R.; Pall, S.; Smith, J.C.; Hess, B.; Lindahl, E. GROMACS: High performance molecular simulations through multi-level parallelism from laptops to supercomputers. *SoftwareX* **2015**, *1–2*, 19–25. [CrossRef]

40. Essmann, U.; Perera, L.; Berkowitz, M.L.; Darden, T.; Lee, H.; Pedersen, L.G. A smooth particle mesh Ewald method. *J. Chem. Phys.* **1995**, *103*, 8577–8593. [CrossRef]

41. Hess, B. P-LINCS: A Parallel Linear Constraint Solver for Molecular Simulation. *J. Chem. Theory Comput.* **2008**, *4*, 116–122. [CrossRef]

42. Miyamoto, S.; Kollman, P.A. SETTLE: An analytical version of the SHAKE and RATTLE algorithm for rigid water models. *J. Comput. Chem.* **1992**, *13*, 952–962. [CrossRef]

43. Nosé, S. A unified formulation of the constant temperature molecular dynamics methods. *J. Chem. Phys.* **1984**, *81*, 511–519. [CrossRef]

44. Hoover, W.G. Canonical dynamics: Equilibrium phase-space distributions. *Phys. Rev. A* **1985**, *31*, 1695–1697. [CrossRef]

45. Parrinello, M.; Rahman, A. Polymorphic transitions in single crystals: A new molecular dynamics method. *J. Appl. Phys.* **1981**, *52*, 7182–7190. [CrossRef]

46. Best, R.B.; Mittal, J. Protein Simulations with an Optimized Water Model: Cooperative Helix Formation and Temperature-Induced Unfolded State Collapse. *J. Phys. Chem. B* **2010**, *114*, 14916–14923. [CrossRef]

47. Best, R.B.; Zheng, W.; Mittal, J. Balanced Protein–Water Interactions Improve Properties of Disordered Proteins and Non-Specific Protein Association. *J. Chem. Theory Comput.* **2014**, *10*, 5113–5124. [CrossRef] [PubMed]

48. Heinig, M.; Frishman, D. STRIDE: A web server for secondary structure assignment from known atomic coordinates of proteins. *Nucleic Acids Res.* **2004**, *32*, W500–W502. [CrossRef] [PubMed]

© 2019 by the authors. Licensee MDPI, Basel, Switzerland. This article is an open access article distributed under the terms and conditions of the Creative Commons Attribution (CC BY) license (http://creativecommons.org/licenses/by/4.0/).

International Journal of
Molecular Sciences

Article

Analysis of Heterodimeric "Mutual Synergistic Folding"-Complexes

Anikó Mentes [†], Csaba Magyar [†], Erzsébet Fichó and István Simon *

Institute of Enzymology, Research Centre for Natural Sciences, Hungarian Academy of Sciences, Magyar Tudósok krt. 2., H-1117 Budapest, Hungary; mentes.aniko@ttk.mta.hu (A.M.); magyar.csaba@ttk.mta.hu (C.M.); ficho.erzsebet@ttk.mta.hu (E.F.)
* Correspondence: simon.istvan@ttk.mta.hu; Tel.: +36-1-3826-710
† These authors contributed equally to the paper.

Received: 13 September 2019; Accepted: 15 October 2019; Published: 16 October 2019

Abstract: Several intrinsically disordered proteins (IDPs) are capable to adopt stable structures without interacting with a folded partner. When the folding of all interacting partners happens at the same time, coupled with the interaction in a synergistic manner, the process is called Mutual Synergistic Folding (MSF). These complexes represent a discrete subset of IDPs. Recently, we collected information on their complexes and created the MFIB (Mutual Folding Induced by Binding) database. In a previous study, we compared homodimeric MSF complexes with homodimeric and monomeric globular proteins with similar amino acid sequence lengths. We concluded that MSF homodimers, compared to globular homodimeric proteins, have a greater solvent accessible main-chain surface area on the contact surface of the subunits, which becomes buried during dimerization. The main driving force of the folding is the mutual shielding of the water-accessible backbones, but the formation of further intermolecular interactions can also be relevant. In this paper, we will report analyses of heterodimeric MSF complexes. Our results indicate that the amino acid composition of the heterodimeric MSF monomer subunits slightly diverges from globular monomer proteins, while after dimerization, the amino acid composition of the overall MSF complexes becomes more similar to overall amino acid compositions of globular complexes. We found that inter-subunit interactions are strengthened, and additionally to the shielding of the solvent accessible backbone, other factors might play an important role in the stabilization of the heterodimeric structures, likewise energy gain resulting from the interaction of the two subunits with different amino acid compositions. We suggest that the shielding of the β-sheet backbones and the formation of a buried structural core along with the general strengthening of inter-subunit interactions together could be the driving forces of MSF protein structural ordering upon dimerization.

Keywords: dehydrons; inter-subunit interactions; intrinsically disordered proteins; ion-pairs; mutual synergistic folding; solvent accessible surface area; stabilization centers

1. Introduction

Mutual synergistic folding (MSF) complexes are a unique subset of intrinsically disordered proteins (IDPs). MSF IDPs can adopt a stable structure during the interaction, without a pre-existing folded partner [1–4]. At the time of the mutual synergistic folding process, the participating IDPs of these complexes synergistically fold into a stable, globular complex. Demarest et al. (2002) investigated the first MSF interaction between the p160 transcriptional coactivator protein ACTR and the tumor suppressor CBP proteins. They found that this MSF complex contains many hydrophobic side-chains and highly specific intermolecular hydrogen bonds, as well as buried intermolecular salt bridges, which help to fold the complex [5]. Since IDPs often have a high net charge, and they have a small content of hydrophobic residues, they are usually not able to form a hydrophobic core [6]. However,

MSF complexes contain more hydrophobic residues, presenting an exception to a general view of IDPs [7,8].

While IDPs mostly have low sequence complexity, MSF complexes are rather heterogeneous, like globular proteins. Furthermore, MSF proteinsare also heterogeneous in amino acid composition similar to globular proteins [8]. The residue-based disorder prediction methods, developed for identifying segments bound to folded proteins, cannot be used for detecting of MSF complexes. Systematic analyses are required to understand and predict these MSF interactions. Nevertheless, this is difficult to implement since a severe weakness of the literature is the little information available about these complexes. At present, the most comprehensive and systematic catalog of MSF complexes is the MFIB (Mutual Folding Induced by Binding) database containing 205 entries [9].

A protein in aqueous solution is only stable when it contains a hydrophobic core buried from water by polar residues. Furthermore, these polar residues shield most of the hydrophobic residues from the solvent. For the first criterion, the protein should contain more residues than a required minimum either as a monomer or as an oligomer. The fulfillment of the second criterion depends on the ratio of the polar and hydrophobic residues because the ratio of the surface and buried residues rapidly decreases by increasing the total number of residues. For a given hydrophilic/hydrophobic ratio, either a long polypeptide chain or oligomerization is needed. MSF proteins fulfill both criteria by oligomerization.

Recently, the physical background of homodimeric MSF complexes from MFIB [7] was analyzed. We identified the residues with solvent accessible main-chain patches (RSAMPs) and studied the "under-wrapped" hydrogen bonds (dehydrons), which are not shielded well enough from solvent [10]. Our results suggested that homodimeric MSF complexes contain more RSAMPs and dehydrons than homodimeric complexes where all the interacting chains are globular in their monomeric form. These properties should contribute to their disordered nature in monomeric form and to their folding in the oligomeric state. In this study, the role of this phenomenon for heterodimeric MSF complexes will be discussed. In the case of heterodimers, the interacting polypeptide chains have different amino acid compositions, which discriminates heterodimers from homodimers. The MFIB database contains, unfortunately, a much lower number of heterodimeric structures when compared to homodimeric ones. Furthermore, there are highly similar proteins among them, which makes redundancy filtering necessary.

2. Results and Discussion

2.1. Sequence-based Analysis

In this study, first, we examined the amino acid composition of the MFIB heterodimeric (MFHE) complexes, which were compared with a globular heterodimeric reference dataset (GLHE), which has similar size distribution for the heterodimeric state (see Figure 1). Note that all GLHE subunits are more than 40 residues away from both axes, while the closest distance of an MFHE chain from the x-axis is less than 20 residues. Also, we will show later (see Figure 5) that the smallest identified globular monomer has 35 residues. In some cases, heterodimeric MSF complexes do not have enough amino acids for creating a hydrophobic core, but in most cases, they have as many residues as globular proteins have, thus other factors might also be responsible for the disordered nature of MFHE proteins.

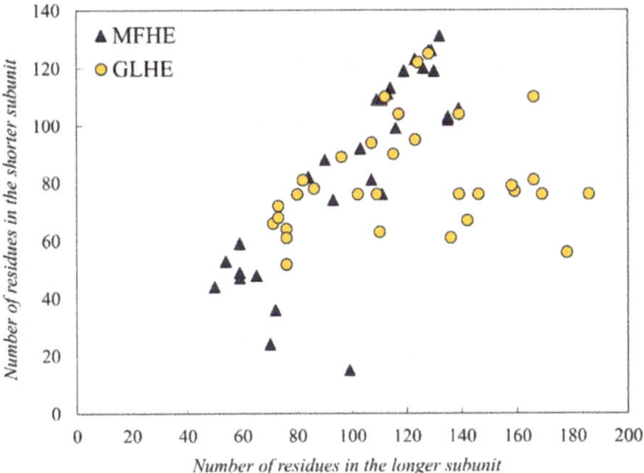

Figure 1. Comparison of the subunit lengths of the Mutual Synergistic Folding (MSF) (MFHE—blue triangles) and globular (globular heterodimeric GLHE—yellow dots) heterodimeric complexes.

Since the beginning of the studies on IDPs, it is known they generally lack hydrophobic residues although alanine has a notably higher content in MFHE complexes compared to GLHE complexes, while the content of other aliphatic residues was similar among the two datasets (see Figure 2A). MFHE complexes have a high net charge, like non-MSF IDPs [11,12].

The amino acid composition of the MFHE and GLHE heterodimers was depicted by a rank-based, indirect gradient analysis method, called Nonmetric MultiDimensional Scaling (NMDS), which creates an ordination based on a distance or dissimilarity matrix, thus it allows decreasing a multidimensional and quantitative, semi-quantitative, qualitative, or mixed variables data set to two dimensions [13]. NMDS demonstrated a separation of MFHE and GLHE complexes and subunits (see Figure 2B,C). The amino acid composition of the subunits, whether globular or MSF complexes are formed, have equal distances from each other as the amino acid compositions of the complexes. Some differences are revealed between the two data sets—the NMDS of the amino acid composition of the MSF heterodimeric complexes showed smaller variation from the globular heterodimeric complexes (see Figure 2C), than the amino acid composition of the MSF subunits from the globular subunits (see Figure 2B). These differences can be explained by the fact that although the amino acid composition of the MSF subunits differs slightly from globular proteins, they are unable to fold into an ordered structure independently. The folding of an MSF subunit requires another partner, which in this case has a different amino acid composition, that could form MSF complexes which have similar amino acid composition than the globular subunits. NMDS also pointed out that MFHE is a diverse group based on their amino acid composition, and these complexes are also clustered according to their structural classes in MFIB [9].

The determination of the amino acids that contribute mainly to the observed difference was revealed by using SIMPER (similarity percentage) analyses. These amino acids were lysine (7.40%; 8.04%), alanine (7.30%; 7.90), leucine (7.14%; 6.64%), glycine (6.86%; 5.83%), arginine (6.39%; 6.70%), and glutamine (6.29%; 6.42%), which values support the similarity of the objects. Mostly aromatic and hydrophobic amino acids cause the amino acid compositions to separate (in slightly different proportions, See Table S1), which case is more common in heterodimeric MSF subunits and complexes if the MSF data were grouped via MFIB for comparison was considered, for the MFIB structural classes (see Figure 2, Table S1), with the exception of glutamine.

Most of the heterodimers from MFIB are histone-type proteins with their high content of lysine and arginine. Acetylated lysine and methylated arginine may interact with proteins containing bromodomains and Tudor domains within the disordered proteins that affect nucleic acid binding and RNA pathways [14].

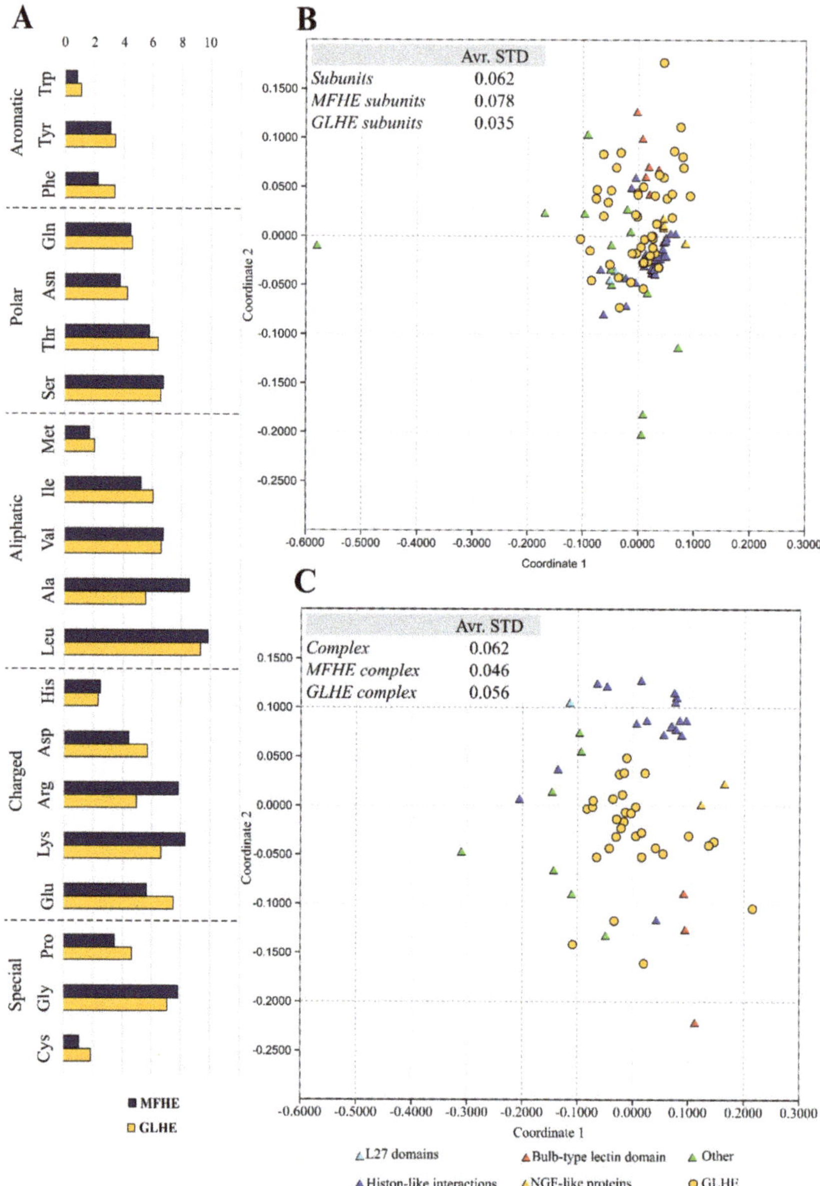

Figure 2. Amino acid composition of the heterodimer datasets, where the types of amino acids were grouped by Mészáros et al. [8] (**A**). The MFHE (triangles) and GLHE (dot) amino acid composition were compared using an indirect gradient analysis method, called Nonmetric Multidimensional Scaling (NMDS), which creates an ordination based on Bray-Curtis distances. In the plot, the objects are protein subunits (**B**) considered separately and complexes (**C**).

The amino acid composition of the homodimeric complexes from MFIB (MFHO), heterodimeric MSF complexes was compared using our small globular protein (SGP) dataset as a standard reference by Kullback-Leibler divergence [15], which measures the extent of the dissimilarity between two probability distributions ($D = \sum_i p_i * \ln \frac{p_i}{q_i}$). MSF heterodimers show about the same similarity to MSF homodimers (D = 1.257) and small globular proteins (D = 1.879), while MSF homodimers are more similar to small globular proteins (D = 0.442). This result is in line with the observation that heterodimeric complexes from MFIB look much more disordered (~20%) than MFIB homodimers (MFHO) (~10%) [7] based on MoRFpred [16] and IUPred [17] results. Some regions of the heterodimeric MFIB complexes are also capable of folding on the surface of a globular protein. Most of these can be found in the DIBS (Disordered Binding Site) database [18]. It is rather rare, but it also shows the elevation of the group inhomogeneity. For example, the cellular tumor antigen p53 protein (UniProt: P04637) is able to establish a coactivator binding domain complex (MFIB: MF2201002, PDB: 2l14) with the CREB-binding MSF protein, although at the same part of the p53 capable to form a transactivator domain complex (DIBS: DI1000009, PDB: 2ly4) with the highly mobile folded B1 protein. We have also found examples of disordered proteins from UniProt (e.g., ID: Q9Y6Q9, Nuclear receptor coactivator 3) which are able to establish an MSF interaction (MF2201001, PDB: 1kbh), and another region is able to form a DIBS interaction (DI1000313, PDB: 3l3x), forming two different types of disordered protein complexes.

It is interesting to note, that a few MFIB homodimers occur in DIBS as ordered interaction partners. For example, the dynein light chain (Tctex-type) protein (UniProt: Q94524), which is disordered in monomeric form based on MFIB (MFIB: MF2110016, PDB: 1ygt), while this homodimeric complex is the ordered part of a DIBS-interaction complex (Cytosolic dynein intermediate chain bound to Tctex-type dynein light chain, DIBS: DI2100002, PDB: 3fm7). An additional example of these multiple structure organizations is the homodimeric S100BEF-hand calcium-binding protein superfamily (MFIB: MF2100013, PDB: 1uwo), which is the ordered component of a DIBS-interaction (RSK1 bound to S100B dimer, DIBS: DI2000012, PDB: 5csf).

Besides the amino acid compositions, other sequential parameters also display differences between GLHE and MFHE. Based on cleverMachine [19] calculations (*p*-value < 0.0001: 56 scale of all 80) and grouped properties results, membrane proteins (*p*-value < 0.0001: 7 scale of 10), nucleic acid binding (*p*-value < 0.0001: 3 scale of 10), disorder propensity (*p*-value < 0.0001: 8 scale of 10), α-helix (*p*-value < 0.0001: 9 scale of 10), β-sheet (*p*-value < 0.0001: 9 scale of 10), aggregation (*p*-value < 0.0001: 8 scale of 10), burial propensity (*p*-value < 0.0001: 10 scale of 10), and hydrophobicity (*p*-value < 0.0001: 2 scale of 10) properties in MFHE are in general stronger than in globular heterodimers (Reference number of the dataset: 196154). While there is no significant difference between the sequences of MFIB homodimers and globular homodimers (GLHO) in most of the properties (exception of some membrane proteins and aggregation scales; *p*-value < 0.0001: 8 scale of all 80) (Reference number of the dataset: 199533).

We analyzed the Pfam database in conjunction with the intermolecular stabilization centers (SCs, see Chapter 2.2. Structure-based analysis) [20] on MFIB heterodimeric and globular heterodimeric complexes (for detailed results, see Table S2). In the MFHE have found 59 Pfam domains in a total of 19 families, while the GLHE have 64 Pfam domains in a total of 37 families. In the case of globular heterodimers, 3 of the 30 complexes have interactions and SCs between the Pfam domains of the monomers, whereas, for MFIB heterodimers much more, at least 15 of the complexes have Pfam domains in which monomers interactions and intermolecular SCs were found. This result confirms that the folding of the MSF proteins is related to their functional role since, in many cases, the two subunits form the biologically relevant unit.

2.2. Structure-based Analysis

In our recent analysis of MSF homodimeric proteins, we found differences in several structural parameters between our dataset and a globular reference dataset. These structural features were investigated including solvent accessibility, hydrogen bonds, stabilization center content, and ion-pairs with an additional investigation of the buried structural core size.

The inter-subunit interface was identified based on the solvent accessible surface area (SASA) calculations. However, an MSF protein subunit in itself does not have an ordered structure, structural properties were also calculated for their monomeric forms, which were created by deleting a polypeptide chain from the heterodimeric PDB structures. This is referred to as their "monomeric structure" hereafter. The all-atom SASA values were calculated for all residues from the heterodimeric and monomeric structures. If the dimeric SASA value was below 20% of the monomeric value, the residue was identified as an interface residue. In the case of the MFIB heterodimeric dataset, 908 interface residues were identified out of the 4615 residues, that is 19.7% of all residues participate in the formation of the interface. In the globular reference heterodimeric dataset 470 interface residues were identified out of the 5155 total residues, i.e., 9.1% of all residues are forming the interface. As a different measure of the interface region, all-atom SASA values were also compared. In MFHE, 27.3% of the total surface area becomes buried upon dimerization, while in GLHE, only 11.6%. This result is in agreement with the finding of Gunasekaran et al., that the per residue interface area is higher in disordered complexes [3] In MSF proteins, the larger interface contact area underlines the importance of inter-subunit interactions, thus inter-subunit interactions were considered hereafter.

Completely buried residues were identified in the MSF and the globular reference heterodimeric datasets using a stricter definition of burial, defining the core of the protein structure shielded from the solvent. We identified all residues, which have less than 10% relative all-atom solvent accessibility in the heterodimeric and monomeric structures, respectively. In MFHE, 10.8% of all residues are buried in monomeric form, while in GLHE this value is 20.9%. If the dimeric structures were analyzed, the values change to 27.7% and 26.3%, respectively. There are significantly fewer residues buried in the monomeric forms of MSF proteins when compared to globular ones. In the dimeric forms, the ratio of buried residues is similar in both cases. Figure 3 shows the number of buried residues in MSF (see Figure 3A) and globular heterodimeric complexes (see Figure 3B).

It can be seen that in the case of MSF heterodimers, there is a more considerable difference between the number of buried residues in the dimeric and monomeric forms, than in the case of globular heterodimers. In the case of globular heterodimers (see Figure 3B), the sum of the number of buried residues in the two monomeric subunits is close to the number of buried residues in the dimeric form. These subunits are ordered by themselves, and they do not need another subunit to help to order their structures. In the case of MSF heterodimers (see Figure 3A), the sum of the number of buried residues in the monomeric forms is lower than in the case of the globular heterodimers and, more importantly, they are much smaller than the number of buried residues in the dimeric form. These polypeptide chains are disordered by themselves, they need the presence of an interacting partner to help in ordering their structures. These protein chains need each other to form a reasonably sized core, needed for a stable, ordered structure.

The secondary structural element content was determined in the heterodimeric structures using the DSSP program [21]. We found that in the MFHE dataset, 43.6% of the residues have the α-helical conformation and only 16.1% of the residues belonged to β-sheets, in the globular heterodimeric dataset, these values were 21.5% and 27.5%, respectively. In the MSF, heterodimeric dataset β-sheets were less abundant than in globular heterodimeric proteins. This will have some consequences in the interpretation of our later results.

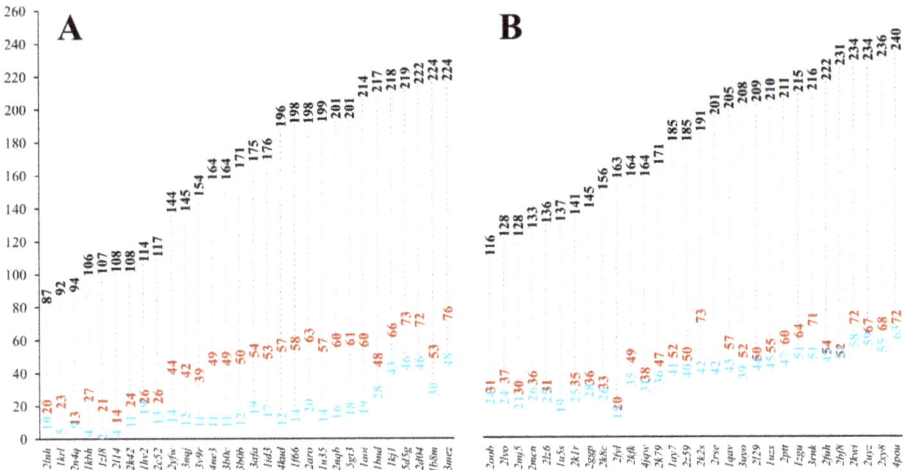

Figure 3. The number of burial residues in MFHE (**A**) and GLHE (**B**) complexes (black: number of all residues in a complex, red: number of buried residues in a heterodimeric complex, blue: sum of numbers of buried residues in the two monomeric subunits. See Figure S1 for the number of buried residues for the homodimeric MFHO and GLHO datasets.

We counted the number of inter-subunit ion-pairs. While there is only a small difference in the number of charged residues between MFHE and GLHE (1224 vs. 1380), the total charge is +320 for all 30 MFHE proteins and –91 for all 30 GLHE proteins. We found only 16 charged residues participating in 8 strong ion-pairs in the MFHE, while 28 residues are participating in 15 ion-pairs in the GLHE dataset. If we also consider weak ion-pairs, these values change to 73 residues participating in 42 ion-pairs for MFHE and 59 residues in 35 ion-pairs for GLHE. This is a 5.25-fold increase for MFHE and only a 2.33-fold increase for GLHE, respectively. Weak ion-pairs, presumably do not contribute to the enthalpic stabilization of the dimers, but probably play a role in the formation of electrostatic complementarity, already observed by Wong et al. in the case of complexes containing IDPs [22] This behavior was unexpected, and further investigation of the role of electrostatic interactions in the stabilization of MSF dimers is planned.

In the case of the MSF homodimers, we found that the main-chain solvent accessibility may play an important role in the stabilization of homodimer structures [8]. We identified residues with solvent accessible main-chain patches (RSAMPs). We have found a total of 161 RSAMPs in the MFHE dataset, and 90 RSAMPs in the GLHE dataset, respectively. There are 2 out of the 30 proteins in the MFHE dataset, which does not contain an RSAMP residue, while there are four such entries in the GLHE dataset. The average RSAMP content was 5.4 per heterodimeric complexes; thus, 17.7% of the interface residues are RSAMPs. In 26 of the 30 globular heterodimeric complexes, the average RSAMP content was 3, thus 19.1% of the interface residues are RSAMPs.

On the one hand, the composition of the RSAMPs of MFIB heterodimers suggested that five types of amino acids (glycine, alanine, isoleucine, leucine, and valine) play a major role in these interactions (see Figure 4). These RSAMP contributing amino acids are mainly hydrophobic, are exposed to the inter-subunit interface. These residues do not contribute to the stabilization of the monomeric form since exposed hydrophobic surfaces are energetically not favorable. However, next to the favorable burial of their main-chain, they might help the formation of the tertiary structure by building sticky hydrophobic patches at the inter-subunit interface. On the other hand, in the case of the globular heterodimer dataset, the two amino acids with the smallest side-chains, glycine and alanine are the most abundant residues under RSAMPs. We investigated the secondary structural distribution of RSAMP, as well. We found that 33.5% of RSAMPs are located in β-sheets and 44.7% in α-helices.

We checked the secondary structural composition of the interface residues, from which RSAMPs are selected. We found that 19.5% of interface residues have β-sheet and 63.9% have α-helical secondary structure. Considering the 3.3-fold higher occurrence of helical secondary structure at the interface, we can conclude that RSAMPs are more abundantly found in β structures, which can be easily broken by disturbing their hydrogen bonding network through interactions with accessible solvent molecules.

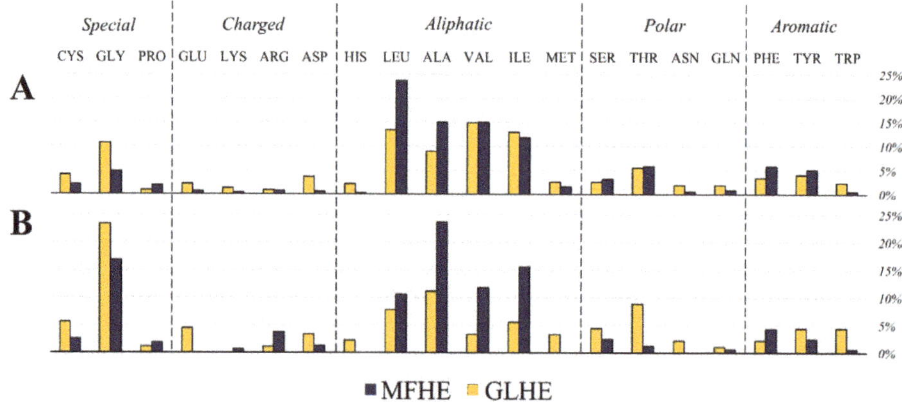

Figure 4. Amino acid composition of the interface (**A**) and the residues with solvent accessible main-chain patches (RSAMPs; (**B**)) of MFHE (blue) and GLHE (yellow) complexes.

We counted the number of inter-subunit hydrogen bonds. We found a total number of 181 H-bonds in the MFHE and only 67 in the GLHE dataset, respectively. This is in agreement with our observation that inter-subunit interactions are of high importance in MSF heterodimers. We calculated the average wrapping of hydrogen bonds [10]. Hydrogen bonds with a low wrapping (dehydrons) are less shielded from the solvent. The average value was 13.8 for the MFHE and 14.6 for the GLHE. Inter-subunit hydrogen bonds are slightly less wrapped in the MSF heterodimers, which also indicates the importance of solvent accessibility.

We also identified inter-subunit stabilization centers in both the MFHE and GLHE datasets. Stabilization centers are special residue pairs, which together with their sequential neighbors, participate in above than average long-range interactions and are believed to contribute to the stabilization of protein structures [23]. The two residues that form a stabilization center are called stabilization center elements (SCEs). In MFHE, the average inter-subunit SCE content was 8.1, and we found at least one inter-subunit SC in 26 of the 30 heterodimers. In GLHE, the average SCE content was 0.5, and we found an inter-subunit SC is only 5 out of the 30 structures.

We investigated if there is a lower size limit for globular proteins, which already bear a buried core structure. Our analysis of monomeric, single-domain globular (SGP) dataset pointed out that proteins with 35 residues are already containing a buried structural core (see Figure 5). Our results, regarding the buried core size of the MFIB heterodimers, indicate that although a couple of polypeptide chains are too small to contain a buried core, this is not a general trend for the MFHE dataset.

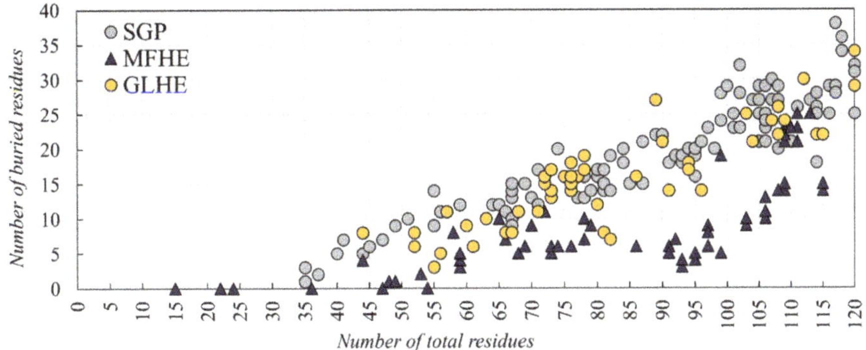

Figure 5. The number of total and buried residues of SGP (grey), GLHE (yellow) and MFHE (blue). For the number of total and buried residues of homodimeric MSF see Figure S2.

3. Conclusions

In our previous article [7], we found that the amino acid composition and sequence properties of MSF homodimers are similar to globular homodimers. However, they have more residues with solvent accessible peptide backbones that make them disordered in monomeric form, but they are ordered in a complex. There are some examples of these interactions in DIBS that prove their ordered nature. According to our results, MFIB heterodimers are less similar to globular proteins than homodimers, based on the calculated sequence and structural features. The MFIB heterodimers like the MFIB homodimers do not lack hydrophobic residues (as non-MSF IDPs), on the contrary, they are enriched in aliphatic residues which would theoretically allow the formation of a hydrophobic core, but in some cases, probably the chain itself is not large enough for the folding.

"Non-MSF" disorder prediction methods identify MFIB protein chains disordered at a short sequence segment which in some cases is confirmed by DIBS. In these DIBS interactions, heterodimeric MFIB subunits could bind to disordered, as well as globular protein regions. Therefore, in the case of heterodimeric MSF complexes, other factors can also affect their disorders than in the case of homodimeric MFIB proteins, because different factors are responsible for order-disordered interactions than for disordered-disordered complexes. This does not exclude that a protein chain can have the capability for both interactions and there has to be another ground why these proteins are unstructured on their own. In most cases of MSF heterodimers, the subunits themselves possibly do not have a low enough "energy" to fold, but the different compositions of the interacting partners may contribute to the stability of the complex. Understanding how sequence and composition and backbone variation affect foldability, will become increasingly crucial in folding protein design methods as more elements are included in the design process [24]. Based on NMDS results, the amino acid composition of the MSF heterodimeric complexes revealed smaller differences from the globular heterodimeric complexes, while the amino acid composition of the subunits showed distant similarity. Aromatic and hydrophobic amino acids are mainly responsible for the separation of the amino acid composition (based on SIMPER analysis) showed on NMDS. The amino acid composition of the MSF subunits is similar to globular proteins, but the MSF subunits together would change the amino acid composition of the complexes for a further reason. The heterodimeric MFIB complexes have a diverse amino acid composition, but they are involved in only a few types of molecular functions, such as DNA or histone binding (based on GO annotations from MFIB), which contributes to functional stability and making improvements in cell interactions [25].

In a recent paper [8], we concluded that MFIB proteins are disordered in monomeric form because they are too small to form a structural core. Our current analyses showed that however there are a couple of MSF protein subunits that do not contain buried residues, this is not a general rule, moreover we found that globular proteins with at least 35 residues already own a buried core (see Figure 5). At the same time, we found that the dimeric structures of MFIB and globular heterodimers contain a similar ratio of buried residues, but in monomeric form the MFIB heterodimers would contain only about half as much buried residues than globular heterodimers. We can conclude that the increased interface area of MFIB heterodimers contributes to the formation of a larger buried core structure. In globular heterodimers, the number of buried residues is increased only by a small margin upon dimerization, while in MFIB heterodimers there is a much larger increase (see Figure 3A). Globular monomers are stable and already own a reasonably sized buried structural core, while MFIB heterodimers are disordered by themselves and they need an interacting partner to form a large enough buried structural core to be stable. According to the structure-based analysis we can deduce that inter-subunit interactions are of high importance in the stabilization of MSF proteins. As in the case of homodimers, shielding of the main-chain from the solvent is an important factor for the stabilization of the heterodimeric structures. Interactions of the main-chain with water molecules might destabilize the secondary structure by breaking the hydrogen bond network, leading to the disruption of the secondary and the tertiary structure. Other interactions, which are identified by our definition of stabilization centers, play an important role in the stabilization of the heterodimeric structures, as well. This is consistent with our results about inter-subunit stabilization centers and Pfam domains, wherein the case of globular heterodimers, a few complexes have interactions and SCs between the Pfam domains of the subunits, whereas for MFIB heterodimers more than half of the complexes have inter-subunit interactions between the Pfam domains of the different chains. This also suggests that the folding of the MSF subunits is related to their functional role.

Though we found that the difference in the number of RSAMPs between MSF and globular proteins is slightly smaller in the case of heterodimers than it was in the case of homodimers, considering the lower β-sheet content of the MFHE dataset, the RSAMP/β-sheet forming residue ratio is correspondingly high in MSF heterodimers and homodimers. We suggest that the shielding of the β-sheet backbones and the formation of a buried structural core together with the general strengthening of inter-subunit interactions together could be the driving forces of MSF protein structural ordering upon dimerization.

Protein folding, the structural organization of proteins in aqueous solution, is realized by monomolecular reactions of intermolecular interactions, even if this is followed later by further macromolecular interactions because of functional or stability reasons. In the case of MSF proteins for the formation of a stable ordered structure intermolecular interactions are needed, therefore it is part of the folding. Opposing the regular folding this is not a monomolecular, but rather a bimolecular reaction, in which the ratio of the participating components and other parameters can be changed. We believe that further experimental and theoretical investigation of the structural organization of MSF proteins can contribute to a more profound understanding of the folding problem.

4. Materials and Methods

There are 49 heterodimeric proteins in the MFIB database. Entries belonging to the "coils and zippers" structural class were excluded, as in the case of homodimers. Since 25 of the 49 heterodimers are histones, filtering of the dataset was necessary to avoid overrepresentation and sequence redundancy of this protein class. Proteins were assigned to the same cluster if their sequence identity was over 90% using the BLASTClust toolkit 2.2.26 [26]. The 2mv7 entry was discarded because it was an outlier due to its fuzzy NMR structure in SASA calculations. One representative structure was kept for the remaining 30 clusters, creating the filtered MFHE dataset (see Table S3). A reference dataset was created of globular heterodimers (GLHE) from the PDBSelect [27] database with a total number of residues less than 240 to match the size distribution of the heterodimer MFIB dataset (see Table S3).

We described the methods in the latest article [7], but briefly: the interface term is used for the contact surface area of the two subunits in the heterodimeric structures. In cases where the term "monomeric structure" is used, calculations were carried out on single polypeptide chains, where the other chain was removed from the PDB files. Residues belonging to the interface region were identified based on solvent accessible surface area (SASA) calculations. All-atom SASA values were calculated using the FreeSASA 2.03 [28] program, residues where the SASA value calculated for the dimeric structure was less than or equal to 20% of the monomeric value, were defined to belong to the interface.

We were looking for residues in the interface that have solvent accessible spots in their main-chain in the monomeric structure, which become buried in the dimeric structures. We identified residues where the main-chain SASA in the dimeric form was less than 20% of the monomeric form value. Only residues with exposed main-chains, with a relative main-chain SASA larger than 0.2 in the monomeric structure, were taken into account. These residues with solvent accessible main-chain patches are called RSAMPs and are believed to be important for structural ordering upon dimerization of the disordered polypeptide chains collected in the MFIB database.

We used an additional Small Globular Protein (SGP) dataset to determine the minimal buried core size of proteins (see Table S3). We collected monomeric single-domain proteins X-ray structures from the PDBSELECT database with less than 120 residues, which do not contain disulfide bonds. Since there was a significant hole in the size distribution of the X-ray structures, monomeric single-domain NMR structures without disulfide bonds were added to the dataset. We excluded rod-like and fuzzy NMR structures using a volume/surface cutoff criterion. Protein volumes were calculated using the ProteinVolume 1.3 program [29].

Secondary structural elements were identified using the DSSP [21] program. Hydrogen bonds were identified using the find_pairs PyMol command using 3.5 Å distance and 45-degree angle criteria [30]. Wrapping of hydrogen bonds was calculated using the dehydron_ter.py program [31]. Stabilization centers (SCs) are special pairs of residues involved in cooperative long-range interactions. The two residues that form a stabilization center are called stabilization center elements (SCEs). SCEs were identified using our SCide server [32]. Ion pairs were defined as pairs of positively and negatively charged residues, with a distance of less than a cutoff value between the charged groups. For strong ion pairs, this value is 4 Å [33], but we introduce additionally, a weak ion-pair definition with a distance cutoff value of 6 Å. Histidine residues were assumed to be neutral in these calculations because of the uncertainty of their protonation states. Ion pairs were identified using our own C++ program. We calculated the total charge of the proteins simply by adding the number of Arg and Lys residues and subtracted the sum of Asp and Glu resides.

Amino acid compositions were determined using MEGA7 software [34]. The amino acid composition of the protein subunits and complexes were visualized in Nonmetric multidimensional scaling (NMDS) in PAST3 [35]. In the plot, one point for each amino acid composition, where close points were more similar in composition (with Bray-Curtis distances). This was followed by a SIMPER analysis (also based on Bray-Curtis distances, in PAST3) to identify those amino acids that contributed most to the observed differences among the type of subunits and complexes. Disorder predictions were revealed by IUPred2A [17] and MoRFpred [16] algorithms.

Supplementary Materials: Supplementary materials can be found at http://www.mdpi.com/1422-0067/20/20/5136/s1. Table S1. Contribution of the amino acids for the observed differences in NMDS by using SIMPER analysis in the subunits (A) and the complexes (B) Table S2. Pfam domains and the intermolecular SCs in globular (A) and MFIB (B) heterodimeric complexes. Table S3. List of PDB entries in the MFHE, GLHE and SGP datasets. Figure S1. The number of burial residues in MFHO (A) and GLHO (B) complexes (black: number of all residues in a complex, red: number of buried residues in a heterodimeric complex, blue: sum of number of buried residues in the two monomeric subunits. Figure S2. The number of total and buried residues of SGP, GLHE, GLHO, MFHE and MFHO.

Author Contributions: Conceptualization, I.S., and C.M.; methodology, A.M., E.F.; software, A.M., E.F., C.M.; validation, A.M., C.M; formal analysis, C.M.; investigation, A.M., E.F.; resources, A.M, E.F.; data curation, A.M., E.F., C.M.; writing—original draft preparation, A.M., C.M., I.S.; writing—review and editing, E.F., A.M., C.M; visualization, A.M.; supervision, I.S.; project administration, I.S.; funding acquisition, I.S.

Funding: This work was financially supported by the National Research, Development and Innovation Office (grant no. K115698). IS was supported by project no. FIEK_16-1-2016-0005 financed under the FIEK_16 funding scheme (National Research, Development and Innovation Fund of Hungary). The work of AM was supported through the New National Excellence Program of the Ministry of Human Capacities (Hungary).

Acknowledgments: The authors acknowledge the support of ELIXIR Hungary (www.elixir-hungary.org).

Conflicts of Interest: The authors declare no conflict of interest. The funders had no role in the design of the study; in the collection, analysis, or interpretation of data; in the writing of the manuscript, or in the decision to publish the results.

Abbreviations

IDP	Intrinsically Disordered Protein
MFIB	Mutual Folding Induced by Binding database
DIBS	Disordered Binding Site database
MFHO	MFIB Homodimeric dataset
MFHE	MFIB Heterodimeric dataset
GLHE	Globular Heterodimeric dataset
NMDS	Non Metric Multidimensional Scaling
RSAMPs	Residues with Solvent Accessible Main-chain Patches
SC/SCE	Stabilization centers/Stabilization center elements
SASA	Solvent Accessible Surface Area
SGP	Small Globular Protein dataset

References

1. Tsai, C.J.; Nussinov, R. Hydrophobic folding units at protein-protein interfaces: Implications to protein folding and to protein-protein association. *Protein Sci.* **1997**, *6*, 1426–1437. [CrossRef] [PubMed]
2. Xu, D.; Tsai, C.J.; Nussinov, R. Mechanism and evolution of protein dimerization. *Protein Sci.* **1998**, *7*, 533–544.
3. Gunasekaran, K.; Tsai, C.J.; Nussinov, R. Analysis of ordered and disordered protein complexes reveals structural features discriminating between stable and unstable monomers. *J. Mol. Biol.* **2004**, *341*, 1327–1341. [CrossRef]
4. Rumfeldt, J.A.; Galvagnion, C.; Vassall, K.A.; Meiering, E.M. Conformational stability and folding mechanisms of dimeric proteins. *Prog. Biophys. Mol. Biol.* **2008**, *98*, 61–84. [CrossRef] [PubMed]
5. Demarest, S.J.; Martinez-Yamout, M.; Chung, J.; Chen, H.; Xu, W.; Dyson, H.J.; Evans, R.M.; Wright, P.E. Mutual synergistic folding in recruitment of CBP/p300 by p160 nuclear receptor coactivators. *Nature* **2002**, *415*, 549–553. [CrossRef] [PubMed]
6. Habchi, J.; Tompa, P.; Longhi, S.; Uversky, V.N. Introducing protein intrinsic disorder. *Chem. Rev.* **2014**, *114*, 6561–6588. [CrossRef]
7. Magyar, C.; Mentes, A.; Fichó, E.; Cserző, M.; Simon, I. Physical Background of the Disordered Nature of "Mutual Synergetic Folding" Proteins. *Int. J. Mol. Sci.* **2018**, *19*, 3340. [CrossRef]
8. Mészáros, B.; Dobson, L.; Fichó, E.; Tusnády, G.E.; Dosztányi, Z.; Simon, I. Sequential, Structural and Functional Properties of Protein Complexes Are Defined by How Folding and Binding Intertwine. *J. Mol. Biol.* **2019**. [CrossRef]
9. Fichó, E.; Reményi, I.; Simon, I.; Mészáros, B. MFIB: A repository of protein complexes with mutual folding induced by binding. *Bioinformatics* **2017**, *33*, 3682–3684. [CrossRef]
10. Fernández, A.; Scott, R. Dehydron: A structurally encoded signal for protein interaction. *Biophys. J.* **2003**, *85*, 1914–1928. [CrossRef]
11. Uversky, V.N.; Gillespie, J.R.; Fink, A.L. Why are "natively unfolded" proteins unstructured under physiologic conditions? *Proteins* **2000**, *41*, 415–427. [CrossRef]
12. Campen, A.; Williams, R.M.; Brown, C.J.; Meng, J.; Uversky, V.N.; Dunker, A.K. TOP-IDP-scale: A new amino acid scale measuring propensity for intrinsic disorder. *Protein Pept. Lett.* **2008**, *15*, 956–963. [CrossRef] [PubMed]
13. Taguchi, Y.H.; Oono, Y. Relational patterns of gene expression via non-metric multidimensional scaling analysis. *Bioinformatics* **2005**, *21*, 730–740. [CrossRef] [PubMed]
14. Bah, A.; Forman-Kay, J.D. Modulation of Intrinsically Disordered Protein Function by Post-translational Modifications. *J. Biol. Chem.* **2016**, *291*, 6696–6705. [CrossRef] [PubMed]

15. Kullback, S.; Leibler, R.A. On information and sufficiency. *Ann. Math. Stat.* **1951**, *22*, 79–86. [CrossRef]

16. Disfani, F.M.; Hsu, W.L.; Mizianty, M.J.; Oldfield, C.J.; Xue, B.; Dunker, A.K.; Uversky, V.N.; Kurgan, L. MoRFpred, a computational tool for sequence-based prediction and characterization of short disorder-to-order transitioning binding regions in proteins. *Bioinformatics* **2012**, *28*, i75–i83. [CrossRef] [PubMed]

17. Mészáros, B.; Erdos, G.; Dosztányi, Z. IUPred2A: Context-dependent prediction of protein disorder as a function of redox state and protein binding. *Nucleic Acids. Res.* **2018**, *46*, W329–W337. [CrossRef]

18. Schad, E.; Fichó, E.; Pancsa, R.; Simon, I.; Dosztányi, Z.; Mészáros, B. DIBS: A repository of disordered binding sites mediating interactions with ordered proteins. *Bioinformatics* **2018**, *34*, 535–537. [CrossRef]

19. Klus, P.; Bolognesi, B.; Agostini, F.; Marchese, D.; Zanzoni, A.; Tartaglia, G.G. The cleverSuite approach for protein characterization: Predictions of structural properties, solubility, chaperone requirements and RNA-binding abilities. *Bioinformatics* **2014**, *30*, 1601–1608. [CrossRef]

20. Dosztányi, Z.; Fiser, A.; Simon, I. Stabilization centers in proteins: Identification, characterization and predictions. *J. Mol. Biol.* **1997**, *272*, 597–612. [CrossRef]

21. Kabsch, W.; Sander, C. Dictionary of protein secondary structure: Pattern recognition of hydrogen-bonded and geometrical features. *Biopolymers* **1983**, *22*, 2577–2637. [CrossRef] [PubMed]

22. Wong, E.T.; Na, D.; Gsponer, J. On the importance of polar interactions for complexes containing intrinsically disordered proteins. *PLoS Comput. Biol.* **2013**, *9*, e1003192. [CrossRef] [PubMed]

23. Magyar, C.; Gromiha, M.M.; Sávoly, Z.; Simon, I. The role of stabilization centers in protein thermal stability. *Biochem. Biophys. Res. Commun.* **2016**, *471*, 57–62. [CrossRef] [PubMed]

24. Saven, J.G. Designing protein energy landscapes. *Chem. Rev.* **2001**, *101*, 3113–3130. [CrossRef]

25. Lee, B.M.; Mahadevan, L.C. Stability of histone modifications across mammalian genomes: Implications for 'epigenetic' marking. *J. Cell. Biochem.* **2009**, *108*, 22–34. [CrossRef]

26. Alva, V.; Nam, S.Z.; Söding, J.; Lupas, A.N. The MPI bioinformatics Toolkit as an integrative platform for advanced protein sequence and structure analysis. *Nucleic Acids. Res.* **2016**, *44*, W410–W415. [CrossRef]

27. Griep, S.; Hobohm, U. PDBselect 1992–2009 and PDBfilter-select. *Nucleic Acids. Res.* **2010**, *38*, D318–D319. [CrossRef]

28. Mitternacht, S. FreeSASA: An open source C library for solvent accessible surface area calculations. *F1000Research* **2016**, *5*, 189. [CrossRef]

29. Chen, C.R.; Makhatadze, G.I. ProteinVolume: Calculating molecular van der Waals and void volumes in proteins. *BMC Bioinforma.* **2015**, *16*, 101. [CrossRef]

30. LLC. *The PyMOL Molecular Graphics System*; Schrodinger Version 1.6; LLC: New York, NY, USA, 2011.

31. Martin, O.A. *Wrappy: A Dehydron Calculator Plugin for PyMOL*; IMASL-CONICET: San Louis, Argentina, 2011.

32. Dosztányi, Z.; Magyar, C.; Tusnády, G.; Simon, I. SCide: Identification of stabilization centers in proteins. *Bioinformatics* **2003**, *19*, 899–900. [CrossRef]

33. Barlow, D.J.; Thornton, M.J. Ion pairs in proteins. *J. Mol. Biol.* **1983**, *168*, 867–885. [CrossRef]

34. Kumar, S.; Stecher, G.; Tamura, K. MEGA7: Molecular Evolutionary Genetics Analysis Version 7.0 for Bigger Datasets. *Mol. Biol. Evol.* **2016**, *33*, 1870–1874. [CrossRef] [PubMed]

35. Hammer, Ø.; Harper, D.A.T.; Ryan, P.D. PAST: Paleontological statistics software package for education and data analysis. *Palaeontol. Electron.* **2002**, *4*, 9.

© 2019 by the authors. Licensee MDPI, Basel, Switzerland. This article is an open access article distributed under the terms and conditions of the Creative Commons Attribution (CC BY) license (http://creativecommons.org/licenses/by/4.0/).

International Journal of
Molecular Sciences

Article

Depicting Conformational Ensembles of α-Synuclein by Single Molecule Force Spectroscopy and Native Mass Spectroscopy

Roberta Corti [1,2], Claudia A. Marrano [1], Domenico Salerno [1], Stefania Brocca [3], Antonino Natalello [3], Carlo Santambrogio [3], Giuseppe Legname [4], Francesco Mantegazza [1], Rita Grandori [3,*] and Valeria Cassina [1,*]

[1]   School of Medicine and Surgery, Nanomedicine Center NANOMIB, University of Milan-Bicocca, 20900 Monza, Italy; r.corti9@campus.unimib.it (R.C.); claudia.marrano@unimib.it (C.A.M.); domenico.salerno@unimib.it (D.S.); francesco.mantegazza@unimib.it (F.M.)
[2]   Department of Materials Science, University of Milan-Bicocca, 20125 Milan, Italy
[3]   Department of Biotechnology and Biosciences, University of Milan-Bicocca, 20126 Milan, Italy; stefania.brocca@unimib.it (S.B.); antonino.natalello@unimib.it (A.N.); carlo.santambrogio@unimib.it (C.S.)
[4]   Scuola Internazionale Superiore di Studi Avanzati, SISSA, 34136 Trieste, Italy; giuseppe.legname@sissa.it
*   Correspondence: rita.grandori@unimib.it (R.G.); valeria.cassina@unimib.it (V.C.)

Received: 11 September 2019; Accepted: 17 October 2019; Published: 19 October 2019

Abstract: Description of heterogeneous molecular ensembles, such as intrinsically disordered proteins, represents a challenge in structural biology and an urgent question posed by biochemistry to interpret many physiologically important, regulatory mechanisms. Single-molecule techniques can provide a unique contribution to this field. This work applies single molecule force spectroscopy to probe conformational properties of α-synuclein in solution and its conformational changes induced by ligand binding. The goal is to compare data from such an approach with those obtained by native mass spectrometry. These two orthogonal, biophysical methods are found to deliver a complex picture, in which monomeric α-synuclein in solution spontaneously populates compact and partially compacted states, which are differently stabilized by binding to aggregation inhibitors, such as dopamine and epigallocatechin-3-gallate. Analyses by circular dichroism and Fourier-transform infrared spectroscopy show that these transitions do not involve formation of secondary structure. This comparative analysis provides support to structural interpretation of charge-state distributions obtained by native mass spectrometry and helps, in turn, defining the conformational components detected by single molecule force spectroscopy.

Keywords: α-synuclein; single molecule force spectroscopy; intrinsically disordered proteins; native mass spectrometry

---

1. Introduction

Intrinsically disordered proteins (IDPs) play crucial regulatory roles in biological systems and lack a specific tertiary structure under physiological conditions [1–4]. Molecular characterization of IDPs requires description of the conformational ensembles populated by the disordered polymers in solution. Single-molecule approaches offer information on dynamic and heterogeneous ensembles, capturing distinct and less populated states, overcoming the limitations of average parameter assessment, intrinsic to bulk methods [5–8].

Usually employed in imaging mode [9,10], atomic force microscopy (AFM) can be used in single-molecule force spectroscopy (SMFS) to characterize the statistical distribution of distinct protein conformers in solution. Indeed, protein unfolding under the action of a pulling force has been demonstrated to characterize the molecular structure of tens of distinct proteins and to distinguish among different conformations induced by ligand binding or mutations [11–13]. In the case of the human, amyloidogenic IDP α-synuclein (AS), at least three major conformational states can be recognized [14–16]: random coil (RC), collapsed states stabilized by weak interactions (WI), and compact conformations stabilized by strong interactions (SI). The SMFS technique has been applied to explore the conformational space populated by the different structures of the protein, revealing distinct conformers of the molecular ensemble and structural effects of point mutations linked to familial Parkinson's disease [4,14–17].

Pure AS in vitro, in the absence of interactors, is largely unstructured at neutral pH, with a small fraction of the population in collapsed states of different compactness, as revealed by NMR spectroscopy [18] and small angle X-ray scattering [19]. A particularly compact, globular state is populated in vivo, as indicated by in-cell NMR in neuronal and non-neuronal mammalian cell types [20]. Dopamine (DA) and epigallocatechin-3-gallate (EGCG) are known to bind AS and redirect the aggregation pathway toward soluble oligomers with different structure and toxicity [21,22].

Native mass spectrometry (native MS) has developed into a central tool for structural biology [23–26]. The analysis of charge states populated by globular and disordered proteins by native MS has shown effects of denaturants [27], stabilizers [28], metal binding [29], and protein–protein interactions [30], just to mention some examples. The application of native MS to free AS in solution reveals multimodal charge-state distributions (CSDs), which are suggestive of a conformational ensemble populated by different conformers, in line with the above-mentioned, in vitro and in vivo evidence [29,31–33]. The charge states obtained by proteins in electrospray have long been recognized as affected by protein compactness at the moment of transfer from solution to gas phase [27,34]. This effect can be rationalized by an influence of protein structure on solvent-accessible surface area [35–37] and apparent gas-phase basicity [38].

A large amount of evidence suggests that the ionization patterns of globular and disordered proteins are similarly affected by conformational properties [23,39–41]. Native MS has described conformational responses of AS to alcohols, pH, and copper binding consistent with NMR and other solution methods [29,33]. Native MS has also suggested that binding of DA and EGCG have distinct structural effects on AS soluble monomers [42,43]. While DA preferentially binds and stabilizes an intermediate form, EGCG promotes accumulation of the most compact AS conformer [42]. This different conformational selectivity could help rationalizing the different structure and toxicity of the resulting oligomers, although the two ligands have similar fibrillation-inhibition effects [42]. Nonetheless, the difficulty to capture IDP compact states by small-angle X-ray scattering and ensemble-optimization method has led to the hypothesis that IDP bimodal CSDs are artifacts resulting from a bifurcated ESI mechanism, rather than distinct components reflecting structural heterogeneity of the original protein sample [44]. The aim of this work is to describe AS conformational ensemble and its response to ligands by orthogonal and highly sensitive biophysical techniques, such as SMFS, in order to test the effect of ligand binding in solution and help interpretation of the available native-MS data on AS and IDPs in general. It is found that, while spectroscopic methods sensitive to secondary structure do not capture these conformational transitions, SMFS and native MS reveal rearrangements of the conformational ensembles, consistent with a loss of structural disorder induced by the ligands.

2. Results

2.1. Single Molecule Force Spectroscopy (SMSF)

The SMFS experiments have been performed on a polyprotein construct containing eight repeats of titin immunoglobulin-like domain (I27) and one grafted AS domain [45–47]. A schematic representation of the polyprotein construct and typical unfolding curves in the absence of ligands are reported in Figure 1A,B.

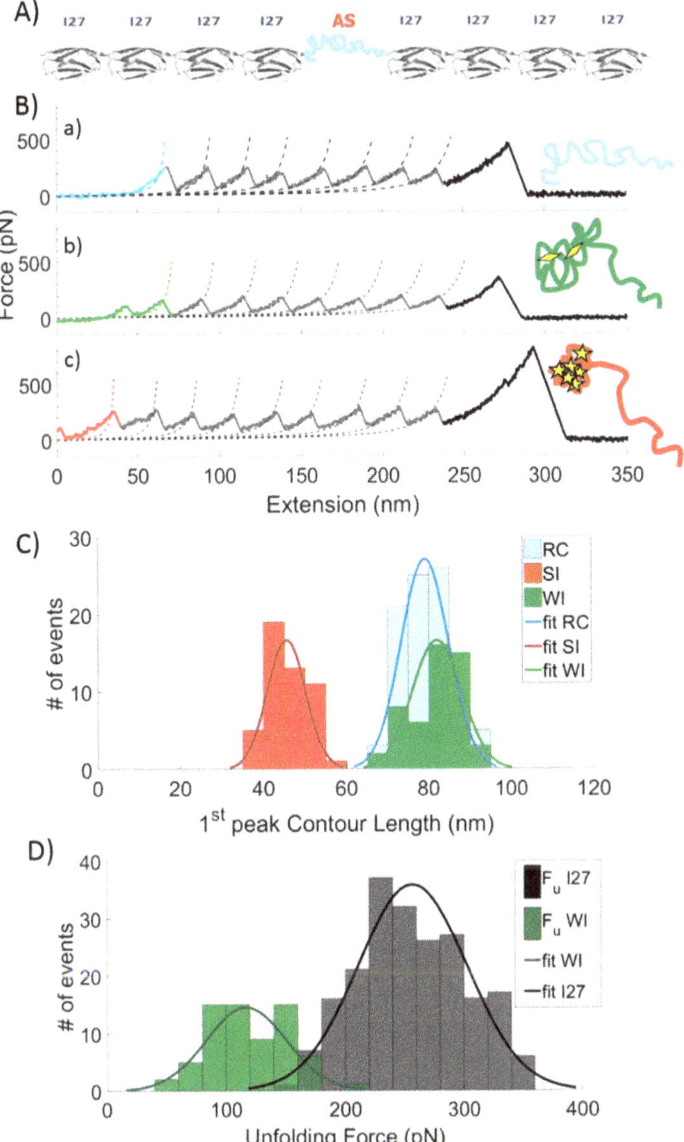

Figure 1. Representative single molecule force spectroscopy (SMFS) recording of α-synuclein (AS) polyprotein and relative statistical analysis. (**A**) Polyprotein construct encompassing the AS full-length polypeptide chain for SMFS experiments. (**B**) Representative force curves of the mechanical unfolding of the polyprotein in distinct conformations stabilized by RC (a), WI (b), and SI (c). Dotted lines are worm-like-chain (WLC) fits to the force-extension curves with free contour length L_C and a fixed persistence length L_p = 0.36 nm (see Figure S2 for raw data). Sketches of AS conformations are shown on the right. Diamonds represent weak interactions stabilizing the AS protein, while stars represent strong interactions. (**C**) Statistical distribution of the contour length of the first peak for RC (L_C = 79 ± 6 nm), WI (L_C = 82 ± 6 nm), and SI (L_C = 46 ± 5 nm) conformations. Solid lines represent the Gaussian fits of the histograms. (**D**) Unfolding force statistical distribution of WI (F_{WI} = 117 ± 34 pN) and I27 modules (F_{I27} = 257 ± 46 pN).

As apparent from Figure 1B, the observed SMFS curves show the typical "sawtooth" pattern, in which the initial part is related to the presence of AS and it is characterized by different mechanical resistances to the unfolding. Each following regular peak is due to the unfolding of an individual I27 domain. Every curve was fitted by means of the worm-like-chain (WLC) model to extract the contour length L_C of each peak (both for I27 and AS) [48]. Consistent with the presence of a heterogeneous conformational ensemble, three distinct patterns can be recognized by analyzing the L_C of the first peak (Figure 1B,C). A first class of curves displays $L_C = 79 \pm 6$ nm (light blue curve, first line of Figure 1B); a second class is characterized by $L_C = 82 \pm 6$ nm and by the presence of at least one small peak before the first regular peak (green curve, second line of Figure 1B); a third class displays $L_C = 46 \pm 5$ nm and it is characterized by the presence of an additional peak whose height is comparable to the one related to an I27 unfolding event (red curve, third line in Figure 1B). In detail, the light blue curves are ascribed to unstructured conformations of AS and classified as random coil (RC), since no additional peak is detected in the first ~80 nm. The green curves display small (one or more) peaks corresponding to an unfolding force ($F_{WI} = 117 \pm 34$ pN) sensibly lower than I27 ($F_{I27} = 257 \pm 46$ pN, Figure 1D, Figure S1 and Table S1). These curves are interpreted as representative of a collapsed state of AS mainly stabilized by weak interactions (WI), characterized by an energy barrier to overcome smaller than the one involved in the I27 unfolding. The third and the latter type of curves, characterized by a shorter L_C is assigned to a collapsed state of AS, mainly stabilized by strong interactions (SI), which presents resistance to unfolding similar to the one shown by the highly mechanostable protein I27. The extension of the first peak (in the curves assigned to the RC conformation) and that of the first of the higher peaks (in the curves assigned to the WI) are all around 80 nm. These peaks occur when AS is completely extended and flanked by eight I27 folded modules. The measured length is due to the contribution of the eight folded I27 modules (a folded module of I27 is 3 nm long, i.e., 3 nm × 8 = 24 nm), the length of eight linkers between each protein module (a linker is 2 aa, i.e., 8 × 0.36 nm × 2 = 5.76 nm) [47], and the length of the completely extended AS (i.e., 140 aa × 0.36 nm = 50.4 nm). By summing all the contributions, one obtains a total extension of 80.16 nm, which is coherent with the measured values. The subset of curves presenting a L_C of the first peak higher than 95 nm, which could be associated with an undesired misfolding event of a I27 module [49], was discarded.

2.2. Effect of DA and EGCG on the Conformational Ensemble

The SMFS measurements were repeated in the presence of either 200 μM DA or 25 μM EGCG (see Figures S3–S6 and Tables S2–S6 for more details). These concentrations were chosen to compare SMFS results with native-MS data [42]. The statistical distributions of AS conformations obtained by SMFS in the presence or absence of ligands are reported in Figure 2A. In solution, at neutral pH and without ligands, AS behaves partially as RC (62% of the molecules) and partially populates collapsed states (~30%, mainly stabilized by WI and ~8%, mainly stabilized by SI), consistent with previously reported SMFS data [14–16]. The addition of either ligand leads to a loss of the RC conformation in favor of the SI conformation, with the most pronounced effect of EGCG (drop of RC from 62% to 36%). The same conditions had been investigated by native MS, showing the presence of intermediate states (I1 and I2), together with random coil (RC) and compact (C) conformations (Figure 2B).

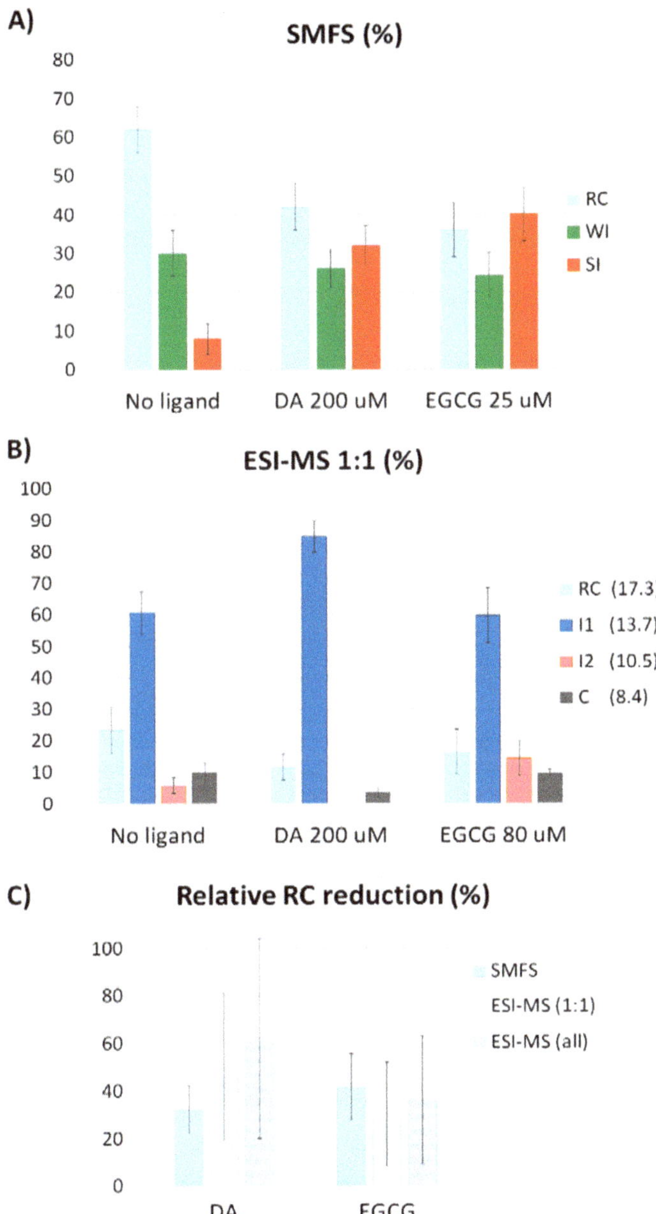

Figure 2. Species distributions as obtained by (**A**) SMFS and (**B**) native MS. The intensity-weighted average charge of the peak envelopes is reported in brackets (i.e., RC = 17.3; I1 = 13.7; I2 = 10.5; C = 8.4). Error bars in panel (A) represent the standard deviation calculated for the normal distribution. Error bars in panel (B) represent the standard deviations from three independent experiments. (**C**) RC reduction in response to ligand binding, relative to the free protein, as obtained by SMFS and native MS, considering the 1:1 protein:ligand complexes (1:1) or the cumulative MS data (all). Error bars in panel (C) represent the propagated standard deviation.

A quantitative comparison between the species distributions obtained by SMFS and native-MS data is shown in Figure 2C. An intrinsic difference between SMFS and MS concerns the discrimination between free and ligand-bound protein molecules, which is possible only by the latter technique. Thus, native-MS data in Figure 2C have been processed by two alternative ways. In one case, only signals of the 1:1 protein:ligand complexes have been considered. This procedure yields more reliable information on the conformational changes induced by ligand binding but is, at the same time, not exactly comparable to the blind molecular selection performed by SMFS. Thus, "cumulative" MS data are also shown (labeled as ESI-MS(all) in Figure 2C), derived by Gaussian fitting of the artificial CSD obtained by the summation of the species-specific CSDs corresponding to the different binding stoichiometries, including the free protein. In either way, the aggregated data for the unstructured (RC) component, represented as relative change from the reference condition of the protein in the absence of ligands, indicate a remarkable loss of the most disordered conformation induced by ligand binding, as assessed by both techniques.

2.3. Comparison to CD and FTIR

For comparison with complementary spectroscopic methods, sensitive to protein secondary structure, far-UV circular dichroism (CD) and Fourier-transform infrared spectroscopy (FTIR) analyses were performed. Representative results are reported in Figure 3. It can be noted that AS spectra in the presence or absence of the ligands, acquired by either technique under the same conditions employed for SMFS experiments, are almost superimposable. Thus, bulk methods probing secondary structure do not capture the conformational changes induced by ligand binding in monomeric AS in solution.

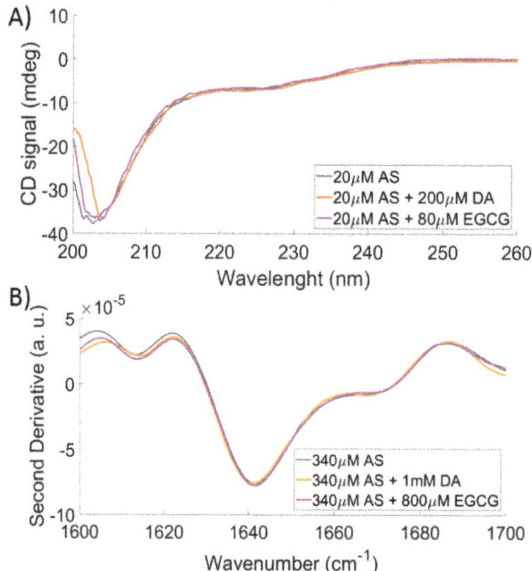

Figure 3. Secondary-structure content as obtained by CD and Fourier-transform infrared spectroscopy (FTIR) techniques. (**A**) Far-UV CD spectra of 20 µM AS in PBS buffer in the absence of ligands (gray), in the presence of 200 µM DA (orange) or 80 µM epigallocatechin-3-gallate (EGCG) (purple). (**B**) Second derivatives in the Amide I region of the FTIR absorption spectra of 340 µM AS in deuterated PBS buffer in the absence of ligands (gray) and in the presence of 1 mM DA (orange) or 800 µM EGCG (purple).

3. Discussion

The results reported here provide direct evidence of the different conformers populated by AS in solution and the structural effects elicited by ligand binding, resulting in a rearrangement of the conformational ensemble [50]. The structural heterogeneity of free AS in solution captured by SMFS is consistent with previous reports by the same approach [14,15], as well as with results from native MS [23,29,33,42], computational simulations [51,52], and chemical crosslinking [53], indicating the presence of at least three different conformational states characterized by different degrees of intramolecular interactions. Furthermore, SMFS is applied here for the first time to probe the effects of the fibrillation inhibitors DA and EGCG on AS conformational properties in solution.

Since the reliability of CSD analysis in the investigation of IDP conformational ensembles by native MS has been questioned [44], the SMFS results obtained in this work are compared to native-MS data. In analogy with SMFS, the CSD analysis of nano-ESI-MS spectra identifies, in addition to the RC component, the presence of three non-RC components, namely the intermediate species I1, I2, and the compact conformation C. Furthermore, both techniques indicate a loss of the most disordered component in response to ligand binding, resulting in the accumulation of the more structured species. Thus, not only the presence of multimodal profiles is confirmed by both techniques, but also a reorganization of the conformational ensemble in the same direction is consistently indicated in the presence of ligands.

Nonetheless, the structural intermediates detected by SMFS and native MS cannot be related in a straightforward way. These discrepancies can be due to the fact that the physical properties detected by the two techniques are different. While SMFS discriminates protein structures according to their mechanical stability under an external tension (quantified by the unfolding force), native MS is affected by structural compactness (quantified by the acquired net charge). Different compaction levels can correspond to similar unfolding force and vice-versa. Accordingly, the WI state, as detected by SMFS, is characterized by a number of variable peaks ranging from 1 to 3 different species, which could be compatible with different AS compaction states. Furthermore, the SMFS instrumental noise, related to the minimum measurable force (around 20pN) limits the minimal detectable unfolding force, below which the less stable AS compact states are counted as RC molecules.

It should also be noted that the conformations with lower unfolding force, as detected by SMFS, could include some components with higher charge-state detected by ESI-MS. This hypothesis can be verified by comparing the two techniques in terms of the response of the RC component to the binding of the ligand. Indeed, upon binding of either ligand there is a compatible trend of loss in such a component, as observed by both techniques, in favor of more compact structures (native MS) or stronger interactions (SMFS). Therefore, interpreting the low-charge components of CSDs as collapsed and partially structured conformational states leads to compatible pictures delivered by SMFS and native MS. Both techniques reveal the presence of partially structured conformers, thus suggesting that the bimodal or multimodal CSDs detected by native MS do not simply reflect artefacts of the ESI mechanism. It is worth pointing out that the conditions employed in this work do not lead neither to AS oxidation (Figure S7) nor to AS oligomerization, which requires incubation at 37 °C, shaking and higher protein concentrations [54], as also indicated by the lack of higher-order aggregates in native-MS spectra [42].

It cannot be ruled out that different ionization and/or transmission efficiency of compact and extended protein ions in native MS might lead to distortions of the apparent molecular ensemble, adding to the difficulties of direct comparison with SMFS data. Indeed, it has been suggested that folded and unfolded molecules could undergo different ESI mechanisms, resulting in different signal yields [55]. However, this effect seems to be protein-specific, since quantitative agreement with solution methods has been observed describing, for instance, the pH-dependent unfolding transition of cytochrome *c* [56]. The underlying mechanism has been identified in the different hydropathy of the exposed regions of normally folded proteins in different conformational states, which could affect their surface activity inside ESI droplets [55]. Such an effect is expected to be much more modest for IDPs, which lack a structured hydrophobic core and whose collapsed conformations are mostly promoted by electrostatic interactions [57]. More systematic, quantitative comparison between native MS and solution methods will be required to further elucidate this point. This first comparative study between a single-molecule technique and CSD analysis by native MS supports the feasibility of combined approaches to describe IDP molecular ensembles.

Based on this study, it seems safe to conclude that SMFS and structural interpretation of CSDs consistently indicate the simultaneous presence of collapsed and partially structured conformers of AS monomer in solution and, most importantly, reveal induced-folding transitions elicited by ligand binding. Furthermore, this study shows that single-molecule protein unfolding can capture changes in AS conformational landscape, induced by variable solution conditions, with remarkable sensitivity and reproducibility. These results indicate that the conformational ensemble depicted by two orthogonal biophysical principles is heterogeneous and reshaped in the same direction by ligand binding.

Another implication of this study is that the AS conformational transitions detected by SMFS under these conditions should not be interpreted in terms of secondary structure formation [14,15]. Indeed, the measured WI and SI components cannot be simply seen as the distinct contributions of van-der-Waals interactions and ordered secondary structure, respectively. In fact, an increase of almost 30% in the SI component, as observed here, would be detected by CD and FTIR spectroscopies, if ascribable to secondary structure. It is conceivable that the AS conformational components detected by SMFS under these conditions differ by contact order, type, and number of interactions, within a picture of similar secondary-structure content. Hence, a new structural interpretation of SMFS data is proposed, in particular for the SI population, differing from the one reported in the literature [14,15], where the SI component was directly associated with the presence of secondary structure.

This comparison points out that the ion-sorting mechanism inherent to MS analyses makes the MS methods more comparable to single-molecules approaches, rather than to bulk spectroscopic techniques, and underscores the importance of multi-technological approach to ensemble characterization. Nevertheless, the WI population detected by SMFS and the intermediate species (I1 and I2) detected by native MS do not necessarily coincide. Actually, two intermediates are detected by MS and only one by SMFS and the WI species found by SMFS does not respond to ligands, while the MS-detected intermediates do. These results indicate that both techniques capture the decrease in structural disorder induced by the ligands, but they describe the partially structured species of the conformational ensemble in different ways. In particular, it seems that the collapsed and partially structured species detected by MS contribute cumulatively to the SI component by SMFS, while the WI component by SMFS does not find correspondence in the MS spectra. These interactions could be too weak to survive the ionization/desolvation step.

4. Materials and Methods

4.1. Cloning, Expression, and Purification of the (I27)4_AS_(I27)4 Polyprotein

In order to obtain a (I27)4_AS_(I27)4 polyprotein, consisting of a single AS molecule, flanked by four repetitions of titin immunoglobulin-like domain (I27) at the N-terminus and at the C-terminus, the cDNA of the human *AS* (NP_000336) was cloned in the pRSet.A(I27)8 expression vector [47], taking advantage of the NheI restriction site placed in the middle of (I27)8 encoding sequence. A mutagenic PCR was performed on the pEGFP_AS vector [58] to delete the start and stop codons and to insert a NheI restriction site at both extremities of the *AS* gene. The PCR was carried out using the Q5®High-Fidelity DNA Polymerase (NEB, cat. #M0491) with the following primers: forward primer 5′ AAAAGCTAGCGATGTATTCATGAAAGGAC 3′, reverse primer 5′ AATTGCTAGCGGCTTCAGGTTCGTAG 3′, (in bold, the NheI restriction site). After sequencing, the pRSet.A (I27)4_AS_(I27)4 vector was used to transform BL21(DE3) *Escherichia coli* cells. Transformed cells were grown in Luria-Bertani medium at 37 °C until they reached an OD600 of 0.4–0.6 and the expression of the polyprotein was induced overnight at 22 °C by the addition of 1 mM IPTG. Cells were subsequently harvested by centrifugation and resuspended in lysis buffer (50 mM Na_2HPO_4, 300 mM NaCl, 10 mM imidazole, 4% Triton™ X-100, and 0.5 mM phenylmethylsulfonyl fluoride) before sonication on ice. The purification was performed by gravity flow column ion metal affinity chromatography (IMAC), taking advantage of the 6× His-tag present at the N-terminus of the polyprotein. The soluble fraction of cell lysate was incubated on Ni+-NTA resin (Roche, cat. #05893682001) for 1 h at 4 °C with gentle agitation. The washing step was carried out in 50 mM Na_2HPO_4, 300 mM NaCl added with 20 mM imidazole, elution was achieved in the same buffer, added with 250 mM imidazole. The presence of the protein in the eluted fractions was verified by SDS-PAGE on a 4–12% polyacrylamide gel (InvitrogenTM, ThermoFisher Scientific, cat. #NW04120BOX) stained with Coomassie Brilliant Blue.

4.2. AFM—Single Molecule Force Spectroscopy

SMFS experiments were carried out on a Nanowizard II (JPK Instruments, Berlin, Germany) at room temperature. Prior to each experiment, every cantilever (Si_3N_4, Bruker MLCT-BIO, Cantilever D, Nominal spring constant k = 0.03 N/m) was individually calibrated using the Equipartition Theorem in the JPK software. Approximately 20 µL of protein (at a concentration of ~2 µM) were deposited onto an evaporated gold coverslip and allowed to adsorb for about 15 min. After this time, 1.8 mL of PBS buffer (pH 7.4, 150 mM) were added to reach an overall protein final concentration of ~20 nM. Constant-velocity, single-molecule pulling experiments were performed at 1 µm/s, with a recorded rate of 4096 Hz. Each experiment was carried out in fresh PBS buffer, to which EGCG (stock diluted in PBS) and DA (stock diluted in acidic MilliQ, pH 4) (stored at 4 °C protected from light) were added to reach the desired final concentration. Each solution was filtered on a filter screen with a porosity of 0.2 µm before each experiment.

4.3. AFM Data Analysis

The resulting force curves were then processed by means of both the JPK-Data Processed software (JPK Instruments, Berlin, Germany) and MATLAB custom-written software. The contour length (L_C) of each peak (both I27 and AS) was calculated by means of WLC fit as a single parameter, while the persistence length (L_P) was kept constant (0.36 nm) [59]. Only curves with a single clear detachment peak, at least seven I27 peaks, and traces with a spurious signal below 45pN in the first 25 nm of the force-extension were considered.

4.4. Native-MS Experiments

Nano-ESI-MS data were taken from Konijnenberg [42]. In particular, nano-ESI-MS spectra were collected after 10-minute incubation of protein-ligand mixtures in 10 mM ammonium acetate, pH 7.4, at a final AS concentration of 20 µM. Quantification from native-MS data was based on Gaussian fitting of CSDs, upon transformation to x = z abscissa axis. The reported values refer to the area of the components obtained for the protein in the absence of ligand and for the 1:1 AS:ligand complexes, from three independent experiments.

4.5. CD and FTIR Experiments

CD and FTIR analyses were performed as previously described [43]. In particular, Far-UV CD spectra of 20 µM AS in PBS buffer were acquired on a J-815 spectropolarimeter (JASCO Corp., Tokyo, Japan) under the following instrumental settings: data pitch, 0.1 nm; scan speed, 20 nm/min; bandwidth, 1 nm; accumulation spectra, 2. A 1-mm path length quartz cuvette was employed. FTIR spectra of 340 µM AS in deuterated PBS buffer were acquired on a Varian 670-IR spectrometer (Varian Australia Pty. Ltd., Mulgrave, VIC, Australia) under the following instrumental settings: resolution, 2 cm^{-1}; scan speed, 25 kHz; scan coadditions, 1000; apodization, triangular; nitrogen-cooled mercury cadmium telluride detector. A temperature-controlled transmission cell with two BaF_2 windows separated by a 100-µm Teflon spacer was employed. Representative spectra from three independent experiments are shown.

5. Conclusions

Single-molecule description of AS conformational ensemble in solution detects differently structured components that are overseen by bulk spectroscopic methods, which probe secondary structure, but are consistent with the different degrees of compactness suggested by CSD analysis. Thus, although ion-mobility studies and molecular-dynamics simulations have shown that IDPs rearrange in the gas phase in a charge-dependent fashion [40], the extent of ionization at the moment of transfer from solution to gas phase, i.e., CSDs, seems to reflect structural heterogeneity in solution rather than ESI artifacts. This correspondence is experimentally established here, independently of assumptions on the underlying ESI mechanism. Combined description by orthogonal biophysical methods can provide valuable constraints for computational simulations of IDP conformational ensembles in the presence or absence of interactors [51].

Supplementary Materials: Supplementary materials can be found at http://www.mdpi.com/1422-0067/20/20/5181/s1.

Author Contributions: Conceptualization, R.G., V.C., D.S., and F.M.; methodology, R.C., V.C., C.A.M., A.N., and C.S.; software, R.C.; investigation, R.C., V.C., C.A.M., A.N., and C.S.; resources, R.G., F.M.; data curation, R.C., V.C., D.S., A.N., and C.S.; writing—original draft preparation, R.C., V.C., C.A.M., R.G., and F.M.; writing—review and editing, V.C., S.B., G.L., R.G., and F.M.; supervision, V.C., D.S., S.B., G.L., R.G., and F.M.

Funding: This research received no external funding.

Acknowledgments: We thank G. Cappelletti and J. Clarke for the kind gift of the DNA plasmids used in this work.

Conflicts of Interest: The authors declare no conflict of interest.

Abbreviations

AS	α-synuclein
C	compact structure detected in native MS
CD	circular dichroism
CSDs	charge state distributions
DA	dopamine
EGCG	epigallocatechin-3-gallate
ESI-MS	electrospray ionization mass spectrometry
F	unfolding force
FTIR	Fourier-transform infrared spectroscopy
I1, I2	Intermediate 1 and 2 detected in native MS
I27	27th titin immunoglobulin-like domain
IDP	intrinsically disordered protein
IDPs	intrinsically disordered proteins
L_C	contour length
L_P	persistence length
native MS	native mass spectrometry
NMR	nuclear magnetic resonance
RC	random coil
SAXS-EOM	small-angle X-ray scattering and ensemble-optimization method
SI	strong interactions
SMFS	single molecule force spectroscopy
WI	weak interactions
WLC	worm-like-chain

References

1. Rezaei-Ghaleh, N.; Parigi, G.; Soranno, A.; Holla, A.; Becker, S.; Schuler, B.; Luchinat, C.; Zweckstetter, M. Local and Global Dynamics in Intrinsically Disordered Synuclein. *Angew. Chem. Int. Ed. Engl.* **2018**, *57*, 15262–15266. [CrossRef] [PubMed]

2. Borgia, A.; Kemplen, K.R.; Borgia, M.B.; Soranno, A.; Shammas, S.; Wunderlich, B.; Nettels, D.; Best, R.B.; Clarke, J.; Schuler, B. Transient misfolding dominates multidomain protein folding. *Nat. Commun.* **2015**, *6*, 8861. [CrossRef] [PubMed]

3. Gruebele, M.; Dave, K.; Sukenik, S. Globular Protein Folding In Vitro and In Vivo. *Annu. Rev. Biophys.* **2016**, *45*, 233–251. [CrossRef] [PubMed]

4. Wright, P.E.; Dyson, H.J. Intrinsically disordered proteins in cellular signalling and regulation. *Nat. Rev. Mol. Cell. Biol.* **2015**, *16*, 18–29. [CrossRef] [PubMed]

5. Oberhauser, A.F.; Marszalek, P.E.; Carrion-Vazquez, M.; Fernandez, J.M. Single protein misfolding events captured by atomic force microscopy. *Nat. Struct. Biol.* **1999**, *6*, 102510–102528. [CrossRef]

6. Rounsevell, R.; Forman, J.R.; Clarke, J. Atomic force microscopy: Mechanical unfolding of proteins. *J. Methods* **2004**, *34*, 100–111. [CrossRef] [PubMed]

7. Ferreon, A.C.M.; Deniz, A.A. Protein folding at single-molecule resolution. *Biochim. Biophys. Acta* **2011**, *1814*, 1021–1029. [CrossRef]

8. Junker, J.P.; Rief, M. Single-molecule force spectroscopy distinguishes target binding modes of calmodulin. *Proc. Natl. Acad. Sci. USA* **2009**, *106*, 14361–14366. [CrossRef]

9. Cassina, V.; Manghi, M.; Salerno, D.; Tempestini, A.; Iadarola, V.; Nardo, L.; Brioschi, S.; Mantegazza, F. Effects of cytosine methylation on DNA morphology: An atomic force microscopy study. *Biochim. Biophys. Acta Gen. Subj.* **2016**, *1860*, 1–7. [CrossRef]

10. Cassina, V.; Seruggia, D.; Beretta, G.L.; Salerno, D.; Brogioli, D.; Manzini, S.; Zunino, F.; Mantegazza, F. Atomic force microscopy study of DNA conformation in the presence of drugs. *Eur. Biophys. J.* **2011**, *40*, 59–68. [CrossRef]

11. Beedle, A.E.M.; Lezamiz, A.; Stirnemann, G.; Garcia-Manyes, S. The mechanochemistry of copper reports on the directionality of unfolding in model cupredoxin proteins. *Nat. Commun.* **2015**, *6*, 7894. [CrossRef] [PubMed]

12. Walder, R.; LeBlanc, M.A.; Van Patten, W.J.; Edwards, D.T.; Greenberg, J.A.; Adhikari, A.; Okoniewski, S.R.; Sullan, R.M.A.; Rabuka, D.; Sousa, M.C.; et al. Rapid Characterization of a Mechanically Labile α-Helical Protein Enabled by Efficient Site-Specific Bioconjugation. *J. Am. Chem. Soc.* **2017**, *39*, 9867–9875. [CrossRef] [PubMed]

13. Garcia-Manyes, S.; Kuo, T.L.; Fernández, J.M. Contrasting the individual reactive pathways in protein unfolding and disulfide bond reduction observed within a single protein. *J. Am. Chem. Soc.* **2011**, *133*, 3104–3113. [CrossRef] [PubMed]

14. Sandal, M.; Valle, F.; Tessari, I.; Mammi, S.; Bergantino, E.; Musiani, F.; Brucale, M.; Bubacco, L.; Samorì, B. Conformational equilibria in monomeric alpha-synuclein at the single-molecule level. *PLoS Biol.* **2008**, *6*, 99–108. [CrossRef]

15. Brucale, M.; Sandal, M.; Di Maio, S.; Rampion, A.; Tessari, I.; Tosatto, L.; Bisaglia, M.; Bubacco, L.; Samorì, B. Pathogenic mutations shift the equilibria of alpha-synuclein single molecules towards structured conformers. *ChemBioChem* **2009**, *10*, 176–183. [CrossRef]

16. Hervàs, R.; Oroz, J.; Galera-Prat, A.; Goñi, O.; Valbuena, A.; Vera, A.M.; Gòmez-Sicilia, A.; Losada-Urzáiz, F.; Uversky, V.N.; Menéndez, M.; et al. Common features at the start of the neurodegeneration cascade. *PLoS Biol.* **2012**, *10*, 1001335. [CrossRef]

17. Zhang, Y.; Hashemi, M.; Lv, Z.; Williams, B.; Popov, K.I.; Dokholyan, N.V.; Lyubchenko, Y.L. High-speed atomic force microscopy reveals structural dynamics of α-synuclein monomers and dimers. *J. Chem. Phys.* **2018**, *148*, 123322. [CrossRef]

18. Stephens, A.D.; Zacharopoulou, M.; Kaminski Schierle, G.S. The Cellular Environment Affects Monomeric α-Synuclein Structure. *Trends Biochem. Sci.* **2019**, *44*, 453–466. [CrossRef]

19. Curtain, C.C.; Kirby, N.M.; Mertens, H.D.; Barnham, K.J.; Knott, R.B.; Masters, C.L.; Cappai, R.; Rekas, A.; Kenche, V.B.; Ryan, T. α-synuclein oligomers and fibrils originate in two distinct conformer pools: A small angle X-ray scattering and ensemble optimisation modelling study. *Mol. Biosyst.* **2015**, *11*, 190–196. [CrossRef]

20. Theillet, F.X.; Binolfi, A.; Bekei, B.; Martorana, A.; Rose, H.M.; Stuiver, M.; Verzini, S.; Lorenz, D.; van Rossum, M.; Goldfarb, D.; et al. Structural disorder of monomeric α-synuclein persists in mammalian cells. *Nature* **2016**, *530*, 45–50. [CrossRef]

21. Zhao, J.; Liang, Q.; Sun, Q.; Chen, C.; Xu, L.; Ding, Y.; Zhou, P. (-)-Epigallocatechin-3-gallate (EGCG) inhibits fibrillation, disaggregates amyloid fibrils of α-synuclein, and protects PC12 cells against alpha-synuclein-induced toxicity. *RSC Adv.* **2017**, *7*, 32508–32517. [CrossRef]

22. Lee, H.J.; Baek, S.M.; Ho, D.H.; Suk, J.E.; Cho, E.D.; Lee, S.J. Dopamine promotes formation and secretion of non-fibrillar alpha-synuclein oligomers. *Exp. Mol. Med.* **2011**, *43*, 216–222. [CrossRef] [PubMed]

23. Santambrogio, C.; Natalello, A.; Brocca, S.; Ponzini, E.; Grandori, R. Conformational Characterization and Classification of Intrinsically Disordered Proteins by Native Mass Spectrometry and Charge-State Distribution. *Proteomics* **2019**, *19*, 1800060. [CrossRef] [PubMed]

24. Lössl, P.; Van de Waterbeemd, M.; Heck, A.J. The diverse and expanding role of mass spectrometry in structural and molecular biology. *EMBO J.* **2016**, *35*, 2634–2657. [CrossRef]

25. Konijnenberg, A.; Butterer, A.; Sobott, F. Native ion mobility-mass spectrometry and related methods in structural biology. *Biochim. Biophys. Acta* **2013**, *1834*, 1239–1256. [CrossRef]

26. Loo, R.R.; Loo, J.A. Salt Bridge Rearrangement (SaBRe) Explains the Dissociation Behavior of Noncovalent Complexes. *J. Am. Soc. Mass Spectrom.* **2016**, *27*, 975–990. [CrossRef]

27. Chowdhury, S.K.; Katta, V.; Chait, B.T. Probing conformational changes in proteins by mass spectrometry. *J. Am. Chem. Soc.* **1990**, *112*, 9012–9013. [CrossRef]

28. Grandori, R.; Matecko, I.; Mayr, P.; Müller, N. Probing protein stabilization by glycerol using electrospray mass spectrometry. *J. Mass Spectrom.* **2001**, *36*, 918–922. [CrossRef]

29. Natalello, A.; Benetti, F.; Doglia, S.M.; Legname, G.; Grandori, R. Compact conformations of α-synuclein induced by alcohols and copper. *Proteins* **2011**, *79*, 611–621. [CrossRef]

30. D'Urzo, A.; Konijnenberg, A.; Rossetti, G.; Habchi, J.; Li, J.; Carloni, P.; Sobott, F.; Longhi, S.; Grandori, R. Molecular basis for structural heterogeneity of an intrinsically disordered protein bound to a partner by combined ESI-IM-MS and modeling. *J. Am. Soc. Mass Spectrom.* **2015**, *26*, 472–481. [CrossRef]

31. Wongkongkathep, P.; Han, J.Y.; Choi, T.S.; Yin, S.; Kim, H.I.; Loo, J.A. Native Top-Down Mass Spectrometry and Ion Mobility MS for Characterizing the Cobalt and Manganese Metal Binding of α-Synuclein Protein. *J. Am. Soc. Mass Spectrom.* **2018**, *29*, 1870–1880. [CrossRef] [PubMed]

32. Testa, L.; Brocca, S.; Santambrogio, C.; D'Urzo, A.; Habchi, J.; Longhi, S.; Uversky, V.N.; Grandori, R. Extracting structural information from charge-state distributions of intrinsically disordered proteins by non-denaturing electrospray-ionization mass spectrometry. *Intrinsic. Disord. Proteins* **2013**, *1*, 25068. [CrossRef] [PubMed]

33. Frimpong, A.K.; Abzalimov, R.R.; Uversky, V.N.; Kaltashov, I.A. Characterization of intrinsically disordered proteins with electrospray ionization mass spectrometry: Conformational heterogeneity of α-synuclein. *Proteins* **2010**, *78*, 714–722. [CrossRef] [PubMed]

34. Verkerk, U.H.; Kebarle, P. Ion-ion and ion-molecule reactions at the surface of proteins produced by nanospray. Information on the number of acidic residues and control of the number of ionized acidic and basic residues. *J. Am. Soc. Mass Spectrom.* **2005**, *16*, 1325–1341. [CrossRef]

35. Testa, L.; Brocca, S.; Grandori, R. Charge-surface correlation in electrospray ionization of folded and unfolded proteins. *Anal. Chem.* **2011**, *83*, 6459–6463. [CrossRef]

36. Kaltashov, I.A.; Mohimen, A. Estimates of protein surface areas in solution by electrospray ionization mass spectrometry. *Anal. Chem.* **2005**, *77*, 5370–5379. [CrossRef]

37. Hall, Z.; Robinson, C.V. Do charge state signatures guarantee protein conformations? *J. Am. Soc. Mass Spectrom.* **2012**, *23*, 1161–1168. [CrossRef]

38. Li, J.; Santambrogio, C.; Brocca, S.; Rossetti, G.; Carloni, P.; Grandori, R. Conformational effects in protein electrospray-ionization mass spectrometry. *Mass Spectrom. Rev.* **2016**, *35*, 111–122. [CrossRef]

39. Natalello, A.; Santambrogio, C.; Grandori, R. Are Charge-State Distributions a Reliable Tool Describing Molecular Ensembles of Intrinsically Disordered Proteins by Native MS? *J. Am. Soc. Mass Spectrom.* **2017**, *28*, 21–28. [CrossRef]

40. Beveridge, R.; Migas, L.G.; Das, R.K.; Pappu, R.V.; Kriwacki, R.W.; Barran, P.E. Ion Mobility Mass Spectrometry Uncovers the Impact of the Patterning of Oppositely Charged Residues on the Conformational Distributions of Intrinsically Disordered Proteins. *J. Am. Chem. Soc.* **2019**, *141*, 4908–4918. [CrossRef]

41. Stuchfield, D.; Barran, P. Unique insights to intrinsically disordered proteins provided by ion mobility mass spectrometry. *Curr. Opin. Chem. Biol.* **2018**, *42*, 177–185. [CrossRef] [PubMed]

42. Konijnenberg, A.; Ranica, S.; Narkiewicz, J.; Legname, G.; Grandori, R.; Sobott, F.; Natalello, A. Opposite Structural Effects of Epigallocatechin-3-gallate and Dopamine Binding to α-Synuclein. *Anal. Chem.* **2016**, *88*, 8468–8475. [CrossRef] [PubMed]

43. Ponzini, E.; De Palma, A.; Cerboni, L.; Natalello, A.; Rossi, R.; Moons, R.; Konijnenberg, A.; Narkiewicz, J.; Legname, G.; Sobott, F.; et al. Methionine oxidation in α-synuclein inhibits its propensity for ordered secondary structure. *J. Biol. Chem.* **2019**, *294*, 5657–5665. [CrossRef] [PubMed]

44. Borysik, A.J.; Kovacs, D.; Guharoy, M.; Tompa, P. Ensemble Methods Enable a New Definition for the Solution to Gas-Phase Transfer of Intrinsically Disordered Proteins. *J. Am. Chem. Soc.* **2015**, *137*, 13807–13817. [CrossRef]

45. Steward, A.; Toca-Herrera, J.L.; Clarke, J. Versatile cloning system for construction of multimeric proteins for use in atomic force microscopy. *J. Protein Sci.* **2002**, *11*, 2179–2183. [CrossRef]

46. Hoffman, T.; Dougan, L. Single molecule force spectroscopy using polyproteins. *Chem. Soc. Rev.* **2012**, *41*, 4781–4796. [CrossRef]

47. Best, R.B.; Brockwell, D.J.; Toca-Herrera, J.L.; Blake, A.W.; Smith, A.; Radford, S.E.; Clarke, J. Force mode atomic force microscopy as a tool for protein folding studies. *J. Anal. Chim. Acta* **2003**, *479*, 87–105. [CrossRef]

48. Bustamante, C.; Marko, J.F.; Siggia, E.D.; Smith, S. Entropic elasticity of lambda-phage DNA. *Science* **1994**, *265*, 1599–1600. [CrossRef]

49. Borgia, M.B.; Borgia, A.; Best, R.B.; Steward, A.; Nettels, D.; Wunderlich, B.; Schuler, B.; Clarke, J. Single-molecule fluorescence reveals sequence-specific misfolding in multidomain proteins. *Nature* **2011**, *474*, 662–665. [CrossRef]

50. Heller, G.T.; Bonomi, M.; Vendruscolo, M. Structural Ensemble Modulation upon Small-Molecule Binding to Disordered Proteins. *J. Mol. Biol.* **2018**, *430*, 2288–2292. [CrossRef]

51. Rossetti, G.; Musiani, F.; Abad, E.; Dibenedetto, D.; Mouhib, H.; Fernandez, C.O.; Carloni, P. Conformational ensemble of human α-synuclein physiological form predicted by molecular simulations. *Phys. Chem. Chem. Phys.* **2016**, *18*, 5702–5706. [CrossRef] [PubMed]

52. Balupuri, A.; Choi, K.E.; Kang, N.S. Computational insights into the role of α-strand/sheet in aggregation of α-synuclein. *Sci. Rep.* **2019**, *9*, 59. [CrossRef] [PubMed]

53. Brodie, N.I.; Popov, K.I.; Petrotchenko, E.V.; Dokholyan, N.V.; Borchers, C.H. Conformational ensemble of native α-synuclein in solution as determined by short-distance crosslinking constraint-guided discrete molecular dynamics simulations. *PLoS Comput. Biol.* **2019**, *15*, e1006859. [CrossRef] [PubMed]

54. Ehrnhoefer, D.E.; Bieschke, J.; Boeddrich, A.; Herbst, M.; Masino, L.; Lurz, R.; Engemann, S.; Pastore, A.; Wanker, E.E. EGCG redirects amyloidogenic polypeptides into unstructured, off-pathway oligomers. *Nat. Struct. Mol. Biol.* **2008**, *15*, 558–566. [CrossRef]

55. Kuprowski, M.C.; Konermann, L. Signal response of coexisting protein conformers in electrospray mass spectrometry. *Anal. Chem.* **2007**, *79*, 2499–2506. [CrossRef]

56. Samalikova, M.; Matecko, I.; Müller, N.; Grandori, R. Interpreting conformational effects in protein nano-ESI-MS spectra. *Anal. Bioanal. Chem.* **2004**, *378*, 1112–1123. [CrossRef]

57. Marsh, J.A.; Forman-Kay, J.D. Sequence determinants of compaction in intrinsically disordered proteins. *Biophys. J.* **2010**, *98*, 2383–2390. [CrossRef]

58. Cartelli, D.; Aliverti, A.; Barbiroli, C.; Santambrogio, C.; Raggi, E.M.; Casagrande, F.V.M.; Cantele, F.; Beltramone, S.; Marangon, J.; De Gregorio, C.; et al. α-Synuclein is a Novel Microtubule Dynamase. *Sci. Rep.* **2016**, *6*, 33289. [CrossRef]

59. Marszalek, P.E.; Lu, H.; Li, H.; Carrion-Vazquez, M.; Oberhauser, A.F.; Schulten, K.; Fernandez, J.M. Mechanical unfolding intermediates in titin modules. *Nature* **1999**, *402*, 100–103. [CrossRef]

© 2019 by the authors. Licensee MDPI, Basel, Switzerland. This article is an open access article distributed under the terms and conditions of the Creative Commons Attribution (CC BY) license (http://creativecommons.org/licenses/by/4.0/).

Article

Sequence and Structure Properties Uncover the Natural Classification of Protein Complexes Formed by Intrinsically Disordered Proteins via Mutual Synergistic Folding

Bálint Mészáros [1,2,3,*], László Dobson [4,5], Erzsébet Fichó [3] and István Simon [3,*]

[1] MTA-ELTE Momentum Bioinformatics Research Group, Department of Biochemistry, Eötvös Loránd University, Pázmány Péter stny 1/c, H-1117 Budapest, Hungary

[2] European Molecular Biology Laboratory, Structural and Computational Biology Unit, Meyerhofstraße 1, 69117 Heidelberg, Germany

[3] Protein Structure Research Group, Institute of Enzymology, RCNS, HAS, Magyar Tudósok krt 2, H-1117 Budapest, Hungary; ficho.erzsebet@ttk.mta.hu

[4] Membrane Protein Bioinformatics Research Group, Institute of Enzymology, RCNS, HAS, Magyar Tudósok krt 2, H-1117 Budapest, Hungary; dobson.laszlo.imre@itk.ppke.hu

[5] Faculty of Information Technology and Bionics, Pázmány Péter Catholic University, Práter u. 50A, H-1083 Budapest, Hungary

* Correspondence: bmeszaros@caesar.elte.hu (B.M.); simon.istvan@ttk.mta.hu (I.S.)

Received: 9 October 2019; Accepted: 30 October 2019; Published: 1 November 2019

Abstract: Intrinsically disordered proteins mediate crucial biological functions through their interactions with other proteins. Mutual synergistic folding (MSF) occurs when all interacting proteins are disordered, folding into a stable structure in the course of the complex formation. In these cases, the folding and binding processes occur in parallel, lending the resulting structures uniquely heterogeneous features. Currently there are no dedicated classification approaches that take into account the particular biological and biophysical properties of MSF complexes. Here, we present a scalable clustering-based classification scheme, built on redundancy-filtered features that describe the sequence and structure properties of the complexes and the role of the interaction, which is directly responsible for structure formation. Using this approach, we define six major types of MSF complexes, corresponding to biologically meaningful groups. Hence, the presented method also shows that differences in binding strength, subcellular localization, and regulation are encoded in the sequence and structural properties of proteins. While current protein structure classification methods can also handle complex structures, we show that the developed scheme is fundamentally different, and since it takes into account defining features of MSF complexes, it serves as a better representation of structures arising through this specific interaction mode.

Keywords: intrinsically disordered protein; IDP; protein–protein interaction; mutual synergistic folding; coupled folding and binding; structural analysis; structure-based classification; fold recognition

---

1. Introduction

Intrinsically disordered proteins (IDPs) are crucial elements of the molecular machinery indispensable for complex life [1,2]. IDPs are parts of regulatory pathways [3], control the cell cycle [4,5], function as chaperones [6,7], and regulate protein degradation [8,9], amongst other functions. In accord, IDPs are typically under tight regulation at several levels [3,10]. While some IDPs fulfill their functions directly through their lack of structure, such as spring-like entropic chains, the majority of disordered proteins interact with other macromolecules, most often other proteins [11]. IDP-mediated interactions are essential for many hub proteins [12,13], and several IDPs serve as interaction scaffolds/platforms for macromolecular assembly [14,15]. Mounting evidence also shows that protein disorder plays a crucial role in the assembly of liquid–liquid phase separated non-membrane-bounded organelles [16].

Depending on the partner protein and the specifics of the interaction, IDPs can bind through several mechanisms. Several IDPs recognize and bind to ordered protein domains, usually through a linear sequence motif [17]. While some IDPs retain their inherent flexibility in the bound form as well [18], in most known cases the complex structure lends itself to standard structure determination methods, such as X-ray crystallography or NMR. These cases of coupled folding and binding have been studied intensively [19–21]. However, IDPs can utilize a fundamentally different molecular mechanism for interaction, through which they reach a folded state as well. Complexes that contain only IDPs as constituent protein chains, without the presence of a previously folded domain, are formed via a process called mutual synergistic folding (MSF) [22]—a much less understood way in which protein folding and binding can merge into a single biophysical process.

A major advancement in the field of IDP interactions in recent years was the development of specialized interaction databases for various mechanisms including coupled folding and binding [23,24], fuzzy complexes [25], mutual synergistic folding [26], and proteins driving liquid–liquid phase separation [27]. Out of these aspects, possibly the most understudied one is mutual synergistic folding, owing to the fact that these are the only interactions where none of the partner proteins have a well-defined structure outside of the complex, forcing us to revise our current approaches used for describing protein structures and complexes. The biological and biophysical properties of these interactions are markedly different from those mediated by other types of proteins. While in other interaction types a stable, folded hydrophobic core is already present in at least one partner, here the folding and binding happen at the same time for all partners. Comparative analysis has not only shown that MSF complexes constitute a separate biologically meaningful class, but also highlighted that these complexes are highly heterogeneous in terms of sequence and structure propreties [28–30].

We now have knowledge of over 140,000 protein structures deposited in the Protein Data Bank (PDB) [31], a major part of which contains several proteins in complex. In each of these cases, the proteins achieve stability either before or upon interacting. A major question is how is stability achieved? Can this be a basis of the definition of biologically meaningful classification? In the case of ordered proteins, current hierarchical classification schemes are rooted in the tertiary protein structures, such as in the case of methods/databases as SCOP (Structural Classification of Proteins) [32] and CATH (Class, Architecture, Topology, Homologous superfamily) [33]. While these methods are extended to classify protein complexes as well, they do not explicitly factor in parameters that describe the interactions or the differences in sequence composition between complexes of similar overall structures. However, in the case of MSF complexes, these differences are defining features, as the interaction is the primary reason for the emergence of the structure itself, and this interaction usually requires highly specialized residue compositions [28]. While other classification methods were developed specifically for protein–protein interactions, they only aim to describe the interface, without taking the overall resulting structure into account [34].

Here we present the first classification method designed to identify biologically relevant types of protein complexes formed via mutual synergistic folding. Our work aims to answer specific questions about the types of MSF complexes based on the currently known more than 200 examples. Are there intrinsic classes of MSF complexes or are all known examples basically unique in terms of sequence and structure? If meaningful groups are definable in an objective way, what are the characteristics of each group in terms of sequence composition and adopted structure? In addition, how is the formation of MSF complexes regulated? Are mechanisms known to be important for other molecular interactions relevant to these complexes as well? If so, are there differences between various MSF groups regarding these regulatory mechanisms and other biologically relevant properties, such as binding strength and subcellular localization?

2. Results

2.1. Sequence-Based Properties Define Four Clusters of Complexes

Complexes formed by mutual synergistic folding were taken from the MFIB (Mutual Folding Induced by Binding) database [26], and each complex has been assigned a feature vector describing the sequence composition of its constituent protein chains. To represent the sequence composition, we use the amino acid grouping previously used for investigating protein–protein complexes involving IDPs [28] (see Data and Methods and Figure 1 for definitions, and Supplementary Table S1 for exact values for all complexes). These vectors were used as input for hierarchical clustering (Supplementary Figure S1) to quantify the sequence-based relationship between various complexes. k-means clustering (Supplementary Figure S2) indicates four as a suitable number of clusters, and, therefore, we use four sequence-based clusters in all subsequent analyses. While this choice is not the only acceptable one based on the k-means results, we aim to have a restricted set of clusters to describe the major types of sequential classes. The main features of the four clusters are shown in Figure 1, while cluster numbers for each complex are shown in Supplementary Table S1.

Figure 1 shows the average sequence compositions of each of the four sequence-based clusters. While clusters were defined based on sequence compositions only, Figure 1 also shows the average heterogeneity of the four clusters, meaning the average normalized difference in sequence composition between the interacting proteins of the complexes (see Data and Methods). Complexes in clusters 1 and 2 are both largely devoid of special residues, including Gly (flexible), Pro (rigid), and Cys (cysteine). Members of these two clusters contain an average fraction of hydrophobic residues; however are slightly depleted in aromatic residues, indicating that π–π interactions are not the dominant source of stability. The most characteristic difference between clusters 1 and 2 is that members of cluster 1 typically contain a high fraction of polar residues, while members of cluster 2 are enriched in charged residues. Also, cluster 1 members are typically formed by proteins with highly different compositions (high heterogeneity values), while cluster 2 members are formed by proteins of very similar compositions.

In contrast, members of clusters 3 and 4 are typically enriched in Gly and Pro and contain a higher-than-average fraction of aromatic residues. Again, polar/charged residue balance is a distinguishing feature, with clusters 3 and 4 showing preferences for polar and charged residues, respectively. Also, similarly to clusters 1 and 2, there is a notable difference in heterogeneity values between clusters 3 and 4: members of clusters 3 and 4 are typically composed of proteins with very similar and different residue compositions, respectively.

		Cluster 1	Cluster 2	Cluster 3	Cluster 4	Average
Number of members		25	60	38	84	-
Average amino acid composition	**Aromatic (FWY)**	0.029	0.049	0.089	0.072	0.064
	Hydrophobic (AILMV)	0.381	0.324	0.287	0.357	0.336
	Flexible (G)	0.024	0.021	0.076	0.056	0.046
	Rigid (P)	0.018	0.005	0.029	0.040	0.026
	Charged (HKRDE)	0.258	0.392	0.239	0.290	0.308
	Polar (NQST)	0.281	0.205	0.254	0.179	0.211
	Cysteine (C)	0.009	0.005	0.027	0.006	0.010
Heterogeneity (average dissimilarity between subunits)		0.162	0.098	0.057	0.117	0.111

Comparison to average:
+30%, +20%, +10%, -10%, -20%, -30%

Figure 1. Average values of sequence features for the four sequence-based clusters. Blue and orange shadings mark values that are over- or under-represented compared with the average of all MSF complexes. Heterogeneity values were not used for cluster definitions.

2.2. Structure-Based Properties Offer A Different Means of Defining Complex Types

The structural properties of the studied complexes were quantified using various features describing secondary structure compositions, various molecular surfaces, and incorporating hydrophobicity measures and atomic contacts (see Supplementary Table S1 and Data and Methods). These structural features were used to describe each complex in the form of a feature vector, and similarly to the analysis of sequence properties, these vectors were input to hierarchical clustering; however, structural features were filtered, and only those that share a modest degree of correlation were kept (see Supplementary Table S2 and Data and Methods for specifics) to avoid bias. The resulting tree is shown in Supplementary Figure S3. In contrast to the sequence-based clustering, k-means within-cluster sum of squares analysis does not indicate any low number of clusters as more optimal than others (Supplementary Figure S4). In order to have a medium number of clusters, we cut the hierarchical tree at a linkage distance that defines five clusters (Supplementary Figure S3), again reflecting our preference to arrive at a moderate number of complex types, to provide a high-level classification scheme. The average values of structural parameters for all five structure classes are shown in Figure 2.

The obtained clusters show distinguishing structural features. Members of cluster 1 incorporate the highest amount of nonhelical secondary structure elements. These complexes heavily rely on a large number of buried hydrophobic residues for stability, and most stabilizing atomic contacts are formed between residues of the same protein, relying less on intermolecular interactions, which tend to be mostly polar in nature.

In contrast, members of cluster 2 adopt mainly helical structures. The stability of these complexes seems to rely more on the interactions formed between the subunits, mostly formed between side chains. The importance of interchain interactions is also reflected in the large relative interface and small relative buried surface areas.

Cluster 3 and 4 complexes exhibit similar features, including a balanced ratio of various secondary structure elements and polar/hydrophobic balance of various molecular surfaces and contacts. For both clusters, interchain contacts rely mostly on side chain–side chain and backbone–backbone contacts. The main difference between the two clusters is the relative role of the interface between the participating proteins. Cluster 3 members have a larger-than-average interface, in terms of both molecular surface and number of contacts, meanwhile cluster 4 complexes have a very restricted interface size, incorporating only a few atomic contacts.

Members of cluster 5 are the most similar to the average in most structural features. There are only weak distinguishing features, including a slightly increased helical content at the expense of extended structural elements, a moderate increase in the role of backbone–side chain interactions in interchain contacts, and the increased ratio of interchain contacts. However, these deviations in average parameter values are modest and—with the exception of the decreased extended structure content—none of them reaches 20% compared to the average values calculated for all complexes.

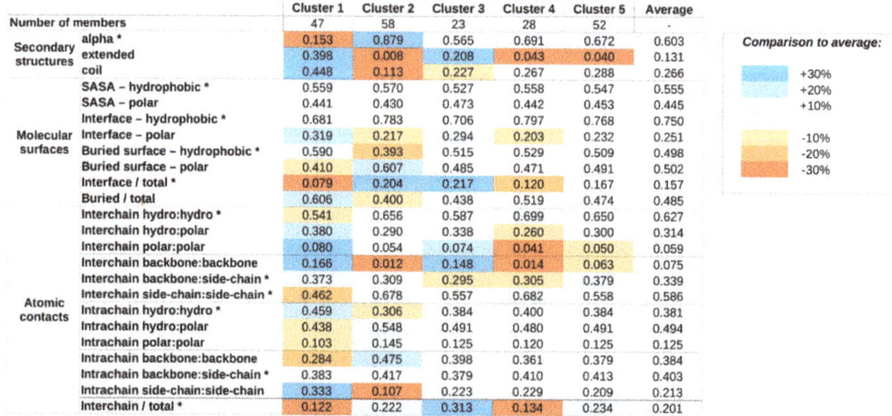

		Cluster 1	Cluster 2	Cluster 3	Cluster 4	Cluster 5	Average
Number of members		47	58	23	28	52	-
Secondary structures	alpha *	0.153	0.879	0.565	0.691	0.672	0.603
	extended	0.398	0.008	0.208	0.043	0.040	0.131
	coil	0.448	0.113	0.227	0.267	0.288	0.266
Molecular surfaces	SASA – hydrophobic *	0.559	0.570	0.527	0.558	0.547	0.555
	SASA – polar	0.441	0.430	0.473	0.442	0.453	0.445
	Interface – hydrophobic *	0.681	0.783	0.706	0.797	0.768	0.750
	Interface – polar	0.319	0.217	0.294	0.203	0.232	0.251
	Buried surface – hydrophobic *	0.590	0.393	0.515	0.529	0.509	0.498
	Buried surface – polar	0.410	0.607	0.485	0.471	0.491	0.502
	Interface / total *	0.079	0.204	0.217	0.120	0.167	0.157
	Buried / total	0.606	0.400	0.438	0.519	0.474	0.485
Atomic contacts	Interchain hydro:hydro *	0.541	0.656	0.587	0.699	0.650	0.627
	Interchain hydro:polar	0.380	0.290	0.338	0.260	0.300	0.314
	Interchain polar:polar	0.080	0.054	0.074	0.041	0.050	0.059
	Interchain backbone:backbone	0.166	0.012	0.148	0.014	0.063	0.075
	Interchain backbone:side-chain *	0.373	0.309	0.295	0.305	0.379	0.339
	Interchain side-chain:side-chain *	0.462	0.678	0.557	0.682	0.558	0.586
	Intrachain hydro:hydro *	0.459	0.306	0.384	0.400	0.384	0.381
	Intrachain hydro:polar	0.438	0.548	0.491	0.480	0.491	0.494
	Intrachain polar:polar	0.103	0.145	0.125	0.120	0.125	0.125
	Intrachain backbone:backbone	0.284	0.475	0.398	0.361	0.379	0.384
	Intrachain backbone:side-chain *	0.383	0.417	0.379	0.410	0.413	0.403
	Intrachain side-chain:side-chain	0.333	0.107	0.223	0.229	0.209	0.213
	Interchain / total *	0.122	0.222	0.313	0.134	0.234	0.201

Comparison to average:

+30%
+20%
+10%
-10%
-20%
-30%

Figure 2. Average values for structure features for the five structure-based clusters. Blue and orange shadings mark values that are over- or under-represented compared to the average of all MSF complexes. SASA—solvent accessible surface area, hydro:hydro—fraction of contacts that are formed between two hydrophobic atoms. Asterisks mark features that were included in the clustering.

2.3. *Defining Interaction Types Based on Sequence and Structure Clusters*

Considering together the previously established sequence- and structure-based clusters, in total 20 types of complexes can be defined (Figure 3). The number of known complexes in possible types shows large variations, with some highly favored ones (e.g., type 2[sequence]/2[structure]) and ones with a single known example (e.g., type 2/1), showing that not all sequence compositions are compatible with all types of adopted structures. In order to arrive at a reasonable number of basic complex types, types with 10 or fewer complexes were either merged with the adjacent sequence clusters or were omitted. As structural differences in general are larger between clusters, types corresponding to different structure clusters were never merged. For structure clusters 1 and 2, only two adjacent sequence clusters were merged, as these contain over 95% and 85% of the complexes, respectively. In contrast, for structure classes 3 and 4, all four sequence clusters were merged, as the distribution of complexes is more even across the sequence space. For structure cluster 5, even a single sequence cluster is enough to capture over 85% of complexes, and thus no merging was employed. This approach yielded five main interaction types, each of which has over 20 complexes. In order to include all known MSF complexes, a sixth pseudo-type was introduced, which contains all structures not compatible with any of the previously described five types (see Supplementary Table S1 for an exhaustive list).

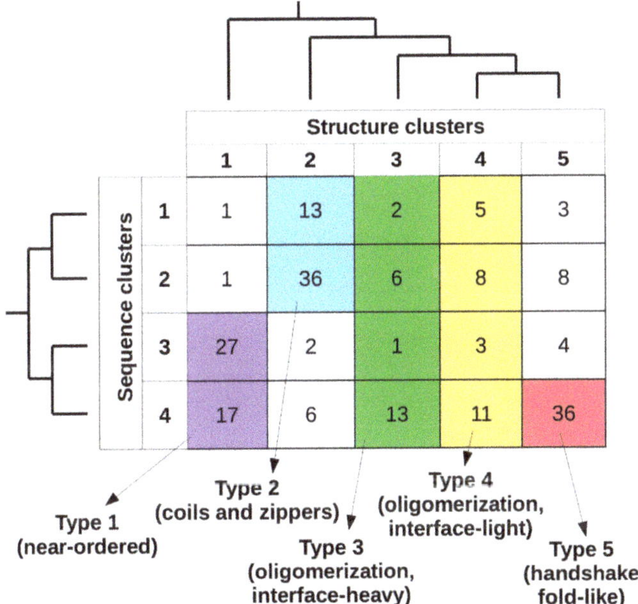

Figure 3. MSF complex types. Colored regions mark separate interaction types considering sequence- and structure-based clusters (vertical and horizontal axes, respectively). The relationship of each sequence-and structure-based cluster taken from the hierarchical clustering (Supplementary Figures S1 and S3) is shown on the corresponding side of the table. Each of the six defined types is assigned a randomly selected color (that is of high contrast), and these are used in later figures to denote the corresponding complex types.

The complex types defined so far are based on structure and sequence features. However, if these types represent biologically meaningful classes, there should be other relevant differences between them in terms of the energetics of the interaction, binding strength, subcellular localization, or the biological regulation of the interaction. In the next chapters, we describe each complex type with biologically important characteristics and assess the potential differences between the members of each class.

2.4. Complex Types Show Characteristic Energetic Properties

From a biological perspective, the strength of association between interacting protein chains and the stability of the resulting complex is of utmost importance. Unfortunately, complexes formed exclusively by IDPs via MSF generally lack targeted measurements concerning thermodynamic and stability parameters. However, low-resolution energy calculations and prediction algorithms can give an indication about the characteristic energetics properties of the uncovered complex types in general. While these methods might have fairly large errors in individual cases, they are well equipped for comparative studies between groups of complexes.

In order to assess the energetic properties of complexes, we employed an energy calculation scheme using low-resolution force fields based on statistical potentials (see Data and Methods). As a reference, energetic properties were calculated for complexes formed exclusively by ordered proteins and complexes formed by an IDP binding to an ordered partner via coupled folding and binding (CFB) (see Data and Methods and Supplementary Tables S3 and S4). Figure 4 shows two types of calculated energies for each complex. On one hand, we calculated the total energy per residue in the whole complex, which reflects the overall stability. On the other hand, we also calculated the fraction of this stabilizing energy coming from intermolecular interactions (i.e., how important the interaction is for stability). In accordance with our expectations, complexes formed by ordered proteins feature strongly bound overall structures, with fairly large negative stabilizing energy/residue. In contrast, CFB complexes in general have less favorable per-residue energies, hinting at their comparatively weakly bound overall structures. However, the energetic feature providing the most recognizable difference between ordered and CFB complexes is the energy contribution of interchain contacts to the overall stability. In the case of ordered complexes, this contribution is fairly limited, as individual subunits have a stable structure on their own. In contrast, if the complex features an IDP, the interaction energy becomes a major contributor to stability (Figure 4a).

While ordered and CFB complexes tend to segregate in this energy space, complexes formed by MSF seem to be more heterogeneous, covering the whole available range of energetic values (Figure 4b). In the case of near-ordered proteins (Type 1), the energies resemble that of ordered complexes, hinting at the borderline ordered nature of the constituent IDPs, with the interaction between subunits playing a minor role. In contrast, coiled-coil-like structures (Type 2) on average have a much less stable complex structure, with interaction playing a substantial role in stability. These complexes resemble IDPs bound to ordered domains, and are expected to include several transient interactions. Other types fall largely between these two extreme cases. Energetics properties of the two types of oligomerization modules (Types 3 and 4) reflect the differences in interface surface area and contact numbers, shown in Figure 2. While the overall stability for both types varies in a very wide range, on average, the contribution of the interaction is higher for interface-heavy complexes (Type 3) than for interface-light ones (Type 4). Handshake-like folds (Type 5) show interesting properties: these complexes are quite stable with only limited variation in the per-residue energies. Yet, they achieve this high stability by relying heavily on the interaction between subunits of the dimer. As opposed to the complexes in Figure 4a, MSF complexes show high overlap in the energy space. This shows that very different structures, with potentially very different sequence compositions, can have similar energetic properties. Also, the high variability of energetic properties within complex types (the main reason for high overlap between different groups) shows that depending on the biological function, similar complexes can be required to have very different stabilities. For example, while several dimeric transcription factors can have similar structures that accommodate DNA-binding, the association and dissociation rates of the dimers (regulating their transcriptional activity) have to adapt to the required expression profiles of the genes they regulate.

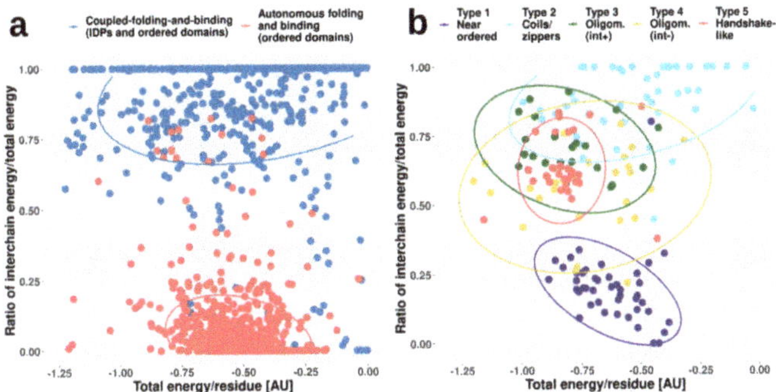

Figure 4. Energetic parameters of various interaction classes. The relative energetic weight of intersubunit interactions in the overall stability (y-axis) as a function of the overall energy per residue (x-axis, measured in arbitrary units, AU) for ordered complexes and complexes formed by coupled folding and binding (**a**), and the five well-defined types of MSF complexes (**b**).

The transient or obligate nature of interactions provides clues about their roles in biological systems. This is at least partially describable through K_d dissociation constants. While there is ample data about K_d values of IDPs binding via CFB to ordered domains [23], these values are largely missing for MSF complexes. In accord, we calculated estimated K_d values for MSF complexes (Supplementary Table S1), with Figure 5 showing the K_d distributions for the six previously defined complex types. In a biological context, actual K_d values can be a nonlinear function of environmental parameters. Unfortunately, this information is largely unknown for most MSF complexes, and such predicted K_d values should be treated with caution and should only be used for comparing group averages, where individual errors can even out. The lowest average K_d values were calculated for complexes with a handshake-like fold (Type 5). The next two types with low K_ds are the near-ordered complexes (Type 1) and interface-heavy oligomerization modules (Type 3). These three types together possibly cover most cases of the interactions where the complex needs to stay stable for an extended period of time, such as histone dimes (Type 5), complexes with enzymatic activity (Type 1) and several transcription factors (Type 3). Coiled-coil-like structures and oligomerization modules with small interfaces in general have a higher K_d, indicating that several transiently bound complexes belong to these types.

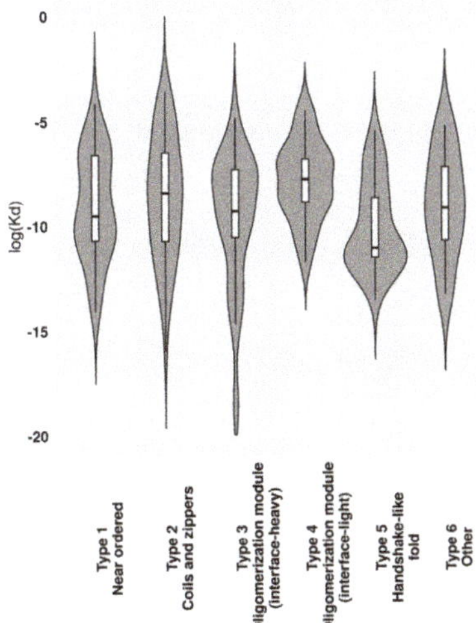

Figure 5. Predicted K_d value distributions for the six types of MSF complexes.

2.5. Interactions Are Heavily Regulated by Several Mechanisms

While the energetics of various interactions can provide clues about their transient/obligatory nature, the regulatory mechanisms can give more direct evidence. For example, while most IDP enzymes (belonging to Type 1) form particularly stable oligomers, indicating an obligate interaction, for example the oligomeric state of superoxide dismutase (SOD1) is known to be controlled by post-translational modification (PTM) serving as an on/off switch [35]; meaning that despite a strong interaction, it is reversible, and the disordered state of the monomers is biologically relevant (Figure 6a). Figure 6a shows additional examples of various regulatory mechanisms of MSF interactions via PTMs. These regulatory steps have already been described in the case of IDPs that bind to ordered domains [36], but have not been studied in the context of IDPs participating in MSF interactions. Apart from the on/off switch exemplified by SOD1, PTMs can control the partner selection of synergistically folding IDPs, such as in the case of another tightly bound complex, formed by H3/H4 histones (Type 5) [37]. PTMs can also tune the affinity of certain interactions, as is the case for the activating p53/CBP interaction (Type 4) [38]. Apart from these mechanisms that directly control the interaction between IDPs, PTMs can have a more indirect effect, modulating the activity of the dimer itself. In the case of the Max dimeric transcription factor, phosphorylation at the N-terminus of the binding region controls the dimer's (Type 4) interaction capacity towards DNA [39]. An even more indirect modulation of function is displayed for the retinoblastoma protein Rb, which in complex with E2F1/DP1 (Type 3) has a strong transcriptional repression activity. Upon methylation, Rb recruits L3MBTL1 [40], which is a direct repressor of transcription via chromatin compaction, augmenting the effect of Rb through a related but separate mechanism extrinsic to the Rb/E2F1/DP1 complex. This way the strength of repression depends on the PTM of the MSF complex, but through an additional protein that is not part of the complex but contributes to the complex function through a parallel mechanism in an indirect way.

To have a more systematic picture of the extent of regulatory mechanisms in MSF interactions, Figure 6b shows the fraction of known MSF complexes with experimentally verified PTM sites (Supplementary Table S5). In total, nearly 30% of studied complexes feature at least one PTM that was experimentally verified in a low-throughput experiment, presenting a regulatory mechanism that is able to directly or indirectly modulate either the interaction itself, or the activity of the resulting complex. The most prevalent PTM is phosphorylation, affecting 22% of complexes, but 10%, 15%, and 5% of MSF complexes contain methylation, acetylation, and ubiquitination sites as well (Figure 6b).

In addition, complex formation can also be regulated through the availability of the subunits participating in the interaction. This availability can depend on the alternative mRNA splicing of the corresponding genes, where certain isoforms lack the binding site (Supplementary Table S6). Also, even if the translated isoform has the binding site, the protein itself can be sequestered by competing interactions with other protein partners (Supplementary Table S7). These mechanisms are present for 11% (alternative splicing) and 16% (competing interactions) of complexes, and together with PTMs, in total 36% of MSF complexes have at least one known regulatory mechanism for modulating the interaction. Furthermore, these regulatory mechanisms often act in cooperation, with seven interactions known to employ PTMs, alternative splicing, and competing interactions as well (Figure 6c).

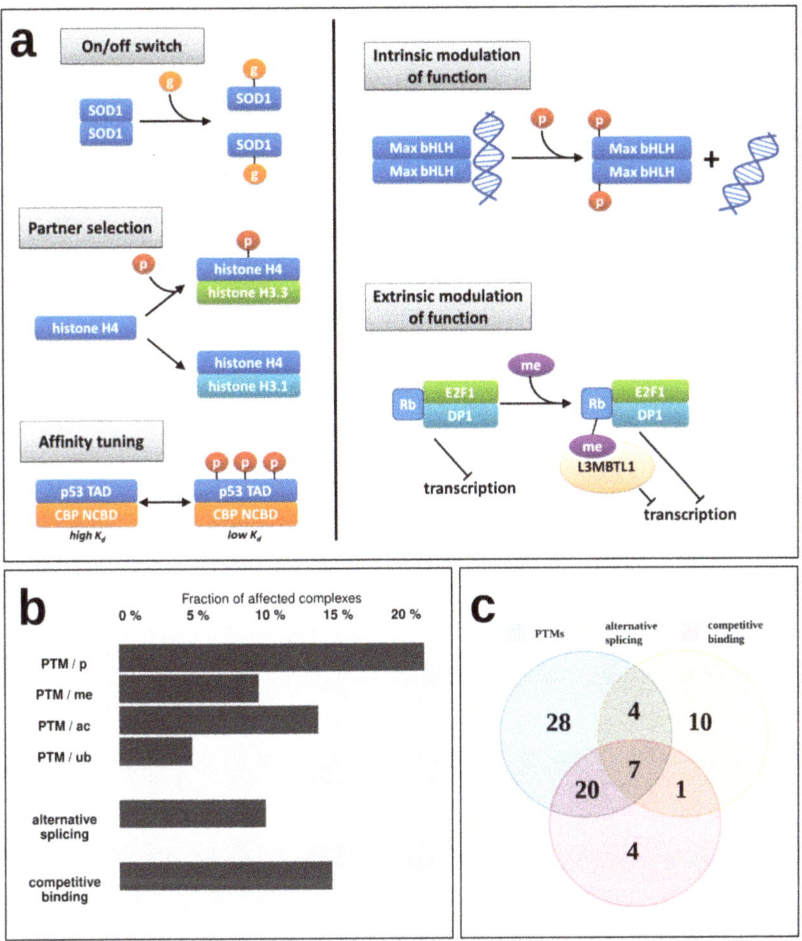

Figure 6. Regulatory mechanisms of MSF complexes. (**a**) examples of regulation and modulation of function through post-translational modifications. p phosphorylation, g—glutathionylation, me—methylation, SOD1—superoxide dismutase, CBP—CREB-binding protein, Rb—retinoblastoma-associated protein. Colored boxes represent interacting chains forming the MSF complexes. (**b**) The fraction of complexes with verified PTM sites, and the fraction of complexes where at least one interactor is regulated via alternative splicing or by competing interactions. (**c**) Number and overlap of MSF complexes affected by the three types of regulatory mechanisms.

2.6. Various Complex Types Show Differential Subcellular Localization

In addition to regulatory mechanisms detailed in the previous chapter, a crucial element in the spatio-temporal control of protein function is subcellular localization [41]. In order to assess this aspect of MSF complexes, and to understand if the defined interaction types have different properties in terms of cellular localization, we used "cellular component" terms from GeneOntology (GO) [42] (see Data and Methods). Various GO terms were condensed into five categories including "Extracellular", "Intracellular", "Membrane", "Nucleus", and "Other" to enable an overview of the differences in localization between the six complex types (Figure 7) (for exact GO terms for each complex see Supplementary Table S8).

The least amount of information is available for Type 1, near-ordered complexes. Albeit GO terms are lacking for most complexes, even the limited annotations highlight that these complexes are able to efficiently function in the extracellular space, which in general is fairly uncommon for IDPs. Coil- and zipper-type helical complexes (Type 2) are somewhat more often attached to the membrane or function in the intracellular space, or in non-nuclear environments, such as the lysosome. In contrast, oligomerization modules (Types 3 and 4) are most prevalent in the nucleus and the intracellular space, which is in line with the function of the high number of transcription factors in these groups. However, modules with a large interface (Type 3) are relatively often found in other compartments, while modules with smaller interfaces (Type 4) also function in the extracellular space. Complexes adopting a handshake-like fold are enriched in histones, which is reflected in their enrichment in the nucleus and the chromatin (classified as "other" in Figure 7). Type 6 complexes are heterogeneous in terms of localization as well, and hence members can be found in all studied localizations to a comparable degree. These preferences in subcellular localization for different complex types reinforce our notion that even though our classification scheme relies on sequence and structure properties alone, the obtained interaction types also have biological meaning.

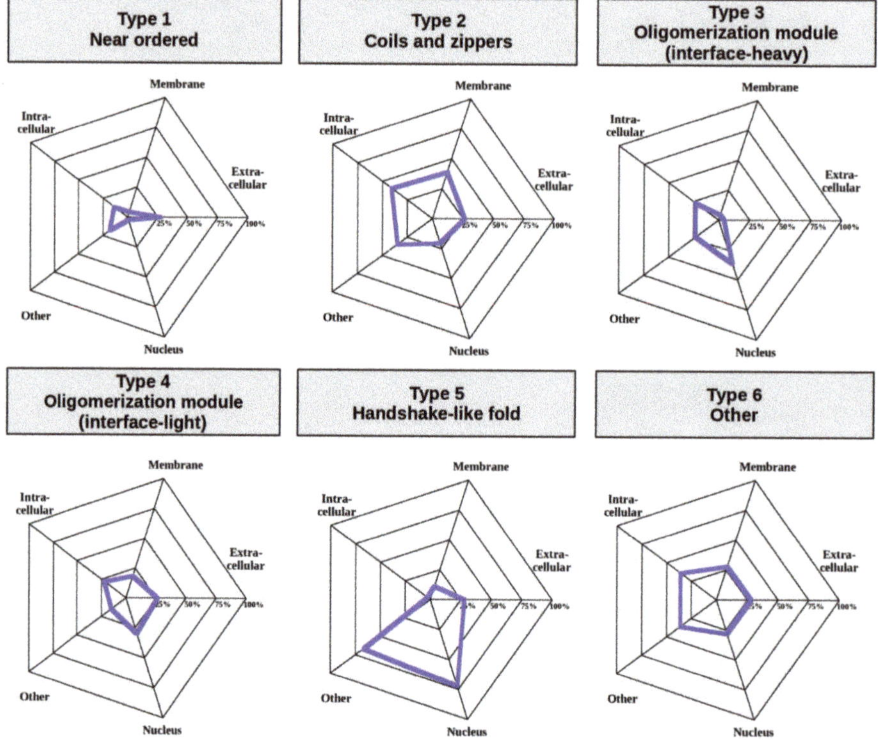

Figure 7. Subcellular localization of MSF complexes belonging to the six types. "Other" contains the "non-membrane-bounded organelle", "secretory granule", "lysosome", "cytoplasmic vesicle lumen", and "transport vesicle" GeneOntology terms.

2.7. The Annotated Catalogue of Complexes Formed via Mutual Synergistic Folding

Considering the previously analyzed features of complexes, averaging the calculated features for the six established interaction types provides the annotated catalogue of MSF interactions (Figure 8). Apart from the main sequential and structural features, Figure 8 also shows example structures, energetic properties, subcellular localization, and the main regulatory mechanisms for each complex type.

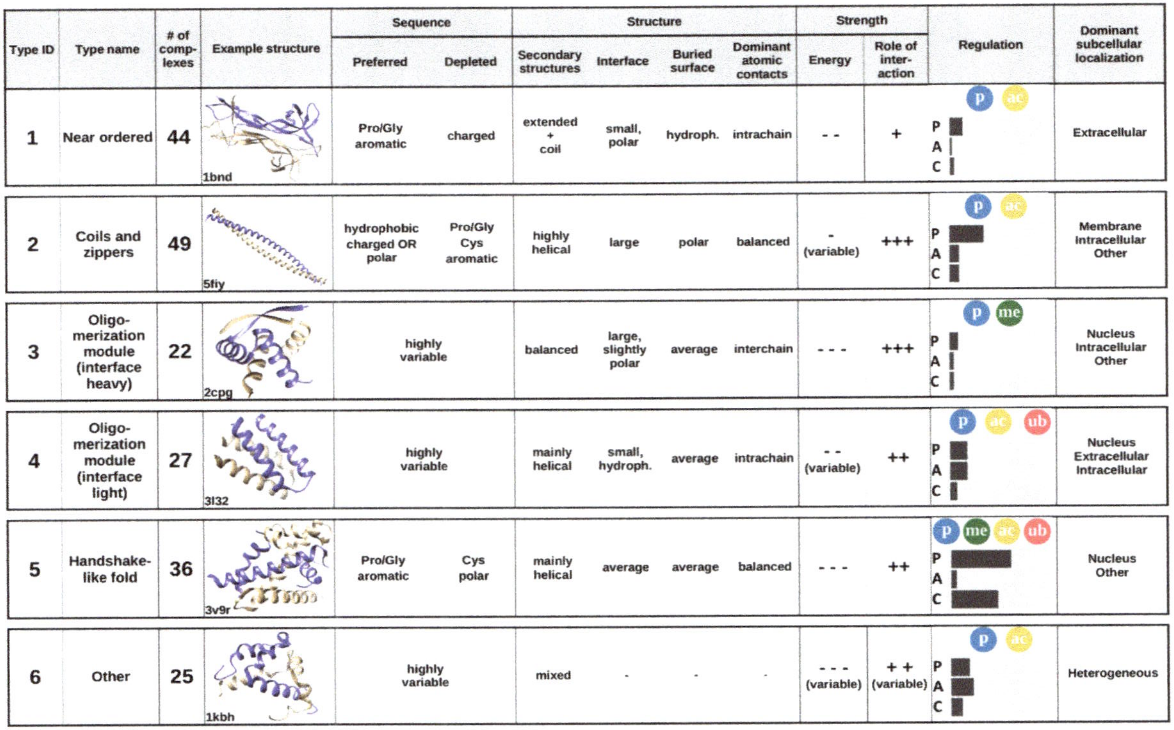

Type ID	Type name	# of comp-lexes	Example structure	Sequence		Structure				Strength		Regulation	Dominant subcellular localization
				Preferred	Depleted	Secondary structures	Interface	Buried surface	Dominant atomic contacts	Energy	Role of inter-action		
1	Near ordered	44	1bnd	Pro/Gly aromatic	charged	extended + coil	small, polar	hydroph.	intrachain	- -	+	P / A / C	Extracellular
2	Coils and zippers	49	5fiy	hydrophobic charged OR polar	Pro/Gly Cys aromatic	highly helical	large	polar	balanced	- (variable)	+++	P / A / C	Membrane Intracellular Other
3	Oligo-merization module (interface heavy)	22	2cpg	highly variable		balanced	large, slightly polar	average	interchain	- - -	+++	P / A / C	Nucleus Intracellular Other
4	Oligo-merization module (interface light)	27	3l32	highly variable		mainly helical	small, hydroph.	average	intrachain	- - (variable)	++	P / A / C	Nucleus Extracellular Intracellular
5	Handshake-like fold	36	3v9r	Pro/Gly aromatic	Cys polar	mainly helical	average	average	balanced	- - -	++	P / A / C	Nucleus Other
6	Other	25	1kbh	highly variable		mixed	-	-	-	- - - (variable)	++ (variable)	P / A / C	Heterogeneous

Figure 8. Annotated types of complexes formed by IDPs, based on sequence and structure features. Horizontal bars in the regulation column show the fraction of complexes in a given group involved in various types of regulatory mechanisms (P—post-translational modifications, A—alternative splicing affecting binding regions, C—competing interactions). Color circles mark the dominant post-translational modification(s) for the group (p—phosphorylation, me—methylation, ac—acetylation, ub—ubiquitination).

The first type of complexes bears a high similarity to ordered protein complexes, and hence are named near ordered. The constituent chains are usually similar, in many cases corresponding to homooligomers, with a high Pro/Gly content and typically only a few charges. The main difference compared to protein complexes formed by ordered proteins is that near ordered subunits are depleted in α-helices [28]. For reaching a stable structure through the interaction, they utilize a large number of intrachain contacts, with inter-subunit interactions through a small polar interface playing only a secondary role in the stability of the complex. This group contains a large number of enzymes, transport proteins, and nerve growth factors, where the exact structure is of utmost importance; however, in contrast to monomeric proteins, the presence of this structure relies on the interaction. This interaction type is mostly regulated through phosphorylation and acetylation of binding site residues. These proteins resemble ordered proteins in their localization as well, with extracellular regions being highly representative.

The second type of complexes contains structures with a high overall similarity, mostly consisting of coiled-coils and zippers, structures composed of parallel interacting helical structures, often stabilized by a restricted set of residues, such as leucines, alanines, or tryptophans. In general, constituent proteins are depleted in residues incompatible with α-helix formation, such as Pro and Gly, and also in aromatic residues. In turn, they are abundant in hydrophobic residues and show an enrichment for either polar or charged residues. The constituent helices usually form a fairly weakly bound system, where the interchain interactions via the relatively large interfaces play a major role. Constituent proteins are able to bury only a small fraction of their polar surfaces. Coiled-coil interactions are often regulated, typically via various types of PTMs, most often through phosphorylation or, to a lesser degree, acetylation. Despite their highly similar structures, complexes in this group convey a large variety of functions, mainly pertaining to regulating transcription and performing membrane-associated biological roles, such as organelle and membrane organization.

The third and the fourth type of complexes are both generic oligomerization modules that can be split according to the importance for the interchain interactions, grouping them as either interface-heavy (Type 3) or interface-light (Type 4) complexes. In both cases, the sequences can be highly variable, and the unifying features are mostly structural. Both types typically have an average-sized relative buried area with balanced hydrophobic/polar composition. However, interface-heavy complexes have a large, slightly polar interface that plays a major role in achieving the tightly bound structures. In contrast, interface-light complexes form a more helical structure and have smaller hydrophobic interfaces that play a more diminished role in achieving the stability of a less tightly bound system. This hints at interface-light complexes being more transient, also supported by the fact that these complexes have a higher number of known regulatory PTMs and are also modulated by alternative splicing. Both type 3 and type 4 complexes preferentially occur in nuclear and intracellular processes, as several of them are ribbon–helix–helix (interface-heavy) or basic helix–loop–helix (interface-light) transcription factors, able to shuttle between the nuclear and the intracellular spaces. In addition to the similarities in subcellular localization, type 4 complexes preferentially occur in the extracellular space, and type 3 complexes in other cell compartments, as well.

The fifth type of complexes typically adopts a handshake-like fold, characteristic of histones and homologous proteins. While these structures are usually largely helical, the interacting proteins often contain a relatively high ratio of prolines and glycines, in addition to the enrichment of aromatic residues. While they are depleted in polar residues, both the interface and the buried surface have a fairly balanced hydrophobic/polar makeup. The complexes are relatively tightly bound, and interchain interactions play a fairly large role in stabilizing the interaction. This type of complex has the highest ratio of both PTMs and competitive interactions, providing a large amount of regulation. In addition, PTMs are highly heterogeneous, containing phosphorylations, acetylations, methylations, and ubiquitinations as well. Members of this cluster primarily serve DNA/chromosome-related functions, and hence are usually located in the nucleus.

While types 1–5 represent well-defined groups with members of clear unifying similarities, the final group serves as an umbrella term for complexes that are not members of any previous structural/sequential class. In accord, these complexes cannot be described by simple characteristic features and are the most sequentially and structurally heterogeneous group. This group contains highly specialized interactions that present unique protein complexes, which are regulated through all three control mechanisms and occur in all studied subcellular localizations.

2.8. Interaction Types Present A Novel Classification of Protein Complexes

The described MSF classification method bears similarity to the approach employed in CATH, as both approaches use a hierarchical classification of PDB structures. However, CATH does not consider interactions and simply relies on the secondary structure elements and their connectivity and arrangement, in contrast to the presented analysis taking into account protein chain interactions too, together with sequence composition features.

Figure 9 shows the studied MSF complexes in both our MSF classification system and in CATH, considering the top two levels ("Class" and "Architecture"). The highest-level CATH definitions, corresponding to "Class", reflect the overall secondary structure element distribution of the structures. In this framework, Type 1 near-ordered complexes mostly occupy the "Mainly Beta" CATH class, while complexes from the other five types mostly fall into the "Mainly Alpha" class or the "Other" class. At the next CATH level, "Architecture", certain MSF type complexes (such as type 2 coils and zippers) are segregated into further subclasses.

Considering "Class" and "Architecture" definitions, there is very little correspondence between the CATH and the new MSF classification. If the two schemes showed a high degree of similarity, the matrix in Figure 9 should be close to a diagonal matrix. In reality, however, off-diagonal elements are large, confirming the novelty of the presented MSF classification scheme.

		CATH classes and architectures							
		Mainly Alpha (1)		Mainly Beta (2)		Alpha Beta (3)		Few Secondary Structures (4)	Other
		Orthogonal bundle (1.10)	Up-down bundle (1.20)	Sandwich (2.60)	Orthogonal prism (2.90)	2-layer sandwich (3.30)	3-layer(aba) sandwich (3.40)	Irregular (4.10)	.
MSF classification	Near-ordered (Type 1)	2	0	11	9	8	4	0	10
	Coils and Zippers (Type 2)	0	31	0	0	0	0	0	18
	Oligomerization (interface-heavy) (Type 3)	8	1	0	0	1	4	1	7
	Oligomerization (interface-light) (Type 4)	11	3	0	0	3	0	3	7
	Hand-shake fold like (Type 5)	29	2	0	0	0	1	0	4
	Other (Type 6)	8	7	0	0	1	0	2	7

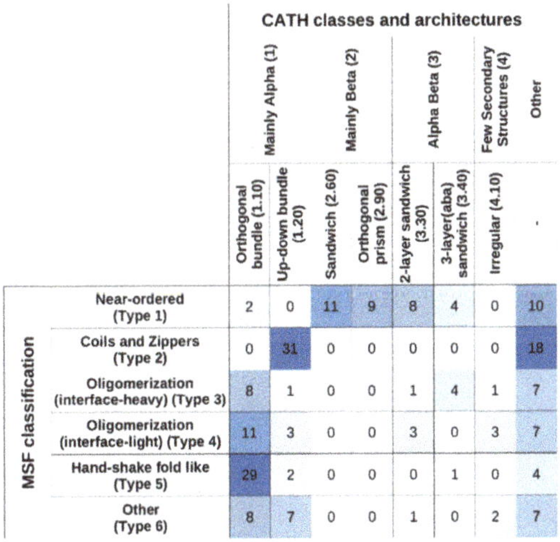

Figure 9. Overlap between CATH and MSF classification.

3. Discussion

Here, we present the first approach aiming at the classification of complex structures formed exclusively by disordered proteins via mutual synergistic folding. We developed and applied a method that can classify these complexes into various types based on sequence- and structure-based properties. The classification scheme takes into account on the one hand, the overall sequence and structure properties of the complex, and on the other hand, the interaction itself, quantifying the role of intra- and intermolecular interactions in relation to the overall contact/surface properties of the structure. As the classification protocol is based on hierarchical clustering, it is freely scalable. Tuning the resolution via changing the number of sequence-based or structure-based clusters, the method can be used to yield any number of types and subtypes. The presented classification is a top-level one highlighting the major types of MSF classes, and this six-way classification scheme will be used to better define MSF complex types in the MFIB [26] database.

While both sequence- and structure-based parameters are taken into account when defining the final complex types, the two sets of descriptors have different roles in the scalability of the method. In our presented approach to defining complex types, the main features are structural properties, while sequence parameters are more descriptive in the sense that they highlight the sequential features needed to be able to fold into a complex of given structural properties (Figure 3). However, sequence features can be used to distinguish subtypes of structure-defined complex types. For example, type 1 near-ordered complexes come in two flavors according to the two sequence clusters they cover (Figures 1 and 3): polar-driven interactions between mostly homodimers, and charge/hydrophobic driven interactions between mostly heterodimers. Also, type 2 complexes (coils and zippers) come in two varieties: relying on polar-driven interactions for heterodimers and charge-driven for homodimers.

In addition to providing a scalable classification scheme, the described method and the defined complex types have biological relevance. The presented complex types have different biological properties; although only information describing the sequence and structure properties were put in, the resulting types show different properties in terms of the energetics and strength of the interactions (Figures 4 and 5), the relevant regulatory processes (Figure 6), and subcellular localization (Figure 7).

The analysis of the energetics properties of the interactions can provide a glimpse into the biophysical details of the binding and folding. The use of low-resolution statistical force fields proved to be a suitable approach to discriminate complexes based on the structural features of constituent chains [28] and to describe the binding of IDPs [43,44]. While complexes of ordered proteins and domain-recognition IDP binding sites have a fairly narrow range in energetics parameters (Figure 4a), complexes formed exclusively by IDPs are more heterogeneous, basically covering the whole range of the energy spectrum (Figure 4b). Furthermore, based on predictions, MSF complexes cover at least 10 orders of magnitude in K_d values (Figure 5). Hence, in terms of binding strength and stability, these complexes have the potential to cover a very wide range of biological functions, overlapping with those of ordered complexes and domain-binding IDPs as well, in agreement with the previous comparative functional analysis of a wide range of interactions [28].

For most known MSF complexes, the resulting structure is instrumental for proper function, such as the coiled-coil structure for the SNAP receptor (SNARE) complex in mediating membrane fusion [45], the dimeric structure for a wide range of transcription factors in precise DNA-binding [46–48], and the proper coordination of catalytic residues for oligomeric enzymes [49,50]. Therefore, for MSF complexes, the interaction de facto switches on the protein function, and hence the precise regulation of the interaction strength is vital in the biological context of these complexes. While structure-based K_d value predictions are informative, in some cases they do not fully describe the interactions. Many MSF complexes are tightly bound, yet they are not necessarily obligate complexes, and their association/dissociation can be under heavy regulation. For example, solely based on K_d values and energetics, type 5 (handshake-like fold) interactions seem to form obligate complexes. However, there are several cases where these interactions do break up in a biological setting, most notably for histones. Histone H4 is able to form dimers with at least eight different H3 variants [51], and it was described that in the case of H3.1 and H3.3, the preference of H4 for these two partners is governed by H4 phosphorylation [37]. The post-translational modifications can enhance complex formation or dissociation in many other cases as well [35]. In addition, competition for the same binding partner and binding site availability as a function of alternative splicing is an additional mechanism for the regulation of the formation of MSF complexes (Figure 6).

Exploring the precise regulatory mechanisms for MSF complexes would be highly informative. Unfortunately, experimental K_d measurements are lacking for the majority of these interactions, and interactions in structural detail have usually been only analyzed in a single PTM state. Therefore, the molecular details and biologically relevant steps of the regulation of these interactions are difficult to assess; but from a biological sense, it is probable that even several low K_d complexes can dissociate rapidly in certain cases. At least some regulatory mechanisms are currently known for about 36% of studied MSF complexes, but the real numbers are bound to be higher. This means that most probably the majority of MSF complexes are not obligate complexes, where the disordered state is physiologically irrelevant, but can exist in both the stable bound state and the disordered unbound state as well, under native conditions. Thus, MSF complexes are integral parts or direct targets of regulatory networks, although the extent of regulation varies with the interaction type considered.

Apart from the studied regulatory mechanisms, additional layers of spatio-temporal regulation can play crucial roles for MSF complexes, similarly to other IDP interactions [41]. An emerging such regulatory mechanism is liquid–liquid phase separation (LLPS). A prime example is the Nck/neuronal Wiskott–Aldrich syndrome protein (N-WASP). N-WASP is known to undergo LLPS when interacting with Nck and nephrin [52], via linear motif-mediated coupled folding and binding. Mutually synergistic folding between the secreted EspFU pathogen protein from enterohaemorrhagic *Escherichia coli* and the autoinhibitory GTPase-binding domain (GBD) in host WASP proteins (MFIB ID:MF2202002, type 5 complex) hijacks the native LLPS-mediated cellular processes [53], showing that competing interactions are not always stoichiometric in nature, and the true extent of MSF regulation is likely to be even more complex than highlighted here.

The difference between complex types in various biological and biophysical properties shows that these type-definitions reflect true biological differences. Apart from being useful for complex classification, the presented method also shows that differences in binding strength, subcellular localization, and regulation are encoded in the sequence and structural properties of proteins. This can be the basis for developing future prediction methods, where these sequence- and structure-based parameters can be used as input for the prediction of biological features of complexes. In addition, the establishment of MSF complex types has direct implications, as knowledge present for a specific complex might be transferable to other complexes of the same type. For example, certain pathological conditions arise through the aggregation of IDPs. A well-known example is transthyretin (TTR) aggregation that can lead to various amyloid diseases, such as senile systemic amyloidosis [54]. Another example from the same near-ordered complex type is the superoxide dismutase SOD1, which is able to form aggregates in amyotrophic lateral sclerosis [55]. While the localization and the biological function of TTR and SOD1 (hormone transport and enzymatic catalysis) are radically different, their potency of malfunctioning (often connected to various mutations) share a high degree of resemblance. On one hand, this marks other type 1 complexes as candidates for toxic aggregation, on the other hand, it indicates that the potential therapeutic techniques for one complex (e.g., CLR01 for TTR) can give clues about potential targeting of other interactions.

Such structural classification approaches can have a high impact on structure research, most importantly in the study of protein structure or evolution, in training and/or benchmarking algorithms, augmenting existing datasets with annotations, and examining the classification of a specific protein or a small set of proteins [56]. Up to date, several structure-based classification approaches have been developed, such as SCOP [32] and CATH [33], which are extended to protein complexes as well. In this sense, previously existing methods are able to classify MSF complexes too. However, the approaches used do not take into account that these structures are only stable in the context of the interaction, and that a certain protein region can adopt fundamentally different structures depending on the interacting partner. The lack of the explicit encoding of parameters describing the properties and importance of the interaction into the classification scheme makes current methods unable to accurately describe the spectrum of MSF complexes, and to date, no such dedicated classification scheme has been proposed. In contrast to previously existing methods that largely encode the same information [57], the presented MSF classification scheme is highly independent (Figure 9), and thus serves as an orthogonal approach capable of properly handling the specific properties of IDP-driven complex formation through mutual synergistic folding.

4. Data and Methods

4.1. Complexes Formed Through Mutual Synergistic Folding (MSF)

MSF complexes were taken from the MFIB database [26]. Two entries, MF2100018 and MF5200001, from the 205 were discarded due to issues with the corresponding PDB structures 1ejp and 1vzj, as constituent chains have an unrealistically low number of interchain contacts. Problems with these two structures are apparent from the high outlier scores and clash scores provided on the PDB server. As the developed classification scheme relies heavily on structural parameters, we opted to leave these two entries out of the calculations. The final list of entries is given in Supplementary Table S1.

4.2. Other Complexes of Ordered and Disordered Proteins

As a reference, two other datasets of protein complexes were used. A set of complexes formed exclusively by ordered single-domain protein interactors was taken from [28]. These 688 complexes (see Supplementary Table S3) are formed via autonomous folding followed by binding, that is, both interacting protein chains adopt a stable structure in their monomeric forms, prior to the interaction. A set of 772 complexes with an IDP interacting with ordered domains was taken from the database of Disordered Binding Sites (DIBS) database [23]. These complexes (see Supplementary Table S4) are formed via coupled folding and binding, where the IDP adopts a stable structure in the context of the interaction.

4.3. Calculating Sequence Features

Similarly to the approach described in [28], the following amino acid groups were used in quantifying sequence composition of proteins: hydrophobic (containing A, I, L, M, V), aromatic (containing F, W, Y), polar (containing N, Q, S, T), charged (containing H, K, R, D, E), rigid (containing only P), flexible (containing only G), and covalently interacting (containing only C). This low-resolution sequence composition at least partially compensates for commonly occurring amino acid substitutions that in most cases do not affect protein structure and function. In all cases, compositions were calculated for the entire complex, including all interacting protein chains. An 8th sequence parameter was used to quantify the compositional difference between subunits. This dissimilarity measure was defined as:
$\Delta_{total} = \sum_{i=1}^{7} \Delta_i$, where Δ_i is the largest composition difference of residue group i between any pair of constituent chains. The average dissimilarities for various sequence-based clusters are shown in Figure 1. For exact sequence composition values for all MSF entries, see Supplementary Table S1.

4.4. Calculating Structure Features

Secondary structure assignment was performed by DSSP [58], using a three-state classification distinguishing helical ('H','G','I'), extended ('B','E'), and irregular ('S','T', unassigned) residues.

Molecular surfaces were calculated using Naccess [59]. Solvent accessible surface area (SASA) was defined by the Nacces absolute surface column. Interface is defined as the increase in SASA as a result of removing interaction partners from the structure. Buried surface was calculated by subtracting interface area and SASA from the sum of standard surfaces of residues in the protein chain. Thus, interface and buried surfaces represent the area that is made inaccessible to the solvent by the partner(s) or by the analyzed protein itself. All calculated areas were split into hydrophobic (H) and polar (P) contributions based on the polarity of the corresponding atom. Polar/hydrophobic assignations were taken from Naccess.

Contacts were defined at the atomic level. Two atoms were considered to be in contact if their distance is shorter than the sum of the two atoms' van der Waals radii plus 1 Angstrom. For exact structural feature values for all MSF entries, see Supplementary Table S1.

4.5. Filtering Features for Clustering

Standard Pearson correlation values were calculated between all sequence and structure features (Supplementary Table S2). If two features show a correlation with an absolute value above 0.7, only one was kept. In each case, we discarded the feature that shows a high correlation with a higher number of other features, or the one with the lower standard deviation. In total, none of the seven sequence parameters were discarded, but 13 out of the 24 structure parameters were omitted from subsequent clustering steps.

4.6. Clustering

Both sequence and filtered structure parameters were used as input for clustering separately. First, hierarchical clustering was done using the scaled features as input, using Euclidean distance and Ward's method (Supplementary Figures S1 and S3). Then, k-means clustering was employed, and the within-groups sum of squares were plotted as a function of the number of clusters (Supplementary Figures S2 and S4). k-means clustering analysis did not provide a clear-cut support for the number of clusters to choose, and hence we opted for choosing a low number of clusters in both cases (four and five in the case of sequence- and structure-based clustering, respectively), that are not in contradiction with the k-means analysis. This choice of cluster numbers reflects our preference for providing an overall high-level classification. Clustering was done using R with the Ward.D2 and k-means packages.

4.7. Energetic Features

Interaction energies for residues were calculated using the statistical potentials described in [60]. These interaction potentials were demonstrated to well describe the energetic features of IDP interactions [43], and are the basis for recognizing them from the sequence [44]. These potentials yield dimensionless quantities in arbitrary units, and hence their absolute values bear no direct physical meaning. However, their signs are accurate, and values below 0 correspond to stabilizing interactions. Furthermore, they can be directly compared, and hence more negative values typically correspond to more stable structures. In each analysis, the total energies were calculated from the residue-level interactions from the entire complex. Two residues were considered to be in interaction if there is at least one heavy atom contact between them. Energetic values are given in Supplementary Tables S1 (for MSF complexes), S3 (for ordered complexes), and S4 (for complexes containing both IDPs and ordered domains).

4.8. Prediction of K_d Values

Dissociation constants for MSF complexes were estimated using the method described in [61]. In each case, the modified PDB structures taken from the MFIB database [26] were used as input. For technical reasons, not all structures yield a K_d value prediction, and thus the number of values used in representing the average per-complex type K_ds (Figure 5) is calculated from fewer values than the actual number of complexes per type. K_d values are listed in Supplementary Table S1.

4.9. Post-Translational Modifications (PTMs), Isoforms and Competitive Binding

Post-translational modifications were taken from the 2 October 2017 version of PhosphoSitePlus [62], PhosphoELM [63], and UniProt [64]. Only PTMs that were identified in low-throughput experiments were used. These were mapped to complex structures using BLAST between UniProt and PDB sequences (Supplementary Table S5). Protein isoforms were taken from the 4 October 2017 version of UniProt (Supplementary Table S6). To determine alternative binding partners for IDPs, all oligomer PDB structures containing the same UniProt region were selected. PDB structures listed as related in the corresponding MFIB entry were removed. Structures containing the same interaction partners as the original complex were also removed (Supplementary Table S7).

4.10. GeneOntology Terms for Assessing Subcellular Localization

Subcellular localization was represented using GeneOntology [42] terms from the cellular_component namespace. Terms attached to complexes in MFIB were mapped to a restricted set of terms, called CellLoc GO Slim, used in previous studies [28] to compare localization of protein–protein interactions. Terms in CellLoc GO Slim were split into five categories: extracellular, intracellular, membrane, nucleus, and other, encompassing other membrane-bounded cellular compartments, such as the lysosome, as well as non-membrane-bounded compartments, such as the chromatin. For CellLoc GO terms attached to MSF complexes, see Supplementary Table S8.

Supplementary Materials: Supplementary materials can be found at http://www.mdpi.com/1422-0067/20/21/5460/s1.

Author Contributions: Conceptualization, B.M., I.S.; Methodology, B.M.; Software, B.M., L.D.; Formal Analysis, B.M., L.D., E.F.; Investigation, B.M., I.S.; Resources, B.M., L.D.; Data Curation, B.M.; Writing—Original Draft Preparation, B.M.; Writing—Review & Editing, B.M., L.D., E.F.; Visualization, B.M.; Supervision, B.M., I.S.; Project Administration, B.M., I.S.; Funding Acquisition, B.M., L.D., I.S.

Funding: This research was funded by the EMBO|EuropaBio fellowship 7544 (B.M.), the UNKP-17-3 new national excellence program of the ministry of human capacities of Hungary (L.D.), the project no. FIEK_16-1-2016-0005 financed under the FIEK_16 funding scheme (National Research, Development and Innovation Fund of Hungary) (I.S.), Hungarian Research and Developments Fund OTKA K115698 (I.S.). Bioinformatic infrastructure was supported by ELIXIR Hungary.

Acknowledgments: The authors express gratitude to Zsuzsanna Dosztányi and Zoltán Gáspári for their comments on the project.

Conflicts of Interest: The authors declare no conflict of interest.

Abbreviations

IDP	Intrinsically Disordered Protein
MSF	Mutual Synergistic Folding
CFB	Coupled Folding and Binding
PTM	Post-Translational Modification
SOD1	Superoxide Dismutase
Rb	Retinoblastoma protein
SCOP	Structural Classification of Proteins
CATH	Class/Architecture/Topology/Homologous superfamily
GO	GeneOntology

References

1.	Dyson, H.J.; Wright, P.E. Intrinsically unstructured proteins and their functions. *Nat. Rev. Mol. Cell Biol.* **2005**, *6*, 197–208. [CrossRef] [PubMed]

2.	Babu, M.M. The contribution of intrinsically disordered regions to protein function, cellular complexity, and human disease. *Biochem. Soc. Trans.* **2016**, *44*, 1185–1200. [CrossRef] [PubMed]

3.	Wright, P.E.; Dyson, H.J. Intrinsically disordered proteins in cellular signalling and regulation. *Nat. Rev. Mol. Cell Biol.* **2015**, *16*, 18–29. [CrossRef] [PubMed]

4.	Galea, C.A.; Wang, Y.; Sivakolundu, S.G.; Kriwacki, R.W. Regulation of cell division by intrinsically unstructured proteins: intrinsic flexibility, modularity, and signaling conduits. *Biochemistry* **2008**, *47*, 7598–7609. [CrossRef]

5.	Fahmi, M.; Ito, M. Evolutionary Approach of Intrinsically Disordered CIP/KIP Proteins. *Sci. Rep.* **2019**, *9*, 1575. [CrossRef]

6.	Tompa, P.; Kovacs, D. Intrinsically disordered chaperones in plants and animals. *Biochem. Cell Biol.* **2010**, *88*, 167–174. [CrossRef]

7. Boczek, E.E.; Alberti, S. One domain fits all: Using disordered regions to sequester misfolded proteins. *J. Cell Biol.* **2018**, *217*, 1173–1175. [CrossRef]

8. He, J.; Chao, W.C.H.; Zhang, Z.; Yang, J.; Cronin, N.; Barford, D. Insights into degron recognition by APC/C coactivators from the structure of an Acm1-Cdh1 complex. *Mol. Cell* **2013**, *50*, 649–660. [CrossRef]

9. Mészáros, B.; Kumar, M.; Gibson, T.J.; Uyar, B.; Dosztányi, Z. Degrons in cancer. *Sci. Signal.* **2017**, *10*. [CrossRef]

10. Gsponer, J.; Futschik, M.E.; Teichmann, S.A.; Babu, M.M. Tight regulation of unstructured proteins: from transcript synthesis to protein degradation. *Science* **2008**, *322*, 1365–1368. [CrossRef]

11. Van der Lee, R.; Buljan, M.; Lang, B.; Weatheritt, R.J.; Daughdrill, G.W.; Dunker, A.K.; Fuxreiter, M.; Gough, J.; Gsponer, J.; Jones, D.T.; et al. Classification of intrinsically disordered regions and proteins. *Chem. Rev.* **2014**, *114*, 6589–6631. [CrossRef] [PubMed]

12. Dosztányi, Z.; Chen, J.; Dunker, A.K.; Simon, I.; Tompa, P. Disorder and sequence repeats in hub proteins and their implications for network evolution. *J. Proteome Res.* **2006**, *5*, 2985–2995. [CrossRef] [PubMed]

13. Hu, G.; Wu, Z.; Uversky, V.N.; Kurgan, L. Functional Analysis of Human Hub Proteins and Their Interactors Involved in the Intrinsic Disorder-Enriched Interactions. *Int. J. Mol. Sci.* **2017**, *18*, 2761. [CrossRef] [PubMed]

14. Cortese, M.S.; Uversky, V.N.; Dunker, A.K. Intrinsic disorder in scaffold proteins: getting more from less. *Prog. Biophys. Mol. Biol.* **2008**, *98*, 85–106. [CrossRef]

15. Snead, D.; Eliezer, D. Intrinsically disordered proteins in synaptic vesicle trafficking and release. *J. Biol. Chem.* **2019**, *294*, 3325–3342. [CrossRef]

16. Harmon, T.S.; Holehouse, A.S.; Pappu, R.V. Differential solvation of intrinsically disordered linkers drives the formation of spatially organized droplets in ternary systems of linear multivalent proteins. *New J. Phys.* **2018**, *20*, 045002. [CrossRef]

17. Davey, N.E.; Van Roey, K.; Weatheritt, R.J.; Toedt, G.; Uyar, B.; Altenberg, B.; Budd, A.; Diella, F.; Dinkel, H.; Gibson, T.J. Attributes of short linear motifs. *Mol. Biosyst.* **2012**, *8*, 268–281. [CrossRef]

18. Tompa, P.; Fuxreiter, M. Fuzzy complexes: polymorphism and structural disorder in protein-protein interactions. *Trends Biochem. Sci.* **2008**, *33*, 2–8. [CrossRef]

19. Sugase, K.; Dyson, H.J.; Wright, P.E. Mechanism of coupled folding and binding of an intrinsically disordered protein. *Nature* **2007**, *447*, 1021–1025. [CrossRef]

20. Wang, Y.; Chu, X.; Longhi, S.; Roche, P.; Han, W.; Wang, E.; Wang, J. Multiscaled exploration of coupled folding and binding of an intrinsically disordered molecular recognition element in measles virus nucleoprotein. *Proc. Natl. Acad. Sci. USA* **2013**, *110*, E3743–E3752. [CrossRef]

21. Shammas, S.L.; Crabtree, M.D.; Dahal, L.; Wicky, B.I.M.; Clarke, J. Insights into Coupled Folding and Binding Mechanisms from Kinetic Studies. *J. Biol. Chem.* **2016**, *291*, 6689–6695. [CrossRef] [PubMed]

22. Demarest, S.J.; Martinez-Yamout, M.; Chung, J.; Chen, H.; Xu, W.; Dyson, H.J.; Evans, R.M.; Wright, P.E. Mutual synergistic folding in recruitment of CBP/p300 by p160 nuclear receptor coactivators. *Nature* **2002**, *415*, 549–553. [CrossRef] [PubMed]

23. Schad, E.; Fichó, E.; Pancsa, R.; Simon, I.; Dosztányi, Z.; Mészáros, B. DIBS: A repository of disordered binding sites mediating interactions with ordered proteins. *Bioinformatics* **2017**. [CrossRef] [PubMed]

24. Fukuchi, S.; Sakamoto, S.; Nobe, Y.; Murakami, S.D.; Amemiya, T.; Hosoda, K.; Koike, R.; Hiroaki, H.; Ota, M. IDEAL: Intrinsically Disordered proteins with Extensive Annotations and Literature. *Nucleic Acids Res.* **2012**, *40*, D507–D511. [CrossRef] [PubMed]

25. Miskei, M.; Antal, C.; Fuxreiter, M. FuzDB: database of fuzzy complexes, a tool to develop stochastic structure-function relationships for protein complexes and higher-order assemblies. *Nucleic Acids Res.* **2017**, *45*, D228–D235. [CrossRef]

26. Fichó, E.; Reményi, I.; Simon, I.; Mészáros, B. MFIB: a repository of protein complexes with mutual folding induced by binding. *Bioinformatics* **2017**, *33*, 3682–3684. [CrossRef]

27. Mészáros, B.; Erdős, G.; Szabó, B.; Schád, É.; Tantos, Á.; Abukhairan, R.; Horváth, T.; Murvai, N.; Kovács, O.P.; Kovács, M.; et al. PhaSePro: the database of proteins driving liquid-liquid phase separation. *Nucleic Acids Res.* **2019**. [CrossRef]

28. Mészáros, B.; Dobson, L.; Fichó, E.; Tusnády, G.E.; Dosztányi, Z.; Simon, I. Sequential, Structural and Functional Properties of Protein Complexes Are Defined by How Folding and Binding Intertwine. *J. Mol. Biol.* **2019**. [CrossRef]

29. Mentes, A.; Magyar, C.; Fichó, E.; Simon, I. Analysis of Heterodimeric "Mutual Synergistic Folding"-Complexes. *Int. J. Mol. Sci.* **2019**, *20*, E5136. [CrossRef]

30. Magyar, C.; Mentes, A.; Fichó, E.; Cserző, M.; Simon, I. Physical Background of the Disordered Nature of "Mutual Synergetic Folding" Proteins. *Int. J. Mol. Sci.* **2018**, *19*, E3340. [CrossRef]

31. wwPDB consortium. Protein Data Bank: the single global archive for 3D macromolecular structure data. *Nucleic Acids Res.* **2019**, *47*, D520–D528. [CrossRef] [PubMed]

32. Chandonia, J.-M.; Fox, N.K.; Brenner, S.E. SCOPe: classification of large macromolecular structures in the structural classification of proteins-extended database. *Nucleic Acids Res.* **2019**, *47*, D475–D481. [CrossRef] [PubMed]

33. Sillitoe, I.; Dawson, N.; Lewis, T.E.; Das, S.; Lees, J.G.; Ashford, P.; Tolulope, A.; Scholes, H.M.; Senatorov, I.; Bujan, A.; et al. CATH: expanding the horizons of structure-based functional annotations for genome sequences. *Nucleic Acids Res.* **2019**, *47*, D280–D284. [CrossRef] [PubMed]

34. Zhao, N.; Pang, B.; Shyu, C.-R.; Korkin, D. Structural similarity and classification of protein interaction interfaces. *PLoS One* **2011**, *6*, e19554. [CrossRef] [PubMed]

35. Redler, R.L.; Wilcox, K.C.; Proctor, E.A.; Fee, L.; Caplow, M.; Dokholyan, N.V. Glutathionylation at Cys-111 induces dissociation of wild type and FALS mutant SOD1 dimers. *Biochemistry* **2011**, *50*, 7057–7066. [CrossRef] [PubMed]

36. Van Roey, K.; Gibson, T.J.; Davey, N.E. Motif switches: decision-making in cell regulation. *Curr. Opin. Struct. Biol.* **2012**, *22*, 378–385. [CrossRef]

37. Kang, B.; Pu, M.; Hu, G.; Wen, W.; Dong, Z.; Zhao, K.; Stillman, B.; Zhang, Z. Phosphorylation of H4 Ser 47 promotes HIRA-mediated nucleosome assembly. *Genes Dev.* **2011**, *25*, 1359–1364. [CrossRef]

38. Lee, C.W.; Ferreon, J.C.; Ferreon, A.C.M.; Arai, M.; Wright, P.E. Graded enhancement of p53 binding to CREB-binding protein (CBP) by multisite phosphorylation. *Proc. Natl. Acad. Sci. USA* **2010**, *107*, 19290–19295. [CrossRef]

39. Bousset, K.; Oelgeschläger, M.H.; Henriksson, M.; Schreek, S.; Burkhardt, H.; Litchfield, D.W.; Lüscher-Firzlaff, J.M.; Lüscher, B. Regulation of transcription factors c-Myc, Max, and c-Myb by casein kinase II. *Cell. Mol. Biol. Res.* **1994**, *40*, 501–511.

40. Saddic, L.A.; West, L.E.; Aslanian, A.; Yates, J.R., 3rd; Rubin, S.M.; Gozani, O.; Sage, J. Methylation of the retinoblastoma tumor suppressor by SMYD2. *J. Biol. Chem.* **2010**, *285*, 37733–37740. [CrossRef]

41. Gibson, T.J. Cell regulation: determined to signal discrete cooperation. *Trends Biochem. Sci.* **2009**, *34*, 471–482. [CrossRef] [PubMed]

42. The Gene Ontology Consortium. The Gene Ontology Resource: 20 years and still GOing strong. *Nucleic Acids Res.* **2019**, *47*, D330–D338. [CrossRef] [PubMed]

43. Mészáros, B.; Tompa, P.; Simon, I.; Dosztányi, Z. Molecular principles of the interactions of disordered proteins. *J. Mol. Biol.* **2007**, *372*, 549–561. [CrossRef] [PubMed]

44. Mészáros, B.; Erdos, G.; Dosztányi, Z. IUPred2A: context-dependent prediction of protein disorder as a function of redox state and protein binding. *Nucleic Acids Res.* **2018**, *46*, W329–W337. [CrossRef] [PubMed]

45. Strop, P.; Kaiser, S.E.; Vrljic, M.; Brunger, A.T. The structure of the yeast plasma membrane SNARE complex reveals destabilizing water-filled cavities. *J. Biol. Chem.* **2008**, *283*, 1113–1119. [CrossRef] [PubMed]

46. Bonvin, A.M.; Vis, H.; Breg, J.N.; Burgering, M.J.; Boelens, R.; Kaptein, R. Nuclear magnetic resonance solution structure of the Arc repressor using relaxation matrix calculations. *J. Mol. Biol.* **1994**, *236*, 328–341. [CrossRef]

47. Madl, T.; Van Melderen, L.; Mine, N.; Respondek, M.; Oberer, M.; Keller, W.; Khatai, L.; Zangger, K. Structural basis for nucleic acid and toxin recognition of the bacterial antitoxin CcdA. *J. Mol. Biol.* **2006**, *364*, 170–185. [CrossRef]

48. Sauvé, S.; Tremblay, L.; Lavigne, P. The NMR solution structure of a mutant of the Max b/HLH/LZ free of DNA: insights into the specific and reversible DNA binding mechanism of dimeric transcription factors. *J. Mol. Biol.* **2004**, *342*, 813–832. [CrossRef]

49. Le Trong, I.; Stenkamp, R.E.; Ibarra, C.; Atkins, W.M.; Adman, E.T. 1.3-A resolution structure of human glutathione S-transferase with S-hexyl glutathione bound reveals possible extended ligandin binding site. *Proteins* **2002**, *48*, 618–627. [CrossRef]

50. Dams, T.; Auerbach, G.; Bader, G.; Jacob, U.; Ploom, T.; Huber, R.; Jaenicke, R. The crystal structure of dihydrofolate reductase from Thermotoga maritima: molecular features of thermostability. *J. Mol. Biol.* **2000**, *297*, 659–672. [CrossRef]

51. Tachiwana, H.; Osakabe, A.; Shiga, T.; Miya, Y.; Kimura, H.; Kagawa, W.; Kurumizaka, H. Structures of human nucleosomes containing major histone H3 variants. *Acta Crystallogr. D Biol. Crystallogr.* **2011**, *67*, 578–583. [CrossRef] [PubMed]

52. Banjade, S.; Wu, Q.; Mittal, A.; Peeples, W.B.; Pappu, R.V.; Rosen, M.K. Conserved interdomain linker promotes phase separation of the multivalent adaptor protein Nck. *Proc. Natl. Acad. Sci. USA* **2015**, *112*, E6426–E6435. [CrossRef] [PubMed]

53. Cheng, H.-C.; Skehan, B.M.; Campellone, K.G.; Leong, J.M.; Rosen, M.K. Structural mechanism of WASP activation by the enterohaemorrhagic E. coli effector EspF(U). *Nature* **2008**, *454*, 1009–1013. [CrossRef] [PubMed]

54. Westermark, P.; Sletten, K.; Johansson, B.; Cornwell, G.G., 3rd. Fibril in senile systemic amyloidosis is derived from normal transthyretin. *Proc. Natl. Acad. Sci. USA* **1990**, *87*, 2843–2845. [CrossRef]

55. Pansarasa, O.; Bordoni, M.; Diamanti, L.; Sproviero, D.; Gagliardi, S.; Cereda, C. SOD1 in Amyotrophic Lateral Sclerosis: "Ambivalent" Behavior Connected to the Disease. *Int. J. Mol. Sci.* **2018**, *19*, E1375. [CrossRef]

56. Fox, N.K.; Brenner, S.E.; Chandonia, J.-M. The value of protein structure classification information-Surveying the scientific literature. *Proteins* **2015**, *83*, 2025–2038. [CrossRef]

57. Hadley, C.; Jones, D.T. A systematic comparison of protein structure classifications: SCOP, CATH and FSSP. *Structure* **1999**, *7*, 1099–1112. [CrossRef]

58. Touw, W.G.; Baakman, C.; Black, J.; te Beek, T.A.H.; Krieger, E.; Joosten, R.P.; Vriend, G. A series of PDB-related databanks for everyday needs. *Nucleic Acids Res.* **2015**, *43*, D364–D368. [CrossRef]

59. Hubbard, S.; Thornton, J. NACCESS Computer Program. 1992. Available online: http://wolf.bms.umist.ac. uk/naccess/ (accessed on 31 October 2019).

60. Dosztányi, Z.; Csizmók, V.; Tompa, P.; Simon, I. The pairwise energy content estimated from amino acid composition discriminates between folded and intrinsically unstructured proteins. *J. Mol. Biol.* **2005**, *347*, 827–839. [CrossRef]

61. Vangone, A.; Bonvin, A.M. Contacts-based prediction of binding affinity in protein-protein complexes. *Elife* **2015**, *4*, e07454. [CrossRef]

62. Hornbeck, P.V.; Zhang, B.; Murray, B.; Kornhauser, J.M.; Latham, V.; Skrzypek, E. PhosphoSitePlus, 2014: mutations, PTMs and recalibrations. *Nucleic Acids Res.* **2015**, *43*, D512–D520. [CrossRef] [PubMed]

63. Dinkel, H.; Chica, C.; Via, A.; Gould, C.M.; Jensen, L.J.; Gibson, T.J.; Diella, F. Phospho.ELM: A database of phosphorylation sites–update 2011. *Nucleic Acids Res.* **2011**, *39*, D261–D267. [CrossRef] [PubMed]

64. UniProt Consortium. UniProt: A worldwide hub of protein knowledge. *Nucleic Acids Res.* **2019**, *47*, D506–D515. [CrossRef] [PubMed]

© 2019 by the authors. Licensee MDPI, Basel, Switzerland. This article is an open access article distributed under the terms and conditions of the Creative Commons Attribution (CC BY) license (http://creativecommons.org/licenses/by/4.0/).

International Journal of
Molecular Sciences

Article

In Silico Study of Rett Syndrome Treatment-Related Genes, *MECP2*, *CDKL5*, and *FOXG1*, by Evolutionary Classification and Disordered Region Assessment

Muhamad Fahmi [1], Gen Yasui [1], Kaito Seki [1], Syouichi Katayama [2], Takako Kaneko-Kawano [2], Tetsuya Inazu [2], Yukihiko Kubota [3] and Masahiro Ito [1,3,*]

[1] Advanced Life Sciences Program, Graduate School of Life Sciences, Ritsumeikan University, Kusatsu, Shiga 525-8577, Japan; gr0343rp@ed.ritsumei.ac.jp (M.F.); sj0048hh@ed.ritsumei.ac.jp (G.Y.); sj0036kf@ed.ritsumei.ac.jp (K.S.)

[2] Department of Pharmacy, College of Pharmaceutical Sciences, Ritsumeikan University, Kusatsu, Shiga 525-8577, Japan; s-kata@fc.ritsumei.ac.jp (S.K.); takanek@fc.ritsumei.ac.jp (T.K.-K.); tinazu@fc.ritsumei.ac.jp (T.I.)

[3] Department of Bioinformatics, College of Life Sciences, Ritsumeikan University, Kusatsu, Shiga 525-8577, Japan; yukubota@fc.ritsumei.ac.jp

* Correspondence: maito@sk.ritsumei.ac.jp

Received: 30 September 2019; Accepted: 5 November 2019; Published: 8 November 2019

Abstract: Rett syndrome (RTT), a neurodevelopmental disorder, is mainly caused by mutations in methyl CpG-binding protein 2 (*MECP2*), which has multiple functions such as binding to methylated DNA or interacting with a transcriptional co-repressor complex. It has been established that alterations in cyclin-dependent kinase-like 5 (*CDKL5*) or forkhead box protein G1 (*FOXG1*) correspond to distinct neurodevelopmental disorders, given that a series of studies have indicated that RTT is also caused by alterations in either one of these genes. We investigated the evolution and molecular features of MeCP2, CDKL5, and FOXG1 and their binding partners using phylogenetic profiling to gain a better understanding of their similarities. We also predicted the structural order–disorder propensity and assessed the evolutionary rates per site of MeCP2, CDKL5, and FOXG1 to investigate the relationships between disordered structure and other related properties with RTT. Here, we provide insight to the structural characteristics, evolution and interaction landscapes of those three proteins. We also uncovered the disordered structure properties and evolution of those proteins which may provide valuable information for the development of therapeutic strategies of RTT.

Keywords: Rett syndrome; intrinsically disordered region; phylogenetic profile analysis; post-transcriptional modification; methyl-CpG-binding protein 2; cyclin-dependent kinase-like 5; forkhead box protein G1

---

1. Introduction

Rett syndrome (RTT; OMIM entry #312750) is a rare disease that was first described by Andreas Rett in 1966 [1]. It is characterized by severe impairment such as deceleration of head growth, loss of speech, seizures, ataxia, movement disorder, and breathing disturbance [2]. Alterations in methyl CpG-binding protein (*MECP*)2, an X-linked gene involved in the regulation of RNA splicing and chromatin remodeling, were confirmed in approximately 95% of individuals diagnosed with RTT [3], while the others were confirmed in either cyclin-dependent kinase-like (*CDKL*)5 or forkhead box protein (*FOXG*)1 alterations as atypical cases of RTT [4,5]. The mutations in *MECP2* are generally paternally derived. Thus, this syndrome mainly affects girls, and the age of onset varies from 6 to 18 months [2,6]. Additionally, Rett syndrome can also affect males with severe phenotype and early lethality following the inactivation of the sole X-linked copy of *MECP2* [7]. In a rare case, it can also exist as somatic mosaicism or co-occur with Klinefelter syndrome in males [8,9]. Even though the causative genes have been determined, the infrequent clinical phenotypes yield to the difficulty in diagnosis. Further, diagnosis may be challenging as many of the clinical features overlap with those of other neurological and neurodevelopmental disorders, and mutation in *MECP2*, *FOXG1*, and *CDKL5* can also cause neurodevelopmental disorders distinct from RTT [10]. As a result, subsequent studies have suggested that alterations in either *CDKL5* or *FOXG1* should be classified as a distinct disorder from RTT as the majority of cases showed some differences in clinical features [11–13] Moreover, recent studies have suggested that RTT is a monogenic disorder caused by mutations that alter the functionality of the methyl-CpG-binding domain (MBD) and the NCoR/SMRT interaction domain (NID) in *MECP2* [14–16]. This may simplify the complication of developing a treatment strategy. But, elucidation on the overlapped symptoms between those three proteins comprehensively on the molecular basis also seems necessary as the study about it remains scarce and it may provide meaningful insight, particularly for RTT.

The MeCP2 structure has been determined using various experimental methods, while the structure of FOXG1 has only been investigated by predictions [17,18]. In the case of CDKL5, the structure of the amino-terminal kinase domain has already been identified, but that of the long carboxy-terminal tail has not been clarified [19]. These proteins have been suggested to contain polypeptide segments that are unable to fold spontaneously into three-dimensional structures; the so-called intrinsically disordered regions (IDRs) exist as dynamic ensembles that rapidly interconvert from molten globule (collapsed) to coiled or pre-molten globule (extended) as a result of the relatively flat energy landscapes [20,21]. The different entities of IDRs and ordered regions (displaying tertiary structures in native conditions) are dictated by the amino acid sequence; the former generally lack bulky hydrophobic residues [22]. Proteins are composed of either fully structured or fully disordered regions (with the latter referred to as intrinsically disordered proteins (IDPs) or a combination of the two, which is the case for most eukaryotic proteins [23]. Although protein function has traditionally been elucidated based on a well-defined structure, it is now widely acknowledged that IDRs contribute to diverse functions, which can be classified into six types: entropic chain activity, display site, chaperone, molecular effector, molecular assembler, and molecular scavenger [23–26]. Excluding entropic chain activity, IDRs adopt specific tertiary conformations—at least locally—in order to perform those functions by binding to other proteins, nucleic acids, membranes, and small molecules or responding to changes in their environment [20,27]. Hence, IDR structure varies over time—i.e., it exhibits spatiotemporal heterogeneity. Moreover, long IDRs contain more modification sites than fully ordered regions, and their flexibility provides more opportunities for displaying these sites [28,29]. These features explain how proteins with IDRs or IDPs interact with and are tightly regulated by various factors to ensure that appropriate levels of proteins are available at the right time to minimize the possibility of inappropriate protein–protein interactions [26]. Thus, misfolding and altered availability of proteins with IDRs or IDPs are more likely to be associated with disease states. Given a similarity in those properties, we proposed that a study concerning the link between MeCP2, CDKL5, and FOXG1 disordered structure properties with RTT or RTT-like syndrome collectively is necessary.

Restoring *Mecp2* gene function in an animal model abolished the symptoms of RTT. Growth factor stimulation (e.g., insulin-like growth factor 1) and the activation of neurotransmitter pathways (e.g., β2-adrenergic receptor pathway) can also partially rescue phenotypes of *Mecp2* knockout mice (RTT model mice), suggesting that the disorder is treatable [15,30,31]. In addition to gene therapy, reactivation of an inactivated X chromosome is known to be a new therapeutic method [32,33]. The therapeutic strategies of RTT are under development, and elucidation on this enigmatic disorder needs various points of view to make advances in understanding. Even though RTT has been determined as a monogenic disorder, the complex biological system compels us to necessarily broaden our perspective; moreover, MeCP2 contains an extensive amount of disordered regions which may facilitate binding with multiple partners. Considering several points above, we investigated the evolution and molecular features of MeCP2, CDKL5, and FOXG1 and their binding partners using phylogenetic profiling to gain a better understanding of their similarities. Additionally, we predicted the structural order–disorder propensity and assessed the evolutionary rates per site of MeCP2, CDKL5, and FOXG1 to investigate the relationships between disordered structure and other related properties with RTT.

2. Results

2.1. Structural Order–Disorder Properties of RTT and RTT-like Causing Proteins during Chordate Evolution

We retrieved 97, 113, and 108 chordates sequences of MeCP2, CDKL5, and FOXG1, respectively, and constructed a heat map of the structural order–disorder propensity for each protein of these genes according to aligned sequences and taxonomic position in the phylogenetic tree (Supplementary Table S1 and Figure 1). This analysis was conducted in order to investigate the evolutionary patterns of structural properties. The results showed that all proteins harbored both ordered and disordered regions; by comparing their distribution to domain and non-domain regions, we found that the catalytic domain and non-domain regions of CDKL5 were ordered and disordered, respectively (Figure 1B). While most regions of MeCP2 were predicted to be disordered, some ordered structures were observed in the MBD (Figure 1A). Furthermore, FOXG1 showed a varied distribution of ordered–disordered regions corresponding to domain and non-domain regions, with the former predicted to be fully ordered (Figure 1C). Although insertions and deletions were frequently detected in disordered regions, particularly in MeCP2 and FOXG1 (Figure 1A,C), the structural order–disorder of all proteins showed to be stable in chordates, excluding a few conformational transitions of FOXG1 and CDKL5 in mammals and fishes, respectively. This indicated that the disordered regions of MeCP2, CDKL5, and FOXG1 tend to be functional either as an entropic chain, transient binding site, or permanent binding site in chordates. Additionally, insertions and deletions were frequently detected in disordered regions. This is caused by their flexibility, which makes sequence alignment difficult; a tendency of linear motifs to lie among the flexible disordered regions; and the permutation of functional modules with respect to others during evolution that is possible in disordered regions, such as SUMO modification sites in *Drosophila melanogaster* and human p53 that are located before and after the oligomerization domain, respectively [26,34].

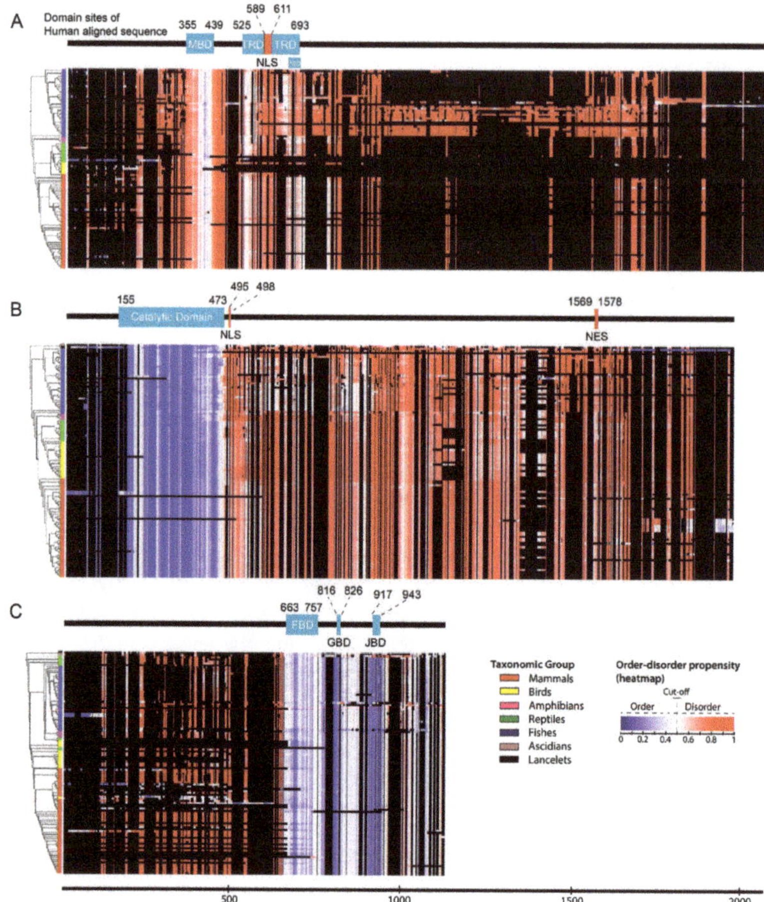

Figure 1. The order–disorder propensity of RTT and RTT-like causing proteins in chordates. Heat maps of the order–disorder propensity were generated according to the taxonomic positions in the phylogenetic tree (rows) and multiple sequence alignment (columns). The heat maps show a color gradient of blue (ordered) to red (disordered), with white as the boundary between the two and black as gaps. Colored boxes between the trees and heat maps indicate the taxonomic group, and bars above the heat maps indicate domain position in the multiple sequence alignment, with light blue and black areas indicating the domain and absence of a domain, respectively. (**A–C**) Heat maps for MeCP2 (**A**), CDKL5 (**B**), and FOXG1 (**C**) are shown. MBD, TRD, NID, FBD, GBD, JBD, NLS, and NES indicate methyl-CpG-binding domain, transcriptional repression domain, NCoR/SMRT interaction domain, forkhead binding domain, Groucho-binding domain, JARID1B binding domain, nuclear localization signal, and nuclear export signal, respectively.

2.2. Rate of Evolution per Site in RTT and RTT-like Causing Proteins

We calculated the evolutionary rates of MeCP2, CDKL5, and FOXG1 in chordates to investigate their relationships with structural features and the distribution of missense point mutations that have previously been suggested to contribute to RTT or RTT-like syndrome. We used the human sequence as a reference and determined standardized evolutionary rate scores (Z scores), with values greater than or less than zero reflecting evolution at a faster and slower than average rate, respectively (Figure 2 and Supplementary Table S2). Evolutionary rates per site showed similar patterns in all proteins, with low rates of evolution more commonly observed in domains and ordered regions; some exceptional cases such as the transcriptional repression domain (TRD) of MeCP2 showed a partial higher rate of amino acid substitution. On the other hand, non-domain regions that were also usually disordered—excluding the ordered region surrounding a domain in FOXG1—typically exhibited a higher evolutionary rate, although some regions with low rates of evolution were nonetheless detected (Figure 2). This was corroborated by the distribution of evolutionary rates for predicted structural order–disorder residues in the three proteins, with disordered residues showing a wide and overlapping distribution that reflected their conservation. The evolutionary rates of ordered and disordered regions are significantly distinct in those three proteins ($p < 2.2e{-}16$ for CDKL5 and FOXG1 and $p < 6.409e{-}08$ for MeCP2, Mann–Whitney U-test; Figure S1).

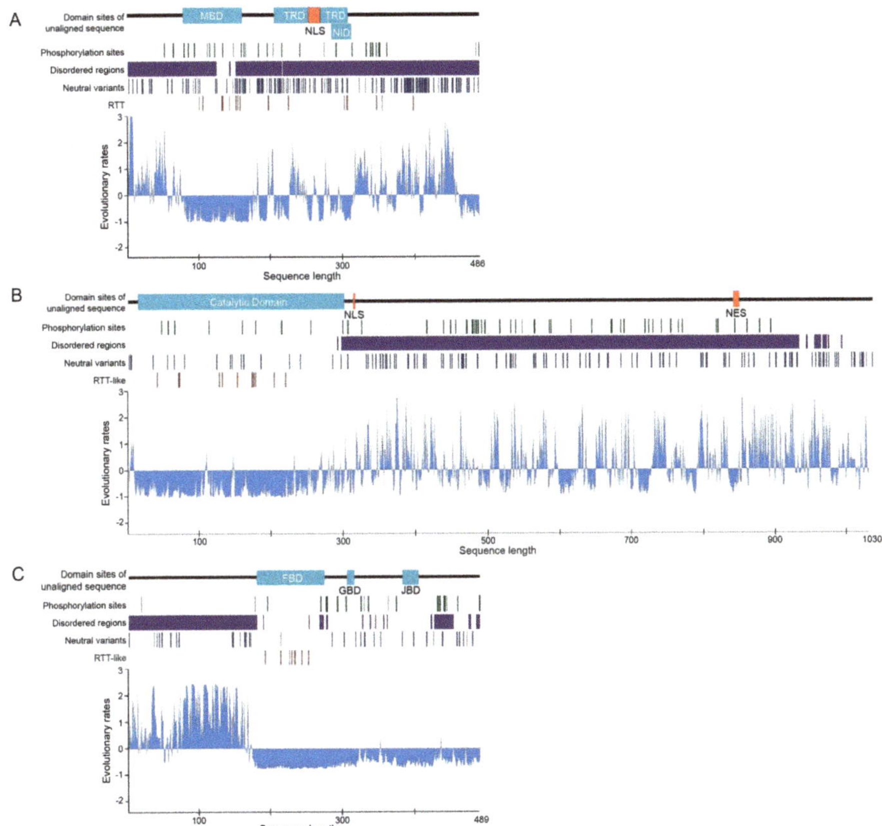

Figure 2. Rate of evolution per site in human RTT-related proteins. (**A–C**) Rates of amino acid substitution in MeCP2 (**A**), CDKL5 (**B**), and FOXG1 (**C**) are shown as blue areas. The bars above charts indicate the position of the domain in the human sequence, with light blue areas indicating the domain and black lines indicating no domain. Conserved phosphorylation sites, disordered region, single nucleotide polymorphisms in the general population, and pathogenic missense point mutation are plotted in green, purple, blue, and red lines, respectively. The x and y axes represent the sequence length and Z score of the evolutionary rates, respectively.

We identified structurally conserved disordered regions, with slowly and rapidly evolving residues reflecting constrained disorder and flexible disorder, respectively [26]. The flexible disorder has a constrained disordered structure despite having rapid evolution of residues; the amino acid substitutions of this property are constrained to residues that confer structural flexibility as the change from structurally disordered to ordered can affect protein function. This type of IDR typically functions as an entropic spring, flexible linker, or spacer without becoming structured and is frequently located outside the domain region [26,35–37]. In contrast, constrained disorder is associated with protein–protein interaction interfaces that adopt a structured conformation or undergo folding upon binding and are thus constrained in terms of sequence, while still requiring flexibility. This module can be present as short linear motifs (SLiMs) or intrinsically disordered domains (IDDs) [26,38]. These regions commonly have secondary structures that may be important for binding and, hence, slowing their evolutionary rates [36,39]. IDDs were observed in the MBD—which was predicted to be partly disordered—and in the TRD and NID of MeCP; it is in accordance with previous reports that structured regions are found only in the MBD, while other regions are extensively disordered [17,18,40]. Most domains with conserved disordered regions are involved in DNA, RNA, and protein binding, which has been demonstrated by those domains of MeCP2 [41]. SLiMs are frequently located outside the domain and may display modification site. In this study, we predicted the constrained disorder regions and conserved phosphorylation sites located outside the domain to be associated with SLiMs, such as the region that spans after the catalytic domain to the C-terminus of human CDKL5.

2.3. Post-Translational Modifications (PTMs)

Phosphorylation is important for modulating the balance of proteins between the bound and unbound states, and previous studies reported that kinases target disordered proteins as many as twice, on average, the number of times they target structured proteins [42,43]. In this study, we predicted PTM (phosphorylation) sites in chordate sequences of MeCP2, CDKL5, and FOXG1 and predicted the conserved human phosphorylation sites to chordates in order to investigate the dynamics of their phosphorylation-related function. We found numerous conserved phosphorylation sites including 60/82 in CDKL5, 30/45 in MeCP2, and all 23 sites in FOXG1 in human (Figure 2 green lines and Supplementary Table S3). Most predicted human phosphorylation sites in MeCP2, CDKL5, and FOXG1 are conserved across chordates and are located in disordered regions; one exception is FOXG1, in which almost half of the phosphorylation sites are located in predicted ordered regions; structural disorder makes such sites accessible for phosphorylation. As PTMs affect the stability, turnover, interaction potential, and localization of proteins within the cell, proteins with disordered regions are more likely to be multifunctional [26]; accordingly, it has shown that MeCP2, CDKL5, and FOXG1 play multiple roles in the molecular basis.

2.4. Disease-Associated Missense Mutation Distribution in the Sequence of RTT and RTT-like Causing Proteins

Plotting missense mutations associated with diseases may yield crucial information on structure–function relationships and the features of the protein. We investigated missense mutations in human MeCP2, CDKL5, and FOXG1 that were previously associated with pathogenic RTT from RettBASE and examined the features of the associated sequences. There were 7, 12, and 18 individual amino acid sites in FOXG1, CDKL5, and MeCP2, respectively, that harbored pathogenic missense mutations associated or previously suggested to be associated with pathogenic RTT (Figure 2 and Supplementary Table S4). When the frequencies were combined with those of cases observed for each mutation, MeCP2 had a higher number of cases (1225) than CDKL5 (30) and FOXG1 (8) (Supplementary Tables S4 and S9). Pathogenic RTT or RTT-like-associated missense mutations were more frequently detected in domain regions for all proteins, and in ordered and slowly evolving regions for MeCP2 and CDKL5 (Supplementary Table S9). On the other hand, many mutation sites in MeCP2 were located close to (or in the case of Ser346Arg and Ser134Cys, overlapped with) phosphorylation sites (Figure 2), although the frequency of cases harboring these mutation sites was low (only one for each).

2.5. Phylogenetic Profiling of RTT and RTT-like Causing Proteins and Their Interaction Partners

We retrieved 240 human proteins interacting with MeCP2, CDKL5, and FOXG1 from BioGRID and UniProt databases (Supplementary Table S5) [44,45]. To illuminate the interconnection of MeCP2, CDKL5, and FOXG1 binding partners as well as their evolutionary relationship, we conducted phylogenetic profiling and cluster analysis of 326 eukaryotes using the retrieved sequences and the sequences of the three proteins, MeCP2, CDKL5, and FOXG1, as queries (Figure 3, Supplementary Table S6). The results showed that the dataset was divided into four clusters, which were defined as Classes 1 to 4. There were 58 conserved proteins in chordates of Class 1, 92 in metazoans of Class 2, 17 in multicellular of Class 3, and 73 in eukaryotes of Class 4. MeCP2 and CDKL5 belonged to Class 1, whereas FOXG1 belonged to Class 2 (Figure 3). FOXG1 and MECP2 showed to have many binding partners that act as a transcription factor or gene expression regulator. In contrast, CDKL5 tend to bind to a fewer number of proteins having functions in regulating cell adhesion, ciliogenesis, and cell proliferation; however, this protein has been shown to interact with MeCP2. As RTT has been determined to occur from the altered functionality of MBD and NID of MECP2, we focused on the widely known binding partners of these domains, such as SIN3 transcription regulator family member A (SIN3A), histone deacetylase (HDAC)1, and nuclear receptor corepressor (NCOR) which play roles as co-repressor complexes. Even though FOXG1 does not directly bind to MeCP2, we found that the binding partners of MeCP2 co-repressor complex are also associated with FOXG1 binding partners that also act as co-repressor complexes such as special AT-rich sequence-binding protein (SATB)2, lysine-specific histone demethylase (KDM)1A, SWI/SNF-related matrix-associated actin-dependent regulator of chromatin subfamily (SMARC)A member 5, A-kinase anchor protein (AKAP)8, of which are ancient proteins within Classes 3 and 4.

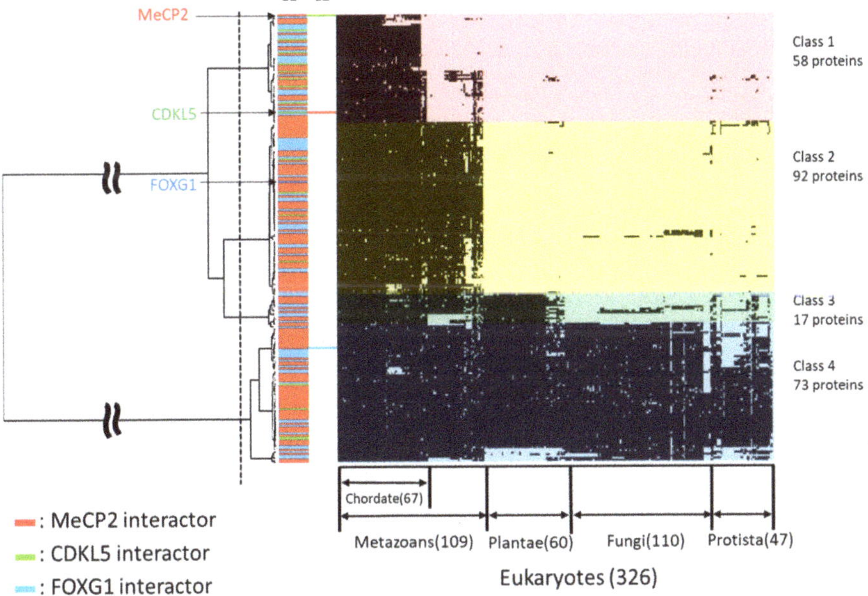

Figure 3. Phylogenetic profiling of MeCP2, CDKL5, and FOXG1 proteins and their interaction partners. The horizontal axis shows 326 eukaryotes for which whole genome sequences are available, and the vertical axis shows 240 human proteins related to RTT. Bar in a1 and a2 shows MeCP2-interactor (red), CDKL5-interactor (green), FOXG1-interactor (blue), respectively. The human orthologous proteins in each species are shown in black. The phylogenetic tree was divided into four clusters (Class 1–4); those conserved across chordates, metazoan, multicellular, and eukaryotes are shown.

2.6. Subcellular Localization and Gene Ontology (GO) Analysis

We predicted the subcellular localization of each protein and GO categories in each class for the evolutionary classification (Figure 4, Supplementary Table S7). Specific GO categories included epigenetic regulation of gene expression, transcriptional regulation, and organogenesis or organ morphogenesis (Figure 4). We confirmed the evolutionary trends of proteins with specific GO categories and their subcellular localization and found that 129 and 48 proteins in Classes 1–4 were expressed in the nucleus only or the nucleus and cytoplasm, respectively. Proteins in Classes 1–4 were represented in the epigenetic regulation of gene expression category, whereas transcriptional regulation was observed only in Classes 1 and 2, and organogenesis and organ morphogenesis were mainly observed in Class 2 (Figure 4).

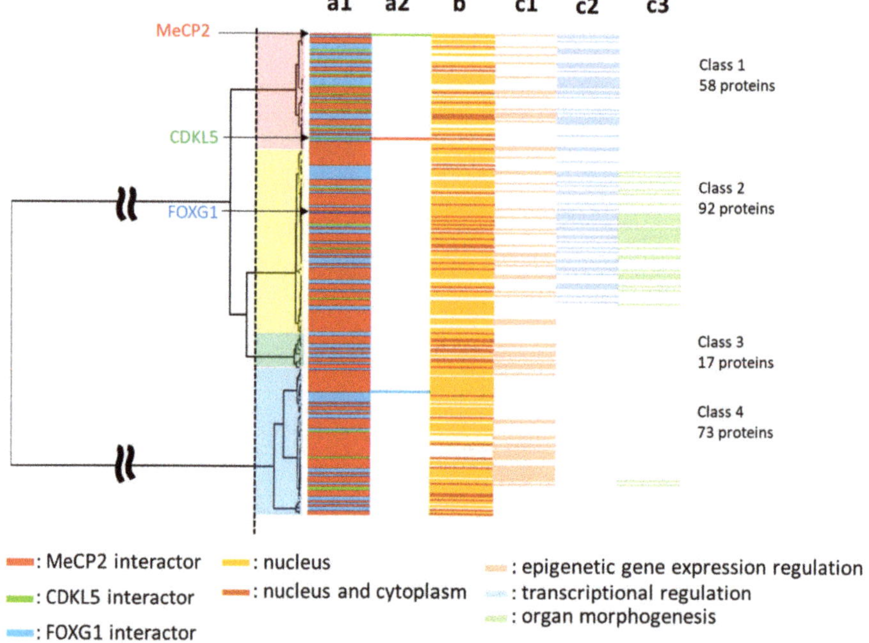

Figure 4. Subcellular localization and specific GO categories of human RTT-related proteins: Phylogenetic trees show interactors, subcellular localization, and specific GO categories for each protein. The vertical axis shows 240 RTT-related proteins, and each bar shows MeCP2-interactor (red), CDKL5-interactor (green), and FOXG1-interactor (blue) (a1 and a2); cellular localization (b); epigenetic regulation of gene expression (c1); transcriptional regulation (c2); and organogenesis (c3).

2.7. Tissue and Organ Localization

Tissue and organ expression data for 237 proteins were extracted from The Human Protein Atlas as transcripts per million (TPM) values [46]. In addition, four proteins were not expressed in the cerebral cortex. Tissues and organs with specific expression were identified using 195 RTT-related human proteins as queries (Figure S2, Supplementary Table S8). There were nine proteins that were specifically expressed in the cerebral cortex including apolipoprotein E, CDKL5, SATB2, spalt-like transcription factor (SALL)1, zinc finger protein (ZNF)483, FOXG1, (sex-determining region Y)-box (SOX)2, homeodomain-interacting protein kinase (HIPK)2, and histone cluster 2 H3 family member A.

3. Discussion

RTT is a progressive postnatal neurodevelopmental disorder; three individual genes, *MECP2*, *CDKL5*, and *FOXG1*, have previously been thought to be the cause of its variants with the altered *MECP2* as the major contributor. Later, it was suggested that RTT is a monogenic disorder caused by either null mutations or mutations that alter the MBD or NID functions of *MECP2* [15,16,47]. MBD and NID facilitate the binding of MeCP2 to modified cytosine in chromatin and recruitment of the NCOR-SMRT complex, respectively; their combination is vital for MeCP2's role as a repressor [48,49]. The altered forms in the other two genes which were previously characterized as variants of RTT were designed as distinct disorders with several overlapping symptoms to RTT. The three proteins have similar extensive amount of disordered regions and play important roles in the brain. The disordered structure itself is a unique property in protein that may contribute to the interaction with a diverse binding partner and the versatility of a protein. While the three proteins may show similar symptoms in the altered form, the investigation on their similarity in the molecular basis remains scarce, particularly on the disordered structure properties and their binding partners. Focusing on RTT, we investigated the evolution of their disordered structures and their binding partners through prediction and phylogenetic profiling, respectively. This approach is important to give an insight into the similarity of biological systems of those proteins structurally and evolutionarily, which may provide useful information for the development of a RTT therapy strategy. RTT itself has attracted considerable attention as its causative protein displays features related to epigenetics and have been shown to have partially or fully disordered structures.

All three proteins have been experimentally determined to play roles and are abundant in the brain, especially the MeCP2_e1 and hCDKL5_1 isoforms [50,51]. It is confirmed by the emergence of neurological impairments in the altered availability or forms of either protein. Through evolutionary analysis and IDRs properties, we provide an additional point of view for that feature. Phylogenetic profiling analysis of MeCP2, CDKL5, and FOXG1 and their interacting proteins showed that 240 molecules formed four clusters—i.e., chordates, metazoans, multicellular, and eukaryotes. Among the three, only *FOXG1* was a member of Class 2, which comprises genes acquired during metazoan evolution, whereas the acquisition of *MECP2* and *CDKL5* was correlated with chordate evolution. The acquisition of *CDKL5* and *MECP2*, and *FOXG1* may contribute to the development of the chordate brain and metazoan nervous system during evolution, respectively. Additionally, order–disorder structure predictions revealed that all three proteins had order–disorder structures that were relatively conserved across chordates. Human MeCP2, CDKL5, and FOXG1 phosphorylation sites were also shown to be relatively conserved to chordates. IDRs properties provide proteins with more interaction areas and PTMs sites, spatiotemporal heterogeneity of structure, and ability to associate and dissociate easily with binding partners. Hence, proteins with long IDRs are likely to have a capacity to bind to many different partners. Accordingly, all three proteins were shown to have multiple binding partners, and FOXG1 and MeCP2 displayed the highest number of partners, some of which were evolutionarily acquired before the metazoan evolved. By cooperating with various proteins partners, particularly the co-repressor complex, FOXG1 or MeCP2 can modulate the expression and suppression of different genes [15,52]. The co-repressor complex itself denotes a conserved mechanism that manifests in diverse forms and may have several functional entities depending on the context in which they are recruited [53]. This indicates the necessity to regulate either FOXG1 or MeCP2 concentration precisely; otherwise, altered availability is likely to be deleterious. Several studies have shown that either overexpression or under-expression of MeCP2 and FOXG1 corresponds to neurological deficits; this phenomenon may not independent from their co-repressor complex that has been showed to play roles in neurogenesis and neuron maturation for FOXG1, and MeCP2, respectively [7,15,52]. On the other hand, CDKL5 binds to a fewer number of proteins that have functions in regulating cell adhesion, ciliogenesis, and cell proliferation. We hypothetically suggest that the amount of CDKL5 binding partners is underestimated since this protein was predicted to have relatively long disordered regions with many constrained disorder features and phosphorylation sites; it also has fewer insertions and deletions than either MeCP2 or FOXG1 along the evolution.

FOXG1 is a transcriptional factor playing an essential role in ventral telencephalon development; it serves as a hallmark of the telencephalon in vertebrates [52,54]. Among the 237 Class 1 or 2 genes, 233 were detected in the cerebral cortex, with nine expressed at a high level (Figure S2). Seven genes were acquired during metazoan evolution, of which four and three encode MeCP2- and FOXG1-interacting molecules, respectively. Since FOXG1 was also acquired during metazoan evolution, acquisition of FOXG1, SATB2, and SALL1 may have played essential roles in development of the neocortex. FOXG1 is transiently expressed in neuronal progenitor cells and regulates their migration to the cortical plate [55]. During this process, FOXG1 expression is upregulated, which contributes to cortical plate development [56]. Similarly, the FOXG1-interacting chromatin remodeling factor SATB2 was found to be expressed in the cortical plate and regulates neocortical development [54,55]. Therefore, it is conceivable that transcriptional co-operation between FOXG1 and SATB2 mediates the laminarization of the neocortex. In support of this possibility, patients with the SATB2 mutation exhibit an RTT-like phenotype [57,58]. There is no direct interplay reported for MeCP2 and FOXG1. The causative regions in the altered form of these proteins that result in the development of RTT or RTT-like disorder exhibited similar functions in regulating the other genes' expression, but likely via a distinct pathway. We suggest that FOXG1 is not a potential target for developing treatment for RTT. However, induced pluripotent stem cell (iPSC)-derived neurons generated from FOXG1+/− patients and patients with MECP2 and CDKL5 mutations reportedly exhibited a similar increase in synaptic cell adhesion protein orphan glutamate receptor δ-1 subunit (GluD1) expression; this result indicates the need for further study to reveal the mechanism of each protein and might be implicated in the clinical symptom overlap among FOXG1-, CDKL5- and MECP2-related syndromes [52,59,60].

CDKL5 belongs to the same molecular pathway of MeCP2. MeCP2 was acquired during chordate evolution; a prerequisite for this step was the acquisition of MeCP2-interacting molecules such as ZNF483, SOX2, HIPK2, and HIST2H2A. The MeCP2 kinase HIPK2 was shown to be required for the induction of apoptotic cell death in neuronal and other cell types via phosphorylation of the MeCP2 N-terminus [61]. Given that CDKL5, another MeCP2 kinase was also acquired during chordate evolution; it is possible that HIPK2 and CDKL5 cooperate to activate MeCP2 during neocortical development. Since apoptotic cell death increased in *Cdkl5* knockout mouse brain, CDKL5 probably has a suppressive function in the apoptosis process in contrast to HIPK2 [62]. Therefore, functional division of their kinases through phosphorylation of MeCP2 is an important issue. Indeed, the CDKL5-interacting domain was shown to be associated with the C-terminus of MeCP2 [63]. Hence, CDKL5 may phosphorylate the carboxy terminus. Thus, both HIPK2 and CDKL5 may activate MeCP2 by phosphorylating different regions of the protein. It has been suggested that MeCP2 also suppresses CDKL5 transcription and that CDKL5 overexpression may also contribute to the typical RTT symptoms [64]. Hence, aiming the catalytic domain of CDKL5 as the key target for developing alternative strategies to treat classical RTT may be essential since its sole impairment resulted in some symptoms that overlapped with those of classical RTT. Additionally, the CDKL5 disordered region, which spans after the catalytic domain to the C-terminus, is suggested to have many SLiMs. The linear motifs theoretically help to determine the various fates of a protein including subcellular localization, stability, and degradation; these motifs are also able to promote recruitment of binding factors and facilitating post-translational modifications [26,38]. Since these motifs typically regulate low-affinity interactions, they can bind to molecules with different structures of similar affinity and facilitate transient-binding, which are favorable properties for drug targets. Accordingly, this region appears to be a potential target for classical RTT treatment. However, this should also consider the expression levels of CDKL5 which are highly modulated spatiotemporally [64,65].

IDRs show unique properties within protein which challenges the traditional viewpoint of the protein structure paradigm. They have differences in residue composition, intramolecular contacts, and functions to ordered regions which cause different evolutionary rates. Generally, they evolve more rapidly than ordered regions, owing to the different accepted point mutations. However, some disordered regions can be highly constrained as they may play crucial roles and have multiple functions; assessing the evolutionary rate of IDRs may thus reveal crucial protein-specific amino acids in the biological system [66]. In this study, we found a unique relationship between evolutionary rates of disordered regions and symptoms of a disease caused by FOXG1. The N-terminus residues of FOXG1 are highly variable and constrained to be disordered, while the residues from FBD to the C-terminus are constrained and contain an ordered structure. It has been reported that mutations in the N-terminal are more likely to be associated with severe phenotypes, and mutations in the C-terminal are associated with milder phenotypes [52]. We reported and predicted a phosphorylation site located in Ser 19 to be conserved in chordates even though it is located among flexible disordered regions; casein kinase 1 (CK1) modifies this site and promotes the nuclear import of FOXG1, which corresponds to neurogenesis in the forebrain [67]. This explains that a flexible disordered region can retain its functional module from phosphorylation, despite harboring numerous insertions and deletions, and that severe phenotypes may result from the altered function of Ser 19 of FOXG1.

Among 236 male testis expressing RTT-related genes, 47 genes expressed at a high level. Because paternal-derived de novo mutation has been shown to affect X-linked MeCP2-related female Rett syndrome [6,68], paternally expressing mutation in these genes may affect the sperm-derived genetic and/or epigenetic inheritance that influence the cause of Rett syndrome in a daughter. Further studies are required to analyze these possibilities.

It is important to remember that the features of structural order–disorder and phosphorylation sites in this study have been inferred using linear sequence predictors and that the sequences and mutation points were retrieved from databases whose data have been collected from studies with various methods. It should be considered that we use canonical isoforms instead of predominant brain isoforms, this option may be able to be applied computationally but should be of concern experimentally. This study provides suggestive or hypothetical conclusions, thus further experimental study is important to verify the findings of this study. Ultimately, the results can still be used and considered as a basis for further identification.

4. Materials and Methods

4.1. Sequence Retrieval, Alignment, and Phylogenetic Analysis of MeCP2, CDKL5, and FOXG1 Proteins

Orthologous sequences of human RTT and RTT-like causing proteins (MeCP2, CDKL5, and FOXG1) in chordates were retrieved from the Kyoto Encyclopedia of Genes and Genomes (KEGG) sequence similarity database (https://www.kegg.jp/kegg/ssdb/) with a Smith–Waterman similarity score threshold of 100 and the bidirectional best hits (best–best hits) option [69]. We primarily used the canonical isoforms MeCP2_e2 and hCDKL_5 instead of those the predominant isoforms in the human brain, MeCP2_e1, and hCDKL5_1. MeCP2_e2 is the most characterized isoform relative to MeCP2_e1, and RettBASE has chosen to name the variants MeCP2_e2 due to historical reason. Variants specific to MeCP2_e1 are still reported in RettBASE with the prefix MeCP2_e1 in the database, but we decided to exclude them in our analysis as we only found one variant that meets our criteria and it cannot be included within the MeCP2_e2 sequence as they differ in the N-terminal region; however, we still reported that variant in our Supplementary Data. CDKL5 has a similar case as MeCP2, but the differences of sequences between hCDKL_5 and hCDKL5_1 are located in the C-terminal region (905–1030 a.a) which does not shift the reported Rett-like variants in the catalytic domain. We selected this option as we primarily collected the RTT and RTT-like variants from RettBASE. The used isoforms do not differ greatly to those predominant brain isoforms. The highest similarity score for each species was used for each of those proteins to minimize redundancy. Datasets were created for each protein and then aligned using MAFFT v.7 (https://mafft.cbrc.jp/alignment/software/) with the iterative refinement method (FFT-NS-i), with a maximum of 1000 iterations [70]. Phylogenetic trees were constructed with the maximum likelihood method using RAxML-HPC2 BlackBox with the RAxML automatic bootstrapping option using Jones, Taylor, and Thornton amino acid substitutions with the + F method and gamma shape parameter (JTT + F + G) model for MeCP2 and CDKL5, and the JTT + G model for FOXG1, which were selected as the best fit models under the Bayesian information criterion (BIC) by ModelTest-NG [71,72]. The outgroup for each tree was selected based on the NCBI Taxonomy Common Tree for the common ancestor within the dataset [73]. Reconstruction of phylogenetic trees and calculation of models were performed in CIPRES Science Gateway (http://www.phylo.org/) [74].

4.2. Structural Order–Disorder Prediction and Secondary Structure Predictions

The structural order–disorder propensity of each protein was predicted using IUPred2A (https://iupred2a.elte.hu/) [75] using the option for long disordered regions. This prediction had values ranging from 0 (strong propensity for an ordered structure) to 1 (strong propensity for a disordered structure), with 0.5 as the cut-off between the propensity for order and disorder. The results for each site of each protein were mapped onto its sequence alignment and taxon position in the phylogenetic tree using iTOL (https://itol.embl.de/) [76].

4.3. Rate of Evolution per Site

We calculated the rate of evolution per site of human CDKL5, FOXG1, and MeCP2 relative to their orthologs using Rate4site (https://m.tau.ac.il/~{}itaymay/cp/rate4site.html) [77]. The aligned sequences of each protein dataset were calculated using the empirical Bayesian principle with the JTT model and 16 discrete categories of the prior gamma distribution. Gaps were treated as missing data, and outputs were standardized as Z scores. The results of the rate of evolution of each residue were then integrated with the structural order–disorder prediction result, and the distribution of the rate of evolution in the structural order and disorder of each protein was evaluated with the Mann–Whitney U-test using R software.

4.4. PTM Prediction

We predicted phosphorylation sites using NetPhos 3.1 (http://www.cbs.dtu.dk/services/NetPhos/) [78] to infer PTM sites conserved between human CDKL5, FOXG1, and MeCP2 sequences and their orthologs. The predictions had values ranging from 0 (strong propensity for obtaining a negative result) to 1 (strong propensity for obtaining a positive result); we used 0.75 as a cut-off to divide the negative and positive results. The prediction results for each sequence were plotted following multiple sequence alignment of each protein dataset. Predicted PTM sites in each dataset were considered as conserved through evolution if they had a positive value according to the 50% majority rule of the amount of sequence in the alignment.

4.5. Point Mutations in MeCP2, CDKL5, and FOXG1

Point mutations in CDKL5, FOXG1, and MeCP2 were identified from RettBASE (http://mecp2.chw.edu.au/) [79]. The amount of mutations variants in general in RettBASE are 929, 298, and 44 for MeCP2, CDKL5, and FOXG1, respectively. We only selected missense mutations that were associated with pathogenic RTT. Additionally, non-pathogenic polymorphisms in the general population for comparison were extracted from the Exome Aggregation Consortium database (http://exac.broadinstitute.org) [80].

4.6. Phylogenetic Profiling and Cluster Analyses of Human MeCP2, CDKL5, and FOXG1 and Their Interacting Proteins

Sequences of human MeCP2, CDKL5, and FOXG1 and their interaction partners identified with BioGRID (https://thebiogrid.org/; release 2019_03) were obtained from the UniProtKB/Swiss-Prot database (https://www.uniprot.org/help/uniprotkb; release 2019_04) and used as the dataset [45,46]. We generated phylogenetic profiles of 326 eukaryotes in the KEGG database (https://www.genome.jp/kegg/) using the dataset as a query [81]. Phylogenetic profiling is a method for detecting the presence or absence of orthologous proteins in a target organism [82]. The presence or absence of proteins homologous to the query in each species was determined using KEGG Ortholog Cluster (https://www.genome.jp/tools/oc/; release 2019_04), this tool uses Smith–Waterman similarity scores of ≥ 150 and symmetric similarity measures to classify the ortholog genes [83]. We suggest that it is a reliable tool to get ortholog data. Profiles were determined based on the Manhattan distance and then clustered using Ward's method [84].

4.7. Protein Expression in Human Tissues

Expression levels of human RTT-related proteins in each tissue were extracted from the Human Protein Atlas (https://www.proteinatlas.org/; release 2019_4) [45] and classified into 37 tissues. The protein expression level was determined using the TPM value, which was corrected for protein expression by gene length. Comparisons of protein expression levels were not shown as a ratio so that proteins with high expression did not skew the results (Equations (1)–(3)). The mean and standard deviation were derived from Equations (1) and (2), and the range was obtained from Equation (3). The range in Equation (3) was taken as the tissue for each of the specifically expressed proteins—i.e., the value was "1" when included in the range of Equation (3) and "0" when it was not included in the expression level of each protein expressed as a percentage. The procedure yielded human protein-specific expression profiles in the context of RTT.

$$\mu = \frac{1}{n}\sum_{i=0}^{n} x_i \tag{1}$$

$$s = \sqrt{\frac{1}{n}\sum_{i=0}^{n}(x_i - \mu)^2} \tag{2}$$

$$\mu + 1.65 \times s < x \tag{3}$$

Here, μ, s, n, and x are the mean, standard deviation, number of samples, and one sample, respectively. The value of 1.65 in Equation (3) is the standard confidence factor for extracting data outside the 90% confidence interval.

4.8. GO Analysis

Specific GO categories in the target protein group were obtained using the Panther tool [85]. Categories with an appearance frequency of $p < 0.05$ were defined as protein group-specific. In this study, we obtained GO categories specific for human proteins related to RTT that were classified based on defined functions.

5. Conclusions

In the last two decades, effort on elucidating RTT has shown a promising trend towards developing a reliable treatment for this disorder. Given a similarity in IDR properties and several overlapping symptoms, we investigated the evolution of MeCP2, CDKL5, and FOXG1 disordered structures and their binding partners through prediction and phylogenetic profiling, respectively. Here, we provided insight to the structural characteristics, evolution and interaction landscapes of those three proteins related to RTT. We suggested that the disordered structures of MECP2, CDKL5, and FOXG1 contribute to the versatility in brain development and may play a crucial role in brain evolution in chordates. We hypothetically suggested that CDKL5 could be a potential target for RTT treatment, particularly by targeting its disordered structure that spans after the catalytic domain to the C-terminus, which shows abundant linear motifs that can bind to molecules with different structures of similar affinity. Finally, this study may provide valuable guidance for experimental research, particularly on the relationship between RTT and disordered regions.

Supplementary Materials: Supplementary materials can be found at http://www.mdpi.com/1422-0067/20/22/5593/s1.

Author Contributions: Conceptualization, M.F., Y.K. and M.I.; methodology, M.F., G.Y., Y.K. and M.I..; software, M.F. and G.Y.; validation, M.F., G.Y. and K.S.; formal analysis, M.F. and G.Y.; investigation, M.F., G.Y., K.S., S.K., T.K.-K., T.I., Y.K. and M.I.; resources, S.K, T.K.-K. and T.I.; data curation, M.F., Y.K. and M.I.; writing—original draft preparation, M.F..; writing—review and editing, S.K., T.K.-K., T.I., Y.K. and M.I.; visualization, M.F., G.Y. and K.S.; supervision, M.I.; project administration, M.I.; funding acquisition, T.I. and M.I.

Funding: This study was supported by the MEXT-supported program for the strategic research foundation at private universities (2015–2019 to T.I) and Takeda Science Foundation.

Acknowledgments: We would like to thank Takahiro Nakamura and Tadasu Shin-I for support and helpful comments.

Conflicts of Interest: The authors declare no competing interest.

Abbreviations

a.a	Amino acids
AKAP8	A-kinase anchor protein 8
APOE	Apolipoprotein E
BioGRID	Biological General Repository for Interaction Datasets
CDKL5	Cyclin-dependent kinase-like 5
CIPRES	Cyberinfrastructure for Phylogenetic Research
CK1	Casein kinase 1
DOI	Digital object identifier
FBD	Forkhead box domain
FOXG1	Forkhead box protein G1
GBD	Groucho-binding domain
GluD1	Glutamate dehydrogenase 1
GO	Gene Ontology
HDAC1	Histone deacetylase 1
HIPK2	Homeodomain-interacting protein kinase 2
IDDs	Intrinsically disordered domains
IDPs	Intrinsically disordered proteins
IDRs	Intrinsically disordered regions
iPSC	Induced pluripotent stem cell
iTOL	Interactive Tree of Life
IUPred	Prediction of Intrinsically Unstructured Proteins
JBD	JARID1B-binding domain
JARID1B	Histone Demethylase Jumonji AT-rich Interactive Domain
JTT	The Jones, Taylor, and Thornton
KDM1A	Lysine-specific histone demethylase 1A
KEGG	Kyoto Encyclopedia of Genes and Genomes
MAFFT	Modified Multiple Alignment Fast Fourier Transform
MBD	Methyl-CpG-binding domain
MeCP2	Methyl-CpG-binding protein 2
NCOR	Nuclear receptor corepressor
NCoR/SMRT	Nuclear receptor co-repressor/silencing mediator of retinoic acid and thyroid hormone receptor
NES	Nuclear export signal
NLS	Nuclear localization signal
NID	NCoR/SMRT interaction domain
OMIM	Online Mendelian Inheritance in Man
PTM	Post-translational modification
RAxML-HPC2	Randomized Axelerated Maximum Likelihood for High-Performance Computing 2
RTT	Rett syndrome
RettBASE	Rett syndrome Variation Database
SALL1	Spalt-like transcription factor 1
SATB2	Special AT-rich sequence-binding protein 2
SIN3A	SIN3 transcription regulator family member A

SLiMs	Short linear motifs
SMARCA5	SWI/SNF-related matrix-associated actin-dependent regulator of chromatin subfamily A member 5
SOX2	SRY-box transcription factor 2
SSDB	Sequence Similarity DataBase
TRD	Transcriptional repression domain
TPM	Transcripts per million
ZNF483	zinc finger protein (ZNF)483

References

1. Rett, A. On a unusual brain atrophy syndrome in hyperammonemia in childhood. *Wien. Med. Wochenschr.* **1966**, *116*, 723–726.
2. Hanefeld, F. The clinical pattern of the Rett syndrome. *Brain Dev.* **1985**, *7*, 320–325. [CrossRef]
3. Laurvick, C.L.; De Klerk, N.; Bower, C.; Christodoulou, J.; Ravine, D.; Ellaway, C.; Williamson, S.; Leonard, H. Rett syndrome in Australia: A review of the epidemiology. *J. Pediatr.* **2006**, *148*, 347–352. [CrossRef]
4. Ariani, F.; Hayek, G.; Rondinella, D.; Artuso, R.; Mencarelli, M.A.; Spanhol-Rosseto, A.; Pollazzon, M.; Buoni, S.; Spiga, O.; Ricciardi, S.; et al. FOXG1 is responsible for the congenital variant of Rett syndrome. *Am. J. Hum. Genet.* **2008**, *83*, 89–93. [CrossRef] [PubMed]
5. Weaving, L.S.; Christodoulou, J.; Williamson, S.L.; Friend, K.L.; McKenzie, O.L.; Archer, H.; Evans, J.; Clarke, A.; Pelka, G.J.; Tam, P.P.; et al. Mutations of CDKL5 cause a severe neurodevelopmental disorder with infantile spasms and mental retardation. *Am. J. Hum. Genet.* **2004**, *75*, 1079–1093. [CrossRef] [PubMed]
6. Trappe, R.; Laccone, F.; Cobilanschi, J.; Meins, M.; Huppke, P.; Hanefeld, F.; Engel, W. MECP2. mutations in sporadic cases of Rett syndrome are almost exclusively of paternal origin. *Am. J. Hum. Genet.* **2001**, *68*, 1093–1101. [CrossRef] [PubMed]
7. Van Esch, H.; Bauters, M.; Ignatius, J.; Jansen, M.; Raynaud, M.; Hollanders, K.; Lugtenberg, D.; Bienvenu, T.; Jensen, L.R.; Gecz, J.; et al. Duplication of the MECP2 region is a frequent cause of severe mental retardation and progressive neurological symptoms in males. *Am. J. Hum. Genet.* **2005**, *77*, 442–453. [CrossRef]
8. Clayton-Smith, J.; Watson, P.; Ramsden, S.; Black, G.C.M. Somatic mutation in MECP2 as a non-fatal neurodevelopmental disorder in males. *Lancet* **2000**, *356*, 830–832. [CrossRef]
9. Ben Zeev, B.; Yaron, Y.; Schanen, N.C.; Wolf, H.; Brandt, N.; Ginot, N.; Shomrat, R.; Orr-Urtreger, A. Rett syndrome: Clinical manifestations in males with MECP2 mutations. *J. Child. Neurol.* **2002**, *17*, 20–24. [CrossRef]
10. Neul, J.L. The relationship of Rett syndrome and MECP2 disorders to autism. *Dialogues Clin. Neurosci.* **2012**, *14*, 253–262.
11. Fehr, S.; Wilson, M.; Downs, J.; Williams, S.; Murgia, A.; Sartori, S.; Vecchi, M.; Ho, G.; Polli, R.; Psoni, S.; et al. The CDKL5 disorder is an independent clinical entity associated with early-onset encephalopathy. *Eur. J. Hum. Genet.* **2013**, *21*, 266–273. [CrossRef] [PubMed]
12. Hector, R.D.; Kalscheuer, V.M.; Hennig, F.; Leonard, H.; Downs, J.; Clarke, A.; Benke, T.A.; Armstrong, J.; Pineda, M.; Bailey, M.E.S.; et al. CDKL5 variants: Improving our understanding of a rare neurologic disorder. *Neurol. Genet.* **2017**, *3*, e200. [CrossRef] [PubMed]
13. Kortum, F.; Das, S.; Flindt, M.; Morris-Rosendahl, D.J.; Stefanova, I.; Goldstein, A.; Horn, D.; Klopocki, E.; Kluger, G.; Martin, P.; et al. The core FOXG1 syndrome phenotype consists of postnatal microcephaly, severe mental retardation, absent language, dyskinesia, and corpus callosum hypogenesis. *J. Med. Genet.* **2011**, *48*, 396–406. [CrossRef] [PubMed]
14. Lyst, M.J.; Ekiert, R.; Ebert, D.H.; Merusi, C.; Nowak, J.; Selfridge, J.; Guy, J.; Kastan, N.R.; Robinson, N.D.; de Lima Alves, F.; et al. Rett syndrome mutations abolish the interaction of MeCP2 with the NCoR/SMRT co-repressor. *Nat. Neurosci.* **2013**, *16*, 898–902. [CrossRef] [PubMed]
15. Lyst, M.J.; Bird, A. Rett syndrome: A complex disorder with simple roots. *Nat. Rev. Genet.* **2015**, *16*, 261–275. [CrossRef]
16. Tillotson, R.; Selfridge, J.; Koerner, M.V.; Gadalla, K.K.E.; Guy, J.; De Sousa, D.; Hector, R.D.; Cobb, S.R.; Bird, A. Radically truncated MeCP2 rescues Rett syndrome-like neurological defects. *Nature* **2017**, *550*, 398–401. [CrossRef]

17. Ghosh, R.P.; Nikitina, T.; Horowitz-Scherer, R.A.; Gierasch, L.M.; Uversky, V.N.; Hite, K.; Hansen, J.C.; Woodcock, C.L. Unique physical properties and interactions of the domains of methylated DNA binding protein 2. *Biochemistry* **2010**, *49*, 4395–4410. [CrossRef]

18. Toth-Petroczy, A.; Palmedo, P.; Ingraham, J.; Hopf, T.A.; Berger, B.; Sander, C.; Marks, D.S. Structured States of Disordered Proteins from Genomic Sequences. *Cell* **2016**, *167*, 158–170. [CrossRef]

19. Canning, P.; Park, K.; Goncalves, J.; Li, C.; Howard, C.J.; Sharpe, T.D.; Holt, L.J.; Pelletier, L.; Bullock, A.N.; Leroux, M.R. CDKL Family Kinases Have Evolved Distinct Structural Features and Ciliary Function. *Cell Rep.* **2018**, *22*, 885–894. [CrossRef]

20. Dunker, A.K.; Lawson, J.D.; Brown, C.J.; Williams, R.M.; Romero, P.; Oh, J.S.; Oldfield, C.J.; Campen, A.M.; Ratliff, C.M.; Hipps, K.W.; et al. Intrinsically disordered protein. *J. Mol. Graph. Model.* **2001**, *19*, 26–59. [CrossRef]

21. Uversky, V.N. Protein folding revisited. A polypeptide chain at the folding-misfolding-nonfolding cross-roads: Which way to go? *Cell Mol. Life Sci.* **2003**, *60*, 1852–1871. [CrossRef] [PubMed]

22. Dyson, H.J.; Wright, P.E. Equilibrium NMR studies of unfolded and partially folded proteins. *Nat. Struct. Biol.* **1998**, *5*, 499–503. [CrossRef] [PubMed]

23. Dunker, A.K.; Babu, M.M.; Barbar, E.; Blackledge, M.; Bondos, S.E.; Dosztanyi, Z.; Dyson, H.J.; Forman-Kay, J.; Fuxreiter, M.; Gsponer, J.; et al. What's in a name? Why these proteins are intrinsically disordered: Why these proteins are intrinsically disordered. *Intrinsically Disord. Proteins* **2013**, *1*, e24157. [CrossRef] [PubMed]

24. Tompa, P. Intrinsically unstructured proteins. *Trends Biochem. Sci.* **2002**, *27*, 527–533. [CrossRef]

25. Tompa, P. The interplay between structure and function in intrinsically unstructured proteins. *FEBS Lett.* **2005**, *579*, 3346–3354. [CrossRef] [PubMed]

26. Van der Lee, R.; Buljan, M.; Lang, B.; Weatheritt, R.J.; Daughdrill, G.W.; Dunker, A.K.; Fuxreiter, M.; Gough, J.; Gsponer, J.; Jones, D.T.; et al. Classification of intrinsically disordered regions and proteins. *Chem. Rev.* **2014**, *114*, 6589–6631. [CrossRef]

27. Uversky, V.N.; Oldfield, C.J.; Dunker, A.K. Intrinsically disordered proteins in human diseases: Introducing the D2 concept. *Annu. Rev. Biophys.* **2008**, *37*, 215–246. [CrossRef]

28. Diella, F.; Haslam, N.; Chica, C.; Budd, A.; Michael, S.; Brown, N.P.; Trave, G.; Gibson, T.J. Understanding eukaryotic linear motifs and their role in cell signaling and regulation. *Front. Biosci.* **2008**, *13*, 6580–6603. [CrossRef]

29. Galea, C.A.; Wang, Y.; Sivakolundu, S.G.; Kriwacki, R.W. Regulation of cell division by intrinsically unstructured proteins: Intrinsic flexibility, modularity, and signaling conduits. *Biochemistry* **2008**, *47*, 7598–7609. [CrossRef]

30. Mellios, N.; Woodson, J.; Garcia, R.I.; Crawford, B.; Sharma, J.; Sheridan, S.D.; Haggarty, S.J.; Sur, M. beta2-Adrenergic receptor agonist ameliorates phenotypes and corrects microRNA-mediated IGF1 deficits in a mouse model of Rett syndrome. *Proc. Natl. Acad. Sci. USA* **2014**, *111*, 9947–9952. [CrossRef]

31. Tropea, D.; Giacometti, E.; Wilson, N.R.; Beard, C.; McCurry, C.; Fu, D.D.; Flannery, R.; Jaenisch, R.; Sur, M. Partial reversal of Rett Syndrome-like symptoms in MeCP2 mutant mice. *Proc. Natl. Acad. Sci. USA* **2009**, *106*, 2029–2034. [CrossRef] [PubMed]

32. Carrette, L.L.G.; Wang, C.Y.; Wei, C.; Press, W.; Ma, W.; Kelleher, R.J., 3rd; Lee, J.T. A mixed modality approach towards Xi reactivation for Rett syndrome and other X-linked disorders. *Proc. Natl. Acad. Sci. USA* **2018**, *115*, E668–E675. [CrossRef] [PubMed]

33. Shah, R.R.; Bird, A.P. MeCP2 mutations: Progress towards understanding and treating Rett syndrome. *Genome Med.* **2017**, *9*, 17. [CrossRef] [PubMed]

34. Mauri, F.; McNamee, L.M.; Lunardi, A.; Chiacchiera, F.; Del Sal, G.; Brodsky, M.H.; Collavin, L. Modification of Drosophila p53 by SUMO modulates its transactivation and pro-apoptotic functions. *J. Biol. Chem.* **2008**, *283*, 20848–20856. [CrossRef]

35. Dyson, H.J.; Wright, P.E. Intrinsically unstructured proteins and their functions. *Nat. Rev. Mol. Cell. Biol.* **2005**, *6*, 197–208. [CrossRef]

36. Fahmi, M.; Ito, M. Evolutionary Approach of Intrinsically Disordered CIP/KIP Proteins. *Sci. Rep.* **2019**, *9*, 1575. [CrossRef]

37. Gsponer, J.; Babu, M.M. The rules of disorder or why disorder rules. *Prog. Biophys. Mol. Biol.* **2009**, *99*, 94–103. [CrossRef]

38. Van Roey, K.; Uyar, B.; Weatheritt, R.J.; Dinkel, H.; Seiler, M.; Budd, A.; Gibson, T.J.; Davey, N.E. Short linear motifs: Ubiquitous and functionally diverse protein interaction modules directing cell regulation. *Chem. Rev.* **2014**, *114*, 6733–6778. [CrossRef]

39. Ahrens, J.; Rahaman, J.; Siltberg-Liberles, J. Large-Scale Analyses of Site-Specific Evolutionary Rates across Eukaryote Proteomes Reveal Confounding Interactions between Intrinsic Disorder, Secondary Structure, and Functional Domains. *Genes* **2018**, *11*, 553. [CrossRef]

40. Wakefield, R.I.; Smith, B.O.; Nan, X.; Free, A.; Soteriou, A.; Uhrin, D.; Bird, A.P.; Barlow, P.N. The solution structure of the domain from MeCP2 that binds to methylated DNA. *J. Mol. Biol.* **1999**, *291*, 1055–1065. [CrossRef]

41. Chen, J.W.; Romero, P.; Uversky, V.N.; Dunker, A.K. Conservation of intrinsic disorder in protein domains and families: II. functions of conserved disorder. *J. Proteome Res.* **2006**, *5*, 888–898. [CrossRef] [PubMed]

42. Grimmler, M.; Wang, Y.; Mund, T.; Cilensek, Z.; Keidel, E.M.; Waddell, M.B.; Jakel, H.; Kullmann, M.; Kriwacki, R.W.; Hengst, L. Cdk-inhibitory activity and stability of p27Kip1 are directly regulated by oncogenic tyrosine kinases. *Cell* **2007**, *128*, 269–280. [CrossRef] [PubMed]

43. Gsponer, J.; Futschik, M.E.; Teichmann, S.A.; Babu, M.M. Tight regulation of unstructured proteins: From transcript synthesis to protein degradation. *Science* **2008**, *322*, 1365–1368. [CrossRef] [PubMed]

44. UniProt, C. The Universal Protein Resource (UniProt) in 2010. *Nucleic Acids Res.* **2010**, *38*, D142–D148. [CrossRef]

45. Chatr-Aryamontri, A.; Oughtred, R.; Boucher, L.; Rust, J.; Chang, C.; Kolas, N.K.; O'Donnell, L.; Oster, S.; Theesfeld, C.; Sellam, A.; et al. The BioGRID interaction database: 2017 update. *Nucleic Acids Res.* **2017**, *45*, D369–D379. [CrossRef]

46. Uhlen, M.; Fagerberg, L.; Hallstrom, B.M.; Lindskog, C.; Oksvold, P.; Mardinoglu, A.; Sivertsson, A.; Kampf, C.; Sjostedt, E.; Asplund, A.; et al. Proteomics. Tissue-based map of the human proteome. *Science* **2015**, *347*, 1260419. [CrossRef]

47. Guy, J.; Alexander-Howden, B.; FitzPatrick, L.; DeSousa, D.; Koerner, M.V.; Selfridge, J.; Bird, A. A mutation-led search for novel functional domains in MeCP2. *Hum. Mol. Genet.* **2018**, *27*, 2531–2545. [CrossRef]

48. Ballestar, E.; Yusufzai, T.M.; Wolffe, A.P. Effects of Rett syndrome mutations of the methyl-CpG binding domain of the transcriptional repressor MeCP2 on selectivity for association with methylated DNA. *Biochemistry* **2000**, *39*, 7100–7106. [CrossRef]

49. Yusufzai, T.M.; Wolffe, A.P. Functional consequences of Rett syndrome mutations on human MeCP2. *Nucleic Acids Res.* **2000**, *28*, 4172–4179. [CrossRef]

50. Mnatzakanian, G.N.; Lohi, H.; Munteanu, I.; Alfred, S.E.; Yamada, T.; MacLeod, P.J.; Jones, J.R.; Scherer, S.W.; Schanen, N.C.; Friez, M.J.; et al. A previously unidentified MECP2 open reading frame defines a new protein isoform relevant to Rett syndrome. *Nat. Genet.* **2004**, *36*, 339. [CrossRef]

51. Williamson, S.L.; Giudici, L.; Kilstrup-Nielsen, C.; Gold, W.; Pelka, G.J.; Tam, P.P.; Grimm, A.; Prodi, D.; Landsberger, N.; Christodoulou, J. A novel transcript of cyclin-dependent kinase-like 5 (CDKL5) has an alternative C-terminus and is the predominant transcript in brain. *Hum. Genet.* **2012**, *131*, 187–200. [CrossRef] [PubMed]

52. Wong, L.C.; Singh, S.; Wang, H.P.; Hsu, C.J.; Hu, S.C.; Lee, W.T. FOXG1-Related Syndrome: From Clinical to Molecular Genetics and Pathogenic Mechanisms. *Int. J. Mol. Sci.* **2019**, *20*, 4176. [CrossRef] [PubMed]

53. Payankaulam, S.; Li, L.M.; Arnosti, D.N. Transcriptional repression: Conserved and evolved features. *Curr. Biol.* **2010**, *17*, R764–R771. [CrossRef] [PubMed]

54. Toresson, H.; Martinez-Barbera, J.P.; Bardsley, A.; Caubit, X.; Krauss, S. Conservation of BF-1 expression in amphioxus and zebrafish suggests evolutionary ancestry of anterior cell types that contribute to the vertebrate telencephalon. *Dev. Genes Evol.* **1998**, *208*, 431–439. [CrossRef]

55. Miyoshi, G.; Fishell, G. Dynamic FoxG1 expression coordinates the integration of multipolar pyramidal neuron precursors into the cortical plate. *Neuron* **2012**, *74*, 1045–1058. [CrossRef]

56. Kumamoto, T.; Toma, K.; Gunadi; McKenna, W.L.; Kasukawa, T.; Katzman, S.; Chen, B.; Hanashima, C. Foxg1 coordinates the switch from nonradially to radially migrating glutamatergic subtypes in the neocortex through spatiotemporal repression. *Cell Rep.* **2013**, *3*, 931–945. [CrossRef]

57. Docker, D.; Schubach, M.; Menzel, M.; Munz, M.; Spaich, C.; Biskup, S.; Bartholdi, D. Further delineation of the SATB2 phenotype. *Eur. J. Hum. Genet.* **2014**, *22*, 1034–1039. [CrossRef]

58. Lee, J.S.; Yoo, Y.; Lim, B.C.; Kim, K.J.; Choi, M.; Chae, J.H. SATB2-associated syndrome presenting with Rett-like phenotypes. *Clin. Genet.* **2016**, *89*, 728–732. [CrossRef]

59. Livide, G.; Patriarchi, T.; Amenduni, M.; Amabile, S.; Yasui, D.; Calcagno, E.; Lo Rizzo, C.; De Falco, G.; Ulivieri, C.; Ariani, F.; et al. GluD1 is a common altered player in neuronal differentiation from both MECP2-mutated and CDKL5-mutated iPS cells. *Eur. J. Hum. Genet.* **2015**, *23*, 195–201. [CrossRef]

60. Patriarchi, T.; Amabile, S.; Frullanti, E.; Landucci, E.; Rizzo, C.L.; Ariani, F.; Costa, M.; Olimpico, F.; Hell, J.W.; Vaccarino, F.M.; et al. Imbalance of excitatory/inhibitory synaptic protein expression in iPSC-derived neurons from FOXG1(+/−) patients and in foxg1(+/−) mice. *Eur. J. Hum. Genet.* **2016**, *24*, 871–880. [CrossRef]

61. Bracaglia, G.; Conca, B.; Bergo, A.; Rusconi, L.; Zhou, Z.; Greenberg, M.E.; Landsberger, N.; Soddu, S.; Kilstrup-Nielsen, C. Methyl-CpG-binding protein 2 is phosphorylated by homeodomain-interacting protein kinase 2 and contributes to apoptosis. *EMBO Rep.* **2009**, *10*, 1327–1333. [CrossRef] [PubMed]

62. Fuchs, C.; Trazzi, S.; Torricella, R.; Viggiano, R.; De Franceschi, M.; Amendola, E.; Gross, C.; Calza, L.; Bartesaghi, R.; Ciani, E. Loss of CDKL5 impairs survival and dendritic growth of newborn neurons by altering AKT/GSK-3beta signaling. *Neurobiol. Dis.* **2014**, *70*, 53–68. [CrossRef] [PubMed]

63. Mari, F.; Azimonti, S.; Bertani, I.; Bolognese, F.; Colombo, E.; Caselli, R.; Scala, E.; Longo, I.; Grosso, S.; Pescucci, C.; et al. CDKL5 belongs to the same molecular pathway of MeCP2 and it is responsible for the early-onset seizure variant of Rett syndrome. *Hum. Mol. Genet.* **2005**, *14*, 1935–1946. [CrossRef] [PubMed]

64. Carouge, D.; Host, L.; Aunis, D.; Zwiller, J.; Anglard, P. CDKL5 is a brain MeCP2 target gene regulated by DNA methylation. *Neurobiol. Dis.* **2010**, *38*, 414–424. [CrossRef] [PubMed]

65. Rusconi, L.; Salvatoni, L.; Giudici, L.; Bertani, I.; Kilstrup-Nielsen, C.; Broccoli, V.; Landsberger, N. CDKL5 expression is modulated during neuronal development and its subcellular distribution is tightly regulated by the C-terminal tail. *J. Biol. Chem.* **2008**, *283*, 30101–30111. [CrossRef] [PubMed]

66. Brown, C.J.; Johnson, A.K.; Dunker, A.K.; Daughdrill, G.W. Evolution and disorder. *Curr. Opin. Struct. Biol.* **2011**, *21*, 441–446. [CrossRef]

67. Regad, T.; Roth, M.; Bredenkamp, N.; Illing, N.; Papalopulu, N. The neural progenitor-specifying activity of FoxG1 is antagonistically regulated by CKI and FGF. *Nat. Cell Biol.* **2007**, *9*, 531–540. [CrossRef]

68. Zhang, Q.; Yang, X.; Wang, J.; Li, J.; Wu, Q.; Wen, Y.; Zhao, Y.; Zhang, X.; Yao, H.; Wu, X.; et al. Genomic mosaicism in the pathogenesis and inheritance of a Rett syndrome cohort. *Genet. Med.* **2019**, *21*, 1330–1338. [CrossRef]

69. Sato, Y.; Nakaya, A.; Shiraishi, K.; Kawashima, S.; Goto, S.; Kanehisa, M. Ssdb: Sequence similarity database in kegg. *Genome Inf.* **2001**, *12*, 230–231. [CrossRef]

70. Katoh, K.; Standley, D.M. MAFFT multiple sequence alignment software version 7: Improvements in performance and usability. *Mol. Biol. Evol.* **2013**, *30*, 772–780. [CrossRef]

71. Stamatakis, A. RAxML-VI-HPC: Maximum likelihood-based phylogenetic analyses with thousands of taxa and mixed models. *Bioinformatics* **2006**, *22*, 2688–2690. [CrossRef] [PubMed]

72. Darriba, D.; Posada, D.; Kozlov, A.M.; Stamatakis, A.; Morel, B.; Flouri, T. ModelTest-NG: A new and scalable tool for the selection of DNA and protein evolutionary models. *Mol. Biol. Evol.* **2019**. [CrossRef] [PubMed]

73. Federhen, S. The NCBI Taxonomy database. *Nucleic Acids Res.* **2012**, *40*, D136–D143. [CrossRef] [PubMed]

74. Miller, M.A.; Pfeiffer, W.; Schwartz, T. Creating the CIPRES Science Gateway for inference of large phylogenetic trees. In Proceedings of the 2010 Gateway Computing Environments Workshop (GCE), New Orleans, LA, USA, 14 November 2010; pp. 1–8.

75. Meszaros, B.; Erdos, G.; Dosztanyi, Z. IUPred2A: Context-dependent prediction of protein disorder as a function of redox state and protein binding. *Nucleic Acids Res.* **2018**, *46*, W329–W337. [CrossRef]

76. Letunic, I.; Bork, P. Interactive Tree of Life (iTOL): An online tool for phylogenetic tree display and annotation. *Bioinformatics* **2007**, *23*, 127–128. [CrossRef]

77. Pupko, T.; Bell, R.E.; Mayrose, I.; Glaser, F.; Ben-Tal, N. Rate4Site: An algorithmic tool for the identification of functional regions in proteins by surface mapping of evolutionary determinants within their homologues. *Bioinformatics* **2002**, *18* (Suppl. 1), S71–S77. [CrossRef]

78. Blom, N.; Gammeltoft, S.; Brunak, S. Sequence and structure-based prediction of eukaryotic protein phosphorylation sites. *J. Mol. Biol.* **1999**, *294*, 1351–1362. [CrossRef]

79. Krishnaraj, R.; Ho, G.; Christodoulou, J. RettBASE: Rett syndrome database update. *Hum. Mutat.* **2017**, *38*, 922–931. [CrossRef]

80. Lek, M.; Karczewski, K.J.; Minikel, E.V.; Samocha, K.E.; Banks, E.; Fennell, T.; O'Donnell-Luria, A.H.; Ware, J.S.; Hill, A.J.; Cummings, B.B.; et al. Analysis of protein-coding genetic variation in 60,706 humans. *Nature* **2016**, *536*, 285–291. [CrossRef]

81. Kanehisa, M.; Furumichi, M.; Tanabe, M.; Sato, Y.; Morishima, K. KEGG: New perspectives on genomes, pathways, diseases and drugs. *Nucleic Acids Res.* **2017**, *45*, D353–D361. [CrossRef]

82. Pellegrini, M.; Marcotte, E.M.; Thompson, M.J.; Eisenberg, D.; Yeates, T.O. Assigning protein functions by comparative genome analysis: Protein phylogenetic profiles. *Proc. Natl. Acad. Sci. USA* **1999**, *96*, 4285–4288. [CrossRef] [PubMed]

83. Nakaya, A.; Katayama, T.; Itoh, M.; Hiranuka, K.; Kawashima, S.; Moriya, Y.; Okuda, S.; Tanaka, M.; Tokimatsu, T.; Yamanishi, Y.; et al. KEGG OC: A large-scale automatic construction of taxonomy-based ortholog clusters. *Nucleic Acids Res.* **2013**, *41*, D353–D357. [CrossRef] [PubMed]

84. Ward, J.H. Hierarchical Grouping to Optimize an Objective Function. *J. Am. Stat. Assoc.* **1963**, *58*, 236–244. [CrossRef]

85. Mi, H.; Muruganujan, A.; Ebert, D.; Huang, X.; Thomas, P.D. PANTHER version 14: More genomes, a new PANTHER GO-slim and improvements in enrichment analysis tools. *Nucleic Acids Res.* **2019**, *47*, D419–D426. [CrossRef] [PubMed]

© 2019 by the authors. Licensee MDPI, Basel, Switzerland. This article is an open access article distributed under the terms and conditions of the Creative Commons Attribution (CC BY) license (http://creativecommons.org/licenses/by/4.0/).

Review

Structural Determinants of the Prion Protein N-Terminus and Its Adducts with Copper Ions

Carolina Sánchez-López [1,†], Giulia Rossetti [2,3,4], Liliana Quintanar [1,*] and Paolo Carloni [2,5,6,*]

[1] Department of Chemistry, Center for Research and Advanced Studies (Cinvestav), 07360 Mexico City, Mexico; magdacarolina29@hotmail.com

[2] Institute of Neuroscience and Medicine (INM-9) and Institute for Advanced Simulation (IAS-5), Forschungszentrum Jülich, Wilhelm-Johnen-Strasse, 52425 Jülich, Germany; g.rossetti@fz-juelich.de

[3] Jülich Supercomputing Center (JSC), Forschungszentrum Jülich, 52428 Jülich, Germany

[4] Department of Oncology, Hematology and Stem Cell Transplantation, Faculty of Medicine, RWTH Aachen University, Pauwelsstraße 30, 52074 Aachen, Germany

[5] Department of Physics and Department of Neurobiology, RWTH Aachen University, 52078 Aachen, Germany

[6] Institute for Neuroscience and Medicine (INM)-11, Forschungszentrum Jülich, 52428 Jülich, Germany

* Correspondence: lilianaq@cinvestav.mx (L.Q.); p.carloni@fz-juelich.de (P.C.)

† Current Address: Instituto de Biología Molecular y Celular de Rosario (IBR-CONICET), Ocampo y Esmeralda, 2000 Rosario, Argentina.

Received: 3 December 2018; Accepted: 18 December 2018; Published: 20 December 2018

Abstract: The N-terminus of the prion protein is a large intrinsically disordered region encompassing approximately 125 amino acids. In this paper, we review its structural and functional properties, with a particular emphasis on its binding to copper ions. The latter is exploited by the region's conformational flexibility to yield a variety of biological functions. Disease-linked mutations and proteolytic processing of the protein can impact its copper-binding properties, with important structural and functional implications, both in health and disease progression.

Keywords: N-terminal prion protein; copper binding; prion disease mutations

---

1. Introduction

Prion diseases or transmissible spongiform encephalopathies (TSEs) are rare neurodegenerative diseases exhibiting symptoms of both cognitive and motor dysfunction, vacuolation of the grey matter in the human central nervous system, neuronal loss, and astrogliosis [1]. A crucial event for the diseases' development is the misfolding of the extracellular, membrane-anchored human prion protein (HuPrP^C) into the fibril-forming isoform called "scrapie" (HuPrP^{Sc}), the major or only component of the infectious particle [2]. This eventually leads to protofibril and fibrillar structures. Accordingly, with the "Protein only hypothesis" by Nobel Laureate Prusiner [3], the feature to undergo induced or spontaneous misfolding depends basically on intrinsic features of the protein. These include the amino acid sequence [4,5] as well as secondary structure elements [6–8], the highly flexible amino terminal region of the protein [9], and posttranslational modification elements [10]. The propensity to form the scrapie form is modulated by a variety of external factors. These include pH [11–13], cofactors like metal ions [14,15], or the presence of proteins [16,17]. Pathogenic mutations (PM) in HuPrP^C are linked to the spontaneous generation of prion diseases [18–21].

HuPrP^C is ubiquitously expressed throughout the body. It is mostly found in the central nervous system. After being synthesized in the rough endoplasmic reticulum, it transits through the Golgi compartment, and it is released to the cell surface where it resides in lipid membrane domains [22]. Though its physiological role is still not clear, HuPrP^C might be involved in neuronal development, cell adhesion, apoptotic events, and cell signaling in the central nervous system. Moreover, HuPrP^C can interact with different neuronal proteins or proteins of the extracellular matrix, as well as with other binders including glycosaminoglycans, nucleic acids, and copper ions [23]. Hence, HuPrP^C has been also proposed as a copper sensing or transport protein [24].

The protein features two signal peptides (1–22 and 232–235, Figure 1), a folded globular domain (GD, residues 125–231), and a naturally unfolded N-terminal tail (N-term_HuPrP^C, hereafter, residues 23–124), which is the focus of this review. The GD consists of two β-sheets (S1 and S2), three α-helices (H1, H2, and H3), one disulfide bond (SS) between cysteine residues 179 and 214, and two potential sites for N-linked glycosylation (green forks in Figure 1) at residues 181 and 197 [25]. H2 and H3 helices linked by the SS-bond constitute the H2 + H3 domain. A glycosylphosphatidylinositol anchor (GPI, in blue in Figure 1) is attached to the C-terminus, which is located on the outside cellular membrane.

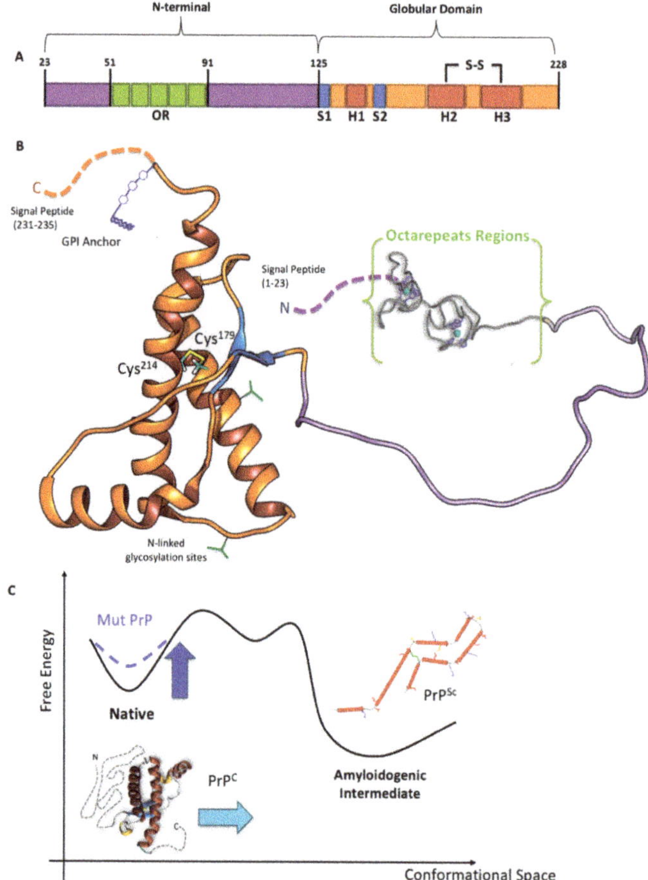

Figure 1. (**A**) Schematic and (**B**) tridimensional view of HuPrP^C. (**C**) Qualitative scheme illustrating the Gibbs free energy change in the conversion from HuPrP^C (left) to HuPrP^{Sc} (right) [26]. The depicted amyloidogenic intermediate is the parallel, in-register β-structure model for the core of recombinant PrP90–231 amyloid fibrils formed in vitro [27], one of the models among others [28–30], whereas the native globular domain (GD) of the HuPrP^C is the nuclear magnetic resonance (NMR) structure by Zahn et al. [25]. Adapted from [31,32].

The HuPrP^C→HuPrP^{Sc} interconversion involves mostly the GD. It may entail increasingly β-stranded intermediate structures [33] (Figure 1C), leading to small aggregates, protofibrils, and finally ordered rigid fibrils [34–38]. Experimental structural information for these is lacking [34–38].

While the structure of the GD of HuPrP^C has been resolved experimentally, the intrinsically disordered nature of the N-term_HuPrP^C has represented a challenge for structural studies. In this paper, the structural properties of the N-term_HuPrP^C are discussed, with a focus on recent insights obtained from computational approaches and on the functional and disease-related implications of copper–N-term_HuPrP^C interactions.

2. The N-Term: Function and Structural Determinants

This naturally unfolded domain contains the major part of the so-called transmembrane domain (termed TM1, comprising roughly residues 112–135) and the preceding "stop transfer effector" (STE, a hydrophilic region containing roughly residues 104–111) [39,40] (Figure 1B). STE and TM1 act in concert to control the co-translational translocation at the endoplasmic reticulum (ER) during the biosynthesis of the protein [41,42].

N-term_HuPrP^C functions as a broad-spectrum molecular sensor [43]. Along with the highly homologous protein from mouse (N-term_MoPrP^C, 93% sequence identity), it interacts with copper ions (see below) and sulphated glycosaminoglycans [44]. In addition, N-term_MoPrP^C interacts with vitronectin [45], the stress-inducible protein 1 (STI1) [46], amyloid-β (Aβ) multimers [47–49], lipoprotein receptor-related protein 1 (LRP1) [50], and the neural cell adhesion molecule (NCAM) [51].

Because experimental structural information on the full-length N-term_HuPrP^C is currently lacking, one has to resort to biocomputing-based predictions. Recently, some of us have used a combination of bioinformatics along with replica-exchange-based Monte Carlo simulation at room temperature, based on a simplified force field, to predict the conformational ensemble on the full-length N-term_MoPrP^C [31,52].

This is expected to be quite similar to that from *Homo sapiens*, given the extremely high sequence identity (93%) with N-term_HuPrP^C [31,52]. Monte Carlo simulations suggest that the N-term_MoPrP^C consists of several regions characterized by different secondary structure elements, consistently with biophysical data [53–57]. Specifically, it contains $19 \pm 8\%$ α-helix, $8 \pm 5\%$ β-sheet, $7 \pm 3\%$ β-bridge, $27 \pm 5\%$ β-turn, $12 \pm 4\%$ bend, $4 \pm 3\%$ 3_{10}-helix, and $1 \pm 1\%$ π-helix. The secondary structure elements are distributed among the N-term in a highly heterogeneous manner (Figure 2A): residues 23-30 are mainly coil/β-turn/bend; residues 31-50 are mainly β-turn/coil/bend/β-bridge; and residues 59-90 form four sequential octarepeat (OR) peptides, with sequence PHGGGWGQ, and are mainly β-turn/coil/bend/β-sheet conformations. In particular, the loop/β-turn conformations in the OR region resemble (backbone RMSD < 2.5 Å) those identified by NMR [57]; residues 89-98 are mainly coil/β-turn/bend/β-sheet; residues 99-117 feature the highest content α-helix of N-term_MoPrP^C regions; and residues 118-125 display a comparable percentage of α-helix and β-turn. Residues 105-125, the "amyloidogenic region", feature transient helical structures (the last eight residues also have a comparable content of beta turn). This is consistent with circular dichroism (CD), nuclear magnetic resonance (NMR), and Fourier transform infrared (FTIR) studies on HuPrP^C fragments [54–56] (Figure 2). The same simulation procedure can be carried out for the known disease-linked mutations (Figure 2).

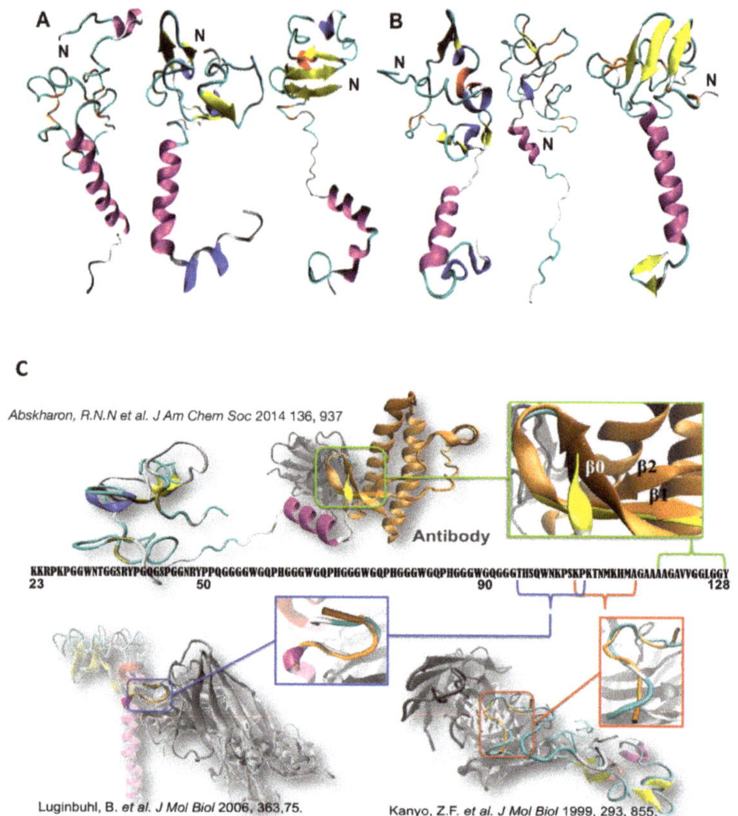

Figure 2. Selected conformations of (**A**) WT N-term_MoPrP^C and (**B**) one PM (N-term_MoPrP^C_Q52P) emerging from molecular simulation [31,52]. These contain transient α-helix (in violet), β-sheet (yellow), β-bridge (orange), β-turn (cyan), 3_{10}-helix (blue), and p-helix (red) elements. (**C**) Superimposition of our conformational ensemble (orange) with available fragments of N-term deposited structures. Readapted from [31,52].

While many PMs in the GD are known to modify significantly the folded structure and to increase its flexibility [58–61], our Monte Carlo calculations suggest that those in the N-term do not impact significantly the global structural properties of the N-term. This finding is consistent with experimental findings showing that PMs in N-term_HuPrP^C do not affect the thermostability or misfolding kinetics of the protein [58,62–64]. On the contrary, our Monte Carlo simulations show that the PMs at the N-term modify local features at the binding sites for known cellular partners, as well as of interdomain interactions. This points to an interference of the PMs with the related physiological functions.

The major differences in the presence of the PMs were observed in the residues binding Cu^{2+} and sulphated GAG (i.e., the OR region and the H110 Cu^{2+}-binding site mouse sequence, H111 in the human sequence). In addition, the PMs affect the SS and the flexibility and increase the hydrophobicity of STE/TM1. The latter contains the putative binding sites for in vivo binding partner proteins such as vitronectin [45] and STI1 [46]. This might affect the biological function of these interactions, which involves the signaling for axonal growth [45] and that for neuroprotection [46], respectively.

The PM Q52P in the OR region, interestingly, affects the flexibility of STE/TM1, while the other six PMs in STE/TM1 also alter the intra-molecular contacts in the OR region suggesting a role played by PMs in altering transient interdomain interactions between the OR region and STE/TM1. Recent studies suggest that N-Term and GD interactions might also serve to regulate the activity and/or toxicity of the PrP^C N-term [65]. Unfortunately, in the reported Monte Carlo study [52], the GD was not taken into account.

The altered local features in STE/TM1 might also impact the interactions of the protein with trans-acting factors in the cytosol and in the ER membrane [66]. This result is consistent with the in vitro data that PMs P101L, P104L, and A116V increase the interactions between MoPrP^C STE/TM1 and a membrane mimetic at pH 7 [67].

3. Copper Binding

Copper ions bind to the N-term of HuPrP^C in vivo [24]. Since the protein is anchored to the neuronal membrane, facing the extracellular space, it is exposed to fluctuations in Cu^{2+} ion concentrations, that can reach 100 μM during synaptic transmission [68]. This represents orders of magnitude larger than the experimentally measured range of binding affinities for Cu(II) ions at the N-term (nM to μM) [69]. Thus, it is plausible that the protein responds to Cu(II) ion concentration changes at the synapse. On the other hand, upon endocytosis, HuPrP^C is exposed to the intracellular reducing environment, where the interaction of its N-term with Cu^+ ions would also be relevant. Two main functions of copper binding to the N-term have been proposed so far: (i) stimulation of HuPrP^C endocytosis [70–73] and (ii) copper sensing associated to cell signaling. Copper-induced endocytosis of HuPrP^C requires its N-term terminus, specifically the octarepeat region, and it might involve conformational alterations of the N-term with the subsequent delivery of copper ions to endosomes. This has led to a proposed role for HuPrP^C in copper transport. However, it is unlikely that HuPrP^C delivers copper efficiently into the cytosol, since high concentrations are needed for copper-induced endocytosis (150–300 μM) [71,73]. On the other hand, HuPrP^C can interact with the human N-methyl-D-aspartate receptors (HuNMDAR) and alpha-amino-3-hydroxy-5-methyl-4-isoxazolepropionic acid receptors (HuAMPAR) involved in synaptic transmission, while both receptors are regulated by Cu ions [74–76]. In the case of HuNMDAR activity, copper binding to HuPrP^C is necessary to regulate the activity of this receptor [75]. Indeed, HuPrP^C and Cu^{2+} are required to inhibit HuNMDAR activity through a mechanism that involves post-translational S-nitrosylation of cysteine residues in HuNMDAR [77]. Overall, these important findings underscore a role for Cu-HuPrP^C interactions in neuroprotective mechanisms, which could be disrupted by other Cu-binding proteins or peptides at the synapse. For instance, human amyloid-β (Aβ) neurotoxicity has recently been linked to its ability to compete for Cu^{2+} ions with HuPrP^C, thereby interfering with the modulation of HuNMDAR activity [75].

The N-term region of HuPrP^C contains six His residues that may serve as anchoring sites for Cu^{2+} ions [78]. The ion binds to different sites within the protein [79–82], which are conserved in mammalian species [83], a fact that underscores its physiological relevance. Four of them are located in the OR region, spanning residues 60-91 with four repeats of the highly conserved octapeptide PHGGGWGQ (Figure 3). Beyond the OR region, two additional His residues, 96 and 111, also act as copper-binding sites in the 92-115 region. Studies on synthetic peptide fragments have suggested that metal coordination modes depend on copper concentration [69], as well as the relative copper:protein ratio and proton concentration [79,80]. At physiological pH, three distinct Cu^{2+} coordination modes have been identified by electron paramagnetic resonance (EPR) [84]. At low Cu:protein ratios, three or four His residues can chelate one metal ion, leading to a multiple histidine Cu-binding mode, named Component 3 (Figure 3). At higher Cu:protein ratios, a species with two His ligands forming a 2N2O equatorial coordination mode is observed (Component 2 in Figure 3). When enough Cu^{2+} is provided to reach a 1:1 ratio for each octapeptide fragment, a species with a 3N1O equatorial coordination mode is formed, named Component 1, where the coordinating residues are as follows: one His imidazole ligand, two deprotonated backbone amide groups, and a carbonyl group from the glycine residues that follow the anchoring His in the sequence (Figure 3) [78,85]. X-ray crystallographic studies of the Cu^{2+} complex with one octapeptide PHGGGWGQ fragment also revealed the participation of a water molecule as an axial ligand, stabilized by hydrogen bonding to the Trp residue [86]. This is the only Cu-binding site fragment that has been characterized so far by X-ray crystallography.

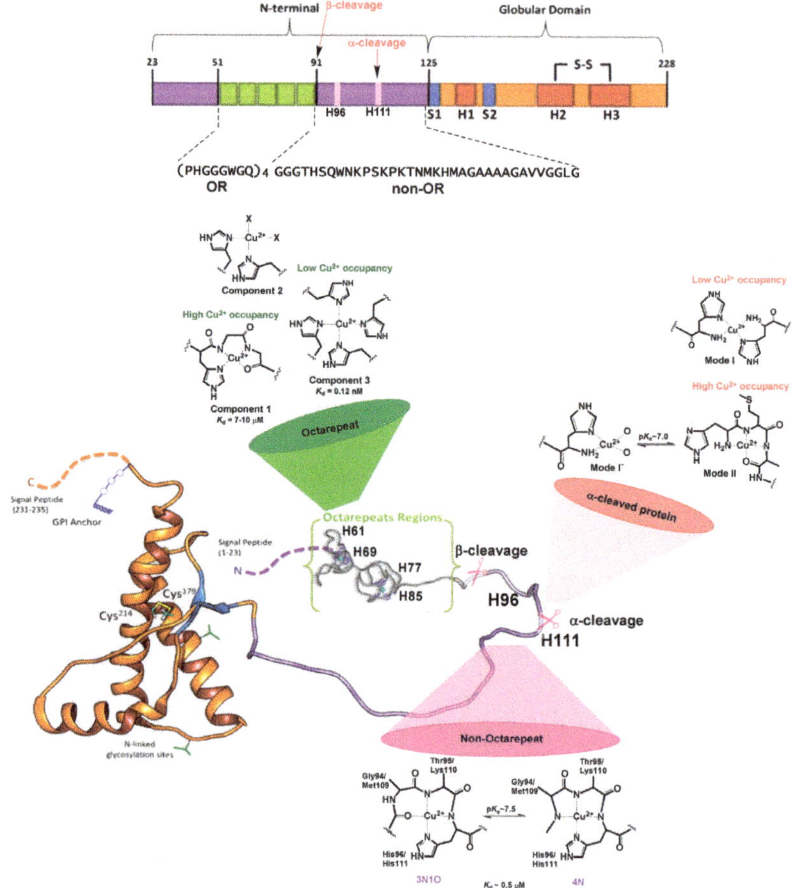

Figure 3. Cu coordination properties of the N-terminal region of human HuPrP^C. The six His residues that act as anchoring sites for Cu ions are highlighted: His61, His69, His77, and His85 in the OR region, and His96 and His111 in the non-OR region. The models for the different Cu^{2+} coordination modes identified at each site at physiological pH are drawn. The impact of α-cleavage processing on the His111 binding site is also shown.

The K_d for the low occupancy multiple-His coordination mode (Component 3) is 0.12 nM, whereas it ranges from 7 to 10 µM for the high-occupancy 3N1O mode (Component 1) [69]. Hence, Cu binding to the OR region displays a negative cooperativity. This is consistent with the formation of intermediate species such as Component 2. Electrochemical studies have determined that the high-occupancy Component 1 species is capable of reducing dioxygen to produce low levels of hydrogen peroxide that may be relevant for cell signaling, whereas the low-occupancy multiple-His mode cannot activate dioxygen at all [87,88].

Outside the OR region, His 96 and 111 act as anchoring sites for Cu^{2+} ions, constituting the closest Cu-binding sites to the amyloidogenic region of HuPrP^C [81,89,90]. An X-ray absorption spectroscopy (XAS) study of a HuPrP (90–231) construct of the protein showed that at low pH Cu ions can coordinate to both His residues in the non-OR region, while at physiological pH only the His111 site is populated [91]. The peptide fragments mostly used to characterize the individual non-OR Cu-binding sites are HuPrP (92-96) and HuPrP (106-115) with sequences GGGTH and KTNMKHMAGA, respectively [85,92]. EPR and other spectroscopic studies have determined that Cu^{2+} coordination is highly pH-dependent in these sites, yielding two different equatorial coordination modes at physiological pH, 3N1O and 4N, related by a pKa value near 7.5 (7.8 for the His 96 site and 7.5 for the His 111 site) [93–96]. In both sites, Cu^{2+} coordination in the 3N1O mode involves the His imidazole group, two deprotonated amide nitrogens, and a carbonyl group from the backbone amide bonds that precede the His residue in the sequence, while a third deprotonated amide group replaces the carbonyl moiety in the 4N equatorial mode (Figure 3). Although the equatorial coordination modes of Cu^{2+} bound to these sites are identical, the presence of two Met residues in the His 111 site provides it with distinct properties, particularly in terms of relative binding affinity, and redox properties. The thioether groups of Met 109 and Met 112 can participate in Cu^{1+} coordination to the PrP(106-115) fragment, as demonstrated by XAS studies, yielding coordination modes where the His111 imidazole ring, the two Met residues, and a backbone carbonyl group stabilize a tetra-coordinated Cu^{1+} species at physiological pH, where Met109 plays a more important role in metal coordination, as compared to Met112 [97]. Anchoring of Cu^{1+} ions by Met residues persists even at low pH values (<5), as those found in endosomes. Thus, the MKHM motif and Cu coordination features of the His111 site assure that the metal ion would still be bound to the protein, even under decreased pH and reducing conditions, such as those encountered upon endocytosis. Additionally, the capability of the His111 site to stabilize both Cu^{2+} and Cu^{1+} makes it a unique site in the N-term region of $HuPrP^C$ that may support redox activity to activate dioxygen.

The relative binding affinity of Cu^{2+} for the non-OR sites has also been studied [98–102]. While a slight preference for Cu^{2+} binding to His111 over His96 has been observed spectroscopically and ascribed to the Met residues nearby, the two sites get loaded simultaneously with K_d values in the range of 0.4 to 0.7 μM at physiological pH [102]. Given the Cu-binding affinity features of the OR region, this implies that upon increasing Cu^{2+} levels, the multiple His species (Component 3) would first form, followed by the population of the non-octarepeat His96/His111 sites, before the OR region is fully loaded to yield the high occupancy mode (Component 1). Overall, the Cu(II) coordination and binding features of the N-term region provide $HuPrP^C$ with the ability to respond to a wide range of Cu concentrations, adopting different metal coordination modes, which in turn may impose different conformations to this unstructured region of the protein.

The conformational flexibility of the N-term_$HuPrP^C$ and the presence of several His residues as Cu anchoring sites provide a platform to accommodate different Cu^{2+} coordination modes as a function of relative metal:protein concentrations. Unlike the static (entatic) Cu active sites of cuproenzymes, where the protein structure imposes restrictions on the metal coordination and geometry, the preferred coordination modes at each Cu-binding site in the flexible N-term domain of $HuPrP^C$ are dictated by the geometric and electronic preferences of the metal ion, which can actually impose metal-induced conformations with potentially different functional implications or a propensity to aggregate. For example, the participation of deprotonated backbone amides in Cu^{2+} coordination, as in the high-occupancy (Component 1) mode of the OR region, inevitably imposes a certain turn in the backbone chain, which is known to yield more compact conformations [81]. Indeed, full loading of Cu^{2+} ions into the OR region yields a conformation where the average Cu–Cu distance is 4 to 7 Å, as determined by EPR [84]. Given the relatively high Cu concentrations needed for endocytosis of $HuPrP^C$ and the Cu-binding affinity features of the OR region, this high-occupancy conformation of the OR region is likely involved in the mechanism of endocytosis.

Recent studies have revealed metal-induced interactions between the N-term and C-term regions of $MoPrP^C$ [103–105]. An NMR study suggested interactions of the region 90-120, containing His96 and His111 Cu-binding sites, with residues in the vicinity of helix-1 (specifically 144-147), while the Cu-loaded OR region may also interact with helix-2, involving residues 174-185 [105]. The latter was confirmed by an elegant NMR and site-directed spin labeling EPR study that provided detailed structural information on how Cu binding at the low occupancy multiple His site (Component 3) in the OR region promotes electrostatic interactions with a highly conserved negatively charged region at helices 2 and 3 of the globular protein [103]. The C-term region of the protein engaged in the interaction with Cu-loaded OR overlaps with the region where neurotoxic PrP antibodies bind, and it also involves acidic residues associated with disease-related mutations, such as E200K. These observations underscore the important role of Cu^{2+} loading into the low-occupancy multiple-His coordination mode in promoting electrostatic interactions between the Cu-bound N-terminal region and the helical C-terminal domain, a stabilizing interaction that is considered to be regulatory for prion conversion [103,105].

On the other hand, in the non-OR region, the two His coordination modes identified at low pH for the PrP(90-231) construct were also found to induce stabilizing interactions with the globular C-terminal domain, whereas Cu^{2+} binding solely to the His111 site induced local beta-sheet structure [91]. Consistently, copper binding to the non-OR sites in the amyloidogenic fragment 90-126 induces a β-sheet-like transition [106]. These observations underscore the important role that Cu binding to the non-OR region may play in amyloid aggregation and prion conversion.

Cu^{2+}-PrP^C interactions and their perturbation by disease-related mutations have been suggested to play a role for Hu/Mo PrP^C aggregation and prion disease progression [107]. Specifically, the GSS-linked Q211P PM [60] (Q212P in Hu numbering) in the HuPrP^C GD can influence the Cu^{2+} binding coordination at H96 and H111 [108], implicating a role of abnormal Cu^{2+} binding in the pathology of PMs in HuPrP^C. As discussed above, the multiple His Cu-binding modes induce stabilizing interactions of the N-term_HuPrP^C with the globular C-terminal domain, whereas Cu^{2+} binding solely to the His111 site induces local beta-sheet structure [91,103,105]. Thus, any perturbation of the local conformation around the Cu-binding sites may have an impact on the stability of the protein. Consistently, structural analysis by molecular simulations of the N-term_HuPrP^C indicates that some disease-linked mutations may affect the local conformation and intramolecular interactions around the Cu^{2+} binding sites, including His111 [31,32], providing a molecular basis to understand their impact on disease progression. Although further studies are needed to understand how Cu binding impacts the folding and conformation of the flexible N-term_HuPrP^C, it is clear that the different Cu^{2+} coordination modes formed at the His anchoring sites can favor distinct metal-induced conformations, while disease-related mutations may also impact the conformation of the N-term region, its Cu-binding properties, and its interactions with the C-terminal globular region of HuPrP^C.

Copper binding may also be affected by a specific post-translational modification, namely the proteolytic processing at specific sites of the N-term region [109]. This includes the following: (i) the β-cleavage of the OR region, leading to the N2 (23-89) and C2 (90-231) fragments [109]. This is induced by reactive oxygen species (ROS) produced in the presence of Cu^{2+} ions [110,111] (It can be also catalyzed by calpain and ADAM8—a member of the A Disintegrin And Metalloproteinase (ADAM) family of enzymes [112,113]). While the N2 fragment may be released, maintaining the Cu^{2+} binding properties of the OR sites as described above, the C2 fragment may remain anchored to the membrane, conserving the His96 and His111 sites, but with a free N-term group at residue 90 [109]. The free NH_2 moiety is expected to change significantly the Cu^{2+} coordination features of these non-OR sites. (ii) The α-cleavage occurs at several sites in the region encompassing residues 109-120, and it is a common feature in a wide range of cell lines [114]. The process, performed by members of the ADAM enzymes [112], increases in the presence of Cu^{2+} ions [115]. This metal also may modulate the relative amount of α-cleavage at each site, possibly by inducing local conformational changes that impact how the protein docks into the protease active site [112,113]. The most described α-cleavage site is located between Lys110 and His111 in the human sequence, producing two fragments N1 (23-110) and C1 (111-231) [109]. The released N1 fragment may still contain the OR region and the His96 site; however, the His111 site may be significantly disturbed, as the cleavage occurs between residues that participate in Cu^{2+} coordination, leaving a His111 with a free NH_2 terminal group at the membrane bound C1 fragment. A recent spectroscopic study determined the impact of α-cleavage processing on Cu^{2+} binding to His111, using a model peptide for the C1 fragment [116]. Indeed, in this fragment His111 and the free NH_2 terminal group act as the main anchoring sites for Cu^{2+}, resulting in coordination modes that are highly dependent on proton and copper concentrations, and are quite different from those characterized for the intact His111 site in the full protein (Figure 3). The Cu-binding affinity features and redox activity of this perturbed His111 site remain to be investigated. It is interesting to note that, while Cu ions can modulate the relative amount of α-cleavage at different sites of the 109-120 region of $HuPrP^C$ [112], the resulting membrane-bound C1 fragments and their Cu-binding properties could in turn determine the metal-induced conformation of the N-term region and its ability to interact with important receptors, such as the HuNMDAR and HuAMPAR [113,117].

4. Conclusions

Recent advances in computational biophysics [31,32,52] have led, for the first time, to a description of the conformational ensemble on the full-length N-term $MoPrP^C$, a fully disordered domain of 125 amino acids, with high similarity to the human domain. This has made it possible to probe the impact of disease-related mutations on the structural properties of this flexible region of the protein.

N-term_$HuPrP^C$ binds copper ions in vivo [24]. It yields a diverse range of Cu coordination modes, each with distinct redox properties and binding affinity features. The Cu-binding properties of the N-term region provide $HuPrP^C$ with the ability to respond to the wide range of Cu concentrations that the protein is exposed to at the synapse, adopting different metal-induced conformations, which in turn may have distinct functional implications. On the other hand, Cu^{2+}-PrP^C interactions and their perturbation by disease-related mutations may play a role in protein aggregation and prion disease progression.

While the interplay between metal ion binding and conformational flexibility in the entire N-term remains to be understood, it is well established that copper displays site-specific effects on its folding, either by promoting stabilizing interactions or inducing conversion to beta-sheet folds. Conversely, molecular simulations suggest that some disease-related mutations may affect the local conformation around the Cu anchoring sites, thus affecting the Cu-binding properties of the N-term_$HuPrP^C$ and the stability of the protein.

Combined computational and experimental studies on the structural impact of Cu^{2+} binding and disease-related mutations at the N-term_$HuPrP^C$, such as those on copper(II)-alpha-synuclein—an intrinsically disordered protein undergoing fibril formation in Parkinson's disease [118]—could advance dramatically our understanding of the functional role of Cu^{2+}-PrP^C interactions in health and disease.

Author Contributions: Conceptualization, all the authors; Writing-Review & Editing, all the authors.

Funding: This research received no external funding.

Acknowledgments: C.S.L. acknowledges SECITI CDMX, Mexico for the postdoctoral fellowship.

Conflicts of Interest: The authors declare no conflict of interest.

References

1. Prusiner, S.B. Molecular biology and pathogenesis of prion diseases. *Trends Biochem. Sci.* **1996**, *21*, 482–487. [CrossRef]
2. Prusiner, S.B. Prions. *Proc. Natl. Acad. Sci. USA* **1998**, *95*, 13363–13383. [CrossRef] [PubMed]
3. Prusiner, S.B. Novel proteinaceous infectious particles cause scrapie. *Science* **1982**, *216*, 136–144. [CrossRef] [PubMed]
4. Gendoo, D.M.A.; Harrison, P.M. Discordant and chameleon sequences: Their distribution and implications for amyloidogenicity. *Protein Sci.* **2011**, *20*, 567–579. [CrossRef] [PubMed]

5. Kuznetsov, I.B.; Rackovsky, S. Comparative computational analysis of prion proteins reveals two fragments with unusual structural properties and a pattern of increase in hydrophobicity associated with disease-promoting mutations. *Protein Sci.* **2004**, *13*, 3230–3244. [CrossRef] [PubMed]

6. Dima, R.; Thirumalai, D. Exploring the propensities of helices in PrP^c to form beta sheet using NMR structures and sequence alignments. *Biophys. J.* **2002**, *83*, 1268–1280. [CrossRef]

7. Dima, R.I.; Thirumalai, D. Probing the instabilities in the dynamics of helical fragments from mouse PrP^C. *Proc. Natl. Acad. Sci. USA* **2004**, *101*, 15335–15340. [CrossRef] [PubMed]

8. Adrover, M.; Pauwels, K.; Prigent, S.; de Chiara, C.; Xu, Z.; Chapuis, C.; Pastore, A.; Rezaei, H. Prion Fibrillization Is Mediated by a Native Structural Element That Comprises Helices H2 and H3. *J. Biol. Chem.* **2010**, *285*, 21004–21012. [CrossRef]

9. Priola, S.A.; Meade-White, K.; Lawson, V.A.; Chesebro, B. Flexible N-terminal region of prion protein influences conformation of protease-resistant prion protein isoforms associated with cross-species scrapie infection in vivo and in vitro. *J. Biol. Chem.* **2004**, *279*, 13689–13695. [CrossRef]

10. Rudd, P.M.; Merry, A.H.; Wormald, M.R.; Dwek, R.A. Glycosylation and prion protein. *Curr. Opin. Struct. Biol.* **2002**, *12*, 578–586. [CrossRef]

11. Abid, K.; Soto, C. The intriguing prion disorders. *Cell. Mol. Life Sci.* **2006**, *63*, 2342–2351. [CrossRef] [PubMed]

12. Swietnicki, W.; Petersen, R.; Gambetti, P.; Surewicz, W.K. pH-dependent stability and conformation of the recombinant human prion protein PrP(90–231). *J. Biol. Chem.* **1997**, *272*, 27517–27520. [CrossRef] [PubMed]

13. Jackson, G.S.; Hosszu, L.L.; Power, A.; Hill, A.F.; Kenney, J.; Saibil, H.; Craven, C.J.; Waltho, J.P.; Clarke, A.R.; Collinge, J. Reversible conversion of monomeric human prion protein between native and fibrilogenic conformations. *Science* **1999**, *283*, 1935–1937. [CrossRef] [PubMed]

14. Choi, C.J.; Kanthasamy, A.; Anantharam, V.; Kanthasamy, A.G. Interaction of metals with prion protein: Possible role of divalent cations in the pathogenesis of prion diseases. *Neurotoxicology* **2006**, *27*, 777–787. [CrossRef] [PubMed]

15. Lehmann, S. Metal ions and prion diseases. *Curr. Opin. Chem. Biol.* **2002**, *6*, 187–192. [CrossRef]

16. Jin, T.; Gu, Y.; Zanusso, G.; Sy, M.; Kumar, A.; Cohen, M.; Gambetti, P.; Singh, N. The chaperone protein BiP binds to a mutant prion protein and mediates its degradation by the proteasome. *J. Biol. Chem.* **2000**, *275*, 38699–38704. [CrossRef]

17. Hachiya, N.S.; Imagawa, M.; Kaneko, K. The possible role of protein X, a putative auxiliary factor in pathological prion replication, in regulating a physiological endoproteolytic cleavage of cellular prion protein. *Med. Hypotheses* **2007**, *68*, 670–673. [CrossRef]

18. Dossena, S.; Imeri, L.; Mangieri, M.; Garofoli, A.; Ferrari, L.; Senatore, A.; Restelli, E.; Baiducci, C.; Fiordaliso, F.; Salio, M.; et al. Mutant Prion Protein Expression Causes Motor and Memory Deficits and Abnormal Sleep Patterns in a Transgenic Mouse Model. *Neuron* **2008**, *60*, 598–609. [CrossRef]

19. Antonyuk, S.V.; Trevitt, C.R.; Strange, R.W.; Jackson, G.S.; Sangar, D.; Batchelor, M.; Cooper, S.; Fraser, C.; Jones, S.; Georgiou, T.; et al. Crystal structure of human prion protein bound to a therapeutic antibody. *Proc. Natl. Acad. Sci. USA* **2009**, *106*, 2554–2558. [CrossRef]

20. Friedman-Levi, Y.; Meiner, Z.; Canello, T.; Frid, K.; Kovacs, G.G.; Budka, H.; Avrahami, D.; Gabizon, R. Fatal Prion Disease in a Mouse Model of Genetic E200K Creutzfeldt-Jakob Disease. *PLoS Pathog.* **2011**, *7*. [CrossRef]

21. Kovacs, G.G.; Trabattoni, G.; Hainfellner, J.A.; Ironside, J.W.; Knight, R.S.G.; Budka, H. Mutations of the prion protein gene phenotypic spectrum. *J. Neurol.* **2002**, *249*, 1567–1582. [CrossRef] [PubMed]

22. Campana, V.; Sarnataro, D.; Zurzolo, C. The highways and byways of prion protein trafficking. *Trends Cell Biol.* **2005**, *15*, 102–111. [CrossRef] [PubMed]

23. Linden, R.; Martins, V.R.; Prado, M.A.M.; Cammarota, M.N.; Izquierdo, I.N.; Brentani, R.R. Physiology of the Prion Protein. *Physiol. Rev.* **2008**, *88*, 673–728. [CrossRef] [PubMed]

24. Brown, D.; Qin, K.; Herms, J.W.; Madlung, A.; Manson, J.C.; Strome, R.; Fraser, P.E.; Kruck, T.; Von Bohlen, A.; Schulz-Schaeffer, W.; et al. The cellular prion protein binds copper in vivo. *Nature* **1997**, *390*, 684–687. [CrossRef] [PubMed]

25. Zahn, R.; Liu, A.Z.; Luhrs, T.; Riek, R.; von Schroetter, C.; Garcia, F.L.; Billeter, M.; Calzolai, L.; Wider, G.; Wuthrich, K. NMR solution structure of the human prion protein. *Proc. Natl. Acad. Sci. USA* **2000**, *97*, 145–150. [CrossRef] [PubMed]

26. Baskakov, I.V.; Legname, G.; Prusiner, S.B. Folding of prion protein to its native α-helical conformation is under kinetic control. *J. Biol. Chem.* **2001**, *276*, 19687–19690. [CrossRef] [PubMed]

27. Surewicz, W.K.; Apostol, M.I. Prion Protein and Its Conformational Conversion: A Structural Perspective. In *Prion Proteins*; Springer: Berlin/Heidelberg, Germany, 2011; Volume 305, pp. 135–167.

28. Govaerts, C.; Wille, H.; Prusiner, S.B.; Cohen, F.E. Evidence for assembly of prions with left-handed beta-helices into trimers. *Proc. Natl. Acad. Sci. USA* **2004**, *101*, 8342–8347. [CrossRef] [PubMed]

29. Cobb, N.J.; Sonnichsen, F.D.; McHaourab, H.; Surewicz, W.K. Molecular architecture of human prion protein amyloid: A parallel, in-register beta-structure. *Proc. Natl. Acad. Sci. USA* **2007**, *104*, 18946–18951. [CrossRef]

30. DeMarco, M.L.; Silveira, J.; Caughey, B.; Daggett, V. Structural properties of prion protein protofibrils and fibrils: An experimental assessment of atomic models. *Biochemistry* **2006**, *45*, 15573–15582. [CrossRef]

31. Rossetti, G.; Carloni, P. Structural Modeling of Human Prion Protein's Point Mutations. *Prog. Mol. Biol. Transl. Sci.* **2017**, *150*, 105–122. [CrossRef]

32. Rossetti, G.; Bongarzone, S.; Carloni, P. Computational studies on the prion protein. *Curr. Top. Med. Chem.* **2013**, *13*, 2419–2431. [CrossRef] [PubMed]

33. Diaz-Espinoza, R.; Soto, C. High-resolution structure of infectious prion protein: The final frontier. *Nat. Struct. Mol. Biol.* **2012**, *19*, 370–377. [CrossRef]

34. Aguzzi, A.; Calella, A.M. Prions: Protein aggregation and infectious diseases. *Physiol. Rev.* **2009**, *89*, 1105–1152. [CrossRef] [PubMed]

35. Aguzzi, A.; O'Connor, T. Protein aggregation diseases: Pathogenicity and therapeutic perspectives. *Nat. Rev. Drug Discov.* **2010**, *9*, 237–248. [CrossRef] [PubMed]

36. Nazabal, A.; Hornemann, S.; Aguzzi, A.; Zenobi, R. Hydrogen/deuterium exchange mass spectrometry identifies two highly protected regions in recombinant full-length prion protein amyloid fibrils. *J. Mass Spectrom.* **2009**, *44*, 965–977. [CrossRef]

37. Sim, V.; Caughey, B. Ultrastructures and strain comparison of under-glycosylated scrapie prion fibrils. *Neurobiol. Aging* **2009**, *30*, 2031–2042. [CrossRef] [PubMed]

38. Stoehr, J.; Weinmann, N.; Wille, H.; Kaimann, T.; Nagel-Steger, L.; Birkmann, E.; Panza, G.; Prusiner, S.B.; Eigen, M.; Riesner, D. Mechanisms of prion protein assembly into amyloid. *Proc. Natl. Acad. Sci. USA* **2008**, *105*, 2409–2414. [CrossRef]

39. Hegde, R.S.; Mastrianni, J.A.; Scott, M.R.; DeFea, K.A.; Tremblay, P.; Torchia, M.; DeArmond, S.J.; Prusiner, S.B.; Lingappa, V.R. A transmembrane form of the prion protein in neurodegenerative disease. *Science* **1998**, *279*, 827–834. [CrossRef]

40. Li, A.M.; Christensen, H.M.; Stewart, L.R.; Roth, K.A.; Chiesa, R.; Harris, D.A. Neonatal lethality in transgenic mice expressing prion protein with a deletion of residues 105–125. *EMBO J.* **2007**, *26*, 548–558. [CrossRef]

41. Ott, C.M.; Akhavan, A.; Lingappa, V.R. Specific features of the prion protein transmembrane domain regulate nascent chain orientation. *J. Biol. Chem.* **2007**, *282*, 11163–11171. [CrossRef]

42. Chakrabarti, O.; Ashok, A.; Hegde, R.S. Prion protein biosynthesis and its emerging role in neurodegeneration. *Trends Biochem. Sci.* **2009**, *34*, 287–295. [CrossRef] [PubMed]

43. Beland, M.; Roucou, X. The prion protein unstructured N-terminal region is a broad-spectrum molecular sensor with diverse and contrasting potential functions. *J. Neurochem.* **2012**, *120*, 853–868. [CrossRef] [PubMed]

44. Silva, J.L.; Vieira, T.C.R.G.; Gomes, M.P.B.; Rangel, L.P.; Scapin, S.M.N.; Cordeiro, Y. Experimental approaches to the interaction of the prion protein with nucleic acids and glycosaminoglycans: Modulators of the pathogenic conversion. *Methods* **2011**, *53*, 306–317. [CrossRef] [PubMed]

45. Hajj, G.N.M.; Lopes, M.H.; Mercadante, A.F.; Veiga, S.S.; da Silveira, R.B.; Santos, T.G.; Ribeiro, K.C.B.; Juliano, M.A.; Jacchieri, S.G.; Zanata, S.M.; et al. Cellular prion protein interaction with vitronectin supports axonal growth and is compensated by integrins. *J. Cell Sci.* **2007**, *120*, 1915–1926. [CrossRef]

46. Zanata, S.M.; Lopes, M.H.; Mercadante, A.F.; Hajj, G.N.M.; Chiarini, L.B.; Nomizo, R.; Freitas, A.R.O.; Cabral, A.L.B.; Lee, K.S.; Juliano, M.A.; et al. Stress-inducible protein 1 is a cell surface ligand for cellular prion that triggers neuroprotection. *EMBO J.* **2002**, *21*, 3307–3316. [CrossRef]

47. Lauren, J.; Gimbel, D.A.; Nygaard, H.B.; Gilbert, J.W.; Strittmatter, S.M. Cellular prion protein mediates impairment of synaptic plasticity by amyloid-beta oligomers. *Nature* **2009**, *457*, 1128–1184. [CrossRef]

48. Nicoll, A.J.; Panico, S.; Freir, D.B.; Wright, D.; Terry, C.; Risse, E.; Herron, C.E.; O'Malley, T.; Wadsworth, J.D.; Farrow, M.A.; et al. Amyloid-beta nanotubes are associated with prion protein-dependent synaptotoxicity. *Nat. Commun.* **2013**, *4*, 2416. [CrossRef]

49. Chen, S.G.; Yadav, S.P.; Surewicz, W.K. Interaction between Human Prion Protein and Amyloid-beta (A beta) Oligomers Role of N-Terminal Residues. *J. Biol. Chem.* **2010**, *285*, 26377–26383. [CrossRef]

50. Parkyn, C.J.; Vermeulen, E.G.; Mootoosamy, R.C.; Sunyach, C.; Jacobsen, C.; Oxvig, C.; Moestrup, S.; Liu, Q.; Bu, G.; Jen, A.; et al. LRP1 controls biosynthetic and endocytic trafficking of neuronal prion protein. *J. Cell Sci.* **2008**, *121*, 773–783. [CrossRef]

51. Schmitt-Ulms, G.; Legname, G.; Baldwin, M.A.; Ball, H.L.; Bradon, N.; Bosque, P.J.; Crossin, K.L.; Edelman, G.M.; DeArmond, S.J.; Cohen, F.E.; et al. Binding of neural cell adhesion molecules (N-CAMs) to the cellular prion protein. *J. Mol. Biol.* **2001**, *314*, 1209–1225. [CrossRef]

52. Cong, X.; Casiraghi, N.; Rossetti, G.; Mohanty, S.; Giachin, G.; Legname, G.; Carloni, P. Role of Prion Disease-Linked Mutations in the Intrinsically Disordered N-Terminal Domain of the Prion Protein. *J. Chem. Theory Comput.* **2013**, *9*, 5158–5167. [CrossRef] [PubMed]

53. Calzolai, L.; Zahn, R. Influence of pH on NMR structure and stability of the human prion protein globular domain. *J. Boil. Chem.* **2003**, *278*, 35592–35596. [CrossRef] [PubMed]

54. Degioia, L.; Selvaggini, C.; Ghibaudi, E.; Diomede, L.; Bugiani, O.; Forloni, G.; Tagliavini, F.; Salmona, M. Conformational Polymorphism of the Amyloidogenic and Neurotoxic Peptide Homologous to Residues-106–126 of the Prion Protein. *J. Biol. Chem.* **1994**, *269*, 7859–7862.

55. Miura, T.; Yoda, M.; Takaku, N.; Hirose, T.; Takeuchi, H. Clustered negative charges on the lipid membrane surface induce beta-sheet formation of prion protein fragment 106–126. *Biochemistry* **2007**, *46*, 11589–11597. [CrossRef] [PubMed]

56. Satheeshkumar, K.S.; Jayakumar, R. Conformational polymorphism of the amyloidogenic peptide homologous to residues 113–127 of the prion protein. *Biophys. J.* **2003**, *85*, 473–483. [CrossRef]

57. Zahn, R. The octapeptide repeats in mammalian prion protein constitute a pH-dependent folding and aggregation site. *J. Mol. Biol.* **2003**, *334*, 477–488. [CrossRef] [PubMed]

58. Van der Kamp, M.W.; Daggett, V. The consequences of pathogenic mutations to the human prion protein. *Protein Eng. Des. Sel.* **2009**, *22*, 461–468. [CrossRef]

59. Rossetti, G.; Cong, X.J.; Caliandro, R.; Legname, G.; Carloni, P. Common Structural Traits across Pathogenic Mutants of the Human Prion Protein and Their Implications for Familial Prion Diseases. *J. Mol. Biol.* **2011**, *411*, 700–712. [CrossRef]

60. Ilc, G.; Giachin, G.; Jaremko, M.; Jaremko, L.; Benetti, F.; Plavec, J.; Zhukov, I.; Legname, G. NMR structure of the human prion protein with the pathological Q212P mutation reveals unique structural features. *PLoS ONE* **2010**, *5*, e11715. [CrossRef]

61. Van der Kamp, M.W.; Daggett, V. Pathogenic Mutations in the Hydrophobic Core of the Human Prion Protein Can Promote Structural Instability and Misfolding. *J. Mol. Biol.* **2010**, *404*, 732–748. [CrossRef]

62. Apetri, A.C.; Surewicz, K.; Surewicz, W.K. The effect of disease-associated mutations on the folding pathway of human prion protein. *J. Biol. Chem.* **2004**, *279*, 18008–18014. [CrossRef] [PubMed]

63. Liemann, S.; Glockshuber, R. Influence of amino acid substitutions related to inherited human prion diseases on the thermodynamic stability of the cellular prion protein. *Biochemistry* **1999**, *38*, 3258–3267. [CrossRef] [PubMed]

64. Swietnicki, W.; Petersen, R.B.; Gambetti, P.; Surewicz, W.K. Familial mutations and the thermodynamic stability of the recombinant human prion protein. *J. Biol. Chem.* **1998**, *273*, 31048–31052. [CrossRef] [PubMed]

65. Evans, E.G.B.; Millhauser, G.L. Copper- and Zinc-Promoted Interdomain Structure in the Prion Protein: A Mechanism for Autoinhibition of the Neurotoxic N-Terminus. In *Prion Protein*; Elsevier: Amsterdam, The Netherlands, 2017; Volume 150, pp. 35–56.

66. Hegde, R.S.; Kang, S.W. The concept of translocational regulation. *J. Cell Biol.* **2008**, *182*, 225–232. [CrossRef] [PubMed]

67. Hornemann, S.; von Schroetter, C.; Damberger, F.F.; Wuthrich, K. Prion Protein-Detergent Micelle Interactions Studied by NMR in Solution. *J. Biol. Chem.* **2009**, *284*, 22713–22721. [CrossRef] [PubMed]

68. Kardos, J.; Kovacs, I.; Hajos, F.; Kalman, M.; Simonyi, M. Nerve endings from rat brain tissue release copper upon depolarization. A possible role in regulating neuronal excitability. *Neurosci. Lett.* **1989**, *103*, 139–144. [CrossRef]

69. Walter, E.D.; Chattopadhyay, M.; Millhauser, G.L. The affinity of copper binding to the prion protein octarepeat domain: Evidence for negative cooperativity. *Biochemistry* **2006**, *45*, 13083–13092. [CrossRef]

70. Lee, K.S.; Magalhaes, A.C.; Zanata, S.M.; Brentani, R.R.; Martins, V.R.; Prado, M.A.M. Internalization of mammalian fluorescent cellular prion protein and N-terminal deletion mutants in living cells. *J. Neurochem.* **2001**, *79*, 79–87. [CrossRef]

71. Pauly, P.C.; Harris, D.A. Copper Stimulates Endocytosis of the Prion Protein. *J. Biol. Chem.* **1998**, *273*, 33107–33110. [CrossRef]

72. Ren, K.; Gao, C.; Zhang, J.; Wang, K.; Xu, Y.; Wang, S.-B.; Wang, H.; Tian, C.; Shi, Q.; Dong, X.-P. Flotillin-1 Mediates PrP^C Endocytosis in the Cultured Cells During Cu^{2+} Stimulation Through Molecular Interaction. *Mol. Neurobiol.* **2013**, *48*, 631–646. [CrossRef]

73. Sumudhu, W.; Perera, S.; Hooper, N.M. Ablation of the metal ion-induced endocytosis of the prion protein by disease-associated mutation of the octarepeat region. *Curr. Biol.* **2001**, *11*, 519–523. [CrossRef]

74. Huang, S.; Chen, L.; Bladen, C.; Stys, P.K.; Zamponi, G.W. Differential modulation of NMDA and AMPA receptors by cellular prion protein and copper ions. *Mol. Brain* **2018**, *11*. [CrossRef] [PubMed]

75. You, H.; Tsutsui, S.; Hameed, S.; Kannanayakal, T.J.; Chen, L.; Xia, P.; Engbers, J.D.T.; Lipton, S.A.; Stys, P.K.; Zamponi, G.W. Aβ neurotoxicity depends on interactions between copper ions, prion protein, and *N*-methyl-D-aspartate receptors. *Proc. Natl. Acad. Sci. USA* **2012**, *109*, 1737–1742. [CrossRef] [PubMed]

76. Stys, P.K.; You, H.; Zamponi, G.W. Copper-dependent regulation of NMDA receptors by cellular prion protein: Implications for neurodegenerative disorders. *J. Physiol.* **2012**, *590*, 1357–1368. [CrossRef]

77. Gasperini, L.; Meneghetti, E.; Pastore, B.; Benetti, F.; Legname, G. Prion Protein and Copper Cooperatively Protect Neurons by Modulating NMDA Receptor Through S-nitrosylation. *Antioxid. Redox Signal.* **2015**, *22*, 772–784. [CrossRef]

78. Aronoff-Spencer, E.; Burns, C.S.; Avdievich, N.I.; Gerfen, G.J.; Peisach, J.; Antholine, W.E.; Ball, H.L.; Cohen, F.E.; Prusiner, S.B.; Millhauser, G.L. Identification of the Cu^{2+} binding sites in the N-terminal domain of the prion protein by EPR and CD spectroscopy. *Biochemistry* **2000**, *39*, 13760–13771. [CrossRef]

79. Millhauser, G.L. Copper Binding in the Prion Protein. *Acc. Chem. Res.* **2004**, *37*, 79–85. [CrossRef]

80. Millhauser, G.L. Copper and the Prion Protein: Methods, Structures, Function, and Disease. *Annu. Rev. Phys. Chem.* **2007**, *58*, 299–320. [CrossRef]

81. Quintanar, L.; Rivillas-Acevedo, L.; Grande-Aztatzi, R.; Gómez-Castro, C.Z.; Arcos-López, T.; Vela, A. Copper coordination to the prion protein: Insights from theoretical studies. *Coord. Chem. Rev.* **2013**, *257*, 429–444. [CrossRef]

82. Wells, M.A.; Jelinska, C.; Hosszu, L.L.; Craven, C.J.; Clarke, A.R.; Collinge, J.; Waltho, J.P.; Jackson, G.S. Multiple forms of copper(II) co-ordination occur throughout the disordered N-terminal region of the prion protein at pH 7.4. *Biochem. J.* **2006**, *400*, 501–510. [CrossRef]

83. Wopfner, F.; Weidenhöfer, G.; Schneider, R.; von Brunn, A.; Gilch, S.; Schwarz, T.F.; Werner, T.; Schätzl, H.M. Analysis of 27 mammalian and 9 avian PrPs reveals high conservation of flexible regions of the prion protein. *J. Mol. Biol.* **1999**, *289*, 1163–1178. [CrossRef] [PubMed]

84. Chattopadhyay, M.; Walter, E.D.; Newell, D.J.; Jackson, P.J.; Aronoff-Spencer, E.; Peisach, J.; Gerfen, G.J.; Bennett, B.; Antholine, W.E.; Millhauser, G.L. The octarepeat domain of the prion protein binds Cu(II) with three distinct coordination modes at pH 7.4. *J. Am. Chem. Soc.* **2005**, *127*, 12647–12656. [CrossRef] [PubMed]

85. Burns, C.S.; Aronoff-Spencer, E.; Legname, G.; Prusiner, S.B.; Antholine, W.E.; Gerfen, G.J.; Peisach, J.; Millhauser, G.L. Copper coordination in the full-length, recombinant prion protein. *Biochemistry* **2003**, *42*, 6794–6803. [CrossRef] [PubMed]

86. Burns, C.S.; Aronoff-Spencer, E.; Dunham, C.M.; Lario, P.; Avdievich, N.I.; Antholine, W.E.; Olmstead, M.M.; Vrielink, A.; Gerfen, G.J.; Peisach, J.; et al. Molecular features of the copper binding sites in the octarepeat domain of the prion protein. *Biochemistry* **2002**, *41*, 3991–4001. [CrossRef] [PubMed]

87. Liu, L.; Jiang, D.; McDonald, A.; Hao, Y.; Millhauser, G.L.; Zhou, F. Copper redox cycling in the prion protein depends critically on binding mode. *J. Am. Chem. Soc.* **2011**, *133*, 12229–12237. [CrossRef] [PubMed]

88. Zhou, F.; Millhauser, G.L. The rich electrochemistry and redox reaacions of the copper sites in the cellular prion protein. *Coord. Chem. Rev.* **2012**, *256*, 2285–2296. [CrossRef] [PubMed]

89. Jackson, G.S.; Murray, I.; Hosszu, L.L.; Gibbs, N.; Waltho, J.P.; Clarke, A.R.; Collinge, J. Location and properties of metal-binding sites on the human prion protein. *Proc. Natl. Acad. Sci. USA* **2001**, *98*, 8531–8535. [CrossRef]

90. Walter, E.D.; Stevens, D.J.; Spevacek, A.R.; Visconte, M.P.; Dei Rossi, A.; Millhauser, G.L. Copper binding extrinsic to the octarepeat region in the prion protein. *Curr. Protein Pept. Sci.* **2009**, *10*, 529–535. [CrossRef]

91. Giachin, G.; Mai, P.T.; Tran, H.N.; Salzano, G.; Benetti, F.; Migliorati, V.; Arcovito, A.; Della Longa, S.; Mancini, G.; D'Angelo, P.; et al. The non-octarepeat copper binding site of the prion protein is a key regulator of prion conversion. *Sci. Rep.* **2015**, *5*, 15253. [CrossRef]

92. Klewpatinond, M.; Davies, P.; Bowen, S.; Brown, D.R.; Viles, J.H. Deconvoluting the Cu(2+) binding modes of full-length prion protein. *J. Biol. Chem.* **2008**, *283*, 1870–1881. [CrossRef]

93. Grande-Aztatzi, R.; Rivillas-Acevedo, L.; Quintanar, L.; Vela, A. Structural models for Cu(II) bound to the fragment 92-96 of the human prion protein. *J. Phys. Chem. B* **2013**, *117*, 789–799. [CrossRef] [PubMed]

94. Hureau, C.; Charlet, L.; Dorlet, P.; Gonnet, F.; Spadini, L.; Anxolabéhère-Mallart, E.; Girerd, J.J. A spectroscopic and voltammetric study of the pH-dependent Cu(II) coordination to the peptide GGGTH: Relevance to the fifth Cu(II) site in the prion protein. *J. Boil. Inorg. Chem.* **2006**, *11*, 735–744. [CrossRef] [PubMed]

95. Hureau, C.; Mathé, C.; Faller, P.; Mattioli, T.A.; Dorlet, P. Folding of the prion peptide GGGTHSQW around the copper(II) ion: Identifying the oxygen donor ligand at neutral pH and probing the proximity of the tryptophan residue to the copper ion. *J. Biol. Inorg. Chem.* **2008**, *13*, 1055–1064. [CrossRef] [PubMed]

96. Rivillas-Acevedo, L.; Grande-Aztatzi, R.; Lomelí, I.; García, J.E.; Barrios, E.; Teloxa, S.; Vela, A.; Quintanar, L. Spectroscopic and electronic structure studies of copper(II) binding to His111 in the human prion protein fragment 106–115: Evaluating the role of protons and methionine residues. *Inorg. Chem.* **2011**, *50*, 1956–1972. [CrossRef] [PubMed]

97. Arcos-López, T.; Qayyum, M.; Rivillas-Acevedo, L.; Miotto, M.C.; Grande-Aztatzi, R.; Fernández, C.O.; Hedman, B.; Hodgson, K.O.; Vela, A.; Solomon, E.I.; et al. Spectroscopic and Theoretical Study of Cu(I) Binding to His111 in the Human Prion Protein Fragment 106–115. *Inorg. Chem.* **2016**, *55*, 2909–2922. [CrossRef] [PubMed]

98. Berti, F.; Gaggelli, E.; Guerrini, R.; Janicka, A.; Kozlowski, H.; Legowska, A.; Miecznikowska, H.; Migliorini, C.; Pogni, R.; Remelli, M.; et al. Structural and dynamic characterization of copper(II) binding of the human prion protein outside the octarepeat region. *Chem. Eur. J.* **2007**, *13*, 1991–2001. [CrossRef] [PubMed]

99. DiNatale, G.; Ösz, K.; Nagy, Z.; Sanna, D.; Micera, G.; Pappalardo, G.; Sóvágó, I.; Rizzarell, E. Interaction of Copper(II) with the Prion Peptide Fragment HuPrP(76–114) Encompassing Four Histidyl Residues within and outside the Octarepeat Domain. *Inorg. Chem.* **2009**, *48*, 4239–4250. [CrossRef] [PubMed]

100. Jones, C.E.; Klewpatinond, M.; Abdelraheim, S.R.; Brown, D.R.; Viles, J.H. Probing copper^{2+} binding to the prion protein using diamagnetic nickel^{2+} and ^{1}H NMR: The unstructured N terminus facilitates the coordination of six copper^{2+} ions at physiological concentrations. *J. Mol. Biol.* **2005**, *346*, 1393–1407. [CrossRef] [PubMed]

101. Nadal, R.C.; Davies, P.; Brown, D.R.; Viles, J.H. Evaluation of copper^{2+} affinities for the prion protein. *Biochemistry* **2009**, *48*, 8929–8931. [CrossRef] [PubMed]

102. Sánchez-López, C.; Rivillas-Acevedo, L.; Cruz-Vásquez, O.; Quintanar, L. Methionine 109 plays a key role in Cu(II) binding to His111 in the 92–115 fragment of the human prion protein. *Inorg. Chim. Acta* **2018**, *481*, 87–97. [CrossRef]

103. Evans, E.G.B.; Pushie, M.J.; Markham, K.A.; Lee, H.-W.; Millhauser, G.L. Interaction between Prion Protein's Copper-Bound Octarepeat Domain and a Charged C-Terminal Pocket Suggests a Mechanism for N-Terminal Regulation. *Structure* **2016**, *24*, 1057–1067. [CrossRef] [PubMed]

104. Spevacek, A.R.; Evans, E.G.B.; Miller, J.L.; Meyer, H.C.; Pelton, J.G.; Millhauser, G.L. Zinc drives a tertiary fold in the prion protein with familial disease mutation sites at the interface. *Structure* **2013**, *21*, 236–246. [CrossRef] [PubMed]

105. Thakur, A.K.; Srivastava, A.K.; Srinivas, V.; Chary, K.V.; Rao, C.M. Copper alters aggregation behavior of prion protein and induces novel interactions between its N- and C-terminal regions. *J. Biol. Chem.* **2011**, *286*, 38533–38545. [CrossRef] [PubMed]

106. Younan, N.D.; Klewpatinond, M.; Davies, P.; Ruban, A.V.; Brown, D.R.; Viles, J.H. Copper(II)-induced secondary structure changes and reduced folding stability of the prion protein. *J. Mol. Biol.* **2011**, *410*, 369–382. [CrossRef] [PubMed]

107. Leal, S.S.; Botelho, H.M.; Gomes, C.M. Metal ions as modulators of protein conformation and misfolding in neurodegeneration. *Coord. Chem. Rev.* **2012**, *256*, 2253–2270. [CrossRef]

108. D'angelo, P.; Della Longa, S.; Arcovito, A.; Mancini, G.; Zitolo, A.; Chillemi, G.; Giachin, G.; Legname, G.; Benetti, F. Effects of the Pathological Q212P Mutation on Human Prion Protein Non-Octarepeat Copper-Binding Site. *Biochemistry* **2012**, *51*, 6068–6079. [CrossRef]

109. Altmeppen, H.C.; Puig, B.; Dohler, F.; Thurm, D.K.; Falker, C.; Krasemann, S.; Glatzel, M. Proteolytic processing of the prion protein in health and disease. *Am. J. Neurodegener. Dis.* **2012**, *1*, 15–31.

110. McMahon, H.E.; Mangé, A.; Nishida, N.; Créminon, C.; Casanova, D.; Lehmann, S. Cleavage of the amino terminus of the prion protein by reactive oxygen species. *J. Biol. Chem.* **2001**, *276*, 2286–2291. [CrossRef]

111. Watt, N.T.; Taylor, D.R.; Gillott, A.; Thomas, D.A.; Perera, W.S.S.; Hooper, N.M. Reactive Oxygen Species-mediated β-Cleavage of the Prion Protein in the Cellular Response to Oxidative Stress. *J. Biol. Chem.* **2005**, *280*, 35914–35921. [CrossRef]

112. McDonald, A.J.; Dibble, J.P.; Evans, E.G.; Millhauser, G.L. A new paradigm for enzymatic control of alpha-cleavage and beta-cleavage of the prion protein. *J. Biol. Chem.* **2014**, *289*, 803–813. [CrossRef]

113. McDonald, A.J.; Millhauser, G.L. PrP overdrive. Does inhibition of alpha-cleavage contribute to PrP^C toxicity and prion disease? *Prion* **2014**, *8*, 183–191. [CrossRef]

114. Oliveira-Martins, J.B.; Yusa, S.; Calella, A.M.; Bridel, C.; Baumann, F.; Dametto, P.; Aguzzi, A. Unexpected tolerance of alpha-cleavage of the prion protein to sequence variations. *PLoS ONE* **2010**, *5*, e9107. [CrossRef] [PubMed]

115. Haigh, C.L.; Lewis, V.A.; Vella, L.J.; Masters, C.L.; Hill, A.F.; Lawson, V.A.; Collins, S.J. PrP^C-related signal transduction is influenced by copper, membrane integrity and the alpha cleavage site. *Cell Res.* **2009**, *19*, 1062–1078. [CrossRef] [PubMed]

116. Sánchez-López, C.; Fernández, C.O.; Quintanar, L. Neuroprotective alpha-cleavage of the human prion protein significantly impacts Cu(II) coordination at its His111 site. *Dalton Trans.* **2018**, *47*, 9274–9282. [CrossRef] [PubMed]

117. Black, S.A.G.; Stys, P.K.; Zamponi, G.W.; Tsutsui, S. Cellular prion protein and NMDA receptor modulation: Protecting against excitotoxicity. *Front. Cell Dev. Biol.* **2014**, *2*. [CrossRef] [PubMed]

118. Villar-Piqué, A.; Lopes da Fonseca, T.; Sant'Anna, R.; Szegö, E.M.; Fonseca-Ornelas, L.; Pinho, R.; Carija, A.; Gerhardt, E.; Masaracchia, C.; Abad Gonzalez, E.; et al. Environmental and genetic factors support the dissociation between α-synuclein aggregation and toxicity. *Proc. Natl. Acad. Sci. USA* **2016**, *113*, 6506–65115. [CrossRef] [PubMed]

© 2018 by the authors. Licensee MDPI, Basel, Switzerland. This article is an open access article distributed under the terms and conditions of the Creative Commons Attribution (CC BY) license (http://creativecommons.org/licenses/by/4.0/).

International Journal of
Molecular Sciences

Review

Modeling of Disordered Protein Structures Using Monte Carlo Simulations and Knowledge-Based Statistical Force Fields

Maciej Pawel Ciemny [1,2], Aleksandra Elzbieta Badaczewska-Dawid [1], Monika Pikuzinska [1], Andrzej Kolinski [1] and Sebastian Kmiecik [1,*]

[1] Faculty of Chemistry, Biological and Chemical Research Center, University of Warsaw, Pasteura 1, 02-093 Warsaw, Poland; maciej.ciemny@fuw.edu.pl (M.P.C.); adawid@chem.uw.edu.pl (A.E.B.-D.); m.pikuzinska@student.uw.edu.pl (M.P.); kolinski@chem.uw.edu.pl (A.K.)
[2] Faculty of Physics, University of Warsaw, Pasteura 5, 02-093 Warsaw, Poland
* Correspondence: sekmi@chem.uw.edu.pl; Tel.: +48-22-55-26-365

Received: 13 December 2018; Accepted: 29 January 2019; Published: 31 January 2019

Abstract: The description of protein disordered states is important for understanding protein folding mechanisms and their functions. In this short review, we briefly describe a simulation approach to modeling protein interactions, which involve disordered peptide partners or intrinsically disordered protein regions, and unfolded states of globular proteins. It is based on the CABS coarse-grained protein model that uses a Monte Carlo (MC) sampling scheme and a knowledge-based statistical force field. We review several case studies showing that description of protein disordered states resulting from CABS simulations is consistent with experimental data. The case studies comprise investigations of protein–peptide binding and protein folding processes. The CABS model has been recently made available as the simulation engine of multiscale modeling tools enabling studies of protein–peptide docking and protein flexibility. Those tools offer customization of the modeling process, driving the conformational search using distance restraints, reconstruction of selected models to all-atom resolution, and simulation of large protein systems in a reasonable computational time. Therefore, CABS can be combined in integrative modeling pipelines incorporating experimental data and other modeling tools of various resolution.

Keywords: coarse-grained; CABS model; MC simulations; statistical force fields; disordered protein; protein structure

---

1. Introduction

There is a growing body of evidence that some proteins act in multiple structural states [1]. It has been demonstrated that the ability of these proteins to switch between distinct structural states may be crucial for their function and regulation [1]. Additionally, a number of key biological functions have been proven to be performed by disordered or partially unstructured proteins [2]. Some proteins fold and obtain their structure only upon binding to their partners, while others form so called "fuzzy complexes" in which both proteins retain a certain degree of disorder [3]. These discoveries modified the core biochemistry principle of "structure determines function". As for now, a consensus has been reached that protein function may be a result of an interplay between protein structure and its dynamics [4,5].

Internal protein motions may be studied both experimentally and with computational methods [6,7]. For example, nuclear magnetic resonance (NMR) spectroscopy is one of the richest sources of information on protein structure and dynamics, especially when accompanied with assisting methods that enhance resolution or provide an additional insight into the dynamics of structures [8]. This approach, however, results in an averaged image of the structural ensemble.

A variety of computational techniques have been developed to assist these challenging experimental studies [7,9]. In the last decades, molecular modeling was dominated by structure-based models or Go-like models (approaches that are biased toward known folded conformations [10,11]). These indeed lead to significant speedup of simulations but may result for example in an unrealistic picture of protein folding, which in reality may also depend on non-native interactions [12–14].

Recent works show that methods combining experimental data and computational approaches may produce the most promising pictures of protein equilibrium dynamics [15,16]. However, the development of these methods poses a number of challenges—both in terms of the validity of the approach and its computationally efficient implementation [17].

Molecular dynamics (MD) has been so far the most widespread computational method for the investigation of protein motions [18]. However, standard all-atom MD implementations are limited to sub-microsecond timescales and may suffer from limited sampling despite recent significant advances in code optimization and hardware [19]. To overcome this problem various MD extensions have been proposed. These extensions include for example replica-exchange MD, meta-dynamics, Markov state models and simulated annealing algorithms [6,20–23].

A number of non-MD sampling methods have also been developed to provide a comprehensive image of protein dynamics using limited computational resources. Of these, Monte Carlo (MC) is perhaps the most commonly used and generally applicable sampling method [11]. Monte Carlo randomly generates conformations and uses an energy-based acceptance criterion that promotes pseudo-trajectory convergence to an energetic minimum. On the expense of losing a direct image of the timescales or kinetics of the ensemble, MC manages to overcome some of the major limitations of MD [24].

Aside from the sampling method, a further extension of effective timescales is possible by using a simplified representation of protein structures to reduce the number of a system's degrees of freedom. The accuracy of the available coarse-grained (CG) models may vary from detailed, almost atomistic representations (Primo [25], Rosetta [26]), medium resolution models (in which a single amino acid is represented by three to five beads: UNRES [27], CABS [28], AWSEM [29], MARTINI [30], PaLaCe [31]), and Scorpion [32]) to significantly simplified models like SURPASS [33,34]. Applications and implementations of these and other CG models are described in detail in a recent review [11].

In addition to the representation and sampling method, the choice of the force field to perform the simulation determines the success of modeling. Traditionally, force fields are divided into two main groups: physics-based, which involve (usually pairwise) interaction terms [35], and those employing a statistical approach; however, most of the successful approaches are usually a mixture of the two. A statistical force field is constructed using the probability of a chosen observable (or a set of observables) in a given ensemble of structures [36]. Early attempts focused on straightforward pairwise contacts [37]; however, with further development, more complex observables were analyzed. This resulted in a generation of knowledge-based force fields, or scores, for various representations, coarse-grained and all-atom: CABS [28], Rosetta [38], DOPE [39], GOAP [40], QUARK [41], Bcl::Score [42] or BACH [36]. Newly developed approaches go a step further and improve the results by combining these methods with experimental data [43,44]. An example of such approach is RosettaEPR [45], which includes distance data from site-directed spin labeling electron paramagnetic resonance experiments. It is generally agreed that statistical force fields frequently allow more accurate scoring than physics-based potentials [11]. The combination of knowledge-based force fields or scores with effective sampling schemes seems to be a promising approach to a number of problems [11], such as protein structure prediction [43,44,46,47], investigation of protein interactions [48] or studies of protein dynamics [17,49–51].

This review briefly describes one of these approaches: an MC-based and knowledge-based interaction scheme for modeling protein–peptide interactions and unfolded states of globular proteins using the CABS coarse-grained protein model. Firstly, the main features of the CABS method will be described, with a focus on their applicability for modeling disordered or unfolded proteins or their fragments. Subsequently, representative case studies will be discussed to provide detailed insights into the modeling results obtained for systems characterized by a varying level of disorder.

2. CABS Dynamics and Interaction Model

Since its development, the CABS model (C-alpha, C-beta and Side chain model) has been applied to a variety of modeling problems, such as protein folding mechanisms [49,50,52–57], protein structure prediction [58–61], protein–peptide docking including large-scale conformational flexibility [62–68] and simulations of near-native fluctuations of globular proteins [69–73]. When combined with careful bioinformatics selection of the generated models, CABS proved to be one of the two most accurate structure prediction tools evaluated in the CASP (Critical Assessment of protein Structure Prediction) experiment [60]. The CABS model uses up to four atoms or pseudo-atoms per residue (see the description below), but outputs protein systems in C-alpha representation only. Therefore, for practical applications, the obtained models need to be reconstructed to all-atom representation. In various multiscale modeling tools discussed below, CABS has been integrated with the MODELLER-based reconstruction procedure [74]. Other reconstruction scenarios are also possible to ensure the best possible quality of local protein structure. This can be realized by combination of different tools for protein backbone reconstruction from the C-alpha trace and side chain reconstruction, like BBQ [75] or SCWRL [76] for example, and optionally further refinement [77].

In this review, we discuss the applicability of the CABS CG model and its knowledge-based statistical force field [28] to the modeling of disordered or unfolded protein states. In the CABS model the polypeptide chain representation is reduced to up to four unified atoms per residue (see Figure 1). These interaction centers represent lattice-confined C-alpha atoms, C-beta atoms, the united side chain pseudo-atom, and additionally, pseudo-atoms representing geometrical centers of peptide bonds needed to define the hydrogen pseudo-bond. An example of a polypeptide chain in CABS representation is presented in Figure 1b. Even though the restriction of the C-alpha trace to the underlying low spacing (0.61 Å [28]) cubic lattice may appear to be a drastic simplification, it is not. Allowing small fluctuations of the C-alpha, C-alpha distance enables hundreds of possible orientations of this pseudo bond, and thereby the resulting model chains do not show any noticeable directional biases. Furthermore, the averaged resolution of the C-alpha traces is acceptable and below 0.5 Å [28]. Additionally, the lattice representation enables pre-calculation of local moves and corresponding changes of interactions, leading to a few times faster simulations in comparison with otherwise equivalent continuous space CG models [11].

The CABS model uses a knowledge-based statistical force field that consists of generic, sequence-independent interaction terms that favor protein-like conformations, and sequence-dependent interaction terms that determine some structural details [11,28,78]. The generic force field terms are derived from general features of polypeptide chains that result in protein-like behavior of the model chains. They account for properties of protein chains such as local stiffness, their biases toward secondary structures and packing compactness. The residue–residue interaction terms are derived from contact geometry statistics derived from folded globular proteins (illustrated in Figure 2a). Nevertheless, the local packing regularities in unfolded states appear to be very similar to that observed in native structures [11,28,33]. Thereby, CABS simulations provided correct pictures of protein folding [49,52–56,60] and flexibility of globular proteins [70,71].

The resulting force field takes a form of a precomputed matrix of contact pseudo-energies, presented schematically in Figure 2b. Additionally, to allow successful modeling of membrane proteins the CABS force field can be extended by introducing effective dielectric constant terms [79].

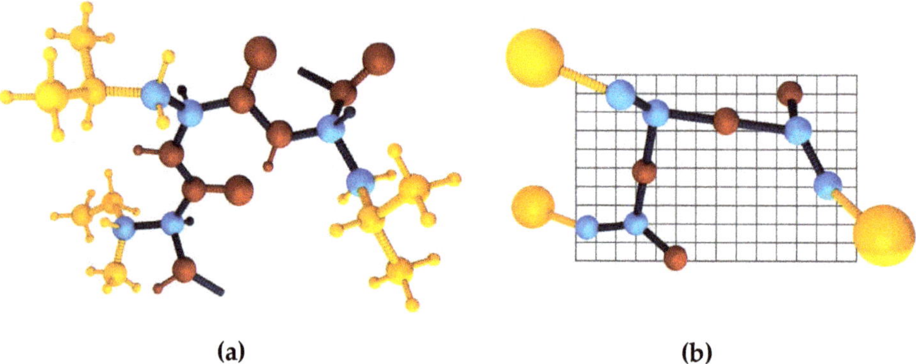

(a)   **(b)**

Figure 1. A three-residue protein fragment in: all-atom (**a**) and CABS model (**b**) representation. The spheres represent atoms: blue, C-alpha and C-beta atoms (the same in both representations); yellow, side chain atoms (one pseudo-atom in CABS); red, atoms involved in the peptide bond (one pseudo-atom in CABS placed in the geometric center of the peptide bond. A single slice (layer) of the lattice that confines the C-alpha trace in the CABS model is also presented.

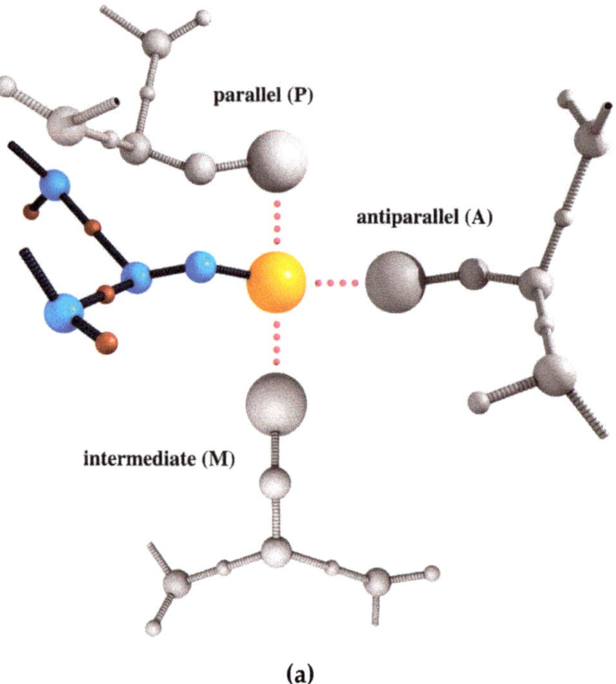

(a)

Figure 2. *Cont.*

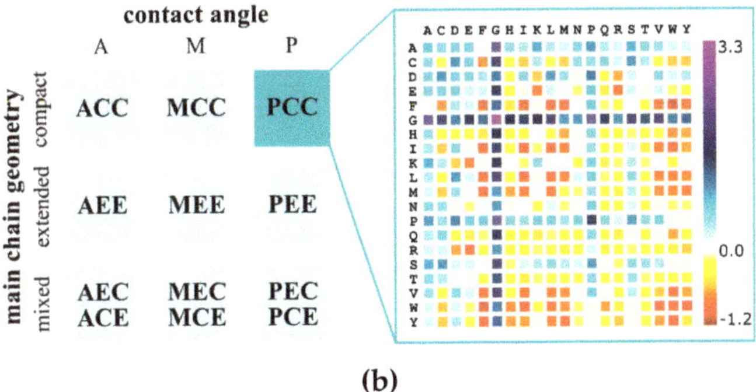

(b)

Figure 2. Key elements of a residue–residue interaction term in the CABS model force field. Panel (**a**) shows three examples of contact geometries in CABS representation: parallel (P), antiparallel (A), and intermediate (M), used to derive contact statistics from experimentally-derived structures of folded globular proteins. Panel (**b**) shows an example matrix of contact energies which depend on the geometry of the contacting pair, main chain geometry (compact (C) or extended (E)) for both amino acids (left part of the panel), and also on the amino acid identities (right part of the panel, the amino acids are represented using the one-letter code). The PCC matrix is presented which shows interaction energies between residues being in parallel orientation (P), where one residue belongs to a compact type of structure (C) and the second one as well (C).

The main difference between CABS and other statistical force fields used in CG models of similar resolution [11] is the context and orientation dependence of side chain interaction pseudo-energy that encodes characteristic patterns observed in globular proteins. For instance, the oppositely charged side chains in single globules mostly contact in an almost parallel fashion (usually on the surface of a globule), while the antiparallel contacts (usually in the buried regions of the protein globule) are very rare. Therefore, in the context dependent force field these antiparallel contacts of oppositely charged residues are treated as repulsive. This way, the CABS force field implicitly incorporates information on the complicated interaction patterns with the solvent (via contact statistics) and its entropic contribution to system thermodynamics [11,28].

Using the mean-force force field derived from folded proteins to simulations of less-structured systems raises justified questions about the validity of this approach in studies of the disordered protein regions. The folding events observed in simulations performed using the CABS force field are consistent with both the experimental data and all-atom MD simulations [49,52,80,81]. Thus, it is hypothesized that unstructured (unfolded, partially unfolded or intrinsically disordered) proteins to a significant extent share similar stabilizing interaction patterns with the patterns observed for their well-structured counterparts [82,83].

The CABS method uses the MC asymmetric Metropolis sampling scheme that governs a set of local motions as well as multi-residue, small distance moves of the C-alpha atoms (see Figure 3). The method uses a replica exchange algorithm with simulated annealing to enhance the sampling of conformational states. The simulation is organized as a set of nested loops, in which the s number of MC steps are organized into the y number of MC cycles, and these in the a number of annealing cycles. Each of the MC steps consists of a per-set number of attempts to perform each of the five standard precomputed moves. The available motions and the details of implementation of the sampling scheme are presented in Figure 3.

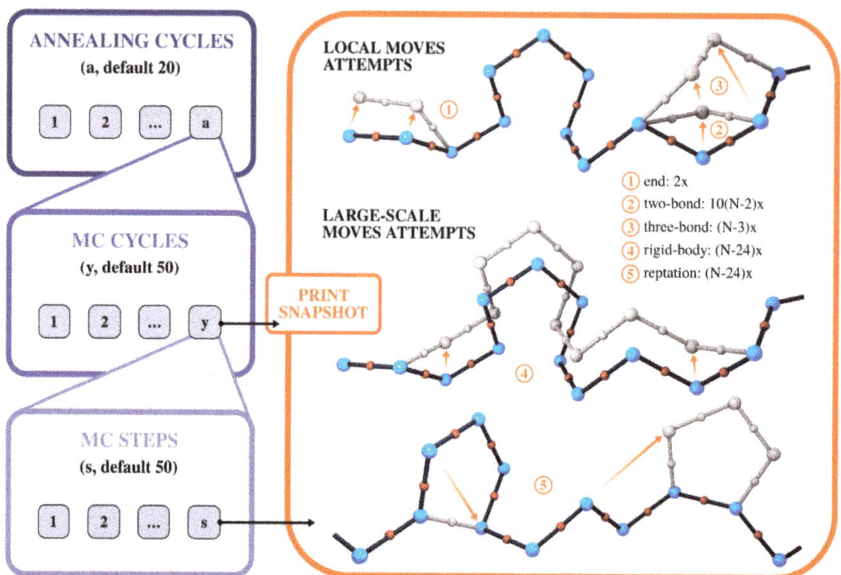

Figure 3. Sampling scheme of the CABS model. Blue panels show implementation details of Monte Carlo (MC) iterations (loops). The orange panel shows all motions that may be performed in a single MC step. The simulation is organized as a set of nested loops, in which the s number of MC steps is organized into the y number of cycles, and these in a annealing cycles (number of *a*, *y* or *s* cycles can be controlled by the user in CABS-flex and CABS-dock standalone packages [72]). In the orange panel, numbers 1 to 5 denote the available moves, presented together with the number of attempts to perform a move in each of the MC steps. The resulting trajectory is comprised of simulation snapshots saved at the end of each MC cycle.

The combination of the key features of CABS—its representation, force field and the scale of the movements used in the MC scheme—makes it suitable for the investigation of protein pseudo-dynamics. As mentioned above, the fine-grained lattice improves sampling efficiency, achieving effective timescales of milliseconds. As compared with MD, this is a considerably broader time range (in the study of flexibility of folded proteins [71] the CABS dynamics was estimated to be around 6×10^3 cheaper in terms of computational cost than the classical MD). The chosen micro-motions allow (via accumulation over simulation steps) cooperative, large-scale motions. The ensemble of structures produced by the CABS method resembles a dynamic ensemble averaged over the effective timescale. Due to the nature of the method, the picture of local dynamics is distorted (on the level of local moves); however, it may be argued (based on the works mentioned above that compared our simulations with experimental data) that the long-time pseudo-dynamics recovers the realistic picture of protein motions averaged over time.

The timescale of the CABS simulations is not a priori defined and depends on the CABS simulation temperature, due to hidden entropic contributions in the force field, accounting for implicit solvent effects and multi-body interactions encoded in the statistical force field. Nevertheless, the effective timescale of MC dynamics can be approximately identified by comparison with MD trajectories from sufficiently long simulations. This comparison was thoroughly discussed previously, and the results were compared to MD results [69] and NMR ensembles [71].

The CABS model is presently used as a simulation engine of a few multiscale modeling tools that merge CABS with models reconstruction to all-atom resolution. Those include the CABS-dock method for flexible protein-peptide docking (available as a web server [62] at http://biocomp.chem.uw.edu.pl/CABSdock and a standalone application [84] at https://bitbucket.org/lcbio/cabsdock/)

(accessed on 30 January 2019). In comparison to other protein–peptide docking tools, reviewed recently [85], CABS-dock offers a unique opportunity for modeling large-scale rearrangements of protein receptor structure during on-the-fly docking of fully flexible peptides. Another CABS-based tool, CABS-flex, enables fast simulations of protein flexibility (available as a web server [73] at http://biocomp.chem.uw.edu.pl/CABSflex and a standalone application [72] at https://bitbucket.org/lcbio/cabsflex/, accessed on 30 January 2019). This approach has been also incorporated as the module in the Aggrescan3D method for prediction of protein aggregation properties (available as a web server [86] at http://biocomp.chem.uw.edu.pl/A3D and a standalone application at https://bitbucket.org/lcbio/aggrescan3D, accessed on 30 January 2019). By using CABS-flex predictions, Aggrescan3D enables predicting the impact of protein conformational fluctuations on aggregation properties. Finally, the CABS model is used in the CABS-fold method for protein structure prediction: in the de novo fashion (from an amino acid sequence only), guided by user-provided templates or user-provided distance restraints (available as a web server [58] at http://biocomp.chem.uw.edu.pl/CABSfold/, accessed on 30 January 2019). The access to CABS-based tools, together with the tools description, is also available from websites of the laboratories: http://biocomp.chem.uw.edu.pl/ and http://lcbio.pl/ (accessed on 30 January 2019).

3. CABS Applications to Simulation of Disordered or Unfolded Proteins

In this section, we review CABS applications to simulations of protein–peptide binding (Section 3.1) and folding of globular proteins (Section 3.2). We briefly discuss modeling results for the binding of three protein–peptide systems and protein folding of one protein system. Figure 4 shows native conformations of these systems determined by X-ray crystallography or NMR. In the figure, they are arranged according to the size of a fully flexible fragment of the modeled system, effective timescales required for a meaningful simulation of their motions, and thus the modeling difficulty: (1) modeling of FxxLF motif peptide docking to an androgen receptor (AR), (2) investigation of binding and folding of an unstructured pKID protein to KIX protein, (3) modeling of p53-derived peptide docking to the MDM2 protein receptor with partially unstructured regions, and (4) simulation of the de novo folding of barnase. The simulations were performed using the CABS-dock method for protein–peptide docking [62] and CABS-flex methodology [72,73] that enable running de novo folding simulations.

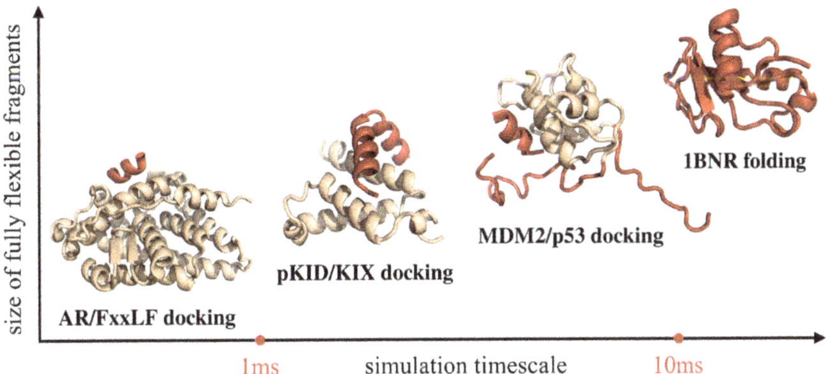

Figure 4. Presentation of the modeling cases discussed in this work. The modeled systems are arranged according to the size of the fully flexible fragment of the modeled system and the effective timescales required to observe their motions. The regions of the systems that were modeled as fully flexible are marked with red, while the regions in which backbone fluctuations were limited to 1 Å RMSD with beige. The presented millisecond values are approximated up to the order of magnitude.

3.1. Protein–Peptide Binding

The CABS-dock method has been extensively tested using the PeptiDB benchmark set of protein–peptide complexes [62,65,87]. One of the benchmark cases is the androgen receptor ligand binding domain (AR) in complex with a peptide with the FxxLF motif [88] (PDB code: 1T7R). To further analyze the interaction details of this complex, we performed blind global docking (using no knowledge about the binding site and peptide conformation) using CABS-dock [62]. As the input we used information on peptide sequence (incorporating the FxxLF motif: SSRFESLFAGEKESR), peptide secondary structure information assigned by the DSSP method [89] and the structure of the AR protein receptor. In this docking study, the peptide structure was simulated as fully flexible, while fluctuations of the protein receptor were limited to small backbone movements around the input structure (around 1 Å). The docking simulation started from random peptide conformations placed in random positions around the receptor structure. During simulation, the peptide remained unstructured until it was bound to the receptor binding site (Figure 5a). The docking simulations provided a set of high-quality models—the best model was characterized by a peptide-RMSD (root-mean-square deviation) value of 1.97 Å—and contact maps in strong agreement with the experimental data. As expected from the experimentally obtained structures and sequence analysis [88] the FxxLF interaction motif residues were most frequently involved in stabilizing hydrophobic interactions with the receptor. These high-frequency contacts are clearly visible in Figure 5a.

The study of the pKID/KIX system [63] involved performing a folding simulation of an intrinsically disordered protein (pKID) and its binding to a well-structured KIX receptor (Figure 5b). According to the experimental studies, the pKID structure is disordered in its unbound form with a slight propensity toward a helix (for detailed description on how one-dimensional secondary structure information is used in the CABS model see [78]). In the complex with the KIX protein, pKID adopts a characteristic conformation of two perpendicular helices that wrap around the receptor. However, most simulation results for the coupled folding and binding of this system published prior to the CABS-based study used models which biased pKID toward its native conformation (see the discussion in [63]). Using our method for studying this system enabled fully flexible treatment of the pKID protein. The obtained results [63] suggested the binding mechanism that involve two encounter complexes and were in well agreement with the available NMR experimental data. The predicted models presented high fractions of native contacts and allowed identification of residues essential for the binding and stabilization of the complex.

In the simulation of MDM2/p53 binding [64], the most challenging task was to adequately model the flexibility of the relatively long, unstructured regions of the protein receptor in addition to the fully flexible peptide [64,90] (Figure 5c). To provide a detailed insight into MDM2/p53 binding, we performed CABS-dock simulations and captured system behavior in agreement with the experimental data [64]. During the simulation, the flexible N- and C- terminal MDM2 fragments remained significantly disordered. The best resulting model was characterized by a peptide-RMSD value of 2.76 Å and 54% of the native contacts while the top ranked model by 3.74 Å and 60%, respectively. During simulations, we observed ensembles of models in which the peptide adopted different conformations loosely bound to the binding site and models in which the N-terminal highly flexible MDM2 fragment was interacting with the binding site. These findings are in agreement with the experimental data suggesting that p53-MDM2 binding is affected by significant rearrangements of the N-terminal MDM2 fragment (see discussion in [64]).

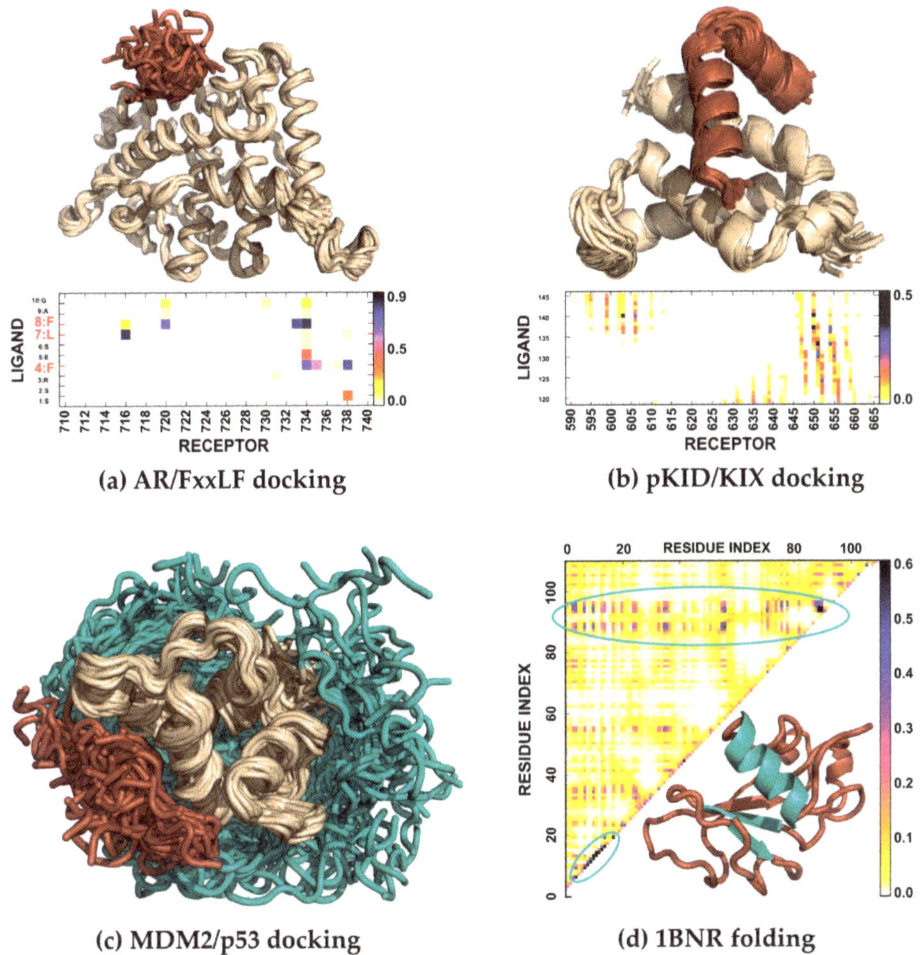

Figure 5. Case studies of modeling disordered or unfolded structures of proteins with CABS based tools. In the figures, red or cyan marks structure fragments simulated as fully flexible (cyan was used to mark regions of interest discussed in the text), while beige marks regions whose motions were confined to small backbone movements (around 1 Å from the input structure). (**a**) Modeling of the dynamics of a flexible peptide representing the FxxLF motif in the proximity of the binding site of AR protein together with an averaged contact map showing frequency of residue–residue contacts during the docking simulation. (**b**) Modeling of coupled folding and binding of the disordered pKID to the KIX domain [63]; the map presents the frequency of contacts of near-native conformations obtained in the simulation. (**c**) Modeling of p53 peptide binding to the MDM2 receptor [64], which includes fully-flexible regions of the protein receptor (shown in cyan) interacting with a fully-flexible peptide (shown in red). (**d**) Modeling of barnase folding [52] in the de novo fashion (using no knowledge about the structure); the map is a residue–residue contact map showing relative contact frequencies in denaturing conditions; the protein fragments that form the folding nucleation site are colored in cyan in the presented folded structure of barnase.

3.2. Folding and Flexibility of Globular Proteins

The CABS model has been applied to de novo simulations of protein folding (using no knowledge about the protein structure) for several model systems that have been extensively studied by experiment and simulation tools. Those studies include barnase [50,52], chymotrypsin inhibitor [50,52], B1 domain of protein G [49,50], B domain of protein A [53], and others [50,54]. The CABS modeling protocol was also extended to enable studies of the chaperonin effect on the folding mechanism [55]. In these works, various parameters have been studied, including residue–residue contact frequency, radius of gyration, residual secondary structure and others. The obtained pictures, which covered protein dynamics from highly denatured states to ensembles close to the folded states, agreed well with available experimental data.

For example, simulation of barnase folding resulted in the adequate reproduction of the folding pathway in strong agreement with NMR data for denatured states and phi-value analysis [52]. The performed simulations show that barnase folding starts with developing a folding nucleation site that consists of protein fragments corresponding to two strands of a beta sheet and one of the helices in the folded structure (presented in Figure 5d). In addition, the characteristic patterns of hydrophobic interactions that are crucial for the initiation and sustenance of folding are in agreement with the experimental data (see discussion in Reference [52], the contact map resulting from these simulations is presented in Figure 5d).

4. Conclusions

The presented case studies review the applications of the CABS model in simulations of disordered or unfolded protein states. As discussed, the method succeeded in capturing the experimentally determined features of the investigated systems, such as binding site localization, key contacts, peptide hot-spot areas, distinctive conformational states of the system, transient encounter complexes and intermediate states in protein folding [49,52,63,64]. Additionally, CABS enables an investigation of fluctuations of globular proteins around the native (input) structure [69–73].

There is a number of tools commonly used for sampling of disordered protein states, which predictions agree with the experimental studies [91–95]. The CABS method is complementary to these and provides a unique approach allowing for effective modeling both ordered and disordered elements of the system. As observed in many previous studies, these features of CABS method allow for providing accurate pictures of folding pathways [49,52–56,60] and near-native dynamics [70,71]. Obviously, due to its coarse-graining, the geometric details are missed, and their reconstructions is approximate [11,28]. The main distinctive feature of CABS method as compared to the available tools is that the ensemble generation is (pseudo-)energy driven and thus may provide some information on the dynamics on the system. This is not the case in the above-mentioned examples of methods based on random-walk [91,92,95].

On the other hand, CABS force field side-chain interactions escape a clear interpretation, which may be a disadvantage compared to physics-based approaches that allow for straightforward and detailed description of each of the terms [93,94].

It is, however, noteworthy that statistical force fields suffer from inherent limitations, depending on the chosen method of derivation. The most commonly discussed challenges include the transferability, solvent interactions and integration of experimental data. Here, we briefly summarize these topics, a detailed discussion of the limitations of this approach, and possible workarounds may be found in review works [11,17]. The transferability of statistical force fields may be limited as they are applicable always to a certain subset of proteins. Therefore, the performance of knowledge-based approaches may be poor for rare or atypical structures, for which appropriate statistics of contact patterns could not be collected. It should also be noted that interactions with solvent are averaged and treated implicitly, which may lead to significant discrepancies if the method is applied to non-standard solvent conditions (such as extreme pH values). The CABS force field is derived assuming averaged effect solvent conditions for folded globular proteins. Therefore, a subtle effect of small molecules, such as pH, cannot be simulated in a strict fashion, although averaged effects (see modeling the chaperonin effect [55]) can be approximately taken into considerations.

One of the most challenging tasks in modeling protein systems is the effective incorporation of sparse experimental data to drive the modeling procedure. In the CABS model, the experimental data may be readily introduced into the simulation as geometry distance restraints and weighted according to their certainty. A thorough discussion of this possibility is presented in the documentation of CABS-based tools for the fast modeling of protein flexibility and protein–peptide docking [66,72,73]. On a similar basis, CABS simulations can be guided by computational predictions from other sources or integrated with other modeling tools of various resolution. Therefore, the CABS model can be incorporated into integrative modeling pipelines that would benefit from its effective sampling scheme. The recently published standalone application and web server tools are available for integration with external pipelines (access links are presented in the last paragraph of Section 2).

Author Contributions: S.K. and A.K. conceptualized this review. M.P. performed the simulations and analyzed the results for the AR/FxxLF system. The review was written by M.P.C., A.E.B-D., A.K. and S.K.

Funding: This research was funded by NCN Poland, grant number MAESTRO2014/14/A/ST6/00088.

Conflicts of Interest: The authors declare no conflict of interest.

Abbreviations

CABS	Cα, Cβ, Side chain model
MC	Monte Carlo
NMR	nuclear magnetic resonance
MD	molecular dynamics
CG	coarse-grained
AR	androgen receptor
DSSP	dictionary of protein secondary structure
RMSD	root-mean-square deviation of atomic positions
PDB	Protein Data Bank
CASP	Critical Assessment of protein Structure Prediction

References

1. Dishman, A.F.; Volkman, B.F. Unfolding the Mysteries of Protein Metamorphosis. *ACS Chem. Biol.* **2018**, *13*, 1438–1446. [CrossRef] [PubMed]
2. Uversky, V.N. Dancing protein clouds: The strange biology and chaotic physics of intrinsically disordered proteins. *J. Biol. Chem.* **2016**, *291*, 6681–6688. [CrossRef] [PubMed]
3. Wright, P.E.; Dyson, H.J. Intrinsically disordered proteins in cellular signalling and regulation. *Nat. Rev. Mol. Cell Biol.* **2015**, *16*, 18–29. [CrossRef] [PubMed]
4. Henzler-Wildman, K.; Kern, D. Dynamic personalities of proteins. *Nature* **2007**, *450*, 964–972. [CrossRef] [PubMed]
5. Vendruscolo, M.; Dobson, C.M. Dynamic visions of enzymatic reactions. *Science* **2006**, *313*, 1586–1587. [CrossRef] [PubMed]
6. Wei, G.; Xi, W.; Nussinov, R.; Ma, B. Protein Ensembles: How Does Nature Harness Thermodynamic Fluctuations for Life? the Diverse Functional Roles of Conformational Ensembles in the Cell. *Chem. Rev.* **2016**, *116*, 6516–6551. [CrossRef]
7. Best, R.B. Computational and theoretical advances in studies of intrinsically disordered proteins. *Curr. Opin. Struct. Biol.* **2017**, *42*, 147–154. [CrossRef]
8. Kay, L.E. NMR studies of protein structure and dynamics. *J. Magn. Reson.* **2011**, *213*, 477–491. [CrossRef]
9. Robustelli, P.; Piana, S.; Shaw, D.E. Developing a molecular dynamics force field for both folded and disordered protein states. *Proc. Natl. Acad. Sci. USA* **2018**, *115*, E4758–E4766. [CrossRef]
10. Bowman, G.R.; Voelz, V.A.; Pande, V.S. Taming the complexity of protein folding. *Curr. Opin. Struct. Biol.* **2011**, *21*, 4–11. [CrossRef]

11. Kmiecik, S.; Gront, D.; Kolinski, M.; Wieteska, L.; Dawid, A.E.; Kolinski, A. Coarse-Grained Protein Models and Their Applications. *Chem. Rev.* **2016**, *116*, 7898–7936. [CrossRef] [PubMed]

12. Zhang, Z.; Chan, H.S. Competition between native topology and nonnative interactions in simple and complex folding kinetics of natural and designed proteins. *Proc. Natl. Acad. Sci. USA* **2010**, *107*, 2920–2925. [CrossRef] [PubMed]

13. Shan, B.; Eliezer, D.; Raleigh, D. The unfolded state of the C-terminal domain of the ribosomal protein L9 contains both native and non-native structure. *Biochemistry* **2009**, *48*, 4707–4719. [CrossRef] [PubMed]

14. Rothwarf, D.M.; Scheraga, H.A. Role of non-native aromatic and hydrophobic interactions in the folding of hen egg white lysozyme. *Biochemistry* **1996**, *35*, 13797–13807. [CrossRef] [PubMed]

15. Cavalli, A.; Montalvao, R.W.; Vendruscolo, M. Using chemical shifts to determine structural changes in proteins upon complex formation. *J. Phys. Chem. B* **2011**, *115*, 9491–9494. [CrossRef] [PubMed]

16. Fu, B.; Kukic, P.; Camilloni, C.; Vendruscolo, M. MD Simulations of Intrinsically Disordered Proteins with Replica-Averaged Chemical Shift Restraints. *Biophys. J.* **2014**, *106*, 481a. [CrossRef]

17. Kar, P.; Feig, M. Recent advances in transferable coarse-grained modeling of proteins. *Adv. Protein Chem. Struct. Biol.* **2014**, *96*, 143–180. [CrossRef]

18. Greener, J.G.; Filippis, I.; Sternberg, M.J.E. Predicting Protein Dynamics and Allostery Using Multi-Protein Atomic Distance Constraints. *Structure* **2017**, *25*, 546–558. [CrossRef]

19. Klepeis, J.L.; Lindorff-Larsen, K.; Dror, R.O.; Shaw, D.E. Long-timescale molecular dynamics simulations of protein structure and function. *Curr. Opin. Struct. Biol.* **2009**, *19*, 120–127. [CrossRef]

20. Bernardi, R.C.; Melo, M.C.R.; Schulten, K. Enhanced sampling techniques in molecular dynamics simulations of biological systems. *Biochim. Biophys. Acta - Gen. Subj.* **2015**, *1850*, 872–877. [CrossRef]

21. Shukla, D.; Hernández, C.X.; Weber, J.K.; Pande, V.S. Markov state models provide insights into dynamic modulation of protein function. *Acc. Chem. Res.* **2015**, *48*, 414–422. [CrossRef] [PubMed]

22. Kolinski, A. Toward more efficient simulations of slow processes in large biomolecular systems: Comment on "Ligand diffusion in proteins via enhanced sampling in molecular dynamics" by Jakub Rydzewski and Wieslaw Nowak. *Phys. Life Rev.* **2017**, *22–23*, 75–76. [CrossRef] [PubMed]

23. Rydzewski, J.; Nowak, W. Ligand diffusion in proteins via enhanced sampling in molecular dynamics. *Phys. Life Rev.* **2017**, *22–23*, 82–84. [CrossRef] [PubMed]

24. Maximova, T.; Moffatt, R.; Ma, B.; Nussinov, R.; Shehu, A. Principles and Overview of Sampling Methods for Modeling Macromolecular Structure and Dynamics. *PLoS Comput. Biol.* **2016**, *12*, e1004619. [CrossRef] [PubMed]

25. Hatherley, R.; Brown, D.K.; Glenister, M.; Bishop, Ö.T. PRIMO: An interactive homology modeling pipeline. *PLoS ONE* **2016**, *11*, e0166698. [CrossRef] [PubMed]

26. Das, R.; Baker, D. Macromolecular Modeling with Rosetta. *Annu. Rev. Biochem.* **2008**, *77*, 363–382. [CrossRef] [PubMed]

27. Czaplewski, C.; Karczyńska, A.; Sieradzan, A.K.; Liwo, A. UNRES server for physics-based coarse-grained simulations and prediction of protein structure, dynamics and thermodynamics. *Nucleic Acids Res.* **2018**, *46*, W304–W309. [CrossRef] [PubMed]

28. Kolinski, A. Protein modeling and structure prediction with a reduced representation. *Acta Biochim. Pol.* **2004**, *51*, 349–371.

29. Davtyan, A.; Schafer, N.P.; Zheng, W.; Clementi, C.; Wolynes, P.G.; Papoian, G.A. AWSEM-MD: Protein structure prediction using coarse-grained physical potentials and bioinformatically based local structure biasing. *J. Phys. Chem. B* **2012**, *116*, 8494–8503. [CrossRef]

30. Marrink, S.J.; Tieleman, D.P. Perspective on the Martini model. *Chem. Soc. Rev.* **2013**, *42*, 6801. [CrossRef]

31. Pasi, M.; Lavery, R.; Ceres, N. PaLaCe: A coarse-grain protein model for studying mechanical properties. *J. Chem. Theory Comput.* **2013**, *9*, 785–793. [CrossRef] [PubMed]

32. Basdevant, N.; Borgis, D.; Ha-Duong, T. Modeling protein-protein recognition in solution using the coarse-grained force field SCORPION. *J. Chem. Theory Comput.* **2013**, *9*, 803–813. [CrossRef] [PubMed]

33. Dawid, A.E.; Gront, D.; Kolinski, A. SURPASS Low-Resolution Coarse-Grained Protein Modeling. *J. Chem. Theory Comput.* **2017**, *13*, 5766–5779. [CrossRef] [PubMed]

34. Dawid, A.E.; Gront, D.; Kolinski, A. Coarse-Grained Modeling of the Interplay between Secondary Structure Propensities and Protein Fold Assembly. *J. Chem. Theory Comput.* **2018**, *14*, 2277–2287. [CrossRef] [PubMed]

35. Lopes, P.E.M.; Guvench, O.; MacKerell, A.D. Current Status of Protein Force Fields for Molecular Dynamics Simulations. In *Molecular Modeling of Proteins*; Humana Press: New York, NY, USA, 2015; pp. 47–71.

36. Cossio, P.; Granata, D.; Laio, A.; Seno, F.; Trovato, A. A simple and efficient statistical potential for scoring ensembles of protein structures. *Sci. Rep.* **2012**, *2*, 351. [CrossRef]

37. Tanaka, S.; Scheraga, H.A. Medium- and Long-Range Interaction Parameters between Amino Acids for Predicting Three-Dimensional Structures of Proteins. *Macromolecules* **1976**, *9*, 945–950. [CrossRef] [PubMed]

38. Tsai, J.; Bonneau, R.; Morozov, A.V.; Kuhlman, B.; Rohl, C.A.; Baker, D. An improved protein decoy set for testing energy functions for protein structure prediction. *Proteins Struct. Funct. Genet.* **2003**, *53*, 76–87. [CrossRef] [PubMed]

39. Shen, M.; Sali, A. Statistical potential for assessment and prediction of protein structures. *Protein Sci.* **2006**, *15*, 2507–2524. [CrossRef]

40. Zhou, H.; Skolnick, J. GOAP: A Generalized Orientation-Dependent, All-Atom Statistical Potential for Protein Structure Prediction. *Biophys. J.* **2011**, *101*, 2043–2052. [CrossRef]

41. Xu, D.; Zhang, Y. Ab initio protein structure assembly using continuous structure fragments and optimized knowledge-based force field. *Proteins Struct. Funct. Bioinforma.* **2012**, *80*, 1715–1735. [CrossRef]

42. Woetzel, N.; Karakaş, M.; Staritzbichler, R.; Müller, R.; Weiner, B.E.; Meiler, J. BCL::Score—Knowledge Based Energy Potentials for Ranking Protein Models Represented by Idealized Secondary Structure Elements. *PLoS ONE* **2012**, *7*, e49242. [CrossRef] [PubMed]

43. Ovchinnikov, S.; Park, H.; Kim, D.E.; Liu, Y.; Wang, R.Y.-R.; Baker, D. Structure prediction using sparse simulated NOE restraints with Rosetta in CASP11. *Proteins Struct. Funct. Bioinforma.* **2016**, *84*, 181–188. [CrossRef] [PubMed]

44. Ovchinnikov, S.; Kim, D.E.; Wang, R.Y.-R.; Liu, Y.; DiMaio, F.; Baker, D. Improved de novo structure prediction in CASP11 by incorporating coevolution information into Rosetta. *Proteins Struct. Funct. Bioinforma.* **2016**, *84*, 67–75. [CrossRef] [PubMed]

45. Hirst, S.J.; Alexander, N.; Mchaourab, H.S.; Meiler, J. RosettaEPR: An integrated tool for protein structure determination from sparse EPR data. *J. Struct. Biol.* **2011**, *173*, 506–514. [CrossRef]

46. Yang, J.; Zhang, W.; He, B.; Walker, S.E.; Zhang, H.; Govindarajoo, B.; Virtanen, J.; Xue, Z.; Shen, H.B.; Zhang, Y. Template-based protein structure prediction in CASP11 and retrospect of I-TASSER in the last decade. *Proteins* **2016**, *84*, 233–246. [CrossRef]

47. Russel, D.; Lasker, K.; Webb, B.; Velázquez-Muriel, J.; Tjioe, E.; Schneidman-Duhovny, D.; Peterson, B.; Sali, A. Putting the Pieces Together: Integrative Modeling Platform Software for Structure Determination of Macromolecular Assemblies. *PLoS Biol.* **2012**, *10*, e1001244. [CrossRef]

48. Rodrigues, J.P.G.L.M.; Bonvin, A.M.J.J. Integrative computational modeling of protein interactions. *FEBS J.* **2014**, *281*, 1988–2003. [CrossRef]

49. Kmiecik, S.; Kolinski, A. Folding pathway of the B1 domain of protein G explored by multiscale modeling. *Biophys. J.* **2008**, *94*, 726–736. [CrossRef]

50. Kolinski, A. Multiscale approaches to protein modeling: Structure prediction, dynamics, thermodynamics and macromolecular assemblies. In *Multiscale Approaches to Protein Modeling: Structure Prediction, Dynamics, Thermodynamics and Macromolecular Assemblies*; Kolinski, A., Ed.; Springer: New York, NY, USA, 2011; pp. 1–355. ISBN 9781441968890.

51. Kmiecik, S.; Kouza, M.; Badaczewska-Dawid, A.E.; Kloczkowski, A.; Kolinski, A. Modeling of Protein Structural Flexibility and Large-Scale Dynamics: Coarse-Grained Simulations and Elastic Network Models. *Int. J. Mol. Sci.* **2018**, *19*, 3496. [CrossRef]

52. Kmiecik, S.; Kolinski, A. Characterization of protein-folding pathways by reduced-space modeling. *Proc. Natl. Acad. Sci. USA* **2007**, *104*, 12330–12335. [CrossRef]

53. Kmiecik, S.; Gront, D.; Kouza, M.; Kolinski, A. From coarse-grained to atomic-level characterization of protein dynamics: Transition state for the folding of B domain of protein A. *J. Phys. Chem. B* **2012**, *116*, 7026–7032. [CrossRef] [PubMed]

54. Kmiecik, S.; Kurcinski, M.; Rutkowska, A.; Gront, D.; Kolinski, A. Denatured proteins and early folding intermediates simulated in a reduced conformational space. *Acta Biochim. Pol.* **2006**, *53*, 131–143. [CrossRef] [PubMed]

55. Kmiecik, S.; Kolinski, A. Simulation of chaperonin effect on protein folding: A shift from nucleation - Condensation to framework mechanism. *J. Am. Chem. Soc.* **2011**, *133*, 10283–10289. [CrossRef] [PubMed]

56. Jamroz, M.; Kolinski, A.; Kmiecik, S. Protocols for efficient simulations of long-time protein dynamics using coarse-grained CABS model. *Methods Mol. Biol.* **2014**, *1137*, 235–250. [CrossRef] [PubMed]

57. Wabik, J.; Kmiecik, S.; Gront, D.; Kouza, M.; Koliński, A. Combining coarse-grained protein models with replica-exchange all-atom molecular dynamics. *Int. J. Mol. Sci.* **2013**, *14*, 9893–9905. [CrossRef] [PubMed]

58. Blaszczyk, M.; Jamroz, M.; Kmiecik, S.; Kolinski, A. CABS-fold: Server for the de novo and consensus-based prediction of protein structure. *Nucleic Acids Res.* **2013**, *41*, W406–W411. [CrossRef] [PubMed]

59. Kmiecik, S.; Jamroz, M.; Kolinski, M. Structure prediction of the second extracellular loop in G-protein-coupled receptors. *Biophys. J.* **2014**, *106*, 2408–2416. [CrossRef]

60. Koliński, A.; Bujnicki, J.M. Generalized protein structure prediction based on combination of fold-recognition with de novo folding and evaluation of models. *Proteins Struct. Funct. Genet.* **2005**, *61*, 84–90. [CrossRef]

61. Jamroz, M.; Kolinski, A. Modeling of loops in proteins: A multi-method approach. *BMC Struct. Biol.* **2010**, *10*. [CrossRef]

62. Kurcinski, M.; Jamroz, M.; Blaszczyk, M.; Kolinski, A.; Kmiecik, S. CABS-dock web server for the flexible docking of peptides to proteins without prior knowledge of the binding site. *Nucleic Acids Res.* **2015**, *43*, W419–W424. [CrossRef]

63. Kurcinski, M.; Kolinski, A.; Kmiecik, S. Mechanism of folding and binding of an intrinsically disordered protein as revealed by ab initio simulations. *J. Chem. Theory Comput.* **2014**, *10*, 2224–2231. [CrossRef] [PubMed]

64. Ciemny, M.P.; Debinski, A.; Paczkowska, M.; Kolinski, A.; Kurcinski, M.; Kmiecik, S. Protein-peptide molecular docking with large-scale conformational changes: The p53-MDM2 interaction. *Sci. Rep.* **2016**, *6*. [CrossRef] [PubMed]

65. Blaszczyk, M.; Kurcinski, M.; Kouza, M.; Wieteska, L.; Debinski, A.; Kolinski, A.; Kmiecik, S. Modeling of protein-peptide interactions using the CABS-dock web server for binding site search and flexible docking. *Methods* **2016**, *93*, 72–83. [CrossRef] [PubMed]

66. Ciemny, M.; Kurcinski, M.; Kozak, K.; Kolinski, A.; Kmiecik, S. Highly flexible protein-peptide docking using cabs-dock. *Methods Mol. Biol.* **2017**, *1561*, 69–94. [CrossRef]

67. Blaszczyk, M.; Ciemny, M.P.; Kolinski, A.; Kurcinski, M.; Kmiecik, S. Protein–peptide docking using CABS-dock and contact information. *Brief. Bioinform.* **2018**, *bby080*. [CrossRef]

68. Ciemny, M.P.; Kurcinski, M.; Blaszczyk, M.; Kolinski, A.; Kmiecik, S. Modeling EphB4-EphrinB2 protein-protein interaction using flexible docking of a short linear motif. *Biomed. Eng. Online* **2017**, *16*, 71. [CrossRef]

69. Jamroz, M.; Orozco, M.; Kolinski, A.; Kmiecik, S. Consistent view of protein fluctuations from all-atom molecular dynamics and coarse-grained dynamics with knowledge-based force-field. *J. Chem. Theory Comput.* **2013**, *9*, 119–125. [CrossRef]

70. Jamroz, M.; Kolinski, A.; Kmiecik, S. CABS-flex: Server for fast simulation of protein structure fluctuations. *Nucleic Acids Res.* **2013**, *41*, W427–W431. [CrossRef] [PubMed]

71. Jamroz, M.; Kolinski, A.; Kmiecik, S. CABS-flex predictions of protein flexibility compared with NMR ensembles. *Bioinformatics* **2014**, *30*, 2150–2154. [CrossRef] [PubMed]

72. Kurcinski, M.; Oleniecki, T.; Ciemny, P.M.; Kuriata, A.; Kolinski, A.; Kmiecik, S. CABS-flex standalone: A simulation environment for fast modeling of protein flexibility. *Bioinformatics* **2018**, *bty685*. [CrossRef] [PubMed]

73. Kuriata, A.; Gierut, A.M.; Oleniecki, T.; Ciemny, M.P.; Kolinski, A.; Kurcinski, M.; Kmiecik, S. CABS-flex 2.0: A web server for fast simulations of flexibility of protein structures. *Nucleic Acids Res.* **2018**, *46*, W338–W343. [CrossRef] [PubMed]

74. Eswar, N.; John, B.; Mirkovic, N.; Fiser, A.; Ilyin, V.A.; Pieper, U.; Stuart, A.C.; Marti-Renom, M.A.; Madhusudhan, M.S.; Yerkovich, B.; Sali, A. Tools for comparative protein structure modeling and analysis. *Nucleic Acids Res.* **2003**, *31*, 3375–3380. [CrossRef] [PubMed]

75. Gront, D.; Kmiecik, S.; Kolinski, A. Backbone building from quadrilaterals: A fast and accurate algorithm for protein backbone reconstruction from alpha carbon coordinates. *J. Comput. Chem.* **2007**, *28*, 1593–1597. [CrossRef] [PubMed]

76. Canutescu, A.A.; Shelenkov, A.A.; Dunbrack, R.L. A graph-theory algorithm for rapid protein side-chain prediction. *Protein Sci.* **2003**, *12*, 2001–2014. [CrossRef] [PubMed]

77. Gront, D.; Kmiecik, S.; Blaszczyk, M.; Ekonomiuk, D.; Koliński, A. Optimization of protein models. *Wiley Interdiscip. Rev. Comput. Mol. Sci.* **2012**, *2*, 479–493. [CrossRef]

78. Kmiecik, S.; Kolinski, A. One-dimensional structural properties of proteins in the coarse-grained cabs model. *Methods Mol. Biol.* **2017**, *1484*, 83–113. [CrossRef] [PubMed]

79. Pulawski, W.; Jamroz, M.; Kolinski, M.; Kolinski, A.; Kmiecik, S. Coarse-grained simulations of membrane insertion and folding of small helical proteins using the CABS model. *J. Chem. Inf. Model.* **2016**, *56*, 2207–2215. [CrossRef]

80. Adhikari, A.N.; Freed, K.F.; Sosnick, T.R. De novo prediction of protein folding pathways and structure using the principle of sequential stabilization. *Proc. Natl. Acad. Sci. USA* **2012**, *109*, 17442–17447. [CrossRef]

81. Adhikari, A.N.; Freed, K.F.; Sosnick, T.R. Simplified protein models: Predicting folding pathways and structure using amino acid sequences. *Phys. Rev. Lett.* **2013**, *111*, 028103. [CrossRef]

82. Konrat, R. NMR contributions to structural dynamics studies of intrinsically disordered proteins. *J. Magn. Reson.* **2014**, *241*, 74–85. [CrossRef]

83. Kmiecik, S.; Wabik, J.; Kolinski, M.; Kouza, M.; Kolinski, A. Coarse-Grained Modeling of Protein Dynamics. In *Computational Methods to Study the Structure and Dynamics of Biomolecules*; Springer: Berlin/Heidelberg, Germany, 2014; Volume 1, pp. 55–79. ISBN 978-3-642-28553-0.

84. Kurcinski, M.; Ciemny, M.P.; Oleniecki, T.; Kuriata, A.; Badaczewska-Dawid, A.E.; Kolinski, A.; Kmiecik, S. CABS-dock standalone: A toolbox for flexible protein-peptide docking. *Bioinformatics* **2019**. submitted.

85. Ciemny, M.; Kurcinski, M.; Kamel, K.; Kolinski, A.; Alam, N.; Schueler-Furman, O.; Kmiecik, S. Protein–peptide docking: Opportunities and challenges. *Drug Discov. Today* **2018**, *23*, 1530–1537. [CrossRef] [PubMed]

86. Zambrano, R.; Jamroz, M.; Szczasiuk, A.; Pujols, J.; Kmiecik, S.; Ventura, S. AGGRESCAN3D (A3D): Server for prediction of aggregation properties of protein structures. *Nucleic Acids Res.* **2015**, *43*, W306–W313. [CrossRef]

87. London, N.; Movshovitz-Attias, D.; Schueler-Furman, O. The Structural Basis of Peptide-Protein Binding Strategies. *Structure* **2010**, *18*, 188–199. [CrossRef] [PubMed]

88. Hur, E.; Pfaff, S.J.; Sturgis Payne, E.; Grøn, H.; Buehrer, B.M.; Fletterick, R.J. Recognition and accommodation at the androgen receptor coactivator binding interface. *PLoS Biol.* **2004**, *2*, E274. [CrossRef] [PubMed]

89. Kabsch, W.; Sander, C. Dictionary of protein secondary structure: Pattern recognition of hydrogen-bonded and geometrical features. *Biopolymers* **1983**, *22*, 2577–2637. [CrossRef] [PubMed]

90. Kussie, P.H.; Gorina, S.; Marechal, V.; Elenbaas, B.; Moreau, J.; Levine, A.J.; Pavletich, N.P. Structure of the MDM2 oncoprotein bound to the p53 tumor suppressor transactivation domain. *Science* **1996**, *274*, 948–953. [CrossRef] [PubMed]

91. Ozenne, V.; Bauer, F.; Salmon, L.; Huang, J.R.; Jensen, M.R.; Segard, S.; Bernadó, P.; Charavay, C.; Blackledge, M. Flexible-meccano: A tool for the generation of explicit ensemble descriptions of intrinsically disordered proteins and their associated experimental observables. *Bioinformatics* **2012**, *28*, 1463–1470. [CrossRef] [PubMed]

92. Feldman, H.J.; Hogue, C.W.V. Probabilistic sampling of protein conformations: New hope for brute force? *Proteins Struct. Funct. Genet.* **2002**, *46*, 8–23. [CrossRef] [PubMed]

93. Vitalis, A.; Pappu, R.V. ABSINTH: A new continuum solvation model for simulations of polypeptides in aqueous solutions. *J. Comput. Chem.* **2009**, *30*, 673–699. [CrossRef] [PubMed]

94. Baul, U.; Chakraborty, D.; Mugnai, M.L.; Straub, J.E.; Thirumalai, D. Sequence effects on size, shape, and structural heterogeneity in Intrinsically Disordered Proteins. *bioRxiv* **2018**, 427476. [CrossRef]

95. Estaña, A.; Sibille, N.; Delaforge, E.; Vaisset, M.; Cortés, J.; Bernadó, P. Realistic Ensemble Models of Intrinsically Disordered Proteins Using a Structure-Encoding Coil Database. *Structure* **2018**. [CrossRef] [PubMed]

© 2019 by the authors. Licensee MDPI, Basel, Switzerland. This article is an open access article distributed under the terms and conditions of the Creative Commons Attribution (CC BY) license (http://creativecommons.org/licenses/by/4.0/).

International Journal of
Molecular Sciences

Review

Modulation of Disordered Proteins with a Focus on Neurodegenerative Diseases and Other Pathologies

Anne H. S. Martinelli [1,†], **Fernanda C. Lopes** [2,3,†], **Elisa B. O. John** [2,3], **Célia R. Carlini** [3,4,5,*] **and Rodrigo Ligabue-Braun** [6,*]

1   Department of Molecular Biology and Biotechnology & Department of Biophysics, Biosciences Institute-IB, (UFRGS), Porto Alegre CEP 91501-970, RS, Brazil; ahsmartinelli@gmail.com
2   Center for Biotechnology, Universidade Federal do Rio Grande do Sul (UFRGS), Porto Alegre CEP 91501-970, RS, Brazil; fernandacortezlopes@gmail.com (F.C.L.); elisabeajohn@gmail.com (E.B.O.J.)
3   Graduate Program in Cell and Molecular Biology, Universidade Federal do Rio Grande do Sul (UFRGS), Porto Alegre CEP 91501-970, RS, Brazil
4   Graduate Program in Medicine and Health Sciences, Pontifícia Universidade Católica do Rio Grande do Sul (PUCRS), Porto Alegre CEP 91410-000, RS, Brazil
5   Brain Institute-InsCer, Laboratory of Neurotoxins, Pontifícia Universidade Católica do Rio Grande do Sul (PUCRS), Porto Alegre CEP 90610-000, RS, Brazil
6   Department of Pharmaceutical Sciences, Universidade Federal de Ciências da Saúde de Porto Alegre (UFCSPA), Porto Alegre CEP 90050-170, RS, Brazil
*   Correspondence: celia.carlini@pucrs.br or celia.carlini@pq.cnpq.br (C.R.C.); rodrigolb@ufcspa.edu.br (R.L.-B.)
†   These authors contributed equally to this work.

Received: 21 December 2018; Accepted: 12 February 2019; Published: 15 March 2019

Abstract: Intrinsically disordered proteins (IDPs) do not have rigid 3D structures, showing changes in their folding depending on the environment or ligands. Intrinsically disordered proteins are widely spread in eukaryotic genomes, and these proteins participate in many cell regulatory metabolism processes. Some IDPs, when aberrantly folded, can be the cause of some diseases such as Alzheimer's, Parkinson's, and prionic, among others. In these diseases, there are modifications in parts of the protein or in its entirety. A common conformational variation of these IDPs is misfolding and aggregation, forming, for instance, neurotoxic amyloid plaques. In this review, we discuss some IDPs that are involved in neurodegenerative diseases (such as beta amyloid, alpha synuclein, tau, and the "IDP-like" PrP), cancer (p53, c-Myc), and diabetes (amylin), focusing on the structural changes of these IDPs that are linked to such pathologies. We also present the IDP modulation mechanisms that can be explored in new strategies for drug design. Lastly, we show some candidate drugs that can be used in the future for the treatment of diseases caused by misfolded IDPs, considering that cancer therapy has more advanced research in comparison to other diseases, while also discussing recent and future developments in this area of research. Therefore, we aim to provide support to the study of IDPs and their modulation mechanisms as promising approaches to combat such severe diseases.

Keywords: intrinsically disordered proteins (IDPs); neurodegenerative diseases; aggregation; drugs; drug discovery

---

1. Introduction

The protein structure–function paradigm was established in the 20th century. The key point of this paradigm is that an ordered (rigid) and unique 3D structure of a protein is an obligatory prerequisite for protein function [1,2]. Nevertheless, recent studies have provided broad and convincing evidences that some proteins do not adopt only one structure, but still are fully functional [3].

The different possible protein conformations are structured (folded), molten globular, pre-molten globular, and unstructured (unfolded) [4].

Since the beginning of the 2000s, a new class of unstructured proteins started to be studied more due to the improvement of techniques to elucidate protein structure. Crystal-structure analysis using X-ray diffraction cannot provide information on unstructured states, with only the absence of electron density in some regions being observed. However, the nuclear magnetic resonance (NMR) technique allowed for the better characterization of these disordered proteins, confirming the flexibility of protein segments that are missing in crystallography experiments [5]. They are defined mainly as intrinsically disordered proteins (IDPs), in spite of some authors defining these proteins as natively denatured [6], natively unfolded [7], intrinsically unstructured [8], and natively disordered proteins [9], among other definitions. We will use the IDP definition to refer to these proteins. The disorder could be also present in some regions of proteins; these regions are named intrinsically disordered regions (IDRs). Intrinsically disordered proteins /IDRs have no single, well-defined equilibrium structure and exist as heterogeneous ensembles of conformers [10].

There are significant differences between the amino acid sequences of IDPs/IDRs in comparison with structured globular proteins and/or domains. These differences are related to amino acid composition, sequence complexity, hydrophobicity, aromaticity, charge, flexibility, type and rate of amino acid substitutions over evolutionary time [11]. Some features of IDPs are the low content of hydrophobic residues and the high load of charged residues [12]. Intrinsically disordered proteins/IDPRs present large hydrodynamic volumes, low content of ordered secondary structure, and high structural heterogeneity. These proteins are very flexible. However, some of them show transitions from the disordered to the ordered state in the presence of natural ligands [10]. The ability of IDPs to return to the highly flexible conformations after performing their biological function, and their predisposition to acquire different conformations according to the environment, are unique properties of IDPs [13] (Figure 1).

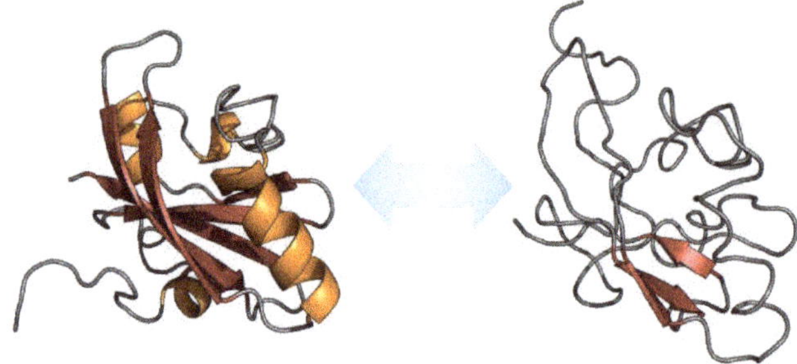

Figure 1. Example of intrinsically disordered proteins (IDP) conformational plasticity. Shown are ordered and disordered extremes in the conformational ensemble described for the photoactive yellow protein from *Halorhodospira halophile* (PDB IDs 3PHY and 2KX6).

It was demonstrated that these IDPs are highly prevalent in many genomes, including humans', and are important in several cellular processes, such as regulation of transcription and translation, cell cycle control, and signaling [3]. It is important to highlight that they are much more common in eukaryotes, in comparison to Eubacteria and Archaea, reflecting the greater importance of disorder-associated signaling and regulation for eukaryotic cells [13]. Intrinsically disordered proteins are present in major disease pathways, such as cancer, amyloidosis, diabetes, cardiovascular, and neurodegenerative diseases. Changes in the environment and/or mutation(s) of IDPs would be expected to affect their normal function, leading to misidentification and missignaling. Consequently, it can result in misfolding and aggregation, which are known to be associated with the pathogenesis of numerous diseases. Some IDPs, such as α-synuclein, tau protein, p53, and BRCA1 are important in neurodegenerative diseases and cancer, being attractive targets for drugs modulating protein–protein interactions. Based on these IDPs and other examples, novel strategies for drug discovery have been developed [11,13]. The ability to modulate the interactions of these proteins offers tremendous opportunities of investigation in chemical biology and molecular therapeutics. Several recent small molecules, such as potential drugs, have been shown to act by blocking protein–protein interactions based on intrinsic disorder of one of the partners [14].

In this review, we will focus on IDPs involved in some neurodegenerative diseases, such as α-synuclein, amyloid β-peptide, and tau protein, while also commenting on cancer associated IDPs, such as p53 and c-Myc, and diabetes-related amylin. In addition, we will summarize the strategies to modulate IDPs action in some diseases and the promising drugs in this field, which are currently more developed for non-neurodegenerative disorders, prompting the need of focusing strategies on IDP-centered drug development for them.

2. Intrinsically Disordered Proteins in Some Diseases

Inside the cell, protein folding is promoted by chaperone machinery that allows the protein to adopt a folded, biologically active form [15]. However, IDPs remain partially or totally unfolded and could cause many neurodegenerative disorders due to some changes in their folding [10]. Neurodegenerative diseases are disorders characterized by progressive loss of neurons associated with deposition of proteins showing altered physicochemical properties in the brain and in peripheral organs. These proteins show misbehavior and disarrangement, affecting negatively their processing, functioning, and/or folding [16,17]. In some of these disorders, there is a conversion of the functional state of specific proteins into an aggregate state that can accumulate as fibrils, causing loss of native function, and consequent gain of a toxic function. The toxicity of these fibrils is caused by disrupting intracellular transport, overwhelming protein degradation pathways, and/or disturbing vital cell functions [16,18]. Misfolding and aggregation of IDPs/IDPRs are especially common in neurodegeneration [16,19,20].

If these misfolded proteins accumulate as deposits of aggregates, they can originate many neurodegenerative diseases such as Alzheimer's, Parkinson's, Huntington's, and prionic diseases, among others [21]. Proteins that accumulate as amyloid fibrils are called amyloidogenic proteins. In order to facilitate the understanding, they can be divided in two groups: 1) proteins that present a well-defined structure with only part of the molecule being disordered, as in the case of prion protein; 2) IDPs like amyloid-β (Aβ), tau and α-synuclein, that show changes in the entire protein [22]. In addition to neurodegenerative diseases, IDPs are also involved in diabetes and different types of cancer. Here, we briefly summarize and cover the general characteristics of some IDPs that can accumulate as fibril aggregates rich in β-structure, and their association with some neurodegenerative diseases, as well as features of cancer- and diabetes-related IDPs.

2.1. α-Synuclein and Parkinson's Disease

Synucleinopathies refer to a group of neurodegenerative diseases, namely Parkinson's disease (PD), dementia with Lewy bodies, and multiple system atrophy, characterized histologically by the presence of inclusions (Lewy bodies and Lewy neurites) composed of aggregated α-synuclein in the central nervous system (CNS) [23,24]. Aggregates containing α-synuclein can be found also in microglia and astrocytes, and in neurons of the peripheral nervous system associated with rarer autonomic diseases. This protein, encoded by the *SNCA* gene located on chromosome 4 region q21, is predominantly expressed in the brain, where it concentrates in nerve terminals. Three isoforms of synuclein, α, β, and γ, are known, but only the α isoform is found in Lewy bodies and neurites. α-synuclein is a single chain with 140 amino acids, and displays 61% identity compared to β-synuclein (134 amino acids), and its sequence contains seven imperfect repeats of eleven amino acids, each with a -KTKEGV- conserved core, separated by nine amino acid residues [25]. Weinreb and co-workers [7] reported, in 1996, the intrinsically disordered nature of α-synuclein in solution, and the protein was found to maintain its disordered state in physiological cell conditions [26]. Upon reversible binding to negatively charged phospholipids, α-synuclein oligomerizes and undergoes structural changes to assume a highly dynamic α-helical conformation while still maintaining partially disordered stretches [27,28].

The physiological role of α-synuclein is still elusive. Mice lacking all three synucleins developed only mild neurodegenerative pathology [29,30]. Lipid-bound α-synuclein accumulates in the plasma membrane of synaptic terminals and synaptic vesicles suggesting a role in neurotransmitter release [31]. The protein has been shown to possess some chaperone activity, interacting with components of the SNARE complex [32] and promoting dilatation of the exocytotic fusion pore [33]. In synucleinopathies, misfolding of lipid-bound α-synuclein occurs leading to β-sheet rich amyloid fibrils, found as the main component of Lewy bodies and Lewy neurites [20,34]. The core of fibrillated protein comprises about 70 amino acids of its repeat region, organized in parallel, in-register β-sheets in a Greek key topology [35].

In contrast to its normal, physiological form, pathological aggregated α-synuclein is extensively phosphorylated at S129 and S87. Other posttranslational modifications present in pathological α-synuclein include nitration, oxidation (for which oxidized by-products of dopamine might contribute) and truncation. Not all of these modifications contribute to accelerate the fibrillation process, since nitration and oxidation decrease fibril formation and stabilize oligomers and protofibrils of α-synuclein. On the other hand, truncated α-synuclein, typically at its C-terminal, shows increased propensity towards fibrillation. In some of the familial forms of PD, point mutations in α-synuclein (E46K, A30P, A53T) alter its propensity to fibrillate [36]. However, about 90% of Parkinson's disease cases are idiopathic [23]. The identification of which α-synuclein species are indeed toxic is yet incomplete and is an intense field of debate. There is a growing perception that soluble oligomeric forms of α-synuclein are the most relevant in terms of toxicity, suggesting that Lewy inclusions might represent a protective response, and that interventions to favor the fibrillation process could be of therapeutic value.

Aggregation of α-synuclein apparently starts in the synapses and the aggregates propagate to nearby neurons through a prion-like mechanism [37,38]. Initial brain structures accumulating intracellular α-synuclein inclusions are the olfactory bulb, glossopharyngeal, and vagal nerves; then the Lewy pathology spreads to other regions of the brain reaching the amygdala and substantia nigra, where it causes death of dopaminergic neurons consequently leading to the motor symptoms characteristic of PD. In the more advanced cases, Lewy bodies and neurites are found in the neocortex, accounting for the cognitive impairment associated to the disease [23,39,40].

2.2. Amyloid β-Peptide, Tau Protein, and Alzheimer's Disease

Alzheimer's disease (AD) is one of the most prevalent neurodegenerative diseases that affects the learning and memory processes beyond the reduction of the brain area, degenerationand death of neurons [41,42]. Diagnosis of AD requires the identification of senile plaques composed by fibril β-amyloid peptides and tangles of tau protein aggregates [41,43]. Amyloid-β (Aβ) peptide is a well-known IDP with several oligomeric forms [44]. Amyloid-β aggregates are formed mainly by peptides containing 39 to 43 amino acids yielded by proteolytic cleavage of amyloid precursor protein (APP) [45–47]. Its aggregated form is significantly linked to Alzheimer's disease, and the generation of Aβ and plaque pathology is linked to the presence of mutations or transport defects related to this protein [48,49].

The amyloid precursor protein (APP) is a transmembrane glycoprotein (type I) that is suggested to be involved in the development of the neurosystem, acting as a cell adhesion molecule [50]. The gene that encodes APP is located in human chromosome 21 [51,52] and this gene yields different isoforms by alternative splicing. Nevertheless, the function of APP is still not understood [43]. The APP proteolytic processing occurs via α, β, and γ-secretase [53]. This process can happen via two pathways: the non-amyloidogenic and the amyloidogenic route (producing toxic $A\beta_{1-40/42}$) [54]. Aβ peptides occur in two major lengths, $A\beta_{1-40}$ and $A\beta_{1-42}$ amino acids, both present in senile plaques [11,46].

Some studies showed that $A\beta_{1-42}$ accumulate as an early event in neuronal dysfunction, acting as seeding in the formation of amyloid plaques [55,56].

The two alloforms, $A\beta_{1-40}$ and $A\beta_{1-42}$, have identical sequences with the exception of two residues in the C-terminus of $A\beta_{1-42}$, causing major differences in conformational behavior, with $A\beta_{1-42}$ being much more folded than $A\beta_{1-40}$ [57]. The amyloid plaques could also be associated to other molecules and metal ions, playing an important role in their assembly and toxicity [58,59]. If some mutations occur in the substrate (APP) or in the γ-secretase regulator proteins (prenisilin-1 and prenisilin-2) it may cause an alteration of APP processing, increasing the levels of $A\beta_{1-42}$ or $A\beta_{1-43}$ peptides formed [60,61]. These mutations are known to be involved in development of early onset AD [62–65].

In order to support the idea that $A\beta$ peptides possess an important role in AD, Simmons and co-workers [66] demonstrated that aggregation of $A\beta$ increased the neurotoxic effect in rat embryonic neuronal cells. Kirkitadze and co-workers [67] studied the $A\beta_{1-40}$ and $A\beta_{1-42}$ oligomerization and assembly into fibrils, showing that the early features of fibril assembly were the increase of intermediates containing α-helix and then their decrease by the assembly of fibrils. Yan and Wang [68] showed that $A\beta_{1-42}$ possesses more tendencies to aggregate in comparison with $A\beta_{1-40}$, and that their C-terminal domain is more rigid.

A structural model for amyloid $A\beta_{1-40}$ using solid state NMR (ssNMR) spectroscopy was proposed. It was found that the first 10 residues are disordered, a β-strand conformation forming β-sheet structure was found between residues 12–24 and 30–40 [69]. After that, other studies were performed with different forms of preparation of the fibrils, with the binding of Cu^{2+}, with mutant forms of the peptide, among others [70–73].

Recently, the peptide $A\beta_{1-42}$ was studied also using ssNMR and in one of the studies it displayed triple parallel β-sheet segments, which is formed by three β-sheets encompassing residues 12–18 (β1), 24–33 (β2), and 36–40 (β3) [74]. Another NMR study of $A\beta_{1-42}$, demonstrated that the fibril core is formed by a dimeric form of the peptide, containing four β-strands in an S-shaped amyloid fold [75]. Wälti and coworkers [76] found similar results: the fibril in dimeric form, forming a double-horseshoe. The different results of these groups were probably due to the differences in the preparation of the fibrils, such as pH, peptide concentration, agitation and ionic strength, as well as the source of the peptide (recombinant or synthetic) [42]. For a detailed review about the structural features of the two peptides, see Reference [42]. Besides the importance of $A\beta_{1-40}$ and $A\beta_{1-42}$, some studies demonstrated that the presence of minor isoforms of $A\beta$ peptides could be involved in aggregation and/or or neurotoxicity [49,77,78], although their effect in AD is not fully understood.

Tau is a microtubule-associated protein initially identified as a protein involved in microtubule (MT) assembly and stabilization [79] and in the axonal transport of proteins [80]. Nowadays, the list of physiological functions of tau has expanded to include diverse roles such as protection against DNA damage and cell signaling [81]. Recent data revealed that tau physiologically interacts with various proteins and subcellular structures, and upon release from neurons, it may even act on other cells, widening the spectrum of its repercussions in health and in diseased states [82].

The single gene encoding the tau protein is present in one copy in the human genome, located in chromosome 17q21 [83,84]. Alternative splicing of this gene can yield six different isoforms of tau with polypeptide chains varying from 352 to 441 amino acids [85,86], all containing either three or four tandem repeats of 31 or 32 amino acid residues, the so-called microtubule binding repeats [81,82]. Tau is composed of 25 to 30% of charged amino acids and contains many proline residues, rendering it full intrinsically disordered. Tau undergoes many types of posttranslational modifications such as phosphorylation, glycosylation, methylation, acetylation, ubiquitinilation, SUMOylation (interaction with Small Ubiquitin-like Modifiers), nitration, among others, which are thought to finely regulate the involvement of the protein in its various biological functions. As a result of "abnormal" phosphorylation, glycosylation, oxidation, truncation or other posttranslational modification [6,82], tau becomes prone to aggregation and forms intracellular deposits, a feature of several neurodegenerative diseases collectively known as "tauopathies". The abnormal tau adopts many transient local foldings among which β-structures of hydrophobic regions, characteristic of neurofibrillary tangles, and paired helical filaments of its microtubule binding domains [87,88] (for a review, see Reference 81]). Tau aggregates can mediate the spreading of the neuropathology to neighboring cells through its paired helical filaments, emerging as a possible target for taoupathy therapies [89]. Oxidation status of tau cysteine residues plays an important role in aggregation. While the formation of intermolecular disulfide bridges aggregates the protein, intramolecular cystine bonds prevent aggregation [90]. Truncation and/or proteolysis of tau yielding lower molecular mass forms of the protein, either in the intracellular or extracellular compartments, were also reported to lead to conformational changes that culminate in toxic, aggregated fibrillar tau [91,92].

In the most common tauopathy, Alzheimer's disease, and in some forms of frontotemporal dementia, the sites of neurodegeneration correlate with deposits of an aberrant hyperphosphorylated tau. All six isoforms of hyperphosphorylated tau are found in tauopathies, resulting in loss of the protein's ability to bind to microtubules and causing disturbance of axonal transport [82]. Phosphorylation of tau may occur in more than 85 putative sites, and distinct kinases and phosphatases are involved in controlling the protein's phosphate content. On the other hand, the glycosylation and/or acetylation status of tau determines its phosphorylation pattern [93].

As a consequence of its disordered nature, tau interacts with a diverse array of partners inside the cell, among which are proteins, small molecules, nucleic acids, and metal ions, with many of these interactions modifying tau's structural properties and biological functions. At least 33 distinct protein partners bind to tau's different domains or motifs, as reviewed in Reference [81]. The multifunctionality of tau resulting from the combination of the wide range of its binding partners and a plethora of posttranslational modifications guarantees its place among true moonlighting proteins [94]. One of such interactions is with the β-amyloid peptide, in a manner that the neurotoxicity of both partners is thought to be reinforced [95,96].

2.3. Prion Protein in Prion Diseases

The term prion was introduced to describe a small proteinaceous agent that was causing neurodegenerative disease in humans and other animals [97]. It was identified as an abnormal form of the prion protein [98,99]. The prion protein (PrP), encoded by the *Prnp* gene, is a glycoprotein, natively found in cells and that could be involved in the maintenance of myelin in neurons among other functions [100–102]. Structural studies of PrP using NMR demonstrated that the N-terminal portion of the recombinant murine PrP is unstructured and flexible, and that the C-terminal portion is globular, containing 3 α-helices and a short anti-parallel β-sheet [103]. A similar structure was found for murine PrP [103], hamster PrP [104], human PrP [105] and bovine Prp [106]. Prions are not considered IDPs per se due to their mixed structural features. Some authors argue in favor of prion-specific classification [107], while others consider them to be IDP-like or IDR-containing proteins [11,108].

In prion diseases, PrP changes are predominantly from an α-helical conformation (PrPC) into a β-sheet-rich structure acquiring a PrPSc form that is misfolded, aggregated and that causes transmissible and fatal neurodegenerative diseases [18,109]. Three kinds of prion diseases have been reported: sporadic, infectious, and hereditary forms, including human disorders like Creutzfeldt–Jakob disease (CJD), Gerstmann–Sträussler–Scheinker disease (GSS), familial atypical dementia, Kuru, and veterinary disorders such as scrapie in sheep, goats, mouse, etc. [110,111]. The basic neurocytological characteristics of these diseases are a progressive vacuolation of neurons and gray matter changing to a spongiform aspect with extensive neuronal loss [112].

Interspecies transmission of prions has been postulated [113], although some interspecific barrier for transmission of PrPSc prions has been established [114]. One factor involved in this barrier could be the difference between the donor and host amino acids sequence [115–117]. A recent study brings new insights on prion replication during species transition [118].

The structural modifications involved in prion propagation and infectivity is the transition of α-helices of PrPc into aggregated β-sheet of PrPSC [109,119]. The presence of PrPSC abnormal form seems to stimulate and serves as template for transition of PrPC into the infectious conformation [120]. Makarava and colleagues [121] reported that prion disease could be induced in wild-type animals by injection of recombinant PrPC fibrils. In order to understand how this transition occurs, Stahl and co-workers performed a study using mass spectrometry and Edman sequencing. They demonstrated that the primary structures of PrPSc were the same as the one predicted for the *PrPC* gene, suggesting that the difference between them is not in RNA modification nor splicing events. In the same study, no covalent modifications were identified in this transition [122]. In another study, using Fourier-transform infrared (FTIR) spectroscopy and circular dichroism (CD), it was shown that PrPC contains around 42% of α-helices in its structure and only 3% of β-sheet content. On the other hand, the modified isoform PrPSc contain a higher content of β-sheet (43%) and a lower content of α-helices (30%) [109].

The PrPSc aggregates present resistance to proteolytic degradation at the C-terminal region, differently from the PrPc normal form [123]. Saverioni and co-workers [124] demonstrated that human PrPSc isolates showed strain-specific differences in their resistance to proteolytic digestion, something that could be linked to aggregate stability. Such aggregates can have heterogeneous sizes [124,125].

When PrPc obtains a β-sheet-rich conformation and misfolded form, it has a tendency to accumulate as amyloid fibers, a useful characteristic for detection and diagnostic of diseases [126,127]. In spite of that, the formation of amyloid plaques is not an obligatory event in prion infectivity [128]. Thinking in a therapeutic target for prion diseases, one approach would be blocking the conversion of PrPC into PrPSc [129].

2.4. p53, c-Myc, and Cancer

Several human diseases, such as cancer, diabetes, and autoimmune disorders, have been found to be associated with deregulation of transcription factors [130]. Carcinogenesis is a multi-step process, resulting in uncontrolled cell growth. Mutations in DNA that lead to cancer disturb these orderly processes by disrupting their regulation. This disruption results in uncontrolled cell division leading to cancer development [131]. Deregulation of multiple transcription factors has been reported in cancer progression. Extensively studied transcription factors that have shown a major role in progression of different types of cancer are p53 and c-Myc, two intrinsically disordered proteins [132,133].

Fifty percent of all human cancer present mutations in *TP53*, and on many other cancers, the function of the p53 protein is compromised. Thus, p53 is a very important target in cancer therapy [134]. Mutations in p53 are found in several types of cancer such as colon, lung, esophagus, breast, liver, brain, reticuloendothelial, and hemopoietic tissues [135]. Additionally, many p53 mutants, instead of losing functions, acquire oncogenic properties, enabling them to promote invasion, metastasis, proliferation, and cell survival [136].

p53 is a key transcription factor involved in the regulation of cell proliferation, apoptosis, DNA repair, angiogenesis, and senescence. It acts as an important defense protein against cancer onset and evolution and is negatively regulated by interaction with the oncoprotein MDM2 (murine double minute 2). In human cancers, the *TP53* gene is frequently mutated or deleted, or the wild-type p53 function is inhibited by high levels of MDM2, leading to the downregulation of tumor suppressive p53 pathways [137–139]. When DNA damage occurs, p53 is activated to promote the elimination or repair of the damaged cells. p53 is phosphorylated by DNA damage response (DDR) kinase, leading to cell cycle arrest, senescence, or apoptosis. In addition, p53 stimulates DNA repair by activating genes encoding components of the DNA repair machinery [140].

Human p53 is a homotetramer of 393 amino acids composed of an intrinsically disordered N-terminal transactivation domain (TAD), followed by a conserved proline-rich domain, a central and structured DNA-binding domain, and an intrinsically disordered C-terminal encoding its nuclear localization signals and oligomerization domain required for transcriptional activity [138,141–147]. Natively unfolded regions account for about 40% of the full-length protein and the disordered regions are extensively used to mediate and modulate interactions with other proteins. Disorder is crucial for p53 function, since its numerous posttranslational modifications are majorly found within the disordered regions [11,146,148]. The full TAD of p53 consists of the N-terminal containing 73 residues and with a net charge of −17, due to its richness in acidic amino acid residues, such as aspartic acid and glutamic acid [143]. The C-terminus, on the other hand, is rich in basic amino acids (mainly lysines) and binds DNA non-specifically [146].

Transactivation domain is a promiscuous binding site for several interacting proteins, including negative regulators as MDM2 and MDM4 [146,149–151]. Transactivation domain is an IDR that undergoes coupled folding and binding when interacting with partner proteins like the E3 ligase, RPA70 (the 70 kDa subunit of replication protein A) and MDM2. p53 forms an amphipathic helix when it binds to the MDM2 in a hydrophobic cleft in its N-terminal domain [137,138,149,152–155]. The p53–MDM2 interaction blocks the binding of p53 to several transcription factors. In addition, MDM2 tags p53 for ubiquitination and consequent degradation by the proteasome and the p53–MDM2 complex tends to be exported from the nucleus, preventing p53 to act as a "cellular gatekeeper" [138,144,156].

The proto-oncogene *c-MYC* encodes a transcription factor that is implicated in various cellular processes such as cell growth, proliferation, loss of differentiation and apoptosis [157]. Elevated or deregulated expression of *c-MYC* has been detected in various human cancers and is frequently associated with aggressive and poorly differentiated tumors. Some of these cancers include breast, colon, cervical, small-cell lung carcinomas, osteosarcomas, glioblastomas, melanoma, and myeloid leukemia [158–160]. c-Myc is a very important protein for understanding and developing therapeutics against cancers and cancer stem cells [161].

c-Myc is an IDP and becomes transcriptionally functional when it forms an heterodimer with its obligate partner Max to assume a coiled-coil structure that recognizes the E-box (enhancer-box)-sequence 5′-CACGTG-3′. The c-Myc N-terminus, its TAD, can activate transcription in mammalian cells when fused to a heterologous DNA-binding domain. The C-terminus of this protein contains a basic-helix-loop-helix-leucine zipper (b-HLH-LZ) domain, and it promotes its interaction with Max, that has the same (b-HLH-LZ) domain, and the sequence-specific DNA binding mentioned above [162–165]. Nuclear magnetic resonance studies of c-Myc disordered region have attributed to it the protein functional plasticity and multiprotein complex formation capacity [166]. Computational and experimental investigations show that c-Myc extensively employs its disorder regions to perform diverse interactions with other partners [161].

It is important to highlight that Max protein is critical for c-Myc's transcriptional activities, both gene activation and repression [162,167]. Considering c-Myc as a target for cancer therapy, one approach to c-Myc inhibition has been to disrupt the formation of this dimeric complex [132]. However, the disruption of c-Myc-Max dimerization is not easy, since both proteins are IDPs and protein–protein interaction involving large flat surface areas are difficult to target with small molecules, such as drugs [168,169].

2.5. Amylin and Diabetes

Diabetes (Type II) is a multifactorial disease characterized by dysfunction of insulin action (insulin resistance) and failure of insulin secretion by pancreatic β-cells [11,170]. One hallmark feature of this disease is the accumulation of amyloid fibrils into pancreatic islets (islets of Langerhans). These amyloid deposits are majority composed by islet amyloid polypeptides (IAPP), also called amylin. Islet amyloid polypeptides are IDPs composed of 37 amino acid residues, co-secreted with insulin by the same pancreatic cells, and its gene is located on chromosome 12 in humans [171–173].

The process of aggregation of IAPP seems to be initiated by interaction of one IAPP monomer to another, progressively leading to the formation of aggregates [174,175]. Analysis of human IAPP using circular dichroism spectroscopy demonstrated that the fibril formation was accompanied by a conformational change of random coil to β-sheet/α-helical structure [176]. These transient conformations were further confirmed by other studies [177–179].

Cytotoxicity of IAPP accumulated as amyloid deposits could be associated with loss of pancreatic β-cells functions and cells apoptosis [180,181]. Recent reviews of computational studies provided mechanistic insights of IAPP structure as monomers and oligomers and their interaction with lipid bilayers in order to understand the IAPP cytotoxicity mediated by membranes [175,182].

3. Strategies for IDP Modulation

Intrinsically disordered proteins can rapidly populate different conformations in solution, usually not assuming a well-defined three-dimensional structure in their native state, as a result of their signature low-sequence complexity and low proportion of bulky hydrophobic amino acids (instead, charged and hydrophilic residues are common) that lead to a flexible, dynamically disordered behavior and larger interaction surface areas than analogous folded regions in globular proteins [183–186]. Considering that IDPs participate in numerous key processes in cell metabolism, it is expected that their activity would be regulated by multiple mechanisms at transcriptional, post-transcriptional, and translational levels, which makes active IDPs accessible in shorter periods compared to structured proteins [187,188]. Some aspects of IDP modulation (mainly for IDPs involved in cell signaling) are covered in this section, arbitrarily grouped in mechanisms that engage in direct structural changes on IDPs in order to achieve stabilization (coupled folding and binding, post-translational modifications), and mechanisms that control IDPs abundance in the cell (mRNA decay, IDP proteasomal degradation, nanny model for stabilization).

3.1. Regulation of IDP Activity through Structural Changes

Intrinsically disordered proteins acting in intracellular pathways contain conserved motifs for interaction with nucleic acids and other proteins, and frequently form low-affinity complexes advantageous to processes like signal transduction [184]. The recognition elements (being often called "SLiMs"—short linear motifs ranging from 3 to 10 amino acids [189–191]) determine a pivotal feature associated with disordered proteins, that is their binding promiscuity, which is carried out through "one-to-many" and "many-to-one" mechanisms [192]. Interestingly, many IDPs are able to adopt ordered structures when interacting with certain targets, characterizing the coupled folding and binding phenomenon. Also, structural polymorphisms can emerge from the IDP conformational landscape in cases where the same disordered protein can assume different defined structures as it binds to different targets. Some of the recognition elements are also targeted for post-translational modifications by regulatory enzymes, enabling disordered-to-ordered transitions in IDPs. However, induced folding in IDPs is not mandatory for activity, as many regions that remain disordered upon partner binding are important to function, constituting "fuzzy" complexes [193,194].

3.1.1. Coupled Folding and Binding

The folding induction by partner interaction is possibly one of the most reported characteristics of IDPs, despite not being a phenomenon absolutely widespread across this class of proteins (considering the fuzzy complexes), nor a completely understood process. There are numerous examples of disordered-to-ordered transitions in proteins implicated in the regulation of gene expression, like transcription factors that assume folded motifs when interacting with DNA. One particular example is the leucine zipper protein GCN4, which presents a basic region that is unstructured in the absence of DNA but becomes a stable helical structure when interacting to its cognate AP-1 site [195]. The transition begins with transient nascent helical forms, observed in the unbound state, that interact with DNA and lead to dramatic structural changes, explained by a reduction of the entropic cost of DNA binding due to restriction of the conformational space accessible to the basic region [8,196]. Thus, the induced folding is usually explained by loss of conformational entropy (from the unbound IDP state) upon target binding and compensatory favorable contributions from reduction in exposed hydrophobic surfaces and enhanced electrostatic interactions [196]. Exhaustive kinetic studies are necessary for the investigation on what order the events of binding and folding occur, and segregate them in schemes of induced fit (IF, when the IDP binds to a partner and then folds) or conformational selection (CS, when the partner only binds to IDPs in a certain conformation) [197]. However, trying to define the coupled folding and binding in two categories may offer a rather simplistic explanation for the IDPs behavior, since the mechanisms can overlap depending upon the system conditions [197,198].

3.1.2. Post-Translational Modifications

Because of their accessibility to modifying enzymes, IDPs are frequent targets for post-translational modifications (PTMs), which expand their functional versatility [185,199]. The occurrence of PTMs causes structural changes on IDPs by affecting their energy landscapes, due to modifications of the physicochemical properties of the primary sequence [185]. Post-translational modifications engage in the addition of chemical functional groups (usually small radicals such as phosphoryl, alkyl, acyl or glycosyl) or involve the direct modification of residues through reactions of oxidation, deimidation, and deamidation. Intrinsically disordered proteins suffering PTMs may have their electrostatic, steric, and hydrophobic properties modified, possibly inducing transformations on the structure due to enhancement/inhibition of contacts of motifs within the IDP chain or with binding partners [185,200–202]. Phosphorylation is one of the most prevalent PTM and constitutes a major regulatory mechanism in various cellular processes involving signal transduction. Replacing a neutral hydroxyl group with a tetrahedral phosphoryl results in new possibilities for intra- and intermolecular electrostatic interactions (e.g., salt bridges and hydrogen bonds). Many other types of PTMs work in a similar fashion (but modulating different chemical properties), providing new forms of interaction that can ultimately cause alterations in IDPs activity, including disordered to ordered transitions.

3.2. Regulation of IDPs Abundance

Intrinsically disordered proteins levels are carefully monitored in the cell, and changes in their abundance are associated with disease, mainly due to defective signal transduction (linked to the occurrence of some cancers [11,203]) and non-specific interactions that generate fibrillar aggregates (present in many neurodegenerative disorders [199,204]). Tight regulation can be achieved in different levels, controlling the half-lives of mRNAs encoding IDPs and the abundance of IDPs themselves. Obviously, there are outliers for these global trends and certain IDPs are present in cells in large amounts or for long periods of time [205]—usually not the IDPs involved in dynamic processes such as cell signaling. Examples include the fibrous muscle protein titin [206,207] and the curious case of tardigrade-specific IDPs, that are constitutively expressed and upregulated in some tardigrade species and are essential for desiccation tolerance [208].

3.2.1. IDP-Encoding mRNAs

A robust study monitoring gene expression in *Saccharomyces cerevisiae* (with similar trends detected for human genes [205]) demonstrated that mRNAs encoding sequences categorized as "highly unstructured" have lower half-lives than mRNAs encoding more structured proteins, having a comparable number of transcription factors regulating them [206]. One of the reasons for the increased mRNA decay is hypothesized to be the short poly(A) tails observed in IDP-encoding mRNAs, that foment RNA degradation pathways. Moreover, other factors related to transcript instability, like the binding of RNA-binding PUF proteins (that usually facilitate deadenylation and subsequent RNA clearance [209,210]) were found to be increased in mRNA coding for disordered proteins [206].

3.2.2. Proteasomal Degradation

Intrinsically disordered proteins undergo proteasomal degradation through two different (but not mutually exclusive) pathways, ubiquitin-dependent (UD) and ubiquitin-independent (UI) [211]. The UD pathway relies on the addition of ubiquitin to the substrate to be degraded, in a process regulated by a series of enzymes and is mediated collectively by the 26S proteasome [212]. It was found that IDPs present a high content of predicted ubiquitination sites [185,205], and, in an analysis of ubiquitinated proteins, there seems to be a correlation between confirmed degradation sites and regions of disorder [213]. Alternatively, the UI pathway is mainly orchestrated by the core of the 20S proteasome, being a default process for degradation of free disordered proteins. Some evidences support the hypothesis that the flexible and extended structures of IDPs (as well as disordered terminal segments in folded proteins) facilitates the interaction with the proteasome, considering that bulky particles have reduced cleavage rates [214,215].

3.2.3. Stabilization through "Nanny" Proteins

As IDPs are usually prone to degradation, there is a need for protein stabilization in some contexts. The ubiquitous enzyme NQO1 functions as a "gatekeeper" of the 20S proteasomes, binding and regulating the degradation of some IDPs, in a mechanism consuming NADH [211,216]. An analogous mechanism characterizes the "nanny" model for IDP protection from cleavage, where there is sequestration of ID segments by interactions with other proteins, resulting in evasion from 20S proteasomal digestion. The binding of the nanny is transient, beginning at the initial stage of IDPs' life cycle, when they are newly synthesized (assuming they are more sensible to digestion in this stage) [154]. Despite the fact that nanny proteins also bind to nascent polypeptide chains, they are not considered chaperones because they do not induce a fixed three-dimensional organization on targets, only assisting on IDP conservation without permanently affecting their disordered structure.

3.3. Modulation of IDPs by Chaperones and Co-Chaperones

Aggregates produced in neurodegenerative diseases have been shown to respond to changes in levels of molecular chaperones, suggesting the possibility of therapeutic intervention and a role for chaperones in disease pathogenesis [217]. The heat shock protein Hsp90 promotes neurodegenerative disorders indirectly [218]. Tau protein accumulation is regulated by a (Hsp90) chaperone system. This chaperone is able to bind Tau, causing a conformational change that allows tau's phosphorylation by glycogen synthase kinase (GSK3β), leading to tau aggregation [219]. The inhibition of this chaperone results in the reduction of tau phosphorylation levels, due to reduction of GSK3β levels [220]. Another approach was performed, the use of a co-chaperone of Hsp90, ATPase homolog 1 (Aha1), this protein is an activator of Hsp90. This approach promoted the increase of the production of aggregated tau in vitro and in mouse model of neurodegenerative disease. Moreover, inhibition of Aha1 reduced tau accumulation in cultured cells. Thus, Aha1 is an interesting target to the treatment of Alzheimer's disease [221].

The Hsp70 is a protein stabilizer, has a cellular protection against neurodegeneration of the central nervous system [218]. Members of the Hsp70 family, such as Hsp70 and Hsc70, bind to misfolded proteins and somehow send them to the lysosome–autophagy pathway or ubiquitin–proteasome system for degradation [222,223]. Folding and degradation of proteins are linked through co-chaperones, such as C-terminus of HSP70-interacting protein (CHIP) and HSJ1 (DNAJB2) [224] which regulate the decisions determining whether misfolded proteins are refolded or degraded. The CHIP is associated with α-synuclein inclusions and act as a co-chaperone, altering its aggregation and enhancing the degradation of the misfolded α-synuclein [225].

Another important chaperone is cyclophilin 40 (CyP40) that is a cis/trans peptidyl-prolyl isomerase (PPIase) and is involved in regulation and orientation of proline residues [226,227]. Tau protein is rich in proline residues and its residues, usually found in β-turns, are involved in tau aggregation propensity [228]. Based on this information, Baker and coworkers [229] demonstrated that CyP40 possess the ability to dissolve amyloids fibrils in vitro. Nuclear magnetic resonance experiments showed that CyP40 acts specifically on proline rich residues performing the disaggregation of tau fibrils and oligomers. This cyclophilin could also interact with others aggregated proteins containing proline, like α-synuclein [229].

4. Known Drugs Acting on IDPs

Despite IDPs abundance in eukaryotes, currently there are no FDA-approved drugs specifically targeting these proteins, only experimental and speculative ones (i.e. drugs that have been evaluated by the United States Food and Drug Administration agency and had their marketing sanctioned). Some experimental drug examples are prevalent in the literature, such as those targeting p53-MDM2, c-Myc-Max, and EWS-Fli1 complexes, while some others are less discussed [230–233]. In this section we provide an overview of the pharmacological modulation of IDPs, neurodegenerative and otherwise.

Intrinsically disordered proteins are normally considered aggregation-avoidant, due to their high proportion of charged residues (as opposed to patches of folding-inducing, hydrophobic residues). Such "non-folding" plasticity is proposed to be advantageous for proteins with multiple partners [234,235]. However, some of the IDPs, as described earlier, are found in conformational diseases and amyloid formation (e.g., in Alzheimer's disease, Parkinson's disease). The suppression of fibril formation, thus, is of therapeutic interest. Drug candidates in this front include molecular tweezers (Table 1), which are ligands designed to bind lysine and arginine specifically, perturbing aggregation [236–238]. These positively-charged residues are prone to interact with negatively-charged regions in the fibril-forming monomers [237]. The SEN1576 compound, a 5-aryloxypyrimidine inhibitor of synaptotoxic Aβ aggregation (Table 1) was shown to be safe and orally bioavailable with good brain penetration [239].

Fragments of amyloid fibrils also served as templates for non-natural amino acid inhibitors of amyloid fibril formation (D-TLKIVW) [240], while the ELN484228 (Table 1) compound was shown to be protective in cell models for vesicular dysfunction via α-Synuclein [235]. Alterations of the neuroleptic agent chlorpromazine allowed for enhanced 20S proteasome activation, inducing degradation of IDPs, such as tau and α-synuclein, but not of structured proteins [241]. These chlorpromazine-derived molecules, despite showing noteworthy potential (even as tools to study the proteasome 20S gate regulation), may interfere with other, physiological, non-pathologic disordered proteins to a still unstudied extent. A naphthoquinone-tryptophan hybrid (NQTrp) (Table 1) was shown to be effective in model systems for tau aggregation [242].

Regarding tumor-associated IDPs, the most commonly mutated gene in human cancers, the tumor suppressor protein p53, is key to cell cycle signaling. It is regulated by binding to various partners, including MDM2 and Taz2 [243–245]. Despite being highly disordered, p53 is not itself the target for the currently screened drug candidates aiming the p53-MDM2 complex. Instead, these ligands aim to occupy the p53-binding site in MDM2. The inhibitors nutlins (Table 1) (*cis*-imidazoline analogs currently in phase-I clinical trials), were shown to be potent against multiple cancerous cell lines, including breast cancer, colorectal cancer, lung cancer, osteosarcoma, prostate cancer, and renal cancer [246–249].

The c-Myc-Max complex involves the IDP transcription factor c-Myc that is activated by binding to Max, being expressed constitutively in various cancer cells [250]. Inhibitor candidates target the disordered c-Myc in this case, including peptidomimetic inhibitors [249,250], the small molecules 10058-F4, 10074-G5 (Table 1), and some others [250–257]. The oncogenic fusion protein EWS-Fli1 is an IDP exclusively present in Ewing's sarcoma [257]. As with c-Myc-Max inhibitors, the small molecule inhibitor YK-4–279 (Table 1) targets the disordered EWS-Fli1 protein directly [257,258].

The AF9-AF4 dimer is found in acute leukemias and is composed by two disordered fusion proteins. The AF9 protein is responsible for turning hematopoietic cells oncogenic [259,260]. The AF4-derived peptide of amino acid sequence PFWT was shown to inhibit AF9 when used in combination with established chemotherapeutic agents [261,262], while some non-peptidic inhibitor candidates have been identified by high-throughput screening [263]. The protein-tyrosine phosphatase 1B (PTP1B), a reticular non-transmembrane enzyme, has been validated as therapeutic target for diabetes, obesity, and breast cancer, due to its role as negative regulator of insulin and leptin signaling [264]. Protein-tyrosine phosphatase 1B has an elongated disordered carboxy-terminus, to which trodusquemine (Table 1) (MSI-1436, a natural product) binds [265]. This aminosterol acts allosterically, stabilizing an inactive form of the enzyme by binding to a non-catalytic disordered site [265].

Table 1. Known drugs acting on IDPs (selected examples).

Compound Name *	Targets	Compound Structure
CLR01 (Molecular tweezers)	Lysine and arginine residues in amyloid proteins	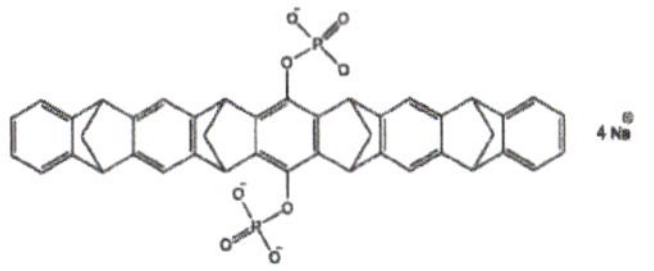
ELN484228	α-Synuclein	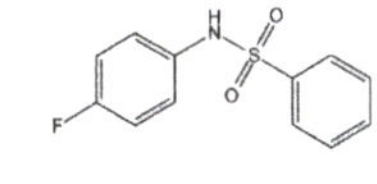
SEN1576	Amyloid β	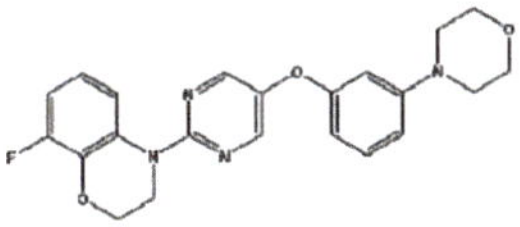
NQTrp	PHF6 (Tau protein)	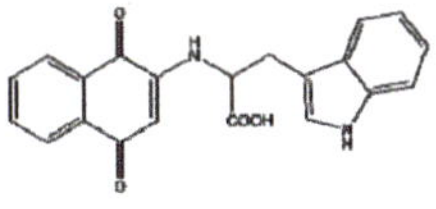
Nutlin-3	p53-MDM2 complex	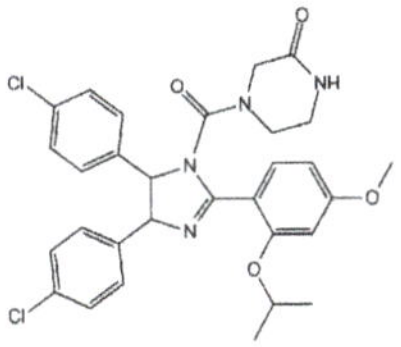
10058-F4	c-Myc-Max complex	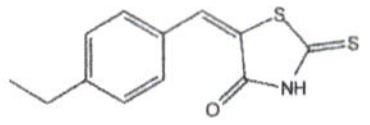
10074-G5	c-Myc-Max complex	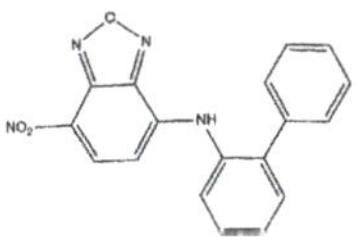
YK-4–279	EWS-Fli1	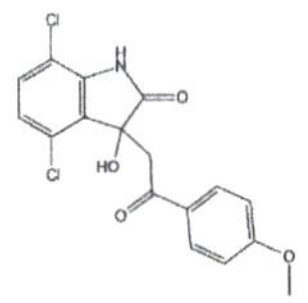
Trodusquemine	PTP1B	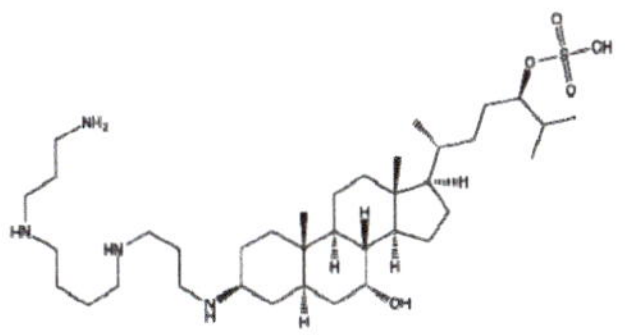

* Unless otherwise specified, these names are directly taken from their lead-compound coding and have no meaning on their own. EWS: Ewing Sarcoma; MDM2: oncoprotein (murine double minute 2); Myc: Myelocytomatosis-transcription factor homolog; NQTrp: naphthoquinone-tryptophan hybrid; PHF6: Tau-derived peptide (amino acid sequence VQIVYK); PTP1B: protein-tyrosine phosphatase 1B.

Inhibitors of α-synuclein aggregation are considered a promising approach. From in vitro studies, a few lead molecules were identified, such as EGCG (epigallocatechin gallate) [266]. Iron is known to induce the aggregation of α-synuclein. Deferiprone is an iron chelator used in thalassemic patients. Two clinical trials have shown a decrease in iron content in the substantia nigra of some PD patients, while a trend for improved motor scores was seen for all enrolled patients [267]. Although copper can induce aggregation of α-synuclein in vitro, levels of copper in the substantia nigra of PD patients are up to 50% lower than that of age-matched controls. The Cu^{2+} complex of diacetylbis-(4-methylthiosemicarbazone), called Cu^{2+}(atsm), showed neuroprotective action in different animal models of PD [268], prompting for a phase I trial. From in vivo studies, promising results were obtained for KYP-2047, an inhibitor of prolyl oligopeptidase, an enzyme shown to interact with α-synuclein [269–271]; the diphenyl-pyrazole compound anle138b, was shown to cross the blood-brain barrier of mice and to reduce aggregation of α-synuclein [269]. The NPT200-11 compound prevented the formation of oligomers of α-synuclein and improved neuropathological symptoms in transgenic mice [272] and it has already been subjected to a phase I clinical trial. Another promising compound, NPT088, is a fusion protein of a general amyloid interaction motif derived from a bacteriophage [273] and a fragment of human immunoglobin, developed by Proclara Biosciences, entered a phase I clinical trial for AD [274]. Phase II clinical trial of NPT088 is expected to include PD patients. Intrabodies, a single chain variable fragment of immunoglobulin expressed intracellularly, have been developed to target oligomeric and fibrillary α-synuclein, conferred neuroprotection, apparently by shifting the dynamics of the aggregation process [275–277]. Addition of a proteasome-addressing sequence to intrabodies targeted pathological forms of α-synuclein to degradation, NbSyn87PEST, directed towards the C-terminal region, and VH14PEST, directed against the NAC hydrophobic interaction domain, effectively degraded α-synuclein in cultured cells [278]. In rats overexpressing wild-type α-synuclein, these proteasome-targeted intrabodies (or nanobodies) decreased the levels of pathological aggregates, increased striatal dopamine levels and improved motor function [279]. Research in this promising field moves to find ways to deliver these compounds in adequate levels in specific areas of the brain, probably by using viral vectors.

Immunotherapies against α-synuclein, based on the evidence of an extracellular pathological protein during spreading of PD to different brain structures, show promising results in animal models [280]. Besides opsonization of the pathological protein for clearance, it is likely that antibodies could block further oligomerization of α-synuclein. Both passive (humanized monoclonal antibodies) and active (vaccine) immunization are being pursued. Pharmaceutical companies have joined the efforts and early clinical trials have been concluded or are under way. A brief description of the more advanced planned immunotherapies follows. A phase I trial was conducted by Roche for PRX002, a monoclonal antibody against the C-terminus of α-synuclein. It was well tolerated and reduced by 96% the levels of serum α-synuclein [281,282]. Affitope PD01A, a synthetic α-synuclein-mimicking peptide developed by Affiris for active immunization, had the first pilot study in 21 PD patients concluded in May 2018. It elicited a specific antibody response and showed good safety and tolerability profiles in a long-term (4 years) outpatient setting. Results of Affitope PD03A phase I clinical trial indicated no severe off-target effects, and a dose-dependent production of antibodies that cross-reacted with the intended α-synuclein epitope. Results from animal studies demonstrated that the antibodies raised against these antigens crossed the blood-brain barrier, decreasing the levels of aggregated α-synuclein, thereby improving motor function [283,284].

Regarding tauopathies, various therapeutic approaches have been tested, aiming to inhibit aggregation of tau, either directly or by preventing its interaction with some partners, and removal of toxic conformers and fibrillated tau [80]. However, the enormous effort put on finding ways to revert or delay the neurodegeneration symptoms associated to fibrillar tau, or to prevent the onset of tauopathies, has been so far unsuccessful, partly due to the intrinsically disordered nature of tau, which hampers drug design based on structural approaches. Small molecules that inhibit tau aggregation in vitro are considered promising leads to anti-tauopathy drugs [285] and the number of new tau inhibitory molecules grows steadily [286]. Nevertheless, there are unanswered questions regarding their effectiveness in vivo and the potential non-specific effects on normal tau physiology that could impact heavily on the CNS. The most studied small inhibitory molecules belong to distinct chemical groups, such as phenotiazines, cyanines, rhodanines, and arylmethines [287,288]. Peptides derived from neuroprotective proteins like NAP (amino acid sequence NAPVSIPQ) and D-SAL (all D-amino acid sequence SALLRSIPA) [289,290], enantiomeric peptides [291], and RNA/DNA aptamers [292] are also attractive components of future anti-tauopathy therapies. Some natural molecules present in cellular medium, such vitamin B_{12} [293] and 8-nitro-$_C$GMP [294], are known to inhibit tau aggregation through oxidation of its cysteine residues. Drugs that bind tau, inducing formation of intramolecular disulfide bonds, such as methylene blue [295] or cinnamon derivatives [296], are potential frames for developing specific tau aggregation inhibitors. Taking into account that accumulation of hyperphosphorylated tau is a hallmark of AD and other neurodegenerative disorders, inhibitors of kinases, particularly of glycogen sintase kinase 3β and of Fyn, a member the Src-family of non-receptor tyrosine kinases, have drawn much attention for their anti-tauopathy potential [80]. Another strategy focuses on dual inhibitors that would interfere on tau aggregability and simultaneously block its interaction with protein partners, particularly kinases [297,298]. Other attempts to develop an anti-tauopathy drug have focused on inhibiting tau interaction with proteases like beta-secretases [299], caspases [300], and calpain [301], and chaperones such as Hsp90 [302], among others.

Tau-targeted immunotherapy began in 2013 [303], and since then a dozen of different types of immunological strategies were subject of clinical trials, including two active immunizations (vaccines) and humanized monoclonal antibodies directed towards distinct tau epitopes aiming passive immunization (reviewed in References [304,305]). These clinical trials are still at early phases and only limited data on the outcomes have been disclosed so far [306,307]. To achieve a successful immunotherapy to treat tauopathies, antibodies should be capable of neutralizing at least one of the many diseased isoforms of tau, either intracellularly or in the extracellular space, and interrupt the processes that lead to tau fibrillation and the neuron-to-neuron spreading. Ideally, the antibodies do not bind to normal tau and are able to cross the blood–brain barrier. In the case of a tau vaccine, the senescence of the immune system of the elderly has to be considered [304]. AADvac1 was conceived as the first vaccine against AD, using as immunogen a tau peptide previously identified to be essential for its pathological aggregation. Active immunization with this peptide elicits antibodies against a stretch of tau's primary sequence (amino acid residues 294–305) and to conformational epitopes as well, targeting mainly extracellular tau, reducing its oligomerization. Tested in different animal models of AD, AADvac1 raised a protective humoral immune response with antibodies that discriminated between normal and pathological tau, reduced the level of neurofibrillary pathology in rat brains and lowered the content of disease-specific hyperphosphorylated tau [308]. Phase 1 clinical trial of AADvac1, conducted in 2013–2015 in patients aged 50–85 years with mild-to-moderate AD immunized weekly for 12 weeks, revealed a favorable safety profile and 29 out of 30 patients given AADvac1 developed an IgG response [309]. After 72 weeks, and booster doses of AADvac1, patients who had developed higher IgG titers showed lower hippocampal atrophy and cognitive decline rates and only mild adverse side effects [310]. A second active immunotherapy against tau has the compound ACI-35 as the immunogen, a peptide containing tau's phospho-epitope pS396/pS404, in a liposome-based formulation able to elicit antibodies against abnormal hyperphosphorylated tau in P301L tau-mice [311].

Attempts of passive immunotherapy utilize humanized monoclonal antibodies, mostly of IgG1 or IgG4 isotypes, which are directed towards stretches of tau's primary sequence known to be involved in the oligomerization of the protein, or to extracellular seeding-capable forms of truncated tau [304]. Phase I clinical trial of ABBV-8E12, one of such humanized monoclonal antibodies [312], revealed a satisfactory safety profile in 30 patients with the progressive supranuclear palsy tauopathy, receiving single doses (2.5 to 50 mg/kg) of the antibody, with no signs of immunogenicity against it [313]. Phase I trials of other two anti-tau antibodies, C2N-8E12 and BMS-986168, were also conducted [307].

Something that is noteworthy, and that demonstrates the close interplay between amyloid β peptide and tau in causing neurodegenerative diseases, therapeutic interventions aimed at one pathology can ameliorate symptoms of the other. This is the case of immunization of triple transgenic AD-like mice with a full-length DNA of amyloid β_{1-42} peptide, which showed a 40% reduction in the brain content of the amyloid β_{1-42} concomitant with a 25–50% decrease of total tau and different phosphorylated tau isoforms [314]. Conversely, passive immunization with antibodies against tau's fragments 6–18 and 184–195 protected triple transgenic AD-like mice by reducing amyloid precursor protein in the CA1 region of hypothalamus and in amyloid plaques [315,316].

5. Status and Challenges in Drug Development for IDPs

The limited number of drugs targeting IDPs currently available (see previous section) may look disappointing, considering the physiological relevance of these proteins. One should be careful, though, to not exaggerate the current lack of IDP-specific drugs as being a reflection of disorder as a limitation for drug development.

The rational drug design strategy has been used successfully since the 1980s [317–319]. It depends on knowledge of the three-dimensional structure of the target protein, based on which ligands (usually inhibitors) are planned with aid of computational tools [320]. By definition, IDPs do not have a single, major conformation, occurring in dynamic conformational ensembles [10], and there is difficulty in using such traditional techniques to design IDP ligands [321]. Hence, most cases of IDP drug development were carried out by experimental screening and not by rational design [322]. Still, detection of IDP "hits" (potential initial drug candidates) through high throughput screening of compounds has been challenging [321] and computational methods achieved some success in predicting good candidates [323].

For IDPs with recognizable/determined metastable structures in their conformational ensembles, such structures could be used for rational drug design. However, IDPs are expected to be promiscuous, acting as hubs for multiple cellular processes [324]. This scenario, described as "protein clouds" [325] has its complexity further increased, with IDP ligands being described as "ligand clouds around protein clouds" [326]. Such roles in protein–protein interactions (PPIs) make IDPs especially interesting as drug targets, but the development of molecules targeting PPIs has been in itself, challenging [327–330].

A large difference is expected between the entropic loss and the enthalpic gain upon binding of a small ligand to an IDP, but some of them were shown to be capable of forming adaptable, specific interfaces for small molecule binding [231,235,321]. Intrinsically disordered proteins are difficult targets, since their interactions with small molecules are weaker and more transient, and the entropic loss is greater, in comparison to structured proteins [331]. The fragment-based drug design approach allows fragments to sample large amounts of chemical space, reducing the number of compounds for screening, with different fragments that bind at different regions of IDPs being able to be linked together via an appropriate linker [331]. Such fragments usually require hydrogen bonds to achieve detectable binding, generating an enthalpic gain that compensates for the entropic loss upon binding of the small molecules, lowering the free energy of the protein upon binding [331].

Still, the over-representation of IDPs in disorders, as summarized by the D^2 concept (for "disorder in disorders") [11] and the D^3 concept (for "disorder in degenerative disorders") [16], points to these proteins as promising therapeutic targets. Attempts to detect potentially druggable cavities in IDPs have identified at least 14 targets that could be subjected to rational drug design [233]. A more general estimate is that 9% of detected cavities may be druggable in IDPs, in comparison to 5% in ordered proteins [233]. These observations are especially interesting, considering that current drugs target around 500 proteins, less than 10% of the estimated potential target list, with very strict classes (such as enzymes and G-protein coupled receptors) accounting for more than 70% of them [230,331,332].

There have been major advances in the detection and prediction of IDP features in protein sequences [293], something that will surely help in the identification of these special drug targets. Major breakthroughs are also being achieved by the combination of experimental methods (especially

NMR and fluorescence techniques) with computational modeling and molecular dynamics simulation of IDPs [294,322]. The latter is one of the few methods that allow for the description of IDPs in their conformational ensemble, instead of a single (or just a few) conformations [132,322,324,333–337].

6. Conclusion and Perspectives

The pharmacological strategies developed so far (and reviewed here) that target IDPs can be separated as binding directly to IDPs and hampering their aggregation by keeping them in the interaction incompetent conformation; interacting with the IDP and promoting the stabilization of non-toxic/ non-amyloidogenic oligometric species; and interacting with the amyloidogenic protein and greatly accelerating its aggregation to minimize the period of toxic oligomer formation [338]. Still, as we described here, there are many pathways acting on IDP control, and these are still unexplored targets for pharmaceutical interference. As the binding mechanisms of IDPs are being better described from a physical chemical standpoint [339,340], it is becoming clear that for candidate molecules to act on IDPs they must deviate from traditional prediction rules for drug-likeness [319,320]. One standout feature of IDP ligands is that they are larger and more three-dimensional than traditional drugs [341]. Adding another layer of complexity to this scenario, some proteins are shown to be conditionally unfolded [342], being disordered only under specific conditions.

Despite being abundant in eukaryotes, in which IDPs have evolutionarily conserved interaction partners [343], the occurrence of disorder in proteins from other organisms is being described. It includes the description of IDPs in Trypanosomatid parasites [344] and in some paramyxoviruses, including measles, Nipah and Hendra viruses [345]. These proteins constitute prospective targets for drug design endeavors, as was also observed for multiple disordered targets in prostate cancer [346]. Furthermore, recent evaluations indicate that IDP-targeted drug development may not be irreconcilable with structure-based drug design [347].

The development of drugs specifically tailored for IDPs is still in its infancy. As with the whole pipeline for drug discovery, there has been continuous progress in this area, and as we proposed in this work, there are many untapped pathways and unexplored targets regarding these proteins. As the biophysical techniques advance to catch up with the diversity of disordered behaviors in proteins, one can expect major developments in this front. Taking the limited but solid cases of success in IDP-specific drug design, we may face a future in which target disorder may be taken as the rule and not the exception.

Author Contributions: Conceptualization, C.R.C. and R.L.-B.; Writing—Original Draft Preparation, A.H.S.M., F.C.L., E.O.J., C.R.C., and R.L.-B.; Writing—Review & Editing, C.R.C. and R.L.-B.; Funding Acquisition, C.R.C.

Funding: The authors of this work have been funded by grants from Coordenação de Aperfeiçoamento de Pessoal de Nível Superior-CAPES and Conselho Nacional de Desenvolvimento Científico e Tecnológico - CNPq.

Conflicts of Interest: The authors declare no conflict of interest.

Abbreviations

AD	Alzheimer's Disease
APP	Amyloid Precursor Protein
CJD	Creutzfeldt–Jakob Disease
CNS	Central Nervous System
CS	Conformational Selection
GSS	Gerstmann–Sträussler–Scheinker Disease
IAPP	Islet Amyloid Polypeptide (Amylin)
IDP	Intrinsically Disordered Protein
IDR	Intrinsically Disordered Region
IF	Induced Fit
MT	Microtubules
NMR	Nuclear Magnetic Resonance
PDB	RCSB Protein Databank

PrP	Prion Protein
PrPSc	Prion Protein, Alternate Conformation
PTM	Post-Translational Modification
SLiMs	Short Linear Motifs
ssNMR	Solid-State Nuclear Magnetic Resonance
TAD	Transactivation Domain (of p53)
UD	Ubiquitin-Dependent
UI	Ubiquitin-Independent

References

1. Dunker, A.K.; Lawson, J.D.; Brown, C.J.; Williams, R.M.; Romero, P.; Oh, J.S.; Oldfield, C.J.; Campen, A.M.; Ratliff, C.M.; Hipps, K.W.; et al. Intrinsically disordered protein. *J. Mol. Graph. Model.* **2001**, *19*, 26–59. [CrossRef]
2. Uversky, V.N. Natively unfolded proteins: A point where biology waits for physics. *Protein Sci.* **2002**, *11*, 739–756. [CrossRef] [PubMed]
3. Wallin, S. Intrinsically disordered proteins: Structural and functional dynamics. *Res. Rep. Biol.* **2017**, *8*, 7–16. [CrossRef]
4. Uversky, V.N. Protein folding revisited. A polypeptide chain at the folding—Misfolding—Nonfolding cross-roads: Which way to go? *Cell. Mol. Life Sci.* **2003**, *60*, 1852–1871. [CrossRef] [PubMed]
5. Dyson, H.J.; Wright, P.E. Intrinsically unstructured proteins and their functions. *Nat. Rev. Mol. Cell Biol.* **2005**, *6*, 197–208. [CrossRef] [PubMed]
6. Schweers, O.; Schönbrunn-Hanebeck, E.; Marx, A.; Mandelkow, E. Structural studies of tau protein and Alzheimer paired helical filaments show no evidence for beta-structure. *J. Biol. Chem.* **1994**, *269*, 24290–24297. [PubMed]
7. Weinreb, P.H.; Zhen, W.; Poon, A.W.; Conway, K.A.; Lansbury, P.T. NACP, a protein implicated in Alzheimer's disease and learning, is natively unfolded. *Biochemistry* **1996**, *35*, 13709–13715. [CrossRef] [PubMed]
8. Wright, P.E.; Dyson, H.J. Intrinsically unstructured proteins: Re-assessing the protein structure-function paradigm. *J. Mol. Biol.* **1999**, *293*, 321–331. [CrossRef] [PubMed]
9. Daughdrill, G.W.; Pielak, G.J.; Uversky, V.N.; Cortese, M.S.; Dunker, A.K. Natively disordered proteins. In *Protein Fold Handbook*; Buchner, J., Kiefhaber, T., Eds.; Wiley VCH: Weinheim, Germany, 2005; pp. 275–357.
10. Uversky, V.N. A decade and a half of protein intrinsic disorder: Biology still waits for physics. *Protein Sci.* **2013**, *22*, 693–724. [CrossRef] [PubMed]
11. Uversky, V.N.; Oldfield, C.J.; Dunker, A.K. Intrinsically Disordered Proteins in Human Diseases: Introducing the D^2 Concept. *Annu. Rev. Biophys.* **2008**, *37*, 215–246. [CrossRef] [PubMed]
12. Uversky, V.N.; Gillespie, J.R.; Fink, A.L. Why are "natively unfolded" proteins unstructured under physiologic conditions? *Proteins Struct. Funct. Bioinform.* **2000**, *41*, 415–427. [CrossRef]
13. Uversky, V.N.; Dunker, A.K. Understanding protein non-folding. *Biochim. Biophys. Acta* **2010**, *1804*, 1231–1264. [CrossRef] [PubMed]
14. Dunker, A.K.; Oldfield, C.J.; Meng, J.; Romero, P.; Yang, J.Y.; Chen, J.W.; Vacic, V.; Obradovic, Z.; Uversky, V.N. The unfoldomics decade: An update on intrinsically disordered proteins. *BMC Genom.* **2008**, *26*, S1. [CrossRef] [PubMed]
15. Hartl, F.U. Molecular chaperones in cellular protein folding. *Nature* **1996**, *381*, 571–580. [CrossRef] [PubMed]
16. Uversky, V.N. The triple power of D^3: Protein intrinsic disorder in degenerative diseases. *Front. Biosci.* **2014**, *19*, 181–258. [CrossRef]
17. Kovacs, G.G. Concepts and classification of neurodegenerative diseases. *Handb. Clin. Neurol.* **2017**, *145*, 301–307. [PubMed]
18. Kransnoslobodtsev, A.V.; Shlyakhtenko, L.S.; Ukraintsev, E.; Zaikova, T.O.; Keana, J.F.W.; Lyubchenko, Y.L. Nanomedicine and protein misfolding diseases. *Nanomedicine* **2005**, *1*, 300–305. [CrossRef] [PubMed]
19. Uversky, V.N. Targeting intrinsically disordered proteins in neurodegenerative and protein dysfunction diseases: Another illustration of the D^2 concept. *Expert Rev. Proteom.* **2010**, *7*, 543–564. [CrossRef] [PubMed]
20. Breydo, L.; Uversky, V.N. Role of metal ions in aggregation of intrinsically disordered proteins in neurodegenerative diseases. *Metallomics* **2011**, *3*, 1163–1180. [CrossRef] [PubMed]

21. Eftekharzadeh, B.; Hyman, B.T.; Wegmann, S. Structural studies on the mechanism of protein aggregation in age related neurodegenerative diseases. *Mech. Ageing Dev.* **2016**, *156*, 1–13. [CrossRef] [PubMed]

22. Eisele, Y.S.; Monteiro, C.; Fearns, C.; Encalada, S.E.; Wiseman, R.L.; Powers, E.T.; Kelly, J.W. Targeting protein aggregation for the treatment of degenerative diseases. *Nat. Rev. Drug Discov.* **2015**, *14*, 759–780. [CrossRef] [PubMed]

23. Goedert, M.; Jakes, R.; Spillantini, M.G. The Synucleinopathies: Twenty Years On. *J. Parkinsons Dis.* **2017**, *7*, S51–S69. [CrossRef] [PubMed]

24. Spillantini, M.; Crowther, R.; Jakes, R.; Hasegawa, M.; Goedert, M. alpha-synuclein in filamentous inclusions of Lewy bodies from Parkinson's disease and dementia with Lewy bodies. *Proc. Natl. Acad. Sci. USA* **1998**, *95*, 6469–6473. [CrossRef] [PubMed]

25. Jakes, R.; Spillantini, M.G.; Goedert, M. Identification of two distinct synucleins from human brain. *FEBS Lett.* **1994**, *345*, 27–32. [CrossRef]

26. Theillet, F.X.; Binolfi, A.; Bekei, B.; Martorana, A.; Rose, H.M.; Stuiver, M.; Verzini, S.; Lorenz, D.; van Rossum, M.; Goldfarb, D.; et al. Structural disorder of monomeric α-synuclein persists in mammalian cells. *Nature* **2016**, *530*, 45–50. [CrossRef] [PubMed]

27. Ferreon, A.C.; Gambin, Y.; Lemke, E.A.; Deniz, A.A. Interplay of alpha-synuclein binding and conformational switching probed by single-molecule fluorescence. *Proc. Natl. Acad. Sci. USA* **2009**, *106*, 5645–5650. [CrossRef] [PubMed]

28. Choi, T.S.; Han, J.Y.; Heo, C.E.; Lee, S.W.; Kim, H.I. Electrostatic and hydrophobic interactions of lipid-associated α-synuclein: The role of a water-limited interfaces in amyloid fibrillation. *Biochim. Biophys. Acta Biomembr.* **2018**, *1860*, 1854–1862. [CrossRef] [PubMed]

29. Spillantini, M.G.; Goedert, M. Neurodegeneration and the ordered assembly of α-synuclein. *Cell Tissue Res.* **2018**, *373*, 137–148. [CrossRef] [PubMed]

30. Greten-Harrison, B.; Polydoro, M.; Morimoto-Tomita, M.; Diao, L.; Williams, A.M.; Nie, E.H.; Makani, S.; Tian, N.; Castillo, P.E.; Buchman, V.L.; et al. αβγ-Synuclein triple knockout mice reveal age-dependent neuronal dysfunction. *Proc. Natl. Acad. Sci. USA* **2010**, *107*, 19573–19578. [CrossRef]

31. Fortin, D.L.; Troyer, M.D.; Nakamura, K.; Kubo, S.; Anthony, M.D.; Edwards, R.H. Lipid rafts mediate the synaptic localization of alpha-synuclein. *J. Neurosci.* **2004**, *24*, 6715–6723. [CrossRef] [PubMed]

32. Burré, J.; Sharma, M.; Südhof, T.C. α-Synuclein assembles into higher-order multimers upon membrane binding to promote SNARE complex formation. *Proc. Natl. Acad. Sci. USA* **2014**, *111*, E4274–E4283. [CrossRef] [PubMed]

33. Logan, T.; Bendor, J.; Toupin, C.; Thorn, K.; Edwards, R.H. α-Synuclein promotes dilation of the exocytotic fusion pore. *Nat. Neurosci.* **2017**, *20*, 681–689. [CrossRef] [PubMed]

34. Gai, W.P.; Yuan, H.X.; Li, X.Q.; Power, J.T.; Blumbergs, P.C.; Jensen, P.H. In situ and in vitro study of colocalization and segregation of alpha-synuclein, ubiquitin, and lipids in Lewy bodies. *Exp. Neurol.* **2000**, *166*, 324–333. [CrossRef] [PubMed]

35. Tuttle, M.D.; Comellas, G.; Nieuwkoop, A.J.; Covell, D.J.; Berthold, D.A.; Kloepper, K.D.; Courtney, J.M.; Kim, J.K.; Barclay, A.M.; Kendall, A.; et al. Solid-state NMR structure of a pathogenic fibril of full-length human α-synuclein. *Nat. Struct. Mol. Biol.* **2016**, *23*, 409–415. [CrossRef] [PubMed]

36. Choi, W.; Zibaee, S.; Jakes, R.; Serpell, L.C.; Davletov, B.; Crowther, R.A.; Goedert, M. Mutation E46K increases phospholipid binding and assembly into filaments of human alpha-synuclein. *FEBS Lett.* **2004**, *576*, 363–368. [CrossRef] [PubMed]

37. Tofaris, G.K.; Goedert, M.; Spillantini, M.G. The Transcellular Propagation and Intracellular Trafficking of α-Synuclein. *Cold Spring Harb. Perspect. Med.* **2017**, *7*, a024380. [CrossRef] [PubMed]

38. Osterberg, V.R.; Spinelli, K.J.; Weston, L.J.; Luk, K.C.; Woltjer, R.L.; Unni, V.K. Progressive aggregation of alpha-synuclein and selective degeneration of Lewy inclusion-bearing neurons in a mouse model of parkinsonism. *Cell Rep.* **2015**, *10*, 1252–1260. [CrossRef] [PubMed]

39. Fusco, G.; Chen, S.W.; Williamson, P.T.F.; Cascella, R.; Perni, M.; Jarvis, J.A.; Cecchi, C.; Vendruscolo, M.; Chiti, F.; Cremades, N.; et al. Structural basis of membrane disruption and cellular toxicity by α-synuclein oligomers. *Science* **2017**, *358*, 1440–1443. [CrossRef] [PubMed]

40. Varela, J.A.; Rodrigues, M.; De, S.; Flagmeier, P.; Gandhi, S.; Dobson, C.M.; Klenerman, D.; Lee, S.F. Optical Structural Analysis of Individual α-Synuclein Oligomers. *Angew. Chem. Int. Ed. Engl.* **2018**, *57*, 4886–4890. [CrossRef] [PubMed]

41. Mattson, M.P. Pathways towards and away from Alzheimer's disease. *Nature* **2004**, *430*, 631–639. [CrossRef] [PubMed]

42. Aleksis, R.; Oleskovs, F.; Jaudzems, K.; Pahnke, J.; Biverstål, H. Structural studies of amyloid-β peptides: Unlocking the mechanism of aggregation and the associated toxicity. *Biochimie* **2017**, *140*, 176–192. [CrossRef] [PubMed]

43. Selkoe, D.J. Alzheimer's disease: Genes, proteins, and therapy. *Physiol. Rev.* **2001**, *81*, 741–766. [CrossRef] [PubMed]

44. Kumari, A.; Rajput, R.; Shrivastava, N.; Somvanshi, P.; Grover, A. Synergistic approaches unraveling regulation and aggregation of intrinsically disordered β-amyloids implicated in Alzheimer's disease. *Int. J. Biochem. Cell Biol.* **2018**, *99*, 19–27. [CrossRef] [PubMed]

45. Kang, J.; Lemaire, H.G.; Unterbeck, A.; Salbaum, J.M.; Masters, C.L.; Grzeschik, K.H.; Multhaup, G.; Beyreuther, K.; Müller-Hill, B. The precursor of Alzheimer's disease amyloid A4 protein resembles a cell-surface receptor. *Nature* **1987**, *325*, 733–736. [CrossRef] [PubMed]

46. Iwatsubo, T.; Odaka, A.; Suzuki, N.; Mizusawa, H.; Nukina, N.; Ihara, Y. Visualization of Aβ42(43) and Aβ40 in senile plaques with end-specific Aβ monoclonals: Evidence that an initially deposited species is Aβ42(43). *Neuron* **1994**, *13*, 45–53. [CrossRef]

47. De Strooper, B.; Saftig, P.; Craessaerts, K.; Vanderstichele, H.; Guhde, G.; Annaert, W.; Von Figura, K.; Van Leuven, F. Deficiency of presenilin-1 inhibits the normal cleavage of amyloid precursor protein. *Nature* **1998**, *391*, 387–390. [CrossRef] [PubMed]

48. Stokin, G.B.; Lillo, C.; Falzone, T.L.; Brusch, R.G.; Rockenstein, E.; Mount, S.L.; Raman, R.; Davies, P.; Masliah, E.; Williams, D.S.; et al. Axonopathy and transport deficits early in the pathogenesis of Alzheimer's disease. *Science* **2005**, *307*, 1282–1288. [CrossRef] [PubMed]

49. Lewis, H.; Beher, D.; Cookson, N.; Oakley, A.; Piggott, M.; Morris, C.M.; Jaros, E.; Perry, R.; Ince, P.; Kenny, R.A.; et al. Quantification of Alzheimer pathology in ageing and dementia: Age-related accumulation of amyloid-beta(42) peptide in vascular dementia. *Neuropathol. Appl. Neurobiol.* **2006**, *32*, 103–118. [CrossRef] [PubMed]

50. Sosa, L.J.; Caceres, A.; Dupraz, S.; Oksdath, M.; Quiroga, S.; Lorenzo, A. The physiological role of the amyloid precursor protein as an adhesion molecule in the developing nervous system. *J. Neurochem.* **2017**, *143*, 11–29. [CrossRef] [PubMed]

51. Goldgaber, D.; Lerman, M.I.; McBride, O.W.; Saffiotti, U.; Gajdusek, D.C. Characterization and chromosomal localization of a cDNA encoding brain amyloid of Alzheimer's disease. *Science* **1987**, *235*, 877–880. [CrossRef] [PubMed]

52. Tanzi, R.E.; Gusella, J.F.; Watkins, P.C.; Bruns, G.A.; St George-Hyslop, P.; Van Keuren, M.L.; Patterson, D.; Pagan, S.; Kurnit, D.M.; Neve, R.L. Amyloid beta protein gene: cDNA, mRNA distribution, and genetic linkage near the Alzheimer locus. *Science* **1987**, *235*, 880–884. [CrossRef] [PubMed]

53. Haass, C. Take five—BACE and the γ-secretase quartet conduct Alzheimer's amyloid β-peptide generation. *EMBO J.* **2004**, *23*, 483–488. [CrossRef] [PubMed]

54. Haass, C.; Schlossmacher, M.G.; Hung, A.Y.; Vigo-Pelfrey, C.; Mellon, A.; Ostaszewski, B.L.; Lieberburg, I.; Koo, E.H.; Schenk, D.; Teplow, D.B.; et al. Amyloid β-peptide is produced by cultured cells during normal metabolism. *Nature* **1992**, *359*, 322–325. [CrossRef] [PubMed]

55. Gouras, G.K.; Tsai, J.; Naslund, J.; Vincent, B.; Edgar, M.; Checler, F.; Greenfield, J.P.; Haroutunian, V.; Buxbaum, J.D.; Xu, H.; et al. Intraneuronal Aβ42 accumulation in human brain. *Am. J. Pathol.* **2000**, *156*, 15–20. [CrossRef]

56. Bitan, G.; Vollers, S.S.; Teplow, D.B. Elucidation of primary structure elements controlling early amyloid β-protein oligomerization. *J. Biol. Chem.* **2003**, *12*, 34882–34889. [CrossRef] [PubMed]

57. Suzuki, N.; Cheung, T.T.; Cai, X.-D.; Odaka, A.; Otvos, L.; Eckman, C.; Golde, T.E.; Younkin, S.G. An increased percentage of long amyloid beta protein secreted by familial amyloid beta protein precursor (beta APP717) mutants. *Science* **1994**, *264*, 1336–1340. [CrossRef] [PubMed]

58. Lovell, M.A.; Robertson, J.D.; Teesdale, W.J.; Campbell, J.L.; Markesbery, W.R. Copper, iron and zinc in Alzheimer's disease senile plaques. *J. Neurol. Sci.* **1998**, *158*, 47–52. [CrossRef]

59. Alexandrescu, A.T. Amyloid accomplices and enforcers. *Protein Sci.* **2005**, *14*, 1–12. [CrossRef] [PubMed]

60. Scheuner, D.; Eckman, C.; Jensen, M.; Song, X.; Citron, M.; Suzuki, N.; Bird, T.D.; Hardy, J.; Hutton, M.; Kukull, W.; et al. Secreted amyloid beta-protein similar to that in the senile plaques of Alzheimer's disease is increased in vivo by the presenilin 1 and 2 and APP mutations linked to familial Alzheimer's disease. *Nat. Med.* **1996**, *2*, 864–870. [CrossRef] [PubMed]

61. Citron, M.; Westaway, D.; Xia, W.; Carlson, G.; Diehl, T.; Levesque, G.; Johnson-Wood, K.; Lee, M.; Seubert, P.; Davis, A.; et al. Mutant presenilins of Alzheimer's disease increase production of 42-residue amyloid beta-protein in both transfected cells and transgenic mice. *Nat. Med.* **1997**, *3*, 67–72. [CrossRef] [PubMed]

62. Mullan, M.; Crawford, F.; Axelman, K.; Houlden, H.; Lilius, L.; Winblad, B.; Lannfelt, L. A pathogenic mutation for probable Alzheimer's disease in the APP gene at the N-terminus of β-amyloid. *Nat. Genet.* **1992**, *1*, 345–347. [CrossRef] [PubMed]

63. Sherrington, R.; Rogaev, E.I.; Liang, Y.; Rogaeva, E.A.; Levesque, G.; Ikeda, M.; Chi, H.; Lin, C.; Li, G.; Holman, K.; et al. Cloning of a gene bearing missense mutations in early-onset familial Alzheimer's disease. *Nature* **1995**, *375*, 754–760. [CrossRef] [PubMed]

64. Selkoe, D.J.; Hardy, J. The amyloid hypothesis of Alzheimer's disease at 25 years. *EMBO Mol. Med.* **2016**, *8*, 595–608. [CrossRef] [PubMed]

65. Lanoiselée, H.M.; Nicolas, G.; Wallon, D.; Rovelet-Lecrux, A.; Lacour, M.; Rousseau, S.; Richard, A.C.; Pasquier, F.; Rollin-Sillaire, A.; Martinaud, O.; et al. APP, PSEN1, and PSEN2 mutations in early-onset Alzheimer disease: A genetic screening study of familial and sporadic cases. *PLoS Med.* **2017**, *14*, e1002270. [CrossRef] [PubMed]

66. Simmons, L.K.; May, P.C.; Tomaselli, K.J.; Rydel, R.E.; Fuson, K.S.; Brigham, E.F.; Wright, S.; Lieberburg, I.; Becker, G.W.; Brems, D.N. Secondary structure of amyloid beta peptide correlates with neurotoxic activity in vitro. *Mol. Pharmacol.* **1994**, *45*, 373–379. [PubMed]

67. Kirkitadze, M.D.; Condron, M.M.; Teplow, D.B. Identification and characterization of key kinetic intermediates in amyloid beta-protein fibrillogenesis. *J. Mol. Biol.* **2001**, *312*, 1103–1119. [CrossRef] [PubMed]

68. Yan, Y.; Wang, C. Aβ42 is More Rigid than Aβ40 at the C Terminus: Implications for Aβ Aggregation and Toxicity. *J. Mol. Biol.* **2006**, *364*, 853–862. [CrossRef] [PubMed]

69. Petkova, A.T.; Ishii, Y.; Balbach, J.J.; Antzutkin, O.N.; Leapman, R.D.; Delaglio, F.; Tycko, R. A structural model for Alzheimer's beta-amyloid fibrils based on experimental constraints from solid state NMR. *Proc. Natl. Acad. Sci. USA* **2002**, *99*, 16742–16747. [CrossRef] [PubMed]

70. Petkova, A.T.; Leapman, R.D.; Guo, Z.; Yau, W.M.; Mattson, M.P.; Tycko, R. Self-propagating, molecular-level polymorphism in Alzheimer's β-amyloid fibrils. *Science* **2005**, *307*, 262–265. [CrossRef] [PubMed]

71. Bertini, I.; Gonnelli, L.; Luchinat, C.; Mao, J.; Nesi, A. A new structural model of Aβ40 fibrils. *J. Am. Chem. Soc.* **2011**, *133*, 16013–16022. [CrossRef] [PubMed]

72. Parthasarathy, S.; Long, F.; Miller, Y.; Xiao, Y.; McElheny, D.; Thurber, K.; Ma, B.; Nussinov, R.; Ishii, Y. Molecular-level examination of Cu^{2+} binding structure for amyloid fibrils of 40-residue Alzheimer's β by solid-state NMR spectroscopy. *J. Am. Chem. Soc.* **2011**, *133*, 3390–3400. [CrossRef] [PubMed]

73. Sgourakis, N.G.; Yau, W.M.; Qiang, W. Modeling an in-register, parallel "Iowa" Aβ fibril structure using solid-state NMR data from labeled samples with Rosetta. *Structure* **2015**, *23*, 216–227. [CrossRef] [PubMed]

74. Xiao, Y.; Ma, B.; McElheny, D.; Parthasarathy, S.; Long, F.; Hoshi, M.; Nussinov, R.; Ishii, Y. Aβ(1–42) fibril structure illuminates self-recognition and replication of amyloid in Alzheimer's disease. *Nat. Struct. Mol. Biol.* **2015**, *22*, 499–505. [CrossRef] [PubMed]

75. Colvin, M.T.; Silvers, R.; Ni, Q.Z.; Can, T.V.; Sergeyev, I.; Rosay, M.; Donovan, K.J.; Michael, B.; Wall, J.; Linse, S.; et al. Atomic Resolution Structure of Monomorphic Aβ42 Amyloid Fibrils. *J. Am. Chem. Soc.* **2016**, *138*, 9663–9674. [CrossRef] [PubMed]

76. Wälti, M.A.; Ravotti, F.; Arai, H.; Glabe, C.G.; Wall, J.S.; Böckmann, A.; Güntert, P.; Meier, B.H.; Riek, R. Atomic-resolution structure of a disease-relevant Aβ(1–42) amyloid fibril. *Proc. Natl. Acad. Sci. USA* **2016**, *113*, E4976–E4984. [CrossRef] [PubMed]

77. Masters, C.L.; Multhaup, G.; Simms, G.; Martins, R.N.; Beyreuther, K. Neuronal origin of a cerebral amyloid: Neurofibriliary tangles of Alzheimer's disease contain the same protein as the amyloid of plaque cores and blood vessels. *EMBO J.* **1985**, *4*, 2757–2763. [CrossRef] [PubMed]

78. Portelius, E.; Bogdanovic, N.; Gustavsson, M.K.; Volkmann, I.; Brinkmalm, G.; Zetterberg, H.; Winblad, B.; Blennow, K. Mass spectrometric characterization of brain amyloid beta isoform signatures in familial and sporadic Alzheimer's disease. *Acta Neuropathol.* **2010**, *120*, 185–193. [CrossRef] [PubMed]

79. Weingarten, M.D.; Lockwood, A.H.; Hwo, S.Y.; Kirschner, M.W. A protein factor essential for microtubule assembly. *Proc. Natl. Acad. Sci. USA* **1975**, *72*, 1858–1862. [CrossRef] [PubMed]

80. Ebneth, A.; Godemann, R.; Stamer, K.; Illenberger, S.; Trinczek, B.; Mandelkow, E. Overexpression of tau protein inhibits kinesin-dependent trafficking of vesicles, mitochondria, and endoplasmic reticulum: Implications for Alzheimer's disease. *J. Cell Biol.* **1998**, *143*, 777–794. [CrossRef] [PubMed]

81. Borna, H.; Assadoulahei, K.; Riazi, G.; Harchegani, A.B.; Shahriary, A. Structure, Function and Interactions of Tau: Particular Focus on Potential Drug Targets for the Treatment of Tauopathies. *CNS Neurol. Disord Drug Targets* **2018**, *17*, 325–337. [CrossRef] [PubMed]

82. Bakota, L.; Ussif, A.; Jeserich, G.; Brandt, R. Systemic and network functions of the microtubule-associated protein tau: Implications for tau-based therapies. *Mol. Cell. Neurosci.* **2017**, *84*, 132–141. [CrossRef]

83. Kosik, K.S.; Joachim, C.L.; Selkoe, D.J. Microtubule-associated protein tau (tau) is a major antigenic component of paired helical filaments in Alzheimer disease. *Proc. Natl. Acad. Sci. USA* **1986**, *83*, 4044–4448. [CrossRef] [PubMed]

84. Andreadis, A.; Brown, W.M.; Kosik, K.S. Structure and novel exons of the human tau gene. *Biochemistry* **1992**, *31*, 10626–10633. [CrossRef] [PubMed]

85. Goedert, M.; Spillantini, M.G.; Jakes, R.; Rutherford, D.; Crowther, R.A. Multiple isoforms of human microtubule-associated protein tau: Sequences and localization in neurofibrillary tangles of Alzheimer's disease. *Neuron* **1989**, *3*, 519–526. [CrossRef]

86. Goedert, M.; Spillantini, M.G.; Potier, M.C.; Ulrich, J.; Crowther, R.A. Cloning and sequencing of the cDNA encoding an isoform of microtubule-associated protein tau containing four tandem repeats: Differential expression of tau protein mRNAs in human brain. *EMBO J.* **1989**, *8*, 393–399. [CrossRef] [PubMed]

87. Goedert, M.; Wischik, C.M.; Crowther, R.A.; Walker, J.E.; Klug, A. Cloning and sequencing of the cDNA encoding a core protein of the paired helical filament of Alzheimer disease: Identification as the microtubule-associated protein tau. *Proc. Natl. Acad. Sci. USA* **1988**, *85*, 4051–4055. [CrossRef] [PubMed]

88. Lee, V.M.; Balin, B.J.; Otvos, L., Jr.; Trojanowski, J.Q. A68: A major subunit of paired helical filaments and derivatized forms of normal Tau. *Science* **1991**, *251*, 675–678. [CrossRef] [PubMed]

89. Santa-Maria, I.; Varghese, M.; Ksiezak-Reding, H.; Dzhun, A.; Wang, J.; Pasinetti, G.M. Paired helical filaments from Alzheimer disease brain induce intracellular accumulation of Tau protein in aggresomes. *J. Biol. Chem.* **2012**, *287*, 20522–20533. [CrossRef] [PubMed]

90. Bhattacharya, K.; Rank, K.B.; Evans, D.B.; Sharma, S.K. Role of cysteine-291 and cysteine-322 in the polymerization of human tau into Alzheimer-like filaments. *Biochem. Biophys. Res. Commun.* **2001**, *285*, 20–26. [CrossRef] [PubMed]

91. Novak, P.; Cehlar, O.; Skrabana, R.; Novak, M. Tau Conformation as a Target for Disease-Modifying Therapy: The Role of Truncation. *J. Alzheimers Dis.* **2018**, *64*, S535–S546. [CrossRef] [PubMed]

92. Florenzano, F.; Veronica, C.; Ciasca, G.; Ciotti, M.T.; Pittaluga, A.; Olivero, G.; Feligioni, M.; Iannuzzi, F.; Latina, V.; Maria Sciacca, M.F.; et al. Extracellular truncated tau causes early presynaptic dysfunction associated with Alzheimer's disease and other tauopathies. *Oncotarget* **2017**, *8*, 64745–64778. [CrossRef] [PubMed]

93. Schedin-Weiss, S.; Winblad, B.; Tjernberg, L.O. The role of protein glycosylation in Alzheimer disease. *FEBS J.* **2014**, *281*, 46–62. [CrossRef] [PubMed]

94. Ligabue-Braun, R.; Carlini, C.R. Moonlighting Toxins: Ureases and Beyond. In *Plant Toxins*; Gopalakrishnakone, P., Carlini, C.R., Ligabue-Braun, R., Eds.; Springer: Dordrecht, The Netherlands, 2015; pp. 199–219.

95. Oliveira, J.M.; Henriques, A.G.; Martins, F.; Rebelo, S.; da Cruz e Silva, O.A. Amyloid-beta Modulates Both AbetaPP and Tau Phosphorylation. *J. Alzheimers Dis.* **2015**, *45*, 495–507. [CrossRef] [PubMed]

96. Ittner, L.M.; Ke, Y.D.; Delerue, F.; Bi, M.; Gladbach, A.; van Eersel, J.; Wölfing, H.; Chieng, B.C.; Christie, M.J.; Napier, I.A.; et al. Dendritic function of tau mediates amyloid-beta toxicity in Alzheimer's disease mouse models. *Cell* **2010**, *142*, 387–397. [CrossRef] [PubMed]

97. Prusiner, S.B. Novel proteinaceous infectious particles cause scrapie. *Science* **1982**, *216*, 136–144. [CrossRef] [PubMed]

98. Prusiner, S.B.; Mckinley, M.P.; Groth, D.F.; Bowman, K.A.; Mock, N.I.; Cochran, S.P.; Masiarz, F.R. Scrapie agent contains a hydrophobic protein. *Proc. Natl. Acad. Sci. USA* **1981**, *78*, 6675–6679. [CrossRef] [PubMed]

99. Prusiner, S.B.; Groth, D.F.; Bolton, D.C.; Kent, S.B.; Hood, L.E. Purification and structural studies of a major scrapie prion protein. *Cell* **1984**, *38*, 127–134. [CrossRef]

100. Westergard, L.; Christensen, H.M.; Harris, D.A. The cellular prion protein (PrP(C)): Its physiological function and role in Disease. *Biochim. Biophys. Acta* **2007**, *1772*, 629–644. [CrossRef] [PubMed]

101. Bremer, J.; Baumann, F.; Tiberi, C.; Wessig, C.; Fischer, H.; Schwarz, P.; Steele, A.D.; Toyka, K.V.; Nave, K.A.; Weis, J.; et al. Axonal prion protein is required for peripheral myelin maintenance. *Nat. Neurosci.* **2010**, *13*, 310–318. [CrossRef] [PubMed]

102. Chakravarty, A.K.; Jarosz, D.F. More than Just a Phase: Prions at the Crossroads of Epigenetic Inheritance and Evolutionary Change. *J. Mol. Biol.* **2018**, *430*, 4607–4618. [CrossRef] [PubMed]

103. Riek, R.; Hornemann, S.; Wider, G.; Glockshuber, R.; Wüthrich, K. NMR characterization of the full-length recombinant murine prion protein, mPrP(23-231). *FEBS Lett.* **1997**, *413*, 282–288. [CrossRef]

104. Donne, D.G.; Viles, J.H.; Groth, D.; Mehlhorn, I.; James, T.L.; Cohen, F.E.; Prusiner, S.B.; Wright, P.E.; Dyson, H.J. Structure of the recombinant full-length hamster prion protein PrP (29-231): The N terminus is highly flexible. *Proc. Natl. Acad. Sci. USA* **1997**, *94*, 13452–13457. [CrossRef] [PubMed]

105. Zahn, R.; Liu, A.; Luhrs, T.; Riek, R.; von Schroetter, C.; Lopez-Garcia, F.; Billeter, M.; Calzolai, L.; Wider, G.; Wuthrich, K. NMR solution structure of the human prion protein. *Proc. Natl. Acad. Sci. USA* **2000**, *97*, 145–150. [CrossRef] [PubMed]

106. Lopez-García, F.L.; Zahn, R.; Riek, R.; Wüthrich, K. NMR structure of the bovine prion protein. *Proc. Natl. Acad. Sci. USA* **2000**, *97*, 8334–8339. [CrossRef] [PubMed]

107. Sabate, R.; Rousseau, F.; Schymkowitz, J.; Ventura, S. What makes a protein sequence a prion? *PLoS Comput. Biol.* **2015**, *11*, e1004013. [CrossRef] [PubMed]

108. Cong, X.; Casiraghi, N.; Rossetti, G.; Mohanty, S.; Giachin, G.; Legname, G.; Carloni, P. Role of Prion Disease-Linked Mutations in the Intrinsically Disordered N-Terminal Domain of the Prion Protein. *J. Chem. Theory Comput.* **2013**, *9*, 5158–5167. [CrossRef] [PubMed]

109. Pan, K.-M.; Baldwin, M.; Nguyen, J.; Gasset, M.; Serban, A.N.A.; Groth, D.; Mehlhorn, I.; Huang, Z.; Fletterick, R.J.; Cohen, F.E. Conversion of alpha-helices into beta-sheets features in the formation of the scrapie prion proteins. *Proc. Natl. Acad. Sci. USA* **1993**, *90*, 10962–10966. [CrossRef] [PubMed]

110. Prusiner, S.B. Molecular biology of prion diseases. *Science* **1991**, *252*, 1515–1522. [CrossRef] [PubMed]

111. Prusiner, S.B. Chemistry and biology of prions. *Biochemistry* **1992**, *31*, 12277–12288. [CrossRef] [PubMed]

112. Brown, P.; Gajdusek, D.C. The Human Spongiform Encephalopathies: Kuru, Creutzfeldt-Jakob Disease, and the Gerstmann-Sträussler-Scheinker Syndrome. In *Transmissible Spongiform Encephalopathies: Current Topics in Microbiology and Immunology*; Chesebro, B.W., Ed.; Springer: Berlin, Germany, 1991; Volume 172, pp. 1–20.

113. Nathanson, N.; Wilesmith, J.; Griot, C. Bovine spongiform encephalopathy (BSE): Causes and consequences of a common source epidemic. *Am. J. Epidemiol.* **1997**, *145*, 959–969. [CrossRef] [PubMed]

114. Pattison, I.H. The relative susceptibility of sheep, goats and mice to two types of the goat scrapie agent. *Res. Vet. Sci.* **1966**, *7*, 207–212. [CrossRef]

115. Scott, M.; Foster, D.; Mirenda, C.; Serban, D.; Coufal, F.; Wälchli, M.; Torchia, M.; Groth, D.; Carlson, G.; DeArmond, S.J.; et al. Transgenic mice expressing hamster prion protein produce species-specific scrapie infectivity and amyloid plaques. *Cell* **1989**, *59*, 847–857. [CrossRef]

116. Prusiner, S.B.; Scott, M.; Foster, D.; Pan, K.-M.; Groth, D.; Mirenda, C.; Torchia, M.; Yang, S.-L.; Serban, D.; Carlson, G.A.; et al. Transgenetic studies implicate interactions between homologous PrP isoforms in scrapie prion replication. *Cell* **1990**, *63*, 673–686. [CrossRef]

117. Bartz, J.C.; McKenzie, D.I.; Bessen, R.A.; Marsh, R.F.; Aiken, J.M. Transmissible mink encephalopathy species barrier effect between ferret and mink: PrP gene and protein analysis. *J. Gen. Virol.* **1994**, *75*, 2947–2953. [CrossRef] [PubMed]

118. Bian, J.; Khaychuk, V.; Angers, R.C.; Fernández-Borges, N.; Vidal, E.; Meyerett-Reid, C.; Kim, S.; Calvi, C.L.; Bartz, J.C.; Hoover, E.A.; et al. Prion replication without host adaptation during interspecies transmissions. *Proc. Natl. Acad. Sci. USA* **2017**, *114*, 1141–1146. [CrossRef] [PubMed]

119. Prusiner, S.B. Molecular biology and pathogenesis of prion diseases. *Trends Biochem. Sci.* **1996**, *21*, 482–487. [CrossRef]

120. Colby, D.W.; Prusiner, S.B. Prions. *Cold Spring Harb. Perspect. Biol.* **2011**, *3*, a006833. [CrossRef] [PubMed]

121. Makarava, N.; Kovacs, G.G.; Bocharova, O.; Savtchenko, R.; Alexeeva, I.; Budka, H.; Rohwer, R.G.; Baskakov, I.V. Recombinant prion protein induces a new transmissible prion disease in wild-type animals. *Acta Neuropathol.* **2010**, *119*, 177–187. [CrossRef] [PubMed]

122. Stahl, N.; Baldwin, M.A.; Prusiner, S.B.; Teplow, D.B.; Hood, L.; Gibson, B.W.; Burlingame, A.L. Structural Studies of the Scrapie Prion Protein Using Mass Spectrometry and Amino Acid Sequencing. *Biochemistry* **1993**, *32*, 1991–2002. [CrossRef] [PubMed]

123. Meyer, R.K.; McKinley, M.P.; Bowman, K.A.; Braunfeld, M.B.; Barry, R.A.; Prusiner, S.B. Separation and properties of cellular and scrapie prion proteins. *Proc. Natl. Acad. Sci. USA* **1986**, *83*, 2310–2314. [CrossRef] [PubMed]

124. Saverioni, D.; Notari, S.; Capellari, S.; Poggiolini, I.; Giese, A.; Kretzschmar, H.A.; Parchi, P. Analyses of protease resistance and aggregation state of abnormal prion protein across the spectrum of human prions. *J. Biol. Chem.* **2013**, *288*, 27972–27985. [CrossRef] [PubMed]

125. Safar, J.; Wille, H.; Itri, V.; Groth, D.; Serban, H.; Torchia, M.; Cohen, F.E.; Prusiner, S.B. Eight prion strains have PrP Sc molecules with different conformations. *Nat. Med.* **1998**, *4*, 1157–1165. [CrossRef] [PubMed]

126. Tzaban, S.; Friedlander, G.; Schonberger, O.; Horonchik, L.; Yedidia, Y.; Shaked, G.; Gabizon, R.; Taraboulos, A. Protease-sensitive scrapie prion protein in aggregates of heterogeneous sizes. *Biochemistry* **2002**, *41*, 12868–12875. [PubMed]

127. DeArmond, S.J.; Sánchez, H.; Yehiely, F.; Qiu, Y.; Ninchak-Casey, A.; Daggett, V.; Camerino, A.P.; Cayetano, J.; Rogers, M.; Groth, D.; et al. Selective neuronal targeting in prion disease. *Neuron* **1997**, *19*, 1337–1348. [CrossRef]

128. Colby, D.W.; Zhang, Q.; Wang, S.; Groth, D.; Legname, G.; Riesner, D.; Prusiner, S.B. Prion detection by an amyloid seeding assay. *Proc. Natl. Acad. Sci. USA* **2007**, *104*, 20914–20919. [CrossRef] [PubMed]

129. Wille, H.; Prusiner, S.B.; Cohen, F.E. Scrapie infectivity is independent of amyloid staining properties of the N-Terminally truncated prion protein. *J. Struct. Biol.* **2000**, *130*, 323–338. [CrossRef] [PubMed]

130. Uversky, V.N.; Davé, V.; Iakoucheva, L.M.; Malaney, P.; Metallo, S.J.; Pathak, R.R.; Joerger, A.C. Pathological unfoldomics of uncontrolled chaos: Intrinsically disordered proteins and human diseases. *Chem. Rev.* **2014**, *114*, 6844–6879. [CrossRef] [PubMed]

131. Mol, P.R. Oncogenes as Therapeutic Targets in Cancer: A Review. *IOSR J. Dent. Med. Sci.* **2013**, *5*, 46–56. [CrossRef]

132. Dunker, A.K.; Uversky, V.N. Drugs for "protein clouds": Targeting intrinsically disordered transcription factors. *Curr. Opin. Pharmacol.* **2010**, *10*, 782–788. [CrossRef] [PubMed]

133. Uversky, V.N. p53 Proteoforms and Intrinsic Disorder: An Illustration of the Protein Structure-Function Continuum Concept. *Int. J. Mol. Sci.* **2016**, *17*, 1874. [CrossRef] [PubMed]

134. Levine, A.J. Targeting therapies for the p53 protein in cancer treatments. *Annu. Rev. Cancer Biol.* **2019**. [CrossRef]

135. Hollstein, M.; Sidransky, D.; Vogelstein, B.; Curtis, C. p53 Mutation Human Cancers. *Science* **1991**, *253*, 49–53. [CrossRef] [PubMed]

136. Muller, P.A.J.; Vousden, K.H. P53 mutations in cancer. *Nat. Cell Biol.* **2013**, *15*, 2–8. [CrossRef] [PubMed]

137. Dawson, R.; Müller, L.; Dehner, A.; Klein, C.; Kessler, H.; Buchner, J. The N-terminal domain of p53 is natively unfolded. *J. Mol. Biol.* **2003**, *332*, 1131–1141. [CrossRef] [PubMed]

138. Kubbutat, M.H.G.; Jones, S.N.; Vousden, K.H. Regulation of p53 stability by Mdm2. *Nature* **1997**, *387*, 299–303. [CrossRef] [PubMed]

139. Nag, S.; Qin, J.; Srivenugopal, K.S.; Wang, M.; Zhang, R. The MDM2-p53 pathway revisited. *J. Biomed. Res.* **2013**, *27*, 254–271. [PubMed]

140. Williams, A.B.; Schumacher, B. p53 in the DNA-damage-repair process. *Cold Spring Harb. Perspect. Med.* **2016**, *6*, a026070. [CrossRef] [PubMed]

141. Cho, Y.; Gorina, S.; Jeffrey, P.D.; Pavletich, N.P. Crystal structure of a p53 tumor suppressor-DNA complex: Understanding tumorigenic mutations. *Science* **1994**, *265*, 346–355. [CrossRef] [PubMed]

142. Clore, G.M.; Ernst, J.; Clubb, R.; Omichinski, J.G.; Kennedy, W.M.P.; Sakaguchi, K.; Appella, E.; Gronenborn, A.M. Refined solution structure of the oligomerization domain of the tumour suppressor p53. *Nat. Struct. Mol. Biol.* **1995**, *2*, 321–333. [CrossRef]

143. Fields, S.; Jang, S.K. Presence of a potent transcription activating sequence in the p53 protein. *Science* **1990**, *249*, 1046–1049. [CrossRef] [PubMed]

144. Haupt, Y.; Maya, R.; Kazaz, A.; Oren, M. Mdm2 promotes the rapid degradation of p53. *Nature* **1997**, *387*, 296–299. [CrossRef] [PubMed]

145. Honda, R.; Tanaka, H.; Yasuda, H. Oncoprotein MDM2 is a ubiquitin ligase E3 for tumor suppressor p53. *FEBS Lett.* **1997**, *420*, 25–27. [CrossRef]

146. Joerger, A.C.; Fersht, A.R. Structural Biology of the Tumor Suppressor p53. *Annu. Rev. Biochem.* **2008**, *77*, 557–582. [CrossRef] [PubMed]

147. Lee, W.; Harvey, T.S.; Yin, Y.; Yau, P.; Litchfield, D.; Arrowsmith, C.H. Solution structure of the tetrameric minimum transforming domain of p53. *Nat. Struct. Mol. Biol.* **1994**, *1*, 877–890. [CrossRef]

148. Uversky, A.V.; Xue, B.; Peng, Z.; Kurgan, L.; Uversky, V.N. On the intrinsic disorder status of the major players in programmed cell death pathways. *F1000Research* **2013**, *2*, 190. [CrossRef] [PubMed]

149. Kussie, P.H.; Gorina, S.; Marechal, V.; Elenbaas, B.; Moreau, J.; Levine, A.J.; Pavletich, N.P. Structure of the MDM2 oncoprotein bound to the p53 tumor suppressor transactivation domain. *Science* **1996**, *274*, 948–953. [CrossRef] [PubMed]

150. Marine, J.-C.; Jochemsen, A.G. Mdmx as an essential regulator of p53 activity. *Biochem. Biophys. Res. Commun.* **2005**, *331*, 750–760. [CrossRef] [PubMed]

151. Schon, O.; Friedler, A.; Bycroft, M.; Freund, S.M.V.; Fersht, A.R. Molecular mechanism of the interaction between MDM2 and p53. *J. Mol. Biol.* **2002**, *323*, 491–501. [CrossRef]

152. Borcherds, W.; Kashtanov, S.; Wu, H.; Daughdrill, G.W. Structural divergence is more extensive than sequence divergence for a family of intrinsically disordered proteins. *Proteins Struct. Funct. Bioinform.* **2013**, *81*, 1686–1698. [CrossRef] [PubMed]

153. Chi, S.W.; Lee, S.H.; Kim, D.H.; Ahn, M.J.; Kim, J.S.; Woo, J.Y.; Torizawa, T.; Kainosho, M.; Han, K.H. Structural details on mdm2-p53 interaction. *J. Biol. Chem.* **2005**, *280*, 38795–38802. [CrossRef] [PubMed]

154. Popowicz, G.M.; Czarna, A.; Rothweiler, U.; Szwagierczak, A.; Krajewski, M.; Weber, L.; Holak, T.A. Molecular basis for the inhibition of p53 by Mdmx. *Cell Cycle* **2007**, *6*, 2386–2392. [CrossRef] [PubMed]

155. Vise, P.D.; Baral, B.; Latos, A.J.; Daughdrill, G.W. NMR chemical shift and relaxation measurements provide evidence for the coupled folding and binding of the p53 transactivation domain. *Nucleic Acids Res.* **2005**, *33*, 2061–2077. [CrossRef] [PubMed]

156. Chene, P. The role of tetramerization in p53 function. *Oncogene* **2001**, *20*, 2611–2617. [CrossRef] [PubMed]

157. Pelengaris, S.; Khan, M.; Evan, G. c-MYC: More than just a matter of life and death. *Nat. Rev. Cancer* **2002**, *2*, 764–776. [CrossRef] [PubMed]

158. Dang, C.V. c-Myc Target Genes Involved in Cell Growth, Apoptosis, and Metabolism. *Mol. Cell. Biol.* **1999**, *19*, 1–11. [CrossRef] [PubMed]

159. Nesbit, C.E.; Tersak, J.M.; Prochownik, E.V. MYC oncogenes and human neoplastic disease. *Oncogene* **1999**, *18*, 3004–3016. [CrossRef] [PubMed]

160. Schlagbauer-Wadl, H.; Griffioen, M.; Van Elsas, A.; Schrier, P.I.; Pustelnik, T.; Eichler, H.; Wolff, K.; Pehamberger, H.; Jansen, B. Influence of Increased c-Myc Expression on the Growth Characteristics of Human Melanoma. *J. Investig. Dermatol.* **1999**, *112*, 332–336. [CrossRef] [PubMed]

161. Kumar, D.; Sharma, N.; Giri, R. Therapeutic interventions of cancers using intrinsically disordered proteins as drug targets: C-myc as model system. *Cancer Inform.* **2017**, *16*. [CrossRef] [PubMed]

162. Amati, B.; Dalton, S.; Brooks, M.W.; Littlewood, T.D.; Evan, G.I.; Land, H. Transcriptional activation by the human c-Myc oncoprotein in yeast requires interaction with Max. *Nature* **1992**, *359*, 423–426. [CrossRef] [PubMed]

163. Blackwell, T.K.; Kretzner, L.; Blackwood, E.M.; Eisenman, R.N.; Weintraub, H. Sequence-specific DNA binding by the c-Myc protein. *Science* **1990**, *250*, 1149–1151. [CrossRef] [PubMed]

164. Blackwood, E.M.; Eisenman, R.N. Max: A helix-loop-helix zipper protein that forms a sequence-specific DNA-binding complex with Myc. *Science* **1991**, *251*, 1211–1217. [CrossRef] [PubMed]

165. Kato, G.J.; Lee, W.M.; Chen, L.L.; Dang, C.V. Max: Functional domains and interaction with c-Myc. *Genes Dev.* **1992**, *6*, 81–92. [CrossRef] [PubMed]

166. Andresen, C.; Helander, S.; Lemak, A.; Farès, C.; Csizmok, V.; Carlsson, J.; Penn, L.Z.; Forman-Kay, J.D.; Arrowsmith, C.H.; Lundström, P.; et al. Transient structure and dynamics in the disordered c-Myc transactivation domain affect Bin1 binding. *Nucleic Acids Res.* **2012**, *40*, 6353–6366. [CrossRef] [PubMed]

167. Mao, D.Y.L.; Watson, J.D.; Yan, P.S.; Barsyte-Lovejoy, D.; Khosravi, F.; Wong, W.W.-L.; Farnham, P.J.; Huang, T.H.-M.; Penn, L.Z. Analysis of Myc bound loci identified by CpG island arrays shows that Max is essential for Myc-dependent repression. *Curr. Biol.* **2003**, *13*, 882–886. [CrossRef]

168. Clausen, D.M.; Guo, J.; Parise, R.A.; Beumer, J.H.; Egorin, M.J.; Lazo, J.S.; Prochownik, E.V.; Eiseman, J.L. In vitro cytotoxicity and in vivo efficacy, pharmacokinetics, and metabolism of 10074-G5, a novel small-molecule inhibitor of c-Myc/Max dimerization. *J. Pharmacol. Exp. Ther.* **2010**, *335*, 715–727. [CrossRef] [PubMed]

169. Raffeiner, P.; Röck, R.; Schraffl, A.; Hartl, M.; Hart, J.R.; Janda, K.D.; Vogt, P.K.; Stefan, E.; Bister, K. In vivo quantification and perturbation of Myc-Max interactions and the impact on oncogenic potential. *Oncotarget* **2014**, *5*, 8869–8878. [CrossRef]

170. Ferrannini, E. Insulin resistance versus insulin deficiency in non-insulin-dependent diabetes mellitus: Problems and prospects. *Endocr. Rev.* **1998**, *19*, 477–490. [CrossRef] [PubMed]

171. Cooper, G.J.; Willis, A.C.; Clark, A.; Turner, R.C.; Sim, R.B.; Reid, K.B. Purification and characterization of a peptide from amyloid-rich pancreases of type 2 diabetic patients. *Proc. Natl. Acad. Sci. USA* **1987**, *84*, 8628–8632. [CrossRef] [PubMed]

172. Westermark, P.; Wernstedt, C.; Wilander, E.; Hayden, D.W.; O'Brien, T.D.; Johnson, K.H. Amyloid fibrils in human insulinoma and islets of Langerhans of the diabetic cat are derived from a neuropeptide-like protein also present in normal islet cells. *Proc. Natl. Acad. Sci. USA* **1987**, *84*, 3881–3885. [CrossRef] [PubMed]

173. Mosselman, S.; Höppener, J.W.; Zandberg, J.; van Mansfeld, A.D.; Geurts van Kessel, A.H.; Lips, C.J.; Jansz, H.S. Islet amyloid polypeptide: Identification and chromosomal localization of the human gene. *FEBS Lett.* **1988**, *239*, 227–232. [CrossRef]

174. Kapurniotu, A. Amyloidogenicity and cytotoxicity of islet amyloid polypeptide. *Biopolymers* **2001**, *60*, 438–459. [CrossRef]

175. Moore, S.J.; Sonar, K.; Bharadwaj, P.; Deplazes, E.; Mancera, R.L. Characterisation of the Structure and Oligomerisation of Islet Amyloid Polypeptides (IAPP): A Review of Molecular Dynamics Simulation Studies. *Molecules* **2018**, *23*, 2142. [CrossRef] [PubMed]

176. Goldsbury, C.; Goldie, K.; Pellaud, J.; Seelig, J.; Frey, P.; Müller, S.A.; Kistler, J.; Cooper, G.J.; Aebi, U. Amyloid fibril formation from full-length and fragments of amylin. *J. Struct. Biol.* **2000**, *130*, 352–362. [CrossRef] [PubMed]

177. Yonemoto, I.T.; Kroon, G.J.; Dyson, H.J.; Balch, W.E.; Kelly, J.W. Amylin proprotein processing generates progressively more amyloidogenic peptides that initially sample the helical state. *Biochemistry* **2008**, *47*, 9900–9910. [CrossRef] [PubMed]

178. Reddy, A.S.; Wang, L.; Singh, S.; Ling, Y.L.; Buchanan, L.; Zanni, M.T.; Skinner, J.L.; de Pablo, J.J. Stable and metastable states of human amylin in solution. *Biophys. J.* **2010**, *99*, 2208–2216. [CrossRef] [PubMed]

179. Qiao, Q.; Bowman, G.R.; Huang, X. Dynamics of an intrinsically disordered protein reveal metastable conformations that potentially seed aggregation. *J. Am. Chem. Soc.* **2013**, *135*, 16092–16101. [CrossRef] [PubMed]

180. Höppener, J.W.; Ahrén, B.; Lips, C.J. Islet amyloid and type 2 diabetes mellitus. *N. Engl. J. Med.* **2000**, *343*, 411–419. [CrossRef] [PubMed]

181. Höppener, J.W.; Lips, C.J. Role of islet amyloid in type 2 diabetes mellitus. *Int. J. Biochem. Cell Biol.* **2006**, *38*, 726–736. [CrossRef] [PubMed]

182. Dong, X.; Qiao, Q.; Qian, Z.; Wei, G. Recent computational studies of membrane interaction and disruption of human islet amyloid polypeptide: Monomers, oligomers and protofibrils. *Biochim. Biophys. Acta Biomembr.* **2018**, *1860*, 1826–1839. [CrossRef] [PubMed]

183. Longhena, F.; Spano, P.; Bellucci, A. Targeting of Disordered Proteins by Small Molecules in Neurodegenerative Diseases. *Handb. Exp. Pharmacol.* **2018**, *245*, 85–110. [PubMed]

184. Babu, M.M.; van der Lee, R.; de Groot, N.S.; Gsponer, J. Intrinsically disordered proteins: Regulation and disease. *Curr. Opin. Struct. Biol.* **2011**, *21*, 432–440. [CrossRef] [PubMed]

185. Wright, P.E.; Dyson, H.J. Intrinsically disordered proteins in cellular signalling and regulation. *Nat. Rev. Mol. Cell Biol.* **2015**, *16*, 18–29. [CrossRef] [PubMed]

186. Bah, A.; Forman-Kay, J.D. Modulation of intrinsically disordered protein function by post-translational modifications. *J. Biol. Chem.* **2016**, *291*, 6696–6705. [CrossRef] [PubMed]

187. Dyson, H.J. Making Sense of Intrinsically Disordered Proteins. *Biophys. J.* **2016**, *110*, 1013–1016. [CrossRef] [PubMed]

188. Van Der Lee, R.; Buljan, M.; Lang, B.; Weatheritt, R.J.; Daughdrill, G.W.; Dunker, A.K.; Fuxreiter, M.; Gough, J.; Gsponer, J.; Jones, D.T.; et al. Classification of intrinsically disordered regions and proteins. *Chem. Rev.* **2014**, *114*, 6589–6631. [CrossRef] [PubMed]

189. Babu, M.M. The contribution of intrinsically disordered regions to protein function, cellular complexity, and human disease. *Biochem. Soc. Trans.* **2016**, *44*, 1185–1200. [CrossRef] [PubMed]

190. Strome, B.; Hsu, I.S.; Li Cheong Man, M.; Zarin, T.; Nguyen Ba, A.; Moses, A.M. Short linear motifs in intrinsically disordered regions modulate HOG signaling capacity. *BMC Syst. Biol.* **2018**, *12*, 75. [CrossRef] [PubMed]

191. Van Roey, K.; Uyar, B.; Weatheritt, R.J.; Dinkel, H.; Seiler, M.; Budd, A.; Gibson, T.J.; Davey, N.E. Short linear motifs: Ubiquitous and functionally diverse protein interaction modules directing cell regulation. *Chem. Rev.* **2014**, *114*, 6733–6778. [CrossRef] [PubMed]

192. Tompa, P. Unstructural biology coming of age. *Curr. Opin. Struct. Biol.* **2011**, *21*, 419–425. [CrossRef] [PubMed]

193. Hu, G.; Wu, Z.; Uversky, V.N.; Kurgan, L. Functional analysis of human hub proteins and their interactors involved in the intrinsic disorder-enriched interactions. *Int. J. Mol. Sci.* **2017**, *18*, 2761. [CrossRef] [PubMed]

194. Tompa, P.; Kovacs, D. Intrinsically disordered chaperones in plants and animals. *Biochem. Cell Biol.* **2010**, *88*, 167–174. [CrossRef] [PubMed]

195. Sharma, R.; Raduly, Z.; Miskei, M.; Fuxreiter, M. Fuzzy complexes: Specific binding without complete folding. *FEBS Lett.* **2015**, *589*, 2533–2542. [CrossRef] [PubMed]

196. Weiss, M.A.; Ellenberger, T.; Wobbe, C.R.; Lee, J.P.; Harrison, S.C.; Struhl, K. Folding transition in the DMA-binding domain of GCN4 on specific binding to DNA. *Nature* **1990**, *347*, 575–578. [CrossRef] [PubMed]

197. Bracken, C.; Carr, P.A.; Cavanagh, J.; Palmer, A.G. Temperature dependence of intramolecular dynamics of the basic leucine zipper of GCN4: Implications for the entropy of association with DNA. *J. Mol. Biol.* **1999**, *285*, 2133–2146. [CrossRef] [PubMed]

198. Shammas, S.L.; Crabtree, M.D.; Dahal, L.; Wicky, B.I.M.; Clarke, J. Insights into coupled folding and binding mechanisms from kinetic studies. *J. Biol. Chem.* **2016**, *291*, 6689–6695. [CrossRef] [PubMed]

199. Hammes, G.G.; Chang, Y.-C.; Oas, T.G. Conformational selection or induced fit: A flux description of reaction mechanism. *Proc. Natl. Acad. Sci. USA* **2009**, *106*, 13737–13741. [CrossRef] [PubMed]

200. DeForte, S.; Uversky, V.N. Order, disorder, and everything in between. *Molecules* **2016**, *21*, 1090. [CrossRef] [PubMed]

201. Xie, H.; Vucetic, S.; Iakoucheva, L.M.; Oldfield, C.J.; Dunker, A.K.; Obradovic, Z.; Uversky, V.N. Functional anthology of intrinsic disorder. 3. Ligands, post-translational modifications, and diseases associated with intrinsically disordered proteins. *J. Proteome Res.* **2007**, *6*, 1917–1932. [CrossRef] [PubMed]

202. Schwalbe, M.; Biernat, J.; Bibow, S.; Ozenne, V.; Jensen, M.R.; Kadavath, H.; Blackledge, M.; Mandelkow, E.; Zweckstetter, M. Phosphorylation of human tau protein by microtubule affinity-regulating kinase 2. *Biochemistry* **2013**, *52*, 9068–9079. [CrossRef] [PubMed]

203. Ou, L.; Ferreira, A.M.; Otieno, S.; Xiao, L.; Bashford, D.; Kriwacki, R.W. Incomplete folding upon binding mediates Cdk4/cyclin D complex activation by tyrosine phosphorylation of inhibitor p27 protein. *J. Biol. Chem.* **2011**, *286*, 30142–30151. [CrossRef] [PubMed]

204. Zeng, Y.; He, Y.; Yang, F.; Mooney, S.M.; Getzenberg, R.H.; Orban, J.; Kulkarni, P. The cancer/testis antigen prostate-associated gene 4 (PAGE4) is a highly intrinsically disordered protein. *J. Biol. Chem.* **2011**, *286*, 13985–13994. [CrossRef] [PubMed]

205. Coskuner-Weber, O.; Uversky, V.N. Insights into the molecular mechanisms of Alzheimer's and Parkinson's diseases with molecular simulations: Understanding the roles of artificial and pathological missense mutations in intrinsically disordered proteins related to pathology. *Int. J. Mol. Sci.* **2018**, *19*, 336. [CrossRef] [PubMed]

206. Edwards, Y.J.K.; Lobley, A.E.; Pentony, M.M.; Jones, D.T. Insights into the regulation of intrinsically disordered proteins in the human proteome by analyzing sequence and gene expression data. *Genome Biol.* **2009**, *10*, R50. [CrossRef] [PubMed]

207. Gsponer, J.; Futschik, M.E.; Teichmann, S.A.; Babu, M.M. Tight regulation of unstructured proteins: From transcript synthesis to protein degradation. *Science* **2008**, *322*, 1365–1368. [CrossRef] [PubMed]

208. Forbes, J.G.; Jin, A.J.; Ma, K.; Gutierrez-Cruz, G.; Tsai, W.L.; Wang, K. Titin PEVK segment: Charge-driven elasticity of the open and flexible polyampholyte. *J. Muscle Res. Cell Motil.* **2005**, *26*, 291–301. [CrossRef] [PubMed]

209. Boothby, T.C.; Tapia, H.; Brozena, A.H.; Piszkiewicz, S.; Smith, A.E.; Giovannini, I.; Rebecchi, L.; Pielak, G.J.; Koshland, D.; Goldstein, B. Tardigrades Use Intrinsically Disordered Proteins to Survive Desiccation. *Mol. Cell* **2017**, *65*, 975.e5–984.e5. [CrossRef] [PubMed]

210. Russo, J.; Olivas, W.M. Conditional regulation of Puf1p, Puf4p, and Puf5p activity alters YHB1 mRNA stability for a rapid response to toxic nitric oxide stress in yeast. *Mol. Biol. Cell* **2015**, *26*, 1015–1029. [CrossRef] [PubMed]

211. Wang, M.; Ogé, L.; Perez-Garcia, M.D.; Hamama, L.; Sakr, S. The PUF protein family: Overview on PUF RNA targets, biological functions, and post transcriptional regulation. *Int. J. Mol. Sci.* **2018**, *19*, 410. [CrossRef] [PubMed]

212. Tsvetkov, P.; Reuven, N.; Shaul, Y. The nanny model for IDPs. *Nat. Chem. Biol.* **2009**, *5*, 778–781. [CrossRef] [PubMed]

213. Inobe, T.; Matouschek, A. Paradigms of protein degradation by the proteasome. *Curr. Opin. Struct. Biol.* **2014**, *24*, 156–164. [CrossRef] [PubMed]

214. Hagai, T.; Azia, A.; Tóth-Petróczy, Á.; Levy, Y. Intrinsic disorder in ubiquitination substrates. *J. Mol. Biol.* **2011**, *412*, 319–324. [CrossRef] [PubMed]

215. Wenzel, T.; Baumeister, W. Conformational constraints in protein degradation by the 20S proteasome. *Nat. Struct. Biol.* **1995**, *2*, 199–204. [CrossRef] [PubMed]

216. Theillet, F.-X.; Binolfi, A.; Frembgen-Kesner, T.; Hingorani, K.; Sarkar, M.; Kyne, C.; Li, C.; Crowley, P.B.; Gierasch, L.; Pielak, G.J.; et al. Physicochemical Properties of Cells and Their Effects on Intrinsically Disordered Proteins (IDPs). *Chem. Rev.* **2014**, *114*, 6661–6714. [CrossRef] [PubMed]

217. Dou, F.; Netzer, W.J.; Tanemura, K.; Li, F.; Hartl, F.U.; Takashima, A.; Gouras, G.K.; Greengard, P.; Xu, H. Chaperones increase association of tau protein with microtubules. *Proc. Natl. Acad. Sci. USA* **2003**, *100*, 721–726. [CrossRef] [PubMed]

218. Wang, H.; Tan, M.S.; Lu, R.C.; Yu, J.T.; Tan, L. Heat shock proteins at the crossroads between cancer and Alzheimer's disease. *BioMed Res. Int.* **2014**, *2014*, 239164. [CrossRef] [PubMed]

219. Tortosa, E.; Santa-Maria, I.; Moreno, F.; Lim, F.; Perez, M.; Avila, J. Binding of Hsp90 to tau promotes a conformational change and aggregation of tau protein. *J. Alzheimers Dis.* **2009**, *17*, 319–325. [CrossRef] [PubMed]

220. Dou, F.; Chang, X.; Ma, D. Hsp90 maintains the stability and function of the tau phosphorylating kinase GSK3β. *Int. J. Mol. Sci* **2007**, *8*, 51–60. [CrossRef]

221. Shelton, L.B.; Baker, J.D.; Zheng, D.; Sullivan, L.E.; Solanki, P.K.; Webster, J.M.; Sun, Z.; Sabbagh, J.J.; Nordhues, B.A.; Koren, J.; et al. Hsp90 activator Aha1 drives production of pathological tau aggregates. *Proc. Natl. Acad. Sci. USA* **2017**, *114*, 9707–9712. [CrossRef] [PubMed]

222. Mayer, M.P. Hsp70 chaperone dynamics and molecular mechanism. *Trends Biochem. Sci.* **2013**, *38*, 507–514. [CrossRef] [PubMed]

223. Young, Z.T.; Rauch, J.N.; Assimon, V.A.; Jinwal, U.K.; Ahn, M.; Li, X.; Dunyak, B.M.; Ahmad, A.; Carlson, G.A.; Srinivasan, S.R.; et al. Stabilizing the Hsp70-Tau complex promotes turnover in models of Tauopathy. *Cell Chem. Biol.* **2016**, *23*, 992–1001. [CrossRef] [PubMed]

224. Westhoff, B.; Chapple, J.P.; van der Spuy, J.; Höhfeld, J.; Cheetham, M.E. HSJ1 is a neuronal shuttling factor for the sorting of chaperone clients to the proteasome. *Curr. Biol.* **2005**, *15*, 1058–1064. [CrossRef] [PubMed]

225. Shin, Y.; Klucken, J.; Patterson, C.; Hyman, B.T.; McLean, P.J. The co-chaperone carboxyl terminus of Hsp70-interacting protein (CHIP) mediates alpha-synuclein degradation decisions between proteasomal and lysosomal pathways. *J. Biol. Chem.* **2005**, *280*, 23727–23734. [CrossRef] [PubMed]

226. Lang, K.; Schmid, F.X.; Fischer, G. Catalysis of protein folding by prolyl isomerase. *Nature* **1987**, *329*, 268–270. [CrossRef] [PubMed]

227. Nigro, P.; Pompilio, G.; Capogrossi, M.C. Cyclophilin A: A key player for human disease. *Cell Death Dis.* **2013**, *4*, e888. [CrossRef] [PubMed]

228. Torbeev, V.Y.; Hilvert, D. Both the cis-trans equilibrium and isomerization dynamics of a single proline amide modulate β2-microglobulin amyloid assembly. *Proc. Natl. Acad. Sci. USA* **2013**, *110*, 20051–20056. [CrossRef] [PubMed]

229. Baker, J.D.; Shelton, L.B.; Zheng, D.; Favretto, F.; Nordhues, B.A.; Darling, A.; Sullivan, L.E.; Sun, Z.; Solanki, P.K.; Martin, M.D.; et al. Human cyclophilin 40 unravels neurotoxic amyloids. *PLoS Biol.* **2017**, *15*, e2001336. [CrossRef] [PubMed]

230. Tsvetkov, P.; Reuven, N.; Shaul, Y. Ubiquitin-independent p53 proteasomal degradation. *Cell Death Differ.* **2010**, *17*, 103–108. [CrossRef] [PubMed]

231. Cheng, Y.; LeGall, T.; Oldfield, C.J.; Mueller, J.P.; Van, Y.Y.J.; Romero, P.; Cortese, M.S.; Uversky, V.N.; Dunker, A.K. Rational drug design via intrinsically disordered protein. *Trends Biotechnol.* **2006**, *24*, 435–442. [CrossRef] [PubMed]

232. Metallo, S.J. Intrinsically disordered proteins are potential drug targets. *Curr. Opin. Chem. Biol.* **2010**, *14*, 481–488. [CrossRef] [PubMed]

233. Wang, J.H.; Cao, Z.X.; Zhao, L.L.; Li, S.Q. Novel strategies for drug discovery based on intrinsically disordered proteins (IDPs). *Int. J. Mol. Sci.* **2011**, *12*, 3205–3219. [CrossRef] [PubMed]

234. Zhang, Y.; Cao, H.; Liu, Z. Binding cavities and druggability of intrinsically disordered proteins. *Protein Sci.* **2015**, *24*, 688–705. [CrossRef] [PubMed]

235. Liu, Z.; Huang, Y. Advantages of proteins being disordered. *Protein Sci.* **2014**, *23*, 539–550. [CrossRef] [PubMed]

236. Zhu, M.; De Simone, A.; Schenk, D.; Toth, G.; Dobson, C.M.; Vendruscolo, M. Identification of small-molecule binding pockets in the soluble monomeric form of the Aβ42 peptide. *J. Chem. Phys.* **2013**, *139*, 035101. [CrossRef] [PubMed]

237. Fokkens, M.; Schrader, T.; Klärner, F.G. A molecular tweezer for lysine and arginine. *J. Am. Chem. Soc.* **2005**, *127*, 14415–14421. [CrossRef] [PubMed]

238. Sinha, S.; Lopes, D.H.; Du, Z.; Pang, E.; Shanmugam, A.; Lomakin, A.; Talbiersky, P.; Tennstaedt, A.; McDaniel, K.; Bakshi, R.; et al. Lysine-specific molecular tweezers are broad-spectrum inhibitors of assembly and toxicity of amyloid proteins. *J. Am. Chem. Soc.* **2011**, *133*, 16958–16969. [CrossRef] [PubMed]

239. O'Hare, E.; Scopes, D.I.; Kim, E.M.; Palmer, P.; Spanswick, D.; McMahon, B.; Amijee, H.; Nerou, E.; Treherne, J.M.; Jeggo, R. Novel 5-aryloxypyrimidine SEN1576 as a candidate for the treatment of Alzheimer's disease. *Int. J. Neuropsychopharmacol.* **2014**, *17*, 117–126. [CrossRef] [PubMed]

240. Prabhudesai, S.; Sinha, S.; Attar, A.; Kotagiri, A.; Fitzmaurice, A.G.; Lakshmanan, R.; Ivanova, M.I.; Loo, J.A.; Klärner, F.G.; Schrader, T.; et al. A novel "molecular tweezer" inhibitor of α-synuclein neurotoxicity in vitro and in vivo. *Neurotherapeutics* **2012**, *9*, 464–476. [CrossRef] [PubMed]

241. Sievers, S.A.; Karanicolas, J.; Chang, H.W.; Zhao, A.; Jiang, L.; Zirafi, O.; Stevens, J.T.; Münch, J.; Baker, D.; Eisenberg, D. Structure-based design of non-natural amino-acid inhibitors of amyloid fibril formation. *Nature* **2011**, *475*, 96–100. [CrossRef] [PubMed]

242. Frenkel-Pinter, M.; Tal, S.; Scherzer-Attali, R.; Abu-Hussien, M.; Alyagor, I.; Eisenbaum, T.; Gazit, E.; Segal, D. Naphthoquinone-Tryptophan Hybrid Inhibits Aggregation of the Tau-Derived Peptide PHF6 and Reduces Neurotoxicity. *J. Alzheimers Dis.* **2016**, *51*, 165–178. [CrossRef] [PubMed]

243. Jones, C.L.; Njomen, E.; Sjögren, B.; Dexheimer, T.S.; Tepe, J.J. Small Molecule Enhancement of 20S Proteasome Activity Targets Intrinsically Disordered Proteins. *ACS Chem. Biol.* **2017**, *12*, 2240–2247. [CrossRef] [PubMed]

244. Joerger, A.C.; Fersht, A.R. The tumor suppressor p53: From structures to drug discovery. *Cold Spring Harb. Perspect. Biol.* **2010**, *2*, a000919. [CrossRef] [PubMed]

245. Li, Z.Y.; Ni, M.; Li, J.K.; Zhang, Y.P.; Ouyang, Q.; Tang, C. Decision making of the p53 network: Death by integration. *J. Theor. Biol.* **2011**, *271*, 205–211. [CrossRef] [PubMed]

246. Huang, Y.Q.; Liu, Z.R. Anchoring intrinsically disordered proteins to multiple targets: Lessons from N terminus of the p53 protein. *Int. J. Mol. Sci.* **2011**, *12*, 1410–1430. [CrossRef] [PubMed]

247. Vassilev, L.T.; Vu, B.T.; Graves, B.; Carvajal, D.; Podlaski, F.; Filipovic, Z.; Kong, N.; Kammlott, U.; Lukacs, C.; Klein, C.; et al. In vivo activation of the p53 pathway by small-molecule antagonists of MDM2. *Science* **2004**, *303*, 844–848. [CrossRef] [PubMed]

248. Tovar, C.; Rosinski, J.; Filipovic, Z.; Higgins, B.; Kolinsky, K.; Hilton, H.; Zhao, X.L.; Vu, B.T.; Qing, W.G.; Packman, K.; et al. Small-molecule MDM2 antagonists reveal aberrant p53 signaling in cancer: Implications for therapy. *Proc. Natl. Acad. Sci. USA* **2006**, *103*, 1888–1893. [CrossRef] [PubMed]

249. Yu, X.; Narayanan, S.; Vazquez, A.; Carpizo, D.R. Small molecule compounds targeting the p53 pathway: Are we finally making progress? *Apoptosis* **2014**, *19*, 1055–1068. [CrossRef] [PubMed]

250. Burgess, A.; Chia, K.M.; Haupt, S.; Thomas, D.; Haupt, Y.; Lim, E. Clinical Overview of MDM2/X-Targeted Therapies. *Front. Oncol.* **2016**, *6*, 7. [CrossRef] [PubMed]

251. Hammoudeh, D.I.; Follis, A.V.; Prochownik, E.V.; Metallo, S.J. Multiple independent binding sites for small molecule inhibitors on the oncoprotein c-Myc. *J. Am. Chem. Soc.* **2009**, *131*, 7390–7401. [CrossRef] [PubMed]

252. Berg, T.; Cohen, S.B.; Desharnais, J.; Sonderegger, C.; Maslyar, D.J.; Goldberg, J.; Boger, D.L.; Vogt, P.K. Small-molecule antagonists of Myc/Max dimerization inhibit Myc-induced transformation of chicken embryo fibroblasts. *Proc. Natl. Acad. Sci. USA* **2002**, *99*, 3830–3835. [CrossRef] [PubMed]

253. Shi, J.; Stover, J.S.; Whitby, L.R.; Vogt, P.K.; Boger, D.L. Small molecule inhibitors of Myc/Max dimerization and Myc-induced cell transformation. *Bioorg. Med. Chem. Lett.* **2009**, *19*, 6038–6041. [CrossRef] [PubMed]

254. Yin, X.Y.; Giap, C.; Lazo, J.S.; Prochownik, E.V. Low molecular weight inhibitors of Myc-Max interaction and function. *Oncogene* **2003**, *22*, 6151–6159. [CrossRef] [PubMed]

255. Zirath, H.; Frenzel, A.; Oliynyk, G.; Segerstrom, L.; Westermark, U.K.; Larsson, K.; Persson, M.M.; Hultenby, K.; Lehtio, J.; Einvik, C.; et al. MYC inhibition induces metabolic changes leading to accumulation of lipid droplets in tumor cells. *Proc. Natl. Acad. Sci. USA* **2013**, *110*, 10258–10263. [CrossRef] [PubMed]

256. Fletcher, S.; Prochownik, E.V. Small-molecule inhibitors of the Myc oncoprotein. *Biochim. Biophys. Acta* **2015**, *1849*, 525–543. [CrossRef] [PubMed]

257. Yu, C.; Niu, X.; Jin, F.; Liu, Z.; Jin, C.; Lai, L. Structure-based Inhibitor Design for the Intrinsically Disordered Protein c-Myc. *Sci. Rep.* **2016**, *6*, 22298. [CrossRef] [PubMed]

258. Erkizan, H.V.; Kong, Y.L.; Merchant, M.; Schlottmann, S.; Barber-Rotenberg, J.S.; Yuan, L.S.; Abaan, O.D.; Chou, T.H.; Dakshanamurthy, S.; Brown, M.L.; et al. A small molecule blocking oncogenic protein EWS-FLI1 interaction with RNA helicase A inhibits growth of Ewing's sarcoma. *Nat. Med.* **2009**, *15*, 750–757. [CrossRef] [PubMed]

259. Hong, S.H.; Youbi, S.E.; Hong, S.P.; Kallakury, B.; Monroe, P.; Erkizan, H.V.; Barber-Rotenberg, J.S.; Houghton, P.; Uren, A.; Toretsky, J.A. Pharmacokinetic modeling optimizes inhibition of the 'undruggable' EWS-FLI1 transcription factor in Ewing Sarcoma. *Oncotarget* **2014**, *5*, 338–350. [CrossRef] [PubMed]

260. Hegyi, H.; Buday, L.; Tompa, P. Intrinsic structural disorder confers cellular viability on oncogenic fusion proteins. *PLoS Comput. Biol.* **2009**, *5*, e1000552. [CrossRef] [PubMed]

261. Huang, Y.Q.; Liu, Z.R. Do intrinsically disordered proteins possess high specificity in protein-protein interactions? *Chem.-Eur. J.* **2013**, *19*, 4462–4467. [CrossRef] [PubMed]

262. Srinivasan, R.S.; Nesbit, J.B.; Marrero, L.; Erfurth, F.; LaRussa, V.F.; Hemenway, C.S. The synthetic peptide PFWT disrupts AF4-AF9 protein complexes and induces apoptosis in t(4;11) leukemia cells. *Leukemia* **2004**, *18*, 1364–1372. [CrossRef] [PubMed]

263. Palermo, C.M.; Bennett, C.A.; Winters, A.C.; Hemenway, C.S. The AF4-mimetic peptide, PFWT, induces necrotic cell death in MV4-11 leukemia cells. *Leuk. Res.* **2008**, *32*, 633–642. [CrossRef] [PubMed]

264. Watson, V.G.; Drake, K.M.; Peng, Y.; Napper, A.D. Development of a high-throughput screening-compatible assay for the discovery of inhibitors of the AF4-AF9 interaction using AlphaScreen technology. *Assay Drug Dev. Technol.* **2013**, *11*, 253–268. [CrossRef] [PubMed]

265. Johnson, T.O.; Ermolieff, J.; Jirousek, M.R. Protein tyrosine phosphatase 1B inhibitors for diabetes. *Nat. Rev. Drug Discov.* **2002**, *1*, 696–709. [CrossRef] [PubMed]

266. Krishnan, N.; Koveal, D.; Miller, D.H.; Xue, B.; Akshinthala, S.D.; Kragelj, J.; Jensen, M.R.; Gauss, C.M.; Page, R.; Blackledge, M.; et al. Targeting the disordered C terminus of PTP1B with an allosteric inhibitor. *Nat. Chem. Biol.* **2014**, *10*, 558–566. [CrossRef] [PubMed]

267. Bieschke, J.; Russ, J.; Friedrich, R.P.; Ehrnhoefer, D.E.; Wobst, H.; Neugebauer, K.; Wanker, E.E. EGCG remodels mature alpha-synuclein and amyloid-beta fibrils and reduces cellular toxicity. *Proc. Natl. Acad. Sci. USA* **2010**, *107*, 7710–7715. [CrossRef] [PubMed]

268. Martin-Bastida, A.; Ward, R.J.; Newbould, R.; Piccini, P.; Sharp, D.; Kabba, C.; Patel, M.C.; Spino, M.; Connelly, J.; Tricta, F.; et al. Brain iron chelation by deferiprone in a phase 2 randomised double-blinded placebo controlled clinical trial in Parkinson's disease. *Sci. Rep.* **2017**, *7*, 1398. [CrossRef] [PubMed]

269. Hung, L.W.; Villemagne, V.L.; Cheng, L.; Sherratt, N.A.; Ayton, S.; White, A.R.; Crouch, P.J.; Lim, S.; Leong, S.L.; Wilkins, S.; et al. The hypoxia imaging agent CuII(atsm) is neuroprotective and improves motor and cognitive functions in multiple animal models of Parkinson's disease. *J. Exp. Med.* **2012**, *209*, 837–854. [CrossRef] [PubMed]

270. Savolainen, M.H.; Richie, C.T.; Harvey, B.K.; Männistö, B.T.; Maguire-Zeiss, K.A.; Myöhänen, T.T. The beneficial effect of a prolyl oligopeptidase inhibitor, KYP-2047, on alpha-synuclein clearance and autophagy in A30P transgenic mouse. *Neurobiol. Dis.* **2014**, *68*, 1–15. [CrossRef] [PubMed]

271. Svarcbahs, R.; Julku, U.H.; Myöhänen, T.T. Inhibition of Prolyl Oligopeptidase Restores Spontaneous Motor Behavior in the α-Synuclein Virus Vector-Based Parkinson's Disease Mouse Model by Decreasing α-Synuclein Oligomeric Species in Mouse Brain. *J. Neurosci.* **2016**, *36*, 12485–12497. [CrossRef] [PubMed]

272. Levin, J.; Schmidt, F.; Boehm, C.; Prix, C.; Bötzel, K.; Ryazanov, S.; Leonov, A.; Griesinger, C.; Giese, A. The oligomer modulator anle138b inhibits disease progression in a Parkinson mouse model even with treatment started after disease onset. *Acta Neuropathol.* **2014**, *127*, 779–780. [CrossRef] [PubMed]

273. Price, D.L.; Koike, M.A.; Khan, A.; Wrasidlo, W.; Rockenstein, E.; Masliah, E.; Bonhaus, D. The small molecule alpha-synuclein misfolding inhibitor, NPT200-11, produces multiple benefits in an animal model of Parkinson's disease. *Sci. Rep.* **2018**, *8*, 16165. [CrossRef] [PubMed]

274. Krishnan, R.; Hefti, F.; Tsubery, H.; Lulu, M.; Proschitsky, M.; Fisher, R. Conformation as the Therapeutic Target for Neurodegenerative Diseases. *Curr. Alzheimer Res.* **2017**, *14*, 393–402. [CrossRef] [PubMed]

275. Levenson, J.M.; Schroeter, S.; Carroll, J.C.; Cullen, V.; Asp, E.; Proschitsky, M.; Chung, C.H.; Gilead, S.; Nadeem, M.; Dodiya, H.B.; et al. NPT088 reduces both amyloid-β and tau pathologies in transgenic mice. *Alzheimers Dement.* **2016**, *2*, 141–155. [CrossRef] [PubMed]

276. Yuan, B.; Sierks, M.R. Intracellular targeting and clearance of oligomeric alpha-synuclein alleviates toxicity in mammalian cells. *Neurosci. Lett.* **2009**, *459*, 16–18. [CrossRef] [PubMed]

277. Bhatt, M.A.; Messer, A.; Kordower, J.H. Can intrabodies serve as neuroprotective therapies for Parkinson's disease? Beginning thoughts. *J. Parkinsons Dis.* **2013**, *3*, 581–591. [PubMed]

278. Emadi, S.; Liu, R.; Yuan, B.; Schulz, P.; McAllister, C.; Lyubchenko, Y.; Messer, A.; Sierks, M.R. Inhibiting aggregation of alpha-synuclein with human single chain antibody fragments. *Biochemistry* **2004**, *43*, 2871–2878. [CrossRef] [PubMed]

279. Butler, D.C.; Joshi, S.N.; Genst, E.; Baghel, A.S.; Dobson, C.M.; Messer, A. Bifunctional Anti-Non-Amyloid Component α-Synuclein Nanobodies Are Protective In Situ. *PLoS ONE* **2016**, *11*, e0165964. [CrossRef] [PubMed]

280. Chatterjee, D.; Bhatt, M.; Butler, D.; De Genst, E.; Dobson, C.M.; Messer, A.; Kordower, J.H. Proteasome-targeted nanobodies alleviate pathology and functional decline in an α-synuclein-based Parkinson's disease model. *NPJ Parkinsons Dis.* **2018**, *4*, 25. [CrossRef] [PubMed]

281. Dehay, B.; Bourdenx, M.; Gorry, P.; Przedborski, S.; Vila, M.; Hunot, S.; Singleton, A.; Olanow, C.; Merchant, K.; Bezard, E.; et al. Targeting alpha-synuclein for treatment of Parkinson's disease: Mechanistic and therapeutic considerations. *Lancet Neurol.* **2015**, *14*, 855–866. [CrossRef]

282. Schenk, D.B.; Koller, M.; Ness, D.K.; Griffith, S.G.; Grundman, M.; Zago, W.; Soto, J.; Atiee, G.; Ostrowitzki, S.; Kinney, G.G. First-in-human assessment of PRX002, an anti-α-synuclein monoclonal antibody, in healthy volunteers. *Mov. Disord.* **2017**, *32*, 211–218. [CrossRef] [PubMed]

283. Jankovic, J.; Goodman, I.; Safirstein, B.; Marmon, T.K.; Schenk, D.B.; Koller, M.; Zago, W.; Ness, D.K.; Griffith, S.G.; Grundman, M.; et al. Safety and Tolerability of Multiple Ascending Doses of PRX002/RG7935, an Anti-α-Synuclein Monoclonal Antibody, in Patients With Parkinson Disease: A Randomized Clinical Trial. *JAMA Neurol.* **2018**, *75*, 1206–1214. [CrossRef] [PubMed]

284. Mandler, M.; Valera, E.; Rockenstein, E.; Weninger, H.; Patrick, C.; Adame, A.; Santic, R.; Meindl, S.; Vigl, B.; Smrzka, O.; et al. Next-generation active immunization approach for synucleinopathies: Implications for Parkinson's disease clinical trials. *Acta Neuropathol.* **2014**, *127*, 861–879. [CrossRef] [PubMed]

285. O'Hara, D.M.; Kalia, S.K.; Kalia, L.V. Emerging disease-modifying strategies targeting alpha-synuclein for the treatment of Parkinson's disease. *Br. J. Pharmacol.* **2018**, *175*, 3080–3089. [CrossRef] [PubMed]

286. Kiss, R.; Csizmadia, G.; Solti, K.; Keresztes, A.; Zhu, M.; Pickhardt, M.; Mandelkow, E.; Toth, G. Structural Basis of Small Molecule Targetability of Monomeric Tau Protein. *ACS Chem. Neurosci.* **2018**, *9*, 2997–3006. [CrossRef] [PubMed]

287. Jouanne, M.; Rault, S.; Voisin-Chiret, A.S. Tau protein aggregation in Alzheimer's disease: An attractive target for the development of novel therapeutic agents. *Eur. J. Med. Chem.* **2017**, *139*, 153–167. [CrossRef] [PubMed]

288. Pickhardt, M.; Neumann, T.; Schwizer, D.; Callaway, K.; Vendruscolo, M.; Schenk, D.; St George-Hyslop, P.; Mandelkow, E.M.; Dobson, C.M.; McConlogue, L.; et al. Identification of Small Molecule Inhibitors of Tau Aggregation by Targeting Monomeric Tau As a Potential Therapeutic Approach for Tauopathies. *Curr. Alzheimer Res.* **2015**, *12*, 814–828. [CrossRef] [PubMed]

289. Baggett, D.W.; Nath, A. The Rational Discovery of a Tau Aggregation Inhibitor. *Biochemistry* **2018**, *57*, 6099–6107. [CrossRef] [PubMed]

290. Shiryaev, N.; Pikman, R.; Giladi, E.; Gozes, I. Protection against tauopathy by the drug candidates NAP (davunetide) and D-SAL: Biochemical, cellular and behavioral aspects. *Curr. Pharm. Des.* **2011**, *17*, 2603–2612. [CrossRef] [PubMed]

291. Ivashko-Pachima, Y.; Gozes, I. NAP protects against Tau hyperphosphorylation through GSK3. *Curr. Pharm. Des.* **2018**, *24*, 3868–3877. [CrossRef] [PubMed]

292. Dammers, C.; Yolcu, D.; Kukuk, L.; Willbold, D.; Pickhardt, M.; Mandelkow, E.; Horn, A.H.; Sticht, H.; Malhis, M.N.; Will, N.; et al. Selection and Characterization of Tau Binding -Enantiomeric Peptides with Potential for Therapy of Alzheimer Disease. *PLoS ONE* **2016**, *11*, e0167432. [CrossRef] [PubMed]

293. Kim, J.H.; Kim, E.; Choi, W.H.; Lee, J.; Lee, J.H.; Lee, H.; Kim, D.E.; Suh, Y.H.; Lee, M.J. Inhibitory RNA Aptamers of Tau Oligomerization and Their Neuroprotective Roles against Proteotoxic Stress. *Mol. Pharm.* **2016**, *13*, 2039–2048. [CrossRef] [PubMed]

294. Rafiee, S.; Asadollahi, K.; Riazi, G.; Ahmadian, S.; Saboury, A.A. Vitamin B12 Inhibits Tau Fibrillization via Binding to Cysteine Residues of Tau. *ACS Chem. Neurosci.* **2017**, *8*, 2676–2682. [CrossRef] [PubMed]

295. Yoshitake, J.; Soeda, Y.; Ida, T.; Sumioka, A.; Yoshikawa, M.; Matsushita, K.; Akaike, T.; Takashima, A. Modification of Tau by 8-Nitroguanosine 3,5-Cyclic Monophosphate (8-Nitro-cGMP): Effects of nitric oxide-linked chemical modification on tau aggregation. *J. Biol. Chem.* **2016**, *291*, 22714–22720. [CrossRef] [PubMed]

296. Sun, W.; Lee, S.; Huang, X.; Liu, S.; Inayathullah, M.; Kim, K.-M.; Tang, H.; Ashford, J.W.; Rajadas, J. Attenuation of synaptic toxicity and MARK4/PAR1-mediated Tau phosphorylation by methylene blue for Alzheimer's disease treatment. *Sci. Rep.* **2016**, *6*, 34784. [CrossRef] [PubMed]

297. George, R.C.; Lew, J.; Graves, D.J. Interaction of Cinnamaldehyde and Epicatechin with Tau: Implications of Beneficial Effects in Modulating Alzheimer's Disease Pathogenesis. *J. Alzheimers Dis.* **2013**, *36*, 21–40. [CrossRef] [PubMed]

298. Gandini, A.; Bartolini, M.; Tedesco, D.; Martinez-Gonzalez, L.; Roca, C.; Campillo, N.E.; Zaldivar-Diez, J.; Perez, C.; Zuccheri, G.; Miti, A.; et al. Tau-Centric Multitarget Approach for Alzheimer's Disease: Development of First-in-Class Dual Glycogen Synthase Kinase 3 beta and Tau-Aggregation Inhibitors. *J. Med. Chem.* **2018**, *61*, 7640–7656. [CrossRef] [PubMed]

299. Llorach-Pares, L.; Nonell-Canals, A.; Avila, C.; Sanchez-Martinez, M. Kororamides, Convolutamines, and Indole Derivatives as Possible Tau and Dual-Specificity Kinase Inhibitors for Alzheimer's Disease: A Computational Study. *Mar. Drugs* **2018**, *16*, 386. [CrossRef] [PubMed]

300. Moussa, C.E. Beta-secretase inhibitors in phase I and phase II clinical trials for Alzheimer's disease. *Expert Opin. Investig. Drugs* **2017**, *26*, 1131–1136. [CrossRef] [PubMed]

301. Mead, E.; Kestoras, D.; Gibson, Y.; Hamilton, L.; Goodson, R.; Jones, S.; Eversden, S.; Davies, P.; O'Neill, M.; Hutton, M.; et al. Halting of Caspase Activity Protects Tau from MC1-Conformational Change and Aggregation. *J. Alzheimers Dis.* **2016**, *54*, 1521–1538. [CrossRef] [PubMed]

302. Rao, M.V.; McBrayer, M.K.; Campbell, J.; Kumar, A.; Hashim, A.; Sershen, H.; Stavrides, P.H.; Ohno, M.; Hutton, M.; Nixon, R.A. Specific calpain inhibition by calpastatin prevents tauopathy and neurodegeneration and restores normal lifespan in tau P301L mice. *J. Neurosci.* **2014**, *34*, 9222–9234. [CrossRef] [PubMed]

303. Blair, L.J.; Sabbagh, J.J.; Dickey, C.A. Targeting Hsp90 and its co-chaperones to treat Alzheimer's disease. *Expert Opin. Ther. Targets* **2014**, *18*, 1219–1232. [CrossRef] [PubMed]

304. Kontsekova, E.; Zilka, N.; Kovacech, B.; Novak, P.; Novak, M. First-in-man tau vaccine targeting structural determinants essential for pathological tau-tau interaction reduces tau oligomerisation and neurofibrillary degeneration in an Alzheimer's disease model. *Alzheimers Res. Ther.* **2014**, *6*, 44. [CrossRef] [PubMed]

305. Novak, P.; Kontsekova, E.; Zilka, N.; Novak, M. Ten Years of Tau-Targeted Immunotherapy: The Path Walked and the Roads Ahead. *Front. Neurosci.* **2018**, *12*, 798. [CrossRef] [PubMed]

306. Shahpasand, K.; Sepehri Shamloo, A.; Nabavi, S.M.; Lu, K.P.; Zhou, X.Z. Tau immunotherapy: Hopes and hindrances. *Hum. Vaccin Immunother.* **2018**, *14*, 277–284. [CrossRef] [PubMed]

307. Cehlar, O.; Skrabana, R.; Kovac, A.; Kovacech, B.; Novak, M. Crystallization and preliminary X-ray diffraction analysis of tau protein microtubule-binding motifs in complex with Tau5 and DC25 antibody Fab fragments. *Acta Crystallogr. Sect. F Struct. Biol. Cryst. Commun.* **2012**, *68*, 1181–1185. [CrossRef] [PubMed]

308. Panza, F.; Solfrizzi, V.; Seripa, D.; Imbimbo, B.P.; Lozupone, M.; Santamato, A.; Tortelli, R.; Galizia, I.; Prete, C.; Daniele, A.; et al. Tau-based therapeutics for Alzheimer's disease: Active and passive immunotherapy. *Immunotherapy* **2016**, *8*, 1119–1134. [CrossRef] [PubMed]

309. Novak, P.; Zilka, N.; Zilkova, M.; Kovacech, B.; Skrabana, R.; Ondrus, M.; Fialova, L.; Kontsekova, E.; Otto, M.; Novak, M. AADvac1, an Active Immunotherapy for Alzheimer's Disease and Non Alzheimer Tauopathies: An Overview of Preclinical and Clinical Development. *J. Prev. Alzheimers Dis.* **2019**, *6*, 63–69. [PubMed]

310. Novak, P.; Schmidt, R.; Kontsekova, E.; Zilka, N.; Kovacech, B.; Skrabana, R.; Vince-Kazmerova, Z.; Katina, S.; Fialova, L.; Prcina, M.; et al. Safety and immunogenicity of the tau vaccine AADvac1 in patients with Alzheimer's disease: A randomised, double-blind, placebo-controlled, phase 1 trial. *Lancet Neurol.* **2017**, *16*, 123–134. [CrossRef]

311. Novak, P.; Schmidt, R.; Kontsekova, E.; Kovacech, B.; Smolek, T.; Katina, S.; Fialova, L.; Prcina, M.; Parrak, V.; Dal-Bianco, P.; et al. FUNDAMANT: An interventional 72-week phase 1 follow-up study of AADvac1, an active immunotherapy against tau protein pathology in Alzheimer's disease. *Alzheimers Res. Ther.* **2018**, *10*, 108. [CrossRef] [PubMed]

312. Theunis, C.; Crespo-Biel, N.; Gafner, V.; Pihlgren, M.; Lopez-Deber, M.; Reis, P.; Hickman, D.; Adolfsson, O.; Chuard, N.; Ndao, D.; et al. Efficacy and Safety of A Liposome-Based Vaccine against Protein Tau, Assessed in Tau.P301L Mice That Model Tauopathy. *PLoS ONE* **2013**, *8*, e72301. [CrossRef] [PubMed]

313. Yanamandra, K.; Jiang, H.; Mahan, T.; Maloney, S.; Wozniak, D.; Diamond, M.; Holtzmanm, D. Anti-tau antibody reduces insoluble tau and decreases brain atrophy. *Ann. Clin. Trans. Neurol.* **2016**, *2*, 278–288. [CrossRef] [PubMed]

314. West, T.; Hu, Y.; Verghese, P.B.; Bateman, R.J.; Braunstein, J.B.; Fogelman, I.; Budur, K.; Florian, H.; Mendonca, N.; Holtzman, D.M. Preclinical and Clinical Development of ABBV-8E12, a Humanized Anti-Tau Antibody, for Treatment of Alzheimer's Disease and Other Tauopathies. *J. Prev. Alzheimers Dis.* **2017**, *4*, 236–241. [PubMed]

315. Rosenberg, R.N.; Fu, M.; Lambracht-Washington, D. Active full-length DNA Aβ_{42} immunization in 3xTg-AD mice reduces not only amyloid deposition but also tau pathology. *Alzheimers Res. Ther.* **2018**, *10*, 115. [CrossRef] [PubMed]

316. Dai, C.L.; Tung, Y.C.; Liu, F.; Gong, C.X.; Iqbal, K. Tau passive immunization inhibits not only tau but also Aβ pathology. *Alzheimers Res. Ther.* **2017**, *9*, 1. [CrossRef] [PubMed]

317. Hol, W.G.J. Protein Crystallography and Computer Graphics—Toward Rational Drug Design. *Angew. Chem.* **1986**, *25*, 767–778. [CrossRef]

318. Roberts, N.A.; Marlin, J.A.; Kinchington, D.; Broadhurst, A.V.; Craig, J.C.; Duncan, I.B.; Galpin, S.A.; Handa, B.K.; Kay, J.; Kröhn, A.; et al. Rational design of peptide-based HIV proteinase inhibitors. *Science* **1990**, *248*, 358–361. [CrossRef] [PubMed]

319. Von Itzstein, M.; Wu, W.Y.; Kok, G.B.; Pegg, M.S.; Dyason, J.C.; Jin, B.; Van Phan, T.; Smythe, M.L.; White, H.F.; Oliver, S.W.; et al. Rational design of potent sialidase-based inhibitors of influenza virus replication. *Nature* **1993**, *363*, 418–423. [CrossRef] [PubMed]

320. Sliwoski, G.; Kothiwale, S.; Meiler, J.; Lowe, E.W., Jr. Computational methods in drug discovery. *Pharmacol. Rev.* **2013**, *66*, 334–395. [CrossRef] [PubMed]

321. Joshi, P.; Vendruscolo, M. Druggability of Intrinsically Disordered Proteins. *Adv. Exp. Med. Biol.* **2015**, *870*, 383–400. [PubMed]

322. Marasco, D.; Scognamiglio, P.L. Identification of inhibitors of biological interactions involving intrinsically disordered proteins. *Int. J. Mol. Sci.* **2015**, *16*, 7394–7412. [CrossRef] [PubMed]

323. Tsafou, K.; Tiwari, P.B.; Forman-Kay, J.D.; Metallo, S.J.; Toretsky, J.A. Targeting Intrinsically Disordered Transcription Factors: Changing the Paradigm. *J. Mol. Biol.* **2018**, *430*, 2321–2341. [CrossRef] [PubMed]

324. Rezaei-Ghaleh, N.; Blackledge, M.; Zweckstetter, M. Intrinsically disordered proteins: From sequence and conformational properties toward drug discovery. *ChemBioChem* **2012**, *13*, 930–950. [CrossRef] [PubMed]

325. Uversky, V.N. Dancing Protein Clouds: The Strange Biology and Chaotic Physics of Intrinsically Disordered Proteins. *J. Biol. Chem.* **2016**, *291*, 6681–6688. [CrossRef] [PubMed]

326. Jin, F.; Yu, C.; Lai, L.; Liu, Z. Ligand clouds around protein clouds: A scenario of ligand binding with intrinsically disordered proteins. *PLoS Comput. Biol.* **2013**, *9*, e1003249. [CrossRef] [PubMed]

327. Bier, D.; Thiel, P.; Briels, J.; Ottmann, C. Stabilization of Protein-Protein Interactions in chemical biology and drug discovery. *Prog. Biophys. Mol. Biol.* **2015**, *119*, 10–19. [CrossRef] [PubMed]

328. Arkin, M.R.; Wells, J.A. Small-molecule inhibitors of protein-protein interactions: Progressing towards the dream. *Nat. Rev. Drug Discov.* **2004**, *3*, 301–317. [CrossRef] [PubMed]

329. Wells, J.A.; McClendon, C.L. Reaching for high-hanging fruit in drug discovery at protein-protein interfaces. *Nature* **2007**, *450*, 1001–1009. [CrossRef] [PubMed]

330. Hopkins, A.L.; Groom, C.R. The druggable genome. *Nat. Rev. Drug Discov.* **2002**, *1*, 727–730. [CrossRef] [PubMed]

331. Drews, J.; Ryser, S. The role of innovation in drug development. *Nat. Biotechnol.* **1997**, *15*, 1318–1319. [CrossRef] [PubMed]

332. Drews, J. Drug discovery: A historical perspective. *Science* **2000**, *287*, 1960–1964. [CrossRef] [PubMed]

333. Li, J.; Feng, Y.; Wang, X.; Li, J.; Liu, W.; Rong, L.; Bao, J. An Overview of Predictors for Intrinsically Disordered Proteins over 2010–2014. *Int. J. Mol. Sci.* **2015**, *16*, 23446–23562. [CrossRef] [PubMed]

334. Heller, G.T.; Aprile, F.A.; Vendruscolo, M. Methods of probing the interactions between small molecules and disordered proteins. *Cell. Mol. Life Sci.* **2017**, *74*, 3225–3243. [CrossRef] [PubMed]

335. Ambadipudi, S.; Zweckstetter, M. Targeting intrinsically disordered proteins in rational drug discovery. *Expert Opin. Drug Discov.* **2016**, *11*, 65–77. [CrossRef] [PubMed]

336. Henriques, J.; Cragnell, C.; Skepö, M. Molecular Dynamics Simulations of Intrinsically Disordered Proteins: Force Field Evaluation and Comparison with Experiment. *J. Chem. Theory Comput.* **2015**, *11*, 3420–3431. [CrossRef] [PubMed]

337. Robustelli, P.; Piana, S.; Shaw, D.E. Developing a molecular dynamics force field for both folded and disordered protein states. *Proc. Natl. Acad. Sci. USA* **2018**, *115*, E4758–E4766. [CrossRef] [PubMed]

338. Uversky, V.N. Unusual biophysics of intrinsically disordered proteins. *Biochim. Biophys. Acta* **2013**, *1834*, 932–951. [CrossRef] [PubMed]

339. Dogan, J.; Gianni, S.; Jemth, P. The binding mechanisms of intrinsically disordered proteins. *Phys. Chem. Chem. Phys.* **2014**, *16*, 6323–6331. [CrossRef] [PubMed]

340. Shirai, N.C.; Kikuchi, M. Structural flexibility of intrinsically disordered proteins induces stepwise target recognition. *J. Chem. Phys.* **2013**, *139*, 225103. [CrossRef] [PubMed]

341. Sammak, S.; Zinzalla, G. Targeting protein-protein interactions (PPIs) of transcription factors: Challenges of intrinsically disordered proteins (IDPs) and regions (IDRs). *Prog. Biophys. Mol. Biol.* **2015**, *119*, 41–46. [CrossRef] [PubMed]

342. Hausrath, A.C.; Kingston, R.L. Conditionally disordered proteins: Bringing the environment back into the fold. *Cell. Mol. Life Sci.* **2017**, *74*, 3149–3162. [CrossRef] [PubMed]

343. Hultqvist, G.; Åberg, E.; Camilloni, C.; Sundell, G.N.; Andersson, E.; Dogan, J.; Chi, C.N.; Vendruscolo, M.; Jemth, P. Emergence and evolution of an interaction between intrinsically disordered proteins. *eLife* **2017**, *6*, e16059. [CrossRef] [PubMed]

344. De Cássia Ruy, P.; Torrieri, R.; Toledo, J.S.; de Souza Alves, V.; Cruz, A.K.; Ruiz, J.C. Intrinsically disordered proteins (IDPs) in trypanosomatids. *BMC Genom.* **2014**, *15*, 1100.

345. Longhi, S. Structural disorder within paramyxoviral nucleoproteins. *FEBS Lett.* **2015**, *589*, 2649–2659. [CrossRef] [PubMed]

346. Russo, A.; Manna, S.L.; Novellino, E.; Malfitano, A.M.; Marasco, D. Molecular signaling involving intrinsically disordered proteins in prostate cancer. *Asian J. Androl.* **2016**, *18*, 673–681. [PubMed]

347. Ruan, H.; Sun, Q.; Zhang, W.; Liu, Y.; Lai, L. Targeting intrinsically disordered proteins at the edge of chaos. *Drug Discov. Today* **2019**, *24*, 217–227. [CrossRef] [PubMed]

© 2019 by the authors. Licensee MDPI, Basel, Switzerland. This article is an open access article distributed under the terms and conditions of the Creative Commons Attribution (CC BY) license (http://creativecommons.org/licenses/by/4.0/).

International Journal of
Molecular Sciences

Review

bHLH–PAS Proteins: Their Structure and Intrinsic Disorder

Marta Kolonko and Beata Greb-Markiewicz *

Department of Biochemistry, Faculty of Chemistry, Wroclaw University of Science and Technology, Wybrzeże Wyspiańskiego 27, 50-370 Wroclaw, Poland
* Correspondence: beata.greb-markiewicz@pwr.edu.pl

Received: 24 June 2019; Accepted: 16 July 2019; Published: 26 July 2019

Abstract: The basic helix–loop–helix/Per-ARNT-SIM (bHLH–PAS) proteins are a class of transcriptional regulators, commonly occurring in living organisms and highly conserved among vertebrates and invertebrates. These proteins exhibit a relatively well-conserved domain structure: the bHLH domain located at the N-terminus, followed by PAS-A and PAS-B domains. In contrast, their C-terminal fragments present significant variability in their primary structure and are unique for individual proteins. C-termini were shown to be responsible for the specific modulation of protein action. In this review, we present the current state of knowledge, based on NMR and X-ray analysis, concerning the structural properties of bHLH–PAS proteins. It is worth noting that all determined structures comprise only selected domains (bHLH and/or PAS). At the same time, substantial parts of proteins, comprising their long C-termini, have not been structurally characterized to date. Interestingly, these regions appear to be intrinsically disordered (IDRs) and are still a challenge to research. We aim to emphasize the significance of IDRs for the flexibility and function of bHLH–PAS proteins. Finally, we propose modern NMR methods for the structural characterization of the IDRs of bHLH–PAS proteins.

Keywords: bHLH–PAS transcription factor; intrinsically disordered region; IDR; C-terminus

1. Introduction to bHLH–PAS Proteins

The basic helix–loop–helix/Per-ARNT-SIM (bHLH–PAS) proteins are a class of transcriptional regulators that commonly occur in living organisms. They play an important role in the regulation of a variety of developmental and physiological events [1]. The maintenance of cellular and systemic oxygen homeostasis is performed by hypoxia-inducible factor 1α (HIF1-α) [2]. In the hypoxia condition, HIF1-α is translocated to the nucleus [3] where it regulates transcription activity related to angiogenesis, cell proliferation/survival, glucose metabolism, and iron metabolism. The incorrect control of the listed processes is fundamental in many diseases, including cancer, strokes, and heart disease [2]. Some bHLH–PAS family members act as receptors for different high and low molecular ligands [1]. The only known small ligand-activated bHLH–PAS protein, aryl hydrocarbon receptor (AHR), is involved in toxin metabolism and binds highly toxic ligands, such as TCDD [4]. The ligated AHR migrates to the nucleus and mediates a wide range of biological responses to poisons. This mediation comprises a wasting syndrome, hepatotoxicity, teratogenesis, and tumor promotion [4]. Overexpression and constitutive activation of the AHR have been observed in various types of tumors [5]. Importantly, the AHR has been described as a critical modulator of host–environment interactions, especially for immune and inflammatory responses [6].

Another interesting example of a bHLH–PAS family member is the single-minded protein (SIM), which plays a significant role during central nerve cord [7] and genital imaginal disc development [8]. As shown, SIM gene mutations contribute to certain dysmorphic features of brain development and also the mental retardation in Down syndrome [9]. Interestingly, SIM overexpression is also associated with breast and prostate cancer [10], which indicates connections between their apparently unrelated signaling pathways.

Members of the bHLH–PAS family were shown to be targets for disease therapy. AHR, highly expressed in multiple organs and tissues, may influence tumorigenesis both by direct effect on the cancer cells and by modulation of the immune system. For this reason, the development of selective AHR modulators active against multiple tumors is a desirable direction of research [11]. Also, targeting of the HIF1-α pathway as a novel cancer therapy is a current project [12]. As AHR was shown to modulate the immune response in the respiratory tract, this protein can be potentially used also as a therapeutic object for the treatment of various inflammatory lung diseases [13,14]. Another member of the family, expressed mainly in the brain, neuronal PAS domain-containing protein 4 (NPAS4) has been proposed as a novel therapeutic target for depression and neurodegenerative diseases [15] and as a component of new stroke therapies [16]. Additionally, NPAS4, whose expression was also detected in the pancreas, was proposed to be a therapeutic target for diabetes [17] and as a treatment during pancreas transplantation [18].

In spite of performing a high diversity of functions, the bHLH–PAS proteins family exhibits a relatively well-conserved domain structure in the N-terminal part of their sequence (Figure 1). The bHLH region contains approximately 60 amino acid (aa) residues and can be divided into two functionally distinctive parts: the basic region responsible for DNA binding (approximately 15 aa), and the neighboring C-terminal HLH region, which takes part in protein dimerization [19]. The PAS domain is located in the central part of the protein and usually comprises about 300 aa residues [1]. It is divided into two structurally conserved regions named PAS-A and PAS-B, which are often connected to a single PAS-associated C-terminal (PAC) motif [20]. The PAS-A and PAS-B regions are separated by a poorly conserved link [1]. The PAS-A region is critical for selecting a dimerization partner and ensuring the specificity of target gene activation [21]. The PAS-B region is usually responsible for sensing diverse exogenous and endogenous signals, and is accompanied by energetic and conformational changes that regulate protein activity [21]. Contrary to conserved domains, the C-termini of bHLH–PAS proteins present significant variability [21] and contain variable transcription activation/repression domains (TAD/RPD) (Figure 1) [22,23]. An example is the mammalian SIM existing in two isoforms: SIM1 and SIM2. Both isoforms present a high amino acid identity in their N-termini (90% identity in the bHLH and PAS regions) and extreme diversity in their C-termini [24]. While SIM1 activates the expression of target genes, SIM2 acts as an inhibitor. Interestingly, the opposite transcriptional effect disappears after the deletion of both SIM1 and SIM2 C-termini, resulting in proteins with a similar activity [25,26]. Moffet and Pelletier [26] demonstrated that a distinct SIM2 C-terminal sequence comprises two repression domains with a high proline/serine and proline/alanine content, respectively. It is a feature of "repressor motifs", which can also be found in a large number of other transcriptional repressors [25,26]. Due to the highly variable amino acid sequence and the lack of predefined domains, C-termini are believed to be responsible for the specific modulation of the functioning of bHLH–PAS proteins and the recognition of partner proteins necessary for their unique action [21].

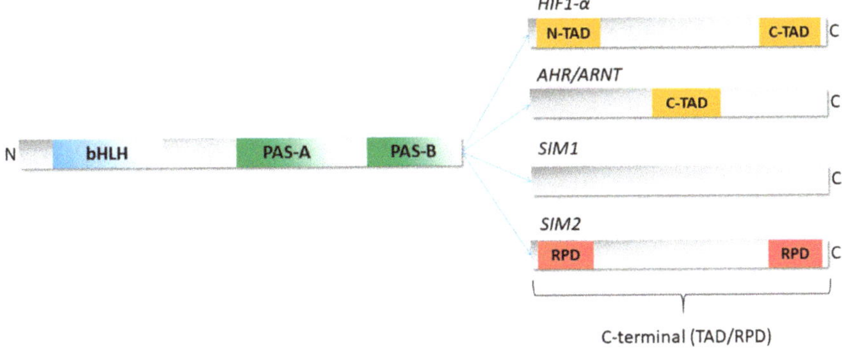

Figure 1. Schematic representation of the bHLH–PAS protein domain structure. The N-terminal part of bHLH–PAS proteins is characterized by the presence of defined domains: bHLH (blue), PAS-A and PAS-B (green). The C-terminal part presents significant diversity and contains variable transactivation/repression domains (TAD/RPD). The C-termini of selected proteins (HIF1-α, AHR/ARNT, SIM1, and SIM2) are presented. Yellow boxes indicate TADs while the red box indicates RPD. Based on [26–29].

Generally, bHLH–PAS proteins can be divided into two classes. While the expression of class I proteins is specifically regulated by diverse physiological states and/or environmental signals [30], class II proteins are expressed continuously and serve as heterodimerization partners for class I members. Only the dimer of the two bHLH–PAS proteins acts as a functional transcription factor complex, regulating the expression of genes under its control [22]. Mammalian bHLH–PAS transcription factors are listed in Table 1.

Table 1. Mammalian class I and class II bHLH–PAS proteins [1,21,30–32].

Class I	Class II	Type of Signal
hypoxia-inducible factors (HIF; HIF1-α, HIF2-α, and HIF3-α)		regulated by hypoxia
aryl hydrocarbon receptor (AHR); aryl hydrocarbon receptor repressor (AHRR)	aryl hydrocarbon receptor nuclear translocator (ARNT), also known as HIF1-β and ARNT2	regulated by xenobiotics
single-minded proteins (SIM1 and SIM2)		developmentally regulated
neuronal PAS domain proteins (NPAS)		developmentally regulated
circadian locomotor output cycles protein kaput (CLOCK)	Circadian rhythm proteins (BMAL1 and BMAL2, also known as ARNTL and ARNTL2)	circadian rhythms

2. bHLH–PAS Protein Conservation between Organisms

bHLH–PAS proteins are highly conserved among different organisms, including vertebrates and invertebrates [33]. Most mammalian representatives possess orthologs in insect species. An example is the *Drosophila melanogaster* TANGO (TGO) protein, which is a homologue of the mammalian class II protein, ARNT [34]. TGO is known as the general dimerization partner for Similar (SIMA), Trachealess (TRH), Single-minded (SIM) protein, Spineless (SS), and Dysfusion (DYS), performing functions equivalent to mammalian ones.

In 2017, the Nobel Prize in Physiology or Medicine was awarded to J. C. Hall, M. Rosbash, and M. W. Young for their discoveries of molecular mechanisms controlling the circadian rhythm in *D. melanogaster*. As shown, the two bHLH–PAS transcription factors CLOCK and CYCLE play a key role as transcriptional activators for *period* (*per*) and *timeless* (*tim*) genes [35,36]. Thanks to the conservation of circadian bHLH–PAS proteins between *D. melanogaster* and mammals [35], the explanation of the fly daily rhythm enabled the understanding of a similar, though much more complicated, process in mammals, controlled by two orthologous to CLOCK/CYCLE heterodimers: CLOCK/BMAL1 and NPAS-2/BMAL1 [22].

In spite of significant similarities, some exceptions between vertebrates and invertebrates can be noticed. The bHLH–PAS transcription factor, Methoprene-tolerant protein (MET), occurs exclusively in insects and to date has no known ortholog in nonarthropod organisms. MET has been recently confirmed as the juvenile hormone (JH) receptor playing a significant role during insect development and maturation [37]. Interestingly, in a few species of insects, like *D. melanogaster* and *Bombyx mori*, there exist the MET paralogs named germ-cell expressed (GCE) and MET2, respectively [38]. MET and GCE participate in modulating JH signaling during *D. melanogaster* development, but their functions are not fully redundant and the proteins exhibit tissue-specific distribution [39]. In turn, the MET2 protein function in *B. mori* is not yet defined [40].

3. Structure of bHLH–PAS Proteins

To date, our knowledge regarding the tertiary structure of bHLH–PAS proteins is limited. All determined structures comprise single isolated domains (PAS-A or PAS-B) or adjacent domains connected with flexible aa chains. C-termini, however, comprising an extensive part of proteins, have not yet been structurally characterized. These regions are not homologous to any described domains and seem to be very disordered. Consequently, it can be seen to be a huge challenge for scientists to determine their structure and combine it with specific protein functions. All bHLH–PAS structures deposited in the Protein Data Bank (PDB) are listed in Table 2 (Nuclear Resonance Magnetism (NMR) structures) and Table 3 (X-ray structures). Most of the listed assemblies correspond to heterodimers.

Table 2. bHLH–PAS protein structures deposited in the PDB obtained with NMR.

Form	Protein	Segment	Organism	PDB ID
monomers	HIF2-α	PAS-B domain	*Homo sapiens*	1P97
	ARNT	PAS-B domain	*Homo sapiens*	1X0O
dimer	HIF-2a:ARNT	PAS-B domains	*Homo sapiens*	2A24

Table 3. bHLH–PAS protein structures deposited in the PDB obtained with X-ray diffraction.

Form	Protein	Segment	Organism	PDB ID
monomers	AHR	PAS-A	*Mus musculus*	4M4X
	ARNT	PAS-B	*Homo sapiens*	2B02
	HIF1-α	PAS-B	*Homo sapiens*	4H6J
dimers	ARNT Homodimer	PAS-B	*Homo sapiens*	4EQ1
	HIF2-α:ARNT	PAS-B	*Homo sapiens*	3F1P
	HIF2-α:ARNT with artificial ligand	PAS-B	*Homo sapiens*	3F1O
	HIF2-α:ARNT	PAS-B	*Homo sapiens*	6D0C
	ARNT/HIF transcription factor/coactivator complex	PAS-B	*Homo sapiens, Mus musculus*	4PKY
	ARNT transcription factor/coactivator complex	PAS-B domain	*Homo sapiens, Mus musculus*	4LPZ
	HIF2-α:ARNT	bHLH; PAS-A; PAS-B	*Mus musculus*	4ZP4

Table 3. *Cont.*

Form	Protein	Segment	Organism	PDB ID
	HIF2-α:ARNT with HRE DNA	bHLH; PAS-A; PAS-B	*Mus musculus*	4ZPK
	HIF1-α:ARNT with HRE DNA	bHLH; PAS-A; PAS-B	*Mus musculus*	4ZPR
	AHR:ARNT	bHLH; PAS-A	*Homo sapiens*	5NJ8
	AHR:ARNT bound to the dioxin response element (DRE)	bHLH; PAS-A	*Homo sapiens, Mus musculus*	5V0L
	AHRR:ARNT	bHLH; PAS-A; PAS-B	*Homo sapiens, Bos taurus*	5Y7Y
	NPAS1:ARNT	bHLH; PAS-A; PAS-B	*Mus musculus*	5SY5
	NPAS3:ARNT in complex with HRE DNA	bHLH; PAS-A; PAS-B	*Mus musculus*	5SY7
	CLOCK-BMAL1	bHLH	*Homo sapiens*	4H10
	CLOCK:BMAL1	bHLH; PAS-A; PAS-B	*Mus musculus*	4F3L

The first step in determining the structure of bHLH–PAS proteins was the isolation and characterization of PAS-B domains from HIF2-α (Figure 2A) [41] and ARNT (Figure 2B) [42]. Both structures were obtained using the NMR technique and presented a fold characteristic for the PAS domain: a five-stranded antiparallel β-sheet flanked by several α-helices [42]. The next step was the crystallization of the isolated PAS-A domain of AHR (Figure 2C) and the PAS-B domains of ARNT (not shown) and HIF1-α (not shown). Interestingly, the tertiary architecture of all structurally characterized PAS domains is very conserved (Figure 2), despite the fact that their primary sequence is highly divergent (sequence identity lower than 20%) [43].

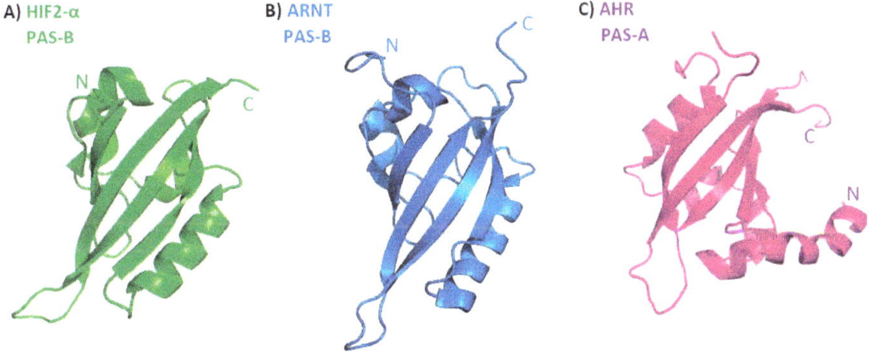

Figure 2. Representation of the PAS fold: a five-stranded antiparallel β-sheet is flanked by several α-helices. (**A**) HIF2-α PAS-B obtained with NMR (PDB 1P97), (**B**) ARNT PAS-B obtained with NMR (PDB 1X0O), (**C**) AHR PAS-A domain obtained with X-ray (PDB 4M4X).

Further experiments led to the cocrystallization of PAS-B domains from the HIF2-α/ARNT heterodimer, which revealed that these two domains form an interaction interface via their β-sheets in an antiparallel form (Figure 3A) [42]. Another measurement covering bHLH domains of BMAL1/CLOCK bound to the DNA defined domain structure and binding properties specifying interactions taking place (Figure 3B) [44]. A typical bHLH domain comprises two long α helices connected by a short loop. The first helix includes the basic domain and interacts with the major groove of the DNA [45]. All presented structures allowed an insight into the organization of bHLH–PAS proteins; however, the structure of the multidomain bHLH–PAS protein was still missing.

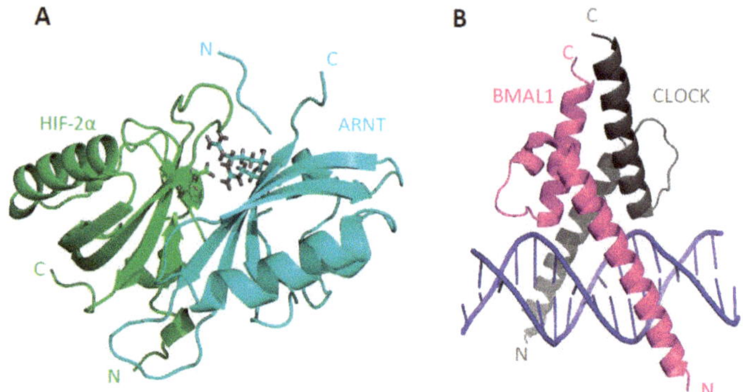

Figure 3. (**A**) HIF2-α PAS-B (green) and ARNT PAS-B (blue) heterodimer (3F1P, [46]). Amino acids creating a salt bridge are marked (HIF2-α E247, ARNT R362, ARNT R379). (**B**) BMAL1 bHLH (magenta) and CLOCK bHLH (grey) domains with E-box DNA (blue) (1H10, [44]).

A turning point was the year 2012, when the first heterodimer comprising the bHLH–PAS-A/PAS-B domains (CLOCK-BMAL1) was crystallized [47] and its structure was resolved (Figure 4A). In 2015, the architecture of two other heterodimers, HIF1-α-ARNT (not shown) and HIF2-α-ARNT (Figure 4B), were obtained [48]. All determined structures present the position of the defined domains in relation to each other in the functional heterodimers. In general, the individual PAS domains are not involved in equal interactions, and the obtained structures are highly asymmetric. Importantly, two groups of heterodimers (based on BMAL-1 or ARNT proteins as a dimerization partner) present separate types of quaternary architecture. All domains in the BMAL-1 group are close spatially to each other (Figure 4A), while ARNT domains do not create intramolecular interactions and can wrap up around a partner protein (Figure 4B) [22,48].

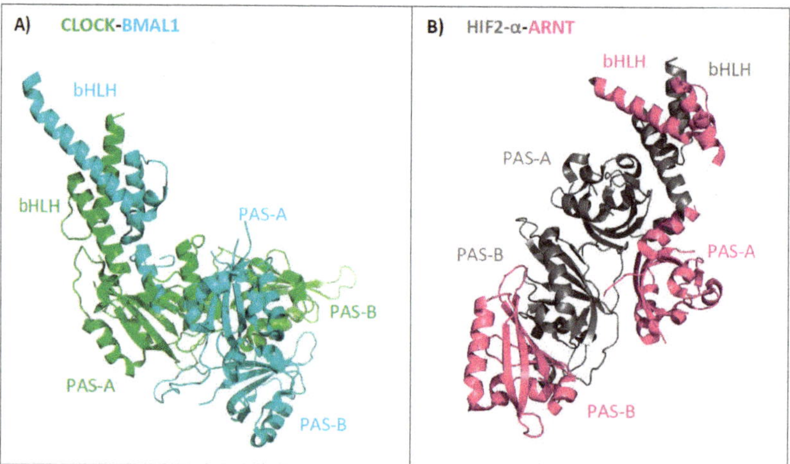

Figure 4. Representatives of the two groups of the bHLH–PAS heterodimers. (**A**) Overall structure of the CLOCK-BMAL1 heterodimer (4f3l, [47]), (**B**) overall structure of the HIF2-α–ARNT heterodimer (4zp4, [48]).

To date, all available structural information concerns mammalian bHLH–PAS proteins. There is almost no information about the structure of proteins derived from other organisms, including invertebrate *D. melanogaster*. It would be interesting to verify evolutionary conservation of the entire bHLH–PAS fold in different organisms. The majority of reported protein structures include defined N-terminal domains, while the structural information about C-terminal regions is still missing and limited to short peptides bound to interacting proteins [22]. An example is a short motif featuring a conserved sequence LIXXL found in *D. melanogaster* MET and GCE, which represents a novel nuclear receptor (NR) box. The Docking models of the MET/GCE NR box associating peptides to the orphan nuclear receptor (FTZ-F1) ligand-binding domain (LBD) revealed their α-helical structure, necessary for hydrophobic interaction [49].

4. Unique Properties of the C-Terminal Domains of bHLH–PAS Proteins as IDRs

While the N-terminal part of bHLH–PAS proteins is responsible for interactions with DNA, ligands/cofactors binding, and heterodimerization, their C-termini are usually responsible for the regulation of the protein and the activity of created complexes [50]. The variability of the amino acid sequence of C-terminal fragments, their transactivation role, and the lack of homology to any described domains prompted us to ask the question about the structural character of these regions and the relationship of their character with the performed function. For a long time, it was believed that spontaneous folding into a well-defined and stable tertiary structure is required for the protein action [51]. However, it is actually known that more than 20–30% of eukaryotic proteins do not have a stable tertiary structure in physiological conditions, but at the same time still perform important biological functions. Such proteins are referred to as intrinsically disordered proteins (IDPs). Simultaneously, over 70% of proteins involved in signal transduction cascades have long intrinsically disordered regions (IDRs). Importantly, the lack of a defined structure is critical for the functionality of IDPs and IDRs [52]. Additionally, the conformational plasticity and elongated shape make them a frequent target of different kinds of post-translational modifications (phosphorylation, acetylation, methylation, and others) that regulate protein activity [53]. IDPs were identified as elements of cellular signaling which control mechanisms and protein interaction networks [54]. IDPs were also shown to take part in disease-related signaling transduction; for example, intrinsically disordered amyloid β-peptides are involved in Alzheimer's disease [55]. Therefore, IDPs can be seen to be targets for drug design strategies.

4.1. In Silico Analyses of Selected bHLH–PAS Proteins

To estimate the occurrence of putative IDRs in bHLH–PAS proteins, we performed in silico analyses of the composition, hydropathy, and sequence complexity of amino acid sequences corresponding to selected proteins. We used the previously described human SIM1 and SIM2, as well as their *D. melanogaster* ortholog, SIM (Figure 5A), representing the class I of the family. To obtain a wider spectrum, we studied other human class I members, AHR, HIF1-α, and CLOCK (Figure 5B), which are engaged in different signal transduction pathways. As mentioned previously, class I proteins dimerize with class II proteins to form a functional complex and are crucial for heterodimer specificity. As each bHLH–PAS class II transcription factor is able to interact with different class I members, we found it to be extremely interesting to perform in silico analysis of the structure of class II members. We chose human ARNT, human BMAL1 (Figure 5C), and, additionally, *D. melanogaster* MET (Figure 5C) as a unique protein with an unknown mammalian homolog. MET can be classified as a class II bHLH–PAS family member based on its ability to not only create heterodimers with its paralog GCE, but also homodimers [56].

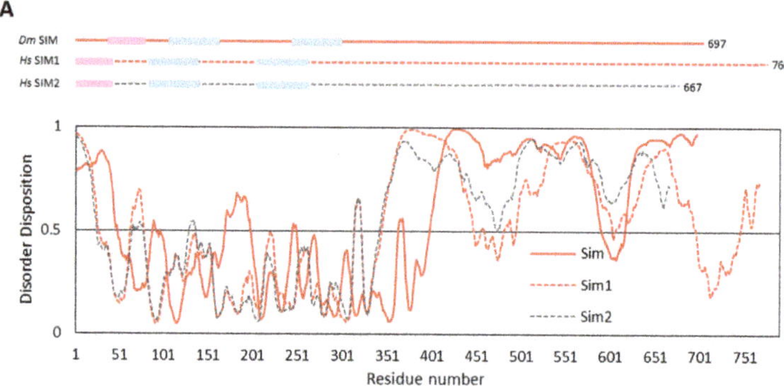

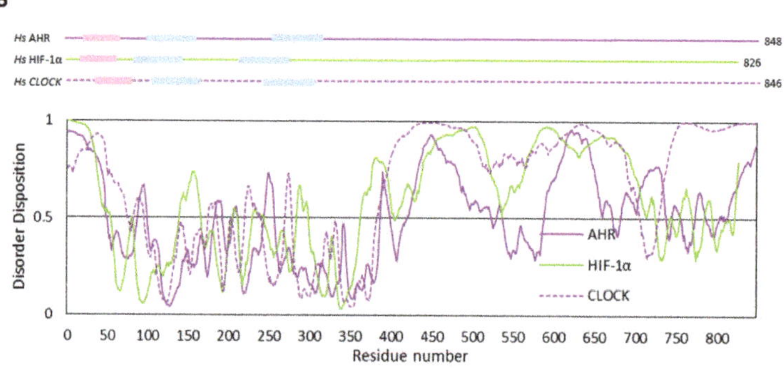

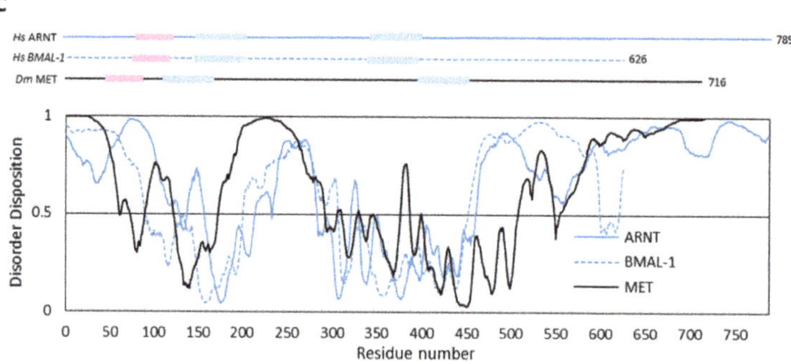

Figure 5. Prediction of intrinsically disordered regions. The top panel presents the domain structure of the analyzed bHLH–PAS proteins. Pink indicates the bHLH domain, whereas blue represents PAS domains. The length of the proteins is marked. The bottom panel presents a prediction of intrinsically disordered regions based on the amino acid sequence of proteins. All calculations were performed using PONDR-VLS2 software [57]. A score over 0.5 indicates disorder. (**A**) The class I proteins: *D. melanogaster* SIM (red line) and its *H. sapiens* orthologs SIM1 (dashed red line) and SIM2 (dashed grey line). (**B**) The class I proteins: *H. sapiens* AHR (violet line), HIF1-α (green line), and CLOCK (violet dashed line). (**C**) The class II proteins: *H. sapiens* ortholog ARNT (blue line), BMAL-1 (blue dashed line), and *D. melanogaster* Met (black line).

We performed in silico analysis using the predictors of intrinsically disordered regions: PONDR-VSL2 [57], PONDR-FIT [58], IUPred [59], and IsUnstruct [60]. Since the results of all the employed predictors were compatible, for the purpose of simplicity, we decided to show only one representative result (PONDR-VSL2) for each protein (Figure 5). All the results (Figure 5) substantiate our hypothesis and indicate the intrinsic character of the long C-termini. It is worth noting that for proteins representing class I (see Figure 5A,B), short ordered fragments in their C-termini are visible. Such fragments are able to act as TAD/RPD or so-called molecular recognition elements (MoREs) [61,62]. The presence of MoREs makes the interactions between partner proteins highly specific and reversible [52]. The presented results revealed some subtle differences in the regions that comprise preserved domains. The structure predicted for class I proteins (Figure 5A,B) is undeniably more ordered, while class II proteins show a marked structure relaxation in their middle part (Figure 5C), which is C-terminally linked to the PAS-A domain responsible for specificity of gene activation by bHLH–PAS proteins [63]. Such a difference explains the ability of class II proteins to serve as an interaction partner for different proteins [22]. The ability of IDRs (and IDPs) to interact with several partners is an undeniable advantage in molecular recognition processes [64]. Importantly, the resulting induced folding may differ depending on the binding partner. For example, a disordered region of p53 protein, a known cell cycle regulator and a tumor suppressor [65], folds into alpha helix or beta strand, depending on the partner protein [66].

4.2. The Impact of Disordered Regions on Protein Function

The flexibility and disorder detected in individual C-termini can be related to the ability of individual bHLH–PAS proteins to perform diverse functions. The differences between SIM1 and SIM2 C-termini, regarding their opposite functions (gene activation/repression), have previously been described. C-terminal regions of two other studied proteins, AHR (class I member) and ARNT (class II member), are characterized by the presence of TADs [67], in which functions are mediated by CBP/p300 and RIP140 coactivators. The C-terminal region of ARNT was additionally proposed to be a crucial activator of the estrogen receptor (ER) [68]. Interestingly, the suppression of AHR activity is also connected with the C-terminus and is mediated by the binding of the small peptide inhibitor [69]. Another repressor of the AHR signaling pathway, AHRR, is distinguished from AHR by the presence of three SUMOylation sites in its C-terminus. As shown, SUMOylation is crucial for full suppressive activity of AHRR [70].

Moreover, the C-terminus of another studied protein, HIF1-α, is characterized by the presence of TADs, and it also interacts with the CBP/p300 coactivator. The C-terminus is additionally responsible for protein stability/degradation and contains sequence motifs influencing subcellular localization: nuclear localization signal (NLS) and nuclear export signal (NES) [71,72].

Another remarkable class I bHLH–PAS protein is CLOCK, comprising a domain with histone transacetylase (HAT) activity in the C-terminus. This domain is responsible for histones acetylation, which affects the transcriptional stimulation of clock-controlled genes. Additional acetylation is performed on the R537 residue of the partner protein, BMAL1. R537 residue is located in the C-terminal part of BMAL1 and its modification facilitates the cryptochrome (CRY1)-mediated repression of specific gene transcription [73]. Importantly, CRY1 competes with the CBP/p300 coactivator for BMAL1 TAD binding, and is not able to bind the C-terminus in the paralog protein, BMAL2. Therefore, C-termini distinguish the circadian functions of these two BMAL paralogs [74].

4.3. Structural Analysis of bHLH–PAS C-Terminal Fragments

To date, the only structurally characterized C-terminal fragment of the bHLH–PAS protein is the *D. melanogaster* MET C-terminus (MET/C) [75]. It was shown by a series of in vitro analyses that MET/C exhibits a highly disordered character and exists in a solution in extended flexible form with predispositions for conformational changes. It is interesting to note that some short secondary motifs in the structure of MET/C have been predicted. Such short ordered fragments can be important during partner recognition and interactions. It was hypothesized that the intrinsic disorder of the C-terminal fragment was indispensable for the functionality of MET due to it modulating the protein's action in a context-specific way. It enables cross-talk between JH signaling and other signaling pathways during *D. melanogaster* development. Previously, it was shown that Met interacts with FTZ-F1 by its C-terminus [49], thereby modulating stage-specific responses to the hormones during *D. melanogaster* metamorphosis [76].

As all the in vitro analyses results obtained for the MET/C [75] were consistent with the in silico studies presented above (Figure 5C), we hypothesize that the disorder character of the bHLH–PAS proteins subfamily C-terminal fragments can be a more common characteristic and also be very important for their functionality. Previously, the importance of the disordered character of regions flanking the bHLH domain of bHLH transcription factors was shown [77–79].

4.4. Structural Analysis of IDPs

While C-terminal regions of the bHLH–PAS family are considered as IDRs, it can be challenging to detect and characterize them. The reason is that IDPs and IDRs do not adopt a single stable structure and the energetically most favorable conformations can be very distinguished [80,81]. The tiny conformational changes can promote IDPs/IDRs aggregation [82]. Additionally, it was shown that IDPs/IDRs can be highly sensitive to proteolysis [83]. Currently, studies focused on the characterization of IDRs and IDPs are rapidly developing, and techniques enabling the study of proteins in solution are still improving.

There are a number of bioinformatics tools allowing primary recognition of disordered proteins. Since IDPs are characterized by the specific aa composition (a low content of hydrophobic and a high content of charged residues [84]), the Composition Profiler [85] is commonly used to compare aa distribution between the studied protein and IDPs (DisProt3.4 database)/globular proteins (PDB S25 database). Additionally, for IDPs and globular proteins distinguishing, the Uversky diagram plotting mean net charge versus mean hydrophobicity is useful [86]. Disorder predictors (like PONDR-VSL2 [57], PONDR-FIT [58], IUPred [59], and IsUnstruct [60] used in this work) allow determining the probability of IDR occurrence utilizing the neural networks, trained on selected sets of ordered and disordered sequences. Another predictor, DynaMine, provides information about protein backbone flexibility [87,88]. IDPs, once purified, can be identified by various experimental methods. First, the underestimated mobility during SDS-PAGE electrophoresis can indicate the extended and elongated shape of the protein [75]. Hydrodynamic analysis comprising Size Exclusion Chromatography (SEC) [89] and Analytical Ultracentrifugation (AUC) are commonly used to determine hydrodynamic properties, like the Stokes radius (R_S), the sedimentation coefficients (s), and the frictional ratios (f/f_0) [90]. The Circular Dichroism (CD) is useful for secondary structures content calculation [91]. All listed techniques allow obtaining preliminary insight into protein structure properties.

One technique commonly used to study the overall shape and structural transitions of biological macromolecules in solution is small-angle X-ray scattering (SAXS) [92]. However, SAXS only provides limited information about the low-resolution overall shape of the molecule, so it is important to combine it with complementary high-resolution methods like NMR that present the local structure [93]. NMR offers unique opportunities that are based on analyzing the deviations from an idealized random coil devoid of any structural propensity [94]. The random coil exhibits characteristic chemical shifts, which are averages of all the possible conformations that amino acids can adopt in a solution. Therefore, NMR chemical shift deviations from random coil values can be used to evaluate the local transient secondary structure of IDPs [80]. The main problem during spectra assignments of IDPs is spectra overlapping (low chemical shifts dispersion) and a significant proton exchange with bulk water that reduces $^1H^N$ signal intensities, which in turn leads to low signal-to-noise ratios [94]. The exchange with water can be reduced by conducting measurements in low temperature or low pH [95]. Low-resolution spectra require the development of a novel NMR technique. Recently, IDP-dedicated methods such as ^{13}C-direct detected experiments, paramagnetic relaxation enhancements (PREs), or residual dipolar couplings (RDCs) have been described [96].

5. Conclusions

The available structure characterization of bHLH–PAS proteins is limited to the relatively well-conserved domains bHLH, PAS-A, and PAS-B. Importantly, all structures deposited in the Protein Data Bank are obtained for mammalian family members, the majority of them being heterodimers. On the other hand, the important parts of bHLH–PAS factors, which comprise their long C-termini, have not yet been structurally characterized. These fragments perform important functions in the specific modulation of protein action and for the recognition of interacting partners.

Performed in silico analysis revealed that the C-termini of representatives of the class I bHLH–PAS protein family members (SIM, SIM1, SIM2, AHR, HIF1-α, and Clock), and also class II (ARNT, BMAL1, and MET), are predicted as intrinsically disordered regions (IDRs) and are not homologous to any described domains. We discussed the known functions of the presented C-termini proteins according to their disorder character. Moreover, we proposed NMR techniques for intrinsically disordered C-termini characterization [94]. We believe that the structural properties of subsequent IDRs predicted in the sequences of bHLH–PAS transcription factors (mainly C-termini) need to be resolved for a full understanding of the way of bHLH–PAS family transcription factors function.

Funding: The work was supported by The National Science Centre (NCN): PRELUDIUM predoctoral grant UMO-2017/27/N/NZ1/01783 and ETIUDA doctoral scholarship UMO-2018/28/T/NZ1/00337, and partially supported by a statutory activity subsidy 0401/0143/18 from the Polish Ministry of Science and Higher Education for the Faculty of Chemistry of Wroclaw University of Science and Technology.

Acknowledgments: We would like to thank Andrzej Ożyhar (Department of Biochemistry, Faculty of Chemistry, Wroclaw University of Science and Technology) for all valuable comments and suggestions.

Conflicts of Interest: The authors declare no conflict of interest.

Abbreviations

Aa	amino acid
AHR	aryl hydrocarbon receptor
AHRR	aryl hydrocarbon receptor repressor
ARNT	aryl hydrocarbon receptor nuclear translocator
bHLH–PAS	helix–loop–helix/Per-ARNT-SIM
CLOCK	circadian locomotor output cycles protein kaput
CRY1	cryptochrome
DYS	dysfusion protein
ER	estrogen receptor
GCE	germ cell-expressed protein
HAT	histone transacetylase
HIF	hypoxia-inducible factors
IDPs	intrinsically disordered proteins
IDRs	intrinsically disordered regions
JH	juvenile hormone
LBD	ligand-binding domain
MET	methoprene-tolerant protein
MoREs	molecular recognition elements
MET/C	C-terminus of the MET protein
NES	nuclear export signal
NLS	nuclear localization signal
NMR	Nuclear Resonance Magnetism
NPAS	neuronal PAS domain-containing proteins
NR	boxes nuclear receptor boxes

PAC	PAS-associated C-terminal motif
PDB	Protein Data Bank
PREs	paramagnetic relaxation enhancements
RDCs	residual dipolar couplings
SAXS	Small-angle X-ray Scattering
SIM	single-minded proteins
SIMA	similar protein
SS	spineless protein
TAD or TRD	transactivation or repression domain
TGO	TANGO protein
TRH	tracheales protein

References

1. Crews, S.T. Control of cell lineage-specific development and transcription by bHLH-PAS proteins. *Genes Dev.* **1998**, *12*, 607–620. [CrossRef] [PubMed]
2. Lee, J.-W.; Bae, S.-H.; Jeong, J.-W.; Kim, S.-H.; Kim, K.-W. Hypoxia-inducible factor (HIF-1)α: Its protein stability and biological functions. *Exp. Mol. Med.* **2004**, *36*, 1–12. [CrossRef] [PubMed]
3. Ema, M.; Hirota, K.; Mimura, J.; Abe, H.; Yodoi, J.; Sogawa, K.; Poellinger, L.; Fujii-Kuriyama, Y. Molecular mechanisms of transcription activation by HLF and HIF1alpha in response to hypoxia: Their stabilization and redox signal-induced interaction with CBP/p300. *EMBO J.* **1999**, *18*, 1905–1914. [CrossRef] [PubMed]
4. Petrulis, J.R.; Kusnadi, A.; Ramadoss, P.; Hollingshead, B.; Perdew, G.H. The hsp90 Co-chaperone XAP2 Alters Importin β Recognition of the Bipartite Nuclear Localization Signal of the Ah Receptor and Represses Transcriptional Activity. *J. Biol. Chem.* **2003**, *278*, 2677–2685. [CrossRef] [PubMed]
5. Xue, P.; Fu, J.; Zhou, Y. The Aryl Hydrocarbon Receptor and Tumor Immunity. *Front. Immunol.* **2018**, *9*, 286. [CrossRef] [PubMed]
6. Neavin, D.; Liu, D.; Ray, B.; Weinshilboum, R. The Role of the Aryl Hydrocarbon Receptor (AHR) in Immune and Inflammatory Diseases. *Int. J. Mol. Sci.* **2018**, *19*, 3851. [CrossRef]
7. Freer, S.M.; Lau, D.C.; Pearson, J.C.; Talsky, K.B.; Crews, S.T. Molecular and functional analysis of Drosophila single-minded larval central brain expression. *Gene Expr. Patterns* **2011**, *11*, 533–546. [CrossRef] [PubMed]
8. Pielage, J.; Steffes, G.; Lau, D.C.; Parente, B.A.; Crews, S.T.; Strauss, R.; Klämbt, C. Novel behavioral and developmental defects associated with Drosophila single-minded. *Dev. Biol.* **2002**, *249*, 283–299. [CrossRef]
9. Chen, H.; Chrast, R.; Rossier, C.; Gos, A.; Antonarakis, S.E.; Kudoh, J.; Yamaki, A.; Shindoh, N.; Maeda, H.; Minoshima, S.; et al. Single–minded and Down syndrome? *Nat. Genet.* **1995**, *10*, 9–10. [CrossRef]
10. Bersten, D.C.; Sullivan, A.E.; Peet, D.J.; Whitelaw, M.L. bHLH–PAS proteins in cancer. *Nat. Rev. Cancer* **2013**, *13*, 827–841. [CrossRef]
11. Safe, S.; Lee, S.O.; Jin, U.H. Role of the aryl hydrocarbon receptor in carcinogenesis and potential as a drug target. *Toxicol. Sci.* **2013**, 1–16. [CrossRef] [PubMed]
12. Masoud, G.N.; Li, W. HIF-1α pathway: Role, regulation and intervention for cancer therapy. *Acta Pharm. Sin. B* **2015**, *5*, 378–389. [CrossRef] [PubMed]
13. Beamer, C.A.; Shepherd, D.M. Role of the aryl hydrocarbon receptor (AhR) in lung inflammation. *Semin. Immunopathol.* **2013**, *35*, 693–704. [CrossRef] [PubMed]
14. Puccetti, M.; Paolicelli, G.; Oikonomou, V.; De Luca, A.; Renga, G.; Borghi, M.; Pariano, M.; Stincardini, C.; Scaringi, L.; Giovagnoli, S.; et al. Towards Targeting the Aryl Hydrocarbon Receptor in Cystic Fibrosis. *Mediat. Inflamm.* **2018**, *2018*, 1601486. [CrossRef] [PubMed]
15. Zhang, Z.; Fei, P.; Mu, J.; Li, W.; Song, J. Hippocampal expression of aryl hydrocarbon receptor nuclear translocator 2 and neuronal PAS domain protein 4 in a rat model of depression. *Neurol. Sci.* **2014**, *35*, 277–282. [CrossRef] [PubMed]
16. Choy, F.C.; Klarić, T.S.; Koblar, S.A.; Lewis, M.D. The Role of the Neuroprotective Factor Npas4 in Cerebral Ischemia. *Int. J. Mol. Sci.* **2015**, *16*, 29011–29028. [CrossRef] [PubMed]
17. Sabatini, P.V.; Lynn, F.C. All-encomPASsing regulation of β-cells: PAS domain proteins in β-cell dysfunction and diabetes. *Trends Endocrinol. Metab.* **2015**, *26*, 49–57. [CrossRef]

18. Speckmann, T.; Sabatini, P.V.; Nian, C.; Smith, R.G.; Lynn, F.C. Npas4 Transcription Factor Expression Is Regulated by Calcium Signaling Pathways and Prevents Tacrolimus-induced Cytotoxicity in Pancreatic Beta Cells. *J. Biol. Chem.* **2016**, *291*, 2682–2695. [CrossRef]

19. Li, X.; Duan, X.; Jiang, H.; Sun, Y.; Tang, Y.; Yuan, Z.; Guo, J.; Liang, W.; Chen, L.; Yin, J.; et al. Genome-Wide Analysis of Basic/Helix-Loop-Helix Transcription Factor Family in Rice and Arabidopsis. *Plant Physiol.* **2006**, *141*, 1167–1184. [CrossRef]

20. Ponting, C.P.; Aravind, L. PAS: A multifunctional domain family comes to light. *Curr. Biol.* **1997**, *7*, R674–R677. [CrossRef]

21. Kewley, R.J.; Whitelaw, M.L.; Chapman-Smith, A. The mammalian basic helix–loop–helix/PAS family of transcriptional regulators. *Int. J. Biochem. Cell Biol.* **2004**, *36*, 189–204. [CrossRef]

22. Wu, D.; Rastinejad, F. Structural characterization of mammalian bHLH-PAS transcription factors. *Curr. Opin. Struct. Biol.* **2017**, *43*, 1–9. [CrossRef] [PubMed]

23. Partch, C.L.; Gardner, K.H. Coactivator recruitment, a new role for PAS domains in transcriptional regulation by the bHLH-PAS family. *J. Cell. Physiol.* **2010**, *223*, 553–557. [CrossRef] [PubMed]

24. Woods, S.L.; Whitelaw, M.L. Differential Activities of Murine Single Minded 1 (SIM1) and SIM2 on a Hypoxic Response Element. *J. Biol. Chem.* **2002**, *277*, 10236–10243. [CrossRef] [PubMed]

25. Moffett, P.; Reece, M.; Pelletier, J. The murine Sim-2 gene product inhibits transcription by active repression and functional interference. *Mol. Cell. Biol.* **1997**, *17*, 4933–4947. [CrossRef] [PubMed]

26. Moffett, P.; Pelletier, J. Different transcriptional properties of mSim-1 and mSim-2. *FEBS Lett.* **2000**, *466*, 80–86. [CrossRef]

27. Vorrink, S.U.; Domann, F.E. Regulatory crosstalk and interference between the xenobiotic and hypoxia sensing pathways at the AhR-ARNT-HIF1α signaling node. *Chem. Biol. Interact.* **2014**, *218*, 82–88. [CrossRef] [PubMed]

28. Tal, R.; Shaish, A.; Bangio, L.; Peled, M.; Breitbart, E.; Harats, D. Activation of C-transactivation domain is essential for optimal HIF-1α-mediated transcriptional and angiogenic effects. *Microvasc. Res.* **2008**, *76*, 1–6. [CrossRef] [PubMed]

29. Fukunaga, B.N.; Probst, M.R.; Reisz-Porszasz, S.; Hankinson, O. Identification of functional domains of the aryl hydrocarbon receptor. *J. Biol Chem.* **1995**, *270*, 29270–29278. [CrossRef]

30. Fribourgh, J.L.; Partch, C.L. Assembly and function of bHLH-PAS complexes. *Proc. Natl. Acad. Sci. USA* **2017**, *114*, 5330–5332. [CrossRef]

31. Hoffman, E.C.; Reyes, H.; Chu, F.F.; Sander, F.; Conley, L.H.; Brooks, B.A.; Hankinson, O. Cloning of a factor required for activity of the Ah (dioxin) receptor. *Science* **1991**, *252*, 954–958. [CrossRef] [PubMed]

32. Andreasen, E.A.; Spitsbergen, J.M.; Tanguay, R.L.; Stegeman, J.J.; Heideman, W.; Peterson, R.E. Tissue-Specific Expression of AHR2, ARNT2, and CYP1A in Zebrafish Embryos and Larvae: Effects of Developmental Stage and 2,3,7,8-Tetrachlorodibenzo-p-dioxin Exposure. *Toxicol. Sci.* **2002**, *68*, 403–419. [CrossRef] [PubMed]

33. Crews, S.T.; Fan, C.-M. Remembrance of things PAS: Regulation of development by bHLH–PAS proteins. *Curr. Opin. Genet. Dev.* **1999**, *9*, 580–587. [CrossRef]

34. Sonnenfeld, M.; Ward, M.; Nystrom, G.; Mosher, J.; Stahl, S.; Crews, S. The Drosophila tango gene encodes a bHLH-PAS protein that is orthologous to mammalian Arnt and controls CNS midline and tracheal development. *Development* **1997**, *124*, 4571–4582. [PubMed]

35. Panda, S.; Hogenesch, J.B.; Kay, S.A. Circadian rhythms from flies to human. *Nature* **2002**, *417*, 329–335. [CrossRef] [PubMed]

36. Paranjpe, D.A.; Sharma, V.K. Evolution of temporal order in living organisms. *J. Circadian Rhythm.* **2005**, *3*, 7. [CrossRef]

37. Li, K.-L.; Lu, T.-M.; Yu, J.-K. Genome-wide survey and expression analysis of the bHLH-PAS genes in the amphioxus Branchiostoma floridae reveal both conserved and diverged expression patterns between cephalochordates and vertebrates. *Evodevo* **2014**, *5*, 20. [CrossRef] [PubMed]

38. Cheng, D.; Meng, M.; Peng, J.; Qian, W.; Kang, L.; Xia, Q. Genome-wide comparison of genes involved in the biosynthesis, metabolism, and signaling of juvenile hormone between silkworm and other insects. *Genet. Mol. Biol.* **2014**, *37*, 444–459. [CrossRef] [PubMed]

39. Abdou, M.; Peng, C.; Huang, J.; Zyaan, O.; Wang, S.; Li, S.; Wang, J. Wnt Signaling Cross-Talks with JH Signaling by Suppressing Met and gce Expression. *PLoS ONE* **2011**, *6*, e26772. [CrossRef]

40. Jindra, M.; Bellés, X.; Shinoda, T. Molecular basis of juvenile hormone signaling. *Curr. Opin. Insect. Sci.* **2015**, *11*, 39–46. [CrossRef]

41. Erbel, P.J.A.; Card, P.B.; Karakuzu, O.; Bruick, R.K.; Gardner, K.H. Structural basis for PAS domain heterodimerization in the basic helix-loop-helix-PAS transcription factor hypoxia-inducible factor. *Proc. Natl. Acad. Sci. USA* **2003**, *100*, 15504–15509. [CrossRef] [PubMed]

42. Card, P.B.; Erbel, P.J.A.; Gardner, K.H. Structural Basis of ARNT PAS-B Dimerization: Use of a Common Beta-sheet Interface for Hetero- and Homodimerization. *J. Mol. Biol.* **2005**, *353*, 664–677. [CrossRef] [PubMed]

43. Henry, J.T.; Crosson, S. Ligand-binding PAS domains in a genomic, cellular, and structural context. *Annu. Rev. Microbiol.* **2011**, *65*, 261–286. [CrossRef] [PubMed]

44. Wang, Z.; Wu, Y.; Li, L.; Su, X.-D. Intermolecular recognition revealed by the complex structure of human CLOCK-BMAL1 basic helix-loop-helix domains with E-box DNA. *Cell Res.* **2013**, *23*, 213. [CrossRef] [PubMed]

45. Jones, S. An overview of the basic helix-loop-helix proteins. *Genome Biol.* **2004**, *5*, 226. [CrossRef] [PubMed]

46. Scheuermann, T.H.; Tomchick, D.R.; Machius, M.; Guo, Y.; Bruick, R.K.; Gardner, K.H. Artificial ligand binding within the HIF2 PAS-B domain of the HIF2 transcription factor. *Proc. Natl. Acad. Sci. USA* **2009**, *106*, 450–455. [CrossRef] [PubMed]

47. Huang, N.; Chelliah, Y.; Shan, Y.; Taylor, C.A.; Yoo, S.-H.; Partch, C.; Green, C.B.; Zhang, H.; Takahashi, J.S. Crystal structure of the heterodimeric CLOCK:BMAL1 transcriptional activator complex. *Science* **2012**, *337*, 189–194. [CrossRef] [PubMed]

48. Wu, D.; Potluri, N.; Lu, J.; Kim, Y.; Rastinejad, F. Structural integration in hypoxia-inducible factors. *Nature* **2015**, *524*, 303–308. [CrossRef]

49. Bernardo, T.J.; Dubrovsky, E.B. The Drosophila juvenile hormone receptor candidates methoprene-tolerant (MET) and germ cell-expressed (GCE) utilize a conserved LIXXL motif to bind the FTZ-F1 nuclear receptor. *J. Biol. Chem.* **2012**, *287*, 7821–7833. [CrossRef]

50. Furness, S.G.B.; Lees, M.J.; Whitelaw, M.L. The dioxin (aryl hydrocarbon) receptor as a model for adaptive responses of bHLH/PAS transcription factors. *FEBS Lett.* **2007**, *581*, 3616–3625. [CrossRef]

51. Mirsky, A.E.; Pauling, L. On the Structure of Native, Denatured, and Coagulated Proteins. *Proc. Natl. Acad. Sci. USA* **1936**, *22*, 439–447. [CrossRef] [PubMed]

52. Tompa, P. Intrinsically unstructured proteins. *Trends Biochem. Sci.* **2002**, *27*, 527–533. [CrossRef]

53. Wright, P.E.; Dyson, H.J. Intrinsically disordered proteins in cellular signalling and regulation. *Nat. Rev. Mol. Cell Biol.* **2014**, *16*, 18–29. [CrossRef] [PubMed]

54. Uversky, V.N. Flexible Nets of Malleable Guardians, Intrinsically Disordered Chaperones in Neurodegenerative Diseases. *Chem. Rev.* **2011**, *111*, 1134–1166. [CrossRef] [PubMed]

55. Ball, K.A.; Wemmer, D.E.; Head-Gordon, T. Comparison of Structure Determination Methods for Intrinsically Disordered Amyloid-β Peptides. *J. Phys. Chem. B* **2014**, *118*, 6405–6416. [CrossRef] [PubMed]

56. Godlewski, J.; Wang, S.; Wilson, T.G. Interaction of bHLH-PAS proteins involved in juvenile hormone reception in Drosophila. *Biochem. Biophys. Res. Commun.* **2006**, *342*, 1305–1311. [CrossRef] [PubMed]

57. Li, X.; Romero, P.; Rani, M.; Dunker, A.K.; Obradovic, Z. Predicting Protein Disorder for N-, C-, and Internal Regions. *Genome Inform.* **1999**, *10*, 30–40.

58. Xue, B.; Dunbrack, R.L.; Williams, R.W.; Dunker, A.K.; Uversky, V.N. PONDR-FIT: A meta-predictor of intrinsically disordered amino acids. *Biochim. Biophys. Acta* **2010**, *1804*, 996–1010. [CrossRef]

59. Dosztanyi, Z.; Csizmok, V.; Tompa, P.; Simon, I. IUPred: Web server for the prediction of intrinsically unstructured regions of proteins based on estimated energy content. *Bioinformatics* **2005**, *21*, 3433–3434. [CrossRef]

60. Lobanov, M.Y.; Sokolovskiy, I.V.; Galzitskaya, O.V. IsUnstruct: Prediction of the residue status to be ordered or disordered in the protein chain by a method based on the Ising model. *J. Biomol. Struct. Dyn.* **2013**, *31*, 1034–1043. [CrossRef]

61. Dziedzic-Letka, A.; Ozyhar, A. Intrinsically disordered proteins. *Postepy Biochem.* **2012**, *58*, 100–109. [PubMed]

62. Vacic, V.; Oldfield, C.J.; Mohan, A.; Radivojac, P.; Cortese, M.S.; Uversky, V.N.; Dunker, A.K. Characterization of Molecular Recognition Features, MoRFs, and Their Binding Partners. *J. Proteome Res.* **2007**, *6*, 2351–2366. [CrossRef] [PubMed]

63. Zelzer, E.; Wappner, P.; Shilo, B.-Z. The PAS domain confers target gene specificity of Drosophila bHLH/PAS proteins. *Genes Dev.* **1997**, *11*, 2079–2089. [CrossRef] [PubMed]

64. Uversky, V.N. Natively unfolded proteins: A point where biology waits for physics. *Protein Sci.* **2002**, *11*, 739–756. [CrossRef] [PubMed]

65. Bell, S.; Klein, C.; Müller, L.; Hansen, S.; Buchner, J. p53 contains large unstructured regions in its native state. *J. Mol. Biol.* **2002**, *322*, 917–927. [CrossRef]

66. Riley, T.; Sontag, E.; Chen, P.; Levine, A. Transcriptional control of human p53-regulated genes. *Nat. Rev. Mol. Cell Biol.* **2008**, *9*, 402–412. [CrossRef] [PubMed]

67. Jain, S.; Dolwick, K.M.; Schmidt, J.V.; Bradfield, C.A. Potent transactivation domains of the Ah receptor and the Ah receptor nuclear translocator map to their carboxyl termini. *J. Biol. Chem.* **1994**, *269*, 31518–31524. [PubMed]

68. Brunnberg, S.; Pettersson, K.; Rydin, E.; Matthews, J.; Hanberg, A.; Pongratz, I. The basic helix-loop-helix-PAS protein ARNT functions as a potent coactivator of estrogen receptor-dependent transcription. *Proc. Natl. Acad. Sci. USA* **2003**, *100*, 6517–6522. [CrossRef]

69. Ren, L.; Thompson, J.D.; Cheung, M.; Ngo, K.; Sung, S.; Leong, S.; Chan, W.K. Selective suppression of the human aryl hydrocarbon receptor function can be mediated through binding interference at the C-terminal half of the receptor. *Biochem. Pharmacol.* **2016**, *107*, 91–100. [CrossRef]

70. Kawajiri, K.; Fujii-Kuriyama, Y. The aryl hydrocarbon receptor: A multifunctional chemical sensor for host defense and homeostatic maintenance. *Exp. Anim.* **2017**, *66*, 75–89. [CrossRef]

71. Kallio, P.J.; Okamoto, K.; Brien, S.O.; Carrero, P.; Makino, Y.; Tanaka, H.; Poellinger, L. Signal transduction in hypoxic cells: Inducible nuclear translocation and recruitment of the CBP/p300 coactivator by the hypoxia-inducible factor-1 α. *EMBO J.* **1998**, *17*, 6573–6586. [CrossRef] [PubMed]

72. Mylonis, I.; Chachami, G.; Paraskeva, E.; Simos, G. Atypical CRM1-dependent nuclear export signal mediates regulation of hypoxia-inducible factor-1alpha by MAPK. *J. Biol. Chem.* **2008**, *283*, 27620–27627. [CrossRef] [PubMed]

73. Doi, M.; Hirayama, J.; Sassone-Corsi, P. Circadian regulator CLOCK is a histone acetyltransferase. *Cell* **2006**, *125*, 497–508. [CrossRef] [PubMed]

74. Xu, H.; Gustafson, C.L.; Sammons, P.J.; Khan, S.K.; Parsley, N.C.; Ramanathan, C.; Lee, H.-W.; Liu, A.C.; Partch, C.L. Cryptochrome 1 regulates the circadian clock through dynamic interactions with the BMAL1 C terminus. *Nat. Struct. Mol. Biol.* **2015**, *22*, 476–484. [CrossRef] [PubMed]

75. Kolonko, M.; Ożga, K.; Hołubowicz, R.; Taube, M.; Kozak, M.; Ożyhar, A.; Greb-Markiewicz, B. Intrinsic Disorder of the C-Terminal Domain of Drosophila Methoprene-Tolerant Protein. *PLoS ONE* **2016**, *11*, e0162950. [CrossRef] [PubMed]

76. Broadus, J.; McCabe, J.R.; Endrizzi, B.; Thummel, C.S.; Woodard, C.T. The Drosophila beta FTZ-F1 orphan nuclear receptor provides competence for stage-specific responses to the steroid hormone ecdysone. *Mol. Cell* **1999**, *3*, 143–149. [CrossRef]

77. Fuxreiter, M.; Simon, I.; Bondos, S. Dynamic protein–DNA recognition: Beyond what can be seen. *Trends Biochem. Sci.* **2011**, *36*, 415–423. [CrossRef]

78. Guo, X.; Bulyk, M.L.; Hartemink, A.J. Intrinsic disorder within and flanking the DNA-binding domains of human transcription factors. *Biocomputing* **2012**, 104–115. [CrossRef]

79. Roschger, C.; Schubert, M.; Regl, C.; Andosch, A.; Marquez, A.; Berger, T.; Huber, C.G.; Lütz-Meindl, U.; Cabrele, C. The Recombinant Inhibitor of DNA Binding Id2 Forms Multimeric Structures via the Helix-Loop-Helix Domain and the Nuclear Export Signal. *Int. J. Mol. Sci.* **2018**, *19*, 1105. [CrossRef]

80. Eliezer, D. Biophysical characterization of intrinsically disordered proteins. *Curr. Opin. Struct. Biol.* **2009**, *19*, 23–30. [CrossRef]

81. Jensen, M.R.; Zweckstetter, M.; Huang, J.; Blackledge, M. Exploring Free-Energy Landscapes of Intrinsically Disordered Proteins at Atomic Resolution Using NMR Spectroscopy. *Chem. Rev.* **2014**, *114*, 6632–6660. [CrossRef] [PubMed]

82. Levine, Z.A.; Larini, L.; LaPointe, N.E.; Feinstein, S.C.; Shea, J.-E. Regulation and aggregation of intrinsically disordered peptides. *Proc. Natl. Acad. Sci. USA* **2015**, *112*, 2758–2763. [CrossRef] [PubMed]

83. Tolkatchev, D.; Plamondon, J.; Gingras, R.; Su, Z.; Ni, F. Recombinant Production of Intrinsically Disordered Proteins for Biophysical and Structural Characterization. In *Instrumental Analysis of Intrinsically Disordered Proteins*; John Wiley & Sons, Inc.: Hoboken, NJ, USA, 2010; pp. 653–670.

84. Dunker, A.K.; Lawson, J.D.; Brown, C.J.; Williams, R.M.; Romero, P.; Oh, J.S.; Oldfield, C.J.; Campen, A.M.; Ratliff, C.M.; Hipps, K.W.; et al. Intrinsically disordered protein. *J. Mol. Graph. Model.* **2001**, *19*, 26–59. [CrossRef]

85. Vacic, V.; Uversky, V.N.; Dunker, A.K.; Lonardi, S. Composition Profiler: A tool for discovery and visualization of amino acid composition differences. *BMC Bioinform.* **2007**, *8*, 211. [CrossRef] [PubMed]

86. Uversky, V.N.; Gillespie, J.R.; Fink, A.L. Why are "natively unfolded" proteins unstructured under physiologic conditions? *Proteins* **2000**, *41*, 415–427.

87. Cilia, E.; Pancsa, R.; Tompa, P.; Lenaerts, T.; Vranken, W.F. From protein sequence to dynamics and disorder with DynaMine. *Nat. Commun.* **2013**, *4*. [CrossRef] [PubMed]

88. Cilia, E.; Pancsa, R.; Tompa, P.; Lenaerts, T.; Vranken, W.F. The DynaMine webserver: Predicting protein dynamics from sequence. *Nucleic Acids Res.* **2014**, *42*, W264–W270. [CrossRef]

89. Uversky, V.N. Size-exclusion chromatography in structural analysis of intrinsically disordered proteins. *Methods Mol. Biol.* **2012**, *896*, 179–194. [CrossRef]

90. Schuck, P. Size-distribution analysis of macromolecules by sedimentation velocity ultracentrifugation and lamm equation modeling. *Biophys. J.* **2000**, *78*, 1606–1619. [CrossRef]

91. Greenfield, N.; Fasman, G.D. Computed circular dichroism spectra for the evaluation of protein conformation. *Biochemistry* **1969**, *8*, 4108–4116. [CrossRef]

92. Kikhney, A.G.; Svergun, D.I. A practical guide to small angle X-ray scattering (SAXS) of flexible and intrinsically disordered proteins. *FEBS Lett.* **2015**, *589*, 2570–2577. [CrossRef] [PubMed]

93. Kachala, M.; Valentini, E.; Svergun, D.I. Application of SAXS for the Structural Characterization of IDPs. In *Intrinsically Disordered Proteins Studied by NMR Spectroscopy*; Springer: Cham, Switzerland, 2015; pp. 261–289.

94. Konrat, R. NMR contributions to structural dynamics studies of intrinsically disordered proteins. *J. Magn. Reson.* **2014**, *241*, 74–85. [CrossRef] [PubMed]

95. Solyom, Z.; Schwarten, M.; Geist, L.; Konrat, R.; Willbold, D.; Brutscher, B. BEST-TROSY experiments for time-efficient sequential resonance assignment of large disordered proteins. *J. Biomol. NMR* **2013**, *55*, 311–321. [CrossRef] [PubMed]

96. Kosol, S.; Contreras-Martos, S.; Cedeño, C.; Tompa, P. Structural Characterization of Intrinsically Disordered Proteins by NMR Spectroscopy. *Molecules* **2013**, *18*, 10802–10828. [CrossRef] [PubMed]

© 2019 by the authors. Licensee MDPI, Basel, Switzerland. This article is an open access article distributed under the terms and conditions of the Creative Commons Attribution (CC BY) license (http://creativecommons.org/licenses/by/4.0/).

International Journal of
Molecular Sciences

Review

The Significance of the Intrinsically Disordered Regions for the Functions of the bHLH Transcription Factors

Aneta Tarczewska and Beata Greb-Markiewicz *

Department of Biochemistry, Faculty of Chemistry, Wroclaw University of Science and Technology, Wybrzeże Wyspiańskiego 27, 50-370 Wroclaw, Poland; aneta.tarczewska@pwr.edu.pl
* Correspondence: beata.greb-markiewicz@pwr.edu.pl

Received: 30 September 2019; Accepted: 22 October 2019; Published: 24 October 2019

Abstract: The bHLH proteins are a family of eukaryotic transcription factors regulating expression of a wide range of genes involved in cell differentiation and development. They contain the Helix-Loop-Helix (HLH) domain, preceded by a stretch of basic residues, which are responsible for dimerization and binding to E-box sequences. In addition to the well-preserved DNA-binding bHLH domain, these proteins may contain various additional domains determining the specificity of performed transcriptional regulation. According to this, the family has been divided into distinct classes. Our aim was to emphasize the significance of existing disordered regions within the bHLH transcription factors for their functionality. Flexible, intrinsically disordered regions containing various motives and specific sequences allow for multiple interactions with transcription co-regulators. Also, based on in silico analysis and previous studies, we hypothesize that the bHLH proteins have a general ability to undergo spontaneous phase separation, forming or participating into liquid condensates which constitute functional centers involved in transcription regulation. We shortly introduce recent findings on the crucial role of the thermodynamically liquid-liquid driven phase separation in transcription regulation by disordered regions of regulatory proteins. We believe that further experimental studies should be performed in this field for better understanding of the mechanism of gene expression regulation (among others regarding oncogenes) by important and linked to many diseases the bHLH transcription factors.

Keywords: bHLH; IDP; IDR; LLPS; disorder prediction; LLPS prediction; transcription; phase separation

---

1. Introduction

The bHLH (basic Helix-Loop-Helix) proteins are the important family of transcription factors (TFs) present in all eukaryotes: from yeasts [1,2] and fungi [3] to plants [4] and metazoans [5–10]. All family members contain the HLH domain responsible for dimerization [11]. This domain is usually preceded by a stretch of basic residues which enable DNA binding [12]. The bHLH TFs recognize tissue-specific enhancers containing E-box sequences which regulate expression of a wide range of genes involved in cell differentiation and development [13].

Currently, a few independent classification systems of the bHLH proteins exists: evolutionary classification based on the phylogenetic studies of the bHLH proteins, which classify the bHLH family members into six A-F classes [7,8,14], and a new one based on the complete amino acid sequence analyses, classifying the bHLH proteins into six clades without assumptions about gene function [15]. Contrary to the previous methods, natural method of classification proposed by Murre [12], which divides the bHLH proteins into seven classes, is based on the presence of additional domains, expression patterns and performed transcriptional function [10]. For purposes of clarity, some attempts to revise and systematize different classification systems were undertaken [16]. In this review we present classification of bHLH proteins according to Murre [12], with some short description of presented classes (Table 1).

Table 1. Classification of bHLH proteins based on [5,7,8,10,12,14,16].

Structural Motif Dimerization	Representative Members	Short Description
class I (E proteins)/ group A		
bHLH, homo- and heterodimerization	Vertebrate: E12, E47 [17], HEB [18,19], TCF4 [20] Invertebrate: Daughterless	transcription activators, ubiquitous expression, neurogenesis, immune cell development, sex development, gonadogenesis
class II/ group A		
bHLH, preferred heterodimerization with class I partners	Vertebrate: MYOD, Myogenin, MYF5-6, Ngn1-3, ATOH, NeuroD, NDRF, MATH, MASH, ASCL1 [21], TAL1/SCL [22], OLIG1-3 [23] Invertebrate: TWIST [24], AS-C	transcription activators, tissue specific expression, muscle development, neuro-genesis, generation of autonomic and olfactory neurons, development of granule neurons and external germinal layer of cerebellum, oligodendrocyte development, specification of blood lineage and maturation of several hematopoietic cells, pancreatic development
class III/ group B		
bHLH-LZ	Vertebrate: MYC [25], USF, TFE3, SREBP1-2 *Drosophila*: MYC Plants: MYC2	transcription activators/represors, oncogenic transformation, apoptosis, cellular differentiation, proliferation, cholesterol-mediated induction of the low-density lipoprotein receptor, jasmonate signaling (plants)
class IV/ group B		
bHLH, heterodimerisation with each other and MYC proteins	Vertabrate: MAD, MAX [26], MXI1 *Drosophila*: MNT, MAX	transcription regulators lacking transactivation domain (TAD)
class V/ group D		
HLH (no basic region)	Vertebrate: ID1-4 [27] Invertebrate:EMC	negative transcription regulators of class I and II (group A) proteins, no DNA binding, regulation by sequestration.
class VI/ group B		
bHLH-O, (presence of proline in basic region)	Vertebrate: HES, HEY1-3 [28], STRA13, HERP1-2 [29] *Drosophila*: HAIRY [30], E(spI)	negative transcription regulators interacting with corepressors (Groucho); neurogenesis, vasculogenesis, mesoderm segmentation, myogenesis, T lymphocyte development, cardiovascular development and homeostasis; effectors of Notch signalling [28]; in *Drosophila*: regulation of differentiation, anteroposterior segmentation and sex determination
class VII/ group C - subclass I		
bHLH-PAS, heterodimerization with subclass II	Vertebrate: AHR [31], HIF1-3α [32], SIM1-2 [33], CLOCK [34], NPAS1-4 [35–39] *Drosophila*: MET [40], GCE, SIMA, TRH	transcription regulation in response to physiological and environmental signals: xenobiotics, hypoxia, development, circadian rhytms
class VII/ group C - subclass II		
bHLH-PAS, homo- and heterodimerization with subclass I	Vertebrate: ARNT [41], ARNT2, BMAL1, BMAL2 *Drosophila*: TANGO, CYCLE	general partners for subclass I bHLH-PAS proteins

Both class I (known as E proteins) and class II of the bHLH TFs do not possess domains additional to the bHLH. Contrary to the class I which is expressed in many tissues, the class II proteins expression is tissue specific. Members of the class II are dimerization partners for the class I transcription factors. Class III comprises proteins possessing Leucine-zipper (LZ) motif in addition to the bHLH. Important members of the class III are proteins belonging to the Myc subfamily, which regulate oncogenic transformation, apoptosis, and cellular differentiation. To class IV belong MAD and MAX which can dimerize with MYC and regulate its activity. Also, MAD/MAX are able to create homo- and heterodimers with each other. Although these TFs do not possess transcription activation domain (TAD), MAD/MAX dimers can influence the transcription in a differentiated way. Class V contains transcriptional inhibitors ID1-3 which are not able to bind DNA and act by the other bHLH proteins sequestration. Interestingly, the fourth member of this class- ID4 function as inhibitor of ID1-3 [42]. Class VI comprise proteins containing additional Orange domain adjacent C-terminally to the bHLH domain (bHLH-O). Transcription factors from the described classes perform regulatory function in various developmental processes including cells differentiation and maintaining pluripotency. For this reason they are often linked to cancer development. Class VII comprise transcription factors which possess PAS (Period-Aryl hydrocarbon receptor nuclear translocator-Single minded) domain located C-terminally to the bHLH domain. PAS domain is crucial for the bHLH-PAS proteins specifity [43]. Structurally, the C-terminal PAS domain is often associated with PAC (C-terminal to PAS) motif [44,45]. bHLH-PAS transcription factors are responsible for sensing environmental signals like the presence of xenobiotics (AHR), hypoxia (HIF) or setting of circadian rhythms of organism (CLOCK, CYCLE, BMAL). The members of subclass II of bHLH-PAS TFs -ARNT proteins are general dimerization partners of the subclass I members.

2. The Role of the bHLH Proteins in Transcription

The regulation of genes expression by multiple transcription factors, cofactors and chromatin regulators establish and maintains a specific state of a cell. Inaccurate regulation of transmitted signals can results in diseases and severe disorders [46]. Therefore, transcription requires balanced orchestration of adjustable complexes of proteins. A key regulator of transcription is Mediator, a multi-subunit Mediator complex which interacts with RNA polymerase II (Pol II), and coordinates the action of numerous co-activators and co-repressors [47–50]. Function of the Mediator is conserved in all eukaryotes, though, the individual subunits have diverged considerably in some organisms [51,52].

Up to date, for some bHLH family representatives, interactions with subunits of the Mediator and/or chromatin remodeling histone acetyltransferases/deacyltransferase, were reported. In plants, the Mediator complex is a core element of transcription regulation important for their immunity [53]. It was shown, that in *Arabidopsis thaliana* important jasmonate signaling and resistance to fungus *Botrytis cinerea*, is dependent on the interaction between MED25 subunit of the Mediator and MYC2 [54–56], and interaction of MED8 subunit of the Mediator with FAMA belonging to the bHLH family [57]. Sterol regulatory element binding proteins (SREBPs) the class II bHLH TFs (Table 1) are transcription activators critical for regulation of cholesterol and fatty acid homeostasis in animals. It was shown that human SREBPs bind CBP/p300 acetyltransferase [58] and MED15 subunit of the Mediator to activate target genes [59]. Also yeast Ino2 was shown to bind MED15 subunit of the Mediator tail [60].

The representative of class II TFs TAL1 (Table 1) is required for the specification of the blood lineage and maturation of several hematopoietic cells. TAL1/SCL is considered as a master TF delineating the cell fate and the identity of progenitor and normal hematopoietic stem cells (HSCs). It regulates other hematopoietic TFs thus has a potential for cell reprogramming [22]. TAL1 also binds CBP/p300 acetyltransferase [61,62]. Similarly MyoD—a myogenic regulatory factor which controls skeletal muscle development binds CBP and recruits histone acetyltransferase to activate myogenic program [63]. Cao et al. showed that of MyoD modify the myoblasts chromatin structure and accessibility [64]. ASCL1 (class II, Table 1) was shown to be a pioneer factor which promotes chromatin accessibility and enables chromatin binding by others TFs [65]. Recently, also AHR (bHLH-PAS, Table 1) was suggested to be a pioneer factor which regulates DNA methylation during embryonic developments in unknown way [66]. In clear cell renal cell carcinoma (ccRCC), the most frequent mutation causes the von Hippel-Lindau (VHL) tumor suppressor inactivation leading to genome-wide enhancer and super-enhancer remodeling. This process is mediated by the interaction of HIF2α and HIF1β (bHLH-PAS, Table 1) with histone acetyltransferase p300 [67]. CLOCK, the other bHLH-PAS subfamily member (Table 1) was shown to mediate histone acetylation in a circadian time-specific manner [68].

Interestingly, the bHLH-O proteins members (class VI, Table 1) HEY proteins can function as transcription repressors as well as transcription activators. They were shown to bind directly DNA and interact with histone deacetylases and other TFs [28,69]. On the other hand, gene activation by HEY is regulated in an indirect way. Multiple HEY binding sites located downstream and close to the transcriptional start site, resulted in a hypothesis that HEY influence the pausing/elongation switch of Pol II [70]. Interestingly, though most of TFs stimulate transcription initiation, MYC (class III, Table 1) was shown to stimulate transcription elongation by recruitment of the elongation factor [71]. The presented studies indicate that the crucial role of the bHLH proteins in maintaining transcriptional regulation of important developmental (e.g., cell differentiation) and oncogenic pathways is dependent on the multiple interactions with basal transcriptional machinery.

3. The bHLH Transcription Factors as IDPs

Intrinsically disordered proteins (IDPs) discovered in 1990s obliterate the paradigm derived from Anfinsen's work, stating that functional proteins must possess a well-defined, ordered, three dimensional structure [72]. Currently it is known, that a large number of proteins is perfectly functional or even multifunctional in a disordered state in which a polypeptide chain undergoes rapid conformational fluctuations [73–76]. Intrinsic disorder can be spread throughout the whole polypeptide chain, or it can be limited to intrinsically disordered regions (IDRs) of various length, which are accompanied by well folded domains [77]. The unique properties of disordered proteins originate from their unusual amino acids composition [78]. IDPs/IDRs are depleted in order promoting amino acid residues (hydrophobic, aromatic, aliphatic side chains). In contrast, they possess unusually high content of charged and hydrophilic amino acid residues [79–81]. As a consequence, disordered polypeptide chains have extremely high net charge and low hydrophobicity [82]. IDPs are pliable and highly dynamic molecules of interconvertible conformations. They may completely or almost completely lack the regular secondary structures. However, the content of secondary structure may also be quite significant and molecules can exist in a molten globule state [83–85]. Various in silico analyses indicated that the proportion of disordered proteins is drastically higher in eukaryotes comparing to prokaryotes [86]. This disproportion reflect the complexity of signaling pathways in which IDPs/IDRs play a crucial role [87]. Due to the flexible and dynamic nature, IDPs/IDRs can form fuzzy complexes, adopting various conformations [88]. According to this, one IDP can form multiple interactions with various partners. Due to a large accessibility of particular residues in a disordered chain, the interaction pattern can be easily modified by posttranslational modifications [89]. For that reason IDPs/IDRs often serve as molecular hubs, modulators and sensors of cellular signals [85].

bHLH TFs are responsible for a control of developmental processes like retinal development, proliferation of progenitors, neurogenesis and gliogenesis. Importantly, this is due to a direct interaction between bHLH TFs and interaction of bHLH TFs with homeodomain factors which create complexes that bind to the specific promoters [90,91]. Transcription of muscle-specific genes during skeletal muscle development is also dependent on the interactions between specific bHLH TFs: MyoD, Myogenin, Myf5 and MRF4 with ubiquitously expressed bHLH E-proteins (E12, E47, TCF4, HEB). Interestingly, it was shown that MyoD interacts with two isoforms of HEB: HEBα and HEBβ. which regulate differentially transcriptional activity of MyoD not only on different, but also on the same promoter [92]. Also interesting is the ability of ID4 to recruit multiple ID proteins to assemble higher order complexes. ID4 restores DNA binding by E47 protein even in the presence of repressing ID1 and ID2. Additionally, the ID proteins can interact with non-bHLH partners expanding regulatory network of ID4 [42]. As a consequence, the ID proteins are proposed as a 'hub' for coordination of multiple cancer events [27]. These examples illustrate the possibility of bHLH TFs to interact with many partners in differentiated way. We suggest that these is related to the disordered character of the bHLH proteins. This hypothesis is substantiated by some experimental studies. Neurogenic bHLH transcriprion factor Neurogenin 2 (Ngn2) was shown to possess long IDR which phosphorylation regulates the activity of the protein [93]. Interestingly, though the bHLH domain was considered as a stable, well ordered structure, partially disordered character of this domain was presented for NeuroD [94], MYC and MAX [95]. We performed in silico analyses to predict the presence of intrinsic disorder and get an insight into the degree of flexibility of bHLH proteins representing all established classes (see Table 1): hHEB (class I), hMYOD (class II), hMYC and atMYC2 (class III) (Figure 1); hMAD1 and hMAX (class IV), hID4 (class V), hHES (class VI) (Figure 2); hAHR, hHIF-1α, hCLOCK and hARNT (class VII) (Figure 3). We used PONDR-VLXT [96,97], http://www.pondr.com/ for the disorder prediction and DynaMine [98,99], http://dynamine.ibsquare.be/submission/ for prediction of the flexibility of proteins backbone.

A representative of the class I, human HEB shows a high content of predicted as disordered and flexible sequences. The only highly ordered/rigid region appears between 577–630 aa which comprise the bHLH domain (Figure 1A). Based on prediction results, we assume HEB as IDP. Also hMyoD, the class II TFs presents a high content of flexible IDRs especially in the C-terminal part of the protein (Figure 1B). As the representatives of the class III we have chosen hMYC (Figure 1C) (for which partial disorder of the bHLH domain was experimentally documented [95]) and *Arabidopsis thaliana* MYC2 (Figure 1D). For both proteins the presence of flexible IDRs was predicted, though they locations were different.

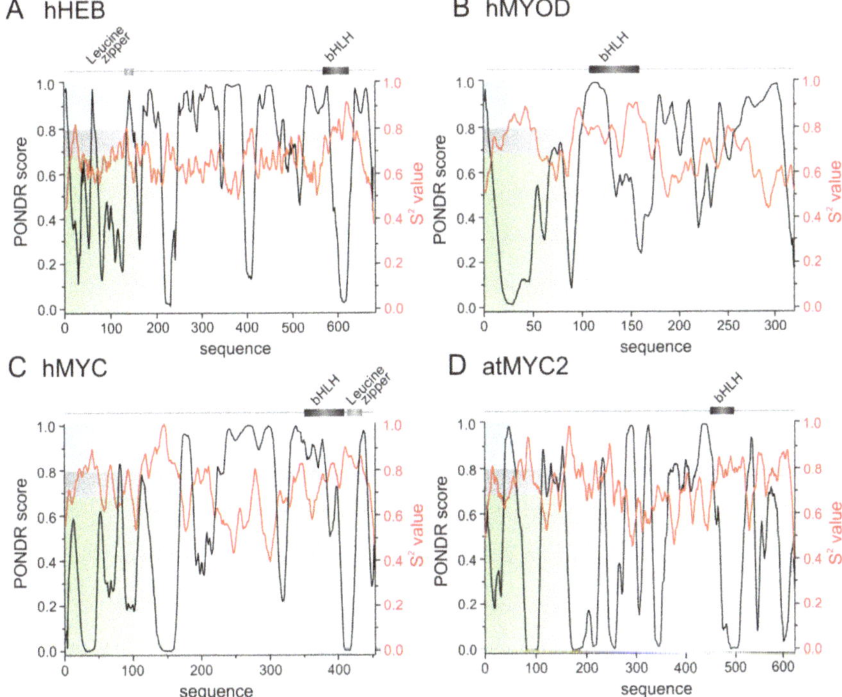

Figure 1. Prediction of intrinsically disordered regions. The top panel presents the domain structure of the analyzed bHLH proteins. Dark grey rectangle indicates the position of bHLH domain, the light grey Leucine zipper. The bottom panel presents a prediction of intrinsically disordered and flexible regions based on the amino acid sequence of proteins. Prediction were performed using PONDR-VLXT (left Y axis) and DynaMine (right Y axis) software. For PONDR prediction, a score above 0.5 indicates disorder. For DynaMine, a S^2 value above 0.8 (blue zone) indicates rigid conformation, 0.69-0.8 (grey zone) is context dependent and a value below 0.69 (green zone) indicates flexible conformation. (**A**) class I human HEB [Q99081], (**B**) class II human MYOD [P15172], (**C**) class III human MYC [P01106-2] and (**D**) *Arabidopsis thaliana* MYC2 [Q39204].

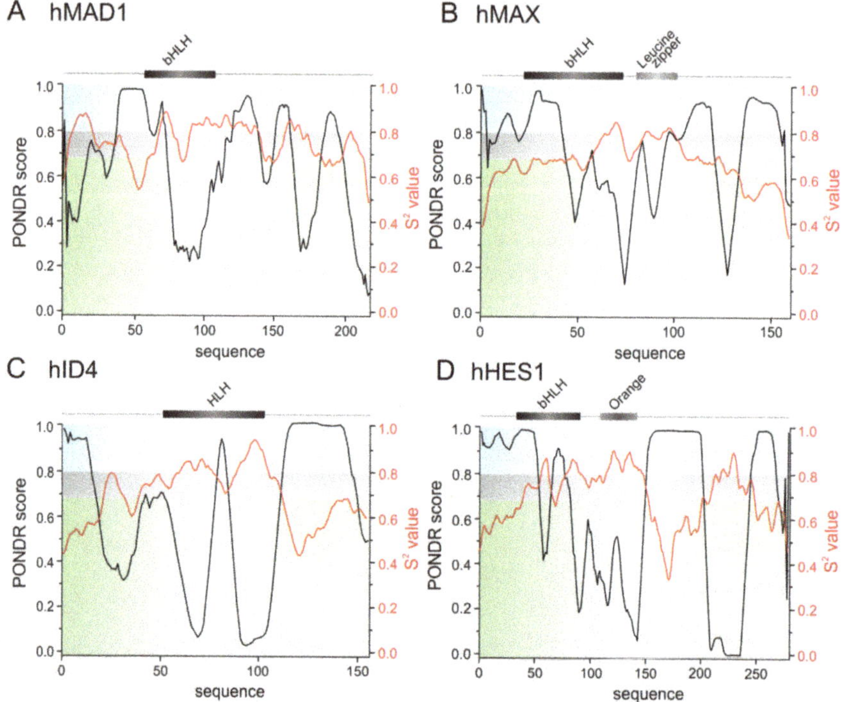

Figure 2. Prediction of intrinsically disordered regions. The top panel presents the domain structure of the analyzed bHLH proteins. Dark grey rectangle indicates the bHLH domain, light grey indicates Leucine zipper or Orange domain. The bottom panel presents a prediction of intrinsically disordered and flexible regions, based on the amino acid sequence of proteins. Predictions were performed using PONDR-VLXT (left Y axis) and DynaMine (right Y axis) software. For PONDR prediction, a score above 0.5 indicates disorder. For Dynamine, a S^2 value above 0.8 (blue zone) indicates rigid conformation, 0.69–0.8 (grey zone) is context dependent and a value below 0.69 (green zone) indicates flexible conformation. (**A**) class IV human MAD [Q9Y6D9] and (**B**) human MAX [P61244], (**C**) class V human ID4 [P47928], (**D**) class VI human HES1 [Q14469].

The representative of the class IV, human MAD1 also shows high content of predicted as disordered and flexible sequences (Figure 2A). Interestingly IDRs of hMAX which belongs to the same class IV are located in the N- and C- protein termini, while the middle part is predicted as possessing more rigid structure (Figure 2B). Also, ID4 belonging to the class V of transcriptional inhibitors presents flexible IDR in the C-terminal part of protein and a shorter one in the N-terminal part (Figure 2C). In addition to similarly located the N- and C-terminal IDRs in the class VI member, human HES1 analysis shows high flexibility/disorder in the central part of protein (Figure 2D).

The class VII proteins comprise the bHLH-PAS subfamily, which additionally to the bHLH domain possess a PAS domain responsible for ligands and co-factors binding. Importantly, their C-termini are usually responsible for the regulation of the protein and created complexes activity [100]. Human AHR, HIF1-α, and CLOCK belong to the subclass I of specialized factors, while human ARNT (the subclass II) is one of the general partners which dimerize with the subclass I proteins and is important for their activity. In contrast to the hAHR, for which relatively short IDRs were predicted within the middle, the N- and the C-terminal part of the protein (Figure 3A), other bHLH-PAS members contain longer IDRs which comprise most of the C-terminal half of proteins and are predicted as highly flexible (hHIF-1α, Figure 3B; hCLOCK, Figure 3C; hARNT, Figure 3D).

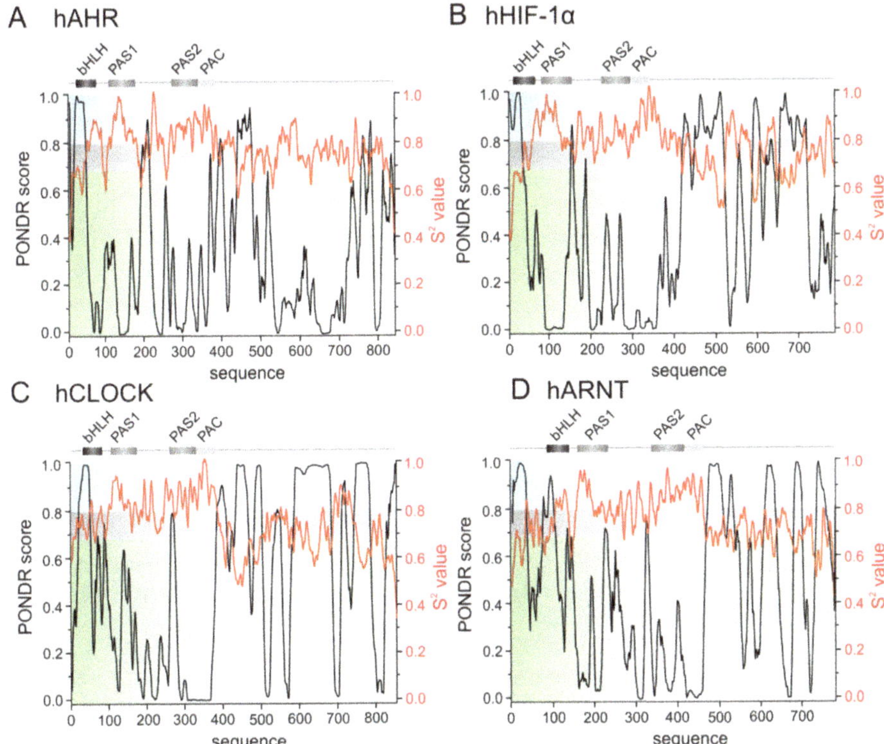

Figure 3. Prediction of intrinsically disordered regions of the class VII bHLH-PAS proteins. The top panel presents the domain structure of the analyzed bHLH–PAS proteins. Dark grey rectangle indicates the bHLH domain, light grey indicates PAS/PAC domains. The bottom panel presents a prediction of intrinsically disordered and flexible regions based on the amino acid sequence of proteins. Prediction were performed using PONDR-VLXT (left Y axis) and DynaMine (right Y axis) software. For PONDR prediction, score above 0.5 indicate disorder. For Dynamine, a S^2 value above 0.8 (blue zone) indicates rigid conformation, 0.69–0.8 (grey zone) is context dependent and a value below 0.69 (green zone) indicates flexible conformation. (**A**) human AHR [P35869], (**B**) human HIF-1α [Q16665], (**C**) human CLOCK [O08785], (**D**) human ARNT [P27540].

To date, the only report, concerning the structure of the full-length bHLH protein is the mentioned study showing Neurogenin as IDP [93]. Based on the presented predictions and our own experience with expression of the selected bHLH proteins (not published), we assume that this is due to the relatively high content of IDRs. This makes overexpression and purification process extremely difficult because of propensity to aggregation and high sensitivity to proteases.

4. The Role of IDPs in Maintaining/Creation of LLPS

Over the last decade, since the pioneering work regarding physical nature of P-bodies was published by Hyman and co-workers [101], many molecular biologists and biophysicists have focused on the significance of spontaneous thermodynamically driven liquid-liquid phase separation (LLPS) in biological systems. LLPS leads to formation of dense, liquid condensates that stably coexist in diluted phase [101,102]. At the molecular level it was shown that LLPS is forced by multiple weak and transient interactions which engage IDPs/IDRs [101,103–106]. Repetitively distributed within IDRs highly charged regions of opposite charges, short motifs such as YG/S-, FG-, RG-, GY-, KSPEA-, SY- and Q/N-rich regions form multivalent interactions between condensate components [107]. A model for the condensate formation and composition proposes that some proteins act as the scaffolds, while others as the clients. The scaffolds are the modular proteins which contain repeated motives that enable heterotypical scaffold-scaffold interaction. As they undergo spontaneous LLPS they are essential for the structural integrity of a condensate [108,109]. Directly interacting sequences called stickers are usually multivalent, whereas the interval sequences which separate stickers, called spacers are responsible for the properties of a condensate [110]. Highly charged and flexible IDRs are in fact frequently identified as scaffolds [108,111]. The clients participate into the condensates by binding to the free, unoccupied scaffold sites [108]. A growing number of evidences indicate that LLPS constitute a fundamental mechanism to compartmentalize the intracellular space. LLPS form the functional centres for biochemical reactions in cytoplasm and membrane-surrounded organelles including nucleus.

The structural and functional organisation of the interior of the nucleus was believed to rely solely on the rigid insoluble nuclear matrix [112]. The rich in A and T DNA sequences known as scaffold/matrix associated regions (S/MARs) attach to nuclear matrix and organise chromatin into higher-order structures which comprise distinct loops and functional units attached to the matrix [113]. That concept is now giving way to a new concept, were dynamic, spontaneously formed condensates, such as nucleolus, splicing speckles, Cajal bodies, PML bodies are the key structural and functional components of the nuclear interior. The barrier-free character of liquid condensates allows for rapid exchange of their components with surrounding so they form an ideal environment for biochemical reactions. On the other hand, nuclear condensates have a stable inert, well-defined structure and can be purified by biochemical methods [114]. It was shown, that the concentration of nucleolar components is close to saturation [115]. It means that small changes in the nucleus can drive spontaneous LLPS. In fact association/dissociation events of nuclear condensates regulate many processes related to gene expression [116] including chromatin structure organisation [117], RNA processing [118], ribosome biogenesis [119]. Importantly, LLPS was shown to be involved in formation of some functional condensates that regulate genes transcription [76,120–122].

5. The Transcription Regulation and LLPS

The genes transcription process require tight regulation to ensure physiological balance of the cell. Knowledge regarding the mechanism of transcription is quite advanced, however some aspects of regulation remains unexplored. Recent findings indicate that regulatory mechanism may tightly depends on the spontaneous LLPS. Transcription of tissue specific gene is initiated at the specific genome regions called super-enhancers (SE). SE first described in embryonic stem cells (ESC) [123] are dense multicomponent assemblies different from typical enhancers [124]. Recently Hnisz [125] performed computational simulation to obtain the probable explanation for typical features of SE. Simulations led to conclusion that formation, activity and unique properties of SE such as sensitivity to concentration of its components, sensitivity to posttranslational modifications, extremely high frequency bursting [126–128] may originate from the fact that SE are liquid condensates assembled/disassembled via spontaneous LLPS [125]. Hnisz and co-workers were the first who point connection and strong dependence between the regulation of transcription initiation at SE and LLPS. Although not experimentally proven, the model serves as the conceptual framework for further research. Recently, Sabari et al. [121] showed that largely disordered BRD4 and MED1 subunit of the Mediator are in close spatial proximity to one another within SE in murine ESC and co-localised puncta show characteristic features of phase separated condensates Moreover, MED1 condensates can incorporate BRD4 and Pol II from nuclear extract [121]. MED1 subunit interacts also with other major pluripotency TFs e.g., OCT-4 [129] and estrogen receptor (ER) [130] forming liquid-like puncta at SE of the key pluripotency genes [121,122]. MED1 condensates depends on the OCT-4 occupancy [122], which are crucial for initiation of tissue specific genes transcription at SE [122,131]. In vitro analyses pointed that formation of MED1-OCT4 liquid condensates occurs via the electrostatic interactions and involves acidic residues enriched in disordered activation domain of the OCT-4 [122]. Interestingly, ER interact with the MED1 subunit by LXXLL motif [132] which is located in the ordered ligand binding domain.

This interaction is regulated by estrogen what means that not only disordered-disordered regions interaction but also disordered-ordered regions interactions play a role in transcription regulation forced by LLPS [122]. Wu et al. [120] showed that largely disordered transcription co-activator TAZ protein forms liquid condensates in vitro and *in vivo*. TAZ condensates compartmentalize DNA binding cofactor TEAD4 and other components of transcription initiation machinery including BRD4, MED1 and CDK9. Importantly, deletion mutant, that is not able to undergo spontaneous LLPS cannot initiate transcription though is able to bind TAZ partners such TEAD4.

Importantly, there are some evidences that not only the initiation, but also the elongation of transcription depends on LLPS. For the transcription elongation essential is hyper-phosphorylation of the YSPTSPS consensus sequence which is repeated multiple times in the disordered C-terminal domain (CTD) of Pol II [133–136]. pTEFb which begins the elongation phase consists of CDK9 kinase associated with cyclin T1 (CycT1). Lu with co-workers [76] concentrated on the function of the lengthy C-terminal IDR of CycT1 in regulation of CDK9 activity. They revealed that a histidine-rich domain (HRD) located in the IDR of CycT1 (residues 480–550) is directly involved in the regulation of the kinase activity [76]. Interestingly, HRD is present also in some other kinases, for example Dyrk1A which phosphorylates CTD of Pol II. Importantly, a homologues kinase Dyrk3 was shown to be responsible for disassembly of stress granules [137] and other cellular condensates during cell division [138]. In vitro studies using a set of recombinant IDRs of the CycT1 and Dyrk1A revealed that the regions can undergo phase separation in a HRD dependent manner. HRD was shown to form condensates which compartmentalize the kinases and the substrate what enables efficient reactions resulting in the hyper-phosphorylation of the CTD of Pol II [76]. Interestingly, the CTD of Pol II can undergo spontaneous LLPS in vitro only in a non-phosphorylated state. The weak CTD-CTD interaction keeps the enzymes molecules in hubs within nucleoplasm. Phosphorylation change the interaction pattern allowing CTD to engage in new multivalent interactions with selected partners [139]. These results indicate that LLPS allows for the condensation of cofactors, that in turn triggers posttranslational modifications leading to the reorganization of the condensate components. Pol II escapes from the promoter site and enables the entry into active elongation stage [76].

Currently not much is known about proteins responsible for formation of the condensates which are important for transcription regulation. The question still remains unanswered which proteins are the scaffolds and which are the clients. Importantly, also not much is known about the involvement of the bHLH TFs in the LLPS process, though they are key players involved in many important cell differentiation and organisms development pathways. As we discussed in previous section, bHLH proteins possess long IDRs which could interact with different partners and be engaged in LLPS. This hypothesis is substantiated by an experimental verification of MyoD possibility to create LLPS [122], and discussed in previous section possibility of some bHLH TFs to interact with the Mediator subunits or other elements of the mechanism which modifies the chromatin accessibility. Interestingly, regulation of circadian clock by BMAL1 comprises binding of CBP, which occurs in discrete nuclear foci. This led to a hypothesis that formation of nuclear bodies containing BMAL1/CBP provides transcriptionally active sites of target genes, like *Per1-2* [34]. Taking the above into consideration, we asked the question if the ability to undergo LLPS is a more general property of the bHLH TFs. As we got positive results for the previously performed prediction of disorder, which was shown to be important for LLPS initiation [76,121,122], we decided to perform in silico analyses to predict if members of the bHLH family comprise putative sequences able to create liquid condensates. We used catGranule program, (http://service.tartaglialab.com/update_submission/216885/dd56e32a89) for computational analyses of the putative propensity to undergo LLPS [140] for the bHLH proteins representing all established classes (see Table 1). Prediction results showed that hHEB (class I), hMyoD (class II), hMYC and 84atMYC2 (class III) (Figure 4) contain sequences with a positive score of propensity to LLPS formation. Interestingly, proteins from the class IV regulators which do not possess TAD: hMAD1 and hMAX, similarly like transcription repressors: hID4 (class V) and hHES (class VI) present very low or even negative score within the whole protein sequence (Figure 5). bHLH-PAS transcription factors representing the class VII, hAHR, hHIF-1α, hCLOCK and hARNT were predicted as containing some sequences with high propensity score (Figure 6). Especially interesting is the observation that the transcription repressors show a very low propensity score to undergo LLPS in contrast to the transcription activators such as hHEB or atMYC2. It is possible that the bHLH repressors inhibit transcription by preventing spontaneous phase separation required to form a complete initiation complex. This hypothesis is substantiated by the observation for TAZ mutants [120], discussed in the previous section.

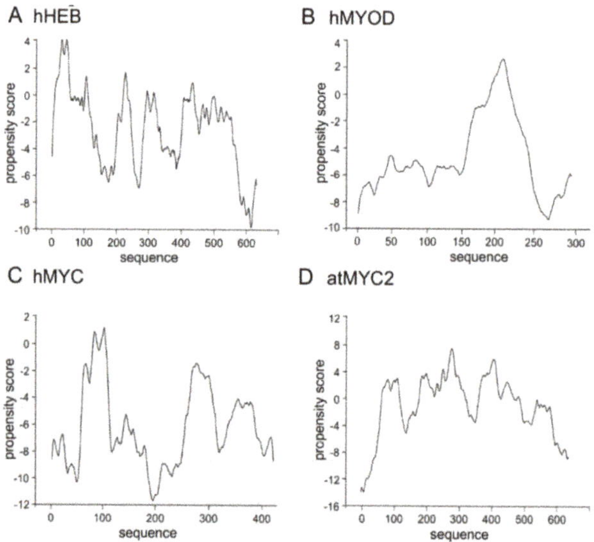

Figure 4. Prediction of propensity of LLPS formation. (**A**) class I human HEB [Q99081], (**B**) class II human MYOD [P15172], (**C**) class III human MYC [P01106-2] and (**D**) *Arabidopsis thaliana* MYC2 [Q39204].

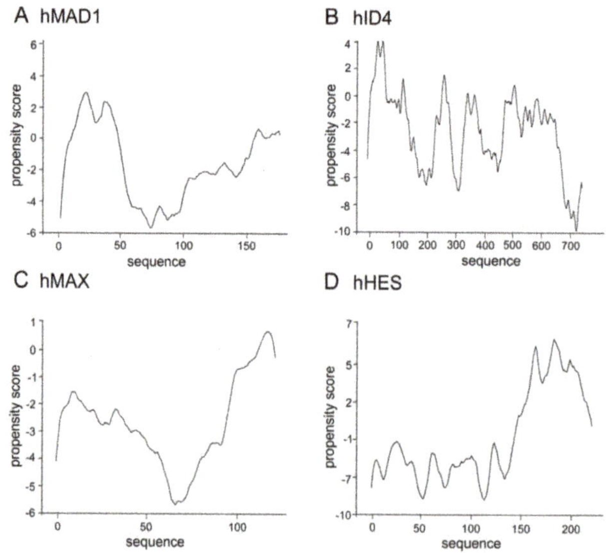

Figure 5. Prediction of propensity of LLPS formation. (**A**) class IV human MAD [Q05195] and (**B**) human MAX [P61244], (**C**) class V human ID4 [P47928], (**D**) class VI human HES1 [Q14469].

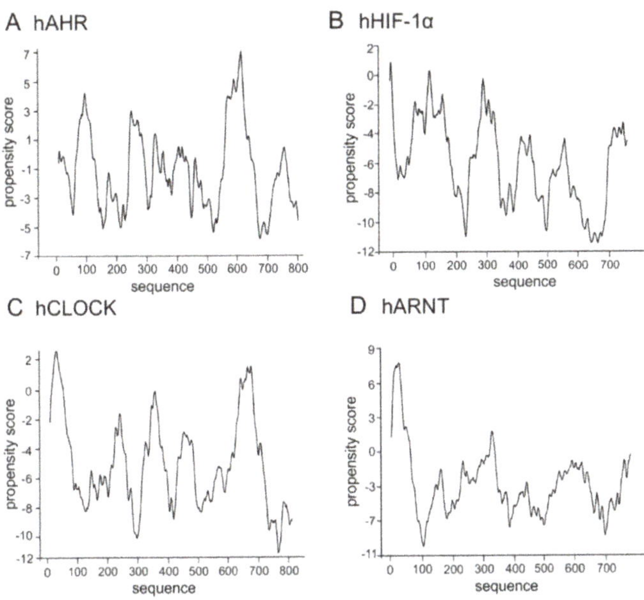

Figure 6. Prediction of propensity of LLPS formation for bHLh-PAS proteins. (**A**) human AHR [P35869], (**B**) human HIF-1α [Q16665], (**C**) human CLOCK [O08785], (**D**) human ARNT [P27540].

As the range of the propensity score is not determined precisely, as a control we performed catGranule prediction for proteins known to create LLPS: nucleophosmin (Figure 7A) and estrogen receptor (Figure 7B) which are deposited in the recently published PhaSePro database (https://phasepro. elte.hu) [141].

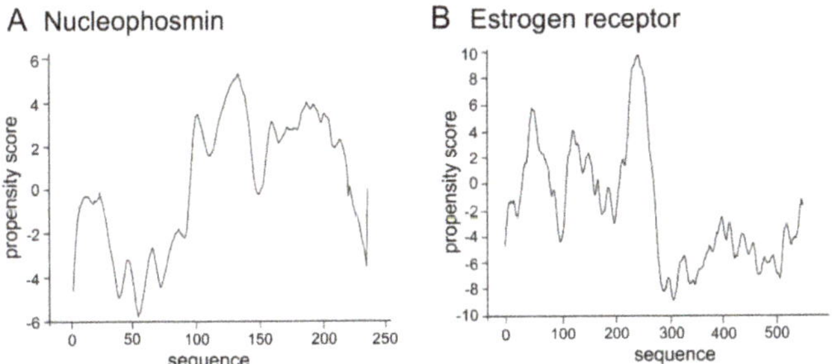

Figure 7. Prediction of propensity of LLPS formation for representative LLPS-enabled proteins. (**A**) nucleophosmin [P06748], (**B**) estrogen receptor [P03372].

Results of performed in silico analyses in comparison to the control show that the selected bHLH proteins have regions that might be involved in multivalent interaction leading to formation of liquid condensates. What would be their role in condensates formation and how would mutations and wrong dimerization/interaction influence formation of the bHLH TFs containing condensate remains a puzzle, however we believe that such an important family of TFs engaged in the crucial pathways and related to many severe disorders like cancer should be the subject of research in this field.

6. Concluding Remarks and Future Perspectives

In eukaryotic cells, regulation of transcription is a dynamic process which requires very precise temporal and spatial coordination of proteins assembling functional complexes. The bHLH family comprises a large group of TFs which utilize conserved DNA binding domain to interact with DNA, but also additional, often disordered domains and motives that allows formation of complex interacting network with various transcription co-factors. It is possible that flexible disordered regions of the bHLH proteins play a role in formation of liquid condensates via LLPS and contribute in this way to regulation of transcription process. Up to date however, there is a lack of experimental evidences. Also recently published PhaSePro database for LLPS does not contain any bHLH TF [141]. We believe that this is due to difficulties with the experimental studies of the bHLH proteins mentioned previously and we expect that some bHLH proteins will be appended in future.

Presented in the previous section predictions may give a hint about the link between LLPS by the bHLH proteins and transcription regulation. This raise a question about functional relevance of this discrepancy between family members. An interesting observation is the predicted low propensity score to form LLPS in the case of transcriptional repressors in contrast to proteins acting as activators. This raise a question about the functional relevance of this discrepancy between family members. Importantly, connection between LLPS and transcription regulation is not limited to the direct interaction between transcription regulators at the active transcription sites. LLPS form nuclear bodies, that maintain, store and modify transcription regulators. Examples include nuclear speckles, polyleukemia bodies, nucleolus, histone locus and others [142]. Within LLPS-formed condensates proteins can undergo acetylation/deacetylation or sumoylation, proteasome-dependent degradation and other posttranslational modifications that influence their functionality [143–145]. Importantly, barrier-free character of these phase separated condensates allows shuttling of its component between the condensates and nucleoplasm, and whenever needed molecules can be recruited from these compartments to the active transcriptionally sites. The discovery that LLPS which is well known in polymer chemistry can play an important role in molecular biology has definitely brought us closer to understanding the cell functionality and regulation of fundamental cellular processes such as transcription. However, our understanding and detailed knowledge is still residual. Many important questions regarding a LLPS concept in transcription regulation remain without answer. We do not know, which components drive association/dissociation events at the active sites. Which molecules serves as a scaffold conditioning formation of liquid condensates and which are just clients. How the type of client molecules influence the function of the phase separated condensates? Also, we do not know which factors and in which way alter LLPS leading to the pathological processes. What would be the role of the bHLH TFs in a condensates formation, and how mutations and incorrect dimerization/interaction of these proteins would impact formation and function of condensates? These questions, as well as many other ones await experimental verification. We believe that such important family of transcription factors which is engaged in crucial pathways and related to many severe diseases like cancer and neurodegenerative disorders, should be the subject of further intensive studies.

Author Contributions: A.T. and B.G.-M. wrote the paper.

Funding: The work was supported by a subsidy from the Polish Ministry of Science and Higher Education for the Wroclaw University of Science and Technology, Faculty of Chemistry.

Acknowledgments: The authors apologize to investigators whose contributions were not cited more extensively because of space limitations.

Conflicts of Interest: The authors declare no conflict of interest.

Abbreviations

AHR	Aryl hydrocarbon receptor
AS-C	Achaete scute complex
ARNT	Aryl hydrocarbon receptor nuclear translocator
bHLH	Helix–loop–helix
ccRCC	Clear cell renal cell carcinoma
CLOCK	Circadian locomotor output cycles protein kaput
CTD	C-terminal domain
CycT1	Cyclin T1
EMC	Extramacrochaetae
E(spl)	Enhancer of split
ER	Estrogen receptor
ESC	Embryonic stem cells
GCE	Germ cell-expressed protein
GRO	Groucho
HAT	Histone transacetylase
HIF	Hypoxia-inducible factor
HSCa	Hematopoietic stem cells
HRD	histidine reach domain
ID	Inhibitor of DNA binding
IDPs	Intrinsically disordered proteins
IDRs	Intrinsically disordered regions
LLPS	liquid-liquid phase separation
LZ	Leucine zipper motif
MET	Methoprene-tolerant protein
MXI1	Max interacting protein
Ngn2	Neurogenin
NPAS	Neuronal PAS domain-containing protein
PAS	Period-arylhydrocarbon nuclear translocator-single minded domain
Pol II	RNA polymerase II
SE	Super-enhancer
SIM	Single-minded protein
SIMA	Similar protein
S/MARs	Scaffold/matrix associate regions
SREBP	Sterol-responsive element-binding protein
TAD	Transactivation domain
TAZ	Tafazzin
TFs	Transcription factors
TRH	Tracheless protein
USF	Upstream stimulatory factor
VHL	Von Hippel-Lindau tumor suppressor

References

1. Robinson, K.A. A network of yeast basic helix-loop-helix interactions. *Nucleic Acids Res.* **2000**, *28*, 4460–4466. [CrossRef] [PubMed]
2. Robinson, K.A. SURVEY AND SUMMARY: Saccharomyces cerevisiae basic helix-loop-helix proteins regulate diverse biological processes. *Nucleic Acids Res.* **2000**, *28*, 1499–1505. [PubMed]
3. Sailsbery, J.K.; Atchley, W.R.; Dean, R.A. Phylogenetic analysis and classification of the fungal bHLH domain. *Mol. Biol. Evol.* **2012**, *29*, 1301–1318. [CrossRef] [PubMed]
4. Pires, N.; Dolan, L. Origin and diversification of basic-helix-loop-helix proteins in plants. *Mol. Biol. Evol.* **2010**, *27*, 862–874.
5. Massari, M.E.; Murre, C. Helix-loop-helix proteins: Regulators of transcription in eucaryotic organisms. *Mol. Cell. Biol.* **2000**, *20*, 429–440. [CrossRef]

6. Simionato, E.; Ledent, V.; Richards, G.; Thomas-Chollier, M.; Kerner, P.; Coornaert, D.; Degnan, B.M.; Vervoort, M. Origin and diversification of the basic helix-loop-helix gene family in metazoans: Insights from comparative genomics. *BMC Evol. Biol.* **2007**, *7*, 33. [CrossRef]

7. Ledent, V.; Paquet, O.; Vervoort, M. Phylogenetic analysis of the human basic helix-loop-helix proteins. *Genome Biol.* **2002**, *3*, 1–18. [CrossRef]

8. Wang, Y.; Chen, K.; Yao, Q.; Zheng, X.; Yang, Z. Phylogenetic Analysis of Zebrafish Basic Helix-Loop-Helix Transcription Factors. *J. Mol. Evol.* **2009**, *68*, 629–640.

9. Sailsbery, J.K.; Dean, R.A. Accurate discrimination of bHLH domains in plants, animals, and fungi using biologically meaningful sites. *BMC Evol. Biol.* **2012**, *12*, 154.

10. Murre, C. Helix–loop–helix proteins and the advent of cellular diversity: 30 years of discovery. *Genes Dev.* **2019**, *33*, 6–25.

11. Murre, C.; McCaw, P.S.; Baltimore, D. A new DNA binding and dimerization motif in immunoglobulin enhancer binding, daughterless, MyoD, and myc proteins. *Cell* **1989**, *56*, 777–783. [CrossRef]

12. Murre, C.; Bain, G.; van Dijk, M.A.; Engel, I.; Furnari, B.A.; Massari, M.E.; Matthews, J.R.; Quong, M.W.; Rivera, R.R.; Stuiver, M.H. Structure and function of helix-loop-helix proteins. *Biochim. Biophys. Acta (BBA) Gene Struct. Expr.* **1994**, *1281*, 129–135. [CrossRef]

13. Ephrussi, A.; Church, G.; Tonegawa, S.; Gilbert, W. B lineage–specific interactions of an immunoglobulin enhancer with cellular factors in vivo. *Science* **1985**, *227*, 134–140. [CrossRef] [PubMed]

14. Atchley, W.R.; Fitch, W.M. A natural classification of the basic helix-loop-helix class of transcription factors. *Proc. Natl. Acad. Sci. USA* **1997**, *94*, 5172–5176. [CrossRef] [PubMed]

15. Skinner, M.K.; Rawls, A.; Wilson-Rawls, J.; Roalson, E.H. Basic helix-loop-helix transcription factor gene family phylogenetics and nomenclature. *Differ. Res. Biol. Divers.* **2010**, *80*, 1–8. [CrossRef]

16. Jones, S. An overview of the basic helix-loop-helix proteins. *Genome Biol.* **2004**, *5*, 226. [CrossRef]

17. Wöhner, M.; Tagoh, H.; Bilic, I.; Jaritz, M.; Poliakova, D.K.; Fischer, M.; Busslinger, M. Molecular functions of the transcription factors E2A and E2-2 in controlling germinal center B cell and plasma cell development. *J. Exp. Med.* **2016**, *213*, 1201–1221. [CrossRef]

18. Yi, S.; Yu, M.; Yang, S.; Miron, R.J.; Zhang, Y. Tcf12, A Member of Basic Helix-Loop-Helix Transcription Factors, Mediates Bone Marrow Mesenchymal Stem Cell Osteogenic Differentiation In Vitro and In Vivo. *Stem Cells* **2017**, *35*, 386–397. [CrossRef]

19. Li, Y.; Brauer, P.M.; Singh, J.; Xhiku, S.; Yoganathan, K.; Zúñiga-Pflücker, J.C.; Anderson, M.K. Targeted Disruption of TCF12 Reveals HEB as Essential in Human Mesodermal Specification and Hematopoiesis. *Stem Cell Rep.* **2017**, *9*, 779–795. [CrossRef]

20. Quednow, B.B.; Brzózka, M.M.; Rossner, M.J. Transcription factor 4 (TCF4) and schizophrenia: Integrating the animal and the human perspective. *Cell. Mol. Life Sci.* **2014**, *71*, 2815–2835. [CrossRef]

21. Huang, C.; Chan, J.A.; Schuurmans, C. Proneural bHLH Genes in Development and Disease. In *Current Topics in Developmental Biology*; Academic Press: Cambridge, MA, USA, 2014; Volume 110, pp. 75–127.

22. Tan, T.K.; Zhang, C.; Sanda, T. Oncogenic transcriptional program driven by TAL1 in T-cell acute lymphoblastic leukemia. *Int. J. Hematol.* **2019**, *109*, 5–17. [CrossRef] [PubMed]

23. Choudhury, S. Genomics of the OLIG family of a bHLH transcription factor associated with oligo dendrogenesis. *Bioinformation* **2019**, *15*, 430–438. [CrossRef] [PubMed]

24. Bouard, C.; Terreux, R.; Honorat, M.; Manship, B.; Ansieau, S.; Vigneron, A.M.; Puisieux, A.; Payen, L. Deciphering the molecular mechanisms underlying the binding of the TWIST1/E12 complex to regulatory E-box sequences. *Nucleic Acids Res.* **2016**, *44*, 5470–5489. [CrossRef] [PubMed]

25. Whitfield, J.R.; Beaulieu, M.-E.; Soucek, L. Strategies to Inhibit Myc and Their Clinical Applicability. *Front. Cell Dev. Biol.* **2017**, *5*, 10. [CrossRef] [PubMed]

26. Amati, B.; Land, H. Myc-Max-Mad: A transcription factor network controlling cell cycle progression, differentiation and death. *Curr. Opin. Genet. Dev.* **1994**, *4*, 102–108. [CrossRef]

27. Lasorella, A.; Benezra, R.; Iavarone, A. The ID proteins: Master regulators of cancer stem cells and tumour aggressiveness. *Nat. Rev. Cancer* **2014**, *14*, 77–91. [CrossRef]

28. Weber, D.; Wiese, C.; Gessler, M. Hey bHLH transcription factors. In *Current Topics in Developmental Biology*; Academic Press: Cambridge, MA, USA, 2014; Volume 110, pp. 285–315.

29. Iso, T.; Kedes, L.; Hamamori, Y. HES and HERP families: Multiple effectors of the notch signaling pathway. *J. Cell. Physiol.* **2003**, *194*, 237–255. [CrossRef]

30. Saha, T.T.; Shin, S.W.; Dou, W.; Roy, S.; Zhao, B.; Hou, Y.; Wang, X.-L.; Zou, Z.; Girke, T.; Raikhel, A.S. Hairy and Groucho mediate the action of juvenile hormone receptor Methoprene-tolerant in gene repression. *Proc. Natl. Acad. Sci. USA* **2016**, *113*, E735–E743. [CrossRef]

31. Wright, E.J.; Pereira De Castro, K.; Joshi, A.D.; Elferink, C.J. Canonical and non-canonical aryl hydrocarbon receptor signaling pathways. *Curr. Opin. Toxicol.* **2017**, *2*, 87–92. [CrossRef]

32. Semenza, G.L. Hypoxia-inducible factors in physiology and medicine. *Cell* **2012**, *148*, 399–408. [CrossRef]

33. Moffett, P.; Pelletier, J. Different transcriptional properties of mSim-1 and mSim-2. *FEBS Lett.* **2000**, *466*, 80–86. [CrossRef]

34. Lee, Y.; Lee, J.; Kwon, I.; Nakajima, Y.; Ohmiya, Y.; Son, G.H.; Lee, K.H.; Kim, K. Coactivation of the CLOCK-BMAL1 complex by CBP mediates resetting of the circadian clock. *J. Cell Sci.* **2010**, *123*, 3547–3557. [CrossRef] [PubMed]

35. Teh, C.H.L.; Lam, K.K.Y.; Loh, C.C.; Loo, J.M.; Yan, T.; Lim, T.M. Neuronal PAS domain protein 1 is a transcriptional repressor and requires arylhydrocarbon nuclear translocator for its nuclear localization. *J. Biol. Chem.* **2006**, *281*, 34617–34629. [CrossRef] [PubMed]

36. Ohsawa, S.; Hamada, S.; Kakinuma, Y.; Yagi, T.; Miura, M. Novel function of neuronal PAS domain protein 1 in erythropoietin expression in neuronal cells. *J. Neurosci. Res.* **2005**, *79*, 451–458. [CrossRef]

37. Gilles-Gonzalez, M.-A.; Gonzales, G. Signal transduction by heme-containing PAS-domain proteins. *J. Appl. Physiol.* **2004**, *96*, 774–783. [CrossRef] [PubMed]

38. Kamnasaran, D. Disruption of the neuronal PAS3 gene in a family affected with schizophrenia. *J. Med Genet.* **2003**, *40*, 325–332. [CrossRef] [PubMed]

39. Ooe, N.; Saito, K.; Mikami, N.; Nakatuka, I.; Kaneko, H. Identification of a Novel Basic Helix-Loop-Helix-PAS Factor, NXF, Reveals a Sim2 Competitive, Positive Regulatory Role in Dendritic-Cytoskeleton Modulator Drebrin Gene Expression. *Mol. Cell. Biol.* **2003**, *24*, 608–616. [CrossRef]

40. Li, M.; Mead, E.A.; Zhu, J. Heterodimer of two bHLH-PAS proteins mediates juvenile hormone-induced gene expression. *Proc. Natl. Acad. Sci. USA* **2011**, *2011*, 638–643. [CrossRef]

41. Brunnberg, S.; Pettersson, K.; Rydin, E.; Matthews, J.; Hanberg, A.; Pongratz, I. The basic helix-loop-helix-PAS protein ARNT functions as a potent coactivator of estrogen receptor-dependent transcription. *Proc. Natl. Acad. Sci. USA* **2003**, *100*, 6517–6522. [CrossRef]

42. Sharma, P.; Chinaranagari, S.; Chaudhary, J. Inhibitor of differentiation 4 (ID4) acts as an inhibitor of ID-1, -2 and -3 and promotes basic helix loop helix (bHLH) E47 DNA binding and transcriptional activity. *Biochimie* **2015**, *112*, 139–150. [CrossRef]

43. Zelzer, E.; Wappner, P.; Shilo, B.-Z. The PAS domain confers target gene specificity of Drosophila bHLH/PAS proteins. *Genes Dev.* **1997**, *11*, 2079–2089. [CrossRef] [PubMed]

44. Cusanovich, M.A.; Meyer, T.E. Photoactive yellow protein: A prototypic PAS domain sensory protein and development of a common signaling mechanism. *Biochemistry* **2003**, *42*, 4759–4770. [CrossRef] [PubMed]

45. Mö Glich, A.; Ayers, R.A.; Moffat, K. Structure and Signaling Mechanism of Per-ARNT-Sim Domains. *Struct. Fold. Des.* **2009**, *17*, 1282–1294. [CrossRef] [PubMed]

46. Roeder, R.G. Transcriptional regulation and the role of diverse coactivators in animal cells. *FEBS Lett.* **2005**, *579*, 909–915. [CrossRef]

47. Ansari, S.A.; Morse, R.H. Mechanisms of Mediator complex action in transcriptional activation. *Cell. Mol. Life Sci.* **2013**, *70*, 2743–2756. [CrossRef]

48. Conaway, R.C.; Conaway, J.W. Function and regulation of the Mediator complex. *Curr. Opin. Genet. Dev.* **2011**, *21*, 225–230. [CrossRef]

49. Poss, Z.C.; Ebmeier, C.C.; Taatjes, D.J. The Mediator complex and transcription regulation. *Crit. Rev. Biochem. Mol. Biol.* **2013**, *48*, 575–608. [CrossRef]

50. Quevedo, M.; Meert, L.; Dekker, M.R.; Dekkers, D.H.W.; Brandsma, J.H.; van den Berg, D.L.C.; Ozgür, Z.; van IJcken, W.F.J.; Demmers, J.; Fornerod, M.; et al. Mediator complex interaction partners organize the transcriptional network that defines neural stem cells. *Nat. Commun.* **2019**, *10*, 2669. [CrossRef]

51. Malik, S.; Roeder, R.G. The metazoan Mediator co-activator complex as an integrative hub for transcriptional regulation. *Nat. Rev. Genet.* **2010**, *11*, 761–772. [CrossRef]

52. Malik, N.; Agarwal, P.; Tyagi, A. Emerging functions of multi-protein complex Mediator with special emphasis on plants. *Crit. Rev. Biochem. Mol. Biol.* **2017**, *52*, 475–502. [CrossRef]

53. An, C.; Mou, Z. The function of the Mediator complex in plant immunity. *Plant Signal. Behav.* **2013**, *8*, e23182. [CrossRef] [PubMed]

54. Chen, R.; Jiang, H.; Li, L.; Zhai, Q.; Qi, L.; Zhou, W.; Liu, X.; Li, H.; Zheng, W.; Sun, J.; et al. The Arabidopsis Mediator Subunit MED25 Differentially Regulates Jasmonate and Abscisic Acid Signaling through Interacting with the MYC2 and ABI5 Transcription Factors. *Plant Cell* **2012**, *24*, 2898–2916. [CrossRef] [PubMed]

55. An, C.; Li, L.; Zhai, Q.; You, Y.; Deng, L.; Wu, F.; Chen, R.; Jiang, H.; Wang, H.; Chen, Q.; et al. Mediator subunit MED25 links the jasmonate receptor to transcriptionally active chromatin. *Proc. Natl. Acad. Sci. USA* **2017**, *114*, E8930–E8939. [CrossRef] [PubMed]

56. Liu, Y.; Du, M.; Deng, L.; Shen, J.; Fang, M.; Chen, Q.; Lu, Y.; Wang, Q.; Li, C.; Zhai, Q. Myc2 regulates the termination of jasmonate signaling via an autoregulatory negative feedback loop[open]. *Plant Cell* **2019**, *31*, 106–127. [CrossRef]

57. Li, X.; Yang, R.; Chen, H. The arabidopsis thaliana mediator subunit MED8 regulates plant immunity to botrytis cinerea through interacting with the basic helix-loop-helix (bHLH) transcription factor FAMA. *PLoS ONE* **2018**, *13*, e0193458. [CrossRef]

58. Oliner, J.D.; Andresen, J.M.; Hansen, S.K.; Zhou, S.; Tjian, R. SREBP transcriptional activity is mediated through an interaction with the CREB-binding protein. *Genes Dev.* **1996**, *10*, 2903–2911. [CrossRef]

59. Yang, F.; Vought, B.W.; Satterlee, J.S.; Walker, A.K.; Jim Sun, Z.Y.; Watts, J.L.; DeBeaumont, R.; Mako Saito, R.; Hyberts, S.G.; Yang, S.; et al. An ARC/Mediator subunit required for SREBP control of cholesterol and lipid homeostasis. *Nature* **2006**, *442*, 700–704. [CrossRef]

60. Pacheco, D.; Warfield, L.; Brajcich, M.; Robbins, H.; Luo, J.; Ranish, J.; Hahn, S. Transcription Activation Domains of the Yeast Factors Met4 and Ino2: Tandem Activation Domains with Properties Similar to the Yeast Gcn4 Activator. *Mol. Cell. Biol.* **2018**, *38*, e00038-18. [CrossRef]

61. Lazrak, M.; Deleuze, V.; Noel, D.; Haouzi, D.; Chalhoub, E.; Dohet, C.; Robbins, I.; Mathieu, D. The bHLH TAL-1/SCL regulates endothelial cell migration and morphogenesis. *J. Cell Sci.* **2004**, *117*, 1161–1171. [CrossRef]

62. Huang, S.; Qiu, Y.; Stein, R.W.; Brandt, S.J. p300 functions as a transcriptional coactivator for the TAL1/SCL oncoprotein. *Oncogene* **1999**, *18*, 4958–4967. [CrossRef]

63. Puri, P.L.; Avantaggiati, M.L. p300 is required for MyoD-dependent cell cycle arrest and muscle-specific gene transcription arrest in the G 0 phase, irreversible exit from the cell cycle. *EMBO J.* **1997**, *16*, 369–383. [CrossRef] [PubMed]

64. Cao, Y.; Yao, Z.; Sarkar, D.; Lawrence, M.; Sanchez, G.J.; Parker, M.H.; MacQuarrie, K.L.; Davison, J.; Morgan, M.T.; Ruzzo, W.L.; et al. Genome-wide MyoD Binding in Skeletal Muscle Cells: A Potential for Broad Cellular Reprogramming. *Dev. Cell* **2010**, *18*, 662–674. [CrossRef] [PubMed]

65. Raposo, A.A.S.F.; Vasconcelos, F.F.; Drechsel, D.; Marie, C.; Johnston, C.; Dolle, D.; Bithell, A.; Gillotin, S.; van den Berg, D.L.C.; Ettwiller, L.; et al. Ascl1 coordinately regulates gene expression and the chromatin landscape during neurogenesis. *Cell Rep.* **2015**, *10*, 1544–1556. [CrossRef] [PubMed]

66. Ko, C.-I.; Puga, A. Does the aryl hydrocarbon receptor regulate pluripotency? *Curr. Opin. Toxicol.* **2017**, *2*, 1–7. [CrossRef] [PubMed]

67. Yao, X.; Tan, J.; Lim, K.J.; Koh, J.; Ooi, W.F.; Li, Z.; Huang, D.; Xing, M.; Chan, Y.S.; Qu, J.Z.; et al. VHL deficiency drives enhancer activation of oncogenes in clear cell renal cell carcinoma. *Cancer Discov.* **2017**, *7*, 1284–1305. [CrossRef] [PubMed]

68. Doi, M.; Hirayama, J.; Sassone-Corsi, P. Circadian regulator CLOCK is a histone acetyltransferase. *Cell* **2006**, *125*, 497–508. [CrossRef] [PubMed]

69. Harada, A.; Ohkawa, Y.; Imbalzano, A.N. Temporal regulation of chromatin during myoblast differentiation. *Semin. Cell Dev. Biol.* **2017**, *72*, 77–86. [CrossRef]

70. Heisig, J.; Weber, D.; Englberger, E.; Winkler, A.; Kneitz, S.; Sung, W.-K.; Wolf, E.; Eilers, M.; Wei, C.-L.; Gessler, M. Target gene analysis by microarrays and chromatin immunoprecipitation identifies HEY proteins as highly redundant bHLH repressors. *PLoS Genet.* **2012**, *8*, e1002728. [CrossRef]

71. Rahl, P.B.; Lin, C.Y.; Seila, A.C.; Flynn, R.A.; McCuine, S.; Burge, C.B.; Sharp, P.A.; Young, R.A. c-Myc Regulates Transcriptional Pause Release. *Cell* **2010**, *141*, 432–445. [CrossRef]

72. Anfinsen, C.B. Principles that govern the folding of protein chains. *Science* **1973**, *181*, 223–230. [CrossRef]

73. Uversky, V.N. A decade and a half of protein intrinsic disorder: Biology still waits for physics. *Protein Sci. A Publ. Protein Soc.* **2013**, *22*, 693–724. [CrossRef] [PubMed]

74. Dyson, H.J.; Wright, P.E. Intrinsically unstructured proteins and their functions. *Nat. Rev. Mol. Cell Biol.* **2005**, *6*, 197–208. [CrossRef] [PubMed]

75. Dunker, A.K.; Lawson, J.D.; Brown, C.J.; Williams, R.M.; Romero, P.; Oh, J.S.; Oldfield, C.J.; Campen, A.M.; Ratliff, C.M.; Hipps, K.W.; et al. Intrinsically disordered protein. *J. Mol. Graph. Model.* **2001**, *19*, 26–59. [CrossRef]

76. Lu, H.; Yu, D.; Hansen, A.S.; Ganguly, S.; Liu, R.; Heckert, A.; Darzacq, X.; Zhou, Q. Phase-separation mechanism for C-terminal hyperphosphorylation of RNA polymerase II. *Nature* **2018**, *558*, 318. [CrossRef] [PubMed]

77. Uversky, V.N.; Dunker, A.K. Understanding protein non-folding. *Biochim. Biophys. Acta (BBA) Proteins Proteom.* **2010**, *1804*, 1231–1264. [CrossRef]

78. Dosztanyi, Z.; Csizmok, V.; Tompa, P.; Simon, I. The pairwise energy content estimated from amino acid composition discriminates between folded and intrinsically unstructured proteins. *J. Mol. Biol.* **2005**, *347*, 827–839. [CrossRef]

79. Campen, A.; Williams, R.; Brown, C.; Meng, J.; Uversky, V.; Dunker, A. TOP-IDP-Scale: A New Amino Acid Scale Measuring Propensity for Intrinsic Disorder. *Protein Pept. Lett.* **2008**, *15*, 956–963. [CrossRef]

80. Williams, R.M.; Obradovi, Z.; Mathura, V.; Braun, W.; Garner, E.C.; Young, J.; Takayama, S.; Brown, C.J.; Dunker, A.K. The protein non-folding problem: Amino acid determinants of intrinsic order and disorder. *Pac. Symp. Biocomput.* **2001**, *2001*, 89–100.

81. Vacic, V.; Uversky, V.N.; Dunker, A.K.; Lonardi, S. Composition Profiler: A tool for discovery and visualization of amino acid composition differences. *BMC Bioinform.* **2007**, *8*, 211. [CrossRef]

82. Uversky, V.N.; Gillespie, J.R.; Fink, A.L. Why are "natively unfolded" proteins unstructured under physiologic conditions? *Proteins Struct. Funct. Genet.* **2000**, *41*, 415–427. [CrossRef]

83. Dunker, A.K.; Obradovic, Z. The protein trinity—Linking function and disorder. *Nat. Biotechnol.* **2001**, *19*, 805. [CrossRef] [PubMed]

84. Tompa, P. The interplay between structure and function in intrinsically unstructured proteins. *FEBS Lett.* **2005**, *579*, 3346–3354. [CrossRef] [PubMed]

85. Van Der Lee, R.; Buljan, M.; Lang, B.; Weatheritt, R.J.; Daughdrill, G.W.; Dunker, A.K.; Fuxreiter, M.; Gough, J.; Gsponer, J.; Jones, D.T.; et al. Classification of intrinsically disordered regions and proteins. *Chem. Rev.* **2014**, *114*, 6589–6631. [CrossRef] [PubMed]

86. Peng, Z.; Yan, J.; Fan, X.; Mizianty, M.J.; Xue, B.; Wang, K.; Hu, G.; Uversky, V.N.; Kurgan, L. Exceptionally abundant exceptions: Comprehensive characterization of intrinsic disorder in all domains of life. *Cell. Mol. Life Sci.* **2014**, *72*, 137–151. [CrossRef]

87. Wright, P.E.; Dyson, H.J. Intrinsically disordered proteins in cellular signalling and regulation. *Nat. Rev. Mol. Cell Biol.* **2015**, *16*, 18–29. [CrossRef]

88. Tompa, P.; Fuxreiter, M. Fuzzy complexes: Polymorphism and structural disorder in protein-protein interactions. *Trends Biochem. Sci* **2008**, *33*, 2–8. [CrossRef]

89. Iakoucheva, L.M.; Radivojac, P.; Brown, C.J.; O'Connor, T.R.; Sikes, J.G.; Obradovic, Z.; Dunker, A.K. The importance of intrinsic disorder for protein phosphorylation. *Nucleic Acids Res.* **2004**, *32*, 1037–1049. [CrossRef]

90. Hatakeyama, J.; Kageyama, R. Retinal cell fate determination and bHLH factors. *Semin. Cell Dev. Biol.* **2004**, *15*, 83–89. [CrossRef]

91. Sölter, M.; Locker, M.; Boy, S.; Taelman, V.; Bellefroid, E.J.; Perron, M.; Pieler, T. Characterization and function of the bHLH-O protein XHes2: Insight into the mechanism controlling retinal cell fate decision. *Development* **2006**, *133*, 4097–4108. [CrossRef]

92. Parker, M.H.; Perry, R.L.S.; Fauteux, M.C.; Berkes, C.A.; Rudnicki, M.A. MyoD Synergizes with the E-Protein HEB To Induce Myogenic Differentiation. *Mol. Cell. Biol.* **2006**, *26*, 5771–5783. [CrossRef]

93. McDowell, G.S.; Hindley, C.J.; Lippens, G.; Landrieu, I.; Philpott, A. Phosphorylation in intrinsically disordered regions regulates the activity of Neurogenin2. *BMC Biochem.* **2014**, *15*, 24. [CrossRef] [PubMed]

94. Aguado-Llera, D.; Goormaghtigh, E.; de Geest, N.; Quan, X.-J.; Prieto, A.; Hassan, B.A.; Gómez, J.; Neira, J.L. The basic helix-loop-helix region of human neurogenin 1 is a monomeric natively unfolded protein which forms a "fuzzy" complex upon DNA binding. *Biochemistry* **2010**, *49*, 1577–1589. [CrossRef] [PubMed]

95. Panova, S.; Cliff, M.J.; Macek, P.; Blackledge, M.; Jensen, M.R.; Nissink, J.W.M.; Embrey, K.J.; Davies, R.; Waltho, J.P. Mapping Hidden Residual Structure within the Myc bHLH-LZ Domain Using Chemical Denaturant Titration. *Structure* **2019**, *27*, 1537–1546. [CrossRef] [PubMed]

96. Romero, P.; Obradovic, Z.; Dunker, A.K. Sequence Data Analysis for Long Disordered Regions Prediction in the Calcineurin Family. *Genome Inform. Workshop Genome Inform.* **1997**, *8*, 110–124.

97. Li, X.; Romero, P.; Rani, M.; Dunker, A.; Obradovic, Z. Predicting Protein Disorder for N-, C-, and Internal Regions. *Genome Inform. Workshop Genome Inform.* **1999**, *10*, 30–40.

98. Cilia, E.; Pancsa, R.; Tompa, P.; Lenaerts, T.; Vranken, W.F. From protein sequence to dynamics and disorder with DynaMine. *Nat. Commun.* **2013**, *4*, 2741. [CrossRef]

99. Cilia, E.; Pancsa, R.; Tompa, P.; Lenaerts, T.; Vranken, W.F. The DynaMine webserver: Predicting protein dynamics from sequence. *Nucleic Acids Res.* **2014**, *42*, W264–W270. [CrossRef]

100. Furness, S.G.B.; Lees, M.J.; Whitelaw, M.L. The dioxin (aryl hydrocarbon) receptor as a model for adaptive responses of bHLH/PAS transcription factors. *FEBS Lett.* **2007**, *581*, 3616–3625. [CrossRef]

101. Brangwynne, C.P.; Eckmann, C.R.; Courson, D.S.; Rybarska, A.; Hoege, C.; Gharakhani, J.; Jülicher, F.; Hyman, A.A. Germline P granules are liquid droplets that localize by controlled dissolution/condensation. *Science* **2009**, *324*, 1729–1732. [CrossRef]

102. Molliex, A.; Temirov, J.; Lee, J.; Coughlin, M.; Kanagaraj, A.P.; Kim, H.J.; Mittag, T.; Taylor, J.P. Phase Separation by Low Complexity Domains Promotes Stress Granule Assembly and Drives Pathological Fibrillization. *Cell* **2015**, *163*, 123–133. [CrossRef]

103. Nott, T.J.; Petsalaki, E.; Farber, P.; Jervis, D.; Fussner, E.; Plochowietz, A.; Craggs, T.D.; Bazett-Jones, D.P.; Pawson, T.; Forman-Kay, J.D.; et al. Phase Transition of a Disordered Nuage Protein Generates Environmentally Responsive Membraneless Organelles. *Mol. Cell* **2015**, *57*, 936–947. [CrossRef] [PubMed]

104. Elbaum-Garfinkle, S.; Kim, Y.; Szczepaniak, K.; Chen, C.C.H.; Eckmann, C.R.; Myong, S.; Brangwynne, C.P. The disordered P granule protein LAF-1 drives phase separation into droplets with tunable viscosity and dynamics. *Proc. Natl. Acad. Sci. USA* **2015**, *112*, 7189–7194. [CrossRef] [PubMed]

105. Lai, J.; Koh, C.H.; Tjota, M.; Pieuchot, L.; Raman, V.; Chandrababu, K.B.; Yang, D.; Wong, L.; Jedd, G. Intrinsically disordered proteins aggregate at fungal cell-to-cell channels and regulate intercellular connectivity. *Proc. Natl. Acad. Sci. USA* **2012**, *109*, 15781–15786. [CrossRef] [PubMed]

106. Mitrea, D.M.; Cika, J.A.; Guy, C.S.; Ban, D.; Banerjee, P.R.; Stanley, C.B.; Nourse, A.; Deniz, A.A.; Kriwacki, R.W. Nucleophosmin integrates within the nucleolus via multi-modal interactions with proteins displaying R-rich linear motifs and rRNA. *eLife* **2016**, *5*, e13571. [CrossRef] [PubMed]

107. Brangwynne, C.P.; Tompa, P.; Pappu, R.V. Polymer physics of intracellular phase transitions. *Nat. Phys.* **2015**, *11*, 899–904. [CrossRef]

108. Banani, S.F.; Rice, A.M.; Peeples, W.B.; Lin, Y.; Jain, S.; Parker, R.; Rosen, M.K. Compositional Control of Phase-Separated Cellular Bodies. *Cell* **2016**, *166*, 651–663. [CrossRef]

109. Ditlev, J.A.; Case, L.B.; Rosen, M.K. Who's In and Who's Out—Compositional Control of Biomolecular Condensates. *J. Mol. Biol.* **2018**, *430*, 4666–4684. [CrossRef]

110. Harmon, T.S.; Holehouse, A.S.; Rosen, M.K.; Pappu, R.V. Intrinsically disordered linkers determine the interplay between phase separation and gelation in multivalent proteins. *eLife* **2017**, *6*, e30294. [CrossRef]

111. Posey, A.E.; Holehouse, A.S.; Pappu, R.V. Phase Separation of Intrinsically Disordered Proteins. In *Methods in Enzymology*; Elsevier: Amsterdam, The Netherlands, 2018; ISBN 9780128156490.

112. Berezney, R.; Coffey, D.S. Identification of a nuclear protein matrix. *Biochem. Biophys. Res. Commun.* **1974**, *60*, 1410–1417. [CrossRef]

113. Linnemann, A.K.; Platts, A.E.; Krawetz, S.A. Differential nuclear scaffold/matrix attachment marks expressed genes. *Hum. Mol. Genet.* **2009**, *18*, 645–654. [CrossRef]

114. Engelke, R.; Riede, J.; Hegermann, J.; Wuerch, A.; Eimer, S.; Dengjel, J.; Mittler, G. The quantitative nuclear matrix proteome as a biochemical snapshot of nuclear organization. *J. Proteome Res.* **2014**, *13*, 3940–3956. [CrossRef] [PubMed]

115. Weber, S.C.; Brangwynne, C.P. Inverse size scaling of the nucleolus by a concentration-dependent phase transition. *Curr. Biol.* **2015**, *25*, 641–646. [CrossRef] [PubMed]

116. Berry, J.; Weber, S.C.; Vaidya, N.; Haataja, M.; Brangwynne, C.P. RNA transcription modulates phase transition-driven nuclear body assembly. *Proc. Natl. Acad. Sci. USA* **2015**, *112*, E5237–E5245. [CrossRef] [PubMed]

117. Gibson, B.A.; Doolittle, L.K.; Jensen, L.E.; Gamarra, N.; Redding, S.; Rosen, M.K. Organization and Regulation of Chromatin by Liquid-Liquid Phase Separation. *bioRxiv* **2019**. [CrossRef]

118. Duronio, R.J.; Marzluff, W.F. Coordinating cell cycle-regulated histone gene expression through assembly and function of the Histone Locus Body. *RNA Biol.* **2017**, *14*, 726–738. [CrossRef]

119. Mitrea, D.M.; Kriwacki, R.W. Phase separation in biology; Functional organization of a higher order Short linear motifs—The unexplored frontier of the eukaryotic proteome. *Cell Commun. Signal.* **2016**, *14*, 1–20. [CrossRef]

120. Wu, T.; Lu, Y.; Gutman, O.; Lu, H.; Zhou, Q.; Henis, Y.I.; Luo, K. Phase separation of TAZ compartmentalizes the transcription machinery to promote gene expression. *bioRxiv* **2019**. [CrossRef]

121. Sabari, B.R.; Dall'Agnese, A.; Boija, A.; Klein, I.A.; Coffey, E.L.; Shrinivas, K.; Abraham, B.J.; Hannett, N.M.; Zamudio, A.V.; Manteiga, J.C.; et al. Coactivator condensation at super-enhancers links phase separation and gene control. *Science* **2018**, *361*, eaar3958. [CrossRef]

122. Boija, A.; Klein, I.A.; Sabari, B.R.; Dall'Agnese, A.; Coffey, E.L.; Zamudio, A.V.; Li, C.H.; Shrinivas, K.; Manteiga, J.C.; Hannett, N.M.; et al. Transcription Factors Activate Genes through the Phase-Separation Capacity of Their Activation Domains. *Cell* **2018**, *175*, 1842–1855.e16. [CrossRef]

123. Chen, X.; Xu, H.; Yuan, P.; Fang, F.; Huss, M.; Vega, V.B.; Wong, E.; Orlov, Y.L.; Zhang, W.; Jiang, J.; et al. Integration of External Signaling Pathways with the Core Transcriptional Network in Embryonic Stem Cells. *Cell* **2008**, *133*, 1106–1117. [CrossRef]

124. Whyte, W.A.; Orlando, D.A.; Hnisz, D.; Abraham, B.J.; Lin, C.Y.; Kagey, M.H.; Rahl, P.B.; Lee, T.I.; Young, R.A. Master transcription factors and mediator establish super-enhancers at key cell identity genes. *Cell* **2013**, *153*, 307–319. [CrossRef] [PubMed]

125. Hnisz, D.; Shrinivas, K.; Young, R.A.; Chakraborty, A.K.; Sharp, P.A. A Phase Separation Model for Transcriptional Control. *Cell* **2017**, *169*, 13–23. [CrossRef] [PubMed]

126. Mansour, M.R.; Abraham, B.J.; Anders, L.; Berezovskaya, A.; Gutierrez, A.; Durbin, A.D.; Etchin, J.; Lee, L.; Sallan, S.E.; Silverman, L.B.; et al. An oncogenic super-enhancer formed through somatic mutation of a noncoding intergenic element. *Science* **2014**, *346*, 1373–1377. [CrossRef] [PubMed]

127. Brown, J.D.; Lin, C.Y.; Duan, Q.; Griffin, G.; Federation, A.J.; Paranal, R.M.; Bair, S.; Newton, G.; Lichtman, A.H.; Kung, A.L.; et al. Nf-kb directs dynamic super enhancer formation in inflammation and atherogenesis. *Mol. Cell* **2014**, *56*, 219–231. [CrossRef] [PubMed]

128. Lovén, J.; Hoke, H.A.; Lin, C.Y.; Lau, A.; Orlando, D.A.; Vakoc, C.R.; Bradner, J.E.; Lee, T.I.; Young, R.A. Selective inhibition of tumor oncogenes by disruption of super-enhancers. *Cell* **2013**, *153*, 320–334. [CrossRef] [PubMed]

129. Apostolou, E.; Ferrari, F.; Walsh, R.M.; Bar-Nur, O.; Stadtfeld, M.; Cheloufi, S.; Stuart, H.T.; Polo, J.M.; Ohsumi, T.K.; Borowsky, M.L.; et al. Genome-wide Chromatin Interactions of the Nanog Locus in Pluripotency, Differentiation, and Reprogramming. *Cell Stem Cell* **2013**, *12*, 699–712. [CrossRef] [PubMed]

130. Manavathi, B.; Samanthapudi, V.S.K.; Gajulapalli, V.N.R. Estrogen receptor coregulators and pioneer factors: The orchestrators of mammary gland cell fate and development. *Front. Cell Dev. Biol.* **2014**, *12*, 34. [CrossRef]

131. Kagey, M.H.; Newman, J.J.; Bilodeau, S.; Zhan, Y.; Orlando, D.A.; Van Berkum, N.L.; Ebmeier, C.C.; Goossens, J.; Rahl, P.B.; Levine, S.S.; et al. Mediator and cohesin connect gene expression and chromatin architecture. *Nature* **2010**, *467*, 430. [CrossRef]

132. Heery, D.M.; Kalkhoven, E.; Hoare, S.; Parker, M.G. A signature motif in transcriptional co-activators mediates binding to nuclear receptors. *Nature* **1997**, *387*, 733. [CrossRef]

133. Cagas, P.M.; Corden, J.L. Structural studies of a synthetic peptide derived from the carboxyl-terminal domain of RNA polymerase II. *Proteins Struct. Funct. Bioinform.* **1995**, *21*, 149–160. [CrossRef]

134. Portz, B.; Lu, F.; Gibbs, E.B.; Mayfield, J.E.; Rachel Mehaffey, M.; Zhang, Y.J.; Brodbelt, J.S.; Showalter, S.A.; Gilmour, D.S. Structural heterogeneity in the intrinsically disordered RNA polymerase II C-terminal domain. *Nat. Commun.* **2017**, *8*, 15231. [CrossRef] [PubMed]

135. Corden, J.L.; Cadena, D.L.; Ahearn, J.M.; Dahmus, M.E. A unique structure at the carboxyl terminus of the largest subunit of eukaryotic RNA polymerase II. *Proc. Natl. Acad. Sci. USA* **1985**, *82*, 7934–7938. [CrossRef] [PubMed]

136. Dahmus, M.E. Reversible Phosphorylation of the C-terminal Domain of RNA Polymerase II. *J. Biol. Chem.* **1996**, *271*, 19009–19012. [CrossRef]

137. Wippich, F.; Bodenmiller, B.; Trajkovska, M.G.; Wanka, S.; Aebersold, R.; Pelkmans, L. Dual specificity kinase DYRK3 couples stress granule condensation/ dissolution to mTORC1 signaling. *Cell* **2013**, *152*, 791–805. [CrossRef] [PubMed]

138. Rai, A.K.; Chen, J.X.; Selbach, M.; Pelkmans, L. Kinase-controlled phase transition of membraneless organelles in mitosis. *Nature* **2018**, *559*, 211. [CrossRef] [PubMed]

139. Boehning, M.; Dugast-Darzacq, C.; Rankovic, M.; Hansen, A.S.; Yu, T.; Marie-Nelly, H.; McSwiggen, D.T.; Kokic, G.; Dailey, G.M.; Cramer, P.; et al. RNA polymerase II clustering through carboxy-terminal domain phase separation. *Nat. Struct. Mol. Biol.* **2018**, *25*, 833. [CrossRef] [PubMed]

140. Bolognesi, B.; Lorenzo Gotor, N.; Dhar, R.; Cirillo, D.; Baldrighi, M.; Tartaglia, G.G.; Lehner, B. A Concentration-Dependent Liquid Phase Separation Can Cause Toxicity upon Increased Protein Expression. *Cell Rep.* **2016**, *16*, 222–231. [CrossRef] [PubMed]

141. Mészáros, B.; Erdős, G.; Szabó, B.; Schád, É.; Tantos, Á.; Abukhairan, R.; Horváth, T.; Murvai, N.; Kovács, O.P.; Kovács, M.; et al. PhaSePro: The database of proteins driving liquid–liquid phase separation. *Nucleic Acids Res.* **2019**. [CrossRef]

142. Darling, A.L.; Liu, Y.; Oldfield, C.J.; Uversky, V.N. Intrinsically Disordered Proteome of Human Membrane-Less Organelles. *Proteomics* **2018**, *18*, 1700193. [CrossRef]

143. Hyman, A.A.; Weber, C.A.; Jülicher, F. Liquid-Liquid Phase Separation in Biology. *Annu. Rev. Cell Dev. Biol.* **2014**, *30*, 39–58. [CrossRef]

144. Uversky, V.N. Protein intrinsic disorder-based liquid–liquid phase transitions in biological systems: Complex coacervates and membrane-less organelles. *Adv. Colloid Interface Sci.* **2017**, *239*, 97–114. [CrossRef] [PubMed]

145. Shin, Y.; Brangwynne, C.P. Liquid phase condensation in cell physiology and disease. *Science* **2017**, *357*, eaaf4382. [CrossRef] [PubMed]

 © 2019 by the authors. Licensee MDPI, Basel, Switzerland. This article is an open access article distributed under the terms and conditions of the Creative Commons Attribution (CC BY) license (http://creativecommons.org/licenses/by/4.0/).

International Journal of
Molecular Sciences

Review

The Role of Post-Translational Modifications in the Phase Transitions of Intrinsically Disordered Proteins

Izzy Owen and Frank Shewmaker *

Department of Biochemistry, Uniformed Services University of the Health Sciences, Bethesda, MD 20814, USA;
izzy.owen@usuhs.edu
* Correspondence: fshewmaker@usuhs.edu; Tel.: +1-301-295-3527

Received: 11 October 2019; Accepted: 2 November 2019; Published: 5 November 2019

Abstract: Advances in genomics and proteomics have revealed eukaryotic proteomes to be highly abundant in intrinsically disordered proteins that are susceptible to diverse post-translational modifications. Intrinsically disordered regions are critical to the liquid–liquid phase separation that facilitates specialized cellular functions. Here, we discuss how post-translational modifications of intrinsically disordered protein segments can regulate the molecular condensation of macromolecules into functional phase-separated complexes.

Keywords: liquid–liquid phase separation; intrinsically disordered regions; post-translational modifications; membraneless organelles

---

1. Introduction to Liquid–Liquid Phase Separation and Membraneless Organelles

Cells contain crowded molecular environments hosting discrete functions that must be separated within time and space. Membrane-less compartments resulting from liquid–liquid phase separation (LLPS) are increasingly being recognized as mechanisms for organizing cellular activities. These distinct regions may be referred to as biomolecular condensates or membrane-less organelles (MLOs). As the names suggest, these organelles are not encapsulated in a membrane, yet contain enriched sets of specific macromolecules. Thus, LLPS is the biologically regulated process by which specific macromolecular components are concentrated into a specific MLO.

MLOs contain proteins, and frequently nucleic acids, and are dynamic in size (generally submicrometer), formation, and composition [1]. They behave like liquid droplets, capable of fusing, deforming, and rearranging [2]—all while being solvated in the larger aqueous environment of the cell. The macromolecular components of MLOs have a higher affinity for each other than for surrounding molecules, allowing for separation from the bulk solution by demixing, thus forming two co-existing liquid states with differing concentrations of particular solutes [3].

The network of multivalent interactions within an MLO is not ordered like a conventional protein complex [4–6]. The interactions are typically characterized as non-static and more dynamic, with less specificity and weaker binding than the forces that hold macromolecular complexes—such as the proteasome or ribosome–into rigid stoichiometric structures [2]. For example, a ribosome consists of large and small subunits with more-or-less specific quaternary arrangement of components that together form a large macromolecular machine. Interactions in MLOs are thought to be less specific, with greater fluctuation of molecular contacts and stoichiometry. The plasticity of interactions may permit these organelles to react more dynamically to specific cellular conditions.

Numerous distinct functional MLOs have been characterized, and their many unique protein constituents have been previously reviewed [7,8]. A recently developed database of nearly 3000 non-redundant LLPS-associated proteins suggests that many MLOs have yet to be fully characterized [9]. Of the MLOs that have been characterized, their diversity and ubiquity is remarkable. MLOs have been observed in cytoplasm and nucleoplasm, and also in canonical membrane-enclosed organelles like mitochondria or chloroplasts [10]. Most commonly, MLOs are linked to specific functions involving ribonucleic acid, such as germ granules [11]. Pathological examples have also been proposed, such as the cytoplasmic inclusion bodies (IBs) within which measles viral RNA is replicated [12]. MLOs may exist transiently, like stress granules (SGs), which are stalled translation complexes that form upon cellular stress [13]. Alternatively, MLOs can have a more persistent presence, like the nucleolus, which is a constant site of ribosome production in the nucleus [7,14]. MLOs may also form in response to spatial necessity, such as neuronal RNA granules, which function in transport of mRNAs from dendrite bodies to distant synapses [15,16].

MLOs may also have roles in the pathogenesis of many diseases, particularly neurodegenerative disorders [17]. For example, many proteins linked to amyotrophic lateral sclerosis (ALS) and frontotemporal dementia (FTD) can undergo LLPS and accumulate within MLOs [18]. Mutations in these proteins not only cause disease but can alter LLPS and the physical properties of the phase-separated state [19,20]. It is hypothesized that aberrant irreversible phase transitions may result in proteinaceous neuronal inclusions that lead directly to cellular dysfunction [2].

2. Intrinsically Disordered Regions Facilitate LLPS

An interesting feature of proteins that undergo LLPS is they frequently contain long segments that lack well-defined three-dimensional structure. These segments are typically termed intrinsically disordered regions (IDRs), or intrinsically disordered proteins (IDPs), because they have no single equilibrium structure; instead, they exist as broad structural (or population) ensembles or they exchange between multiple conformations rapidly. IDRs are usually defined as being approximately 30 amino acids or longer [21], and their distinguishing characteristic is a relative paucity of hydrophobic amino acids to drive folding into a narrow conformational landscape.

The sequence composition of IDRs can vary, but is commonly disproportionately represented by only a few amino acids (i.e., low-complexity). Some low-complexity sequences are called yeast prion-like because they are compositionally very similar to the domains that enable certain yeast proteins to form self-propagating amyloid fibers. Yeast prion domains (and prion-like domains (PrLDs)) are usually very rich in hydrophilic amino acids (e.g., asparagine, glutamine, serine, and tyrosine). Other low-complexity sequences may disproportionately contain charged amino acids, such as the arginine/glycine repeats (RGG or GRG), which occur in several IDRs within liquid phase-separating proteins. Repeating (or spatially distributed) motifs of a subset of amino acids are also common to IDRs [6]. IDRs of MLO-forming proteins are also enriched in amino acids that can form π–π interactions, in which induced electrostatic interactions occur between sp^2 hybridized atoms [22]. These interactions can also involve the backbone amide bonds, which are accessible due to the non-folded arrangement of IDRs.

The significance of IDRs in liquid phase-separating proteins is they enable diverse networks of transient interactions with moderate affinities (i.e., reversible, due partly to entropic penalties of IDRs adopting binding conformations). Relative to folded domains, IDRs have greater accessible conformational space and flexibility for forming molecular contacts. The frequent presence of repetitive motifs can enable numerous low-affinity interactions with the potential for high-avidity binding [23]. IDRs can therefore support multivalent interactions, meaning they can form multiple molecular contacts with a potential variety of binding partners. Thus, an MLO may emerge from a continuous network of IDRs forming inter-protein (or RNA-protein) multivalent contacts.

3. Post-Translational Modifications of Intrinsically Disordered Domains Can Govern LLPS

The lack of secondary structure makes IDRs especially susceptible to post-translational modifications (PTMs) [24]. In fact, IDRs are disproportionately modified post-translationally relative to the entire proteome [25–27]. A variety of PTMs can alter IDR charge, hydrophobicity, size, and structure.

These changes may occur through additions of functional groups (e.g., phosphoryl, methyl, acyl, glycosyl, alkyl, etc.), or subtler chemical changes such as oxidation, deimidation, and deamidation [28].

There are many biological examples of PTMs serving as on/off switches, where they regulate a cellular event, such as protein signaling, localization, and degradation. In the case of IDRs and phase separation, PTMs can similarly have on/off functions by altering the nature of intermolecular contacts that support MLO formation or dissolution [29] (Table 1). Here, we discuss examples of PTMs and IDRs in proteins that undergo functional phase separation in cells. We evaluate the hypothesis that the combination of IDR multivalency and the capacity to be extensively modified results in reversible networks of interactions that can be regulated by specific cellular cues (Figure 1).

Table 1. Examples of post-translational modifications (PTMs) of intrinsically disordered regions (IDRs) altering the liquid–liquid phase separation (LLPS) of proteins. The underlined proteins have multiple PTMs that affect the phase separation propensity. Arrows indicate if PTMs promote (↑) or inhibit (↓) LLPS. C-terminal domain (CTD), N-terminal domain (NTD), prion-like domain (PrLD), arginine-glycine-glycine (RGG), stress granules (SGs), amyotrophic lateral sclerosis (ALS), and frontal temporal dementia (FTD).

PTM	Protein Example	Region Modified	Proposed Effects of PTM on LLPS (↓↑)		Type of MLO	Disease Link
Serine/Threonine Phosphorylation	FMRP	CTD IDR	↑		Neuronal granules	Fragile X syndrome
	TIAR-2	CTD PrLD	↑	Increases electrostatic interactions	SGs	
	Phosphoprotein	Internal IDR	↑		Inclusion bodies	Measles
	tau	Internal IDR	↑		SGs	Alzheimer's disease
	MEG-3	Internal IDR	↓		P granule	
	FUS	NTD PrLD	↓	Introduces electrostatic repulsion	SGs, nuclear paraspeckles	ALS, FTD
	TDP-43	NTD domain & CTD IDR	↓		SGs	ALS, FTD
Arginine Methylation	(SDMA) LSM4	49 aa, NTD RGG domain	↑	Changes hydrophobicity and H-bonding	Processing bodies	
	(ADMA) hnRNPA2	CTD IDR at RGG sites	↓		SGs	ALS, FTD
	RAP55A	36 aa, NTD RGG domain	↓	Changes hydrophobicity and H-bonding	SGs, Processing bodies	Primary biliary cirrhosis
	FUS	41 aa, CTD RGG domain	↓		SGs, nuclear paraspeckles	ALS, FTD
Arginine Citrullination	FUS	RGG domain	↓	Disrupts charge-charge interactions	SGs, nuclear paraspeckles	ALS, FTD
Lysine Acetylation	DDX3X	NTD IDR	↓	Disrupts cation–π interactions	SGs	Intellectual disability
	tau	Internal IDR	↓		SGs	Alzheimer's disease
Lysine Ribosylation	hnRNPA1	Glycine-rich region	↑	Increases multivalency	SGs	ALS, FTD

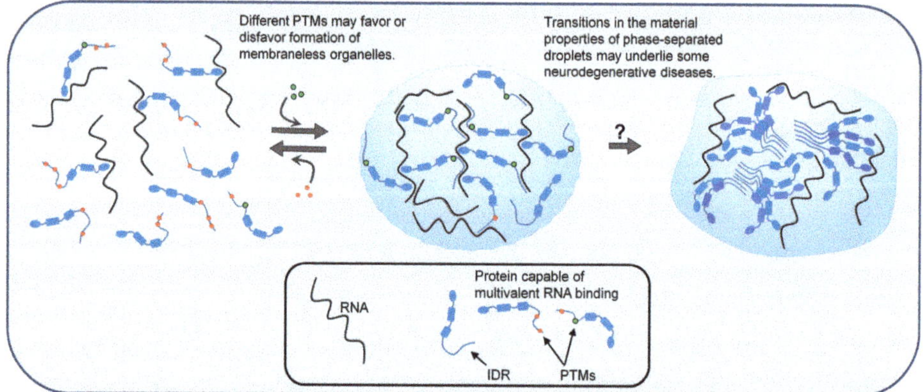

Figure 1. Liquid–liquid phase separation (LLPS) of biopolymers, such as proteins and RNA, is a mechanism by which cells organize their contents into specific functional structures called membraneless organelles (MLOs). Post-translational modifications (PTMs) of intrinsically disordered proteins can influence LLPS and thus regulate the formation and dissolution of MLOs. The figure depicts different patterns of PTMs favoring dispersed or condensed states. Changes in the material properties of liquid-phase separated granules are hypothesized to cause some neurodegenerative diseases. According to this hypothesis, droplets lose their liquid (reversible) properties and adopt more rigid (less reversible) internal structures, which may be glass-like, or in some cases, may have solid amyloid-like structures. These irreversible phase states may have gain-of-function toxicity to neurons.

3.1. Serine/Threonine/Tyrosine Phosphorylation

Phosphorylation is the covalent attachment of a phosphoryl group to an amino acid hydroxyl group. The phosphoryl group is negatively charged, so its addition changes a polar, uncharged residue to a negatively charged amino acid. Serine is the most commonly phosphorylated residue, followed by threonine and tyrosine [30]. The addition of charges to macromolecules may promote certain charge–charge interactions that drive complex coacervation (phase separation of oppositely charged polymers) [31]. Alternatively, addition of phosphates may cause charge repulsion or steric hindrance, thus inhibiting phase separation [2]. Depending on the protein context, the phosphate modification of amino acids can either favor or disfavor phase separation.

Serine/threonine phosphorylation has been shown to promote phase separation of IDPs such as fragile X mental retardation protein (FMRP) [32], TIA-1/TIAL RNA binding protein homolog (TIAR-2) [33], and microtubule-binding protein tau [34]. FMRP has 12 serine residues within its C-terminal IDR (aa 445–632) that have been identified as targets of casein kinase II (CKII). In vitro phosphorylation by CKII results in an increase in the negative charge densities throughout this IDR, increasing the propensity for multivalent electrostatic interactions and promoting phase separation [32]. TIAR-2 also contains a C-terminal intrinsically disordered PrLD that facilitates its LLPS into cytosolic granules [33]. The PrLD of TIAR-2 was shown to be serine phosphorylated when expressed in mechanosensory neurons. Ten serine residues in the intrinsically disordered PrLD were predicted as phospho-sites using NetPhos3.1 [35]. Expression of a non-phosphorylatable (S→A) TIAR-2 mutant (at 10, 8, or 2 serine residues) showed significantly less granule formation in the axons of neurons when compared to wild type. Alternatively, phosphomimetic (S→E) TIAR-2 mutants showed similar levels of granule formation when compared to wild type. These data suggest phosphorylation of serine residues promotes LLPS and formation of TIAR-2 positive granules in neurons of *C. elegans* [33].

PTMs also affect virally encoded proteins and their LLPS capabilities in cells. Measles virus phosphoprotein is a 507-amino acid virally encoded protein composed of multiple IDRs. Measles virus phosphoprotein and nucleoprotein undergo LLPS to form IBs. Phosphoprotein is phosphorylated at multiple sites, but Serine 86 and Serine 151—both of which are in IDRs—have been identified as regulatory sites for IB formation. Mutation or inhibition of phosphorylation at these two sites results in irregular and small IBs [12].

Examples of serine/threonine phosphorylation that disrupt LLPS include maternal-effect germline proteins (MEGs) in P granules of *C. elegans* [36], fused in sarcoma (FUS) [37–39] and TAR DNA-binding protein 43 (TDP-43) [40]. Proper segregation of P granules in zygotes of *C. elegans* requires the expression of MEG proteins [41]. Interestingly, two of the MEG family proteins (MEG-1 and MEG-3) are phosphorylated within their IDRs by a regulatory kinase (MBK-2) [36]. MBK-2 activity and counteractive phosphatase (PPTR-1) activity on MEGs is required for P granule disassembly and formation, respectively [36].

FUS and TDP-43 are frequently studied proteins because their phase separation in vivo has been linked to amyotrophic lateral sclerosis (ALS) [42]. A current hypothesis is that MLOs containing these proteins may promote their stochastic conversion into solid, pathological aggregates [43]. FUS contains an intrinsically disordered N-terminal PrLD, which is necessary and sufficient to drive LLPS [5]. The ~160 amino acid PrLD has 32 putative phosphorylation sites, 12 of which have been identified as PIKK family kinase consensus sites [44]. Phosphomimetic substitution (S/T→E) at 6 or 12 PIKK consensus sites diminishes FUS's ability to phase separate and form fibrillar aggregates in vitro [37]. In cells, a decrease in cytoplasmic aggregation is also observed upon increase in phosphomimetic substitution, suggesting a potential therapeutic target for disrupting pathological aggregate formation [37]. TDP-43 has a C-terminal PrLD, which is multiphosphorylated and aggregated in ALS motor neurons [45]. Two phosphorylation sites in the PrLD, Serine 409 and 410, identified in samples from frontotemporal lobar dementia patients were shown to regulate TDP-43 cytoplasmic granule formation. Phosphomimetic substitution (S→D) at Serine 409 and 410 showed a significant reduction in the number of cells containing TDP-43 puncta [46]. Interestingly, TDP-43 phase separation is regulated by PTMs in both an IDR and a structured domain. A single phosphorylation event in its N-terminal structured domain at Serine 48 is sufficient to suppress its LLPS in vitro and in cells [40]. Serine 48 is conserved in most species evaluated, including flies, mice, and humans [40].

There are two kinases, SKY1 and DYRK3, that have the ability to phase separate into SGs and, in the stages of recovery following a stress response, phosphorylate proteins containing IDRs, resulting in dissolution of the granules [47,48]. SKY1 is a yeast protein kinase with a PrLD that enables its recruitment into SGs. In SGs, SKY1 phosphorylates NLP3 at Serine 441, which is located in its serine-arginine rich C-terminal IDR [49]. This phosphorylation event promotes SG dissolution [48]. Similarly, DYRK3 (human homolog of MBK-2) was shown to phase separate into SGs via its intrinsically disordered N-terminal domain. Aside from regulation of SGs, DYRK3 was identified as a factor that controls phase separation and dissolution of several condensates containing IDPs during mitosis [50]. This kinase is interesting because it has broad-specificity and is generally proline-directed. Some proteins sensitive to DYRK3 inhibition, all of which contain IDRs, include splicing-speckle marker SC35, SG marker PABP, and pericentriolar-material protein PCM1. DYRK3 expression results in dissolution of these granules during mitosis, whereas a kinase-dead mutant or inhibition of DYRK3 results in the persistence of granules [50].

For some proteins, phosphorylation can be a driver or inhibitor of condensate formation, but there are instances where it is not clearly binary. Tau441 contains numerous IDRs and putative serine/threonine phosphorylation sites throughout the protein [51]. In experiments performed with bacterially produced recombinant full-length tau441 and molecular crowding agents, LLPS was driven mostly by electrostatic intermolecular interactions. There was no requirement for phosphorylation [52]. However, in a different study, phosphorylation was found to be required to initiate tau441 LLPS in vitro [34]. Of importance, there are 22 phospho-sites analyzed in this study, 15 of the sites are located in IDRs. Interestingly, LLPS of p-tau was dependent on hydrophobic interactions [34], whereas unphosphorylated tau LLPS was more dependent on ionic interactions [52]. These examples suggest the biophysical mechanism driving LLPS can be altered by the post-translational state of the protein.

3.2. Arginine Methylation

Methylation of arginine residues is important for regulating phase separation and recruitment of proteins into MLOs. The side chain of arginine contains a positively charged guanidinium head group that can be multiply methylated. This reaction is catalyzed by protein methyltransferases (PRMTs) using a donor methyl group from S-adenosyl methionine (SAM). Arginine methylation does not change the charge of the side group, but instead alters its volume, charge distribution, hydrophobicity, and potential for hydrogen bonding [53].

Including unmethylated arginine, there are four differential arginine methylation patterns. Arginines can be monomethylated (MMA) or dimethylated (DMA); arginine dimethylation can exist as symmetrical dimethylation (SDMA) or asymmetrical dimethylation (ADMA). Symmetric dimethylation occurs when two methyl groups are added to the two different nitrogen atoms within the guanidino group, whereas asymmetric dimethylation is when two methyl groups are added to the same nitrogen [54]. These reactions are catalyzed by different methyltransferases. PRMTs are grouped into three subtypes based on their catalytic activity. Type I PRMTs include PRMT1, 2, 3, 4, 6, and 8, while type II PRMTs include PRMT5 and 9 and there is one type III methyltransferase, PRMT7 [55,56]. All three classes of PRMTs have the ability to catalyze MMA reactions, but the DMA reactions that occur subsequent to this reaction are specific to type I and II enzymes. Type I PRMT enzymes catalyze the reaction of MMA to ADMA and type II enzymes catalyze MMA to SDMA. PRMT enzymes target arginine residues within glycine-arginine-glycine (GRG) or arginine-glycine-glycine (RGG) sequences [57], which are preferred sites of arginine methylation [58], and are frequently encoded as multiple repeats within low-complexity regions [6].

The hydrogen bond potential of the head group is the same for SDMA and ADMA, but the location of the methyl group can alter the orientation of hydrogen bonding. Additionally, SDMA and ADMA have different electrostatic surface potentials to the head group, resulting in shifting of charge [59]. The hydrophobicity of the head group is also modified upon methylation. Arginine hydrophobicity incrementally increases following the addition of MMA, ADMA, and SDMA, respectively [53]. Hydrophobic residues are generally located within the folded core of proteins, so within the context of an IDR, these changes may have profound effects on the propensity to fold or bind other macromolecules.

The guanidinium electrons are delocalized into π orbitals, enabling interactions via π-stacking [54]. Arginines can form cation-π interactions, which occur between the positively charged guanidinium group and the available electrons of the π orbital of aromatic rings [60]. These cation-π interactions have been shown to drive protein condensate formation [61]. Methylation of arginine has been shown to both favor and disfavor phase separation in certain contexts. ADMA in RGG motifs has been shown to disrupt favorable interactions and thus disrupt phase separation of IDPs such as dead-box helicase 4 (DDX4) [6], heterogeneous nuclear ribonucleoproteins A2 (hnRNPA2) [62], RAP55A [63,64], and FUS [65]. SDMA within RGG motifs, however, has been shown to drive phase separation. In the case of U6 snRNA-associated Sm-like protein (LSM4), SDMA is necessary for phase separation and processing body formation [66]. PRMT5 catalyzes SDMA of LSM4 at multiple arginine residues in its intrinsically disordered C-terminal RGG-containing domain. Mutation of the arginine residues in this domain or knockdown of PRMT5 diminish SDMA and processing body formation in cells [66].

YTH domain-containing family (YTHDF) of proteins contain numerous IDRs, which allows them to undergo LLPS. These proteins are found in cytoplasmic phase-separated SGs, P-bodies, and neuronal RNA granules [67]. In vitro experiments show that enzymatic modifications drive phase separation, but interestingly, the modifications occur to mRNA, not the protein. Methylation of adenosine in RNA, specifically the formation of N^6-methyladenosine (m^6A), is the most commonly modified nucleotide in mRNA. This modification increases mRNA's multivalency and seeds phase separation of YTHDF proteins in vitro [67]. The protein has the ability to form droplets in solution without the presence of m^6A, but only at much higher concentrations. Importantly, the mRNA that seeded droplet formation of YTHDF was multimethylated; unmethylated or singly methylated mRNA did not show droplet formation [67].

3.3. Arginine Citrullination

Citrullination is another PTM that occurs to arginine residues. Instead of the addition of a functional group, the arginine side chain undergoes an oxidation (or deimination) reaction. In this reaction, peptidylarginine deiminases (PADs) catalyze the oxidation of an imine group (=NH), forming a ketone group (=O) [60,68]. This modification removes the positive charge, leaving a neutrally charged amino acid. Interestingly, the consensus site for PADs are the same RG/RGG motifs that are common to many RNA-binding and phase separating proteins [69]. Citrullination of FUS by PAD4 was shown to diminish FUS recruitment to SGs [69]. PAD4 knockout in mouse embryonic fibroblasts showed a greater amount of FUS sequestration into SGs than when PAD4 was overexpressed in these cells, suggesting citrullination hinders FUS phase separation. Cation-π interactions between arginine residues and the π orbitals of tyrosine residues modulate FUS phase separation [60], but when citrullination occurs, the positive charge of the arginine side chain is removed, disrupting cation-π interactions and disrupting FUS phase separation in vitro [60].

3.4. Lysine Acetylation

Similar to citrullination, acetylation neutralizes the positive charge of an amino acid. Lysine residues contain a positively charged amino head group that can be neutralized by addition of an acetyl group; this not only changes lysine's charge state but also increases its hydrophobicity [70]. Acetyl groups are enzymatically added via acetyltransferases and removed by deacetylases [71]. Acetylation has been shown to disrupt simple coacervation of DDX3X (dead box RNA helicase 3) in vitro [72]. DDX3X has two IDRs: at the C-terminus and the N-terminus. Analysis of an acetylome dataset identified several acetylated lysines in the N-terminal IDR that play a role in DDX3X incorporation into SGs [72]. DDX3X is a substrate of acetyltransferase CREB-binding protein (CBP) and histone deacetylase 6 (HDAC6). To better understand the role of acetylation of DDX3X and SG incorporation, acetyl mimetic (K→Q) constructs and acetyl-dead (K→R) were constructed and expressed in DDX3X knock-out cell lines. Expression of acetyl-dead DDX3X (or inhibition of CBP) increased SG volume, whereas expression of acetyl-mimetic mutant (or inhibition of HDAC6) decreased SG volume.

The formation of SG has been proposed as a two-step process. First a stable core structure is formed, which is followed by the recruitment of IDPs into an outer shell structure [73]. The increase in volume is an important step in SG maturation. Of significance to this growth mechanism, the interaction partners of the acetyl-dead and acetyl mimetic DDX3X mutants were different. The non-acetylated DDX3X interacts with numerous SG components, whereas the acetyl mimetic loses its capacity for interactions with SG proteins, thus showcasing how lysine acetylation can be used to regulate MLO maturation [72].

Another protein that is lysine acetylated is tau, which is of particular interest since its solid-phase aggregation in neurons is linked to Alzheimer's disease [74]. Tau is an IDP, and like DDX3X, its ability to phase separate is disrupted by lysine acetylation [75]. Ferreon et al. found that recombinant tau, when incubated with enzymatically-active p300 histone acetyltransferase (HAT), becomes hyperacetylated (ac-tau) [75]. This acetylation (removal of positive charges) was observed to disfavor LLPS, which is consistent with the previous observations of Wegmann et al. that phosphorylation (addition of negative charges) promotes tau phase separation [34]. Using mass spectrometry analysis, 15 acetylation sites were identified, 8 of which are located in IDRs. Tau readily undergoes LLPS in vitro in low-salt conditions, but ac-tau was unable to form droplets under the same conditions. The neutralization of charged residues was concluded to disrupt electrostatic interactions required for tau LLPS [52,75]. Interestingly, tau also phase separates into SGs [76]. Ukmar-Godec et al. showed that tau association into SGs is altered by the acetylation state of the lysines. Unmodified full length tau441 readily associated with SGs following proteasome inhibition by MG132. Consistent with the in vitro findings above, acetylation of tau strongly reduced the association of the protein with SGs in HeLa cells [77]. Lastly, Ac-tau also showed decreased solid-phase aggregation propensity and reduced thioflavin-t reactivity, which indicates less propensity to form amyloid-like solid aggregates.

This suggests acetyltransferases and deacetylases are potential therapeutic targets for prevention of pathological tau aggregation [75].

3.5. Poly(ADP-Ribosylation)

Poly(ADP-ribosyl)ation or PARylation is a reversible covalent addition of multiple NAD-derived ADP-ribose (ADPr) molecules to a protein [78]. ADPr units can be added to glutamate, aspartate, lysine, arginine, or serine residues by poly(ADP-ribose) polymerases (PARPs) and removed by PAR glycohydrolases (PARGs) [78]. Aside from the physical addition of ADPr units, polyADP-ribose (PAR) molecules are freely synthesized polymers that can modify phase separation of some IDPs [79]. PAR is a multivalent, anionic, nucleic acid-mimicking (similar to RNA) biopolymer that can be bound by phase separating proteins [2]. Cellular stress conditions and DNA damage have been shown to cause an upregulation of PAR synthesis [79]. PAR, PARPs, and PARGs have all been shown to play a regulatory role in SG dynamics [80].

The SG component hnRNPA1 contains both a PAR-binding domain and a PARylation site at Lysine 298 within a glycine rich IDR. PARylation at hnRNPA1 Lysine 298 is important for hnRNPA1 nucleocytoplasmic shuttling, a necessary step for localization to SGs following cellular stress [81]. Interestingly, like numerous other proteins, hnRNPA1 contains a PAR-binding motif (PBM). In vitro experiments showed hnRNPA1 phase separation increasing in response to increased PAR concentration in solution. Mutating the hnRNPA1 PBM resulted in no phase separation in the presence of PAR, implying this interaction is domain specific. TDP-43 also contains a PBM and co-phase separates with hnRNPA1 in vitro and in SGs [81,82]. In vitro, PAR binding via hnRNPA1 PBM is necessary for the co-phase separation of TDP-43 and hnRNPA1 in low-salt concentrations. In cells, both TDP-43 and hnRNPA1 need functional PBMs to localize to SGs, highlighting the role of PAR in protein–protein interaction and phase transition [81,82].

4. Membraneless Organelles and Neurodegenerative Diseases

A connection between MLOs and neurodegenerative disease has been widely observed [83]. Specifically, many proteins that are genetically or histopathologically linked to neurodegeneration are also found in neuronal MLOs [18]. Likewise, proteins with intrinsically disordered PrLDs are notoriously disproportionately linked to neurodegenerative disease [84], and many of these proteins are both capable of undergoing LLPS and frequently found within inclusions of diseased neurons [85].

Why do the same proteins that functionally undergo LLPS appear to adopt pathological meso-scale aggregates in cells? A leading hypothesis is proteins within MLOs may undergo additional transitions into oligomeric species, solid-phase aggregates [86], or droplet-like structures with dramatically different material properties [17] (Figure 1). For example, expression of an ALS-linked TDP-43 mutant results in an MLO that is more viscous and resistant to solvation, suggesting it has a stabilized internal structure [19]. Similarly, in vitro, ALS-mutant FUS can transition from a droplet state into a solid aggregate more rapidly than wild-type FUS [42]. Once formed, such aggregates are thought to be detrimental to cell function and contribute to neuronal degeneration. Possibly, the high concentration of specific proteins within MLOs may potentiate these stochastic, irreversible phase transitions. Since disease-linked proteins like tau and TDP-43 are hyper- and multi-phosphorylated, respectively, within neuronal cytoplasmic inclusions, it is possible the PTMs are facilitating solid-phase transitions; alternatively, the PTMs may simply mark failed attempts at solubilization.

In the case of many IDRs, the abundance of hydrophilic amino acids and lack of stable tertiary structures may facilitate solid-phase transitions into highly ordered amyloid conformations. Amyloid is a well-ordered, filamentous polymeric state composed of a single protein species, much like a one-dimensional crystal [87]. It usually consists of polypeptides aligning in parallel in-register beta sheets [88] and is notoriously difficult to solubilize. MLOs may provide an environment in which some enriched IDR-containing proteins can stochastically adopt amyloid-like conformations, thus explaining the presence of certain MLO-linked proteins in pathological neuronal inclusions. Examples include tau

(Alzheimer's disease), TDP-43 (ALS), and FUS (frontotemporal dementia). Importantly, crystal-like arrangements would be disrupted by PTMs occurring within the structural core of amyloid [89]; thus, targeting specific modifying enzymes could offer a viable therapeutic strategy for neurodegenerative disorders that feature solid-phase inclusions.

5. Future Directions

Experimentation with MLOs is frequently focused on a few protein species. However, in vivo, the entire repertoire of macromolecules within individual biocondensates remains largely unknown. Additionally, for any given MLO, specific protein components can exhibit a broad array of PTMs, thus making it difficult to dissect which modifications are altering LLPS or perhaps serving other non-structural functional roles. Going forward, a major challenge will be to determine the precise relationship between MLOs and disease processes. For example, many human viruses encode proteins with PrLDs [90]. Given what we know about this type of protein domain, it is possible many viruses exploit LLPS during replication and infection, yet antiviral drugs do not specifically target LLPS mechanisms. In the case of neurodegenerative diseases, the pathological connections between aberrant phase transitions and neuronal death are not fully understood, and there are many non-unifying hypotheses. No drugs specifically target phase-separation processes in any neurodegenerative disease. Yet, given the almost complete lack of drugs for treating these diseases, manipulating the enzymes that regulate biocondensation may provide a new target paradigm.

Author Contributions: This manuscript was written by I.O. and F.S.

Funding: This work is supported by the National Institute of General medical Sciences (NIGMS) of the National Institutes of Health (NIH) under Award Number R35GM119790.

Acknowledgments: We thank Debra Yee and Hala Wyne for reviewing and editing our text.

Conflicts of Interest: The authors declare no conflict of interest. The funders had no role in the design of the study; in the collection, analyses, or interpretation of data; in the writing of the manuscript, or in the decision to publish the results.

Abbreviations

ADMA	Asymmetric dimethyl arginine
ADPr	NAD-derived ADP-ribose
IBs	Inclusion bodies
IDP	Intrinsically disordered protein
IDR	Intrinsically disordered region
LCD	Low-complexity domain
LLPS	Liquid–liquid phase separation
MMA	Monomethyl arginine
MLO	Membrane-less organelle
PTM	Post-translational modification
PrLD	Prion-like domain
SDMA	Symmetric dimethyl arginine
SG	Stress granule
RGG/RGR	Arginine/glycine repeats

References

1. Darling, A.L.; Liu, Y.; Oldfield, C.J.; Uversky, V.N. Intrinsically Disordered Proteome of Human Membrane-Less Organelles. *Proteomics* **2018**, *18*, e1700193. [CrossRef]
2. Shin, Y.; Brangwynne, C.P. Liquid phase condensation in cell physiology and disease. *Science* **2017**, *357*, eaaf4382. [CrossRef]
3. Hyman, A.A.; Weber, C.A.; Jülicher, F. Liquid-Liquid Phase Separation in Biology. *Annu. Rev. Cell Dev. Biol.* **2014**, *30*, 39–58. [CrossRef]

4.	Murthy, A.C.; Dignon, G.L.; Kan, Y.; Zerze, G.H.; Parekh, S.H.; Mittal, J.; Fawzi, N.L. Molecular interactions underlying liquid-liquid phase separation of the FUS low-complexity domain. *Nat. Struct. Mol. Biol.* **2019**, *26*, 637–648. [CrossRef]

5.	Burke, K.A.; Janke, A.M.; Rhine, C.L.; Fawzi, N.L. Residue-by-Residue View of In Vitro FUS Granules that Bind the C-Terminal Domain of RNA Polymerase II. *Mol. Cell* **2015**, *60*, 231–241. [CrossRef]

6.	Nott, T.J.; Petsalaki, E.; Farber, P.; Jervis, D.; Fussner, E.; Plochowietz, A.; Craggs, T.D.; Bazett-Jones, D.P.; Pawson, T.; Forman-Kay, J.D.; et al. Phase transition of a disordered nuage protein generates environmentally responsive membraneless organelles. *Mol. Cell* **2015**, *57*, 936–947. [CrossRef]

7.	Uversky, V.N. Intrinsically disordered proteins in overcrowded milieu: Membrane-less organelles, phase separation, and intrinsic disorder. *Curr. Opin. Struct. Biol.* **2017**, *44*, 18–30. [CrossRef]

8.	Courchaine, E.M.; Lu, A.; Neugebauer, K.M. Droplet organelles? *EMBO J.* **2016**, *35*, 1603–1612. [CrossRef]

9.	You, K.; Huang, Q.; Yu, C.; Shen, B.; Sevilla, C.; Shi, M.; Hermjakob, H.; Chen, Y.; Li, T. PhaSepDB: A database of liquid-liquid phase separation related proteins. *Nucleic Acids Res.* **2019**. [CrossRef]

10.	Antonicka, H.; Shoubridge, E.A. Mitochondrial RNA Granules Are Centers for Posttranscriptional RNA Processing and Ribosome Biogenesis. *Cell Rep.* **2015**, *10*, 920–932. [CrossRef]

11.	Marnik, E.A.; Updike, D.L. Membraneless organelles: P granules in Caenorhabditis elegans. *Traffic* **2019**, *20*, 373–379. [CrossRef]

12.	Zhou, Y.; Su, J.M.; Samuel, C.E.; Ma, D. Measles Virus Forms Inclusion Bodies with Properties of Liquid Organelles. *J. Virol.* **2019**, *93*, e00948-19. [CrossRef]

13.	Molliex, A.; Temirov, J.; Lee, J.; Coughlin, M.; Kanagaraj, A.P.; Kim, H.J.; Mittag, T.; Taylor, J.P. Phase separation by low complexity domains promotes stress granule assembly and drives pathological fibrillization. *Cell* **2015**, *163*, 123–133. [CrossRef]

14.	Drino, A.; Schaefer, M.R. RNAs, Phase Separation, and Membrane-Less Organelles: Are Post-Transcriptional Modifications Modulating Organelle Dynamics? *BioEssays* **2018**, *40*, 1800085. [CrossRef]

15.	Sephton, C.F.; Yu, G. The function of RNA-binding proteins at the synapse: Implications for neurodegeneration. *Cell. Mol. Life Sci.* **2015**, *72*, 3621–3635. [CrossRef]

16.	Krichevsky, A.M.; Kosik, K.S. Neuronal RNA granules: A link between RNA localization and stimulation-dependent translation. *Neuron* **2001**, *32*, 683–696. [CrossRef]

17.	Nedelsky, N.B.; Taylor, J.P. Bridging biophysics and neurology: Aberrant phase transitions in neurodegenerative disease. *Nat. Rev. Neurol.* **2019**, *15*, 272–286. [CrossRef]

18.	Ryan, V.H.; Fawzi, N.L. Physiological, Pathological, and Targetable Membraneless Organelles in Neurons. *Trends Neurosci.* **2019**, *42*, 693–708. [CrossRef]

19.	Gopal, P.P.; Nirschl, J.J.; Klinman, E.; Holzbaur, E.L.F. Amyotrophic lateral sclerosis-linked mutations increase the viscosity of liquid-like TDP-43 RNP granules in neurons. *Proc. Natl. Acad. Sci. USA* **2017**, *114*, E2466–E2475. [CrossRef]

20.	Boeynaems, S.; Bogaert, E.; Kovacs, D.; Konijnenberg, A.; Timmerman, E.; Volkov, A.; Guharoy, M.; De Decker, M.; Jaspers, T.; Ryan, V.H.; et al. Phase Separation of C9orf72 Dipeptide Repeats Perturbs Stress Granule Dynamics. *Mol. Cell* **2017**, *65*, 1044–1055.e5. [CrossRef]

21.	Van der Lee, R.; Buljan, M.; Lang, B.; Weatheritt, R.J.; Daughdrill, G.W.; Dunker, A.K.; Fuxreiter, M.; Gough, J.; Gsponer, J.; Jones, D.T.; et al. Classification of Intrinsically Disordered Regions and Proteins. *Chem. Rev.* **2014**, *114*, 6589–6631. [CrossRef] [PubMed]

22.	Vernon, R.M.; Chong, P.A.; Tsang, B.; Kim, T.H.; Bah, A.; Farber, P.; Lin, H.; Forman-Kay, J.D. Pi-Pi contacts are an overlooked protein feature relevant to phase separation. *eLife* **2018**, *7*, e31486. [CrossRef] [PubMed]

23.	Tompa, P.; Davey, N.E.; Gibson, T.J.; Babu, M.M. A Million Peptide Motifs for the Molecular Biologist. *Mol. Cell* **2014**, *55*, 161–169. [CrossRef] [PubMed]

24.	Dyson, H.J. Expanding the proteome: Disordered and alternatively-folded proteins. *Q. Rev. Biophys.* **2011**, *44*, 467–518. [CrossRef] [PubMed]

25.	Holt, L.J.; Tuch, B.B.; Villen, J.; Johnson, A.D.; Gygi, S.P.; Morgan, D.O. Global analysis of Cdk1 substrate phosphorylation sites provides insights into evolution. *Science* **2009**, *325*, 1682–1686. [CrossRef]

26.	Iakoucheva, L.M.; Radivojac, P.; Brown, C.J.; O'Connor, T.R.; Sikes, J.G.; Obradovic, Z.; Dunker, A.K. The importance of intrinsic disorder for protein phosphorylation. *Nucleic Acids Res.* **2004**, *32*, 1037–1049. [CrossRef]

27. Collins, M.O.; Yu, L.; Campuzano, I.; Grant, S.G.N.; Choudhary, J.S. Phosphoproteomic Analysis of the Mouse Brain Cytosol Reveals a Predominance of Protein Phosphorylation in Regions of Intrinsic Sequence Disorder. *Mol. Cell. Proteom.* **2008**, *7*, 1331–1348. [CrossRef]

28. Li, Q.; Shortreed, M.R.; Wenger, C.D.; Frey, B.L.; Schaffer, L.V.; Scalf, M.; Smith, L.M. Global Post-Translational Modification Discovery. *J. Proteome Res.* **2017**, *16*, 1383–1390. [CrossRef]

29. Bah, A.; Forman-Kay, J.D. Modulation of Intrinsically Disordered Protein Function by Post-translational Modifications. *J. Biol. Chem.* **2016**, *291*, 6696–6705. [CrossRef]

30. Hofweber, M.; Dormann, D. Friend or foe-Post-translational modifications as regulators of phase separation and RNP granule dynamics. *J. Biol. Chem.* **2019**, *294*, 7137–7150. [CrossRef]

31. Pak, C.W.; Kosno, M.; Holehouse, A.S.; Padrick, S.B.; Mittal, A.; Ali, R.; Yunus, A.A.; Liu, D.R.; Pappu, R.V.; Rosen, M.K. Sequence Determinants of Intracellular Phase Separation by Complex Coacervation of a Disordered Protein. *Mol. Cell* **2016**, *63*, 72–85. [CrossRef] [PubMed]

32. Tsang, B.; Arsenault, J.; Vernon, R.M.; Lin, H.; Sonenberg, N.; Wang, L.Y.; Bah, A.; Forman-Kay, J.D. Phosphoregulated FMRP phase separation models activity-dependent translation through bidirectional control of mRNA granule formation. *Proc. Natl. Acad. Sci. USA* **2019**, *116*, 4218–4227. [CrossRef] [PubMed]

33. Andrusiak, M.G.; Sharifnia, P.; Lyu, X.; Wang, Z.; Dickey, A.M.; Wu, Z.; Chisholm, A.D.; Jin, Y. Inhibition of Axon Regeneration by Liquid-like TIAR-2 Granules. *Neuron* **2019**, *104*, 290–304. [CrossRef] [PubMed]

34. Wegmann, S.; Eftekharzadeh, B.; Tepper, K.; Zoltowska, K.M.; Bennett, R.E.; Dujardin, S.; Laskowski, P.R.; MacKenzie, D.; Kamath, T.; Commins, C.; et al. Tau protein liquid–liquid phase separation can initiate tau aggregation. *EMBO J.* **2018**, *37*, e98049. [CrossRef] [PubMed]

35. Blom, N.S.; Gammeltoft, S.; Brunak, S. Sequence and structure-based prediction of eukaryotic protein phosphorylation sites. *J. Mol. Biol.* **1999**, *294*, 1351–1362. [CrossRef]

36. Wang, J.T.; Smith, J.; Chen, B.-C.; Schmidt, H.; Rasoloson, D.; Paix, A.; Lambrus, B.G.; Calidas, D.; Betzig, E.; Seydoux, G. Regulation of RNA granule dynamics by phosphorylation of serine-rich, intrinsically disordered proteins in *C. elegans*. *eLife* **2014**, *3*, e04591. [CrossRef]

37. Monahan, Z.; Ryan, V.H.; Janke, A.M.; Burke, K.A.; Rhoads, S.N.; Zerze, G.H.; O'Meally, R.; Dignon, G.L.; Conicella, A.E.; Zheng, W.; et al. Phosphorylation of the FUS low-complexity domain disrupts phase separation, aggregation, and toxicity. *EMBO J.* **2017**, *36*, 2951–2967. [CrossRef]

38. Lin, Y.; Currie, S.L.; Rosen, M.K. Intrinsically disordered sequences enable modulation of protein phase separation through distributed tyrosine motifs. *J. Biol. Chem.* **2017**, *292*, 19110–19120. [CrossRef]

39. Rhoads, S.N.; Monahan, Z.T.; Yee, D.S.; Leung, A.Y.; Newcombe, C.G.; O'Meally, R.N.; Cole, R.N.; Shewmaker, F.P. The prionlike domain of FUS is multiphosphorylated following DNA damage without altering nuclear localization. *Mol. Biol. Cell* **2018**, *29*, 1786–1797. [CrossRef]

40. Wang, A.; Conicella, A.E.; Schmidt, H.B.; Martin, E.W.; Rhoads, S.N.; Reeb, A.N.; Nourse, A.; Montero, D.R.; Ryan, V.H.; Rohatgi, R.; et al. A single N-terminal phosphomimic disrupts TDP-43 polymerization, phase separation, and RNA splicing. *EMBO J.* **2018**, *37*, e97452. [CrossRef]

41. Seydoux, G. The P Granules of *C. elegans*: A Genetic Model for the Study of RNA–Protein Condensates. *J. Mol. Biol.* **2018**, *430*, 4702–4710. [CrossRef] [PubMed]

42. Patel, A.; Lee, H.O.; Jawerth, L.; Maharana, S.; Jahnel, M.; Hein, M.Y.; Stoynov, S.; Mahamid, J.; Saha, S.; Franzmann, T.M.; et al. A Liquid-to-Solid Phase Transition of the ALS Protein FUS Accelerated by Disease Mutation. *Cell* **2015**, *162*, 1066–1077. [CrossRef] [PubMed]

43. March, Z.M.; King, O.D.; Shorter, J. Prion-like domains as epigenetic regulators, scaffolds for subcellular organization, and drivers of neurodegenerative disease. *Brain Res.* **2016**, *1647*, 9–18. [CrossRef] [PubMed]

44. Rhoads, S.N.; Monahan, Z.T.; Yee, D.S.; Shewmaker, F.P. The Role of Post-Translational Modifications on Prion-Like Aggregation and Liquid-Phase Separation of FUS. *Int. J. Mol. Sci.* **2018**, *19*, 886. [CrossRef] [PubMed]

45. Neumann, M.; Kwong, L.K.; Lee, E.B.; Kremmer, E.; Flatley, A.; Xu, Y.; Forman, M.S.; Troost, D.; Kretzschmar, H.A.; Trojanowski, J.Q.; et al. Phosphorylation of S409/410 of TDP-43 is a consistent feature in all sporadic and familial forms of TDP-43 proteinopathies. *Acta Neuropathol.* **2009**, *117*, 137–149. [CrossRef] [PubMed]

46. Brady, O.A.; Meng, P.; Zheng, Y.; Mao, Y.; Hu, F. Regulation of TDP-43 aggregation by phosphorylation and p62/SQSTM1. *J. Neurochem.* **2011**, *116*, 248–259. [CrossRef]

47. Wippich, F.; Bodenmiller, B.; Trajkovska, M.G.; Wanka, S.; Aebersold, R.; Pelkmans, L. Dual Specificity Kinase DYRK3 Couples Stress Granule Condensation/Dissolution to mTORC1 Signaling. *Cell* **2013**, *152*, 791–805. [CrossRef]

48. Shattuck, J.E.; Paul, K.R.; Cascarina, S.M.; Ross, E.D. The prion-like protein kinase Sky1 is required for efficient stress granule disassembly. *Nat. Commun.* **2019**, *10*, 3614. [CrossRef]

49. Gilbert, W.; Siebel, C.W.; Guthrie, C. Phosphorylation by Sky1p promotes Npl3p shuttling and mRNA dissociation. *RNA* **2001**, *7*, 302–313. [CrossRef]

50. Rai, A.K.; Chen, J.-X.; Selbach, M.; Pelkmans, L. Kinase-controlled phase transition of membraneless organelles in mitosis. *Nature* **2018**, *559*, 211–216. [CrossRef]

51. Mair, W.; Muntel, J.; Tepper, K.; Tang, S.; Biernat, J.; Seeley, W.W.; Kosik, K.S.; Mandelkow, E.; Steen, H.; Steen, J.A. FLEXITau: Quantifying Post-translational Modifications of Tau Protein in Vitro and in Human Disease. *Anal. Chem.* **2016**, *88*, 3704–3714. [CrossRef] [PubMed]

52. Boyko, S.; Qi, X.; Chen, T.-H.; Surewicz, K.; Surewicz, W.K. Liquid-liquid phase separation of tau protein: The crucial role of electrostatic interactions. *J. Biol. Chem.* **2019**, *294*, 11054–11059. [CrossRef] [PubMed]

53. Evich, M.; Stroeva, E.; Zheng, Y.G.; Germann, M.W. Effect of methylation on the side-chain pKa value of arginine. *Protein Sci.* **2016**, *25*, 479–486. [CrossRef] [PubMed]

54. Lorton, B.M.; Shechter, D. Cellular consequences of arginine methylation. *Cell Mol. Life Sci.* **2019**, *76*, 2933–2956. [CrossRef] [PubMed]

55. Morales, Y.; Cáceres, T.; May, K.; Hevel, J.M. Biochemistry and regulation of the protein arginine methyltransferases (PRMTs). *Arch. Biochem. Biophys.* **2016**, *590*, 138–152. [CrossRef]

56. Fulton, M.D.; Brown, T.; Zheng, Y.G. The Biological Axis of Protein Arginine Methylation and Asymmetric Dimethylarginine. *Int. J. Mol. Sci.* **2019**, *20*, 3322. [CrossRef]

57. Thandapani, P.; O'Connor, T.R.; Bailey, T.L.; Richard, S. Defining the RGG/RG Motif. *Mol. Cell* **2013**, *50*, 613–623. [CrossRef]

58. Boisvert, F.-M.; Chenard, C.A.; Richard, S. Protein Interfaces in Signaling Regulated by Arginine Methylation. *Sci. Signal.* **2005**, *2005*, 2. [CrossRef]

59. Fuhrmann, J.; Clancy, K.W.; Thompson, P.R. Chemical Biology of Protein Arginine Modifications in Epigenetic Regulation. *Chem. Rev.* **2015**, *115*, 5413–5461. [CrossRef]

60. Qamar, S.; Wang, G.; Randle, S.J.; Ruggeri, F.S.; Varela, J.A.; Lin, J.Q.; Phillips, E.C.; Miyashita, A.; Williams, D.; Ströhl, F.; et al. FUS Phase Separation Is Modulated by a Molecular Chaperone and Methylation of Arginine Cation-π Interactions. *Cell* **2018**, *173*, 720–734. [CrossRef]

61. Wang, J.; Choi, J.-M.; Holehouse, A.S.; Lee, H.O.; Zhang, X.; Jahnel, M.; Maharana, S.; Lemaitre, R.; Pozniakovsky, A.; Drechsel, D.; et al. A Molecular Grammar Governing the Driving Forces for Phase Separation of Prion-like RNA Binding Proteins. *Cell* **2018**, *174*, 688–699. [CrossRef] [PubMed]

62. Ryan, V.H.; Dignon, G.L.; Zerze, G.H.; Chabata, C.V.; Silva, R.; Conicella, A.E.; Amaya, J.; Burke, K.A.; Mittal, J.; Fawzi, N.L. Mechanistic View of hnRNPA2 Low-Complexity Domain Structure, Interactions, and Phase Separation Altered by Mutation and Arginine Methylation. *Mol. Cell* **2018**, *69*, 465–479. [CrossRef] [PubMed]

63. Yang, W.-H.; Yu, J.H.; Gulick, T.; Bloch, K.D.; Bloch, D.B. RNA-associated protein 55 (RAP55) localizes to mRNA processing bodies and stress granules. *RNA* **2006**, *12*, 547–554. [CrossRef] [PubMed]

64. Matsumoto, K.; Nakayama, H.; Yoshimura, M.; Masuda, A.; Dohmae, N.; Matsumoto, S.; Tsujimoto, M. PRMT1 is required for RAP55 to localize to processing bodies. *RNA Biol.* **2012**, *9*, 610–623. [CrossRef]

65. Hofweber, M.; Hutten, S.; Bourgeois, B.; Spreitzer, E.; Niedner-Boblenz, A.; Schifferer, M.; Ruepp, M.-D.; Simons, M.; Niessing, D.; Madl, T.; et al. Phase Separation of FUS Is Suppressed by Its Nuclear Import Receptor and Arginine Methylation. *Cell* **2018**, *173*, 706–719. [CrossRef]

66. Arribas-Layton, M.; Dennis, J.; Bennett, E.J.; Damgaard, C.K.; Lykke-Andersen, J. The C-Terminal RGG Domain of Human Lsm4 Promotes Processing Body Formation Stimulated by Arginine Dimethylation. *Mol. Cell. Biol.* **2016**, *36*, 2226–2235. [CrossRef]

67. Ries, R.J.; Zaccara, S.; Klein, P.; Olarerin-George, A.; Namkoong, S.; Pickering, B.F.; Patil, D.P.; Kwak, H.; Lee, J.H.; Jaffrey, S.R. m6A enhances the phase separation potential of mRNA. *Nature* **2019**, *571*, 424–428. [CrossRef]

68. Anzilotti, C.; Pratesi, F.; Tommasi, C.; Migliorini, P. Peptidylarginine deiminase 4 and citrullination in health and disease. *Autoimmun. Rev.* **2010**, *9*, 158–160. [CrossRef]

69. Tanikawa, C.; Ueda, K.; Suzuki, A.; Iida, A.; Nakamura, R.; Atsuta, N.; Tohnai, G.; Sobue, G.; Saichi, N.; Momozawa, Y.; et al. Citrullination of RGG Motifs in FET Proteins by PAD4 Regulates Protein Aggregation and ALS Susceptibility. *Cell Rep.* **2018**, *22*, 1473–1483. [CrossRef]

70. Patel, J.; Pathak, R.R.; Mujtaba, S. The biology of lysine acetylation integrates transcriptional programming and metabolism. *Nutr. Metab.* **2011**, *8*, 12. [CrossRef]

71. Drazic, A.; Myklebust, L.M.; Ree, R.; Arnesen, T. The world of protein acetylation. *Biochim. Biophys. Acta Proteins Proteom.* **2016**, *1864*, 1372–1401. [CrossRef] [PubMed]

72. Saito, M.; Hess, D.; Eglinger, J.; Fritsch, A.W.; Kreysing, M.; Weinert, B.T.; Choudhary, C.; Matthias, P. Acetylation of intrinsically disordered regions regulates phase separation. *Nat. Chem. Biol.* **2019**, *15*, 51–61. [CrossRef] [PubMed]

73. Jain, S.; Wheeler, J.R.; Walters, R.W.; Agrawal, A.; Barsic, A.; Parker, R. ATPase-Modulated Stress Granules Contain a Diverse Proteome and Substructure. *Cell* **2016**, *164*, 487–498. [CrossRef] [PubMed]

74. Barghorn, S.; Davies, P.; Mandelkow, E. Tau Paired Helical Filaments from Alzheimer's Disease Brain and Assembled in Vitro Are Based on β-Structure in the Core Domain. *Biochemistry* **2004**, *43*, 1694–1703. [CrossRef]

75. Ferreon, J.C.; Jain, A.; Choi, K.-J.; Tsoi, P.S.; MacKenzie, K.R.; Jung, S.Y.; Ferreon, A.C. Acetylation Disfavors Tau Phase Separation. *Int. J. Mol. Sci.* **2018**, *19*, 1360. [CrossRef]

76. Brunello, C.A.; Yan, X.; Huttunen, H.J. Internalized Tau sensitizes cells to stress by promoting formation and stability of stress granules. *Sci. Rep.* **2016**, *6*, 30498. [CrossRef]

77. Ukmar-Godec, T.; Hutten, S.; Grieshop, M.P.; Rezaei-Ghaleh, N.; Cima-Omori, M.-S.; Biernat, J.; Mandelkow, E.; Söding, J.; Dormann, D.; Zweckstetter, M. Lysine/RNA-interactions drive and regulate biomolecular condensation. *Nat. Commun.* **2019**, *10*, 2909. [CrossRef]

78. Alemasova, E.E.; Lavrik, O.I. Poly(ADP-ribosyl)ation by PARP1: Reaction mechanism and regulatory proteins. *Nucleic Acids Res.* **2019**, *47*, 3811–3827. [CrossRef]

79. Altmeyer, M.; Neelsen, K.J.; Teloni, F.; Pozdnyakova, I.; Pellegrino, S.; Grøfte, M.; Rask, M.-B.D.; Streicher, W.; Jungmichel, S.; Nielsen, M.L.; et al. Liquid demixing of intrinsically disordered proteins is seeded by poly(ADP-ribose). *Nat. Commun.* **2015**, *6*, 8088. [CrossRef]

80. Leung, A.K.L.; Vyas, S.; Rood, J.E.; Bhutkar, A.; Sharp, P.A.; Chang, P. Poly(ADP-ribose) regulates stress responses and microRNA activity in the cytoplasm. *Mol. Cell* **2011**, *42*, 489–499. [CrossRef]

81. Duan, Y.; Du, A.; Gu, J.; Duan, G.; Wang, C.; Gui, X.; Ma, Z.; Qian, B.; Deng, X.; Zhang, K.; et al. PARylation regulates stress granule dynamics, phase separation, and neurotoxicity of disease-related RNA-binding proteins. *Cell Res.* **2019**, *29*, 233–247. [CrossRef] [PubMed]

82. McGurk, L.; Gomes, E.; Guo, L.; Mojsilovic-Petrovic, J.; Tran, V.; Kalb, R.G.; Shorter, J.; Bonini, N.M. Poly(ADP-Ribose) Prevents Pathological Phase Separation of TDP-43 by Promoting Liquid Demixing and Stress Granule Localization. *Mol. Cell* **2018**, *71*, 703–717. [CrossRef] [PubMed]

83. Aguzzi, A.; Altmeyer, M. Phase Separation: Linking Cellular Compartmentalization to Disease. *Trends Cell Biol.* **2016**, *26*, 547–558. [CrossRef] [PubMed]

84. An, L.; Harrison, P.M. The evolutionary scope and neurological disease linkage of yeast-prion-like proteins in humans. *Biol. Direct* **2016**, *11*, 32. [CrossRef]

85. Harrison, A.F.; Shorter, J. RNA-binding proteins with prion-like domains in health and disease. *Biochem. J.* **2017**, *474*, 1417–1438. [CrossRef]

86. Li, Y.R.; King, O.D.; Shorter, J.; Gitler, A.D. Stress granules as crucibles of ALS pathogenesis. *J. Cell Biol.* **2013**, *201*, 361–372. [CrossRef]

87. Tycko, R. Amyloid polymorphism: Structural basis and neurobiological relevance. *Neuron* **2015**, *86*, 632–645. [CrossRef]

88. Shewmaker, F.; McGlinchey, R.P.; Wickner, R.B. Structural Insights into Functional and Pathological Amyloid. *J. Biol. Chem.* **2011**, *286*, 16533–16540. [CrossRef]

89. Hu, Z.-W.; Ma, M.-R.; Chen, Y.-X.; Zhao, Y.-F.; Qiang, W.; Li, Y.-M. Phosphorylation at Ser8 as an intrinsic regulatory switch to regulate the morphologies and structures of Alzheimer's 40-residue β-amyloid (Aβ40) fibrils. *J. Biol. Chem.* **2017**, *292*, 8846. [CrossRef]

90. Tetz, G.; Tetz, V. Prion-like Domains in Eukaryotic Viruses. *Sci. Rep.* **2018**, *8*, 8931. [CrossRef]

© 2019 by the authors. Licensee MDPI, Basel, Switzerland. This article is an open access article distributed under the terms and conditions of the Creative Commons Attribution (CC BY) license (http://creativecommons.org/licenses/by/4.0/).

International Journal of
Molecular Sciences

Concept Paper

Pathogens and Disease Play Havoc on the Host Epiproteome—The "First Line of Response" Role for Proteomic Changes Influenced by Disorder

Erik H. A. Rikkerink

The New Zealand Institute for Plant & Food Research Ltd., 120 Mt. Albert Rd., Private Bag 92169, Auckland 1025, New Zealand; erik.rikkerink@plantandfood.co.nz; Tel.: +64-9-925-7157

Received: 8 February 2018; Accepted: 7 March 2018; Published: 8 March 2018

Abstract: Organisms face stress from multiple sources simultaneously and require mechanisms to respond to these scenarios if they are to survive in the long term. This overview focuses on a series of key points that illustrate how disorder and post-translational changes can combine to play a critical role in orchestrating the response of organisms to the stress of a changing environment. Increasingly, protein complexes are thought of as dynamic multi-component molecular machines able to adapt through compositional, conformational and/or post-translational modifications to control their largely metabolic outputs. These metabolites then feed into cellular physiological homeostasis or the production of secondary metabolites with novel anti-microbial properties. The control of adaptations to stress operates at multiple levels including the proteome and the dynamic nature of proteomic changes suggests a parallel with the equally dynamic epigenetic changes at the level of nucleic acids. Given their properties, I propose that some disordered protein platforms specifically enable organisms to sense and react rapidly as the first line of response to change. Using examples from the highly dynamic host-pathogen and host-stress response, I illustrate by example how disordered proteins are key to fulfilling the need for multiple levels of integration of response at different time scales to create robust control points.

Keywords: intrinsically disordered proteins; epiproteome; disordered protein platform; molecular recognition feature; post-translational modifications; physiological homeostasis; stress response; RIN4; p53; molecular machines

1. Introduction

Survival of both individuals and a species is predicated, in no small measure, on their ability to respond to a changing environment. Faced with the challenge of drastic changes, organisms have stark options of fight or flight. Flight comes with its own series of challenges (including adapting to a new environment, or competing with others that have already occupied the new niche). Either way there is a strong evolutionary imperative to acquire an ability to adapt. Adaptation is likely to require response at different time scales from immediate (at the level of the individual) to geological scale (at the level of species and genus). Rapid response can be both a benefit and a cost, as a quick change in direction can sometimes prove to be detrimental in the fullness of time and might, therefore, be likely to favor rapid responses that are also readily reversible. Critical decision points used by the organism to drive response in a particular direction need to be robustly integrated into the core physiology of the cell. In this review I argue in favor of the broader interpretation of the term epiproteome to encapsulate the concepts that (1) changes at the post-translational level are ideally placed to respond in real time and that (2) flexible proteins displaying significant disorder are ideal platforms that can be decorated with post-translational changes and used to integrate responses that potentially have competing impacts on cellular resources.

The term epiproteome was first coined by Dai and Rasmussen [1] to refer to proteomic changes directly associated with epigenetic modifications, namely histone acetylation. Some researchers argue that histone modifications are part of epigenetics, although others argue that their lack of heritability means they should not be included in that term. A search for epiproteome/epiproteomics in PubMed Central yields references to post-translational modification (PTM) changes in histones and a small number that use the term in a wider sense to refer to other PTM [2]. Below I argue in favor of the broad interpretation that includes all PTM.

I suggest that an understanding of the epiproteome (i.e., changing alternative post-translational protein states) in combination with the critical nodal positions occupied by disordered proteins, provides a new basis to comprehend the hypervariable PTM theatre. Its features enable integration of multiple post-translational signals to match the demands of a flexible response. The best examples of hypervariable theatres of response to stress are the battle between hosts and their pathogens and/or their changing environment. Epiproteomic changes offer the host an elegant real-time control of its responses. Unfortunately, this also makes the PTM theatre an Achilles heel, able to be exploited by pathogens. Arguably this explains why so much of a pathogens weaponry appears to be enzymatic and focused on the PTM level of host organisation [3].

2. Review

Below I address five key points or questions that address the key demands on a highly integrated cellular stress control point, namely: (1) A broad interpretation of the concept of the epiproteome; (2) How an organism can communicate between (and marshal) sets of proteins that need to respond to stress while also integrating the, on occasion conflicting, demands of distinct but simultaneous stresses; (3) Are there some key exemplars in plants and animals that point towards solutions to meet the demanding challenge of multiple stresses? (4) What are the characteristics required for a node that can successfully integrate response to simultaneous challenges? (5) How can multiple diverse signals be coordinated in real-time to deliver a coherent response?

2.1. Why Use the Broad Interpretation of the Term Epiproteome?

Our concepts of how the molecular machinery in a cell operates have changed radically over the last three decades. One-dimensional models of static proteins acting alone to promote a particular enzymatic step have been superseded by an understanding that proteins typically act as parts of molecular nano-machines in complexes, and are dynamically controlled by a combination of their own intrinsic flexibility [4], their micro-environment, location and their interactions with their partners. We know that robust control of cellular processes needs to occur at multiple levels [5] that can include modifications of chromatin, transcription, post-transcription, translation and PTM. Epigenetic changes and their role in host plasticity have been widely discussed over the last decade [6,7], including their role in responding to challenges such as pathogens [8]. More recently epitranscriptomic changes have become a topic of renewed interest [9]. It is timely therefore to focus on the epiproteome and the key role that PTM could play in coordinated cellular response to pathogens and other stresses. There are of course thousands of papers referring to specific post-translational modifications or similar terms. Initially epiproteomics referred simply to changes in the specific proteome associated with DNA epigenetic changes [1]. And indeed it is still sometimes used in this way now [10]. Only recently have papers used the term to refer to the sum of all post-translational changes in all proteins or a subset such as the redox or cysteine epiproteome [2,11]. The term epiproteomics evokes a parallel with the temporal nature of epigenetics that is entirely appropriate, and perhaps even central, to the importance of the PTM level of control. Therefore below the term epiproteome is applied in a much more general way to any post-translational modification of any protein in any protein complex. PTM can lead to a large ensemble of forms of the components of the proteome existing in a dynamic state within a cell. Unfortunately, the research tools required to analyse PTM states properly are still expensive to run and hence our current picture of the dynamics of these states is inadequate. PTM alternative states are however likely to be more significant than random noise and there are numerous individual cases where this is confirmed.

2.2. Marshalling a Diverse Set of Responding Proteins More or Less in Unison

When combined with the concept of the role of intrinsic disorder in signalling, the significance of epiproteomic changes are placed in a new light. Coherently controlled epiproteomic changes would have the potential to alter the response of many proteins simultaneously, whether they be members of the same protein complex or dispersed complexes that need to be coordinated with each other. Individual PTM changes have been shown to play key roles in a number of different properties including the formation or dissolution of protein interactions [12], the conformation of a protein [13], the membrane localisation of a protein [14] or the inactivation [15] or degradation [16] of a protein. When such changes are driven by a significant change in physiology of the entire cell, such as its redox potential or pH, there is significant scope for matching coordinated changes in the protein complexes within the cell that need to react to the new state. Thus, the epiproteome is uniquely positioned to play a vital role in marshalling multi-protein responses. There are some known examples of this in situations of stress in nature already. A plant pathogen effector was recently shown to acetylate several proteins that interact with each other in a complex [17]. Another example is the cell signalling that results from electrophilic oxidized lipid products, so called reactive lipid species (RLS), that can react with the amino acids cysteine, lysine and histidine because of their nucleophilic nature. RLS effects on signalling events are largely restricted to the modification of cysteine residues in proteins. RLS induced modifications appear to participate in multiple physiological processes including inflammation, induction of antioxidants and even cell death through the modification of signalling proteins [18].

2.3. Animal and Plant Exemplars: p53 and RIN4

The p53 transcription factor in mammals is best known as a target for cancer therapy and understanding the interaction between stress and cancer but is also associated with facets of aging and microbial responses [19]. Additionally, p53 is a target for microbial manipulation by both viruses and bacteria [20]. The p53 protein interacts with a remarkable array of partners and a key characteristic that allows p53 to act as such a key node/hub is believed to be its disordered characteristics [21]. A particularly pertinent facet of disorder in this discussion is its accessibility to PTM events. In highly structured proteins, only a minority of surface exposed residues and short flexible disordered loops are available for PTM. In disordered proteins/regions, the majority of residues are exposed and provide a readily available platform for epiproteomic modification. Disorder is particularly common in regulatory proteins such as transcription factors (TF) in both animals and plants [22]. The TF p53 has a typical platform with a number of disordered regions (often highly charged) attached to a more structured domain that interacts with DNA. The number of proteins that p53 interacts with, the roles played by epiproteome changes in p53 that link with protein–protein interactions, and the types of PTM events have grown into a very complex interacting network [23,24]. The p53 platform epitomizes the plasticity of disordered proteins and the vital importance of epiproteomic changes to their flexible response. PTM changes are clustered in and around Molecular Recognition Features (MoRFs)—short semi-ordered segments within a largely unstructured backbone that drive interactions with multiple protein partners [25] (see Figure 1).

Disorder-associated properties of p53 have likely also played a significant role in the evolution of this protein family. In a recent study of the evolution of p53 and related proteins in metazoans, Joerger et al. [26] suggest that mutations which stabilized formation of tetrameric p53 forms early in the evolution of vertebrates may have freed up the C-terminal region to adopt a disordered structure. Disorder then may have allowed the C-terminus to undergo numerous PTM and evolve the ability to interact with multiple partner regulatory proteins. They suggest that this disorder assisted evolutionary path allowed p53 to acquire many novel somatic functions by rewiring signalling pathways. Different parts of the p53 disordered regions have been shown to possess significant variation in divergence rates [27]. Indeed, it has been argued for some time that disordered regions in general have novel properties and show increased rates of mutation that suggest they are often under diversifying selection pressure and may be important to allow organisms to adapt [28–31].

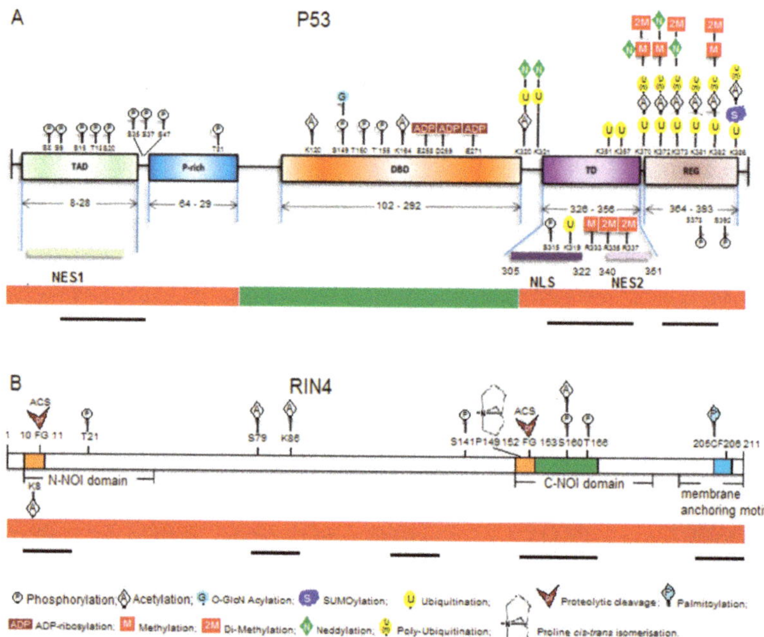

Figure 1. Multiple PTMs cluster in the Molecular Recognition Features within the disordered regions in p53 (**A**) and RIN4 (**B**). Consensus disordered regions are indicated by the red bar at the bottom while ordered regions are indicated by the green bar, putative MoRFs within the disordered regions are indicated by black bars as determined by Uversky (2016; p. 53, [23]) or Sun et al. (2014; RIN4, [32]) by application of disorder prediction programs. PTMs for p53 are a modified form of those identified by Gu and Zhu (2012) [24]. TAD: Transactivation domain; P-rich: proline-rich domain; DBD: DNA binding domain; TD: tetramerization domain; REG: C-terminal regulatory domain; NES1/2: N-terminal (1) and C-terminal (2) nuclear export sequences; NLS: nuclear localization sequence; N-NOI: N-terminal Nitrate induced domain; C-NOI: C-terminal Nitrate induced domain; ACS: AvrRpt2 cleavage site.

The nature of the role played by RPM1-Interacting protein 4 (RIN4) in plant defence is an enigma. That its role is important is hard to question, as RIN4 is targeted directly or indirectly by a number of plant pathogen effectors [32,33]. Moreover the activity of effectors result in several different epiproteomic changes to RIN4 including phosphorylation [34,35], proteolytic cleavage [36], proline isomerisation [37] and acetylation [17]. We have suggested that, like p53, RIN4 is largely intrinsically disordered and is therefore a viable platform for multiple PTM events [32]. As for p53, RIN4 has MoRFs that correlate with conserved motifs and sites of critical importance within RIN4 that are targeted for epiproteomic modification by pathogens and/or the cell itself, or are juxtaposed to modified residues (Figure 1). Two known examples of how RIN4 works are illustrated by recent research [33,37]. Chung and colleagues suggest that two phosphorylation sites within RIN4 are in competition with each other and may be responsible for driving RIN4 in the direction of either innate (molecular pattern triggered) immunity or effector triggered immunity. The authors suggest that RIN4 is a 'phospho-switch' and I note that one of these sites (T166*p*-phosphorylated) sits in the middle of one of the MoRFs identified by Sun et al. [32], while the other sits near the boundary of this MoRF (S141*p*). Li and colleagues [37] identify a proline isomerisation site at P149 in the same MoRF that is, in turn, influenced by phosphorylation at T166. The T166*p* epiproteomic variant has a reduced affinity for the ROC1 enzyme that drives a *cis* to *trans* isomerisation at P149. These first examples of multiple proteomic forms illustrate how RIN4 constitutes the most compelling example yet of a plant protein playing a parallel role to that of p53, as a platform that appears primed to 'collect' epiproteomic signals.

2.4. Characteristics Required for an Integrated Response to Simultaneous Challenges

In order to be able to integrate responses, a hub must be capable of multiple interactions with various partners and collect signatures from various input pathways that can then be coherently interpreted. A high number of flexible and reversible interactions, and ability to make subtle changes to the equilibrium between the various states of control, would constitute a further advantage for such a hub. As these requirements match key characteristics of disordered regions it has been recognised for some time that such hub proteins are highly enriched for disorder [38]. Epiproteomic modification to MoRFs or neighbouring sites in a disordered platform could either block interactions, block other PTM changes at the same site (e.g., acetylation of a serine residue sometimes phosphorylated), or change the charge profile in disordered regions that then changes the dynamic of how (and/or whether) a MoRF interacts with a specific partner. While a degree of subtlety is important sometimes, a hard on-off switch will be important at other times. Disordered proteins can also undergo major conformational switches and even these can be linked to epiproteomic changes by adding larger modifying groups (e.g., glutathionylation, or AMPylation), isomerisation events around critical prolines, or by targeting the entire protein for proteolytic degradation for example.

In a recent analysis of human cells, Chavez and colleagues [39] used novel protein cross-linking methods combined with mass spectrometry to directly identify PTM decorated proteins that are physically associated with each other in complexes. New software advances have also enabled data analysis to focus on cross-linked peptides [40]. In cross-linking analyses distance constraints can be imposed by the type of chemical linker arm used, while addition of biotin groups permits enrichment for cross-linked fragments (e.g., by using avidin-mediated affinity capture technologies). Although the majority of cross-linked peptides identified were derived from homo-dimer interactions, acetylated and methylated peptides from core histone proteins participating in hetero-dimers were particularly common in this analysis. Almost half of the cross-linked histone peptides were found to contain at least one PTM event. Histones are known carry a number of highly significant PTM events. The multiple cases of linkages found between specific peptides increases the likelihood that these have biological relevance in terms of the protein interaction zones between the partners. Interestingly many of the cross-linked peptides with PTM contained modified lysine or arginine residues (residues that are also particularly enriched in disordered regions of proteins). Cross-linking sites were common in the disordered N- and C-termini of histones. In fact, it has been known for more than two decades that the histone tails are the sites where some of the most significant PTM takes place and that these modifications play key roles in the formation or dissolution of chromatin remodelling complexes. These tails serve as recognition sites for chromatin assembly as well as the assembly of the multi-component transcription machinery [41]. The largely positively charged disordered N-terminal tail also contributes to inter-nucleosome binding by contacting an acidic patch in the structured component of histone H2A/H2B dimers to influence histone stacking [42].

The cross-linking analysis allowed Chavez and colleagues [39] to build a significant interactome network map and highlights the importance of the combination of disordered regions and PTM to interactions in such networks, the hub position occupied by histones and the importance of their lysine/arginine rich disordered tails to drive their ability to organize into multi-component complexes. Other biophysical methods of experimentation can also provide indications of how closely associated proteins are in vitro or in vivo. Hydrogen-deuterium exchange (HDX) provides a measure of how exposed different parts of proteins are to PTM [43]. Changes in HDX patterns upon binding with partners can indicate likely interaction zones in protein complexes and were initially used to map antibody binding sites [44]. Other techniques like Förster (fluorescence) resonance energy transfer (FRET) also lend themselves to analysing protein disorder. For example, Vassall et al. [45] used FRET measurements to analyse the order-to-disorder transition of the myelin basic protein (MBP). MBP is largely disordered in aqueous conditions but forms alpha helical recognition fragments upon binding to membranes and its protein partners. The MBP FRET studies, when combined with other tools used to probe structural transition in largely disordered proteins (such as circular dichroism and the membrane-mimetic solvent trifluoroethanol), yielded some surprising results. The data suggested that an intermediate conformational form between disorder and alpha helical state is in fact more compact than the alpha helical form (the latter would normally be expected to have more compactness). This longer form may provide a better bridge across to its complexing protein partners as well as facilitating faster binding to the membrane. Disorder-associated characteristics possessed by histone hubs allow them to integrate epigenetic marks with downstream modifications in mRNA expression response and transfer signals between the epigenetics and transcriptomics levels of response. The disorder properties of MBP on the other hand allow MBP to peripherally attach itself to the cytoplasmic membrane as well as interact with both cytoskeletal proteins like actin and signalling proteins that respond to Ca^{2+}-triggered protein cascades.

One of the ways that cells coordinate their response to changing situation such as stress is to form recognizable sub-cellular organelles. Examples include stress granules (SG), processing bodies (P-bodies) and nuclear stress bodies. Such organelles do not contain membranes, a factor that differentiates them from permanent cellular compartments like the ER, nucleus and mitochondria. Functional organelles must be able to keep interacting with their surrounding liquid environment and yet they must have an ability to form an interphase boundary with this environment. In a recent review Uversky [46] suggests that disorder can provide a crucial component required for forming this liquid-to-liquid interphase. Examples of this are the role that the RNA-binding protein TIA-1 plays to promote assembly of SG through its disordered domains and the disordered regions of a number of the RNA-binding proteins found in human and yeast stress granules. The latter were found to be able to undergo liquid-liquid phase transition in vitro on their own, or when combined with RNA [47]. The phase separated droplets promoted by this organisation can then also recruit other proteins with disordered regions. Furthermore mutations in the key disordered regions or PTM sites involved in regulation can then lead to aberrant fibers or granules that may then contribute to neurodegenerative conditions.

2.5. How Can Multiple Diverse Signals Be Coordinated in Real-Time?

Responses need to be organised at both the temporal and spatial levels. An important biological question is how can organisms create control points that match such elaborate requirements? Significant PTM changes can be very rapid with response times measured in minutes as opposed to hours or even days for many other types of regulation responses [48]. Rapid response makes this level of regulation ideal for responding in real time to challenges perceived by the organisms. A successful reaction to stress is dynamic and requires both sequential, temporal and spatial separation of components and the ability to be nimble in response. The high degree of sophistication required by a successful response is elegantly matched with the opportunities offered by disordered platforms to rapidly integrate PTM signals through multiple MoRFs, multiple targeted PTM sites and reversible as well as competing PTM changes at particular sites. Moreover, PTM changes can also be spatially compartmentalised by limiting where matching substrates and enzymatic functions are co-expressed. As discussed above compartmentalisation can even be aided or driven by the ability of disordered proteins to contribute to phase transition in examples like stress bodies.

Importantly many PTM changes are reversible involving balancing modifications such as phosphorylation/dephosphorylation or acetylation/de-acetylation. Pathogens in turn interfere with PTM processes by developing modifications that can compete with these changes, e.g., phospholyase reactions that break a unique phospho-threonine bond in a protein kinase activation site and make this site un-available for re-phosphorylation [15]. The very properties that make PTM changes so dynamic also make this level of response technically very demanding to illustrate. In order to capture such dynamic potential, sampling time needs to be adjusted to a much finer timescale than commonly used. In addition, techniques that can capture protein associations in real time and are not affected by their readily reversible nature (such as cross-linking techniques) will be required. This will need to be matched with detection techniques sensitive enough to identify any PTM, yet robust enough to be able to scan across complex proteomes. Physical and software enrichment strategies that can overcome the challenge of these limitations in concert with much more sensitive mass spectrometry instrumentation have recently become available. I suggest that disordered protein regions in particular have properties that indicate they are likely to feature prominently in these novel analyses in the near future. Their dynamic ability to change their binding partnerships and to be decorated by multiple PTM events, as illustrated by the examples of p53, RIN4 and histones presented above, suggest that this is one of the major reasons that disorder has become such a common feature of proteins in complex multi-cellular organisms. Indeed this fits with the proposal that disorder was a key enabler on the road to multi-cellular lifestyles [31]. The ability of disordered regions to sense the physiological milieu in which they find themselves by a combination of PTM events, charge profiles and electrostatic interactions suggests that sensing change in this milieu is a specific biological niche that disordered proteins occupy.

3. Conclusions

The animal and plant exemplars, p53 and RIN4, show some key similarities. I suggest that their disordered platform is specifically designed to integrate diverse signals that arrive via alternate post-translational changes inside (or sometimes in close proximity to) MoRFs. PTM changes have a great deal of flexibility and can be very rapid and reversible (e.g., phosphorylation and de-phosphorylation) and the term epiproteomics evokes the dynamic nature of these changes. In addition to reversibility, PTM sites can be; locked in competitive battles (e.g., phosphorylation and acetylation [24]), display competition between sites (as illustrated by the RIN4 phospho-switch concept [33]), result in more subtle shifts in equilibrium (e.g., by changing the charge profile and flexibility of the environment around a MoRF), or result in drastic conformational changes suited to acting as a molecular on/off switch (e.g., by proline isomerisation or multiple phosphorylations). I suggest that the main role of the p53 and RIN4 (and probably many other) proteins containing large disordered domains is to act as sensors and integrators of stress signals from multiple distinct sources via changes to the epiproteome. Moreover, this could explain why examples such as RIN4 and p53 play such important roles in plant and animal disease respectively.

Acknowledgments: This research was funded by Discovery Science Grants to EHAR on intrinsically disordered proteins (DS-1166 and DS-2002) from The New Zealand Institute for Plant & Food Research Ltd. (Auckland, New Zealand). The author thanks Joanna Bowen and Xiaolin Sun for editorial suggestions.

Conflicts of Interest: The author declares no conflict of interest.

References

1. Dai, B.; Rasmussen, T.P. Global epiproteomic signatures distinguish embryonic stem cells from differentiated cells. *Stem Cells* **2007**, *25*, 1567–2574. [CrossRef] [PubMed]
2. Go, Y.M.; Jones, D.P. The redox proteome. *J. Biol. Chem.* **2013**, *288*, 26512–26520. [CrossRef] [PubMed]
3. Block, A.; Alfano, J.R. Plant targets for *Pseudomonas syringae* type III effectors: Virulence targets or guarded decoys? *Curr. Opin. Microbiol.* **2011**, *14*, 39–46. [CrossRef] [PubMed]
4. Uversky, V.N.; Dunker, A.K. Controlled chaos. *Science* **2008**, *322*, 1340–1341. [CrossRef] [PubMed]
5. Payne, J.L.; Wagner, A. Mechanisms of mutational robustness in transcriptional regulation. *Front. Genet.* **2015**, *6*, 322. [CrossRef] [PubMed]
6. Riddihough, G.; Zahn, L.M. What is epigenetics? *Science* **2010**, *330*, 611. [CrossRef] [PubMed]

7. Cortijo, S.; Wardenaar, R.; Colomé-Tatché, M.; Gilly, A.; Etcheverry, M.; Labadie, K.; Caillieux, E.; Hospital, F.; Aury, J.M.; Wincker, P.; et al. Mapping the epigenetic basis of complex traits. *Science* **2014**, *343*, 1145–1148. [CrossRef] [PubMed]

8. Gómez-Díaz, E.; Jordà, M.; Peinado, M.A.; Rivero, A. Epigenetics of host–pathogen interactions: The road ahead and the road behind. *PLoS Pathog.* **2012**, *8*, e1003007. [CrossRef] [PubMed]

9. Gokhale, N.S.; Horner, S.M. RNA modifications go viral. *PLoS Pathog.* **2017**, *13*, e1006188. [CrossRef] [PubMed]

10. Zheng, Y.; Huang, X.; Kelleher, N.L. Epiproteomics: Quantitative analysis of histone marks and codes by mass spectrometry. *Curr. Opin. Chem. Biol.* **2016**, *33*, 142–150. [CrossRef] [PubMed]

11. Go, Y.-M.; Chandler, J.D.; Jones, D.P. The cysteine proteome. *Free Radic. Biol. Med.* **2015**, *84*, 227–245. [CrossRef] [PubMed]

12. Nishi, H.; Hashimoto, K.; Panchenko, A.R. Phosphorylation in protein-protein binding: Effect on stability and function. *Structure* **2011**, *19*, 1807–1815. [CrossRef] [PubMed]

13. Andreotti, A.H. Native state proline isomerisation: An intrinsic molecular switch. *Biochemistry* **2003**, *42*, 9515–9524. [CrossRef] [PubMed]

14. Resh, M.D. Covalent lipid modifications of proteins. *Curr. Biol.* **2013**, *23*, R431–R435. [CrossRef] [PubMed]

15. Li, X.; Lin, H.; Zou, Y.; Zhang, J.; Long, C.; Li, S.; Chen, S.; Zhou, J.M.; Shao, F. The phosphothreonine lyase activity of a bacterial type III effector family. *Science* **2007**, *315*, 1000–1003. [CrossRef] [PubMed]

16. Rosebrock, T.R.; Zeng, L.R.; Brady, J.J.; Abramovitch, R.B.; Xiao, F.M.; Martin, G.B. A bacterial E3 ubiquitin ligase targets a host protein kinase to disrupt plant immunity. *Nature* **2007**, *448*, 370–374. [CrossRef] [PubMed]

17. Lee, J.; Manning, A.J.; Wolfgeher, D.; Jelenska, J.; Cavanaugh Keri, A.; Xu, H.; Fernandez, S.M.; Michelmore, R.W.; Kron, S.J.; Greenberg, J.T. Acetylation of an NB-LRR plant immune-effector complex suppresses immunity. *Cell Rep.* **2015**, *13*, 1670–1682. [CrossRef] [PubMed]

18. Higdon, A.; Diers, A.R.; Oh, J.Y.; Landar, A.; Darley-Usmar, V.M. Cell signalling by reactive lipid species: New concepts and molecular mechanisms. *Biochem. J.* **2012**, *442*, 453–464. [CrossRef] [PubMed]

19. Maclaine, N.J.; Hupp, T.R. The regulation of p53 by phosphorylation: A model for how distinct signals integrate into the p53 pathway. *Aging* **2009**, *1*, 490–502. [CrossRef] [PubMed]

20. Zaika, A.I.; Wei, J.; Noto, J.M.; Peek, R.M. Microbial regulation of p53 tumor suppressor. *PLoS Pathog.* **2015**, *11*, e1005099. [CrossRef] [PubMed]

21. Oldfield, C.J.; Meng, J.; Yang, J.Y.; Yang, M.Q.; Uversky, V.N.; Dunker, A.K. Flexible nets: Disorder and induced fit in the associations of p53 and 14-3-3 with their partners. *BMC Genom.* **2008**, *9* (Suppl. 1), S1. [CrossRef] [PubMed]

22. Sun, X.; Rikkerink, E.H.A.; Jones, W.T.; Uversky, V.N. Multifarious roles of intrinsic disorder in proteins illustrate its broad impact on plant biology. *Plant Cell* **2013**, *25*, 38–55. [CrossRef] [PubMed]

23. Uversky, V.N. P53 proteoforms and intrinsic disorder: An illustration of the protein structure–function continuum concept. *Int. J. Mol. Sci.* **2016**, *17*, 1874. [CrossRef] [PubMed]

24. Gu, B.; Zhu, W.-G. Surf the post-translational modification network of p53 regulation. *Int. J. Biol. Sci.* **2012**, *8*, 672–684. [CrossRef] [PubMed]

25. Cheng, Y.G.; Oldfield, C.J.; Meng, J.W.; Romero, P.; Uversky, V.N.; Dunker, A.K. Mining alpha-helix-forming molecular recognition features with cross species sequence alignments. *Biochemistry* **2007**, *46*, 13468–13477. [CrossRef] [PubMed]

26. Joerger, A.C.; Wilcken, R.; Andreeva, A. Tracing the evolution of the p53 tetramerization domain. *Structure* **2014**, *22*, 1301–1310. [CrossRef] [PubMed]

27. Xue, B.; Brown, C.J.; Dunker, A.K.; Uversky, V.N. Intrinsically disordered regions of p53 family are highly diversified in evolution. *Biochim. Biophys. Acta* **2013**, *1834*, 725–738. [CrossRef] [PubMed]

28. Dunker, A.K.; Garner, E.; Guilliot, S.; Romero, P.; Albrecht, K.; Hart, J.; Obradovic, Z.; Kissinger, C.; Villafranca, J.E. Protein disorder and the evolution of molecular recognition: Theory, predictions and observations. *Pac. Symp. Biocomput.* **1998**, *3*, 473–484.

29. Brown, C.J.; Johnson, A.K.; Dunker, A.K.; Daughdrill, G.W. Evolution and disorder. *Curr. Opin. Struct. Biol.* **2011**, *21*, 441–446. [CrossRef] [PubMed]

30. Nilsson, J.; Grahn, M.; Wright, A.P. Proteome-wide evidence for enhanced positive Darwinian selection within intrinsically disordered regions in proteins. *Genome Biol.* **2011**, *12*, R65. [CrossRef] [PubMed]

31. Romero, P.R.; Zaidi, S.; Fang, Y.Y.; Uversky, V.N.; Radivojac, P.; Oldfield, C.J.; Cortese, M.S.; Sickmeier, M.; LeGall, T.; Obradovic, Z.; et al. Alternative splicing in concert with protein intrinsic disorder enables increased functional diversity in multicellular organisms. *Proc. Natl. Acad. Sci. USA* **2006**, *103*, 8390–8395. [CrossRef] [PubMed]

32. Sun, X.; Greenwood, D.R.; Templeton, M.D.; Libich, D.S.; McGhie, T.K.; Xue, B.; Yoon, M.; Cui, W.; Kirk, C.A.; Jones, W.T.; et al. The intrinsically disordered structural platform of the plant defence hub protein RIN4 provides insights into its mode of action in the host-pathogen interface and evolution of the NOI protein family. *FEBS J.* **2014**, *281*, 3955–3979. [CrossRef] [PubMed]

33. Chung, E.H.; El-Kasmi, F.; He, Y.; Loehr, A.; Dangl, J.L. A plant phosphoswitch platform repeatedly targeted by type III effector proteins regulates the output of both tiers of plant immune receptors. *Cell Host Microbe* **2014**, *16*, 484–494. [CrossRef] [PubMed]

34. Mackey, D.; Holt, B.F.; Wiig, A.; Dangl, J.L. RIN4 interacts with *Pseudomonas syringae* type III effector molecules and is required for RPM1-mediated resistance in *Arabidopsis*. *Cell* **2002**, *108*, 743–754. [CrossRef]

35. Liu, J.; Elmore, J.M.; Lin, Z.-J.D.; Coaker, G. A receptor-like cytoplasmic kinase phosphorylates the host target RIN4, leading to the activation of a plant innate immune receptor. *Cell Host Microbe* **2011**, *9*, 137–146. [CrossRef] [PubMed]

36. Axtell, M.J.; Chisholm, S.T.; Dahlbeck, D.; Staskawicz, B.J. Genetic and molecular evidence that the *Pseudomonas syringae* type III effector protein AvrRpt2 is a cysteine protease. *Mol. Microbiol.* **2003**, *49*, 1537–1546. [CrossRef] [PubMed]

37. Li, M.; Ma, X.; Chiang, Y.H.; Yadeta, K.A.; Ding, P.; Dong, L.; Zhao, Y.; Li, X.; Yu, Y.; Zhang, L.; et al. Proline isomerization of the immune receptor-interacting protein RIN4 by a cyclophilin inhibits effector-triggered immunity in *Arabidopsis*. *Cell Host Microbe* **2014**, *16*, 473–483. [CrossRef] [PubMed]

38. Haynes, C.; Oldfield, C.J.; Ji, F.; Klitgord, N.; Cusick, M.E.; Radivojac, P.; Uversky, V.N.; Vidal, M.; Iakoucheva, L.M. Intrinsic disorder is a common feature of hub proteins from four eukaryotic interactomes. *PLoS Comput. Biol.* **2006**, *2*, e100. [CrossRef] [PubMed]

39. Chavez, J.D.; Weisbrod, C.R.; Zheng, C.; Eng, J.K.; Bruce, J.E. Protein interactions, post-translational modifications and topologies in human cells. *Mol. Cell. Proteom.* **2013**, *12*, 1451–1467. [CrossRef] [PubMed]

40. Weisbrod, C.R.; Chavez, J.D.; Eng, J.K.; Yang, L.; Zheng, C.; Bruce, J.E. In vivo protein interaction network identified with novel real-time chemical cross-linked peptide identification strategy. *J. Proteome Res.* **2012**, *12*, 1569–1579. [CrossRef] [PubMed]

41. Luger, K.; Richmond, T.J. The histone tails of the nucleosome. *Curr. Opin. Genet. Dev.* **1998**, *8*, 140–146. [CrossRef]

42. Chen, Q.; Yang, R.; Korolev, N.; Liu, C.F.; Nordenskiöld, L. Regulation of nucleosome stacking and chromatin compaction by the histone H4 N-terminal tail–H2A acidic patch interaction. *J. Mol. Biol.* **2017**, *429*, 2075–2092. [CrossRef] [PubMed]

43. Sheerin, D.J.; Buchanan, J.; Kirk, C.; Harvey, D.; Sun, X.; Spagnuolo, J.; Li, S.; Liu, T.; Woods, V.A.; Foster, T.; et al. Inter- and intra-molecular interactions of *Arabidopsis thaliana* DELLA protein RGL1. *Biochem. J.* **2011**, *435*, 629–639. [CrossRef] [PubMed]

44. Paterson, Y.; Englander, S.W.; Roder, H. An antibody binding site on cytochrome c defined by hydrogen exchange and two-dimensional NMR. *Science* **1990**, *249*, 755–759. [CrossRef] [PubMed]

45. Vassall, K.A.; Jenkins, A.D.; Bamm, W.; Haruaz, G. Thermodynamic analysis of the disorder-to-α-helical transition of 18.5-kDa myelin basic protein reveals an equilibrium intermediate representing the most compact conformation. *J. Mol. Biol.* **2015**, *427*, 1977–1992. [CrossRef] [PubMed]

46. Uversky, V.N. Protein intrinsic disorder-base liquid-liquid phase transitions in biological systems: Complex coacervates and membrane-less organelles. *Adv. Colloid Interface Sci.* **2017**, *239*, 97–114. [CrossRef] [PubMed]

47. Lin, Y.; Protter, D.S.; Rosen, M.K.; Parker, R. Formation and maturation of phase-separated liquid droplets by RNA-binding proteins. *Mol. Cell* **2015**, *60*, 208–219. [CrossRef] [PubMed]

48. Yosef, N.; Regev, A. Impulse control: Temporal dynamics in gene transcription. *Cell* **2011**, *144*, 886–896. [CrossRef] [PubMed]

© 2018 by the author. Licensee MDPI, Basel, Switzerland. This article is an open access article distributed under the terms and conditions of the Creative Commons Attribution (CC BY) license (http://creativecommons.org/licenses/by/4.0/).